HORTICULTURAL SCIENCE

Apple blossoms. [Courtesy International Harvester Corporation.]

HORTICULTURAL SCIENCE

Fourth Edition

Jules Janick

Purdue University

W. H. Freeman and Company

New York

Library of Congress Cataloging-in-Publication Data

Janick, Jules, 1931–
Horticultural science.

Includes bibliographies and index.
1. Horticulture. I. Title.
SB318.J35 1986 635 85-20521
ISBN 0-7167-1742-5

Printed in the United States of America

1 2 3 4 5 6 7 8 9 0 HL 4 3 2 1 0 8 9 8 7 6

To Shirley and Peter and Robin

Contents

Preface

Horticulture is concerned with those plants whose cultivation brings rewards, whether in the form of profits or personal pleasure, sufficient to warrant the expenditure of intensive effort. This art—which entails great skill and timing—has an ancient tradition. But modern horticulture integrates many natural phenomena with advanced technology, and so becomes a scientific discipline in its own right. The primary purpose of this textbook is to examine the scientific concepts on which horticulture is based. A comprehension of the science gives meaning and scope to the art and makes possible the improvement of centuries-old practices.

After an introductory chapter, "Horticulture and Human Affairs," the biology of horticulture is presented in Part I. Although horticulture is concerned with the interaction of humans and plants, plants must first be considered as part of their natural community. The aim of horticultural practices is to produce a growing plant; to understand the biological rationale behind these practices students must have a knowledge of plant relationships, structure, growth, development, and reproduction. The five chapters in Part I provide this knowledge to students who have not been formally introduced to plant science. For those who have already studied biology, these chapters will serve as a quick review. Students who are simultaneously studying botany or biology should consider these chapters variations on a theme.

Part II, "Horticultural Environment," considers the environment of the growing plant. Each chapter discusses a major component of the environment: soil (the storehouse of water and nutrients), water, radiant energy in the form of light and heat, and air.

Part III deals with the technology of horticulture. The progress and survival of humanity depends on our skill in managing the productivity of the plants that sustain us or, in the case of ornamental plants, make life worth living. Techniques will be presented as general horticultural practices, but their relation to individual crops will also be stressed. This treatment should make the information more meaningful to students and allow them to apply techniques to a wider range of crops.

Part IV describes the industry of horticulture, analyzing individual crops and their locations. The distinguishing characteristics and special problems of the industry are emphasized. Horticultural crop species are discussed in traditional groupings, such as fruits, vegetables, and ornamentals. The book concludes with a discussion of the esthetics of horticulture.

The fourth edition of this text includes various changes in horticultural science that have taken place since the third edition was published in 1979. Some innovations and techniques are covered for the first time in this edition, while others are given expanded treatment. Graphs and tables have been updated, and figures added.

The organization of the fourth edition has been improved from that of previous editions. The three parts (biology, technology, and industry of horticulture) and 15 chapters of the third edition have been expanded to four parts and 22 chapters. The greatest change is the consolidation of scattered discussions of the horticultural environment into the five new chapters ("Soil," "Water," "Light," "Temperature," and "Air") forming Part II. In addition, material on cell and plant reproduction has been brought together in a new chapter entitled "Reproduction" in Part I ("Horticultural Biology"), and separate chapters are devoted to nutrition and to pruning and training in Part III ("Horticultural Technology"). This rearrangement and expansion should facilitate classroom presentation.

The fourth edition, like the first three, has been designed primarily for the beginning horticulture student but also for those whose interests may be only incidentally associated with horticulture. The text assumes no great familiarity with botany or plant science, and its 22 chapters can be covered adequately in a semester. The skills associated with horticulture can be reviewed in a laboratory in a sequence similar to that used in this text. Key references at the end of each chapter encourage further study.

It is a pleasure to acknowledge past and present colleagues who have been generous with their time, their information, and their support of various editions of this text. Among the many are T. E. Boudreaux, K. M. Brink, E. D. Carpenter, N. W. Dana, N. W. Desrosier, H. C. Dostal, Dominic Durkin, F. H. Emerson, E. R. Emino, W. C. Fonteno, W. H. Gabelman, A. T. Guard,

P. M. Hasegawa, C. E. Hoxsie, Jermone Hull, Jr., K. W. Johnson, J. L. Johnson, A. C. Leopold, N. W. Marty, C. A. Mitchell, C. R. Parks, C. L. Pfeiffer, J. R. Shay, I. Spear, E. C. Stevenson, G. Tereshkovich, R. B. Tukey, J. F. Tuite, G. F. Warren, G. E. Wilcox, and M. Workman.

At W. H. Freeman and Company, special thanks are due to Linda Chaput, James A. Dodd, Susan Moran, Robin A. Hessel, and Elizabeth Szabla for their enthusiastic support of the fourth edition. Last, I wish to express my appreciation to Harvey McCaleb, editor emeritus, for his many years of friendship.

October 1985 Jules Janick

1 Horticulture and Human Affairs

The origins of horticulture are intimately associated with the history of humanity. The term *horticulture*, which is probably of relatively recent origin, first appeared in written language in the seventeenth century.[1] The word is derived from the Latin *hortus*, garden, and *colere*, to cultivate. The concept of the culture of gardens (Anglo-Saxon *gyrdan*, to enclose) as distinct from the culture of fields—that is, **agriculture**—is a medieval concept, indicative of the practices in temperate Europe during the Middle Ages. Agriculture now refers broadly to the technology of raising plants and animals. **Horticulture** is that part of plant agriculture concerned with so-called garden crops, as contrasted with **agronomy** (field crops, mainly grains and forages) and **forestry** (forest trees and products). The relation of these disciplines to the rest of science and technology is portrayed in Figure 1-1.

Horticulture deals with an enormous number of plants. Garden crops traditionally include fruits, vegetables, and all the plants grown for ornamental purposes, as well as spices and medicinal plants. Many horticultural products are utilized in the living state and are thus highly perishable; constituent water is

[1]The first known use of the word *horticultura* is in Peter Lauremberg's treatise of that name, written in 1631. *Horticulture* is first mentioned in English by E. Phillips in *The New World of English Words*, London, 1678.

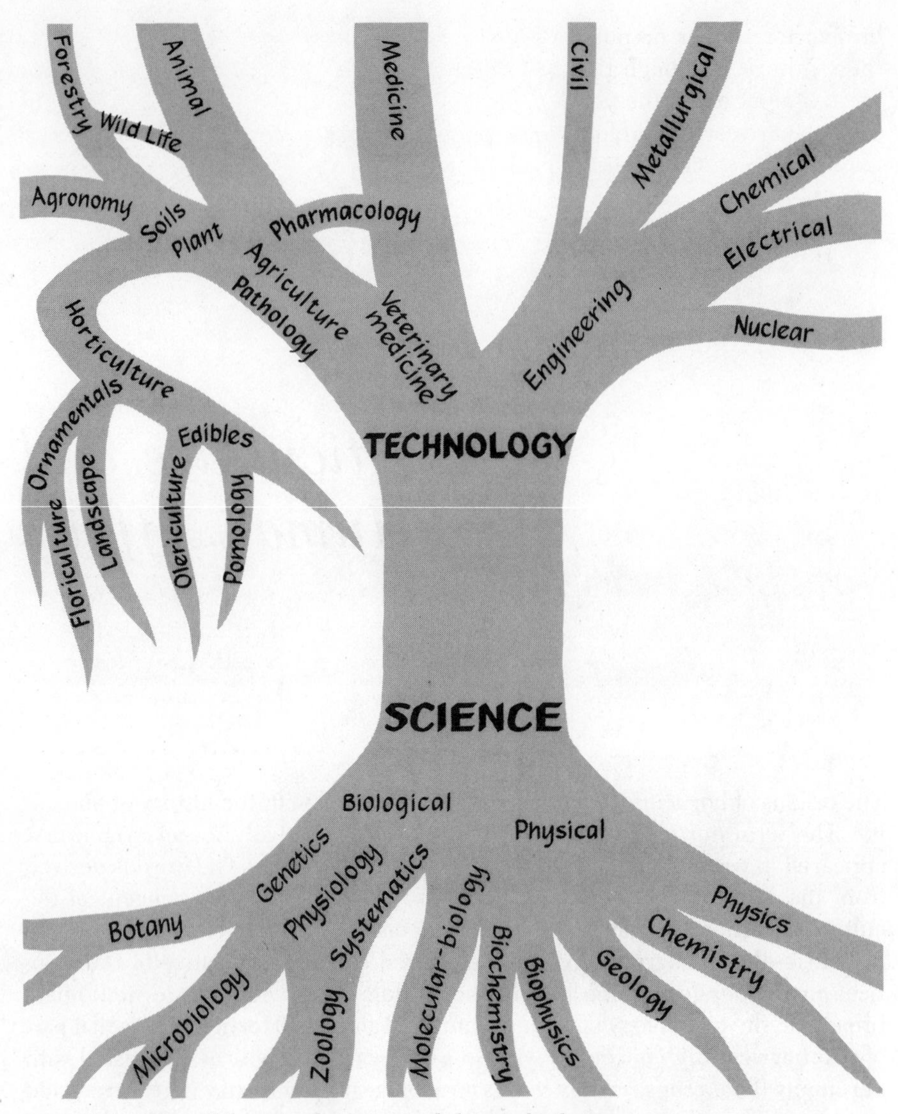

FIGURE 1-1. A tree of the natural sciences and their technologies.

essential to their quality. In contrast, the products of agronomy and forestry are often utilized in the nonliving state and usually contain a high percentage of dry matter. Custom has delineated the boundary line for some crops; for example, tobacco and, in some locations, potatoes are considered agronomic crops in the United States. In the main, however, horticulture deals with crops that are intensively cultivated—that is, plants that are of high enough value to warrant a large input of capital, labor, and technology per unit area of land. Pine trees grown for timber or pulp are an example of extensive forestry, rather than horticulture: Although the value per acre of pine stands may be large if harvested all at once, the yearly increment of value is relatively small (50 to 100 dollars per year). Therefore, the economic importance of our forest industries is largely a function of the tremendous area involved. When pine trees are grown

A

B

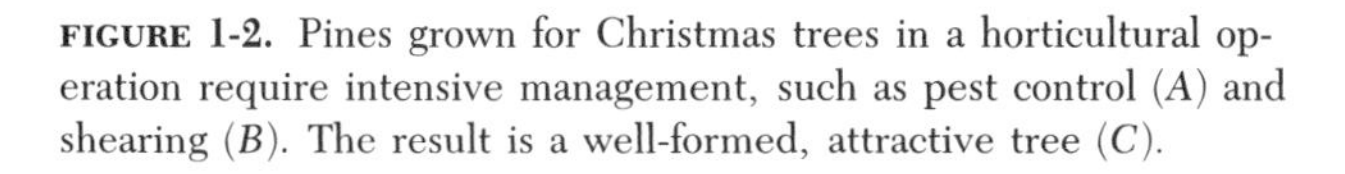

FIGURE 1-2. Pines grown for Christmas trees in a horticultural operation require intensive management, such as pest control (*A*) and shearing (*B*). The result is a well-formed, attractive tree (*C*).

C

more intensively, as are Christmas trees (Figure 1-2), they are often considered a horticultural crop. Pine trees grown as nursery stock for use in ornamental plantings assume sufficient value to justify large expenditures for fertilization, pruning, harvesting, and marketing, and are therefore a true horticultural crop. (The value of good nursery land may be equivalent to that of suburban real estate!) Similarly, the presence of the *sugary* gene in maize, an agronomic crop, increases its value enough to warrant using more intensive cultural methods (the use of better seed—always single-cross hybrids—spray programs, and programmed harvesting procedures) and transforms it into a horticultural crop, sweet corn.

Horticulture thus can be defined as the branch of agriculture concerned with intensively cultured plants directly used by people for food, for medicinal purposes, or for esthetic gratification. The industry is usually subdivided according to the kinds of products and the uses to which they are put. The production of edibles is represented by **pomology** (fruit crops) and **olericulture** (vegetable crops); the production of ornamentals is represented by **floriculture** (flowers and foliage plants) and **landscape horticulture** (trees, shrubs, and ground covers). These terms are not mutually exclusive. For example, many edible plants (apples) are used as ornamentals, and many plants often classed as ornamentals (poppy, pyrethrum) have pharmacological and industrial uses (Figure 1-3).

FIGURE 1-3. The dried flowers of pyrethrum (*Chrysanthemum cinerariaefolium*) are the source of a natural insecticide. Ornamental forms (*C. coccineum*) are known as painted daisies. [Photograph courtesy E. R. Honeywell.]

The esthetic use of plants is a unique feature of horticulture, distinguishing it from other agricultural activities. It is this aspect of horticulture that has led to its universal popularity. In the United States ornamental horticulture has undergone a renaissance brought about by an increased standard of living coincidental with the development of suburban living. The satisfaction of this bent in the American family has created an expanding industry out of ornamental horticulture, whose practice had formerly been confined to well-to-do fanciers.

Horticulture is an ancient art, and many of its practices have been empirically derived. However, modern horticulture—indeed, all modern agriculture—has become intimately associated with science, which has served not only to provide the methods and resources to explain the art but has also become the guiding force for its improvement and refinement. Horticulture will never become wholly a science, nor is this particularly desirable. Its curious mixture of science (botany to physics), technology, and esthetics makes horticulture a refreshing discipline that has continually absorbed people's interest and challenged their ingenuity. The rise of the world environmental movement in the 1960s and 1970s resulted in a virtual explosion of interest in plants and gardening, an interest that shows no sign of waning. The science of horticulture still remains the dynamic influence in the proper use and understanding of the horticultural art, and it is with this phase of horticulture that this book is largely concerned.

ECONOMIC POSITION OF HORTICULTURE

It is difficult to ascertain the precise position of any large and diverse industry in our economy. This is particularly true of horticulture, which involves not only the many facets of production, but the added increments of processing, service, and maintenance. For example, ornamentals such as woody perennials are not consumed but are invested in plantings, which increase in value with the passage of time. The value of this wealth is ordinarily not taken into consideration until we become painfully aware of it through the tolls taken by severe weather or through the encroachment of concrete and steel. The replacement of large trees and shrubs is usually economically prohibitive but not necessarily horticulturally impossible.

Agriculture is the largest industry in the United States, and commercial horticulture represents a significant portion of American agriculture. In the 1980s the total U.S. agricultural output provided the nation's farmers an aggregate gross income of over $150 billion per year (Figure 1-4). The upward spiraling of production inputs (machinery, fertilizers, pesticides) reflects the increasing technology inherent in today's agriculture.

Agriculture has increased in production to keep up with a growing population's demand for a high level of nutrition and a high standard of living. Hor-

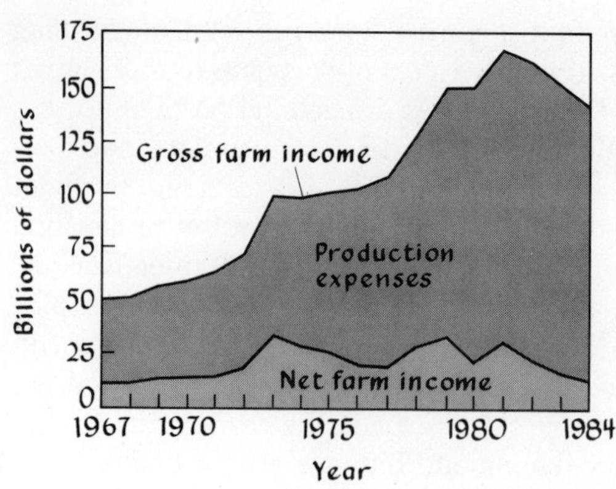

FIGURE 1-4. Gross income from agriculture in the United States. Note that production expenses have more than tripled from 1967 to 1984. In 1980 agricultural income to the farmer exceeded $150 billion. The personal consumption of food in 1982 in the United States was $349 billion ($255 billion at home and $94 billion away from home). Of this about $51 billion represents retail value of imported food. [Data from Economic Research Service, USDA.]

ticulture's percentage share in this expanding industry has been relatively stable over the past 75 years (Figure 1-5). In 1981, when horticulture represented 12.8 percent of total agricultural receipts, cash receipts were $3.5 billion for greenhouse and nursery crops, $6.5 billion for fruits and nuts, and $8.4 billion for vegetable crops, making a total horticultural value of $18.3 billion. The grower's share of the horticultural food dollar is about 24 percent, and this value is probably even lower for greenhouse and nursery crops. Thus, on the retail level horticulture represents at least a $68-billion industry. In the United States about 40 percent by weight of the food consumed consists of horticultural products. In view of this fact alone, one may expect horticulture to maintain its increasing importance in our lives and in our economy.

These figures should not imply that each facet of the horticultural industry has shared or will share equally in this increase. The fortunes of individual crops in the United States (Figure 1-6) reflect the changing habits and preferences of the American people and the technological changes in the food industry. For example, a trend in the consumption of fresh vegetables has been toward an increase in the per capita consumption of those vegetables used in salads (lettuce, tomato) and a decrease in the consumption of starchy root crops.

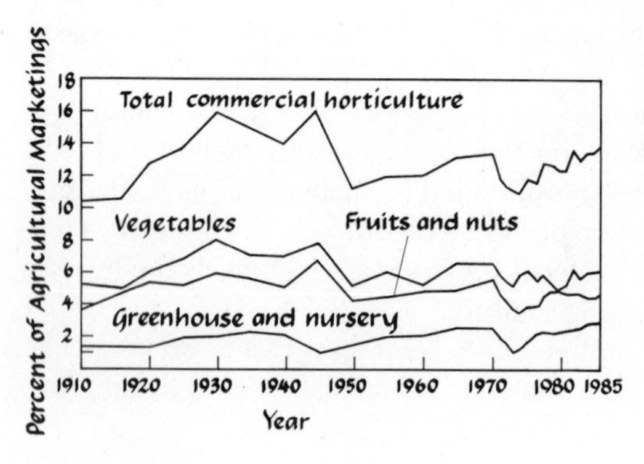

FIGURE 1-5. Horticulture's share of the total agricultural market in the United States from 1910 to the present has been relatively constant. [Data from Economic Research Service, USDA.]

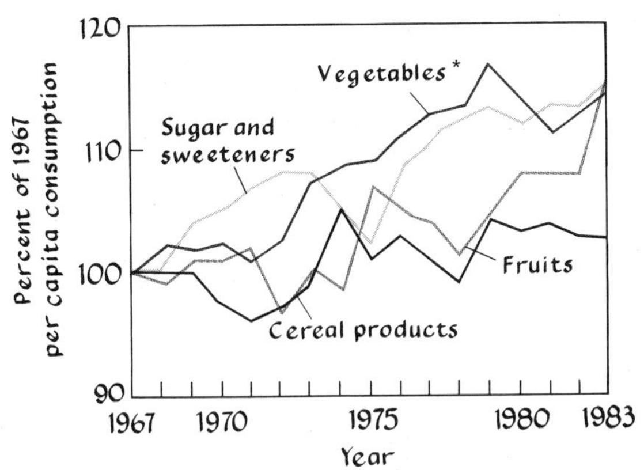

A

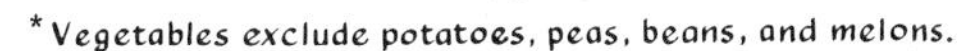

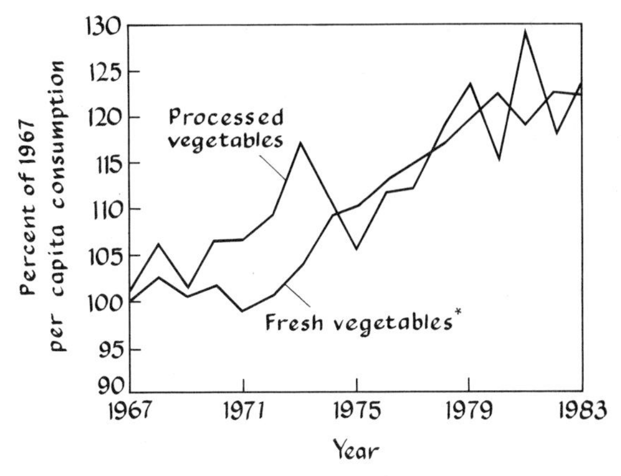

B *Excludes fresh potatoes, dried beans, and dried peas.

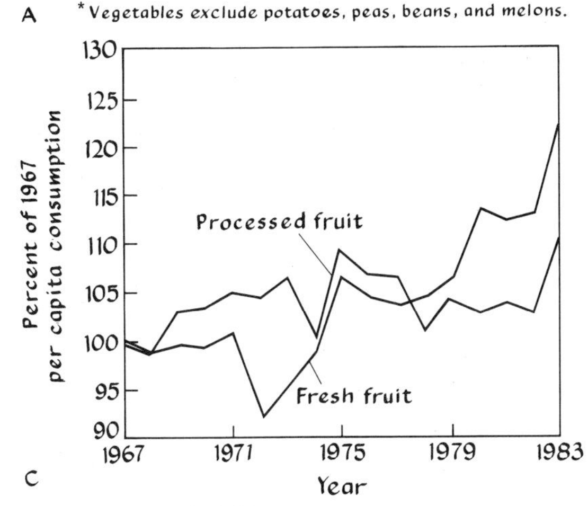

C

FIGURE 1-6. Changing patterns of vegetable and fruit consumption and production in the United States. Note, for example, the increasing per capita consumption of fruits and vegetables (*A*, *B*, *C*). The increase in sugar and sweeteners may be explained by the increasing popularity of soft drinks. Note also that the fortunes of different vegetables (*D*, *E*, *F*) varies greatly. Per capita potato consumption has stabilized at over 110 pounds with more than half attributed to processed potatoes (*G*) (*next page*). The increasing consumption of fruits is due both to increasing per capita consumption of citrus and noncitrus fruit (*H*, *I*). [Data from Economic Research Service, USDA.]

CHANGE IN PER CAPITA
FRESH VEGETABLE CONSUMPTION

Total pounds 1978-1980 Change since 1970-1972

Lettuce* 26.7 15%

Onions† 11.8 11%

Tomatoes 13.2 10%

Cabbage 8.8 -1%

Carrots 6.0 -3%

Sweet corn 7.3 -6%

* Lettuce includes escarole.
† Onions includes about 3 pounds of dehydrated onions.

D

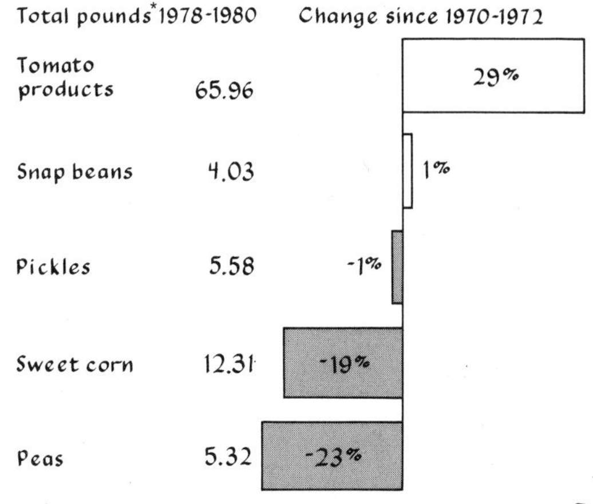

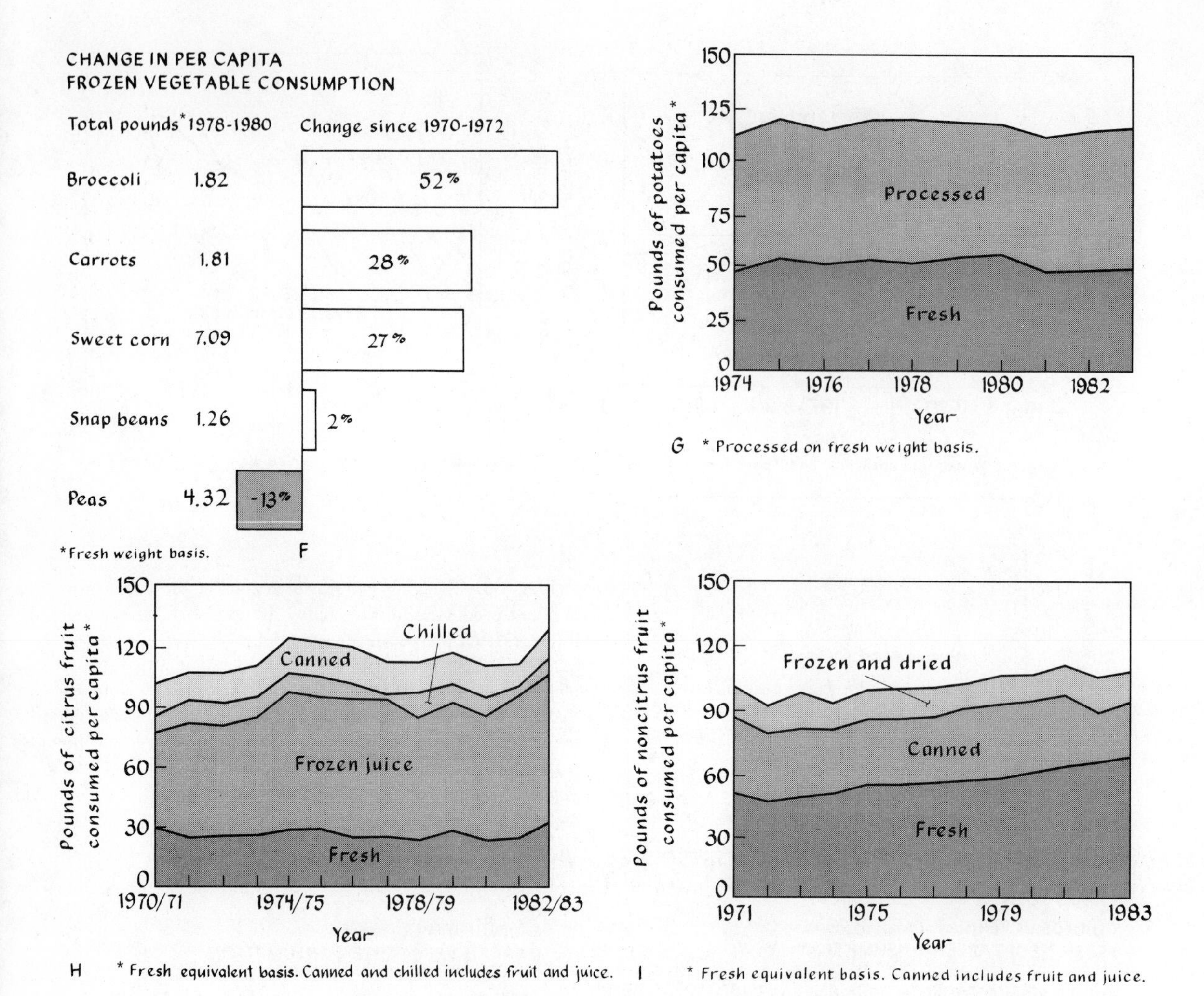

Consumption of potatoes has stabilized because of the increasing consumption of processed potatoes. Fruit consumption per capita has gradually increased, with citrus fruits gaining at the expense of apples. The consumption of processed fruits and vegetables has stabilized but, on a fresh-weight-equivalent basis, still accounts for more than half of the total consumed.

A HISTORICAL PERSPECTIVE

Along with the discovery of fire, the "invention" of agriculture represents the most significant achievement in human civilization. In primitive societies based on food gathering or hunting, each individual must be totally involved with the urgencies of securing sustenance. Notwithstanding the systematic and efficient

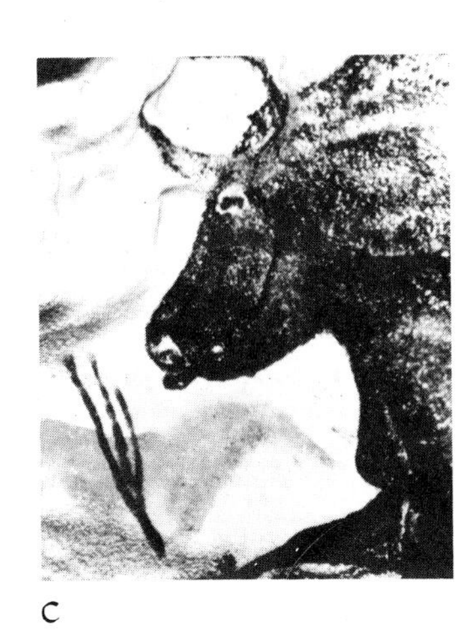

FIGURE 1-7. Hunting and gathering in the Stone Age. Upper Paleolithic cave paintings (*ca.* 30,000 to 15,000 B.C.E.) from Castellon, Spain, portray hunting of stags (*A*) and gathering of honey (*B*). A close-up of a beautifully wrought cave painting of a bison from Lascaux in southwestern France (*C*) shows what appears to be a representation of plants. Could this be the first plant illustration? [(*A*) and (*B*) from C. Singer, E. J. Holmyard, and A. R. Hall (eds.), *History of Technology,* vol. 1, Oxford University Press, 1954, courtesy The Metropolitan Museum of Art. (*C*) from P. J. Ucko and A. Rosenfeld, *Paleolithic Cave Art,* Weidenfeld & Nicolson, 1967.]

organization of certain food-gathering societies, each adult is pressed into continual activity. Abundance proves to be temporary and exceptional. The limiting factor in the development of primitive societies becomes the availability and dependability of a food supply.

People have been food collectors for the great part of human existence (Figure 1-7). Food production by the cultivation of edible plants and the domestication of animals is of relatively recent origin, dating back 7000 to 10,000 years to what is known as the Neolithic Age. Only through the gradual development of a system of agriculture could the accumulation of surpluses become a regular occurrence. The immediate reward of a surplus is the release from food production of people who can contribute to society in other ways: The gradual development of agriculture, with its increased efficiency and dependability, encouraged the development of other new classes of specialists—artisans, clerks, priests. The standard of living of a people increases as the activities of its specialists increase. Table 1-1 broadly dates the progress of civilization in terms of advances in agriculture and horticulture.

Selection of Edible Plants

The origin of civilization can be traced to the discovery that it is possible to assure oneself of a plentiful food supply by planting seed. Rapidly growing vegetables and cereals, which produce a crop within a season, must have been

TABLE 1-1.
Dating the past.

Developments in horticulture and plant agriculture	Years ago	Historical era
Gene transfer	Today	
Gene synthesis		
Genetic code deciphered		Space exploration
Structure of the gene (DNA) discovered		
Isolation of phytochrome		
Mechanical harvesting		
Plastic films		
Radiation preservation		
Polyploid and mutation breeding		
Organic phosphate pesticides		
Chemical weed control		
Tissue culture		Controlled release of atomic energy
Auxin research		
Respiration cycle discovered		
Plant virus studies		
Hybrid corn		
Photoperiodism discovered		
Gasoline-powered tractor		
Concept of essential elements		Successful air flight
Bailey's *Cyclopedia of Horticulture*		
Plant nutrition investigations	100	
Beginnings of agricultural chemistry		
U.S. Agricultural Experiment Stations		
Morrill (Land Grant College) Act		
Mendel discovers laws of heredity		American Civil War
Origins of plant pathology		
Modern plow		
Reaper		
Canning of food		American Revolution
Gardens of Versailles		
Discovery of microscope		
Importation of plant species		Discovery of America
Rebirth of botanical sciences		Beginnings of modern science
Monastery gardens	1000	Norman conquest of England
Herbals		Dark Ages
Roman gardens		Roman civilization
Legume rotation		Birth of Christ
Botanical works of Theophrastus		Golden Age of Greece
Fruit cultivars		
Hanging gardens of Babylon		
Grafting		
Irrigation		Egyptian civilization
Domestication of crop plants		
Beginnings of agriculture	10,000	Neolithic Age

A B C

FIGURE 1-8. Domestication of crop plants in the New World. These pre-Columbian ceramic jars, which are about 3000 years old, combine human and animal forms with the forms of plant crops: (*A*) squash, (*B*) potato, (*C*) peanut. [From the University Museum, University of Pennsylvania.]

the first plants cultivated (Figure 1-8). The technology involved in cultivating nut or fruit trees, for example, is considerably intricate and time consuming; as a result, these edibles were probably gathered from the wild. Even today in the United States some food is gathered from indigenous uncultivated plants. For example, Maine's blueberry industry depends upon such a source. The cultivation of cereals and the domestication of animals led to a permanent agriculture, which provided the medium for the growth of an advanced civilization. Although it is not known when the cultivation of plants first took place, it is known that the bulk of our present-day food plants were selected by the people of many lands long before recorded history. The ways in which wild plants were transformed to their present cultivated forms are often obscure, and the original ancestors of many of our crop plants cannot be traced. The same, of course, is true for our domestic animals. Our debt to primitive peoples is enormous.

Somewhere in the now dry highlands of the Indus, Tigris, Euphrates, or Nile Rivers, the technology we call agriculture was conceived. In about 3000 years, the primitive existence of Stone Age peoples was transformed to the full-fledged urban cultures of Egypt and Sumeria. By this time the date, fig, olive, onion, and grape, the backbone of ancient horticulture, had been brought under cultivation, and the technological base necessary to ensure a productive agriculture—land preparation, irrigation, and pruning—had been discovered. Five thousand years ago, field cultivation with a plow pulled by oxen had replaced cultivation with the hoe; by the time of the flowering of Egyptian culture, agriculture and horticulture were established disciplines (Figure 1-9).

FIGURE 1-9. Ancient Egyptian horticulture as depicted by its contemporary artists. (*A*) A primitive hoe (*right*) made from a forked branch and a more developed form (*left*) with hafted wooden blade (Middle Kingdom, Egypt, 2375 to 1800 B.C.E.). (*B*) Land preparation and cultivation (from a tomb at Beni Hasan, *ca.* 1900 B.C.E.). (*C*) Agricultural implements and land reclamation (from a tomb at Thebes, *ca.* 1420 B.C.E.). (*D*) Irrigating a vegetable garden (*ca.* 1900 B.C.E.). (*E*) Harvesting figs (from a tomb at Beni Hasan, *ca.* 1900 B.C.E.). (*F*) Irrigating a date palm using a *shaduf*—a mechanical aid that uses a pole weighted on one end with a clay counterweight (from a tomb at Thebes, *ca.* 1500 B.C.E.) (*G*) Harvesting grapes from an arbor and treading to express the juice (from a tomb at Thebes, *ca.* 1500 B.C.E.). (*H*) Formal gardens containing doum palms, data palms, acacias, and other trees and shrubs (from a tomb at Thebes, *ca.* 1450 B.C.E.). [(*A*) to (*F*) and (*H*) from C. Singer, E. J. Holmyard, and A. R. Hall (eds.), *History of Technology*, vol. 1, Oxford University Press, 1954, courtesy The Metropolitan Museum of Art. (*G*) from M. Gothein, *A History of Garden Art*, vol. 1, Dutton, 1928.]

E

F

G

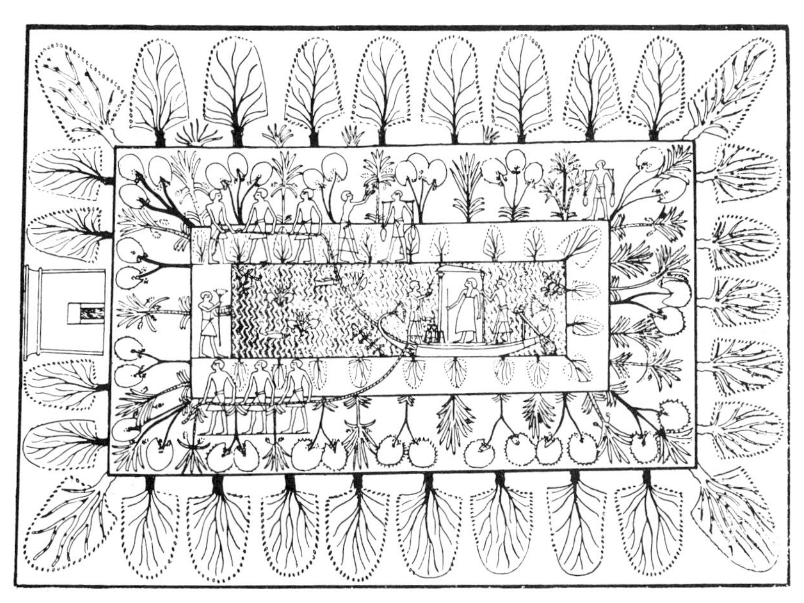

H

A

FIGURE 1-10. Gathering lilies (*A*) and expressing their oil for perfume (*B*) in Ancient Egypt (*ca.* 400 B.C.E.). [From C. Singer, *History of Technology*, vol. 1, Oxford University Press, 1954.]

Egypt and the Fertile Crescent

Roughly 7000 to 8000 years ago, the people living in the Nile valley, which had been inhabited for at least 20,000 years, developed the beginnings of agriculture. In Egypt, humanity's rendezvous with civilization began. By 3500 B.C.E. (before the common era) Egypt had established a centralized government with Memphis as its capital, and, by 2800 B.C.E., it had developed to a high enough level of civilization to support such monumental engineering projects as the pyramids.

The great accomplishment of Egyptian agriculture was systematic irrigation through hydraulic engineering. Among the notable horticultural achievements of the Egyptians were their formal gardens, which were complete with pools and were cared for continuously by gardeners; the creation of a spice and perfume industry (Figure 1-10); and the development of a **pharmacopoeia**—a collection of drug and medicinal plants (Figure 1-11).

The Egyptians cultivated a great number of fruits, including date, grape, olive, fig, banana, lemon, and pomegranate, as well as such vegetables as cucumber, artichoke, lentil, garlic, leek, onion, lettuce, mint, endive, chicory, radish, and various melons. In addition, many other plants (such as papyrus, castor bean, and date palm) were cultivated for fiber, oil, and other "industrial" uses. The Egyptians also developed the various technologies associated with the food industry—baking, fermenting, drying, and pottery making.

B

Egyptian influence lasted an incredible 35 centuries. The Egyptians produced, in addition to their great technology, a magnificent art and a complex, if bewildering, theology. By the time Egypt had become a Roman province (3 B.C.E.), its influence had greatly permeated the ancient world—an influence that is still felt today.

East of Egypt, the ancient cultures of Mesopotamia—Babylonia and Assyria—added to the technology of horticulture such innovations as irrigated

FIGURE 1-11. Plants and seeds brought back from Syria by Thothmes III, as carved on the temple walls at Karnak, Egypt (about 1450 B.C.E.). [From C. Singer, *History of Technology*, vol. 1, Oxford University Press, 1954.]

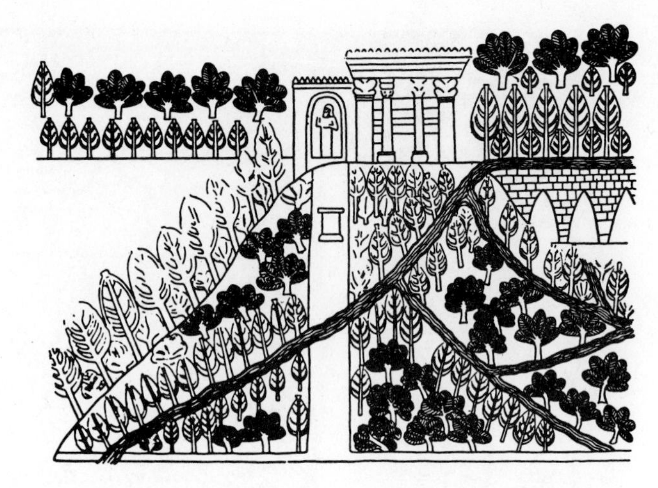

FIGURE 1-12. A royal Assyrian park, watered by streams from an arched aqueduct. From a relief in the palace at Nineveh (seventh century B.C.E.). [From C. Singer, *History of Technology*, vol. 1, Oxford University Press, 1954.]

terraces, gardens, and parks (Figures 1-12 and 1-13). Irrigation canals lined with burnt brick with asphalt-sealed joints helped keep 10,000 square miles under cultivation, which in 1800 B.C.E. fed more than 15 million people. By 700 B.C.E., an Assyrian herbal was compiled that contained the names of more than 900 plants, including 250 vegetable, drug, and oil crops. By the second century B.C.E., a treatise on agriculture consisting of 28 volumes was compiled by Mago, who lived in the ancient city-state of Carthage in North Africa. The Old Testament prophecy that all nations "shall beat their swords into plowshares and their spears into pruning hooks" (Isaiah 2:4, *ca.* 800 B.C.E.) attests also to the importance of agriculture and horticulture in that period.

Greece

The Greeks, while adding only modestly to practical agriculture (Figure 1-14), did devote considerable attention to botany. Although the botanical writings of Aristotle (384–322 B.C.E.) are lost, a portion of the writings of his most famous

FIGURE 1-13. Darius hunting in a grove of palms (fourth century B.C.E.?). [From M. Gothein, *History of Garden Art*, vol. 1, J. M. Dent & Sons, 1928.]

FIGURE 1-14. Harvesting olives. From a Greek black-figured vase (sixth century B.C.E.). [From C. Singer, *History of Technology*, vol. 2, Oxford University Press, 1958.]

student, Theophrastus of Eresos, still exists. These writings cover various botanical subjects, including taxonomy, physiology, and natural history. His books, *History of Plants* and *Causes of Plants*, treat such topics as classification, propagation, geographic botany, forestry, horticulture, pharmacology, viticulture, plant pests, and flavors and odors. Theophrastus, now referred to as the Father of Botany, influenced botanical thinking until the seventeenth century—which reflects both the genius of the Greeks and the stagnation of science that was to come about in the Dark and Middle Ages.

Later Hellenistic botanists included Attalos III of Pergamon (king from 138 to 133 B.C.E.); and, in the following century, Mithridates VI of Pontos, who was much interested in toxicology and poisonous plants; Cratevas (Krateus), physician to Mithridates and author of a treatise on roots as well as an illustrated herbal; and Nicholas of Damascus, author of the great botanical treatise *De Plantis*. Although justifiably best known for their botanical studies, the Greeks also wrote on agriculture. The Roman soldier-encyclopedist Varro (116–27 B.C.E.) lists more than 50 Greek authorities for his agricultural treatise *Res Rusticae*. The first-century herbal of Pedanius Dioscorides (*De Materia Medica*) remained the standard botanical-medical reference up to medieval times (see Figure 2-14). Every botanist up to the sixteenth century was either a translator of, or a commentator on, Dioscorides.

There are some who trace the downfall of Greece not primarily to wars or internal decay but to a gradual erosion brought about by the effects of increasing population on declining resources. The loss of a sound agricultural base was due to a combination of factors, among which were shallow soil, poor conservation practices, and perhaps the reluctance of the Greek mind to seriously consider the mundane problems of agriculture. With this loss, the urban centers of Greece, weakened by the intercity struggles and rivalries that culminated in the 27-year Peloponnesian Wars, eventually crumbled before the armies of Macedonia.

Rome

The enigma of the thousand-year history of Rome (roughly 500 B.C.E. to 500) is not that Rome fell but that it held together as long as it did, in spite of its being a conglomeration of diverse peoples and lands. Unlike the Greeks, the Romans were extremely interested in practical agriculture. It was a vital part of the economy; the largest single group of producers was agricultural. On this firm foundation, Rome rose to glory and, according to some, declined along with its eroded soils.

Roman agriculture had many chroniclers. The earliest was Cato (234–149 B.C.E.), who wrote of farming and useful gardening in his *De Agri Cultura*. Varro, cited earlier, produced a longer commentary on agriculture, which was followed by Columella's 12-book treatise (*ca.* 50). It is from these sources, and from the *Georgics* of Virgil (70–19 B.C.E.), the writings of Pliny the Younger (62–116), and especially the *Natural History* and *Palladius* of Pliny the Elder (23–79), that the agrarian history of Rome has been reconstructed.

The Romans were great borrowers. Although they produced little that was new, they did make great improvements upon what they borrowed. The horticultural technology used by the Romans (much of which can be traced to earlier sources, especially Greece and Egypt) was refined, codified, and made more practical. Their agricultural writings mention grafting and budding, the use of many kinds of fruits and vegetables, legume rotation, fertility appraisals, and even cold storage of fruit. Mention can be found of a prototype greenhouse, or *specularium*, constructed of mica and used for vegetable forcing.

In many respects, the well-to-do Roman was a modern type, civilized and urbane, yet the Roman remained bound to the land through business. His problems were largely managerial ones: the care and handling of slaves, the management of income properties, the vagaries of profit and loss. The typical Roman—soldier, farmer, voluptuary—is strikingly similar to a recent counterpart—namely, the aristocrat of the antebellum South.

It was in Rome that ornamental horticulture was first developed to a high level. From the beginnings of Roman history hereditary estates ranging in size from 1 to 4 acres were referred to as gardens (*hortus*) rather than farms (*fundus*). Early Rome has been described as a market place serving a hamlet of truck gardeners. With the conquest of new lands came the development of large slave plantations, which eventually led to free tenancy and estates, and finally to a manorial system. The great fortunes of Rome were invested in farm land. The good life was that of a gentleman farmer; the sign of wealth was the country estate.

The dwelling on an estate reflected the wealth of its owner. The prosperous Roman had a little place in the country, a *suburbanum*. It included fruit orchards, in which grew apples, pears, figs, olives, pomegranates, and flower gardens, with lilies, roses, violets, pansies, poppies, iris, marigolds, snapdragons, and asters. The mansions of the wealthy were quite splendid. Formal gardens were enclosed by frescoed walls and were amply endowed with statuary and fountains, trellises, flower boxes, shaded walks, terraces, topiary ("bush sculpture"), and even heated swimming pools. The rule was luxury; the desired effect was extravagance.

Rome was largely a parasitic empire based on borrowed culture, slave labor, and stolen goods. This was not destined to last, however, for in the fifth century, the Roman Empire disintegrated, and Europe took a giant step backward to the village.

Horticulture and Classical Antiquity

Our cultural heritage in art, literature, and ethics is largely traceable to Greek and Roman influences. According to the great historian of science Charles Singer, this has resulted in an overemphasis on the importance of the technology of these cultures, which tended to be lower than that of the more ancient cultures of Egypt and Mesopotamia, from which they were derived. The rise of both Greece and Rome was similar in some respects to the rise of the Huns and the Goths—a victory of vigorous barbarism over worn-out but advanced cultures. The story of civilization and technology is not the steady upward climb of the past 600 years. The technology of horticulture is a good example of this. One is hard pressed for examples of progress made by the Greeks or Romans that are comparable to those of ancient Egypt. Significant advance was to await the Renaissance.

A specific example of this delay concerns the state of knowledge on the role of sex in plants. The cultivation of the date palm and fig in Mesopotamia clearly shows that Mesopotamian horticulturists understood the function of the nonbearing staminate plants of the date palm (Figure 1-15) and that they recognized the principle underlying caprification of the fig (the use of the wild caprifig, which shelters a parasitic wasp that carries pollen from the caprifig to the pistillate blossoms of the edible fig). Theophrastus was aware of this ancient concept, but this information was virtually lost until the Dutch botanist Camerarius in 1694 experimentally proved the sexual nature of plants. Similarly, the ancient arts of graftage and irrigation, part of the basic technology of horticulture, were not improved until very recently. Advances in technology that came about during the early medieval period are largely traceable to Eastern sources—China, Islam, and the Byzantine Empire.

Medieval Horticulture

After the fall of Rome, during the so-called Dark Ages, some horticultural art and technology survived in monastic gardens. Gardening became an integral part of monastic life, providing food, ornament, and medicines. Many fruit and vegetable strains were preserved, and some of them were even improved. The gardener (*hortulanus* or *gardinarius*) became a regular officer of the monastery.

FIGURE 1-15. Winged guardian spirit of the Assyrians pollinating blossoms of the date palm. From the Palace of King Ashur-Nasir-Pal II at Nimrud, Iraq (ninth century B.C.E.). [Courtesy Museum of Fine Arts, Boston.]

The few botanical and horticultural writings of this period were, for the most part, compilations, and most were based largely on Pliny's *Natural History*. Many centuries passed before the horticultural technology of the Romans was equaled.

Revived interest in the art of horticulture began in Italy with the Renaissance. As feudalism gave way to trade, producing a real rise in the standard of living, garden culture again began to be practiced widely. By the thirteenth and fourteenth centuries, orchards and gardens were common outside of monasteries (Figure 1-16). As meat became more important in the medieval diet, gardens became important as a source of spices and condiments. The horticultural revival spread from Italy to France and then to England. The *Maison Rustique*, published by Charles Estienne (1504–1564) and John Liebault, is a delightful sourcebook of late medieval horticulture. The section on the apple, quoted in the following paragraphs,[2] illustrates such practices as fertilization, graftage, pruning, breeding, dwarfing, transplanting, insect control, girdling to promote flowering, harvesting, processing, and culinary and medicinal utilization.

[2]From *Maison Rustique, or, The Countrey Farme; Compiled in the French Tongue by Charles Stevens (Estienne), and John Liebault, Doctors of Physicke* (1616 edition, augmented by Gervase Markham).

FIGURE 1-16. Representation of a medieval garden.

OF THE APPLE-TREE

The Apple-tree

The Apple-tree which is most in request, and the most precious of all others, and therefore called of *Homer*, the Tree with the goodly fruit, groweth any where, and in as much as it loveth to have the inward part of his wood moist and sweatie, you must give him his lodging in a fat, blacke, and moist ground; and therefore if it be planted in a gravelly and sandie ground, it must be helped with watering, and batling with dung and smal mould in the time of Autumne. It liveth and continueth in all desireable good estate in the hills and mountaines where it may have fresh moisture, being the thing that it searcheth after, but even there it must stand in the open face of the South. Some make nurceries of the pippins sowne, but and if they be not afterward removed and grafted, they hold not their former excellencie: it thriveth somewhat more when it is set of braunches or shoots: but then also the fruit proveth late and of small value: the best is to graft them upon wild Apple-trees, Plum-trees, Peach-trees, Peare-trees, Peare-plum-trees, Quince-trees, and especially upon Peare-trees, whereupon grow the Apples, called Peare-maines, which is a mixture of two sorts of fruits: as also, when it is grafted upon Quince-trees, it bringeth forth the Apples, called Apples of Paradise, as it were sent from heaven in respect of the delicatenesse of their cote, and great sweetnesse, and they are a kind of dwarffe Apples, because of their stocke the Quince-tree, which is but of a smal stature.

The Apple loveth to be digged twice, especially the first yeare, but it needeth no dung, and yet notwithstanding dung and ashes cause it prosper better, especially the dung of Sheepe, or for lesse charges sake, the dust which in Sommer is gathered up in the high waies. You must many times set at libertie the boughes which intangle themselves one within another; for it is nothing else but aboundance of Wood, wherewith it being so replenished and bepestred, it becometh mossie, and bearing lesse fruit. It is verie subject to be eaten and spoyled of Pismires and little wormes, but the remedie is to set neere unto it the Sea-onion: or else if you lay swines dung at the roots, mingled with mans urine, in as much as the Apple-tree doth rejoyce much to be watered with urine. And to the end it may beare fruit aboundantly, before it begin to blossome, compasse his stocke about, and tie unto it some peece of lead taken from some spout, but when it beginneth to blossome, take it away. If it seeme to be sicke, water it diligently with urine, and to put to his root Asses dung tempered with water. Likewise, if you will have sweet Apples, lay to the roots Goats dung mingled with mans water. If you desire to have red Apples, graft an Apple-tree upon a blacke Mulberrie-tree. If the Apple-tree will not hold and beare his fruit till it be ripe, compasse the stocke of the Apple-tree a good foot from the roots upward, about with a ring of a lead, before it begin to blossome, and when the apples shall begin to grow great, then take it away.

Gathering of Apples

Apples must be gathered when the moone is at the full, in faire weather, and about the fifteenth of September, and that by hand without any pole or pealing downe: because otherwise the fruit would be much martred, and the young siences broken or bruised, and so the Apple-tree by that meanes should be spoyled of his young wood which would cause the losse of the Tree. See more of the manner of gathering of them in the Chapter next following of the Peare-tree: and as for the manner of keeping of them, it must be in such sort as is delivered hereafter.

You shall thaw frozen Apples if you dip them in cold water, and so restore them to their naturall goodnesse. There is a kind of wild Apple, called a Choake-apple, because they are verie harsh in eating, and these will serve well for hogges to eat. Of these apples likewise you may make verjuice if you presse them in a Cyderpress, or if you squeese them under a verjuice milstone.

Vinegar

Vinegar is also made after this manner: You must cut these apples into gobbets, and leave them in their peeces for the space of three dayes, then afterward cast them into a barrell with sufficient quantitie of raine water, or fountaine water, and after that stop the vessell, and so let it stand thirtie daies without touching of it. And then at the terme of those daies you shall draw out vinegar, and put into them againe as much water as you have drawne out vinegar. There is likewise made with this sort of Apples a kind of drinke, called of the Picardines, Piquette, and this they use in steed of Wine. Of other sorts of Apples, there is likewise drinke made, which is called Cyder, as we shall declare hereafter.

An Apple cast into a hogshead full of Wine, if it swim, it sheweth that the Wine is neat: but and if it sinke to the bottome, it shewes that there is Water mixt with the Wine.

Neat wine
Mingled wine

Infinit are the sorts and so the names of Apples comming as well of natures owne accord without the helpe of man, as of the skill of man, not being of the race of the former: in everie one of which there is found some speciall qualitie, which others have not: but the best of all the rest, is the short shanked apple, which is marked with spottings, as tasting and smelling more excellently than any of all the other sorts. And the smell of it is so excellent, as that in the time of the plague there is nothing better to cast upon the coales, and to make sweet perfumes of, than the rinde thereof. The short stalked Apple hath yet furthermore one notable qualitie: for the kernells being taken out of it, and the place fill up with Frankincense, and the hole joyned and fast closed together, and so rosted under hot embers as that it burne not, bringeth an after medicine or remedie to serve when all other fayle, to such as are sicke of a pleurisie, they having it given to eat: sweet apples doe much good against melancholicke affects and diseases, but especially against the pleurisie: for if you roast a sweet apple under the ashes, and season it with the juice of licorice, starch and sugar, and after give it to eat evening and morning two houres before meat unto one sicke of the pleurisie, you shall helpe him exceedingly.

The rise of landscape architecture is one indication of the values of the Renaissance. Gardening became formalized. The design of gardens became as important as the design of the structures they surrounded (Figure 1-17). The peak of Renaissance horticulture is to be found in the magnificent gardens of LeNôtre (1613–1700), the most notable being those at the Palace of Versailles, which was built for Louis XIV. Building this prodigious chateau and landscaping the countryside around it took approximately 25 years and employed thousands of workers. The gardens required the engineering of tremendous irrigation projects capable of supplying as many as 1400 jets of water. In one year (1688) 25,000 trees were purchased. Ancient gardens had been more than equaled; modern gardens have not surpassed it.

The New World

The discovery of the New World in 1492—a convenient if inaccurate date to assign to the beginning of the modern age—was inspired by a search for a new route to the spice-rich Orient. The early conquistadores found, in the mixture

FIGURE 1-17. Astronomical observatory (Arcis Uraniburgi) of Tycho Brahe (1546–1601) at the Danish island of Hveen. [Courtesy Oliver Dunn, from Joan Blaeu, *Grooten Atlas*, 1664–1665.]

of advanced and Stone Age cultures that they encountered in the New World, an agricultural technology that was to have a profound influence on the history of the rest of the world (Figure 1-18). The horticultural contributions of the New World include many new vegetables (maize, potato, tomato, sweetpotato, squash, pumpkin, peanut, kidney bean, and lima bean); fruits and nuts (cranberry, avocado, Brazil nut, cashew, black walnut, pecan, and pineapple); and other important crops (chocolate, vanilla, wild rice, chili, quinine, cocaine, and tobacco). Primitive people in the New World had brought under cultivation practically all of the indigenous plants we now use.

The discovery of America was the most spectacular result of the era of exploration. The broadening of trade routes greatly stimulated horticultural progress. The transplantation of plant species from the Old World to the New World, and vice versa, marks the beginning of our great horticultural industries. The bulb industry in Holland, the cacao industry of Africa, and the banana and coffee industries of Central and South America can all be traced to those importations of plant species.

A

B

FIGURE 1-18. New World agriculture, from an illustrated calendar by Felipe Guamán Poma de Ayafa—a Peruvian of Incan-Spanish ancestry—who drew it in the 1580s to present to the King of Spain as part of a treatise on Inca life. The basic implements were a hand hoe (*A*) and a foot hoe (*taclla*) (*B*).

The Beginnings of Experimental Science

Science is a method of inquiry whose aim is the organization of information. In the ancient world, technology was the parent of science; only recently has science become the parent of technology. The speculative use of information built up through observation and experimentation became a powerful force during the Middle Ages. The discoveries of da Vinci (1452–1519), Galileo (1564–1642), and Newton (1642–1727) in astronomy and the physical sciences represent the flowering of this "new" development.

Renewed interest in botanical studies accompanied the revival of learning, as evidenced by the increasing appearance of herbals—descriptive treatises of plant material (Figure 1-19). The development of an appetite for scholarly works coincided with the European invention, in the middle of the fifteenth century, of printing from movable type (printing from wood blocks was known earlier in China), which had tremendous impact on the spread of ideas.

As a result of this approach, the seventeenth century saw a rebirth of botanical studies. Fundamental studies in plant anatomy and morphology were initiated by Marcello Malpighi (1628–1694) and Nehemiah Grew (1641–1712). The discovery of "cells" in cork by Robert Hooke (1635–1703) initiated the study of cytology, which was destined to reunite botany and zoology. The roots of genetics can be traced to the experimental studies of the Dutch botanist Rudolph Jacob Camerarius (1665–1721), who demonstrated sexuality in plants, and to the later hybridization experiments of J. G. Koelreuter (1733–1806). Interest in plant classification was revived, and, as the number of known plants increased, a series of attempts were made to formulate a workable system of classification.

FIGURE 1-19. Title page of Thomas Johnston's 1633 edition of John Gerard's *Herball* engraved by John Payne. John Gerard (*lower center*) holds a potato stem in flower and fruit. Homage is paid to the Roman goddess Ceres (growing vegetation) and Pomona (fruit trees) along with the "ancients": Theophrastus of Eresos (*ca.* 372–288 B.C.E.), "Father of Botany," and Pedanius Dioscorides of Anazarbos, who flourished during the second half of the first century and whose *Re materia medica* was the authoritative work on plant medicinals for 1500 years. The Latin quotation from Genesis 1:29 is translated as "Behold, I have given you every herb bearing seed." [From B. Henrey, *British Botanical and Horticultural Literature Before 1800*, Oxford University Press, 1975.]

It remained for Linnaeus (1707–1778) to develop a workable method based on the structure of reproductive parts. The beginnings of plant physiology were stimulated by Harvey's discovery of blood circulation in animals in 1628. Not until the eighteenth century, however, were fundamental studies in physiology undertaken, such as the investigations by Stephen Hale (1677–1761) of the movement of sap and the work of Joseph Priestly (1733–1804) on the production of oxygen by plants.

The history of the plant sciences becomes meaningful only when the significance of the fundamental discoveries is understood. The history of botany and experimental horticulture in the last 150 years is in a sense the subject of the following chapters. Similarly, the modern history of the horticultural industry, which represents the accumulated technology and science of many lands, cannot be stated briefly. For the study of this, the reader must investigate particular crops and particular countries. This subject will be discussed briefly, however, in Part IV, "Horticultural Industry."

Influence of Twentieth-Century Technology

The most remarkable feature of agriculture in the United States today is that the increased production of the last 75 years has taken place in spite of a decreasing acreage and a shrinking farm population. In 1910, each farmer produced for himself and 7 others; in 1982 each farmer "supported" 78 people (Figure 1-20). This increase in efficiency is the result of improved technology. The magnitude of this increase in efficiency has been such that many of the recent agricultural problems in the United States are a result of overproduction.

Technological change can be defined as the change in production resulting directly from the use of new knowledge. In general, technological change is measured by the change in output (production) per unit input of land, labor, and capital within a given period of time. The rate of technological change is not constant, because scientific progress and its application to agriculture do not ordinarily occur as a steady flow.

In agriculture, technological change may produce savings in either labor or capital. The classic agricultural inventions—the plow, the reaper, and the tractor—have reduced labor. They substitute capital for labor, which does not necessarily increase yield *per se*. The technological improvement of labor-saving devices results in capital savings. Technological changes as a result of genetic gain, improved nutrition, or irrigation bring about an increase in yield per unit of land and become capital improvements, saving land and reducing expenses.

The increase in United States agricultural production from 1880 to 1920 was largely a reflection of the increase in expenditures—more land and more

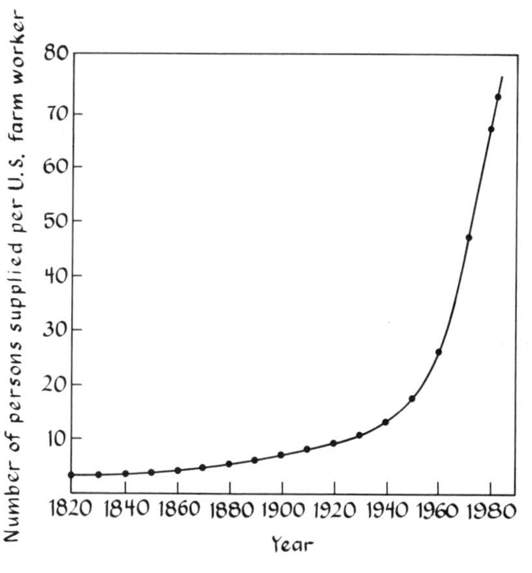

FIGURE 1-20. The number of persons supplied by each U.S. farm worker. In 1820 each farm worker supplied himself and 3 other persons; in 1980 the farm worker supplied himself and 75 other people: 51 in the United States and 24 abroad. Note the dramatic increase since 1940. [Data from Economic Research Service, USDA.]

labor. Returns per acre remained relatively constant. The replacement of horse power by the gasoline engine in the 1920s was the first great step in the twentieth-century scientific revolution. Hundreds of thousands of acres formerly used for feed for draft animals were made available for other purposes. As a result, farm acreage and the farm labor force, which had been expanding before that time, were stabilized during the decade after World War I.

The proliferation of genetic investigations in the quarter century following the rediscovery in 1900 of the revolutionary paper on inheritance by Gregor Mendel (1822–1884) began to yield technological advances in the form of improved plant cultivars. The development of hybrid maize was the most spectacular of these achievements. These improved genetic stocks in combination with the increased use of inorganic fertilizers accounted for a large part of the tremendous increase in production necessitated by World War II.

In the late 1940s agriculture was stimulated by a whole new set of technological advances made possible by basic research in the preceding decades. Agricultural chemicals in the form of weed killers, organic fungicides, and organic insecticides quickly followed the spectacular commercial success of the broadleaf weed killer, 2,4-D. The effects of these improvements were to increase yields per unit area as well as to conserve labor. Mechanization increased in the 1950s to include even "chore" jobs; and recently, supplemental irrigation in the eastern United States (which, of course, has long been used in the West) has permitted the use of additional fertilizer. The net result has been a steady increase in agricultural efficiency. This trend does not appear to be changing (Figure 1-21).

The horticultural industry has followed this trend in agriculture very closely. The average yield per acre of the California tomato processing industry provides a striking example of this pattern. Technological improvements in the form of genetic gain in addition to improved fertilization and irrigation practices have more than quadrupled the average yield per acre within four decades, as shown in Figure 1-22.

In the late 1950s as a result of the political fallout from Sputnik I, the first space satellite launched by the Soviet Union, the U.S. federal government began to support basic scientific research on a very large scale. Although agricultural research was not the major recipient of research funds, increases in agricultural research were carried along by the rising tide of generally increased research fundings. Much general biological research contributed to the rapid advances in understanding the actions of herbicides and growth regulators, research that had been initiated by military interest in biological warfare. In retrospect, studies performed during this period had a major impact on agriculture and horticulture.

Although the dollar amount of research funds provided to support applied agricultural research has not diminished, the percentage of such funds has declined sharply over the years. In essence a cycle is being repeated: The

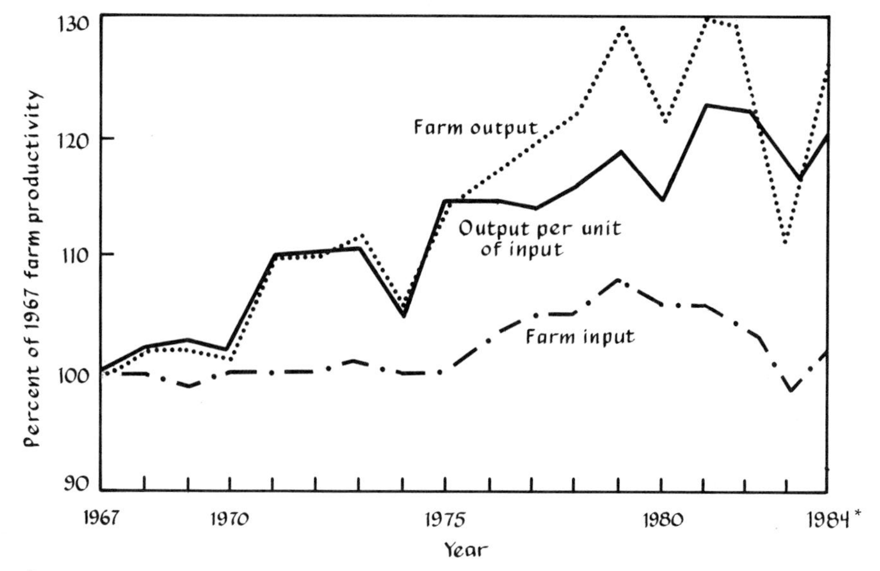

FIGURE 1-21. (*A*) U.S. farm productivity and efficiency, as measured in terms of output per unit of input, has continued to increase since 1967. (*B*) Crop production has climbed faster than population, leaving large quantities in excess of domestic use and available for export. [Data from Economic Research Service, USDA.]

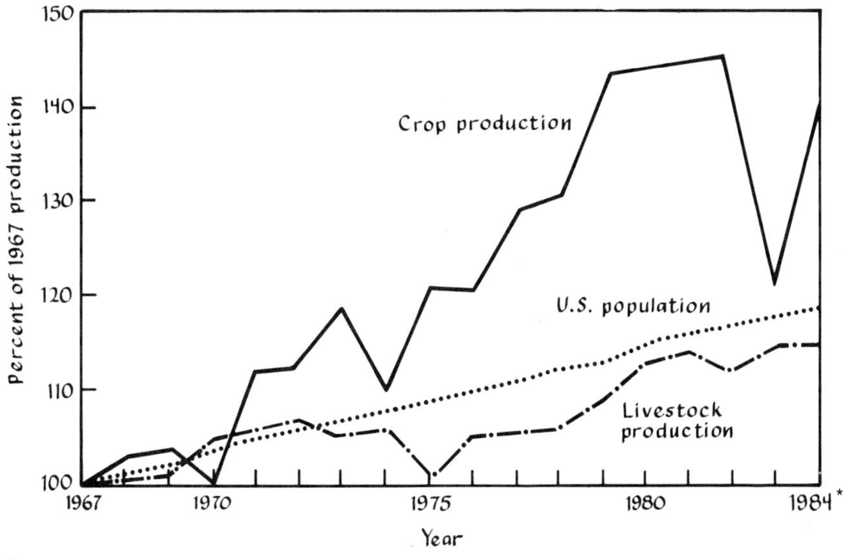

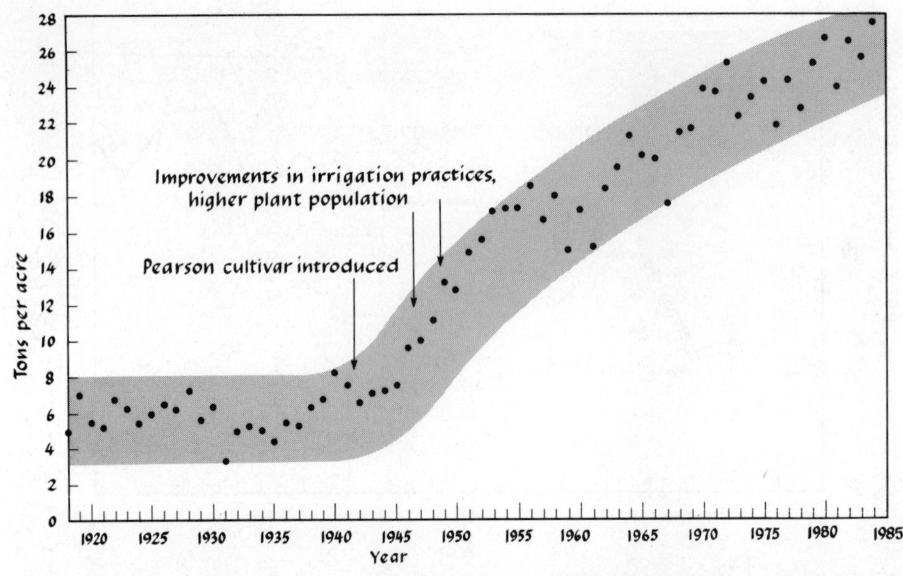

FIGURE 1-22. Yield per acre of processing tomatoes in California. [Data from Agricultural Marketing Service, USDA.]

freedom from agricultural problems has enabled a concentrated effort in the "luxury" areas of research. Present studies concentrated in genetics ("genetic engineering") cannot but have a profound effect on the future technologies of crop production.

The Ecology Movement

In the 1960s two approaching crises impinged on world consciousness. The first crisis was an emerging world food shortage caused by unchecked population increase in the tropical areas of the world. The second, first brought into prominence by Rachel Carson's *Silent Spring* (1962), was the pollution resulting from technology itself. The two problems were soon shown to be part of a single issue—the distribution of the earth's renewable and nonrenewable resources. Ironically, as we raced to conquer the moon, we discovered the earth. Whereas our plant-dominated planet had traditionally been thought of as "Mother Earth," the source of bounty and plenty, the new concept, that of a "Spaceship Earth," stressed the finiteness of earth's resources. The "new" earth was seen as if from the vantage of an observer on the moon: small and isolated, yet beautiful and precious as never before (Figure 1-23). It soon became apparent that there was indeed a limit to earth's resources and that increasing population

A

B

FIGURE 1-23. (*A*) Spaceship earth as viewed from the moon. (*B*) Seedlings of cabbage growing on moon fines. [Courtesy NASA.]

acted as a growing divisor of the world's wealth. Worst of all was the realization that the narrow application of technology, geared only to short-term increases in production, was leading at best to a fouling of the nest and at worst to mass poisoning.

One social consequence was the rise of the ecology movement. Conceived in fear and born in crisis, the ecology movement inspired religious fervor and struck a responsive chord in the United States during the late 1960s and early 1970s. Although its message was strident and its thrust sometimes veered uncomfortably toward anti-intellectualism and antiscience, the ecology movement proved itself to be a force to be reckoned with. The banning of DDT, its *cause celebre*, was the most notable of its victories. The ecology movement spread to the field of social problems and displayed concern for the displaced agricultural worker, the human jetsam of agricultural progress.

The ecology movement had an unforeseen effect on horticulture. As strident and politically conscious voices directed their fury against pesticides and herbicides, young people caught up in the movement, rejecting the materialism of their depression-reared parents, found the land in general and gardening in particular. Vegetarianism was rediscovered; yogurt was in, junkfood out. The new buzz words *organic* and *natural* (soon to be shamefully exploited by the consumer-oriented food and cosmetic industries) replaced the epithets *science* and *technology*. At U.S. colleges and universities agricultural enrollment—particularly in horticulture—soared. The land grant colleges—formerly scorned by the sophisticated as little more than "cow colleges"—were transformed from sleepy state schools to major research universities. The roaring expansion, ignited by the "GI Bill" which gave returning World War II veterans unequalled educational opportunities and fueled by increasing federal support for research and education, exploded under the impact of student population increases brought about by the postwar baby boom.

The clamor over pollution has made agricultural technologists in the United States more aware of the long-range implications of their action. The feeling now exists that technological progress combined with statesmanship can reverse the trend toward environmental deterioration and lead to not only sustained but increased agricultural production.

Although agricultural technology may have been scorned in the United States when surpluses were a big problem, it was eagerly adapted in the poor, underdeveloped countries of the tropics and subtropics. In the late 1960s a dramatic increase in the production of major food crops occurred in the developing world. A central breakthrough occurred in plant breeding—the creation of high-yielding, fertilizer-responsive cultivars of wheat and rice suitable for the tropics. At the same time more emphasis was placed on the use of agricultural technology in developing countries, such as India and Mexico. Increased use of technology in rural areas plus the "miracle" grains introduced what is known as the "green revolution" and advanced a number of tropical countries a small step toward self-sufficiency in foodstuffs. Progress in population control has also been encouraging in at least some countries.

The Energy Crisis

In the 1970s a new shock wave reverberated on world consciousness—the energy crisis. The Arab oil embargo of 1973, coinciding with a quadrupling of crude oil prices, initiated a general economic disorientation (Figure 1-24). The

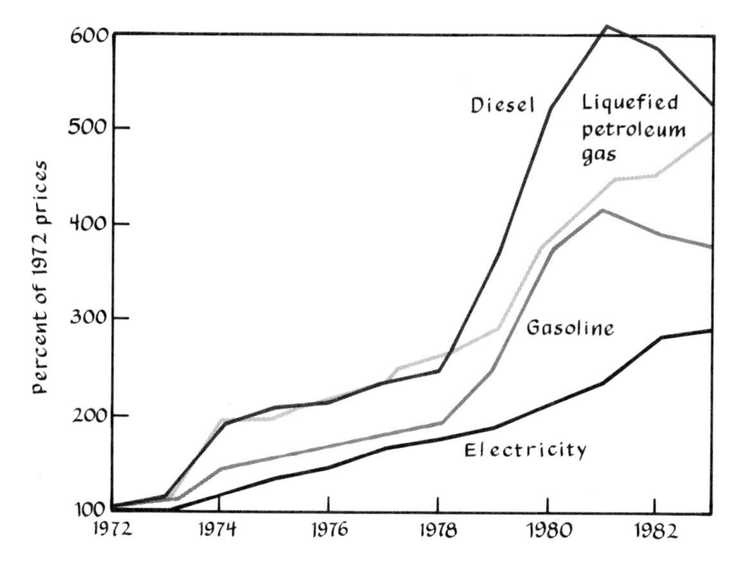

FIGURE 1-24. Changes in energy prices paid by farmers. The dramatic rise in world energy prices after 1973 had tremendous impact on world economics as well as agriculture. The effect was especially traumatic to greenhouse growers who had switched from coal, whose prices remained stable, to oil. [Data from Economic Research Service, USDA].

energy crises jolted the horticultural industries—especially those portions that were oil-dependent. The greenhouse industry, which had completed the conversion from coal to oil, found itself in near panic. The world, already divided into rich and poor, East and West, free economy and controlled economy, now was divided once again into oil exporters and oil importers. As oil prices rose, a massive redistribution of wealth and power occurred whose consequences are still unknown and unpredictable.

The energy crisis, like the ecology movement, had a destabilizing effect on agriculture, especially in underdeveloped countries, because the cost of many of the new technological products—chiefly fertilizer and pesticides—turned out to be directly related to the cost of energy. Those who had been disturbed about presumed deleterious effects of modern agriculture on the environment were joined by new voices concerned about the high energy requirements of high-technology agriculture. For the first time progress itself was questioned and continued progress was labeled a dangerous myth. A new concept called intermediate, or appropriate, technology was offered as a counterforce. Appropriate technology was defined as a level of technology between primitive and advanced, but one that was at the same time appropriate to get the job done, to conserve resources, to preserve long-term human values, and to maximize human happiness!

It is still too early to decide if intermediate technology is merely a nostalgic yearning to return to what is perceived as the simpler ways of the past. To those who have experienced hard times it is difficult to glorify what appears to be a return to primitive living and subsistence agriculture. It also seems foolhardy, however, to deny the argument of declining resources as energy costs spiral upward. And those who are caught in the frustrations of a complex modern society find it difficult to scoff, without some pangs of self-doubt, at the idea that perhaps "small is beautiful," a motto proposed by E. F. Shumacher, one of the prophets of intermediate technology.

However, it turns out that energy consumption by agriculture in the United States is trivial when compared to energy consumption by the rest of our society. Although the total food industry consumes one-sixth of all energy used in the United States, actual food production accounts for only a fifth of that energy use, or only 3 percent of the U.S. total. It is interesting to note that it takes more than twice as much energy for home food preparation as for food production. In this light it is difficult to see how the U.S. farmer can have much of an impact on the energy crisis, and it seems likely that a reduction in food production by conserving energy on the farm may indeed aggravate our current problems. However, horticultural industries do use a disproportionate share of the energy used by agriculture, and as a result the energy crisis continues to have a major impact on horticulture. The variables in the energy equation still remain unknown quantities.

RECENT CHANGES AND FUTURE TRENDS

The improvement in horticultural technology can be expected to continue in the developed (i.e., wealthy) countries. In the past 40 years the main increase in agricultural productivity in the United States has come about from a continual reduction in labor input, a trend that shows no sign of abatement (Figure 1-25). In agriculture and horticulture this has been achieved by the combination of a mechanical and a chemical revolution. This combination has eliminated most of the back-breaking labor associated with land preparation, cultivation, weed control, and harvest. For example, during the 1960s teams of horticulturists and agricultural engineers attacked the problems of mechanically harvesting fruits and vegetables. By the 1980s the list of mechanically harvested crops included processing tomatoes, pod vegetables, bulb crops, grapes, nursery stock, nuts, tart cherries, and even processing raspberries (Figure 1-26). To be sure some crops resisted mechanical harvesting, particularly fruits for fresh consumption (apples, peaches, strawberries), but mechanical aids, bulk containerization, and training to alter plant architecture have greatly simplified these operations. And interestingly enough, the incentive to completely eliminate all hand harvest labor has waned—even as the task appeared within reach. With many areas finding it difficult to achieve constant full employment some have felt it socially undesirable to replace or at best displace hand labor in concentrated agricultural areas. For example, complete mechanical harvesting of head lettuce in California is feasible, yet a combination of social and political concerns and increased efficiency of the harvesting crews and the processing system have in essence side-stepped the rush to full mecha-

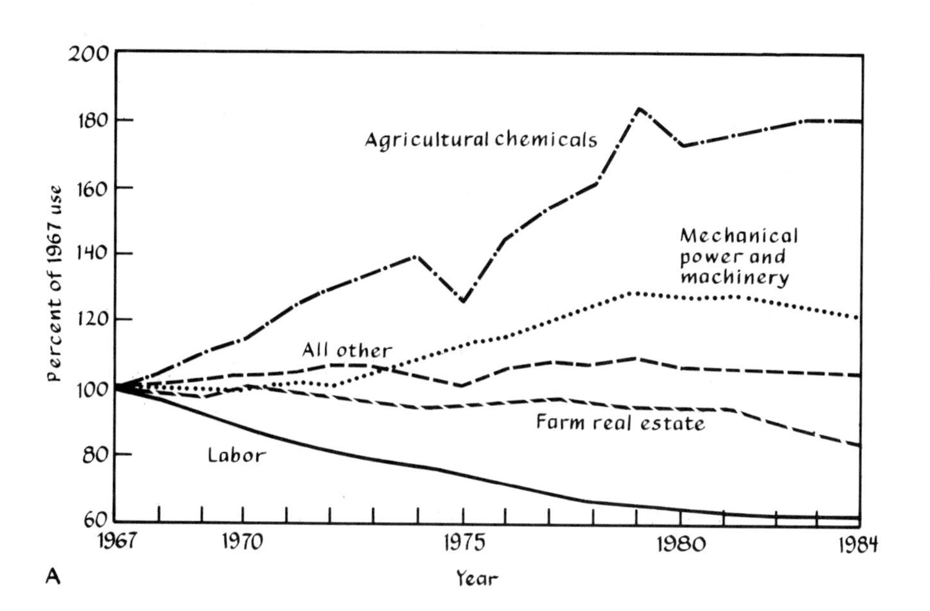

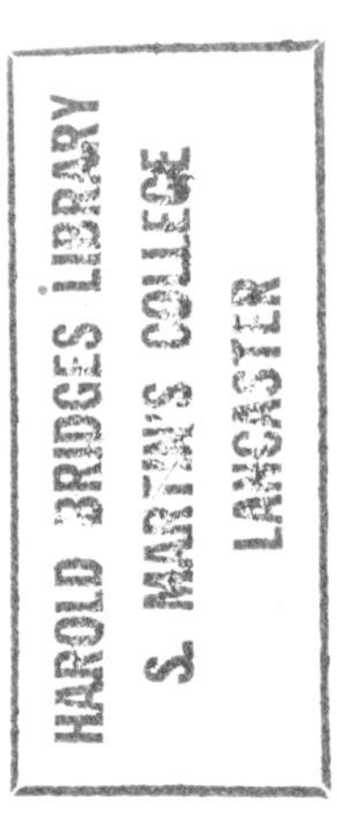

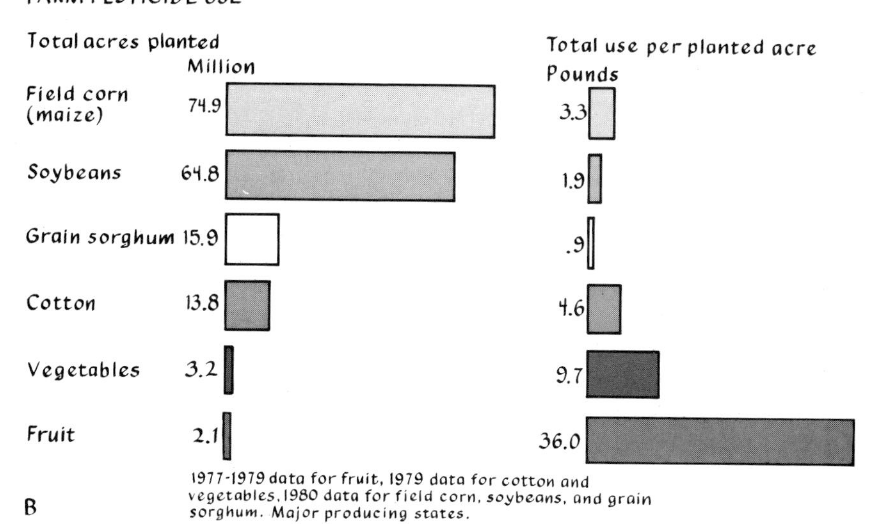

FIGURE 1-25. Use of selected farm inputs. (*A*) Farm labor inputs have declined steadily since 1967, while farm real estate area has remained relatively constant. Mechanical power and machinery has increased by nearly a third, and agricultural chemicals by over two-thirds. (*B*) Horticulture accounts for a disproportionate share of technological inputs. For example, vegetable production uses three times and fruit crop production ten times as much pesticide as field corn (maize). [Data from Economic Research Service, USDA.]

FIGURE 1-26. Harvesting red raspberries in Oregon. [Photograph courtesy Lloyd W. Martin and F. J. Lawrence.]

nization. The same is true with regard to mechanical harvesting of citrus crops in Florida.

The chemical revolution has completely transformed agriculture. (The agricultural chemical industry is now larger than the pharmaceutical industry!) Today's large-scale crop production—particularly of horticultural crops—is inconceivable without the inputs of chemical fertilizers, pesticides, and herbicides. Yet, the problems (and expense) of these materials have altered the philosophy of chemical use in order to reduce the undesirable effects associated with indiscriminate use. For example, instead of seeking a purely chemical solution to pest control a system referred to as integrated pest management (IPM) seeks to incorporate and coordinate various avenues of control, including genetic resistance, insect avoidance, biological control, and the minimum use of pesticides achieved by monitoring pest population levels and the precision application of pesticides only when necessary. The new wrinkle in the chemical revolution—a technology that is very much a horticultural innovation—has been the widespread use of growth-regulating materials. At present the use of growth regulators has become standard practice for many horticultural industries.

For example, the upsurges in productivity and in quality found in the Washington State apple industry have been based on new spur strains, new cultivars, and long-term controlled-atmosphere storage of apples in combination with the chemical control of plant productivity and fruit quality. The chemical component includes chemical thinning (blossom and fruit removal) agents to avoid biennial bearing; chemical bioregulants, such as daminozide, to control vegetative growth, delay watercore, and increase fruit firmness; ethylene-generating substances to increase soluble solids and reduce astringency; a mixture of gibberellin and cytokinin to improve fruit shape; ethoxyquin and diphenylamine to control scald, a storage defect; and new fungicides to control storage molds. (Other uses of growth regulators in the apple industry include defoliation of nursery stock, rooting promotors, and materials to prevent premature fruit drop.) The use of growth-regulating chemicals is particularly extensive with floriculture crops and is expanding rapidly to vegetable crops.

What of the future? The next technological breakthrough likely will be a genetic revolution brought about by new techniques in the manipulation of genetic information. Advances in recombinant DNA and cellular techniques have opened up new avenues for genetic modification. While traditional plant improvement based on the creation of variable population and selection has been an essential part of agricultural progress, the new genetic manipulation of DNA has the potential for advances of another order of magnitude. It is clear that we have entered a new era of plant improvement. It is difficult to predict precisely what this research will yield, but without a doubt these new techniques auger for fundamental changes in agriculture.

Many of the improvements in the past have been made possible by the organization of research essential to agriculture. The system of agricultural experiment stations authorized by the Morrill Act of 1862 has provided a steady supply of basic and applied research. The trend may be accelerating because funding for agriculture has become more diverse. At present the private sector in the United States spends more for research than does the public sector. Agricultural research is also carried out by a network of international crop centers. Finally, today's basic research expenditures in biology have tremendous implications for tomorrow's agriculture.

It is unfortunate that these comforting predictions of expanding technology are not applicable to all parts of the world. Many developing countries with low rates of agricultural productivity are experiencing rapid population increases, so that agricultural productivity per capita has been stagnant in spite of increasing rates of agricultural productivity (Figure 1-27). A combination of population control and the extension of advanced agricultural systems to these areas must be accomplished to fulfill our hopes for a better world for all people.

Horticulture has a special place in agriculture because it deals with an enormous diversity of plant life and because it brings a unique perspective to science and technology. Although horticulture is concerned with plants that nourish our bodies, it is also concerned with plants that delight our palate, enrich our senses, soothe our nerves, and free our spirits. Horticulture is an integral part of the world of "high tech," but, as important, horticulture is essential to make our world a place fit for humans.

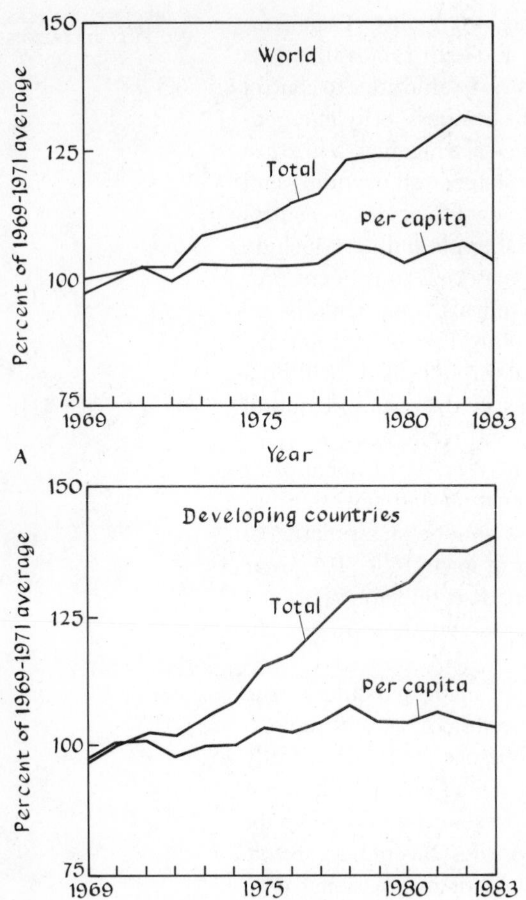

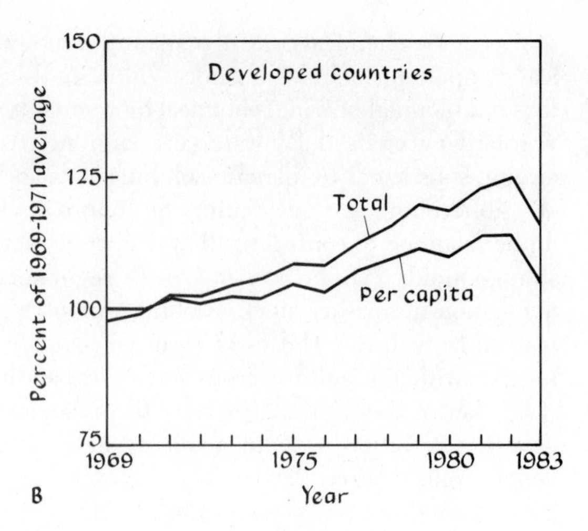

FIGURE 1-27. (*A*) Changes in world agricultural production. World production of agricultural commodities over the past 15 years has increased at an annual compound rate of about 2.2 percent, but only 0.4 percent on a per capita basis. (*B*) Changes in agricultural production of developed countries. (*C*) Changes in agricultural production of developing countries. Production in the developing countries as a group has been growing at about 2.9 percent annually, but the growth in population has kept production per capita from rising and has thus continued at nearly the same level. [Data from Economic Research Service, USDA.]

Selected References

Arber, A., 1938. *Herbals, Their Origin and Evolution: A Chapter in the History of Botany, 1470–1670*, 2d ed. Cambridge, Eng.: Cambridge University Press. (The definitive work on the subject.)

Bailey, L. H., 1947. *The Standard Cyclopedia of Horticulture*. New York: Macmillan. (Probably the best work on horticulture ever assembled. The first edition in 1900 was to compile a "complete record of North American horticulture as it [existed] at the close of the nineteenth century." The work was revised in 1914. Its most valuable use today is for its treatment of plant materials.)

SELECTED REFERENCES

Berrall, J. S., 1966. *The Garden: An Illustrated History.* New York: Viking. (A pictorial romp through garden history.)

Gothein, M. L., 1928. *A History of Garden Art.* New York: Dutton. (Authoritative work on the subject, in two volumes.)

Gras, N. S. B., 1940. *A History of Agriculture,* 2d ed. New York: F. S. Crofts. (An authoritative treatment.)

Harland, J. R., 1975. *Crops and Man.* Madison, Wis.: American Society of Agronomy. (A well-written review of crop origins.)

Hendrick, U. P., 1950. *A History of Horticulture in America to 1860.* New York: Oxford University Press. (A chronicle of early United States horticulture.)

Henrey, B., 1975. *British Botanical and Horticultural Literature before 1800.* New York: Oxford University Press. (A history and bibliography of botanical and horticultural books from the sixteenth, seventeenth, and eighteenth centuries, in three volumes.)

Huxley, A., 1978. *An Illustrated History of Gardening.* New York: Paddington Press. (The newest entry in the world of garden history; profusely illustrated.)

Hyams, E., 1971. *A History of Gardens and Gardening.* New York: Praeger. (The garden, especially in the landscape sense, from the Stone Age to the present.)

Leonard, J. N., 1973. *The First Farmers.* New York: Time-Life Books. (A popular account of the origins of agriculture, profusely illustrated.)

Reed, H. S., 1942. *A Short History of the Plant Sciences.* Waltham, Mass.: Chronica Botanica Co. (An excellent review.)

Simmonds, N. W., 1976. *Evolution of Crop Plants.* Cambridge, Mass.: Harvard University Press. (Origins on a crop-by-crop basis.)

Singer, C., E. J. Holmyard, and A. R. Hall (eds.), 1954–1958. *A History of Technology.* London: Oxford University Press. (A monumental work, in five volumes, on science and technology up to 1900. Agriculture and related fields are well documented.)

van den Muijzenberg, E. W. B., 1980. *A History of Greenhouses.* Wageningen, The Netherlands: Institute for Agricultural Engineering. (A fascinating history of controlled-environment structures.)

Von Hagen, V. W., 1961. *The Ancient Sun Kingdoms of the Americas.* Cleveland: World Publishing. (A popular history with much technological information on the Aztec, Maya, and Inca civilizations.)

Wright, R., 1938. *The Story of Gardening.* Garden City, N.Y.: Garden City Pub. Co. (A popular history of horticulture and gardening.)

Corn seedling. [Courtesy International Harvester Corporation.]

HORTICULTURAL BIOLOGY *I*

Classification 2

by Arthur H. Westing

Since ancient times people have named and categorized the many plants that surround them and upon which they are dependent for their very existence. One can readily surmise that the earliest classifications simply divided plants into the harmful ones and the useful ones (a division of value to this day!). Additionally, plants were probably divided according to their uses. Such classifications thus met the need of organizing what must otherwise have been a bewildering array of objects. Practical systems of this sort are, of course, perfectly valid, provided they are logically conceived and consistent, and therefore capable of predictive use.

Often superimposed on the practical systems of classification are those based upon growth habit or other gross physiological characteristics. Thus plants can be characterized as being **succulent (herbaceous)** or **woody.** Succulent seed plants processing self-supporting stems are known as **herbs.** A plant whose stem requires support for upright growth may be a climbing or trailing plant; if nonwoody, such a plant is known as a **vine,** whereas a woody plant of such habit is correctly called a **liana.** The self-supporting woody plants are either **shrubs** or **trees,** the trees being characterized by a single central axis and the shrubs by several more or less upright stems. Trees are usually taller than shrubs. Occasionally, the distinction between trees and shrubs may be obscured by environmental conditions or by horticultural "training."

Plants that are leafless during a portion of the year (ususally the winter) are referred to as **deciduous;** those whose leaves persist the year round are known as **evergreens.** Evergreens actually may lose their leaves annually, but not until a new set of leaves has developed. The deciduous habit is often associated with temperate regions; the persistent, with tropical regions.

Another classification of obvious importance to the horticulturist is based on life span and divides plants into annuals, biennials, and perennials. The **annual** plant is one that normally completes its entire life cycle during a single growing season. Spinach, lettuce, and petunia are examples of annuals.

The **biennial** plant normally completes its life cycle during a period of two growing seasons. During the first summer its growth is entirely vegetative, the plant often being low in form—a so-called rosette. The winter following the first growing season provides the low temperatures necessary to stimulate this type of plant to "bolt"—to send up a seed stalk during the second growing season, and then to flower and set fruit. Celery, parsnip, and evening primrose are among the biennials. Because certain biennials such as carrots or beets are grown for their overwintering storage organs, they are harvested as annuals at the end of the first growing season. Similarly, if the climate is mild enough, an annual such as spinach can be planted in the fall and harvested the following spring, and thus, although not requiring a period of low temperature, it can be grown in much the same way that biennials are grown.

The **perennial** plant grows year after year, often taking many years to mature. Unlike annuals and biennials, the perennial does not necessarily die after flowering. Although herbaceous plants are found in all three categories, woody plants are usually perennial. The many fruit trees, as well as the ornamental shrubs and trees, are examples known to all (Figure 2-1). Asparagus, rhubarb, and our various bulb crops are among the herbaceous perennials, whose above-ground parts are killed each year in temperate regions but whose roots remain alive to send up new shoots in the spring. When a subtropical perennial (such as tomato, eggplant, or coleus) is grown in a temperate region, it cannot survive the relatively cold winters, and under such conditions it is treated as an annual. An interesting situation exists with respect to the genus *Rubus* (the raspberries and other brambles) in which the roots are perennial and the shoots are biennial (Figure 2-2).

Plants can also be classified according to their temperature tolerances. Thus, when horitculturists speak of **tender** plants and **hardy** plants, they are referring to the plant's ability to withstand low winter temperatures. For woody plants the additional distinction is sometimes made between so-called **wood hardiness** and **flower-bud hardiness.** The former refers to the winter cold-resistance of the plant as a whole; the latter, to the ability of the flower buds to survive low winter temperatures. For example, even though apricot trees could survive in many parts of the United States, their culture is restricted to California because of their limited flower-bud hardiness. Similarly, the ginkgo can be grown as an ornamental in central Canada but cannot "flower" (that is, produce

FIGURE 2-1. Blue spruce (*Picea pungens;* family Pinaceae, order Coniferales, class Gymnospermae). This majestic tree is found scattered primarily on the middle and upper slopes of the Rocky Mountains. Mature trees are often 100 feet tall and live for more than 500 years. The several available cultivars are prized highly as ornamentals, owing not only to their beautiful habit and foliage but also to their ability to withstand drought and extremes of temperature. [Photograph by J. C. Allen & Son.]

strobili) and set fruit there. (It must be borne in mind that temperate-zone plants "harden off" in the fall to become far more cold-resistant in winter than in summer.)

Plants are sometimes also classified according to their temperature requirements during the growing season. For example, peas are a typical cool-season crop, whereas tomatoes are a typical warm-season crop. This characteristic is sometimes related to seed-germination requirements.

The landscape architect may wish to classify plants according to their habitat or site preferences. In addition to recognizing the obvious desert and aquatic types, the landscape architect must know whether an ornamental plant is best suited to moist or dry sites, to sunny or shady conditions, or to acid or alkaline soils.

FIGURE 2-2. Black raspberry (*Rubus occidentalis;* family Rosaceae, subclass Dicotyledoneae). Raspberries are noted for their fruits, equally delicious raw or as preserves. There are numerous cultivars of this native North American shrub and its various related species. Although the roots of the raspberry are perennial, its shoots are biennial. Members of this genus are often referred to as brambles, particularly in England. Crosses with the American red raspberry, *Rubus idaeus* var. *strigosus*, have produced the purple raspberry sometimes referred to as *Rubus neglectus*. [Photograph by J. C. Allen & Son; harvested berries of 'Black Hawk' raspberry courtesy Irvin Denison, Iowa State University.]

A HORTICULTURAL PLANT CLASSIFICATION

Elaborations of the ancient plant classifications, which were based upon the uses to which plants were put, are still the most important ones to the horticulturist. Plants are conveniently separated into those that are edible, those that serve as sources of drugs or spices, those that are of ornamental value, and so forth. Although almost any intensively cultured plant might be considered within the province of horticulture, the primary focus is on the various traditional "garden" plants. The grains, for example, are traditionally excluded from horticulture and are considered field crops.

The horticulturist divides the **edible** garden plants into **vegetables** and **fruits.** Generally considered as **vegetables** are those herbaceous plants of which some portion is eaten, either cooked or raw, during the principal part of the meal. Common examples are artichoke (edible flower, Figure 2-3), sweetpotato (edible root, Figure 2-4), spinach (edible leaf), asparagus (edible stem), beet (edible root), cauliflower (edible flower), eggplant (edible fruit), and pea (edible seed). **Fruits,** on the other hand, are the plants from which a more or less succulent fruit or closely related structure is commonly eaten as a dessert or snack. Fruit plants are most often perennial and are usually woody. Whereas those of the temperate zones are primarily deciduous, the tropical and subtrop-

FIGURE 2-3. Artichoke (*Cynara scolymus;* family Compositae, subclass Dicotyledoneae). The fleshy flower head of the artichoke is cooked as a vegetable and is considered a delicacy by many. This perennial herb is native to the Mediterranean region and is grown as a field crop in parts of Europe and the United States. The artichoke illustrated is not to be confused with the Jerusalem artichoke (*Helianthus tuberosus*), another vegetable plant in the same family grown for its edible tuber. [Photograph by J. C. Allen & Son.]

FIGURE 2-4. Sweetpotato (*Ipomoea batatas;* family Convolvulaceae, subclass Dicotyledoneae). The edible portion of this vine is the tuberlike root, which is cooked and eaten as a vegetable. Sweetpotatoes originated in tropical America but are now a common starchy food throughout the warmer regions of the world and are a truck crop of some importance in the southern United States. The many cultivars fall into two general types, those having dry, mealy flesh and those having soft, moist flesh. Sweetpotatoes are often known as yams in the southern United States but are not to be confused with the true yam (*Dioscorea* spp., family Dioscoreaceae, subclass Monocotyledoneae), another very important starchy, tuberlike vegetable grown in the tropics. [Photograph by J. C. Allen & Son.]

ical plants are usually evergreen. Fruits borne on trees are termed **tree fruits,** among which are the apple, pear, cherry (Figure 2-5), orange, papaya (Figure 2-6), and date (Figure 2-7).

Fruits borne on low-growing plants such as shrubs, lianas, and some herbs are known as **small fruits**—not necessarily because the fruits are small but because the plants are of small stature. (The preferred term in England is **soft fruits.**) Examples of small, or soft, fruits are the blackberry, cranberry, grape, raspberry, and strawberry. **Nuts** are edible seeds enclosed in a hard shell and borne on trees or shrubs. Nut crops are often considered along with fruit crops by horticulturists; familiar examples are the cashew, pecan, and walnut. The peanut, which is the seed of a herbaceous plant, is not a true nut in the horticultural sense but is considered an oilseed.

No precise distinction can be made between the terms *fruit* and *vegetable*. Although the definitions given above hold true for most edible plants, especially those grown in the north temperate regions, the ancient and popular origins of these terms have resulted in certain anomalies. Thus, when the edible portion of a herbaceous plant is stem, leaf, or root, there is seldom any question that one is dealing with a vegetable. However, rhubarb, with its edible petiole, is considered a fruit in some parts of the world because of its use as a dessert. When the edible portion of a herbaceous plant corresponds to the botanical fruit, the situation is often more confusing. Thus, the banana and the pineapple are considered fruits, whereas the melon, tomato, cucumber, squash, and pepper are variously regarded in culinary usage as vegetables or fruits, depending upon national or even local tradition. As a result of a question

FIGURE 2-5. Sour cherry (*Prunus cerasus;* family Rosaceae, subclass Dicotyledoneae). Cherries have been an important fruit crop in Europe and in Asia since the beginnings of agriculture. Hundreds of cultivars of this species and the closely related sweet cherry (*P. avium*) are cultivated throughout the temperate portions of the world. Because of their blossoms, cherry trees also rank among our most valued ornamentals. Some species of cherry provide an important cabinet wood that is known for its high silky luster and great beauty. [Photograph by J. C. Allen & Son.]

FIGURE 2-6. Papaya (*Carica papaya;* family Caricaceae, subclass Dicotyledoneae). The treelike semiherb that bears this delicious fruit is a native of tropical America that is now cultivated pantropically. Papayas have become exceedingly popular in Hawaii and elsewhere. They are mostly eaten raw but are also either boiled as vegetables or pickled. The fruits are a commercial source of the enzyme papain. [Photograph by J. C. Allen & Son.]

of import duties the tomato was legally established in this country as a vegetable by a U.S. Supreme Court decision of 1893! It must be recognized that the terms *fruit, nut,* and *vegetable* may have different meanings depending on whether the context is botanical, horticultural, culinary, or common parlance.

There are other groups of plants that are consumed but do not fit into the popular classification of vegetables, fruits, or nuts. These include crops such as cacao, coffee, and tea (**beverage plants**) and culinary herbs, flavorings, and tropical spices, grouped under the broad terms of **herbs** and **spices.** Herbaceous plants grown in the temperate regions for flavoring or essences are called culinary herbs; tropical plants that provide aromatic or fragrant products are referred to as spices.

Plants used for their ornamental value are commonly separated into **flower** or **landscape plants.** Flowers (often herbaceous) are primarily grown for their blossoms (Figure 2-8) but are occasionally grown for their showy leaves (Figure

FIGURE 2-7. Date palm (*Phoenix dactylifera;* family Palmae, subclass Monocotyledoneae). Date palms have been cultivated for their fruits along the Tigris and Euphrates rivers for more than 4000 years. Apparently originally native to the Near East, date palms are now widely cultivated throughout the warmer regions of the world, including parts of California and Arizona. Dates are the principal food crop from Iran to the Arabian Peninsula and all throughout North Africa. The trees, often 100 feet tall, are also locally important for their wood as well as for their sugary sap. [Photograph by J. C. Allen & Son.]

FIGURE 2-8. Lady slipper (*Calceolaria herbeohybrida;* family Scrophulariaceae, subclass Dicotyledoneae). Lady slippers, or calceolarias, are admired for their showy pouch-shaped flowers, which come in a variety of colors. These herbaceous flowers are native to tropical America. The lady slipper illustrated is not to be confused with the orchids of the same name (*Cypripedium* spp.; family Orchidaceae, subclass Monocotyledoneae). [Photograph by J. C. Allen & Son.]

FIGURE 2-9. Caladium (*Caladium bicolor;* family Araceae, subclass Monocotyledoneae). Caladium is a deciduous herbaceous perennial native to tropical America. It is grown as a foliage plant because of its beautifully and variously patterned leaves. [Photograph by J. C. Allen & Son.]

2-9). They are often separated according to life span: The petunia (an annual), the foxglove (a biennial), and the peony (a perennial) are examples. **Bulbs** and **corms** (crocus, gladiolus, narcissus, and tulip) are a special class of perennial flower crops. Tropical ornamentals grown primarily for indoor use in temperate areas, such as philodendron and sanseveria, are treated separately as **foliage plants** even though they are grown as landscape plants in the tropics.

Landscape, or nursery, plants are most often produced for permanent outdoor plantings. They are often classified according to use, form, and growth habit. The horticulturist recognizes **lawn** and **turf plants** (herbaceous perennials); **ground covers** (either herbaceous or woody perennials, Figure 2-10); **deciduous** and **evergreen shrubs,** such as virburnum (Figure 2-11) and juniper

FIGURE 2-10. Periwinkle (*Vinca minor;* family Apocynaceae, subclass Dicotyledoneae). Periwinkle, also known as myrtle, is an often cultivated, trailing, evergreen herb. It makes a hardy ground cover that is especially suited to moist shady areas. The plant illustrated is a particularly attractive variegated cultivar that has solitary blue flowers. [Photograph by J. C. Allen & Son.]

FIGURE 2-11. Arrowwood viburnum (*Viburnum dentatum;* family Caprifoliaceae, subclass Dicotyledoneae). These handsome woody perennials are among our most important ornamental shrubs. They are noted for their attractive clusters of flowers and fruits, as well as for their foliage. There are about two dozen viburnum species of horticultural importance. The arrowwood illustrated here is a native of the United States and prefers moist sites. The tough, pliant shoots were once used to make arrows. [Photograph by J. C. Allen & Son.]

FIGURE 2-12. Pfitzer juniper (*Juniperus chinensis* "Pfitzeriana"; family Cupressaceae, order Coniferales, class Gymnospermae). This beautiful evergreen shrub from China forms a broad pyramid with its spreading branches and nodding branchlets. The grayish-green leaves are needle-shaped on juvenile plants and scalelike on mature ones, a widespread phenomenon in this family. This shrub, which may be an interspecific hybrid, is one of the rare gymnospermous polyploids. The blue berrylike fruits of junipers are used to give gin its characteristic flavor. Seen climbing the wall in the background is Boston ivy *Parthenocissus tricuspidata,* a deciduous liana. [Photograph by J. C. Allen & Son.]

(Figure 2-12); and **deciduous** and **evergreen trees,** such as birch (Figure 2-13) and spruce (Figure 2-1).

A number of plants that provide the raw materials for various manufactured products are known as **industrial crops.** Many are often treated as horticultural crops, especially when of tropical origin. These include **drugs** and **medicinals,** such as digitalis and quinine; **oilseeds,** particularly tropical and subtropical trees and shrubs, such as jojoba, oilpalm, and tung; and **extractives** and **resins,** such as chickle, rubber and slash pine.

Of course, the various categories here described are not all mutually exclusive. Thus a grape vine, when cultured for its edible berry, is considered to be a small fruit, but when grown in a garden to cover a trellis or arbor it is considered to be an ornamental. A walnut tree can be grown for its nuts, or as an ornamental, or for its highly prized wood. A hazel shrub may be considered to be a crop plant (filberts) by the commercial grower, an ornamental by the gardener, or an undesirable or weed species by the forester.

FIGURE 2-13. Weeping birch (*Betula pendula;* family Betulaceae, subclass Dicotyledoneae). The slender, gracefully drooping branches and the white papery bark explain the extensive use of the weeping birch as a lawn tree. Numerous cultivars are available. Tannins for the leather industry have been extracted from the bark in Europe, where it is native. [Photograph by J. C. Allen & Son.]

Classification

The horticultural plant classification described here (and outlined on the next page) is an operational one based upon the use to which we put the plants. As such it is not only an overlapping classification but may often be an empirical and arbitrary one as well. Moreover, it is a classification that is subject to change over the years as we discover new uses for plants and abandon old ones. The fundamental value of this classification, of course, is that it groups economic plants according to the similarities of their cultural requirements as well as according to the similarities of their utilization and marketing.

A HORTICULTURAL PLANT CLASSIFICATION

EDIBLES

Vegetables

PLANTS GROWN FOR AERIAL PORTIONS

- Cole crops (broccoli, cabbage, cauliflower)
- Legumes or pulse crops (bean, pea)
- Solanaceous fruit crops (*Capsicum* pepper, eggplant, tomato)
- Vine crops or cucurbits (cucumber, melon, squash, pumpkin)
- Greens or pot herbs (chard, dandelion, spinach)
- Mushrooms (*Agaricus, Lentinus*)
- Other vegetables (asparagus, okra, sweet corn)

PLANTS GROWN FOR UNDERGROUND PORTIONS

- Root crops
  - Temperate (beet, carrot, radish, turnip)
  - Tropical (cassava, sweet potato, taro, yam)
- Tuber crops (Jerusalem artichoke, potato)
- Bulb and corm crops (garlic, onion, shallot)

Fruits

TEMPERATE (DECIDUOUS)

- Small fruits
  - Berries (blueberry, cranberry, strawberry)
  - Brambles (blackberry, raspberry)
  - Vines (grape, kiwifruit)
- Tree fruits
  - Pome fruits (apple, pear, quince)
  - Stone fruits (apricot, cherry, peach, plum)

SUBTROPICAL AND TROPICAL (EVERGREEN)

- Herbaceous and vine fruits (banana, papaya, passion fruit, pineapple)
- Tree fruits
  - Citrus (grapefruit, lemon, lime, mandarin, orange)
  - Noncitrus (avocado, date, fig, mango, mangosteen)

Nuts

- Temperate (almond, chestnut, filbert, pecan, pistachio)
- Tropical (Brazil nut, cashew, macadamia)

Beverage crops

- Seed (cacao, coffee)
- Leaf (maté, tea)

Herbs and spices
- Culinary herbs (dill, rosemary, sage)
- Flavorings (peppermint, spearmint)
- Tropical spices (cinnamon, clove, nutmeg, pepper)

ORNAMENTALS

Flower, bedding, and foliage plants
- Annuals (marigold, petunia, zinnia)
- Biennials (English daisy, foxglove)
- Perennials (daylily, delphinium, iris, peony, rose)
  - Bulbs and corms (crocus, gladiolus, narcissus, tulip)

Landscape (nursery)
- Lawn and turf (bermudagrass, bluegrass, fescue, perennial ryegrass)
- Ground covers and vines (English ivy, Japanese spurge, myrtle)
- Evergreen shrubs and trees
  - Broadleaf (holly, rhododendron)
  - Narrowleaf (fir, juniper, yew)
- Deciduous shrubs (dogwood, forsythia, lilac, viburnum)
- Deciduous trees (ash, crabapple, magnolia, sugar maple)

INDUSTRIAL

Drugs and medicinals (digitalis, quinine)

Oil seeds (jojoba, oilpalm, tung)

Extractives and resins (Scotch pine, Pará rubber tree)

A SCIENTIFIC PLANT CLASSIFICATION

People are by nature methodical creatures, and they have always attempted to discover natural order in the universe. Their physical environment they early classified into its "elements"—air, earth, water, and fire. All living things were classed as either animal or vegetable, and the vegetable or plant kingdom was subdivided in many ways. As long ago as 300 B.C.E., the Greek philosopher Theophrastus classified all plants into annuals, biennials, and perennials according to their life spans, and into herbs, shrubs, and trees according to their growth habits. He further divided the trees according to their deciduous or evergreen natures and their various branching habits.

During the 20 centuries following the time of Theophrastus, a multitude of classifications and innumerable lists of useful plants ("herbals") were accepted and discarded (Figure 2-14). It was not until the middle of the eighteenth century that a Swedish physician named Carl von Linné (but better known by the Latinized version of his name, Linnaeus) revolutionized the fields of plant and animal classification, or taxonomy. His labors earned him the title "father of taxonomy." He catalogued organisms by groups—some large, some small—that depended upon structural or morphological similarities and differences.

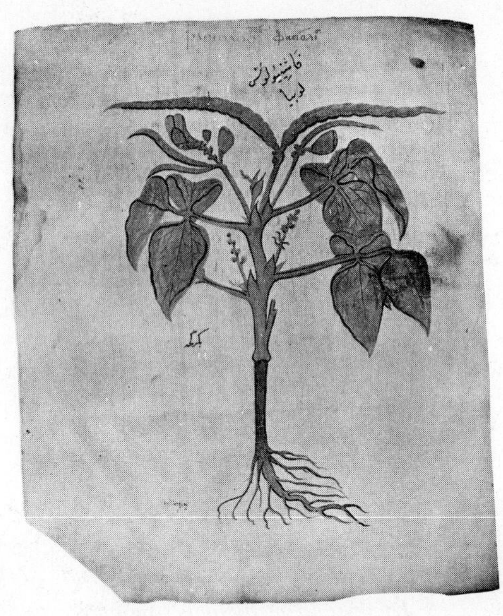

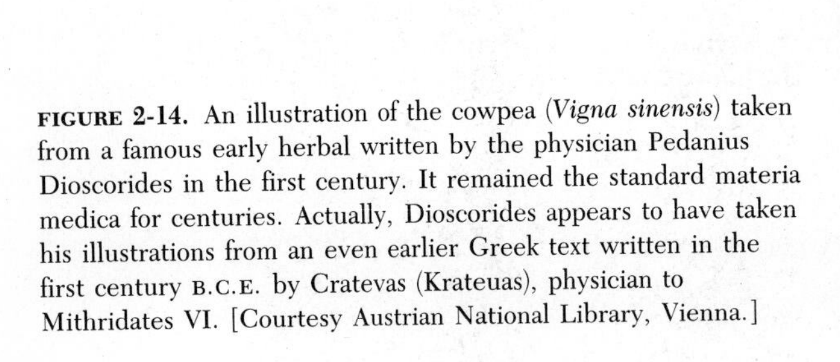

FIGURE 2-14. An illustration of the cowpea (*Vigna sinensis*) taken from a famous early herbal written by the physician Pedanius Dioscorides in the first century. It remained the standard materia medica for centuries. Actually, Dioscorides appears to have taken his illustrations from an even earlier Greek text written in the first century B.C.E. by Cratevas (Krateuas), physician to Mithridates VI. [Courtesy Austrian National Library, Vienna.]

He recognized the value of basing the taxonomic criteria for plants on the morphology of their sexual or reproductive parts—the plant organs least likely to be influenced by environmental conditions. Linnaeus singlehandedly described and assigned names to more than 1300 different plants (and to as many or more animals) from the far reaches of the earth. It was he who brought order out of chaos by standardizing a worldwide system of nomenclature. Linnaeus has been described rightfully as the greatest cataloguer and classifier of all time.

The next important advance in plant classification came in the middle of the nineteenth century, with the publication of Charles Darwin's *Origin of Species*. The principles of evolution propounded in this book had perhaps nowhere a more profound impact than on the fundamentals of taxonomy. Darwin's concept of evolution finally provided a natural framework upon which to hang a scheme of classification. According to this concept, all plants on earth today are more or less closely related taxonomically according to their proximity on the family tree of plant evolution. Thus the classification is based upon the notion that genetic relationships exist between all plants, and that present-day plants are, through successive generations, the offspring of ancient ancestral plants. Of course, the possibility exists that the plant kingdom owes its origin to more than one primordial organism. It is further assumed that there has occurred throughout the history of the earth, and is still occurring, an evolution of plant characteristics that has brought about increasing complexity in structure and genetic organization.

To the taxonomist, or systematist, falls the task of reconstructing these evolutionary connections—a task that is simple in theory, but that is formidable (and in large part impossible) in actual practice. The task is an immense one because of the countless numbers of now extinct and largely unknown plants. Paradoxically, however, it is the very discontinuities resulting from these extinctions that make it possible for the taxonomist to establish and delimit various categories of plants. Without them there probably would be an essentially unclassifiable continuum stretching from the lowest plant to the highest. The work of classification is further complicated by the occasional hybridizations that reticulate the family tree and by different patterns of evolution. The approach systematists take when they attempt to determine evolutionary lines and group relationships can perhaps best be described as scientific sleuthing. They base much of their case on morphological and anatomical comparisons and glean as much information as possible from distributional or geographic evidence. Of fundamental importance, however, is fossil plant (or paleobotanical) evidence: Each new fossil found fills another small gap in the family tree. The recently developed methods of determining the ages of sedimentary rocks and the plant remains preserved in them have been of inestimable help in this regard.

Particularly among closely allied groups, the modern taxonomist also leans heavily on cytological, genetic, ecological, and even physiological and biochemical evidence. For example, some decades ago the German botanist Karl Mez demonstrated the feasibility of determining plant-group affinities by comparing their constituent proteins. Borrowing standard serological techniques, he would sensitize a rabbit by injecting into its bloodstream the proteins from one plant. He then determined the degree to which this plant was related to another from the degree to which the newly developed antibodies in the rabbit's blood reacted to the proteins of the second plant. The less pronounced the reaction, the more distantly related the plants were assumed to be. More recently the taxonomy of the pines has been verified, and to some degree clarified, on the basis of their oleoresin chemistry. Chromatography, a powerful chemical technique whereby very complex mixtures are separated into their component parts by passing them through selectively adsorbent materials, has become an indispensable adjunct to biochemical taxonomy. Additionally, the electronic computer permits taxonomists to compare simultaneously and with relative ease several dozen morphological or other characteristics of allied plant groups in order to assist them in determining taxonomic affinities. Imaginative research in a variety of related fields and a subsequent correlation of the information constitute the difficult task that confronts the modern taxonomist.

One important attempt at a natural classification (although it was based primarily on floral morphology) was made toward the close of the nineteenth century by the German botanist August Eichler. This approach was elaborated by Adolph Engler, who with his collaborator Karl Prantl published a 20-volume classification of all the plants then known. Although much of the information that has since come to light indicates that Engler's arrangement of the plants is phylogenetically incorrect in many respects, it is still used by most botanists in the United States and elsewhere in the world. A widespread conversion to more recent systems is unlikely at this time, for none of them is sufficiently improved to warrant such a major effort. The classification of the plant kingdom that follows, however, has been revised to conform to the latest information available.

THE PLANT KINGDOM

The plant **kingdom** is, first of all, separated into about a dozen major **phyla,** or **divisions.** The most advanced (that is, most recently evolved) division is the one with which the horticulturist is directly concerned. This is the division that contains the so-called higher plants—those with roots, stems, leaves, and a vascular or tracheary system (the source of the name Tracheophyta). The remaining divisions contain almost no horticultural crop plants as such (edible mushrooms are an exception) and are therefore primarily of indirect interest to us. They are responsible, however, for many of our crop diseases. Included among these lower divisions are such diverse plants as the mosses, algae, fungi, bacteria, and slime molds. (In a recently developed system of classification, many of the lower plants are separated from the plant kingdom on the basis of their cellular organization. In this system, the fungi are given a kingdom of their own, the blue-green algae and bacteria are put together in a separate kingdom, and many of the other algae are placed in yet another kingdom with the protozoans and the slime molds.)

A synopsis of the traditional plant categories, or **taxa,** is given below. Each taxon is subdivided by the one below it. All categories need not be used, but the sequence must not be altered. Intermediate subdivisions are frequently made and designated by the prefix *sub-*. The categories from kingdom to family are called the major taxa; those below family are called the minor taxa.

Kingdom
- Division
  - Class
    - Order
      - Family
        - Genus (*pl.* genera)
          - Species (*abbr.* sp., *pl.* spp.)
            - Variety (*abbr.* var.)
              - Form (*abbr.* f.)
                - Individual

The division Tracheophyta is divided by many into about a dozen **classes,** including the horticulturally important Filicinae (or ferns), Gymnospermae (or cycads, ginkgoes, taxads, and conifers), and Angiospermae (the many flowering plants). Figure 2-15, constructed from current information, represents a possible family tree of the living vascular plants. The evolutionary lines, or phylogenetic histories, within the Filicinae and Gymnospermae have been defined comparatively well, primarily because a relatively complete fossil record exists for these classes. The phylogeny of the Angiospermae, on the other hand, is unfortunately still much more of a mystery, partially owing to the great scarcity of paleobotanical information.

The gymnosperms, represented by less than 700 living species, are primarily evergreen trees of the temperate zones of the world. They characteristically have naked seeds, which are usually borne on cones, and often have narrow or needlelike leaves. Gymnosperms provide lumber, pulp, turpentine, rosin, and some edible seeds, and are often highly prized as ornamental plants.

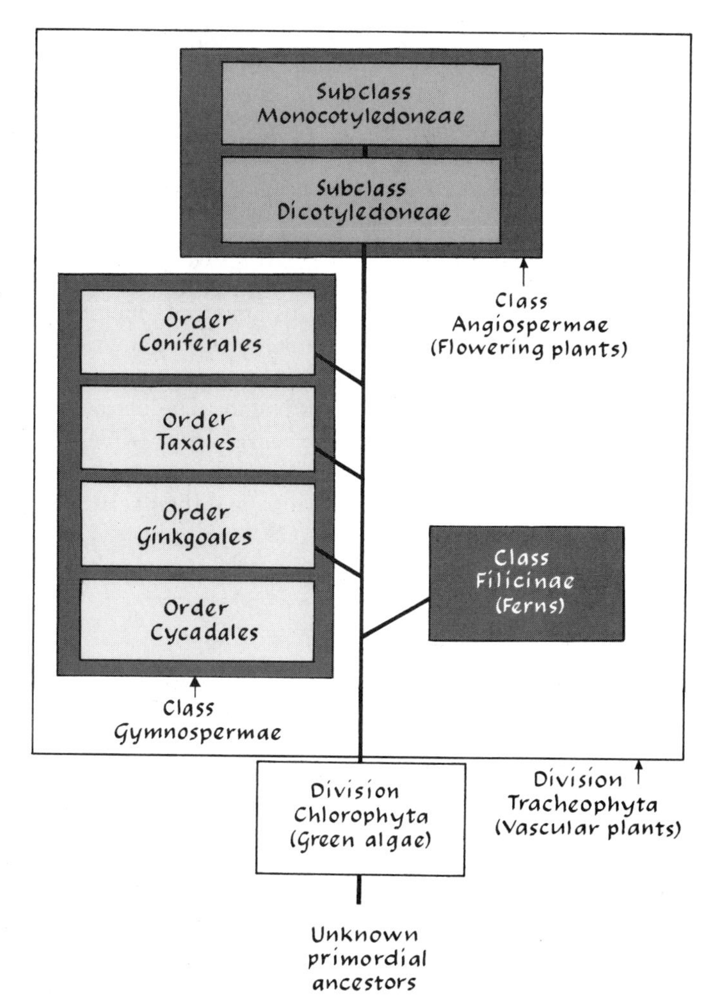

FIGURE 2-15. An abridged family tree of the living vascular plants. The solid lines indicate direct lines of descent. The higher the relative position of a group on the family tree, the more recent its emergence in the evolution of plants.

Classification

The angiosperms, numbering some 250,000 species, are worldwide in their distribution and are found in almost all conceivable habitats. Economically, they are the most important class of plants. They are our primary source of food and beverage, shelter and clothing, paper and rubber, oils and spices. The angiosperms characteristically have seeds that are fully enclosed in a fruit, and they often have broad leaves.

Although most classes are divided directly into **orders** (the Cycadales, Ginkgoales, Taxales, and Coniferales previously mentioned are orders of gymnosperms), the class Angiospermae is first divided into two major subclasses, the Dicotyledoneae (or dicots) and the Monocotyledoneae (or monocots). The dicots are characterized by two cotyledons ("seed leaves") in their seedling stage, usually by flower parts in fours or fives or multiples of these numbers, and often by reticulate leaf venation; the monocots are characterized by one cotyledon, by flower parts in threes or multiples thereof, and often by parallel leaf venation. The dicots, which number about 200,000, include most of our broad-leafed herbs, shrubs, and trees. The 50,000 or so monocots are grouped into such orders as the Liliales (of lilies), the Palmales (or palms), and the Graminales (or grasses and sedges). The grasses and palms together provide the main source of food for most peoples on earth, and the main source of shelter for many.

Orders are split in turn into groups knows as **families.** The family is frequently encountered in taxonomic studies because it is often small enough to permit relatively easy study of the natural relatonships among its members. The horticulturist will often note structural and cultural similarities among the genera within a family. Well-known families among the monocots are the Orchidaceae (or orchids) and the Gramineae (or grasses). Notable families among the dicots are the Leguminosae (or legumes), the Cucurbitaceae (which includes the gourds and melons), the Umbelliferae (which includes carrots and celery), and the Rosaceae (roses, pome fruits, stone fruits, and their relatives).

SOME WELL-KNOWN FAMILIES OF HORTICULTURAL INTEREST

GYMNOSPERMS

PINACEAE—PINE FAMILY

The pine family consists of 9 genera and 210 species of conifers, mostly evergreen trees and shrubs, widely distributed but largely in temperate regions of the Northern Hemisphere. Cultivated genera include *Abies* (fir), *Picea* (spruce), *Pinus* (pine), and *Tsuga* (hemlock).

ANGIOSPERMS

Monocotyledons

AMARYLLIDACEAE—AMARYLLIS FAMILY

Closely related to the Liliaceae, the amaryllis family consists of about 90 genera and 1200 species found principally in the temperate and warm regions of South America, South Africa, and the Mediterranean region. Plants have tunicate bulbs or corms. Horticultural food crops include onion, garlic, and chives—all species of *Allium.* Other cultivated genera include *Amaryllis, Clivia,* and *Narcissus.*

GRAMINEAE (POACEAE)—GRASS FAMILY

The grass family, one of the largest in the plant kingdom, is subdivided into as many as 27 tribes with about 700 genera and 7000 species distributed worldwide. Most are herbaceous plants of low stature, but a few are woody (notably bamboos). The family includes both annuals (many cereal grains) and perennials (turfgrasses). Root systems are fine and fibrous, and stems (culms) are cylindrical and generally hollow. The leaves, which are composed of a sheath and a blade, are parallel-veined; they grow from the base, allowing regrowth after plants are clipped or mowed. The inflorescence is composed of groups of flowers called spikelets, which are variously arranged (in spikes, panicles, or racemes). Flowers are usually perfect, small, and inconspicuous. The fruit is usually a seedlike grain (caryopsis). The grass family is of paramount importance to agriculture, containing all of the cereal grains (rice, wheat, barley, oats, corn, and sorghum) and many forage plants. Horticultural plants include sweet corn (*Zea mays*), bluegrass (*Poa*), and fescues (*Festuca*).

LILIACEAE—LILY FAMILY

With about 240 genera and 3000 species distributed worldwide, the lily family consists mainly of herbaceous plants with specialized food storage organs (rhizomes, bulbs, or fleshy roots). The only horticultural food crop of importance is asparagus (*Asparagus officinalis*). There are many ornamentals, including about 90 *Lilium* species (lilies), *Aloe*, *Aspidistra*, *Hemerocallis* (daylily), *Trillium*, and *Tulipa*.

ORCHIDACEAE—ORCHID FAMILY

Among the largest of the families of flowering plants, the orchid family contains 600 to 800 genera and 17,000 to 30,000 species. Orchids are found mainly in tropical forests, where they are often epiphytes living high in the branches of large trees. Some species are saprophytic, containing no chlorophyll but obtaining their nourishment from dead organic matter. Terrestrial forms have coarse roots with sparse branching. The flowers, which are insect pollinated, are usually large and gaudy, and many are bizarrely shaped. *Vanilla planifolia* is economically important as a source of the flavoring vanilla and many other orchids are important as a source of cut flowers (*Cattleya*, *Dendrobium*, *Vanda*, *Cymbidium*, *Phalaenopsis*) or as "conservatory" plants raised by fanciers and hobbyists.

PALMAE (ARACACEAE)—PALM FAMILY

The palm family is subdivided into 8 tribes, about 210 genera, and more than 4000 species, which are distributed in the tropics and subtropics. Most palms are unbranched woody shrubs or trees with large persistent leaves in a terminal tuft. Flowers are small, usually in a long branched panicle; fruits are berrylike, drupelike (a fleshy fruit with a single pit), or nutlike. Some palms are cultivated for their fruits or nuts, including the date palm (*Phoenix dactylifera*), the coconut palm (*Cocos nucifera*), and the oil palm (*Elaeis guineensis*). Many are popular landscape ornamentals, such as the royal palm (*Roystonia regia*).

Dicotyledons

CACTACAE—CACTUS FAMILY

The cactus family comprises 50 to 220 genera and 800 to 2000 species of fleshy, succulent herbs, shrubs, and trees, most of them native to the drier regions of North and South America. The fleshy stems, often spiny, may function as leaves. Flowers are often large and brightly colored. Many cacti are cultivated as ornamentals, and a few are of minor significance as food crops.

CHENOPODIACEAE—GOOSEFOOT FAMILY

A small group of about 75 genera and 500 species, the goosefoot family consists of annual or perennial herbs or shrubs often found growing in saline soils. It includes common spinach (*Spinacia oleracea*) and beet (*Beta vulgaris*), which includes both the red garden type and the large white sugar beet. The saltbrush (*Atriplex*) is sometimes grown for forage in desert regions.

COMPOSITAE (ASTERACEAE)—SUNFLOWER FAMILY

Perhaps the largest plant family, with 1000 genera and more than 23,000 species, the sunflower family comprises about a tenth of all species of flowering plants. Most of the members of the family are herbaceous; and, although the family is widely distributed, most of its members are best adapted to temperate and cool climates. The inflorescence is typically a head composed of florets on a short or broadened axis. Edible plants in this family include lettuce (*Lactuca sativa*), Jerusalem artichoke (*Helianthus tuberosus*), sunflower (*Helianthus annuus*), endive (*Cichorium intybus*), salsify (*Tragopogon porrifolius*), globe artichoke (*Cynara scolymus*), and dandelion (*Taraxacum officinale*). The family is rich in ornamentals, including such well-known genera as *Aster*, *Chrysanthemum*, *Calendula*, *Dahlia*, *Gerbera*, *Tagetes* (marigold), and *Zinnia*.

CRUCIFERAE (BRASSICACEAE)—MUSTARD FAMILY

The mustard family consists of about 350 genera and 3200 species of pungent herbs indigenous to temperate and cold climates. Garden vegetables in this family include cabbage, broccoli, Brussels sprouts, cauliflower, Chinese cabbage, kohlrabi, turnip, and mustard (all variant forms and species of *Brassica*), and radish (*Raphanus sativa*). Ornamentals include *Alyssum*, sweet allysum (*Lobularia*), stocks (*Matthiola*), and candytuft (*Iberis*). Shepherd's purse (*Capsella bursa-pastoris*) and wild mustard (*Brassica* sp.) are common temperate-zone weeds.

CUCURBITACEAE—GOURD FAMILY

The gourd family consists of about 100 genera and 500 to 750 species. The plants in this family have tendrils, leaves that are often rough to the touch, and large fleshy fruits with many seeds. Food plants include watermelon (*Citrullus lanatus*), melons (*Cucumis melo*), cucumber (*Cucumis sativus*), pumpkin, summer squash and marrow (*Cucurbita pepo*), and winter squash and pumpkin (*Cucurbita maxima*).

ERICACEAE—HEATH FAMILY

The heath family consists of about 70 genera and 1900 species, usually shrubs, widely distributed in acid soils. Cultivated genera include *Rhododendron* (rhododendron and azalea) and *Vaccinium* (blueberry and cranberry).

EUPHORBIACEAE—SPURGE FAMILY

The tropical and temperate herbs, trees, and shrubs that constitute the spurge family nearly all contain milky sap, or latex, which will exude from cut surfaces. Many tropical species are adapted to dry areas. There are about 280 genera and 7300 species. The largest genus, *Euphorbia*, includes the poinsettia (*E. pulcherrima*) and the "crown of thorns" (*E. milii*). Economically important crops include the Pará rubber tree (*Hevea brasiliensis*), cassava (*Manihot esculenta*), and castor bean (*Ricinus communis*).

LAURACEAE—LAUREL FAMILY

The laurel family consists of 47 genera and 2000 to 2500 species of aromatic woody plants, most of which are tropical. Well-known examples are the laurel tree (*Laurus nobilis*), cinnamon (*Cinnamomum zeylanicum*), camphor (*C. camphora*), and avocado (*Persea americana*).

LEGUMINOSAE (FABACEAE)—PEA FAMILY

The pea family, a widely distributed family of about 600 genera and 12,000 species, is one of the three largest in the plant kingdom. It is divided into three subfamilies: Faboideae, with butterflylike flowers; Caesalpinoideae, with irregular showy flowers; and Mimosoideae, with dense flower heads with many stamens. Herbaceous plants are most common in temperate climates, but trees, shrubs, and vines predominate in the tropics. The leaves are commonly pinnate (featherlike, with leaflets arranged in two rows along a common axis), and the fruit is a flattened, dry, dehiscent pod called a legume, which gives the family its name. Many members of the family accumulate large amounts of nitrogenous substances in their tissues as a result of a symbiotic association with nitrogen-fixing bacteria that live in nodules on their roots. The high protein content makes many legumes important as sources of food and forage. Beans (*Phaseolus*, *Vicia*, and *Vigna*), peas (*Pisum sativum*), soybean (*Glycine max*), and peanuts (*Arachis hypogaea*) are all important as human food, while alfalfa (*Medicago sativa*) and clover (*Trifolium* spp.) are important forage plants. Other members of the family yield gums, such as the famous gum arabic (*Acacia senegal*), and dyes (*Cassia* and *Haematoxylon* spp.). Various tree legumes are used for lumber or as ornamentals: redbud (*Cercis canadensis*), Kentucky coffee-tree (*Gymnocladus dioica*), and honeylocust (*Gleditsia tricanthos*).

ROSACEAE—ROSE FAMILY

The rose family consists of about 100 genera and 3000 species, most of which are perennial herbs, shrubs, and trees. Flower parts (petals, sepals, and stamens) are commonly in multiples of five. The family is sometimes divided into three subfamilies: Rosoideae (rose subfamily), which contains roses (*Rosa* spp.), raspberries and blackberries (*Rubus* spp.), and strawberries (*Fragaria* spp.); Amygdaloidea (peach subfamily), which contains the genus *Prunus* (peach, cherry, apricot, almond, and plum); and Maloideae (apple subfamily), which includes apple (*Malus*), pear (*Pyrus*), quince (*Cydonia*), and hawthorn (*Crataegus*).

RUTACEAE—RUE FAMILY

The rue family comprises about 150 genera and 1500 species, most of them tropical and subtropical aromatic trees and shrubs. Notable are the *Citrus* species (sweet orange, mandarin, lemon, lime, grapefruit, citron, and related hybrids) and rue (*Ruta graveolens*), a pungent, shrubby herb of temperate climates.

SOLANACEAE—NIGHTSHADE FAMILY

There are about 90 genera and 2200 species in the nightshade family, most of them native to South America. *Solanum* is the largest genus with about 1700 species. Many species contain poisonous alkaloids, such as solanine, nicotine, and atropine. Economically important plants include potato (*Solanum tuberosum*), tomato (*Lycopersicon esculentum*), eggplant (*Solanum melongena*), chili peppers (*Capsicum* spp.), tobacco (*Nicotiana tabacum*), and petunia (*Petunia hybrida*).

UMBELLIFERAE (APIACEAE)—CARROT FAMILY

The carrot family consists mostly of north-temperate aromatic herbs, nearly all of them annual or biennial. The 250 genera and 2500 species are characterized by clusters of many small flowers borne in flat-topped umbels or heads. Many are important food or spice plants, including carrot (*Daucus carota*), parsnip (*Pastinaca sativa*), celery (*Apium graveolens*), parsley (*Petroselinum crispum*), anise (*Pimpinella anisum*), caraway (*Carum carvi*), coriander (*Coriandrum sativum*), fennel (*Foeniculum vulgare*), and dill (*Anethum graveolens*). The family also contains various poisonous plants, such as the infamous poison hemlock (*Conium maculatum*).

The Minor Plant Categories

The categories mentioned thus far—the divisions, classes, orders, and families—are known as the major taxa. They are of great evolutionary or phylogenetic importance, but it is the somewhat less arbitrary categories below the family—the minor categories— with which the horticulturist is most concerned. They are, in decreasing order of magnitude (or genetic diversity), the **genus, species, variety, form,** and **individual.** Each of these categories will now be discussed in turn.

The **genus** (the category below family) is a group (taxon) in which the member plants have much in common morphologically. It is usually a small enough group that all of its members can be brought together, and their genetic, cytological, and other relationships studied. It is for this reason the most intensively studied taxon. The generic concept is an old one. Even the ancients were familiar with the oaks (*Quercus*), the roses (*Rosa*), the tulips (*Tulipa*), the pines (*Pinus*), and the maples (*Acer*). Since many of our genera were established long before the time of Darwin, they were based primarily on readily apparent morphological characteristics. As a result of their diverse and frequently popular (that is, nonscientific) origins, these genera are often most annoyingly of different levels of genetic complexity. Modern taxonomists strive to base their systematic studies not merely on the traditional morphological criteria but on genetic and other experimental evidence as well. As a matter of fact, when possible, a taxonomist's main criterion for including or not including a population of plants within a particular genus is a genetic one that can be experimentally verified. The members of such a genus are, by their definition, at least somewhat capable of crossing among themselves, and are absolutely incapable of crossing with the plants of any other genus. A genus of this sort, bounded by total genetic incompatibility, is referred to as a **comparium.** The white oak and red oak groups within the genus *Quercus* (a traditional genus) could each be considered a comparium. A synopsis of such modern taxonomic categories as the comparium is presented below, each category being subdivided by the one underneath it.

Comparium: Integrity maintained entirely by genetic barriers
(Often equivalent to a traditional genus)

Cenospecies: Integrity maintained by genetic barriers reinforced by ecological barriers
(Often equivalent to a traditional genus or species)

Ecospecies: Integrity maintained by ecological barriers reinforced by genetic barriers
(Often equivalent to a traditional species)

Ecotype: Integrity maintained entirely by ecological barriers
(Often equivalent to a traditional subspecies, variety, or form)

The **species** (the category below genus) is made up of plants that often exhibit many more morphological similarities than do the members of the genus. The members of a species have been referred to as coming from like parents and producing like progeny. They often constitute an exclusive interbreeding population. It was long held that all species had been created originally just as they look today.

The species has always been the basic unit of all taxonomic work. It is the fundamental category upon which Linnaeus based his system of nomenclature. He and his students established that each species be identified by two names, the generic and the specific. Examples of this binomial system of nomenclature are *Quercus alba*, the scientific name for white oak, and *Cornus florida*, that of flowering dogwood. It should be emphasized that a specific epithet is invariably preceded by a generic name.

Throughout the world the genus and species are given in Latin using the Roman alphabet (the genus is capitalized; the species should not be capitalized; and both are italicized). The name (or abbreviation of a name) that often follows such a binomial designation is that of the person who first named and described the species—for example, *Spinacia oleracea* Linnaeus (spinach). The naming of plants is governed by the International Code of Botanical Nomenclature, now strictly adhered to by the botanists of every nation. The details of the Code were worked out in Vienna at the turn of the century by botanists from all over the world. Further international congresses have been held from time to time to resolve minor conflicts and to revise the Code where necessary. Here is an example of worldwide cooperation to be envied by our national leaders!

Modern taxonomists (whose interest is to identify evolutionary units and whose system of classification of the minor taxa is based on genetics and ecology to a far greater extent than was the system of their traditional counterparts) have sometimes found it necessary to make use of the concept of the **cenospecies** and that of its subdivision, the **ecospecies.** Taxonomists thus divide the genetically isolated comparium into one or more cenospecies. Each cenospecies encompasses a group of similar plants separated from other cenospecies almost entirely by genetic barriers. Cenospecies do not cross under natural conditions but can be made to cross. The offspring of such a cross are sterile and are often unable even to survive.

The ecospecies maintains its integrity through a combination of ecological and genetic mechanisms. The genetic barriers between ecospecies are somewhat less effective than those found between cenospecies, but these barriers are reinforced by spatial separation that results from the differing ecological requirements of related ecospecies.

Upon investigation, some traditional genera, and even some traditional species, turn out to be equivalent to cenospecies. The one or more ecospecies that make up a cenospecies may often be found the equivalent of the traditional species.

A subclassification of the traditional species is the **variety** (or **botanical variety**—not to be confused with the cultivated variety, or cultivar, which will be discussed in the next section). When one or more of the populations of plants making up a species is sufficiently different in appearance from the remaining members of the species, it is often given varietal status. A variety is designated by a trinomial, such as *Juniperus communis* var. *depressa* (prostrate common juniper). As soon as a variety is established, the remaining typical members of that species are designated by a trinomial formed by repeating the specific name—*Juniperus communis* var. *communis*, for the example just given. A population deviating not quite enough to be called a variety is sometimes called a race or, more properly, a **form** (or **forma**). Two forms of the attractive ornamental *Taxus cuspidata* (Japanese yew) are *Taxus cuspidata* f. *densa* (an erect form) and *Taxus cuspidata* f. *nana* (a spreading form) (Figure 2-16). On the other hand, a group deviating perhaps somewhat more from the species norm than a variety may be ranked as a **subspecies.** (It must be added that in practice the fine distinctions among subspecies, varieties, and forms are sometimes lost.)

Two varieties of a single species, through isolation and subsequent divergent evolution, may eventually differ enough to be considered as two species. If, however, the barrier between the two groups breaks down, they may again

A

B

FIGURE 2-16. Japanese Yew (*Taxus cuspidata;* family Taxaceae, order Taxales, class Gymnospermae). The dark green foliage and small scarlet fruits make this hardy evergreen shrub from the Orient a favorite ornamental. (*A*) An erect form (*T. cuspidata* f. *densa*); (*B*) a spreading form (*T. cuspidata* f. *nana*). There are numerous other yews, some of which attain tree proportions. Yew wood makes good archery bows. [Photograph by J. C. Allen & Son.]

converge. The taxonomist is often hard pressed to decide whether to combine or split two such groups.

The modern systematist who prefers to revise the classification of plant populations on the basis of experimentation with particular emphasis on genetic and ecological considerations finds it necessary, as we have learned, to introduce such new concepts as the comparium, cenospecies, and ecospecies. Each ecospecies may be further divided into **ecotypes.** The ecotype is thus a substitute for the traditional subspecies, variety, or form. It is fully capable of crossing with other ecotypes within its ecospecies, but it is separated from them by differing habitat preferences, which make for different geographic distributions. Ecotypic variation can often be expected to occur in species of wide latitudinal extent. If, on the other hand, the ecospecies exhibits a continuous gradient of characters from one end of its range to the other, this is referred to as **clinal,** or **ecoclinal,** variation.

Normally self-pollinating plants such as tomatoes or peas maintain inbred lines that become stabilized after several generations and thenceforth breed true. Such populations of plants, even more genetically homogeneous than ecotypes, are referred to as **biotypes.**

A special situation exists in *Citrus, Mangifera* (mango), some species of *Rubus* (brambles), *Poa* (bluegrass), and several other plants. These plants outwardly seem to depend upon sexual propagation. In reality, although pollination may occur and may trigger fruit production, fertilization occurs only occasionally. Most new plants are produced from seeds that developed asexually from diploid ovular tissue, thereby bypassing the usual meiotic division. This process is called **apomixis,** and the genetically similar lines thus maintained in these plants are referred to as **apomicts.**

The horticulturist is continuously on the lookout for an individual horticultural plant that, by some quirk of genetic recombination or chance mutation, is especially desirable. To perpetuate this valuable selection or "**sport,**" which might not breed true if left to its own sexual devices, the horticulturist may propagate it vegetatively—that is, create a **clone.** The clone is a group of plants all of which arose from a single individual (the **ortet**) through some means of vegetative or asexual propagation. Each member of a clone, which is technically know as a **ramet,** has a genetic makeup identical with the ortet, or clonal progenitor. The 'Redhaven' peach, the 'Delicious' apple, the 'Russet Burbank' potato, and the 'Thompson Seedless' grape are all examples of clones.

It should perhaps be reemphasized that of the traditional taxa in general use today most are based strictly on floral morphology, some on profound ignorance, and only a handful on sound principles and experimental evidence. *Prunus* (cherry, plum), *Rubus* (raspberries and other brambles), *Vitis* (grape), *Rosa* (rose), and *Crataegus* (hawthorn) are, for example, just a few of the horticulturally important genera requiring extensive revision on the basis of modern taxonomic experimentation. The field is wide open for those with imagination and broad scientific interests.

Cultivar: The Cultivated Variety

The cultivar is the entity that is of most relevance to horticulturists. In fact, cultivars in a sense can be considered the very keystone of horticulture. For horticulturists the term *apple* is vague. The 'Delicious' apple, on the other hand, represents a very specific type of apple tree with a characteristic fruit that has a unique combination of flavor, shape, and behavior.

The term **cultivar,** a contraction of the term *cultivated variety*, refers to a named group of plants within a cultivated species that is distinguishable by a character or group of characters and that maintains its identity when propagated either asexually or sexually. Examples include the 'Jonathan' apple, the 'Tendercrop' snapbean, the 'Danish Ballhead' cabbage, and the 'Golden Cross' sweet corn. (Note that the cultivar name is capitalized, printed in nonitalic type, and set off by single quotation marks.)

Cultivar has replaced the older term *variety*, which unfortunately has a double meaning. In the taxonomic sense *variety* (i.e., a *botanical variety* or *varietas*) is a subclass of a species with a distinct difference in appearance. The names of botanical varieties are always in Latin, and usage of this term is governed by the Botanical Code of Nomenclature. In the horticultural or agricultural sense, *variety* indicates a *cultivated variety* as defined above. To avoid confusion the term *cultivar* has replaced *variety* in the *horticultural* sense, and the term *botanical variety* has replaced *variety* in the *botanical* sense.

There are several categories of cultivars based on their mode of reproduction. Thus, biologically speaking, a cultivar can be one of several entities distinguished by its method of propagation. These include clonal cultivars, pure-line cultivars, open-pollinated cultivars, and hybrid cultivars.

Clonal cultivars are asexually propagated clones. Thus, such cultivars of potato, strawberry, and rose as the 'Russet Burbank' potato, the 'Tioga' strawberry, and the 'Peace' rose are unique genotypes produced by vegetative propagation (tubers in the potato, runner plants in the strawberry, and bud grafts in the rose). They are not normally propagated by seed because, as these crops are typically heterozygous and cross-pollinating, seed propagation would result in genetic segregation, and the unique characteristics of the clone would be lost via recombination and segregation. (The 'Russett Burbank' potato does not even produce viable seeds.)

Variation in clones due to genetic changes (mutations) are called strains or, in horticultural parlance, **sports.** When such a variant can be demonstrated to be sufficiently different from the parent cultivar to render it worthy of a new name, it is regarded as a distinct new cultivar. Thus, more than 100 genetic sports of 'Delicious' apple have been described. Some of these strains have been given new cultivar names that stress their origin from 'Delicious' apple, such as 'Starking Delicious,' 'Starkrimson Delicious,' 'Richared Delicious,' and 'Redchief Delicious.'

In some plants clonal propagation can be made by seed that is produced vegetatively. This is known as apomixis. 'Kentucky' bluegrass (*Poa pratensis*), a favorite for temperate lawns as well as a prime pasture species, is largely apomictic. Many types of citrus cultivars, especially rootstocks, are normally propagated via apomictic seed.

Pure-line cultivars are inbred lines of either self- or cross-pollinating species. Self-pollinating species, such as the pea or tomato, essentially breed true after six generations of selfing because these lines become homozygous (see Chapter 17) and each plant when grown from seed reproduces its genetic constitution in tact. Thus, pure-line cultivars of self-pollinating crops breed true from seed provided care is maintained to eliminate genetic contamination due to seed mixing or to an occasional outcross.

Inbred lines of cross-pollinated crops can be considered as cultivars, but such lines must be artificially selfed (or be vegetatively propagated) to maintain their identity. In cross-pollinated crops inbred lines are usually nonvigorous but are valuable as parents to produce hybrid seed.

Hybrid cultivars are produced by intercrossing genetically divergent parental lines. In some cases different species may be involved. Hybrid cultivars are heterozygous and do not breed true from sexual seeds. However, hybrid cultivars can be either vegetatively propagated, or, in the case of plants nonadapted to vegetative propagation, they can be created by crossing established pure lines. Hybrids that are produced anew each generation from intercrossing two homozygous lines are known as F_1 hybrid cultivars. Other types of hybrid cultivars (three-way crosses, double crosses, topcrosses) will be discussed in Chapter 17.

Open-pollinated cultivars of cross-pollinated crops, such as cabbage or spinach, are produced by a combination of selection and isolation of parental lines. They are seed-propagated populations that are differentiated from other populations by the presence of one or more characters. Open-pollinated populations still maintain variability, but selection maintains the distinguishable characteristic and hence cultivar identity.

Cultivar Identification

In the definition of cultivars there is the implicit assumption that a cultivar be distinguishable from other cultivars by some recognized criteria. The problem of cultivar identification is not a trivial one because many characters that might appear useful for identification show large changes in response to the environment and may also be affected by human judgment. Proper identification is essential to establish "trueness-to-name" in commercial channels, to obtain patent protection, to settle infringement disputes, and to determine synonymy—that is, whether two or more cultivars with different names are in fact the same or different cultivars.

For example, the identification and classification of fruit plant clones has been based on pomological descriptions of the fruit and the plant. Usually the fruit is of paramount importance and is often the sole reliable means of identification. In this connection descriptive pomological terms have been developed for each fruit species based on botanical and horticultural considerations—a specialized field known as **systematic pomology.** For example, fruit descriptions of pome fruits (apple, pear, quince) are based on fruit characters such as basin, calyx, calyx tube, carpels, cavity core, corelines, color, dots, flesh type and color, form, season of ripening, size, skin, stem, and use. Plant descriptions involve buds, bark color, foliage, plant form, hardiness, size, and vigor. New quantitative methods in systematics, complemented by computer technology, have been developed that aid in discrimination and classification—a field known as **numerical taxonomy.**

The identification of nursery stock cultivars from vegetative characters alone has been used to detect misnamed trees in the nursery row. However, the system is only practical with a limited number of cultivars, and the process requires great skill and experience. In general, this technique has proven impractical because of the proliferation of cultivars and the increasing numbers of sports and strains that may differ only by fruit characters.

The problem of cultivar identification has become increasingly important with the advent of plant patenting. A precise, nondisputable, single means of cultivar identification would have many commercial benefits.

Two new approaches to the problem of distinguishing cultivars are the identification of morphological ultrastructure and the determination of chemical taxonomy. Use of the scanning electron microscope has made possible the identification of new morphological details on bud, leaf, and pollen surfaces, but the general applicability of this technique remains unknown. A number of biochemical analysis techniques have been used for botanical systematics. The most promising technique for cultivar identification appears to be the separation of plant proteins by gel electrophoresis, a process referred to as "plant fingerprinting," but cultivar keys have been developed for only a limited number of cultivars.

THE IDENTIFICATION OF PLANTS

When people are confronted with plants unknown to them they may turn for assistance to an appropriate book of plant descriptions. Horticulturists living in the continental United States are fortunate in having at their disposal several regional manuals, as well as two that span the continent. These latter two manuals (see Selected References at the end of this chapter) describe virtually all of the horticultural species capable of surviving under our environmental conditions (including greenhouse conditions). One was compiled by the renowned American horticulturist Liberty Hyde Bailey, the other by the well-known dendrologist Alfred Rehder.

Should the plant in question be an exotic not known to be cultivated in the continental United States, it will not be covered by either Bailey or Rehder.

One can then often learn which manual to turn to by consulting the list compiled by Sidney F. Blake and Alice C. Atwood (see Selected References).

The use of manuals enables horticulturists to identify most plants they are likely to encounter. The possibility always exists, however, that the unknown plant may have been considered to be too rare or too unimportant horticulturally for inclusion, or even that it is a plant new to science. The safest recourse under such circumstances is to send the plant to some botanical institute for identification (or naming!) by a professional taxonomist.

The actual identification of an unknown plant is usually accomplished through the use of the **analytical keys** that are a part of most manuals. A key contains the diagnostic features of all plants listed in the manual arranged in such a manner that all but the correct plant can be rejected. A dichotomous key, the kind most commonly used, is constructed as a series of couplets, each containing a pair of contrasting statements. One examines the first couplet and chooses the statement that fits the unknown plant; this leads to the next correct couplet to use, and that to the next, and so on until the plant is "keyed out." The following is an example of such an analytical key.

AN ANALYTICAL KEY TO THE WALNUTS (*Juglans*)

1A	Leaf scars with a hairy upper fringe; pith not light brown	2
1B	Leaf scars with a glabrous upper edge; pith light brown	3
2A	Pitch dark brown; fruits solitary or in clusters of 2 to 5	Butternut, or white walnut (*J. cinerea*)
2B	Pith violet-brown; fruits in clusters of 12 to 20	Japanese walnut (*J. sieboldiana*)
3A	Leaves glabrous; bark smooth and silvery gray	Persian, or English, walnut (*J. regia*)
3B	Leaves pubescent; bark rough and dark brown	Black walnut (*J. nigra*)

Some keys may be ambiguous with respect to species, especially if an attempt has been made to separate the species by characters other than those upon which the species were originally based. If desired, final verification can be made by comparison with a known example of that species. Such examples can be found as dried and pressed specimens (known as herbarium specimens) in many botanical institutes. Finally, it must be pointed out that in order to get full information on varieties one frequently must consult monographs dealing with the particular plant group in question.

CONCLUSION

The subject of plant classification, although perhaps the most elementary branch of horticulture, is also its most inclusive one. Nothing in horticulture can be discussed in a scientific way without some taxonomic knowledge; furthermore, a classification serves to integrate and to summarize all that we know about our plants, whether morphological, genetic, ecological, or physiological. A knowledge of classification often permits the horticulturist to predict the cultural requirements of a plant. It also helps the horticulturist predict its graft compatibilities and the other plants with which it will hybridize. Finally, it aids horticulturists in their search for and development of new plants of horticultural importance.

Selected References

Bailey, L. H., 1949. *Manual of Cultivated Plants Most Commonly Grown in the Continental United States and Canada*, rev. ed. New York: Macmillan. (This manual provides a means for the identification of 5347, or almost all, cultured tracheophyte species covered by the title. Varieties are not described. Identification is done through keys, descriptions, and some illustrations. There is a glossary.)

Bailey, L. H., E. Z. Bailey, and the Staff of the L. H. Bailey Hortorium, 1976. *Hortus Third: A Concise Dictionary of Plants Cultivated in the United States and Canada.* New York: Macmillan. (This monumental work, whose earlier editions were called *Hortus* [1930] and *Hortus Second* [1941], covers 34,305 families, genera, and species of cultivated plants.)

Blake, S. F., and A. C. Atwood, 1942, 1961. *Geographical Guide to Floras of the World; Annotated List with Special Reference to Useful Plants and Common Plant Names. I: Africa, Australia, North America, South America, and Islands of the Atlantic, Pacific, and Indian Oceans* (USDA Misc. Pub. 401; 1942). *II: Western Europe: Finland, Sweden, Norway, Denmark, Iceland, Great Britain with Ireland, Netherlands, Belgium, Luxembourg, France, Spain, Portugal, Andorra, Monaco, Italy, San Marino, and Switzerland* (USDA Misc. Pub. 797; 1961). Washington, D.C.: U.S. Government Printing Office. (This exhaustive bibliography, containing 9866 titles, is the only one of its kind. It covers over half the world, but does not include Germany, central and eastern Europe, or Asia and associated islands.)

Lawrence, G. H. M., 1951. *Taxonomy of Vascular Plants.* New York: Macmillan. (A standard text in systematic botany. Part I covers both the theoretical and practical aspects of classification, nomenclature, and identification. Part II describes all tracheophyte families known to be native to, or introduced into, the United States. The appendix contains a detailed glossary of botanical terms.)

Porter, C. L., 1967. *Taxonomy of Flowering Plants*, 2d ed. New York: W. H. Freeman and Company. (An excellent elementary text.)

Rehder, A., 1940. *Manual of Cultivated Trees and Shrubs Hardy in North America Exclusive of the Subtropical and Warmer Temperate Regions*, 2d ed. New York: Macmillan. (This manual includes 2550 species with about 2900 varieties and about 580 hybrids, or almost all the cultured trees, shrubs, lianas, and partially woody plants covered by the title. There are keys to the species, a glossary, and species descriptions that include indications of hardiness.)

Sneath, P. H. A., and R. R. Sokal, 1973. *Numerical Taxonomy: The Principles and Practice of Numerical Classification.* New York: W. H. Freeman and Company. (An advanced text in the quantitative approach to taxonomic classification.)

Swain, T. (ed.), 1963. *Chemical Plant Taxonomy.* London: Academic Press. (Chemical and allied techniques for determining relationships among plants.)

3 *Structure*

THE PLANT BODY

Flowering plants, which make up almost the entire range of horticultural interest, show extreme diversity in size and structure. Nevertheless, there are essential similarities: Many structures (for example, the air-borne root of the orchid and the swollen root of the sweetpotato), although superficially very different, are functionally and morphologically related. The flowering plant consists of two basic parts—the **root,** the portion that is normally underground, and the **shoot,** the portion that is normally above ground. The shoot is made up of stems and leaves. The leaves grow from enlarged portions of the stem called **nodes.** The shoot shows several significant modifications. The **flower** may be thought of as a specialized stem with leaves adapted for reproductive functions. **Buds** are miniature leafy or flowering stems. Figure 3-1 illustrates the fundamental plant parts.

The plant body is made up ultimately of microscopic components called **cells,** which, although they are essentially similar, may be structurally and functionally very different. Masses of cells in various combinations and arrangements build up the various morphological structures of the plant.

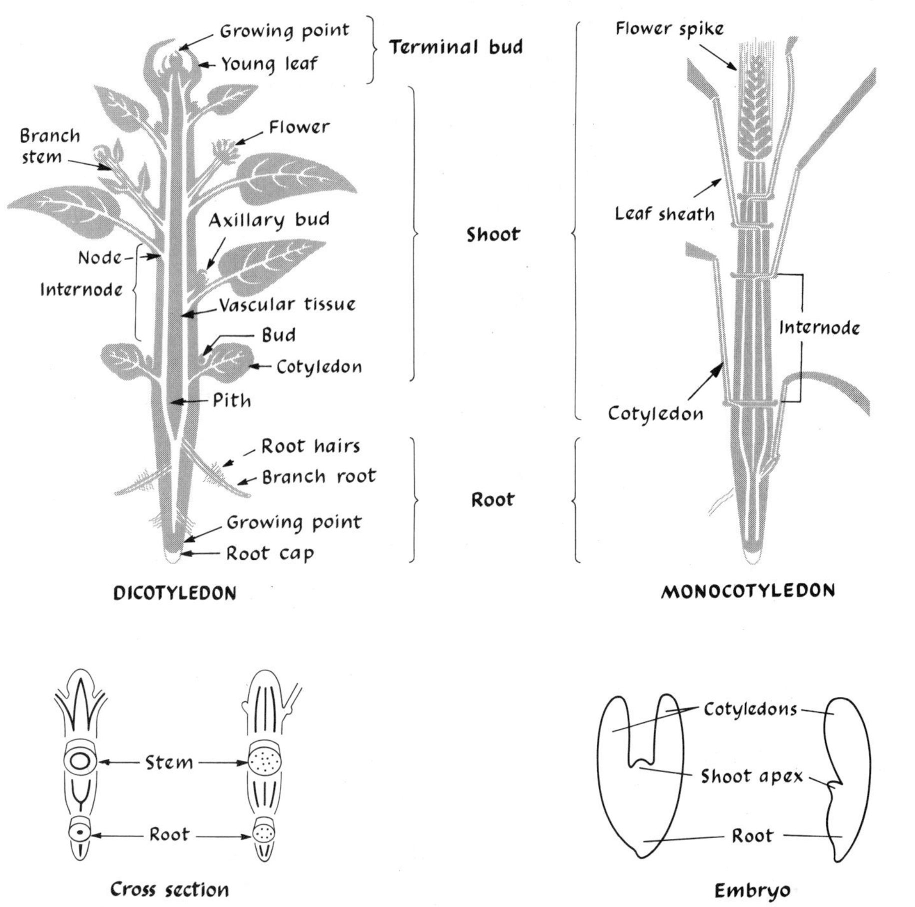

FIGURE 3-1. The fundamental plant parts. [From J. Janick et al., *Plant Science*, 2d ed., W. H. Freeman and Company, copyright © 1974.]

THE CELL AND ITS COMPONENTS

The structural unit of plants, as well as of animals, is the cell. **Cytology,** the study of cells, is concerned with their organization, structure, and function. The concept that the cell is the basic unit of higher organisms is one of the tenets of biology. The complex multicellular organism is an integrated collection of living and nonliving cells. The high degree of synchronization and coordination in the total organism creates an entity that is in effect greater than the sum of its parts.

Plant cells vary in shape from spherical, through polyhedral and ameboidal, to tubular. Most cells are between 0.025 mm and 0.25 mm in size

(0.001 to 0.1 inch), but some cells (long, tubular fibers) are as much as 2 feet long. The concept of the typical, or generalized, plant cell, such as the one in Figure 3-2, is a useful one. The most prominent components visible under the light microscope are the densely staining nucleus and the more or less rigid cell wall that encloses the cytoplasm. The cytoplasm contains a number of structural bodies, or organelles, such as plastids, mitochondria, and vacuoles, and various

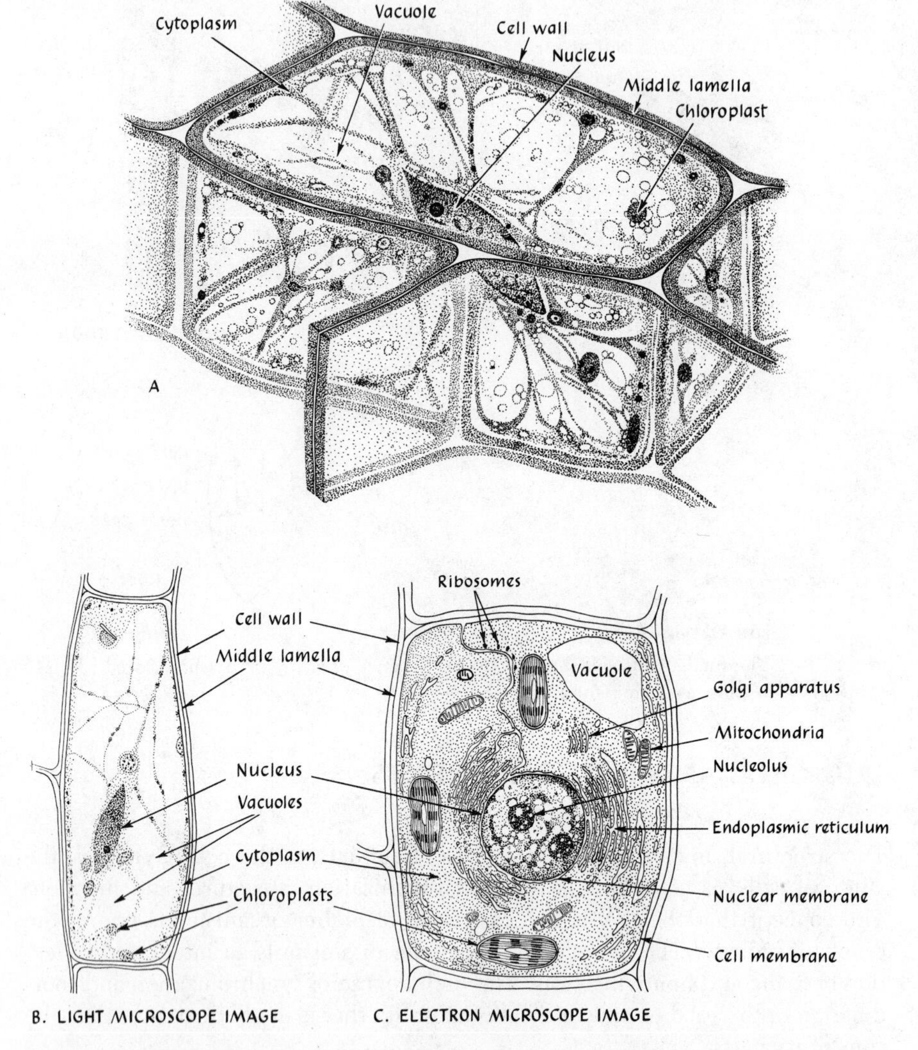

FIGURE 3-2. Three views of a plant cell and its component parts. Recent electron microscope studies have produced a new concept of the ultrastructure of the cell. [Parts *A* and *B* after K. Esau, *Plant Anatomy*, Wiley, 1953. Part *C* after J. Brachet, "The Living Cell," copyright © 1961 by Scientific American, Inc., all rights reserved.]

other entities such as crystals, starch grains, and oil droplets. A distinction is often made between living and nonliving substances in the cell, but such distinctions have little meaning.

The **cytoplasm** (less precisely termed protoplasm) is an extremely complex substance, both physically and chemically. It is composed (by fresh weight) of 85 to 90 percent water; the remaining 10 to 15 percent consists of organic and inorganic substances that are either dissolved (salts and carbohydrates) or in the colloidal state (proteins and fats). The physiological activity of the cell occurs in the cytoplasm. Although the living state appears to be dependent on the cytoplasm within the structural confines of a cell, various so-called vital processes (such as photosynthesis) have recently been found to be possible outside of the cell. The concept of what constitutes a vital process changes as biological techniques increase in sophistication.

Surrounding the cytoplasm is the **plasma membrane,** which lies against the inside of the cell wall. The cell wall is to some degree permeable to all solutes and solvents; the plasma membrane is semipermeable. The plasma membrane is composed of lipoproteins, which account for its elasticity and high permeability to fatty substances. Although it is ultramicroscopic in thickness, the plasma membrane has been viewed in plasmolyzed cells: The electron microscope has revealed that is is not a mere boundary to the cell but has structural definition. The cell appears permeated by internal membranes that, along with the external membrane, are thought to constitute a system called the **endoplasmic reticulum.**

The **nucleus,** a dense, usually spheroidal body, is located within the cytoplasm. The close affinity of the nuclear material for many dyes makes it the most conspicuous feature of a stained cell. The nucleus is, in effect, the control center of the cell, since it contains the **chromosomes** (*chromo*, colored; *soma*, body). The chromosomes contain deoxyribonucleic acid (DNA) in association with protein. The arrangement of the DNA provides genetic information for the cell, in somewhat the same way that punch cards provide information for an electronic computer. The information is relayed to the cytoplasm in the form of RNA (ribonucleic acid), a substance similar to DNA, and it affects the machinery of the cell through the control of protein synthesis. The actual site of protein synthesis is not the nucleus but small particles in the cytoplasm called **ribosomes.** The information relayed by DNA provides the basis for the physiological functioning of the cell and thus determines the metabolic and morphological features of the organism. It is also a major part of the hereditary bridge between generations. About 95 percent of the DNA in the cell is in the nucleus; the other 5 percent is divided between plastids and mitochondria. Just how DNA does all of this will be discussed in Chapter 6 where we focus our attention on plant reproduction.

The plant **cell wall,** one of the structures that distinguishes plant cells from animal cells, is usually thought of as a deposit, or secretion, of the cytoplasm. It is composed of three basic groups of compounds: polysaccharide materials (which are long-chain units formed from such simple sugars as glucose), lignin, and pectin.

The principal polysaccharide in the cell wall is generally **cellulose,** which is an *unbranched* polymer of a few thousand glucose molecules. Other cell-wall polysaccharides are frequently referred to as hemicelluloses, which is a vague, broad term for *branched* chain polysaccharides containing a variety of monosaccharide units, only one of which is glucose, as well as such nonsugar components as protein.

Cellulose and its derivatives are the most abundant materials in the structure of plants. They are highly combustible materials. They are indigestible directly by mammals, including humans, because mammals do not possess an enzyme that is capable of breaking the bonds between the glucose units in cellulose. However, various bacteria do secrete an appropriate enzyme, and those mammals (such as ruminants) that sustain themselves on a diet rich in cellulose are able to do so because these bacteria grow in their digestive tracts. Through fermentation, these bacteria begin the decomposition of the cellulose, with the remaining digestion completed by the host animal.

Lignin consists of a complex mixture of chemically related compounds—specifically, polymers of phenolic acid. The deposition of lignin (lignification) hardens the cellulose walls into an inelastic and enduring material resistant to microbial decomposition. Because lignin causes yellowing in paper, it must be dissolved from any wood pulp used in the manufacture of high-quality papers.

Pectins are important components of the cell wall. They are acidic polysaccharides, specifically water-soluble polymers of galacturonic acid that forms sols and gels with water. The most familiar kind of pectin is that used as a solidifying agent for jams and jellies.

The cell wall is laid down in distinct layers, and its thickness varies greatly with the age and type of cell (Figure 3-3). Three regions of the cell wall are generally distinguished:

1. The **middle lamella** is a pectinacious material associated with the intercellular substance. The slimy nature of rotted fruit is due to the dissolving of the middle lamella by fungal organisms.

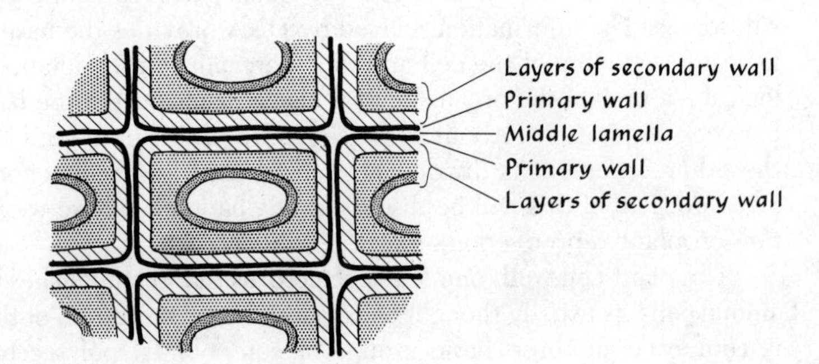

FIGURE 3-3. Diagrammatic cross section of a tracheid, showing the structure of the cell wall. The secondary cell wall deposited inside the primary wall may consist of three layers.

2. The **primary cell wall** is the first wall formed in the developing cell. It is composed largely of cellulose and pectic compounds, but closely related substances and noncellulose compounds may also be present, and it may become lignified. The primary cell wall is the wall of dividing and growing cells; in many cells it is the only wall.
3. The **secondary cell wall** is laid down inside the primary cell wall after the cell has ceased to enlarge. It appears to have a mechanical function and is similar in structure to the primary wall, although it is higher in cellulose. It often contains some lignin.

The cell wall in plants is not continuous. It appears to be pierced by cytoplasmic strands (plasmodesmata) that provide a living connection between cells. Furthermore, thin areas called pits occur in the walls of some cells (Figure 3-4).

Plastids are specialized disk-shaped bodies in the cytoplasm, and they are peculiar to plant cells. They are classified, on the basis of the presence or absence of pigment, into leucoplasts (colorless) or chromoplasts (colored). Leucoplasts occur in mature cells that are not exposed to light, and some types are involved in the storage of starch. Of the colored plastids, those containing chlorophyll (**chloroplasts**) are the most significant, for they are the complete structural and functional unit of photosynthesis, the process by which carbon dioxide and water are transformed to carbon-containing compounds. The reactions result in the formation of starch grains within the chloroplast.

There are about 20 to 100 chloroplasts in each chlorophyllous cell of a typical green leaf. (Mature leaf cells of spinach, however, may contain up to 500

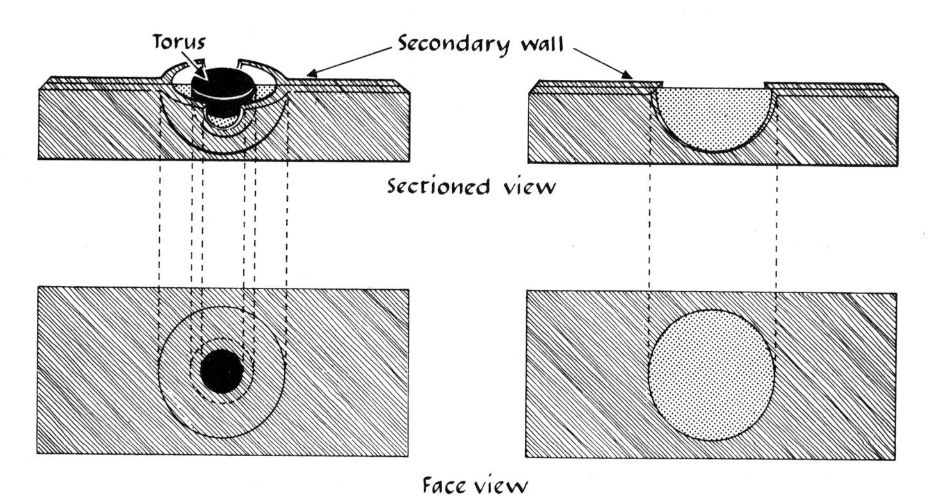

FIGURE 3-4. Pits are thin areas in the cell wall. (*Left*) Bordered pit; (*right*) simple pit. [Adapted from A. J. Eames and L. H. MacDaniels, *An Introduction to Plant Anatomy*, McGraw-Hill, 1947.]

chloroplasts.) Electron microscope photographs have revealed the internal structure of chloroplasts (Figure 3-5). Structural units called **grana,** which resemble stacked coins, contain the chlorophyll and are the receptors of light. The actual transformation of carbon dioxide to carbon-containing compounds occurs in the surrounding material, called **stroma.**

Chloroplasts have a kind of independent existence in the cell. Although they are under the influence of nuclear genes, the chloroplasts are self-replicating and contain DNA; they apparently arise only from preexisting plastids.

Mitochondria are the cell's power centers. They appear as small, dense granules under the light microscope, but electron microscopy has revealed that they have a complex involuted internal structure. The mitochondria are made up of proteins and phospholipids. All of the known functions of mitochondria are related to enzymatic activity connected with oxidative metabolism. This activity occurs through the formation of the energy-carrying substance called adenosine triphosphate (ATP). Mitochondria are self-replicating and contain DNA.

Vacuoles are membrane-lined cavities located within the cytoplasm. They are filled with a watery substance known as the cell sap, which contains a number of dissolved materials—salts, pigments, and various organic metabolic constituents. In actively dividing cells the numerous vacuoles are small; in

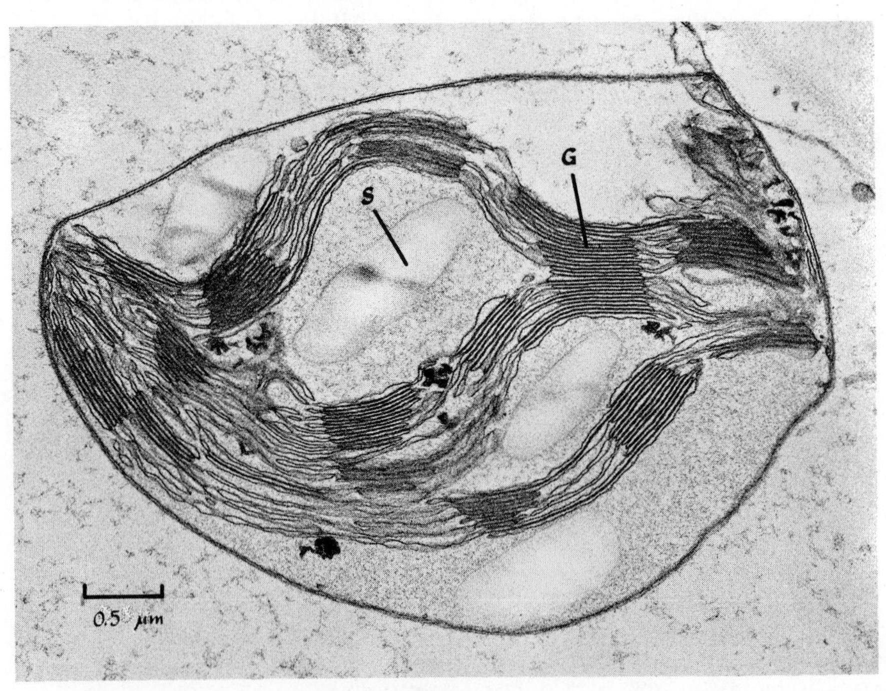

FIGURE 3-5. An electron micrograph of a chloroplast from a tobacco leaf. Note granum (*G*) and starch grain (*S*). [Courtesy W. M. Laetsch.]

mature cells they coalesce into one large vacuole that occupies the center of the cell, pushing the cytoplasm and the nucleus next to the cell wall.

A number of complex materials may be found in the cytoplasm; among these are crystals, starch grains, oil droplets, silica, resins, gums, alkaloids, and many organic substances. Many of these compounds are reserve or waste products of the cell.

TISSUES AND TISSUE SYSTEMS

Although the plant ultimately originates from a single cell (the fertilized egg), the marvels of cell division and differentiation produce an organism composed of many kinds of cells that are structurally and physiologically diverse. It is this difference in cell morphology and cell arrangement that results in the complex variation between plants and within an individual plant.

Plants can be shown to be made up of groups of similar types of cells that are organized in a definite pattern. Continuous, organized masses of similar cells are known as **tissues.** Tissues have been classified in several ways. No universally accepted system exists. The following system is a logical one that retains the customary botanical terms.

- **Meristematic tissue:** actively dividing undifferentiated cells
- **Permanent tissues:** nondividing differentiated cells
  - **Simple tissues:** composed of one type of cell
    - **Parenchyma:** simple thin-walled cells
    - **Collenchyma:** thicker-walled "parenchyma"
    - **Sclerenchyma:** thick-walled supporting cells
  - **Complex tissue:** composed of more than one type of cell
    - **Xylem:** water-conducting tissue
    - **Phloem:** food-conducting tissue

Meristematic Tissue

Meristematic tissue is composed of cells actively or potentially involved in cell division and growth. The meristem not only perpetuates the formation of new tissue but also perpetuates itself. Since many so-called permanent tissues may, under proper stimulation, assume meristematic activity, no strict line of demarcation exists between meristematic and permanent tissue.

Meristematic tissues are located in various portions of the plant (Figure 3-6). Those at the tips of shoots and roots are known as **apical meristems.** The shoot apical meristem is known horticulturally as the growing point. The increase in girth of woody stems results from the growth of lateral meristems,

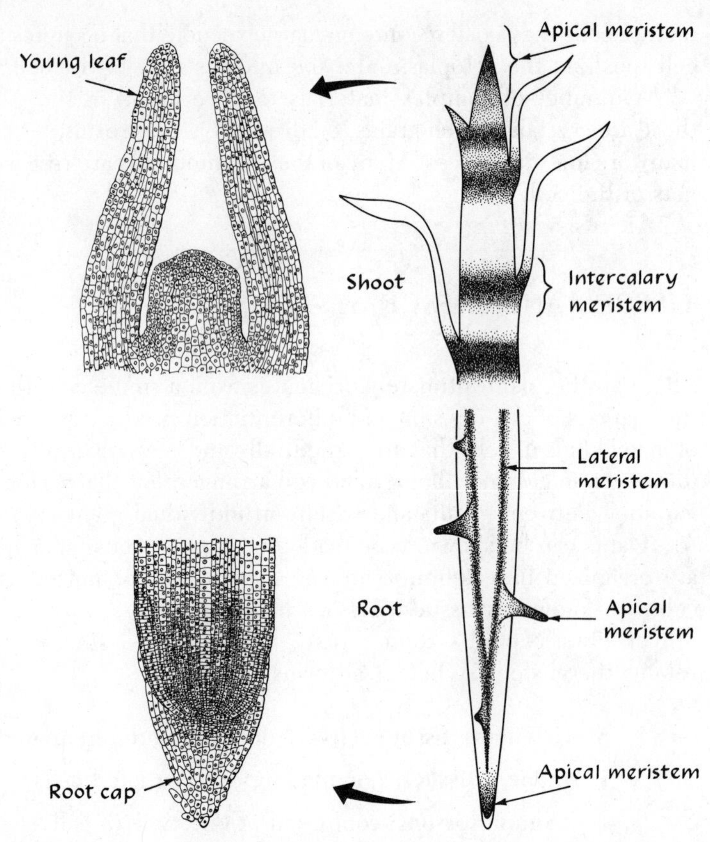

FIGURE 3-6. Diagrammatic longitudinal section of a grass plant, showing the location of the meristems. These shaded areas are the youngest parts of the plant. [Adapted from A. J. Eames and L. H. MacDaniels, *An Introduction to Plant Anatomy*, McGraw-Hill, 1947.]

specifically referred to as the **cambium.** The meristematic regions of grasses become "isolated" near the nodes and are called **intercalary meristems.** Thus, the mowing of lawns does not interfere with the growth of the grass plant because the growing points are not damaged by mowing. Tissues differentiated from apical meristems are referred to as **primary tissues.** Others, especially tissues formed from the cambium, are **secondary tissues.**

Although there are many exceptions, meristematic cells are usually small, often roughly spherical to brick-shaped, and have thin walls with inconspicuous vacuoles. In sections prepared for microscopic examination they appear darkly stained, owing to the small amount of cytoplasm in relation to the nucleus.

Permanent Tissues

Permanent tissues are made up of nondividing differentiated cells derived directly from meristems. They are referred to as **simple tissues** when they are composed of one type of cell (Figure 3-7), and as **complex tissues** when they are mixtures of cell types.

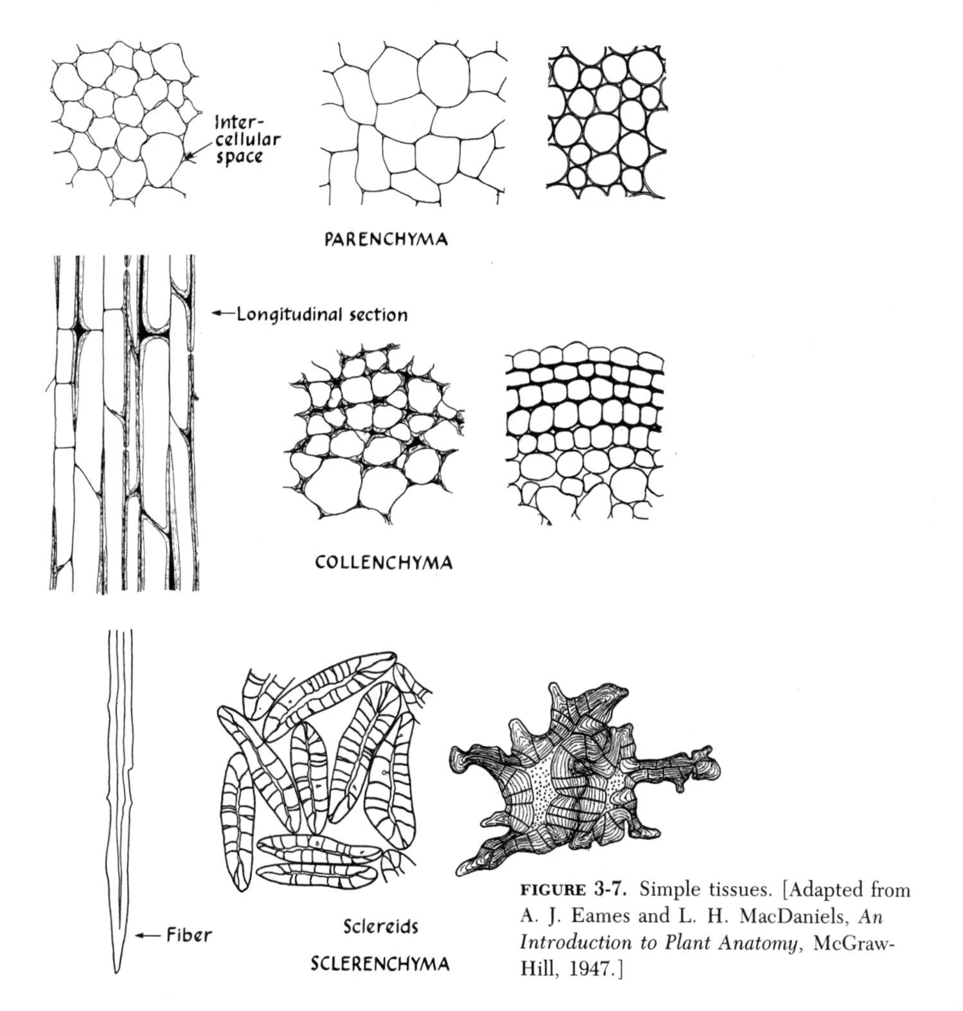

FIGURE 3-7. Simple tissues. [Adapted from A. J. Eames and L. H. MacDaniels, *An Introduction to Plant Anatomy*, McGraw-Hill, 1947.]

Simple Tissues

Simple tissues are of three types: parenchyma, collenchyma, and sclerenchyma.

Parenchyma is relatively undifferentiated, unspecialized vegetative tissue. It makes up a large portion of many plants, such as the fleshy portion of fruits, roots, and tubers.

Collenchyma is tissue characterized by elongated cells with thickened primary walls composed of cellulose and pectic compounds. (It may be thought of as thick-walled parenchyma). This tissue functions largely as mechanical support in early growth. The strands at the outer edge of a celery stalk are collenchyma (Figure 3-8).

Sclerenchyma is tissue composed of especially thick-walled cells that are often lignified. When these cells are long and tapered they are usually referred

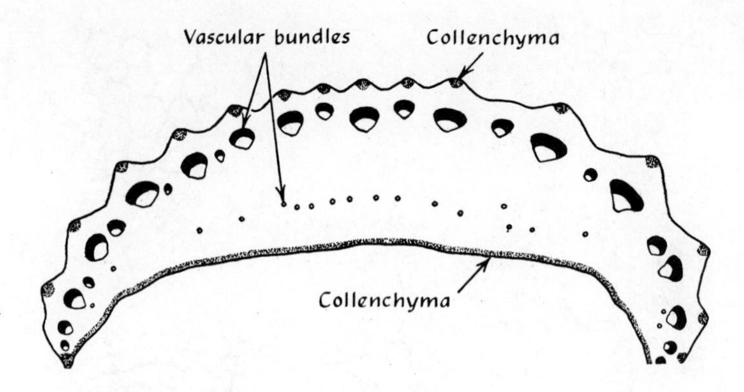

FIGURE 3-8. Cross section of a celery petiole, showing the distribution of collenchyma and vascular bundles. [Adapted from K. Esau, *Plant Anatomy*, Wiley, 1953.]

to as fibers. Others are referred to as sclereids. Clusters of these sclereids, or "stone cells," are responsible for the gritty texture of pears. In masses, sclereids are responsible for the hardness of walnut shells and of peach and cherry pits. Unlike parenchyma and collenchyma, sclerenchyma cells are nonliving when mature.

Complex Tissues

Combinations of simple tissue and specialized tissue form complex tissues, the two major types of which are xylem and phloem.

Xylem, the principal water-conducting tissue, is an enduring tissue that consists of living and nonliving cells. Wood is largely xylem. Herbaceous plants also contain xylem, although the amount is very much less than that contained in "woody" plants. The water-conducting function in xylem is accomplished through specialized nonliving cells called **tracheids.** These are elongated tapered cells with walls that are hard and usually lignified, although not especially thick. The water, in the form of cell sap, moves readily through the empty tracheid, flowing from cell to cell through the numerous pits between them. The cell walls of tracheids are often unevenly thickened or sculptured, which makes these cells easy to distinguish in longitudinal section.

In addition to tracheids, a specialized series of cell members called vessels also functions in water conduction. These are formed from meristematic cells from which the cell contents and end walls have been dissolved. Such cells are lined up end to end, and the series may be many feet long. Xylem typically includes fibers and parenchyma cells (Figure 3-9).

Xylem is formed by differentiation of the apical meristems of root and shoot. In perennial woody plants, secondary xylem is also formed in the familiar annual rings. The spring wood consists of larger cells with thinner walls and appears lighter, or less dense, than the summer wood.

The principal food-conducting tissue is the **phloem.** Its basic components are series of specialized cells called **sieve elements.** It is through these elon-

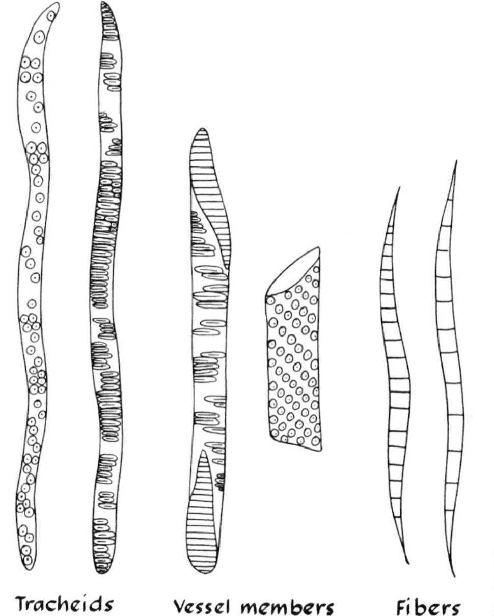

FIGURE 3-9. Tracheids, vessels, and fibers of the xylem. [Adapted from K. Esau, *Plant Anatomy*, Wiley, 1953.]

gated living cells with thin cellulose walls that the food is conducted from one part of the plant to another. Upon maturity the nucleus of the sieve cell disappears. Specialized cells called **companion cells** are in intimate association with the sieve cells (Figure 3-10). In addition to these, fibers and sclereids may be present in phloem. The fibers of hemp and flax are derived from the phloem tissue.

The phloem is formed, as is the xylem, both by the apical meristem and by the cambium. The phloem, however, is not enduring, and the old phloem disintegrates in woody stems. It is protected by special meristematic tissues (**cork cambium**) that produce parenchymatous tissue. The phloem, the corky tissue, and the other incidental tissues constitute **bark.**

ANATOMICAL REGIONS

The tissues that form the various regions of the plant can be classified in terms of structure and function. In much the same way, the bricks, boards, pipe, and wire used in the construction of a house (figuratively speaking, its cells and tissues) can be classified, on the basis of structure and function, as masonry, frame, plumbing system, and wiring system. Since these are interrelated, it is sometimes hard to decide where one region ends and another begins. (Is an electric hot water heater part of the electrical system or the plumbing system?)

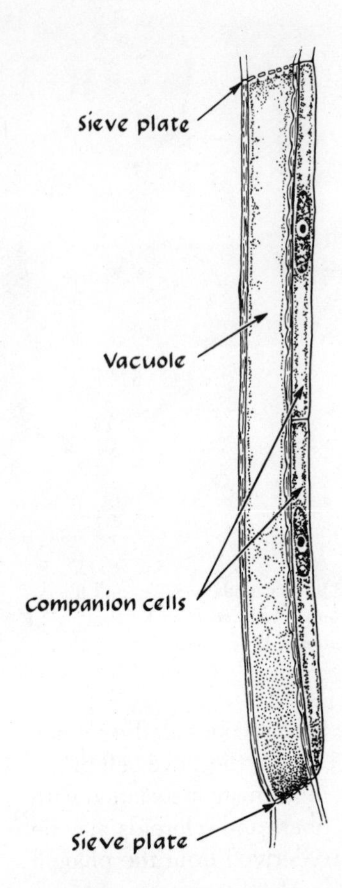

FIGURE 3-10. The sieve elements of the phloem. [Adapted from K. Esau, *Plant Anatomy*, Wiley, 1953.]

Horticultural plants are grossly divided into the vascular system (plumbing), the cortex (frame and insulation), and the epidermis (siding, floor, and roof) (Figure 3-11). The pith, pericycle, endodermis, and secretory glands, however, are components of one or more of these regions.

The Vascular System

The vascular system consists principally of the xylem and phloem tissues. The vascular system serves as the conduction system of the plant, but because it also serves in support it may be compared to both the circulatory and skeletal systems of animals. There are differences between the structural relationships of xylem and of phloem. Typically, the vascular system forms a continuous ring in the stem, in which the inner portion is xylem, surrounding an area of parenchymatous tissue known as the **pith.** The vascular system, however, may be discontinuous, and may appear as a series of strands in longitudinal section and as bundles in cross section (Figure 3-12). This is usual in monocotyledonous plants.

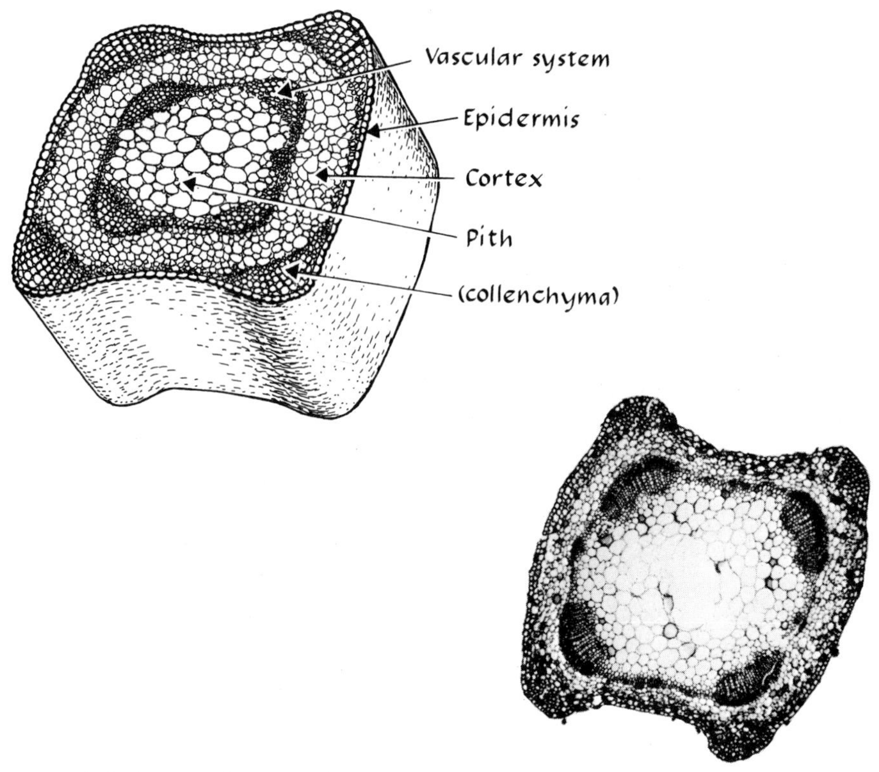

FIGURE 3-11. Anatomical regions of a mint stem.

In roots, the vascular system makes up the core, and the pith is absent. It is separated from the cortex by specialized tissues called the **pericycle** and the **endodermis.** The pericycle encircles the vascular system. It is composed of parenchymatous tissue and is the source of the branch roots and stems that arise from the root. The endodermis is commonly a single sheet of cells separating the vascular system from the cortex, and it is not absolutely clear whether it is part of the vascular system or of the cortex. It appears to have a protective function. The pericycle and endodermis are usually absent in the stem.

Cortex

The cortex is the region between the vascular system and the epidermis. It is made up of primary tissues, predominantly parenchyma. In older woody stems, the formation of cork in the cortex, with the subsequent disintegration of the outer areas, tends to obliterate the cortex as a distinct area. Cork is formed when mature tissue is infiltrated by a waxy substance known as **suberin,** which essentially waterproofs the cell walls. This process is known as **suberization,**

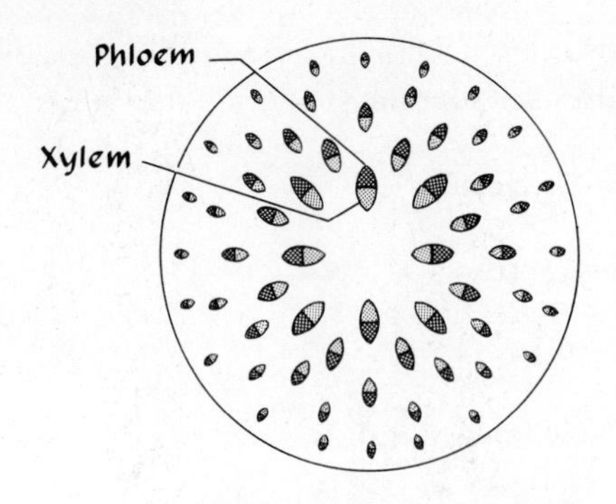

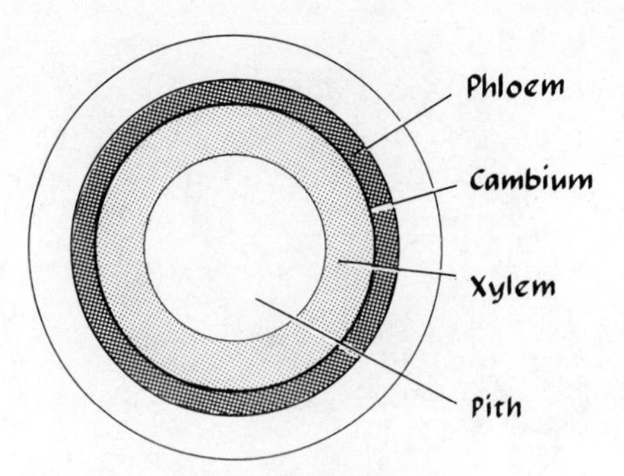

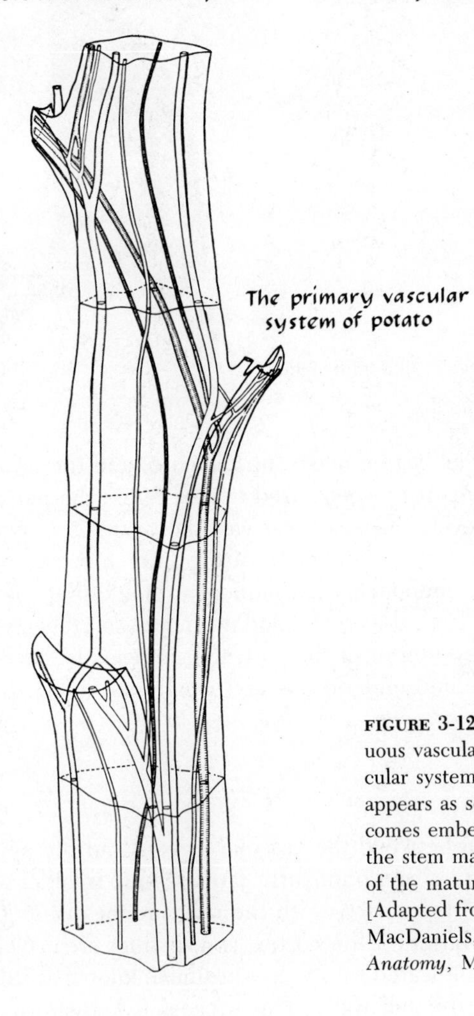

FIGURE 3-12. Discontinuous and continuous vascular systems. The primary vascular system of the potato stem initially appears as separate bundles but becomes embedded in secondary tissue as the stem matures. The vascular system of the mature stem is continuous. [Adapted from A. J. Eames and L. H. MacDaniels, *An Introduction to Plant Anatomy*, McGraw-Hill, 1947.]

and the corky protective sheath that it produces is called the **periderm.** Callus tissue (the "scar tissue" that is formed in response to wounding) may also become corky. The cork industry is based on the large amounts of this tissue produced by wounding of the cork oak (*Quercus suber*).

Differentiated portions of the cortex that form ruptured, rough areas on woody stems and in bark are referred to as **lenticels.** These are "breathing pores": They allow for the exchange of gases between the plant's tissues and the atmosphere. The "dots" on apple skin are also lenticels. Because they penetrate the epidermis, lenticels of fruits can allow for the entrance of decay organisms and act as a point of water loss.

Epidermis

The epidermis is a continuous cell layer that envelops the plant. Except in older stems and roots, in which it may be obliterated, the epidermis sheaths the entire plant. The structure of the epidermis varies, and it may be composed of different kinds of primary tissue. Slightly above the root tips, the epidermal cells form tubelike extensions called root hairs, which function in the absorption of water and inorganic nutrients. Hairs are also found in epidermal cells of the shoot. The velvety feel of rose petals is due to the uneven surface of their upper epidermal cells. The guard cells forming **stomata** are modified epidermal cells (Figure 3-13).

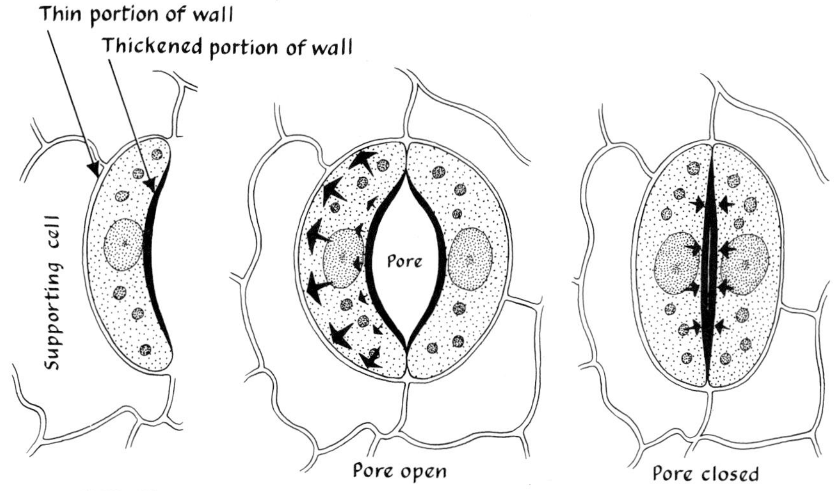

FIGURE 3-13. The stomata are epidermal structures composed of two guard cells that form a pore. The portion of the guard cell wall abutting the pore is thicker than the rest of the cell wall. This causes the pore to open when the guard cells become turgid. [Adapted from J. Bonner and A. W. Galston, *Principles of Plant Physiology*, W. H. Freeman and Company, copyright © 1952.]

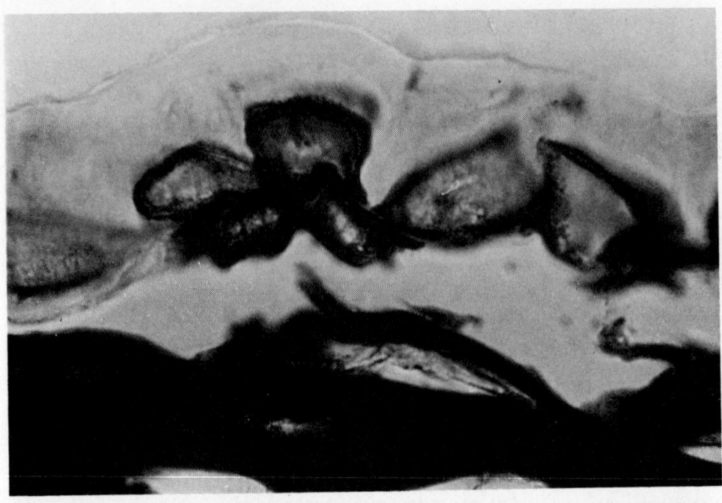

FIGURE 3-14. The epidermis and cuticle of the 'Stayman Winesap' apple fruit. The skin of an apple is several cells thick and consists of epidermis and hypodermis. The latter is composed of compressed layers of cortical cells. The cause of the separation of the epidermis and hypodermis shown in the photograph is unknown. It produces a white patch on the fruit, sometimes referred to as "scarfskin." [Photograph courtesy D. F. Dayton.]

A significant feature of epidermal cell structure is the **cuticle,** a waxy layer that appears on the exposed surface of the cell. The waxy material, **cutin,** acts as a protective covering that prevents the desiccation of inner tissues (Figure 3-14). It is particularly noticeable in such fruits as the apple, nectarine, and cherry, where its accumulation results in a bloom that can be polished to a high gloss. The cuticle is responsible for the tendency of water to bead on leaves and fruits, which, because many disease organisms are water borne, may be very important to the plant's ability to resist disease.

Secretory Glands

A number of morphologically unrelated plant structures—some very simple, others quite complex—secrete or excrete complex metabolic products, often in viscous or liquid form. Many of these substances are of economic importance—for example, resins, rubber, mucilages, and gums.

Secretory functions are carried out by many multicelled, hairlike epidermal appendages called **trichomes** (Figure 3-15). Moreover, complex secretory structures called glands develop from subepidermal and epidermal tissue in various parts of the plant body. The fragrance of flowers is produced by glands called nectaries. Some specialized glands secrete essential oils (see page 132). In citrus fruits such glands are found in the peel, as can be verified by squeezing a portion of the rind next to a flame: Rupturing the glands releases the volatile and inflammable essential oils. Gums, mucilages, and resins are formed

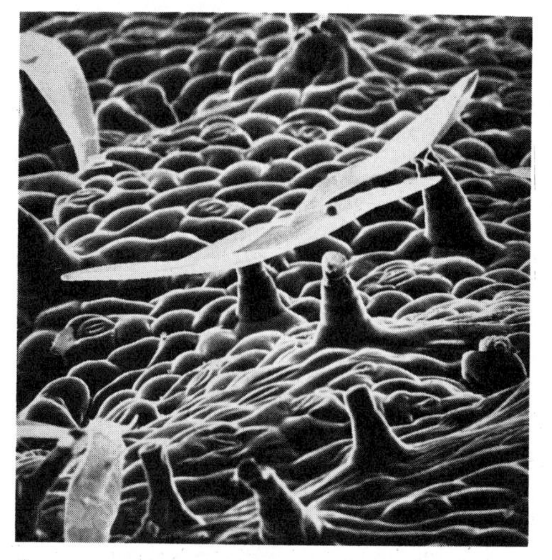

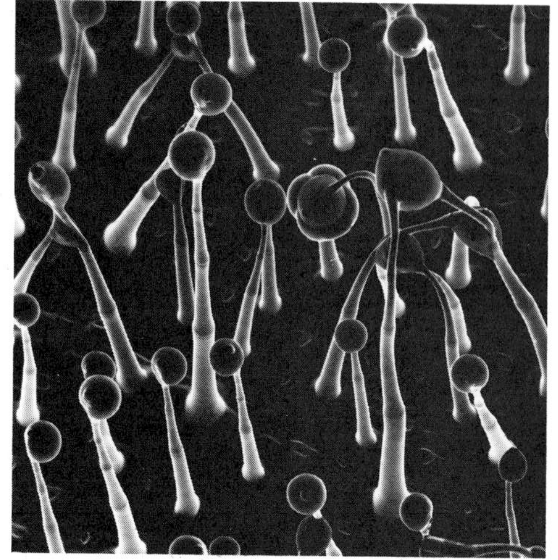

A B

FIGURE 3-15. (*A*) Upper surface of chrysanthemum leaf showing T-shaped trichomes (×144). [Stereoscan electron photomicrograph furnished by C. V. Cutting, Long Ashton Research Station, Bristol, England.] (*B*) Upper surface of the leaf of *Solanum pennellii*, a relative of the potato, showing globular-tipped trichomes. [From C. D. Clayberg, *HortScience* 10:13–15, 1975.

in ducts and canals, intercellular spaces between groups of specialized cells. Often these substances are secreted in response to wounding, as in the stem of peach and cherry.

Latex, a milky, viscous substance containing various materials, particularly gums, is found in some angiosperm families. It may be formed in ordinary parenchyma or in a complex series of branching tubes called laticifers. They are not considered distinct tissue but are associated with other tissues, commonly the phloem. Latex is released under pressure when these cells are punctured. Rubber is a constituent of some latex-forming plants. The content varies greatly, but in the rubber tree (*Hevea brasiliensis*) 40 to 50 percent of the latex is rubber. The function of the latex duct network is not clear, but it is felt that it probably serves as an excretory system for the plant and plays some role in wound healing.

MORPHOLOGICAL STRUCTURES

The Root

The root, although visibly inconspicuous, is a major component of the plant in terms of both function and absolute bulk. It may consist of more than half of the dry weight of the entire plant body. The root is structurally adapted for its major function of absorbing water and nutrients. Owing to its complex branching (which occurs irregularly, rather than at nodes as in stems) and its tip area of root hairs, the root presents a very large surface in intimate contact with the soil. The process of absorption in most plants is carried out mainly in the root hairs, which are constantly renewed by new growth. The growth of the plant is largely limited by the extent of its underground expansion. This vast network of roots anchors the plant and supports the superstructure of food-producing leaves. The older roots may also serve as storage organs for elaborated foodstuffs, as in the sweetpotato and carrot.

The original seedling root, or **primary root,** begins the formation of the root system of the plant by establishing a branching pattern. When the primary root becomes the main root of the plant the network is referred to as a **taproot** system, as in the walnut, carrot, beet, and turnip. In many plants, however, the primary root ceases growth when the plant is still young, and the root system is taken over by new roots that grow adventitiously out of the stem, forming a **fibrous** root system, as is typical of grasses (Figure 3-16). In addition many taprooted plants (for example, apple) form an upper network of fibrous

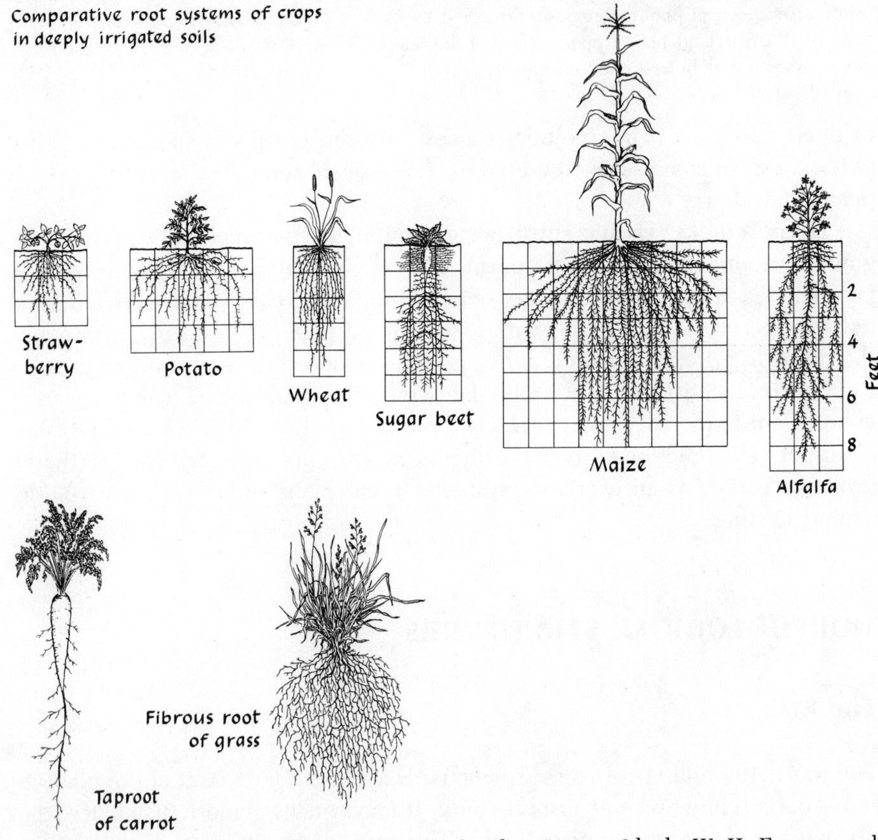

FIGURE 3-16. Root systems. [From J. Janick et al., *Plant Science*, 2d ed., W. H. Freeman and Company, copyright © 1974.]

feeder roots. This permits deep anchorage and a more reliable water supply while providing absorptive capacity in the more fertile upper layers of soil. A fibrous root system may be produced artificially by destroying the taproot. This is accomplished by transplanting or undercutting and is a standard horticultural practice with shrubs and trees. Growers of nursery stock endeavor to build up a fibrous root system concentrated in a "ball" below the plant. This permits even relatively large plants to be successfully transplanted.

In general, plants having a fibrous root system are shallow-rooted in comparison with taprooted plants. Shallow-rooted plants will, of course, be more subject to drought and will show quicker response to variations in fertility treatments.

The morphological structure of the root is shown in Figure 3-17. The arrangement of its vascular system is different from that of the stem. Note the lack of pith and the predominant pericycle and endodermis. A cambium serves to increase the girth of perennial roots, as in the stem.

The major root modifications of horticultural interest are those affecting the storage function. Roots of certain species become swollen and fleshy (Figure 3-18) with stored food in the form of starches and sugars. Some of these storage roots, such as the carrot, sweetpotato, and turnip, are edible. In some plants with storage roots, such as the sweetpotato and the dahlia, the stored food, coupled with the ability of the roots to form adventitious shoot buds, renders the roots important in propagation.

The Shoot

The shoot has been described as a "central axis with appendages." The "central axis"—that is, the stem—supports the food-producing leaves and connects them with the nutrient-gathering roots. The stem is also a storage organ, and in many plants its structure is greatly modified for this function. Young green stems also have a small role in food production because of the chlorophyll they contain. Plants assume extremely varied forms, ranging from a single upright shoot, as in the date palm, to the prostrate-branched "creepers." It is the structure and growth pattern of the stem that determine the form of the plant. Basic structural and anatomical features of herbaceous and woody stems are shown in Figure 3-19.

The upright growth of plants having one active growing point and a rigid stem is considered normal, and our descriptive terms are used to differentiate other growth patterns. Typical shrubby or bushy growth is brought about by the absence of a main truck, or **central leader.** Such growth is characterized by a number of erect or semierect stems, none of which dominates. The distinguishing feature is form rather than size. Similarly, slender and flexible stems that cannot support themselves in an erect position are known as **vines.** Vines will trail unless mechanical support is used to make them grow upright. They may be either herbaceous (morning-glory, pea) or woody (grape).

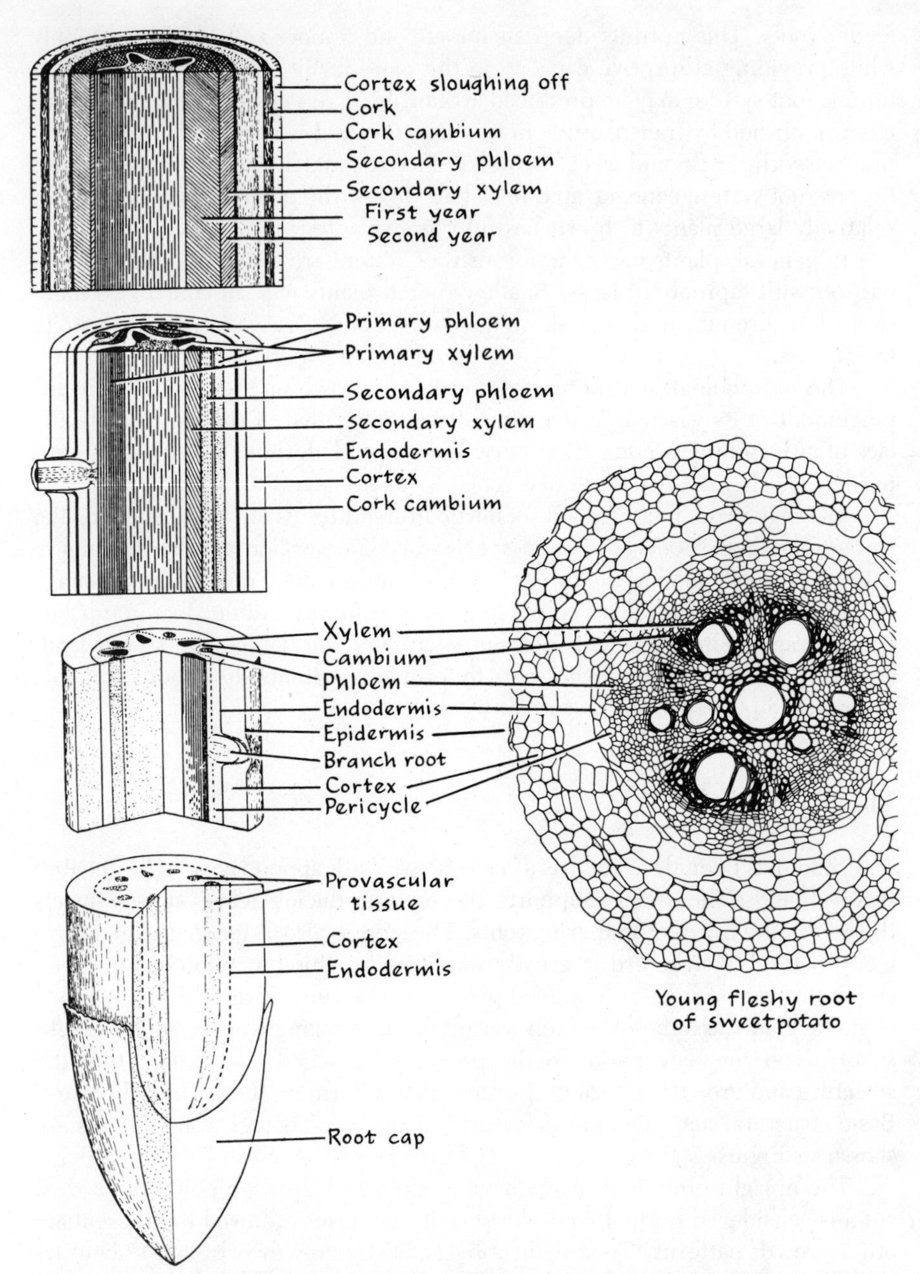

FIGURE 3-17. Diagrammatic sections through a root. [Adapted from P. Weatherwax, *Plant Biology*, Saunders, 1947.]

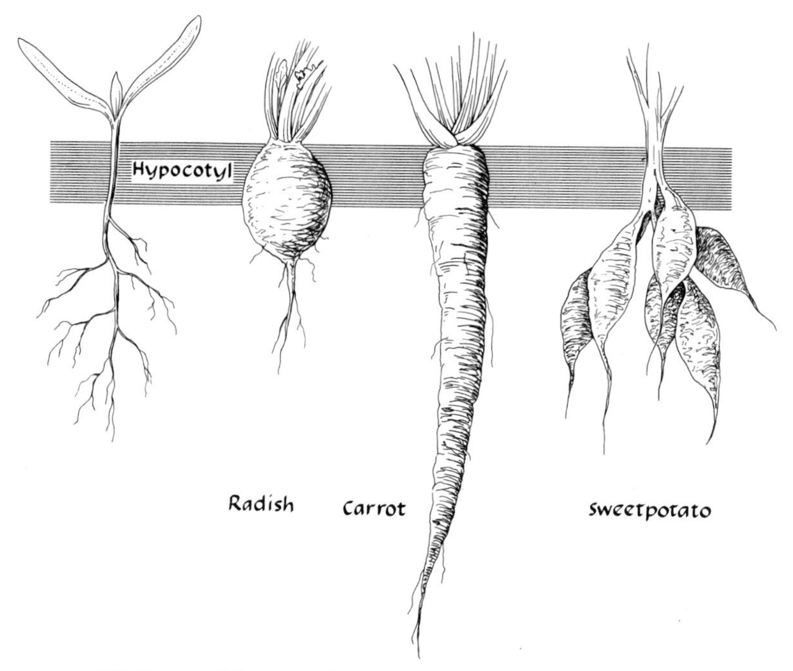

FIGURE 3-18. Root modifications. The swollen roots of the radish and carrot include the hypocotyl, a transition zone between the rudimentary seedling root and shoot.

Buds

The stem is divided into mature regions and actively growing regions in which growth and differentiation take place. An embryonic stem is called a bud. All buds do not grow actively; many exhibit arrested development or dormancy but are nevertheless potential sources of further growth. Although the buds of some plants may be so embedded in the stem tissues as to be relatively inconspicuous, the buds of others may be fairly elaborate, conspicuous structures. The form, structure, and arrangement of buds proves to be a useful guide in describing woody plants even when the leaves are absent, as in winter. Typical buds are shown in Figure 3-20. Growth may originate from a single **terminal** bud or from **lateral** buds that occur in the leaf axis. In addition, buds may be formed in internodal regions of the stems, leaves, or roots, often as a result of injury. These are called **adventitious** buds.

Buds may produce leaves, flowers, or both, and are referred to as **leaf, flower,** or **mixed** buds, respectively. When more than one bud is present at a leaf axis, all but the central or basal bud are called **accessory** buds. The arrangement or topology of buds on a stem is based on the leaf arrangement. If two or more leaves are opposed to each other at the same level, the leaf (and bud) arrangement is said to be **opposite,** or **whorled.** When they are at different

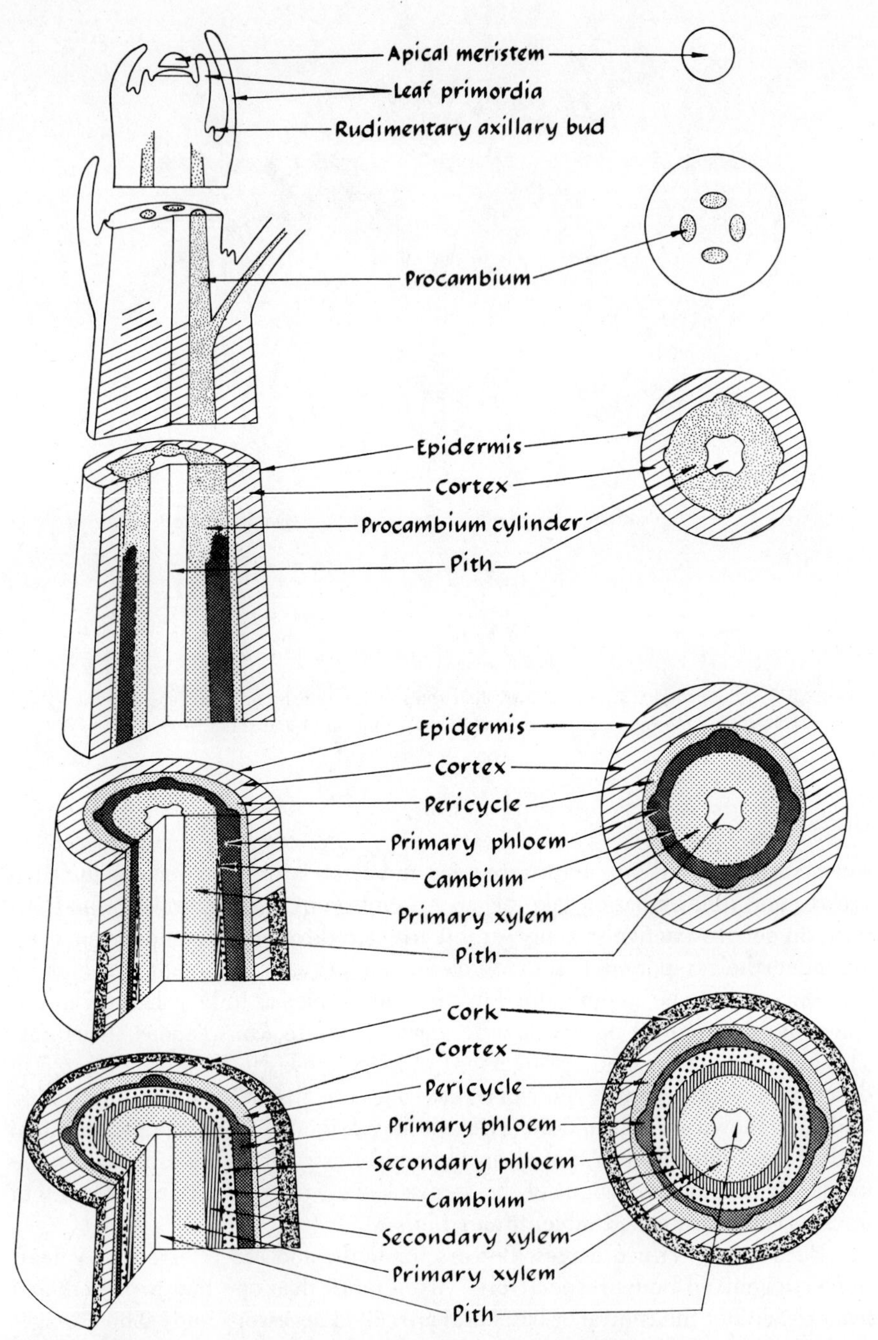

FIGURE 3-19. Diagrammatic sections through a stem. [Adapted from R. M. Holman and W. W. Robbins, *A Textbook of General Botany*, Wiley, 1939.]

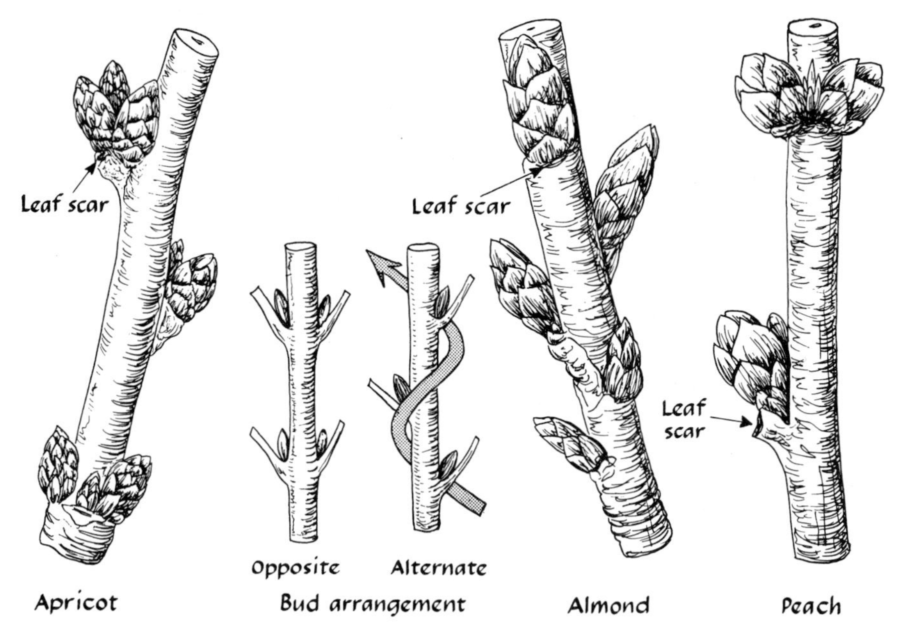

FIGURE 3-20. Dormant fruit and leaf buds of apricot, almond, and peach. The arrangement in each is alternate. [Adapted from Q. B. Zielinski, *Modern Systematic Pomology*, Wm. C. Brown, 1955.]

levels, they are arranged in a spiral and are said to be **alternate** (Figure 3-20). The spiral pattern of leaf arrangement (**phyllotaxy**) is varied and is expressed as a fraction (½, ⅓, ⅖, ⅜), where the numerator is the number of turns to get to a leaf directly above another and the denominator is the number of buds passed. Phyllotaxy has taxonomic significance, since it is often the same throughout a genus or even throughout a whole family.

Stem Modifications

The stem may be greatly modified from a basically cylindrical structure (Figure 3-21). Some of these alterations may appear quite bizarre; yet upon close analysis these modifications can be shown to be basically stemlike in structure; that is, they have nodes, leaves, or similar scalelike structures, and they function in transport or storage. Stem modifications can be divided into above-ground forms (crowns, stolons, spurs) and below-ground forms (bulbs, corms, rhizomes, tubers). Since many stem modifications contain large amounts of stored food, they may be significant in propagation and, as in the tubers of potatoes, important as a source of food.

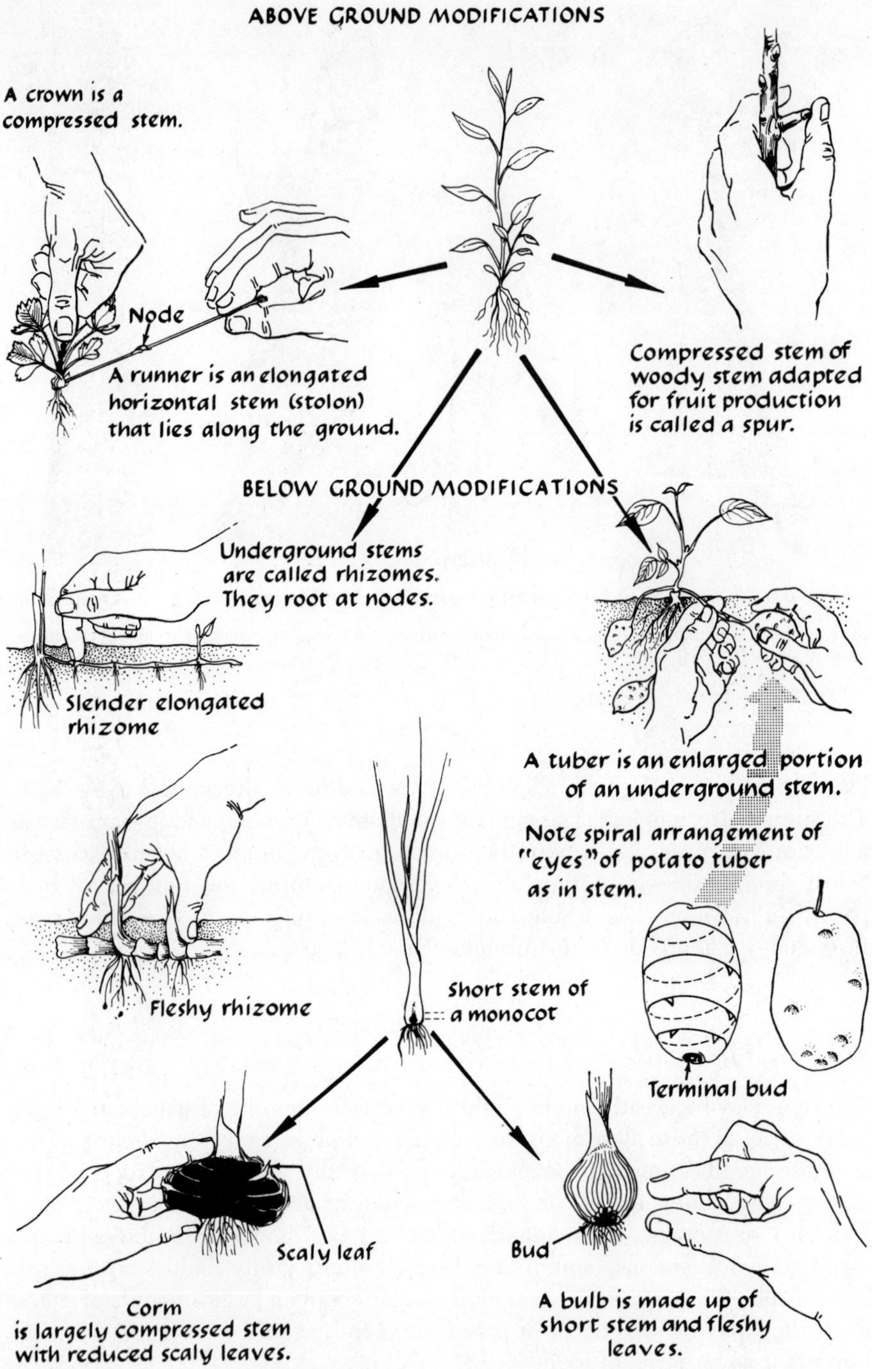

ABOVE GROUND MODIFICATIONS

A crown is a compressed stem.

Node

A runner is an elongated horizontal stem (stolon) that lies along the ground.

Compressed stem of woody stem adapted for fruit production is called a spur.

BELOW GROUND MODIFICATIONS

Underground stems are called rhizomes. They root at nodes.

Slender elongated rhizome

A tuber is an enlarged portion of an underground stem.

Note spiral arrangement of "eyes" of potato tuber as in stem.

Terminal bud

Fleshy rhizome

Short stem of a monocot

Scaly leaf

Bud

Corm is largely compressed stem with reduced scaly leaves.

A bulb is made up of short stem and fleshy leaves.

FIGURE 3-21. Stem modifications. Note that all have nodes and leaflike structures.

A

B

FIGURE 3-22. The crown of a strawberry is a compressed stem (*A*). This can be clearly seen by elongating the stem with gibberellic acid (*B*). Note that the runners are formed from the leaf axils. [Photographs courtesy J. Hull, Jr.]

ABOVE-GROUND STEM MODIFICATIONS. The **crown** of a plant is generally that portion just above and just below ground level.[1] This portion of the plant may be greatly enlarged, as in the bald-cypress (*Taxodium distichum*). Crowns may be thought of as "compressed" stems. The underlying structure of the strawberry crown can be clearly seen by artificially elongating it through treatment with gibberellic acid, as shown in Figure 3-22. Leaves and flowers arise from the crown by buds, as they do in stems. In addition, fleshy buds from crowns may produce a whole new plant, referred to as **crown divisions.** The crown may be modified into a food storage organ, as it is in asparagus.

Short, many-noded, horizontal branches growing out of the crown, bearing fleshy buds or leafy rosettes, are referred to variously as offsets, slips, suckers, and pips. These stem modifications, which can be collectively termed **offshoots,** are important in providing both natural and artificial means of propagation.

Stolons are stems that grow horizontally along the ground. A **runner** is a stolon with long internodes originating at the base or crown of the plant. At some of the nodes, roots and shoots develop. A well-known example of a plant with runners is the cultivated strawberry.

Spurs are stems of woody plants whose growth is restricted. They are characterized by greatly shortened internodes, and are usually attached to normal branches. In mature fruit trees, such as apple, pear, and quince, flowering is largely confined to spurs. Spurs are not irrevocably static and may revert to normal stem growth even after many years of fruiting.

[1]But notice that, in the terminology of forestry, *crown* refers to the branched top of a tree.

BELOW-GROUND STEM MODIFICATIONS. A **bulb** is essentially a compressed modification of the shoot. It consists of a short, flattened, or disk-shaped stem surrounded by fleshy, leaflike structures called scales, which may enclose shoot or flower buds. Bulbs are found only in some monocotyledonous plants. The scales, filled with stored food, may be continuous and form a series of concentric layers surrounding a growing point, as in the onion or tulip (**tunicate bulbs**), or they may form a more or less random attachment to a small portion of stem, as in the Easter lily (**scaly bulbs**). Bulbs commonly grow under the ground or at ground level, although certain bulblike structures (**bulbils**) may be formed on aerial stems in association with flower parts, as they are in some kinds of onions.

Corms are short, fleshy, underground stems having few nodes. The corm is almost entirely stem; the few rudimentary leaves are nonfleshy. The gladiolus and crocus are propagated by corms. Like bulbs, corms are found only in some monocotyledonous plants.

Rhizomes are horizontal undergound stems. They may be compressed and fleshy, as in *Iris*, or slender with elongated internodes, as in turf grasses (such as Bermuda grass). Normally, roots and shoots develop from the nodal regions. Such weeds as quack grass and Canadian thistle are particularly insidious because they spread so rapidly, owing to their natural propagation by rhizomes.

Tubers are greatly enlarged fleshy portions of underground stems. They are typically noncylindrical. (The word *tuber* is derived from a Latin word meaning lump.) The edible portion of the potato is a tuber. The "eyes" arranged in a spiral around the tuber are the buds. Each eye consists of a rudimentary leaf scar and a cluster of buds.

The Leaf

Leaves are the photosynthetic organs upon which higher plants depend for the formation of carbon-containing compounds. The leaf is basically a flat appendage of the stem arranged in such a pattern as to present a large surface for the efficient absorption of light energy. The leaf **blade** is usually attached to the stem by a stalk, or **petiole.** Leaflike outgrowths of the petioles, known as **stipules,** are commonly present.

The anatomical structure of the leaf is shown in Figure 3-23. Note that the vascular system in leaves forms a branching network of **veins.** The veining is typically netlike in dicots and parallel in monocots. The leaf blade, although commonly bilaterally symmetrical, is not radially symmetrical, since it has a distinct upper and lower side. Beneath the upper epidermal layer, which is characterized by heavy deposits of cutin, lie series of elongated closely packed "palisade" cells that are particularly rich in chloroplasts. The irregularly arranged cells beneath the palisade cells produce a spongelike region (**spongy mesophyll**) that provides an area necessary for gaseous exchange in photosynthesis and transpiration. The lower epidermis is interspersed with **stomata** openings in the leaf that permit the exchange of gases and water vapor with the environment.

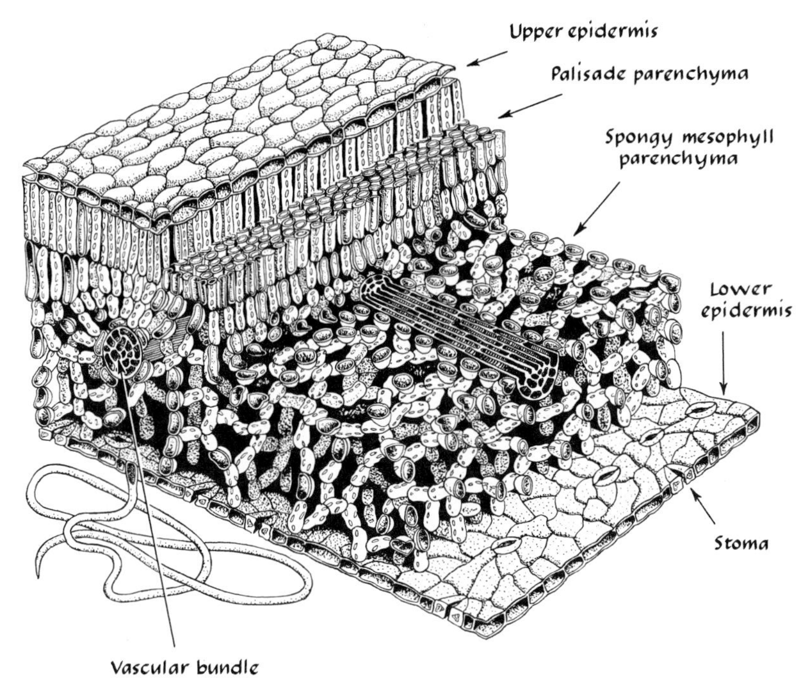

FIGURE 3-23. Structure of an apple leaf.

Leaves of plants vary from the flat thin disks described to the stemlike fleshy structures found in the common house plant *Sansevieria.* The tendrils of peas are modifications of the leaf. Leaves are the edible portions of many plants, such as lettuce, spinach, and cabbage (Figure 3-24), and they are often the chief features of many ornamentals (such as poinsettia), especially when they are rich in red and yellow pigments.

The Flower

The flower shows great variety in structure, composition, and size. The principal flower parts are shown in Figure 3-25.

Sepals (collectively, the **calyx**) enclose the flower in bud. They are usually small, green, leaflike structures below the petals.

Petals (collectively, the **corolla**) are the conspicuous portion of most flowers. They are often highly colored, though rarely with green pigments, and they may contain perfume glands as well as nectar glands that produce a viscous sugary substance. The extremely large, showy flowers of many cultivated ornamentals are the result of rigorous selection.

Stamens are often considered the "male" parts of flowers. Each stamen consists of a pollen-bearing **anther** supported by a **filament.** When the pollen is mature, it is discharged through the ruptured anther wall.

FIGURE 3-24. The cabbage head consists of large fleshy leaves attached to a compressed stem. [Photograph by J. C. Allen & Son.]

The **pistil,** which is often considered the "female" part, consists of an **ovule**-bearing base (or **ovary**) supporting an elongated region (or **style**) whose expanded tip (or surface) is called the **stigma.** The ovule gives rise to the seed. The mature ovary (with or without seeds) becomes the fruit.

The petals and sepals of the flower, as well as the reproductive parts—that is, the stamens and pistils—are essentially modified leaves. The leafy origins of the stamens can be clearly shown in the petaloid anthers or "extra" petals of the cultivated rose (Figure 3-26). These flower parts are borne on an enlarged portion of the flower-supporting stem called the **receptacle.**

Flowers composed of sepals, petals, stamens, and pistils are referred to as **complete** (Figure 3-27). **Incomplete** flowers lack one or more of these parts. For example they may lack stamens (**pistillate** or "female" flowers) or pistils (**staminate** or "male" flowers). Those that contain both stamens and pistils (**perfect, bisexual,** or **hermaphroditic** flowers) may lack calyx or corolla.

Similarly, plants are referred to as **staminate, pistillate,** or **perfect** on the basis of the type of flowers they bear. When both staminate and pistillate flowers occur on the same plant, as in maize, the sex type is **monoecious.** Species in which the sexes are separated into staminate and pistillate plants are **dioecious** (date palm, papaya, spinach, asparagus, hemp). Other combinations of flower types also occur. For example, muskmelons have perfect and staminate flowers on the same plant; this sex type is referred to as **andromonoecious.**

There are many ways in which the flowers are arranged on the plant. The term **inflorescence** refers to a flower cluster. Some of the more common types of inflorescence are diagrammed in Figure 3-28.

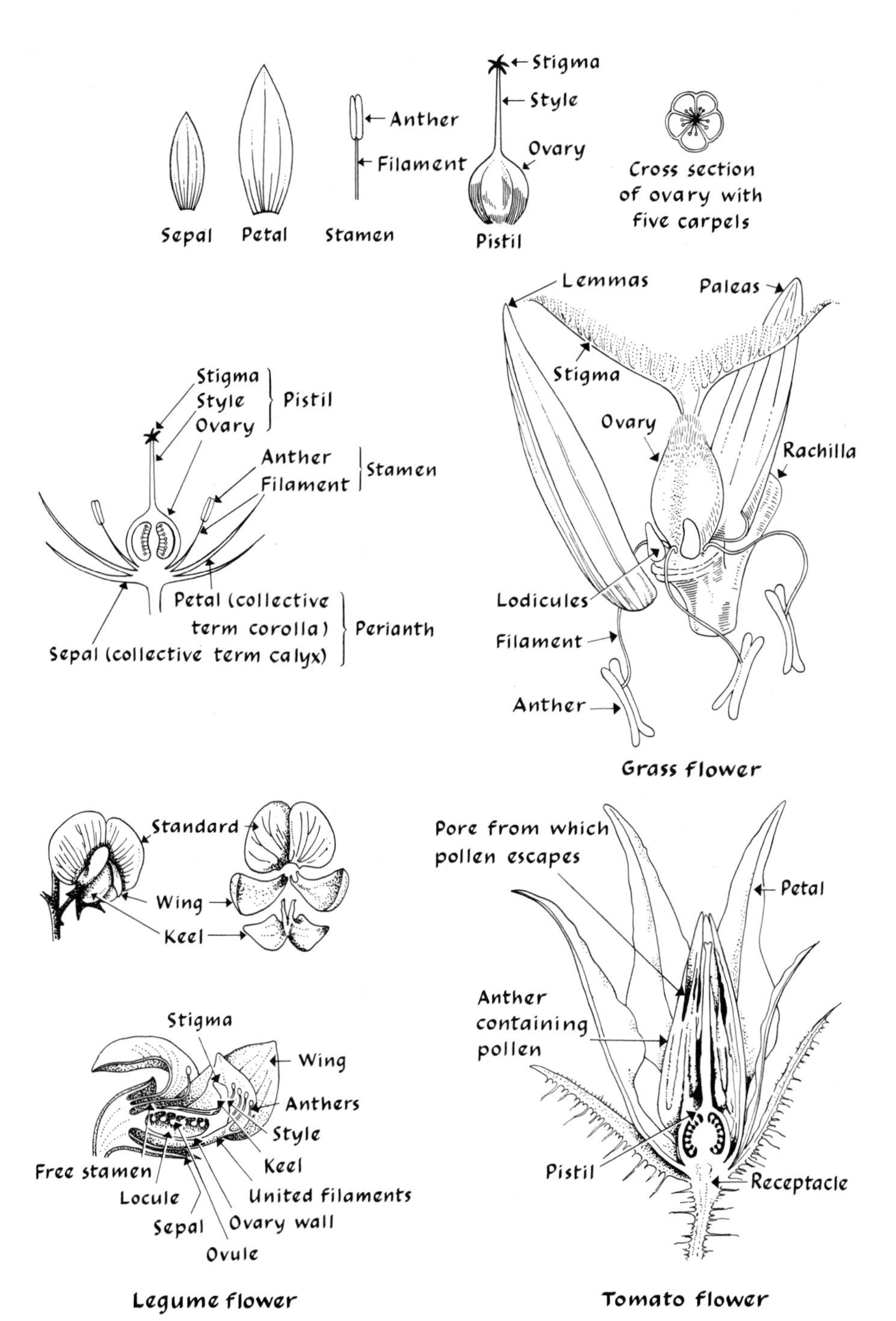

FIGURE 3-25. The structure of flowers. [After J. Janick et al., *Plant Science*, 2d ed., W. H. Freeman and Company, copyright © 1974.

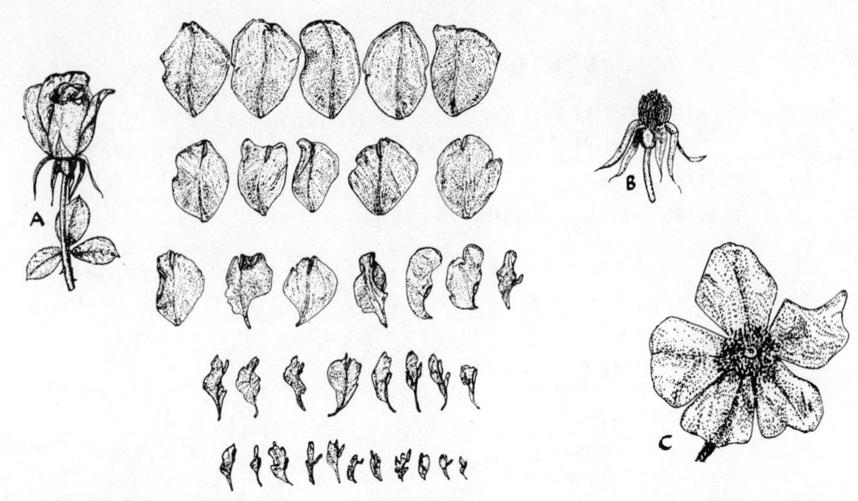

FIGURE 3-26. Roses have five sepals, five petals, numerous stamens or petaloid anthers, and several pistils. A petaloid is a petal-like structure developed from a stamen, forming what is termed semidouble or double flowers. (*A*) Double flowering rose with all five petals intact. From this specimen the five petals and 31 petaloids have been removed progressively inward. Each petaloid shows a rudimentary anther. (*B*) The same specimen after removing all the petals and petaloids to expose the numerous stamens. The number of petaloids depends largely on genetic factors. (*C*) A single rose with five petals and no petaloids. [From E. R. Honeywell, *Roses* (Extension Circular 427), Purdue University.]

FIGURE 3-27. The flower of the lily has all parts and is therefore perfect and complete. When grown commercially, the anthers are removed because the pollen stains the petals. [Photograph by J. C. Allen & Son.]

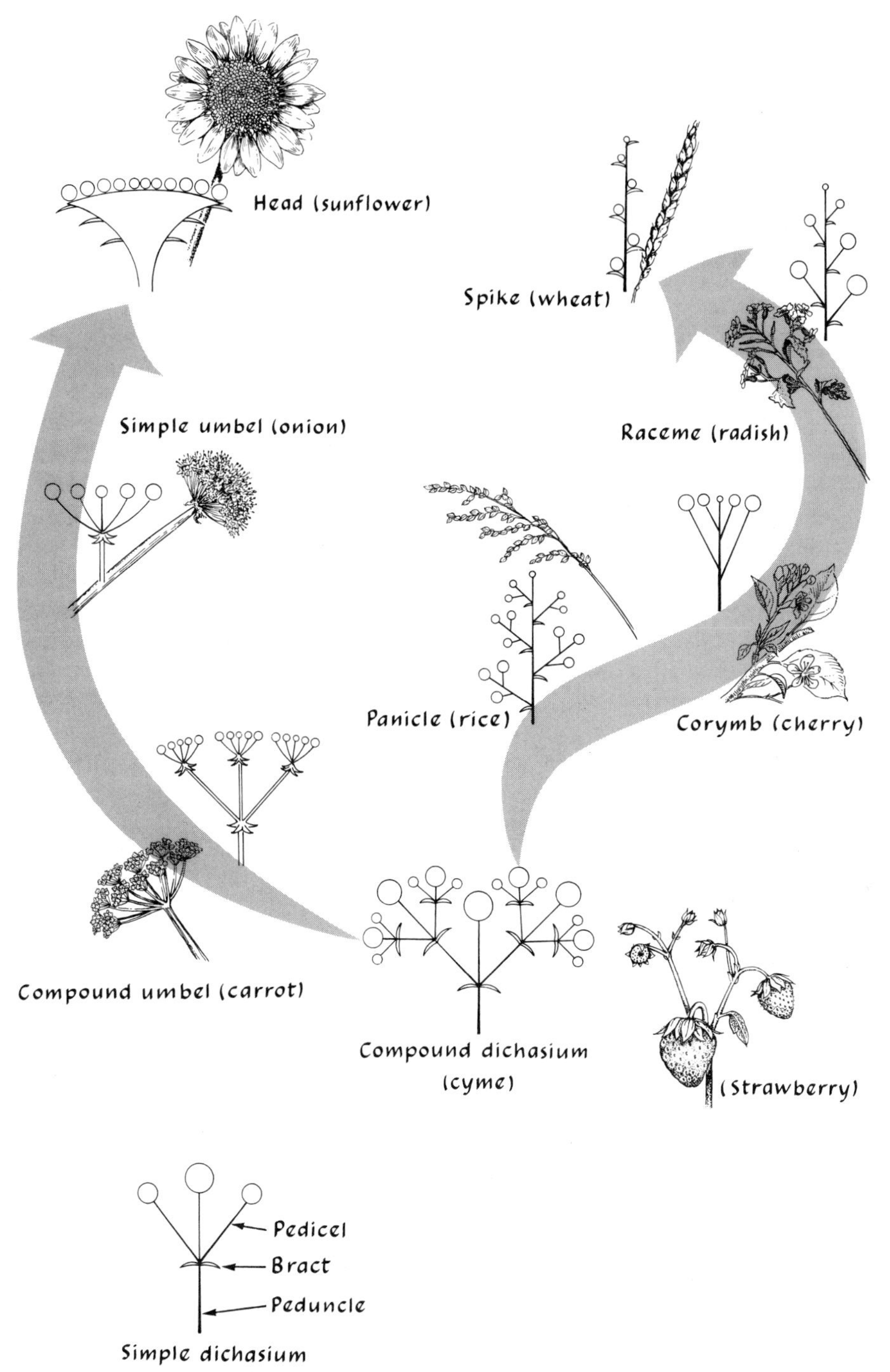

FIGURE 3-28. Common inflorescences. [After J. Janick et al., *Plant Science,* 2d ed., W. H. Freeman and Company, copyright © 1974.]

The Fruit

The botanical term *fruit* refers to the mature ovary and other flower parts associated with it. Thus, it may include the receptacle as well as the withered remnants of the petals, sepals, stamens, and stylar portions of the pistil. It also includes any seeds contained in the ovary.

The structure of the fruit is related to the structure of the flower. Fruits are classified, according to the number of ovaries incorporated in the structure, as **simple, aggregate,** or **multiple** fruits. They may also be classified by the nature and structure of the ovary wall; their ability to split apart when ripe (**dehiscence**), as well as the manner in which this occurs; and the way in which the seed is attached to the ovary.

SIMPLE FRUITS. The majority of flowering plants have fruits composed of a single ovary. These are referred to as simple fruits. In the mature fruit (when the enclosed seed is fully developed) the ovary wall may be **fleshy** (composed of large portions of living succulent parenchyma) or **dry** (made up of nonliving sclerenchyma cells with lignified or suberized walls).

The ovary wall, or **pericarp,** is composed of three distinct layers. These are, from outer to inner layer, the **exocarp, mesocarp,** and **endocarp.** When the entire pericarp of simple fruits is fleshy, the fruit is referred to as a **berry** (not the same as the horticultural term for the edible portion of some "small fruits"). The tomato, grape, and pepper are berry fruits. The muskmelon is a berry (specifically, a **pepo**) with a hard rind made up of exocarp and receptacle tissue (Figure 3-29). Citrus fruits are also berries but of a sort called **hesperidium:** The rind is made up of exocarp and mesocarp, and the edible juicy portion is endocarp.

Simple fleshy fruits having a stony endocarp (such as peach, cherry, plum, and olive) are known as **drupes** (or **stone fruits**). The skin of these fruits is the

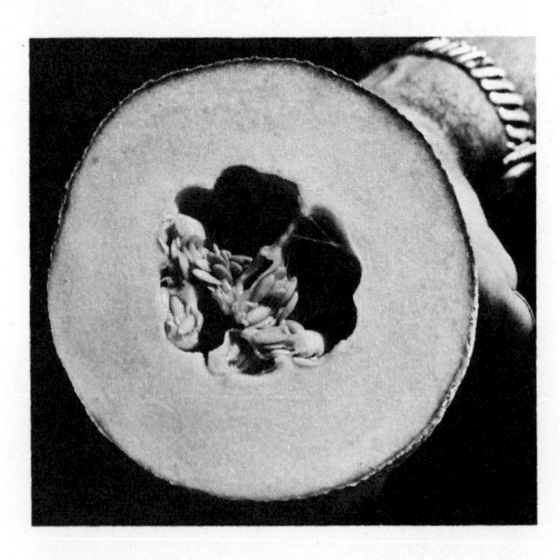

FIGURE 3-29. The muskmelon is a berry fruit called a pepo. The rind is exocarp; the edible flesh is mesocarp. [Photograph by J. C. Allen & Son.]

exocarp; the fleshy edible portion is the mesocarp. Simple fleshy fruits in which the inner portion of the pericarp forms a dry paperlike "core" are known as **pomes** (apple, pear, and quince, for example).

The dry, dehiscent, simple fruits include such types as pods (pea), follicles (milk weed), capsules (jimson weed), or siliques (crucifers). The dry simple fruits that do not dehisce when ripe include achenes (sunflower), caryopses (maize), samaras (maple), schizocarps (carrot), and nuts (walnut). These are diagrammed in Figure 3-30.

AGGREGATE FRUITS. Each aggregate fruit is derived from a flower having many pistils on a common receptacle. The individual fruits of the aggregate may be drupes (stony), as in blackberries, or achenes (that is, one-seeded, dry fruits attached to the receptacle at a single point), as in strawberries. In the strawberry, the fleshy edible portion is the receptacle.

MULTIPLE FRUITS. Each multiple fruit is derived from many separate but closely clustered flowers. Familiar examples of multiple fruits are the pineapple, fig, and mulberry. The beet "seed" is really a multiple fruit.

The Seed

A seed is a miniature plant in an arrested state of development. Most seeds contain a built-in food supply (the orchid seed is an exception). Structurally the seed is a matured ovule, although various parts of the ovary may be incorporated in the **seed coat.** The miniature plant, or embryo, develops from the union of gametes, or sex cells. (The details of the fertilization process will be discussed in Chapter 6, "Reproduction.") By the time the seed is **mature,** the **embryo** is differentiated into a rudimentary shoot (**plumule**), a root (**radicle**), and one or two specialized seed leaves (**cotyledons**). A transition zone between the rudimentary root and shoot is known as the **hypocotyl.** Diagrams of various seeds are shown in Figure 3-31.

The stored food is present in seeds as carbohydrates, fats, and proteins. Seeds are thus a rich source of food as well as of fats and oils for industrial purposes. This stored food may be derived from a tissue called the **endosperm,** which is formed as a result of the fertilization process. The endosperm may produce a specialized region of the mature seed, as in maize (albuminous seed), or it may be absorbed by the developing embryo (exalbuminous seed) and the cotyledons serve as food-storage organs (for example, as in beans and walnuts).

Seeds vary greatly in size and shape. Most plants can be identified by their seeds alone. In addition, great variation exists among seeds of a species. Differences include such things as the presence or absence of spines (spinach), color variation (beans), and the chemical composition of stored food (sugary versus starchy maize).

A. SIMPLE FRUITS

DRY

Dehiscent

Pod of pea

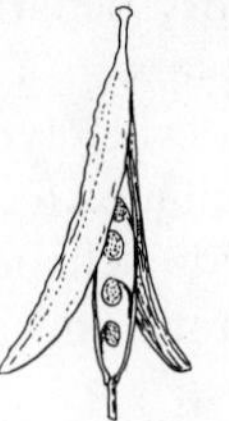
Silique of crucifer

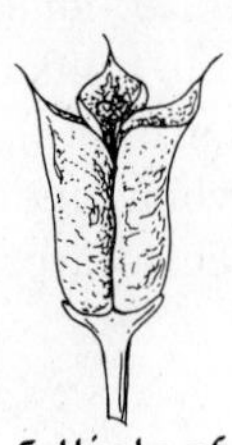
Follicle of larkspur

Capsule of Jimson weed

Nondehiscent

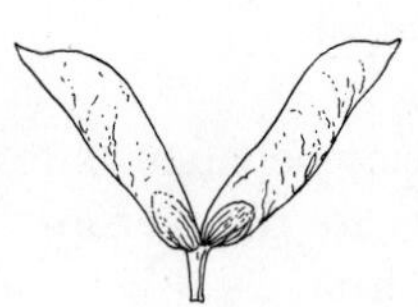
Samara of maple

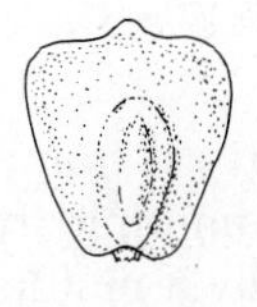
Caryopsis of maize

Achene of sunflower

Schizocarp of carrot

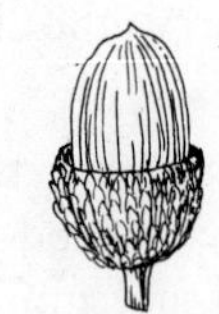
Nut (acorn) of oak

FLESHY

Berry of tomato

Pepo of squash

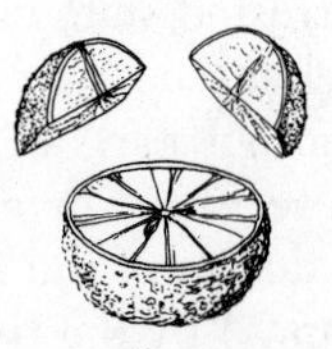
Hesperidium of orange

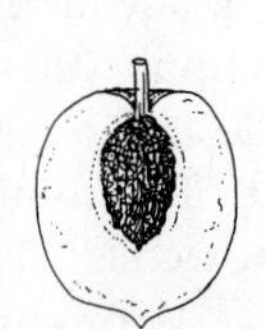
Stone or drupe of peach

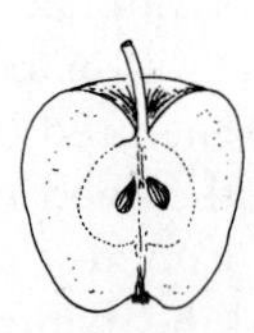
Pome of apple

B. AGGREGATE FRUITS

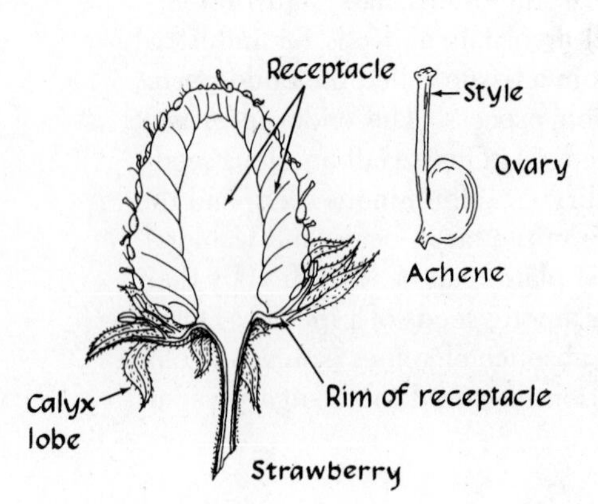

Strawberry

C. MULTIPLE FRUITS

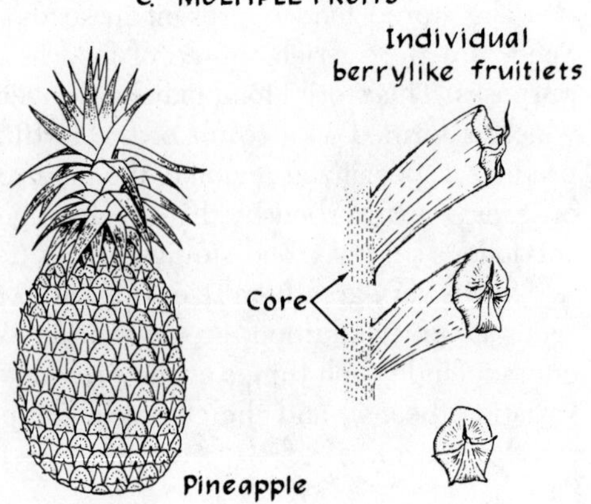

Pineapple

FIGURE 3-30. Various types of fruits. Simple fruits are made up of a single ovary. Aggregate fruits are made up of many pistils on a common receptacle. In the strawberry, the seedlike structures are achenes—small, dry, nondehiscent, one-seeded fruits. Multiple fruits are derived from closely clustered flowers. The individual fruitlets in the pineapple are berrylike. [Adapted from R. M. Holman and W. W. Robbins, *A Textbook of General Botany*, Wiley, 1939.]

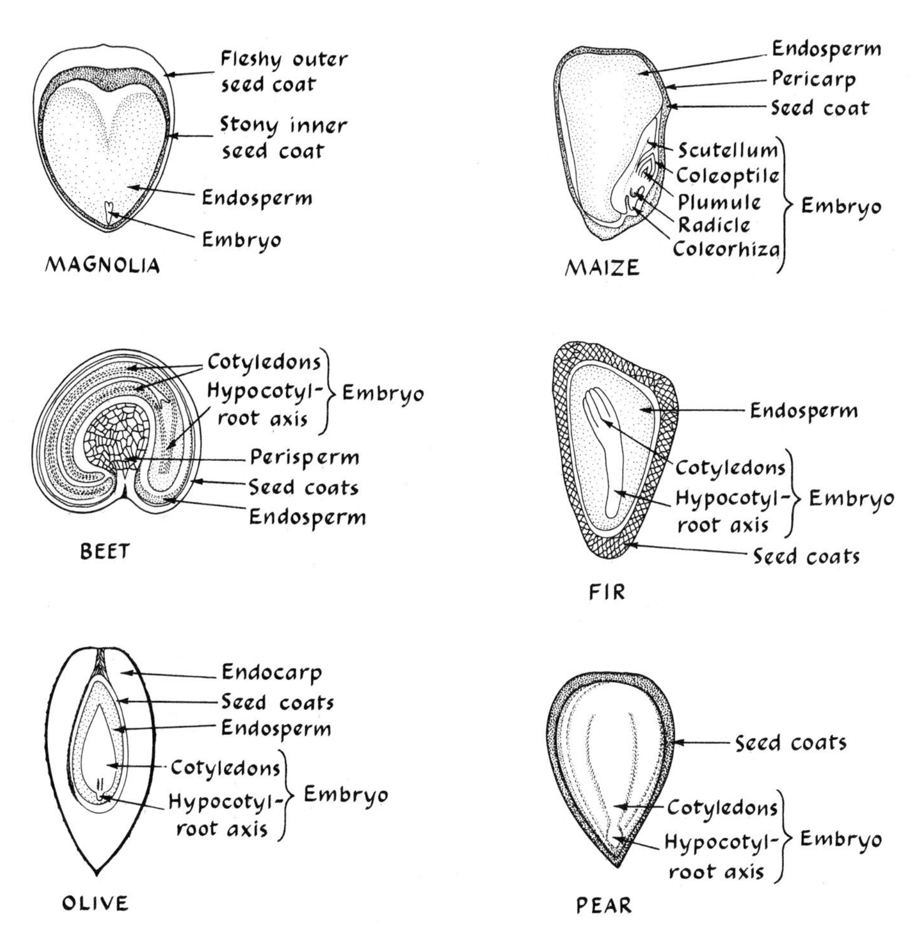

FIGURE 3-31. Structure of seeds and one-seeded fruits. [Adapted from H. T. Hartmann and D. E. Kester, *Plant Propagation: Principles and Practices*, Prentice-Hall, 1959.]

Seed germination refers to the change from the status of arrested development to active growth. The subsequent seeding stage, the interval during which the young foraging plant becomes dependent on its own food manufacturing structures, is diagrammed in Figure 3-32.

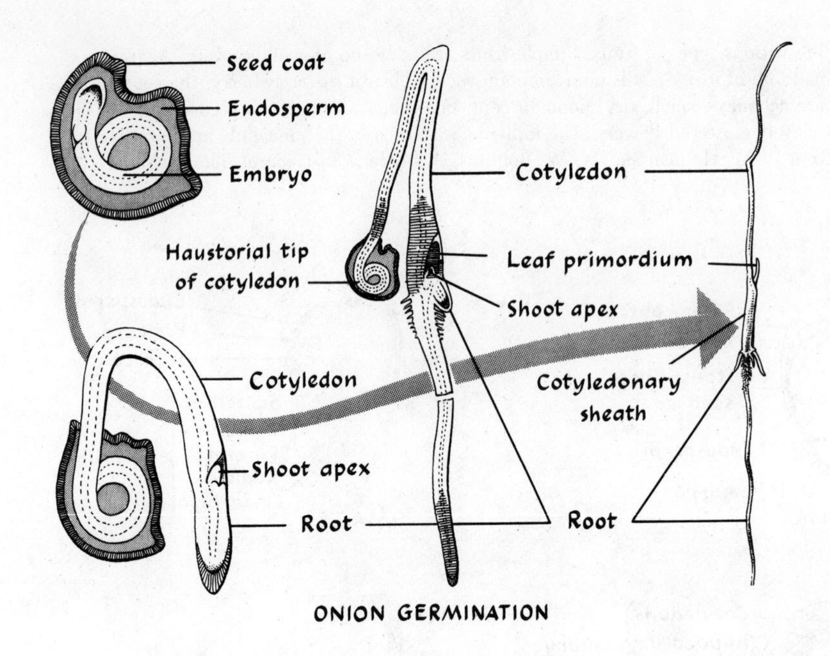

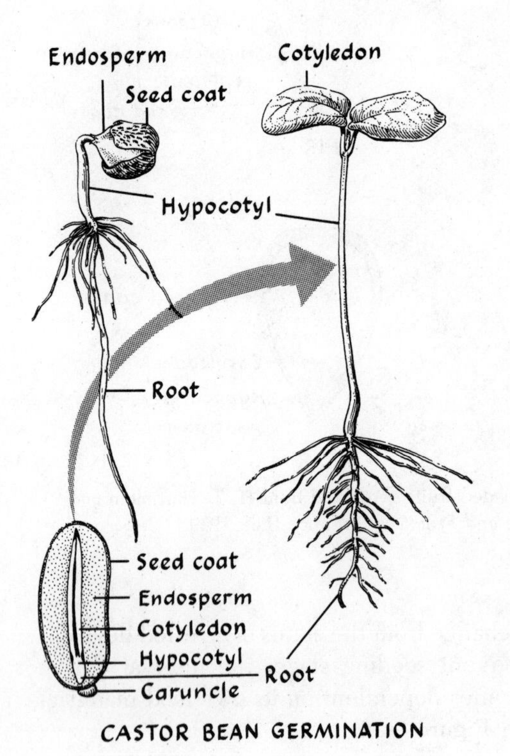

FIGURE 3-32. Seed germination and seedling morphology in onion (a monocot) and castor bean (a dicot). In some dicotyledonous plants the expansion of the hypocotyl elevates the cotyledon above ground (epigeous germination), whereas in others the hypocotyl fails to expand, and the cotyledons remain below ground (hypogeous germination). [Adapted from A. S. Foster and E. M. Gifford, Jr., *Comparative Morphology of Vascular Plants*, 2d ed., W. H. Freeman and Company, copyright © 1974.]

Selected References

Bold, H. C., 1967. *Morphology of Plants*, 2d ed. New York: Harper & Row. (A basic text on the subject.)

Cutler, D. J., 1978. *Applied Plant Anatomy*. New York: Longman. (A useful book on plant anatomy and histology.)

Eames, A. J., and L. H. MacDaniels, 1947. *An Introduction to Plant Anatomy*. New York: McGraw-Hill. (An elementary but authoritative text on vascular plant anatomy.)

Esau, K., 1977. *Anatomy of Seed Plants*, 2d ed. New York: Wiley. (A comprehensive advanced treatise.)

Foster, A. S., and E. M. Gifford, Jr., 1974. *Comparative Morphology of Vascular Plants*, 2d ed. New York: W. H. Freeman and Company. (A physiogenetic approach to plant morphology.)

Heyward, H. E., 1938. *The Structure of Economic Plants*. New York: Macmillan. (This book is especially noted for its treatment of particular species of plants, including corn, onion, beet, radish, pea, celery, sweetpotato, tomato, squash, and lettuce.)

Raven, P. H., R. Evert, and H. Curtis, 1981. *Biology of Plants*, 3d ed. New York: Worth. (A fine introduction to botany.)

Weier, T. C., C. R. Stocking, M. G. Barbou, and T. L. Rost, 1982. *Botany: An Introduction to Plant Biology*, 6th ed. New York: Wiley. (An excellent introductory text.)

4 *Growth and Metabolism*

The growth of a multicellular organism is characterized by increases both in bulk and in complexity. A multicellular organism begins life as a single microscopic cell, the zygote, and may develop into a large individual composed of literally millions of cells arranged into complex histological and morphological structures. Multicellular organisms are complex chemical factories producing the highly specialized substances needed to sustain life.

The events associated with increases in size and in complexity of whole plants can be divided into the interrelated processes of growth and development. **Growth,** in a restricted sense, refers to an irreversible increase in size, reflecting a net increase in protoplasm, brought about by increases both in cell size and in the number of cells. **Development** includes both **morphogenesis,** the creation of pattern and shape as a result of morphological and anatomical development, and **differentiation,** which involves physiological and biochemical specialization.

The increase in the amount of protoplasm that is associated with the growth of an organism is brought about through a series of events in which water, organic chemical compounds, and inorganic salts are transformed into living matter. With respect to whole plants, this process involves the production of carbohydrates (**photosynthesis**), the uptake of water and nutrients, and the synthesis of complex proteins and fats from carbon fragments and inorganic compounds as well as the degradation of those proteins and fats (**metabolism**). The required chemical energy is provided by photosynthesis by day and by **respiration** at night. These physiological processes are functions of both individual cells and multicellular organisms. They are not unrelated to each other any more than the ignition system is unrelated to the compression stroke of the cylinders in the gasoline engine. Nevertheless, their classification into separate processes is conceptually useful. In this chapter we introduce the subject of plant growth, emphasizing growth on a cellular level. In Chapter 5 we will discuss plant development.

THE CELLULAR BASIS OF PLANT GROWTH

Growth is due to increases in cell size and in the number of cells. Each cellular division (**cytokinesis**) produces two cells about half the size of the original cell, each of which grows to the original size before dividing again. Thus, one aspect of growth must be accomplished by cell enlargement. Enlargement of some cells may be extensive. Some leaf cells (pallisade parenchyma) may be 20 times the length and 5 times the width of actively dividing cells of the shoot apex. After pollination, most of the increase in the size of fruits of apple and pear is due to cell enlargement. However, there are limitations to the growth of cells based on the relationship between volume and surface area. (The volume of a sphere increases faster than its surface area.) Further, cells that become very large lose their capacity to divide. Thus, for sufficient growth to occur, cells must divide as well as enlarge after division.

The increase in the length of shoots and roots is due to active cell division at the meristems (the shoot or root tips in dicotyledonous plants) and expansion of new cells as they become separated from the tip. The growth of many fruits is usually due to cell division at the time the ovary is formed and rapid cell enlargement after pollination.

Growth may be measured in various ways. Plant physiologists usually measure growth by increases in dry weight, that is, the weight after water has been driven off. Horticulturalists typically measure growth by increases in plant or organ size or by increases in fresh weight. As it turns out, because growth involves two factors (both the number of cells and the size of cells), there is no single way to measure growth.

Photosynthesis

Photosynthesis is the process in which carbon dioxide and water are transformed, in the presence of light, into carbon-containing, energy-rich, organic compounds. This conversion of light energy into chemical energy is the most significant of the life processes. With few exceptions all of the organic matter in living things is ultimately provided through this sequence of biochemical reactions.

Photosynthesis takes place primarily in the presence of two pigments, chlorophyll *a* and *b*, and, as far as we know, it takes place only in the chloroplasts of living cells. Photosynthesis is a complex series of integrated processes that can be stated in abbreviated form by the following chemical reaction:

$$\underset{\text{Water}}{12H_2O} + \underset{\text{Carbon dioxide}}{6CO_2} + \underset{\text{of chlorophyll}}{\overset{\text{light energy}}{\text{in the presence}}} \longrightarrow \underset{\text{Sugar}}{C_6H_{12}O_6} + \underset{\text{Oxygen}}{6O_2} + \underset{\text{Water}}{6H_2O}$$

The series of photosynthetic reactions can be grouped into a light phase (the reactions that require light) and a dark phase (the reactions that do not require light).

The first step in this series is independent of temperature and consists in the trapping of light energy, which accomplishes the cleavage of the water molecule into hydrogen and oxygen (**photolysis**). The oxygen is released as gaseous molecular oxygen, and the hydrogen is trapped by a hydrogen acceptor, nicotanamide adenine dinucleotide phosphate (NADP). Thus, the liberation of oxygen in photosynthesis is independent of the synthesis of carbohydrates. This step has been referred to as the **Hill reaction** (the NADP serving as the natural Hill reagent). The trapping of light energy by the conversion of adenosine diphosphate (ADP) to adenosine triphosphate (ATP) occurs in a process known as **photophosphorylation.** The combination of the Hill reaction and phosphorylation is known as the **light phase** of photosynthesis.

The conversion of light energy to chemical energy is achieved by the formation of energy carriers such as ATP and $NADPH_2$. ATP has been termed "energy currency" and $NADPH_2$ as "reducing power." ATP is involved in energy transfers in many vital processes of the cell. It is formed from ADP through the addition of a third phosphate group. The captured light energy stored in the third phosphate bond becomes available following the conversion of ATP back to ADP. NADP accepts electrons and hydrogen atoms produced in photolysis and transfers these to other compounds. Indeed, the crucial reactions of photosynthesis are the conversion of ADP to ATP and the reduction of NADP to $NADPH_2$.

The **dark phase** of photosynthesis (also called the **Calvin cycle**) is greatly affected by temperature and has been shown to be independent of light. Essentially the hydrogen atoms from water are transferred by the hydrogen-carrying acceptor ($NADPH_2$) to a low-energy organic acid derived from CO_2 to produce, with the help of the energy of the ATP, a "carbohydrate" of higher energy, from which sugars are formed.

This reduction reaction—that is, the addition of electrons and hydrogen atoms to carbon dioxide—results in the formation of sugar units. The precise pathways by which carbon dioxide is synthesized to sugar are now almost completely known. A substance formed early in this synthesis has been identified as a 3-carbon phosphorus-containing compound, phosphoglyceric acid, two molecules of which eventually give rise to a single 6-carbon sugar and, finally, to starch grains in the chloroplasts, as shown in Figure 4-1.

Photosynthetic efficiency (the rate of net photosynthesis) is equal to the rate of "gross" photosynthesis less any photosynthate that may be lost through respiration:

$$\text{Net photosynthesis} = \text{gross photosynthesis} - \text{respiration}$$

Recent studies have indicated that there are two distinct categories of plants in regard to net photosynthetic rate (Table 4-1). One group, the most typical, fixes carbon from ribulose diphosphate into a 3-carbon acid (phosophoglyceric acid) utilizing the enzyme ribulose-diphosphate carboxylase. Thus, this type of photosynthesis is referred to as the C_3 type. The second group of plants, which have a much higher rate of net photosynthesis, include tropical grasses, sugarcane, and species of *Atriplex* and *Amaranthus* (these two genera include many fast-growing weeds). In this group, CO_2 is first fixed into phosphoenolpyruvate (PEP) to yield 4-carbon acids (such as oxaloacetic, malic, and aspartic acid) and involves the enzyme PEP carboxylase; hence it is known as the C_4 type of photosynthesis. C_4 plants carry out both C_3 and C_4 photosynthesis, whereas C_3 plants lack the C_4 pathway.

Although the C_4 pathway requires slightly more energy than the C_3 pathway, this requirement is offset by other features. The most important advantage is the apparent absence of a type of light-dependent respiration (hence the term **photorespiration**) that is linked to the photosynthetic cycle in C_3 plants. Photorespiration lowers the apparent efficiency of CO_2 assimilation in C_3 plants. The apparent absence of photorespiration in C_4 plants occurs because the photosynthetic affinity for CO_2 is so high that photorespiratory CO_2 is recycled. The process is different from normal ("dark") respiration, which is independent of light. (See the discussion under "Respiration," p. 136.) The occurrence of photorespiration was difficult to detect because it is hard to distinguish the CO_2 fixed in photosynthesis from the CO_2 given off by photorespiration.

The difference in photorespiration between the C_3 and C_4 systems can be demonstrated by enclosing plants in a sealed, illuminated container. As photosynthesis increases, CO_2 is taken up by the plant and fixed, and the effect is to lower the amount normally present in the air. For C_3 plants, a steady state of CO_2 concentration in the air (known as the **compensation point**) is reached when there remains about 50 ppm (parts per million) at 77°F (25°C). This amount is as large as it is because photorespiration releases CO_2. With a C_4 plant, such as maize, which lacks photorespiration, the amount of CO_2 in such a

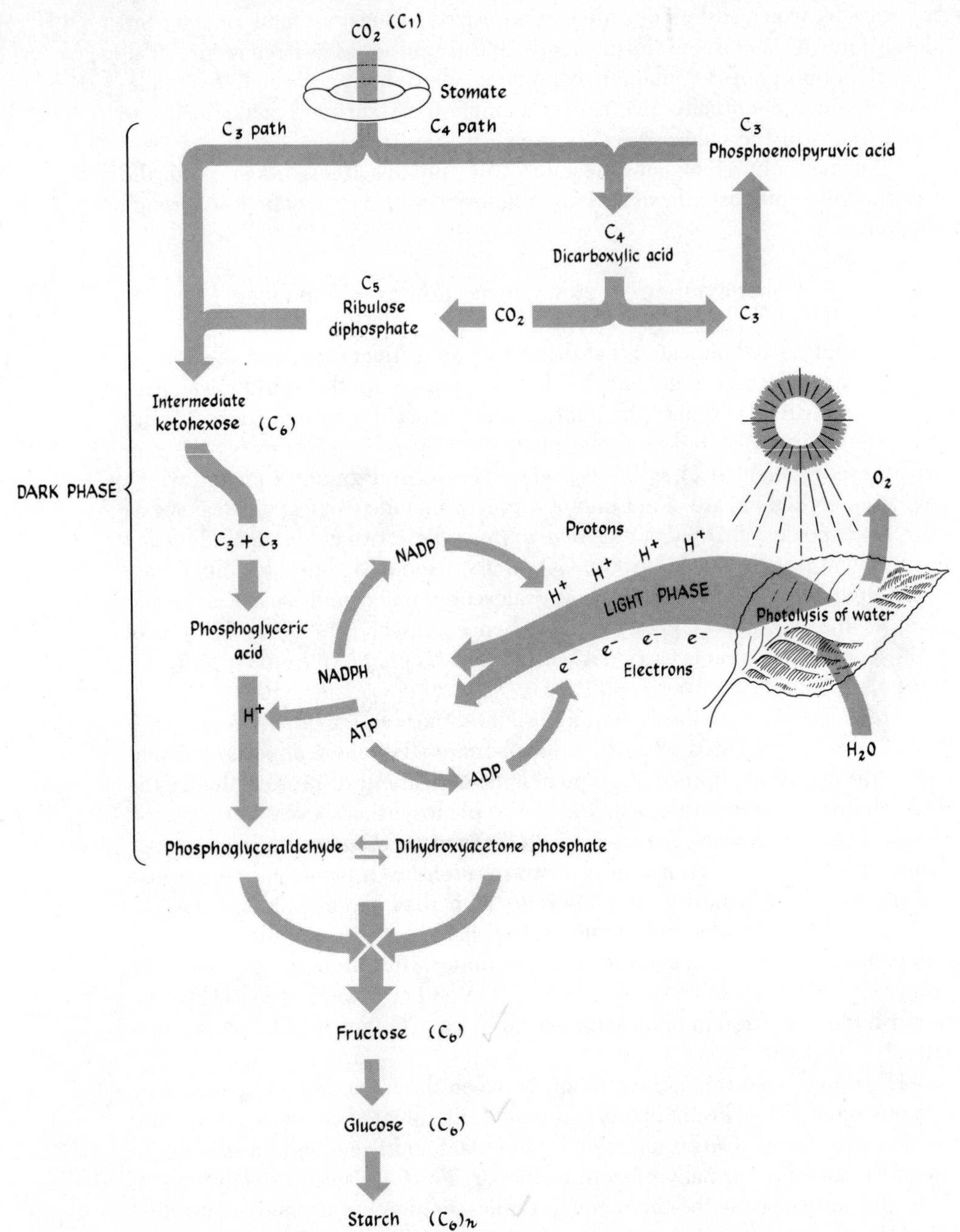

FIGURE 4-1. Photosynthesis. Note the C_3 and C_4 pathways.

TABLE 4-1. Comparative rates of maximum photosynthesis.

Type of plant	Maximum photosynthesis (milligrams CO_2 fixed per square decimeter of leaf surface per hour)
C_3 photosynthesis	
Slow-growing perennials (desert species, orchids)	1–10
Evergreen woody plants	5–15
Deciduous woody plants	15–30
Rapidly growing annuals (wheat, soybean, sugar beet, sunflower)	20–50
C_4 photosynthesis	
Tropical grasses, sugarcane, maize, *Amaranthus*, *Atriplex*	50–90

SOURCE: Jarvis and Jarvis, 1964, *Plant Physiology* 17:645–666.

sealed container diminishes to less than 10 ppm. A soybean plant (C_3) enclosed with a maize plant (C_4) will die because the maize exhausts the CO_2.

The rate of photorespiration increases with temperature faster than the rate of gross photosynthesis. Thus many C_3 plants are nonproductive at high temperatures (77 to 95°F, or 25 to 35°C), whereas C_4 plants such as tropical grasses increase in productivity at these higher temperatures.

Other specific adaptations are associated with the C_4 system. These include enhanced translocation of photosynthetic products, specialized leaf anatomy and chloroplast structure, and a greater ratio of chlorophyll *a* to chlorophyll *b*.

Nutrient Absorption and Translocation

Living organisms may be classified on the basis of their nutritional requirements for organic, or carbon-containing, compounds as either autotrophic or heterotrophic. The basis of plant nutrition is the concept that practically all green plants are **autotrophic** (i.e., self-nourishing) with respect to organic compounds. Green plants with few exceptions can subsist entirely on inorganic materials (nutrients) because they manufacture their own food, a food being defined as any substance from which an organism obtains transformable energy and materials for the maintenance of growth. In contrast, practically all other life forms—animals and nongreen plants—are **heterotrophic** (i.e., not self-nourishing). Their sustenance is derived only from organic matter—ultimately from plants but sometimes with the aid of microorganisms. Although the inorganic mineral constituents supplied to the plant from the soil are often referred to as "plant food," this is not precise terminology.

It should be remembered that the plant is not autotrophic in all stages of its life cycle. Immature embryos (in developing seeds) are heterotrophic and are nourished by the sporophyte. Nourishment to young seedlings is provided in the form of food stored in the cotyledon or endosperm until the development of leaves and roots makes the seedling self-nourishing. Similarly, most cultured tissues must be nourished by a carbon source (typically sucrose) in addition to mineral salts and some vitamins. This will be discussed further on page 346. Interestingly, plants can absorb sugars and other complex substances through their leaves and roots.

Although chemical analysis of plant cells indicates the presence of many different elements, only 16 have been shown to be essential to plant life. The most abundant elements, carbon, hydrogen, and oxygen, are derived largely from carbon dioxide and water. The other 13 elements (iron, potassium, calcium, magnesium, nitrogen, phosphorus, sulfur, manganese, boron, zinc, copper, molybdenum, and chlorine) are derived ultimately from the soil in the form of inorganic salts. (Table 4-2). Plant growth is dependent on the availability of the essential nutrients. Since nutrients and water are ultimately supplied to the cell from the soil, the study of plant nutrition is largely concerned with the biology and chemistry of the soil. (See Chapters 7 and 13.)

TABLE 4-2 The thirteen essential elements derived from the soil.

Essential element	Percentage in representative agricultural soils	Amount (lb/acre)
Fe	3.5	70,000
K	1.5	30,000
Ca	0.5	10,000
Mg	0.4	8000
N	0.1	2000
P	0.06	1000
S	0.05	1000
Mn	0.05	1000
Cl	0.01	200
B	0.002	40
Zn	0.001	20
Cu	0.0005	5
Mo	0.0001	2

SOURCE: J. Bonner and A. W. Galston, 1952, *Principles of Plant Physiology* (W. H. Freeman and Company).

With respect to absorption, the cell can be considered as a mass of protoplasm surrounded by a differentially permeable membrane that permits passage of water and inorganic salts but restrains the passage of most large complex molecules, such as sucrose. Molecules move through a selectively permeable membrane by **diffusion.** The movement of water through such a membrane is referred to as **osmosis** and involves diffusion as well as bulk flow caused by hydrostatic pressure differences. The osmotic movement of molecules can be demonstrated in nonliving closed systems by immersing a differentially permeable membrane that contains sugar-water solution into a solution of pure water (or one with a lesser amount of sugar). The water moves from the solution of high solvent concentration (pure water) to the solution of low solvent concentration (sugar solution) as is illustrated in Figure 4-2. Living cells, however, are able to accumulate certain ions in a manner unaccounted for by diffusion. The cell appears to act as a metabolic pump. This process, known as active uptake, requires energy, which is supplied by respiration. The ability of molecules to move in and out of plant cells is related to the size of the molecules, their solubility in fats, and their ionic charge; membrane permeability is affected by the ionic concentration of the nutrient medium. Monovalent ions (K^+, Na^+, Cl^-) appear to increase the permeability of membranes, whereas polyvalent cations (Ca^{2+} and Mg^{2+}) decrease membrane permeability. Furthermore, different ions interact in their effects on membrane permeability.

Translocation may be defined as the movement of inorganic or organic solutes from one part of the plant to another. The transport of water and solutes in and out of single cells and simple multicellular plants is accomplished largely by diffusion. In higher plants, however, this conduction of solutes is carried out largely in distinct tissue systems. Physiological specialization in multicellular

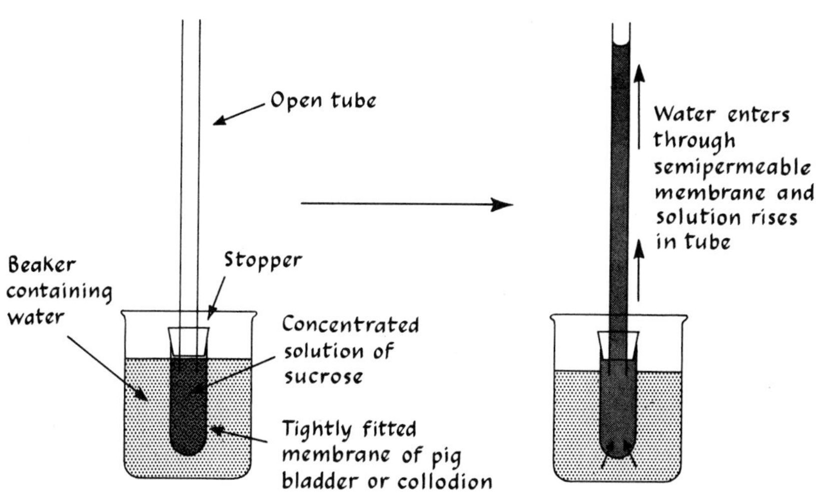

FIGURE 4-2. Diffusion in an artificial osmotic system. [Adapted from J. Bonner and A. W. Galston, *Principles of Plant Physiology*, W. H. Freeman and Company, copyright © 1952.]

plants is made possible because of the rapid, large-scale transport of substances within the plant. This movement is largely a two-way stream, in which water and its solutes move up from the roots through the xylem, and synthesized sugars move out of the leaves to other parts of the plant through the phloem. (Figure 4-3). There is, however, some movement of minerals in the phloem, and the xylem of woody stems functions in the upward movement of organic compounds, especially at certain seasons of the year.

The upward movement of water and solutes in the xylem of higher plants is

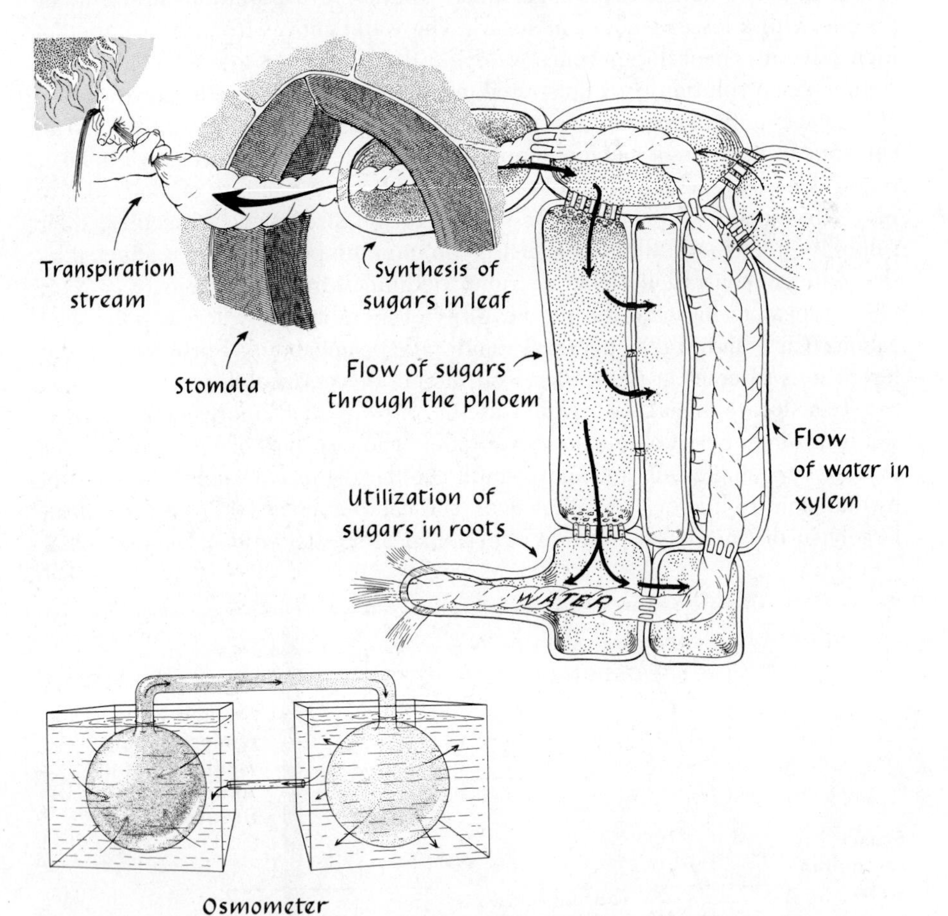

FIGURE 4-3. Diagram of the translocation of water and synthesized sugars in the plant. The upward movement of water through the xylem can be explained on the basis of a tension on the continuous water column in the plant. This tension, produced by the evaporation of water from the leaf (transpiration), is transmitted to the absorbing cells of the root. Sugars synthesized in the leaves move through the sieve tubes of the phloem. Phloem transport is a pressure flow brought about by a high osmotic concentration in the leaf cells and a low concentration in the receiving cells. A model of this system, called the osmometer, is shown at lower left. [Adapted from J. Bonner and A. W. Galston, *Principles of Plant Physiology*, W. H. Freeman and Company, copyright © 1952.]

related in part to **transpiration,** the evaporative loss of water vapor from the leaves through their numerous stomata. As water is lost by the cells, a diffusion-pressure deficit draws the water from the xylem elements, which form large numbers of continuous tubes from roots to leaves. Thus, the tension is transmitted through the entire column to the root cells and results in increased water absorption. The rate of transpiration is affected by the degree to which the stomata are opened and by such environmental factors as temperature and vapor pressure of water in the atmosphere, which affect the rate of water evaporation. The opening of the stomata is a mechanical process regulated by the turgidity of the guard cells (see Figure 3-13).

The movement of sugars occurs principally in the phloem. Phloem transport appears to be accomplished by increased osmotic concentration in the leaf mesophyll cells brought about by the high concentrations of dissolved photosynthates. These sugars then move into the sieve tubes of the phloem by a process that is not clearly understood. The resultant sugar gradient appears to produce a flow, and other substances are swept along with it. The sugars are utilized in the receiving cells for respiration, growth, or storage. There is also evidence of lateral transport between xylem and phloem.

Metabolism

All of the various materials produced in the plant are ultimately derived from the carbon compounds produced by photosynthesis and from the inorganic nutrients and water absorbed from the soil (Figure 4-4). The synthesis (anabolism) and degradation (catabolism) of these organic materials is known as **metabolism.** The degradation of sugars and fats and the release of energy in respiration are examples of catabolic metabolism. The step-by-step elucidation of the pathways in plant metabolism comprises some of the most interesting chapters in the history of biochemistry.

Plants are cultivated for the complex molecules they synthesize. Carbohydrates, proteins, and lipids are the compounds of most concern because they are the major constituents of food. The metabolism of carbohydrates has already been dealt with briefly in the various discussions of photosynthesis. Proteins are second in abundance to carbohydrates as plant constituents (Table 4-3). An understanding of the relations between carbohydrate metabolism and nitrogen metabolism is necessary to a thorough comprehension of plant development.

Nitrogen Metabolism

Although nitrogen accounts for only 1 to 5 percent of plant dry weight, nitrogen-containing substances (typically 16 percent nitrogen) account for 5 to 30 percent. Most of the nitrogen in plants is present in amino acids and proteins. Plants are able to survive in an inorganic world because they can convert inor-

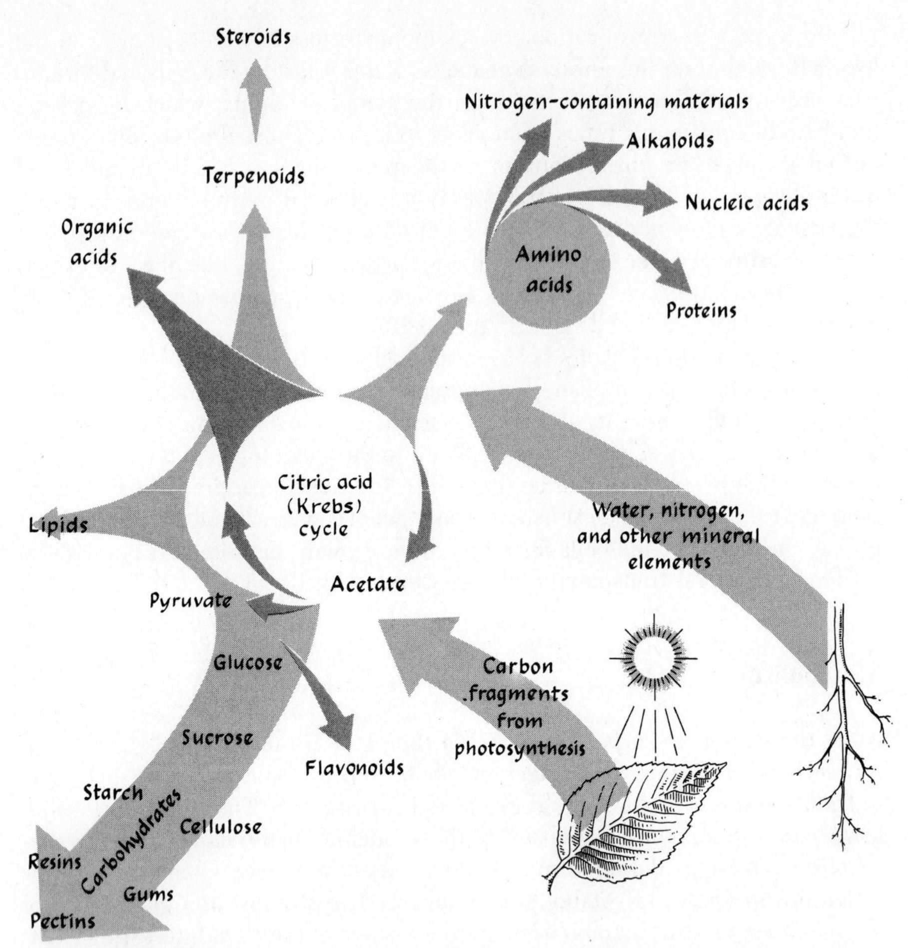

FIGURE 4-4. Metabolic pathways in green plants. [From J. Janick et al., *Plant Science*, 2d ed., W. H. Freeman and Company, copyright © 1974.]

ganic carbon (carbon dioxide from the air) and nitrogen (nitrates and ammonium ions) to organic forms.

Except for the nitrogen-fixing legumes and a few other plants that depend on their bacterial partners, plants are dependent on the nitrate and ammonium ions in the soil solution as a source of nitrogen. Since microorganisms in the soil rapidly oxidize the reduced forms of nitrogen, the most common source of nitrogen taken up by plants is the nitrate ion, NO_3^-. Before combining with organic acids to form amino acids, nitrate is reduced through a series of enzymatic reactions. The reduced nitrogen groups NH_2 and NH_3 combine with the carbon frameworks formed during the oxidation of sugars to form amino acids. The ammonia (NH_3) initially combines with the organic acid α-ketoglutaric acid to form the amino acid glutamic acid, an energy-requiring process. Other

TABLE 4-3 Nutrient constituents of certain plant parts.

Plant part	Percent dry matter	Nutrients as percentage of dry matter			
		Crude protein*	Fat	Carbohydrate	Mineral
Cabbage leaf	15.8	16.5	2.5	62.0	19.0
Beet root	16.4	9.8	0.6	82.9	6.7
Potato tuber	21.2	10.0	0.5	84.0	5.2
Wheat seed	89.5	14.7	2.1	81.0	2.1
Soybean seed	90.0	42.1	20.0	32.8	5.1

*Includes protein and nitrogen-containing compounds.

amino acids are constructed via the transfer of the amino group (NH_2) from this amino acid to other organic acids from the metabolic pool (a process known as **transamination**).

Amino acids join to form proteins through a complex series of reactions regulated by the nucleic acids present in the cell. The primary sites of protein synthesis in plants are the tissues in which new cells are being formed, such as tips of stems and roots, buds, cambium, and developing storage organs. The green leaf is also an important site of protein synthesis, since both the carbon frameworks and the inorganic nitrogen necessary for the formation of amino acids are readily available there. Leaf protein is in a continuous state of turnover. When leaves are placed in the dark, are excised from the plant, or reach the state of senescence, protein breaks down and soluble nitrogen compounds are formed that can be reincorporated into amino acids in other leaves or organs of the plant.

Because the nitrogen metabolism and the carbohydrate metabolism of the plant are closely linked, both processes are markedly affected by gross changes in the environment of the plant, such as heavy nitrogen fertilization, reduced light intensity, or drought. These effects are reflected in altered patterns of growth and development.

Plant Constituents

A great number of organic compounds, many of great complexity, are formed in metabolic processes. The principal organic compounds are carbohydrates, lipids (fats), and proteins. In addition, there are a myriad of other compounds. (Hundreds of different compounds contribute to the flavor of coffee alone!) Some of these metabolites are commonly referred to as essential oils, pigments, and vitamins. These terms, however, are not specific in the chemical sense, and they often refer to mixtures of materials, many of them unrelated. For example, pigments—substances that preferentially absorb certain wavelengths of visible light—differ structurally and chemically. Not all substances produced in the various metabolic pathways are, so far as is known, essential to the plant. Such substances are characteristically produced only by certain plants or plant groups. Many of them (rubber and menthol, for example) are highly prized and constitute the prime reason for the cultivation of the species from which they are derived.

The following sections deal with the various categories of metabolic substances.

Carbohydrates

Carbohydrates consist of a large group of compounds composed of carbon, hydrogen, and oxygen in the ratio of one atom of carbon to two atoms of hydrogen to one atom of oxygen. They were called carbohydrates originally because they incorrectly appeared to early chemists to be "hydrates of carbon" ($C \cdot H_2O$). Carbohydrates make up the bulk of the dry weight of plants. Although there are many kinds, the main carbohydrates of plants are sugars, starches, and celluloses. Animal products contain only small amounts of carbohydrates with the exception of milk, which contains the sugar lactose.

Plants are the prime producers of carbohydrates; animals merely reshift the molecules. Carbohydrates are classified on the basis of their architecture as mono-, oligo-, and polysaccharides. **Monosaccharides** are simple carbon compounds consisting usually of three to seven carbons; in plants the common monosaccharides contain five carbons (pentose sugars, such as sucrose) and six carbons (hexose sugars, such as glucose and fructose) (Figure 4-5). **Disaccharides** are produced from the linkage of two monosaccharides. Thus, maltose is a disaccharide composed of two molecules of glucose; sucrose (ordinary table sugar) is a disaccharide composed of one molecule each of glucose and fructose. Sucrose is the main translocatable form of carbohydrate in plants. Some plants store extensive amounts of sucrose and are cultivated for this purpose (e.g., sugarcane and sugar beet). The linkage of two monosaccharides is called a **condensation reaction** because it releases one molecule of water; the reverse process is called **hydrolysis** since one molecule of water is required. **Oligosaccharides** yield on hydrolysis a discrete number of monosaccharide units. **Polysaccharides** are complexes (high-molecular-weight polymers) that on hydrolysis produce a large, undefined number of monosaccharide units. Branched chains of hundreds of glucose molecules make up **starch** (Figure 4-5), a storage form of carbohydrate found in fleshy tubers, bulbs, and roots. Starch is the principal reserve carbohydrate of many plants. Two forms of starch are known: a straight unbranched chain of glucose molecules linked together (amylose) and a branched chain (amylopectin). Starch is insoluble in water, but in the plant it is readily broken down into soluble sugars by the enzyme **amylase.** Starch is very digestible and is the source of energy of many horticultural foods, such as

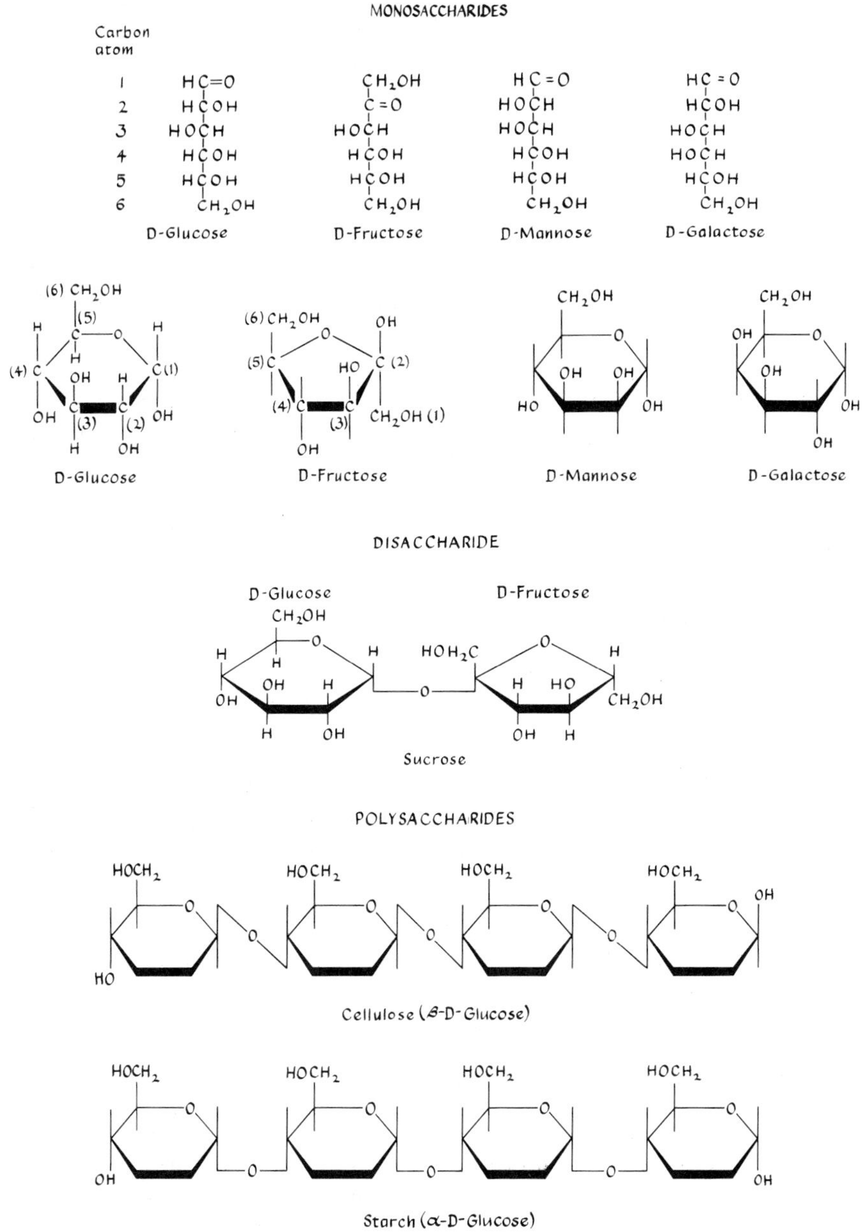

MONOSACCHARIDES

Carbon atom

D-Glucose D-Fructose D-Mannose D-Galactose

D-Glucose D-Fructose D-Mannose D-Galactose

DISACCHARIDE

D-Glucose D-Fructose

Sucrose

POLYSACCHARIDES

Cellulose (β-D-Glucose)

Starch (α-D-Glucose)

FIGURE 4-5. Structure of carbohydrates. The single sugars are found as open-chains (in solution) or in ring configurations. The four 6-carbon, or hexose, sugars (glucose, fructose, mannose, and galactose) occur in all living plants as derivatives or components of polysaccharides, but only glucose and fructose are found in molecular form in the cytoplasm or vacuole. Sugars are transported within most plants as sucrose, a disaccharide formed from glucose and fructose. Starch and cellulose are the most abundant polysaccharides in plants.

potato and sweet corn. The sweetness of sweet corn is due to various genes (principally *sugary*) which act to accumulate sugars at the expense of starch in the developing seed.

Inulin, a polymer of fructose, is the principal reserve carbohydrate of the Jerusalem artichoke (*Helianthus tuberosus*). Inulin has interesting medical uses because, unlike the glucose-containing carbohydrates, it can be metabolized by diabetics.

Cellulose occurs in plants as interwoven and interconnected strands of the cellulose polymer of glucose (Figure 4-5). Cellulose is very insoluble and cannot be broken down into sugars by the human digestive system because the appropriate enzyme is lacking, but cellulose can be broken down by some microorganisms. Some plant-eating animals can utilize cellulose because of the fermentation action of microorganisms in their gastrointestinal tracts. This is accomplished very efficiently within ruminant animals (cattle, sheep, goats) and those species with a large cecum, such as the horse and the rabbit.

Carbohydrate foods are commonly classified on the basis of digestibility into two components: *crude fiber*, the coarse portion that consists largely of cellulose and related substances, and *nitrogen-free extract*, principally starches and sugars. The crude fiber is not easily digested and yields a lesser proportion of its energy than does the nitrogen-free extract. Although crude fiber is of little direct nutritional use to humans, it is now believed to perform important indirect functions, such as stimulating peristaltic action of the intestines. Crude fiber is the basis of many animal rations; the average animal ration is 75 percent carbohydrates.

Carbohydrates are not in themselves essential as a human food. Like other animals, humans can be maintained on a carbohydrate-free diet by substituting lipids as an energy source, as do the Inuit, an arctic people. But carbohydrates in the form of sugars and starches are an inexpensive and convenient source of food energy. About 40 to 45 percent of the calories in the diet of the average American are provided by carbohydrates.

Lipids

Fats and fatlike substances are grouped under the general term *lipids*, substances that are characteristically soluble in organic solvents, such as ether or chloroform. There are three classes of lipids that are important in plants: **true fats** (reserve food material), **phospholipids** (structural materials, particularly membranes), and **waxes** (largely cuticular components).

Fats are esters, that is, unions of organic acids (fatty acids) (Figure 4-6*A*) and certain alcohols, characteristically glycerol, a 3-carbon compound in which each carbon bears a hydroxyl group (Figure 4-6*B*). Fatty acids contain even numbers of carbons (4 to 26 in plant fatty acids) and are linked in long chains, typically unbranched. Fats contain carbon, hydrogen, and oxygen, but, unlike carbohydrates, the proportion of oxygen is very low. The common storage forms of fats are known as triglycerides because three fatty acids are attached to

A

Acid	Composition	Chemical structure	Abbreviated formula*
Lauric	$C_{12}H_{24}O_2$	$CH_3(CH_2)_{10}COOH$	12:0
Myristic	$C_{14}H_{28}O_2$	$CH_3(CH_2)_{12}COOH$	14:0
Palmitic	$C_{16}H_{32}O_2$	$CH_3(CH_2)_{14}COOH$	16:0
Stearic	$C_{18}H_{36}O_2$	$CH_3(CH_2)_{16}COOH$	18:0
Oleic	$C_{18}H_{34}O_2$	$CH_3(CH_2)_7CH{:}CH(CH_2)_7COOH$	18:1
Ricinoleic	$C_{22}H_{34}O_3$	$CH_3(CH_2)_5CHOHCH_2CH{:}CH(CH_2)_7COOH$	18:1
Erucic	$C_{22}H_{42}O_2$	$CH_3(CH_2)_7CH{:}CH(CH_2)_{11}COOH$	22:1
Linoleic	$C_{18}H_{32}O_2$	$CH_3(CH_2)_4CH{:}CHCH_2CH{:}CH(CH_2)_7COOH$	18:2
Linolenic	$C_{18}H_{30}O_2$	$CH_3CH_2CH{:}CHCH_2CH{:}CHCH_2CH{:}CH(CH_2)_7COOH$	18:3
Chaulmoogric	$C_{18}H_{32}O_2$	CH=CH / CH_2—CH_2 >$CH(CH_2)_{12}COOH$	18:1

*Number of carbons:number of double bonds.

B. Glycerol and a triglyceride

H_2COH
|
$HCOH$
|
H_2COH

Glycerol

Hydrocarbon chain of fatty acid

Glycerol —
$H_2C—O—\overset{O}{\overset{\|}{C}}—(CH_2)_n—CH_3$
$HC—O—\overset{O}{\overset{\|}{C}}—(CH_2)_n—CH_3$
$H_2C—O—\overset{O}{\overset{\|}{C}}—(CH_2)_n—CH_3$

Ester linkage

A fat molecule

C. Oleopalmitostearin

$H_2C\text{-}O\text{-}\overset{O}{\overset{\|}{C}}\text{-}(CH_2)_{14}CH_3$ — Hydrocarbon chain of: Palmitic acid (16:0)
$H_2C\text{-}O\text{-}\overset{O}{\overset{\|}{C}}\text{-}(CH_2)_7CH{:}CH(CH_2)_7CH_3$ — Oleic acid (18:1)
$H_2C\text{-}O\text{-}\overset{O}{\overset{\|}{C}}\text{-}(CH_2)_{16}CH_3$ — Stearic acid (18:0)

D. Phospholipid

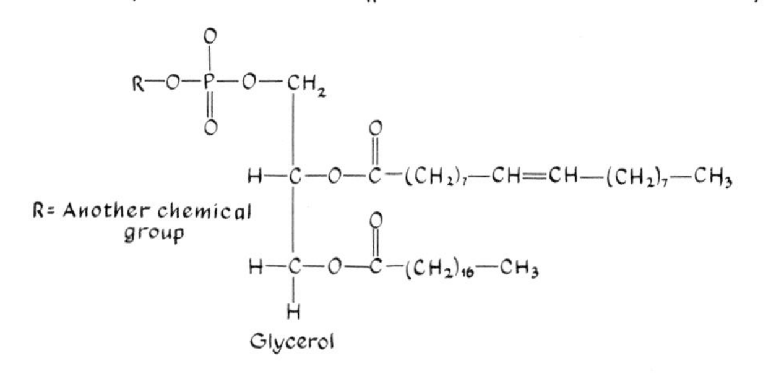

Phospholipids form a film on water

and align themselves in rows in cells to form membranes.

Polar heads

Nonpolar tails

Polar heads

FIGURE 4-6. (*A*) Some of the principal fatty acids of higher plants. (*B*) Glycerol (*left*) and the generalized structure of a fat molecule (*right*). (*C*) Oleopalmitostearin, the principle triglyceride of cocoa butter. (*D*) Phospholipids composed of glycerol, fatty acids, and a phosphate group. [(*D*) is adapted from P. H. Raven, R. F. Evert, and H. Curtis, *Biology of Plants*, 3d ed., Worth, 1981.]

glycerol in an ester linkage. The major triglyceride of cocoa seeds in composed of the fatty acids palmitate, oleate, and stearate (Figure 4-6*C*).

Fats occur in small amounts in all living cells. In plants fats generally accumulate in seeds (Table 4-4) and serve as reserve food for the growing seedling. There is an inverse relation between the amount of carbohydrate and the amount of fat in seeds. The amount of fat in plant seed may be altered by selective breeding. Long-term selection in maize has produced lines with high (greater than 14 percent) and low (less than 2 percent) fat contents. Some fruits, such as olives and avocados, are also high in fat.

The differences in fats are related to the kinds and arrangement of their fatty acids. Fatty acid molecules differ in their length (usually from 4 to 26 carbons) and in the degree of hydrogen saturation—that is, in the number of double bonds between carbon atoms. A fatty acid molecule that contains no double bonds is said to be saturated. The degree of saturation affects their stability and firmness. The distinction between oils and fats is that fats are solid at room temperature, whereas oils are liquid. Oils have large numbers of double bonds; that is, they do not carry all of the hydrogen possible (they are polyunsaturated). Plant oils may be changed to fats by the addition of hydrogen **(hydrogenation).**

Fats are insoluble in water and must be rendered soluble (digested) before they can be utilized by the body: Digestion splits the fatty acids from glycerol. The position at which a fatty acid is attached to the glycerol affects digestibility, because a fatty acid attached at the mid-position is removed last, if at all. Digestibility depends also on the length of the fatty acid molecule; the short-chain fatty acids are more easily digested.

Fats are the most concentrated food energy source and have always been highly valued as food. Some of the unsaturated fatty acids are essential in the human diet. The most important is linoleic acid, an 18-carbon unsaturated fatty acid with reactive double bonds at the ninth and twelfth positions. Fats aid the absorption of certain fat-soluble vitamins and are partly responsible for the feeling of satiety we experience after eating.

TABLE 4-4 Percentage of crude fat in the seeds of various crops.

Crop	Percent crude fat in seed
Coconut	65
Cacao	50–55
Peanut	40–50
Flax (linseed)	30–35
Soybean	15–20
Cotton seed	15–20
Maize	5

In the United States, fat contributes 40 to 50 percent of the calories in the average diet; in some less prosperous parts of the world, fat contributes only about 20 percent. It is not clear with whom the advantage lies, for in some individuals the level of cholesterol—a natural fatty material in the blood—increases to high levels with high intakes of saturated fats. High cholesterol levels have been associated with cardiovascular disorders, but the precise relation between fat intake, cholesterol levels, and cardiovascular disease are not at all clear.

Waxes are a third type of lipid important in plants. Waxes are long-chain fatty acids and long-chain alcohols, often of great complexity. Waxes are found on the surfaces of leaves and fruits. The lipid found in the fruit of jojoba (*Simmondsia chinensis*) is a liquid wax that has properties similar to the properties of sperm whale oil. Leaf waxes of some species are a minor source of commercial waxes. However, waxes provide a critical interface between plants and their aerial environment and thus are very important in horticulture because they affect plant response to foliar application of chemicals.

Phospholipids are composed of a glycerol backbone with two of its three positions filled by fatty acids and the third filled by a phosphorus-containing molecule (Figure 4-6*D*). The phosphate "head" has polarity (opposing charges) but the fatty acid "tails" are nonpolar. The phosphate head is hydrophilic (water loving or soluble in water) and the fatty acid tail is hydrophobic (water fearing or unsoluble in water). The alignment of phospholipids is important in the structure of cellular membranes, which act as barriers and channels to diffusion of various substances including ions. The inner layer of membranes consists of a double row of phospholipids with the phosphate ends aligned outward.

Proteins

Protein (the term applied to proteins collectively) is a major constituent of living material. Next to water, protein is the most abundant substance in our bodies, accounting for about half the dry weight. Protein also makes up the greatest proportion of the dry weight of most plant cells if cellulose is excluded (and lignin in woody plants).

Proteins are complex molecules made up of different combinations of amino acids (Figure 4-7). Amino acids, the basic unit of proteins, are synthesized in plants (and microorganisms) from carbohydrate fragments and from the nitrogen in ammonium ions (NH_4^+). Each of the amino acids contains a carboxyl group (—COOH) and an amino group (—NH_2). Although proteins contain much carbon, hydrogen, and oxygen, nitrogen is their characteristic element. Sulfur is found in two amino acids, methionine and cysteine. Since most proteins average 16 percent nitrogen, the protein content of foods may be estimated by multiplying the nitrogen content by a factor of 6.25.

Although there are many amino acids, only about twenty are found in protein. Amino acids differ in the number of carbon atoms and attached groups. The amino acids combine at their amino and carboxyl groups to form peptide

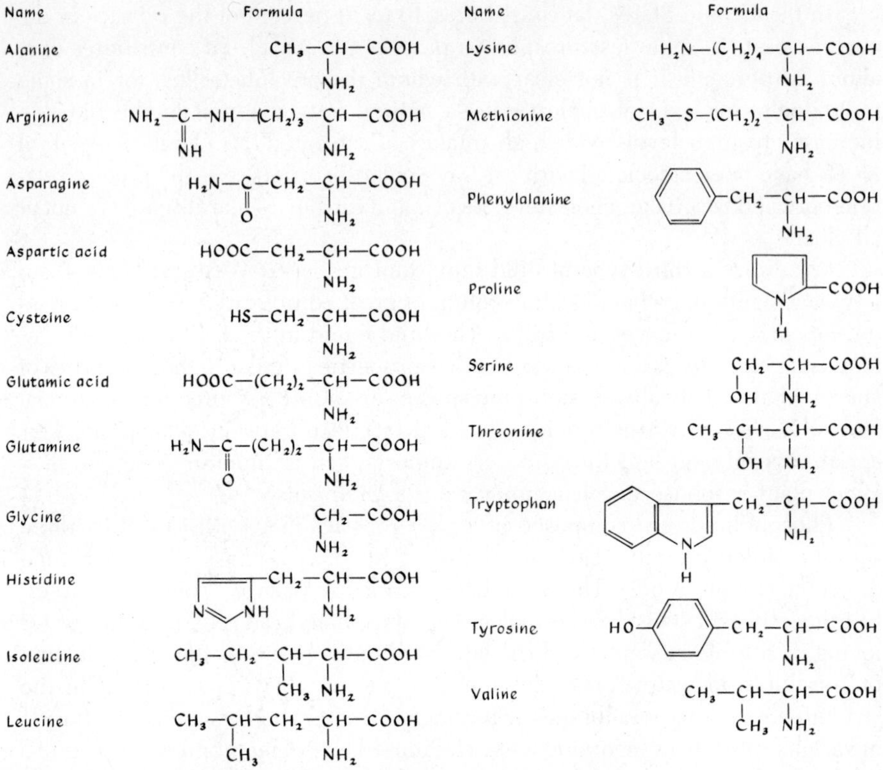

Name	Formula	Name	Formula
Alanine	$CH_3-CH(NH_2)-COOH$	Lysine	$H_2N-(CH_2)_4-CH(NH_2)-COOH$
Arginine	$NH_2-C(=NH)-NH-(CH_2)_3-CH(NH_2)-COOH$	Methionine	$CH_3-S-(CH_2)_2-CH(NH_2)-COOH$
Asparagine	$H_2N-C(=O)-CH_2-CH(NH_2)-COOH$	Phenylalanine	$-CH_2-CH(NH_2)-COOH$
Aspartic acid	$HOOC-CH_2-CH(NH_2)-COOH$	Proline	N(H) $-COOH$
Cysteine	$HS-CH_2-CH(NH_2)-COOH$	Serine	$CH_2(OH)-CH(NH_2)-COOH$
Glutamic acid	$HOOC-(CH_2)_2-CH(NH_2)-COOH$	Threonine	$CH_3-CH(OH)-CH(NH_2)-COOH$
Glutamine	$H_2N-C(=O)-(CH_2)_2-CH(NH_2)-COOH$	Tryptophan	N(H) $-CH_2-CH(NH_2)-COOH$
Glycine	$CH_2(NH_2)-COOH$	Tyrosine	$HO-$ $-CH_2-CH(NH_2)-COOH$
Histidine	N NH $-CH_2-CH(NH_2)-COOH$	Valine	$CH_3-CH(CH_3)-CH(NH_2)-COOH$
Isoleucine	$CH_3-CH_2-CH(CH_3)-CH(NH_2)-COOH$		
Leucine	$CH_3-CH(CH_3)-CH_2-CH(NH_2)-COOH$		

FIGURE 4-7. Structure of the twenty amino acids that occur in proteins.

linkages. A protein can be defined as a group of amino acids held together by peptide linkages. Proteins may be very large. The molecular weight of gliadin, one of the wheat proteins, is 27,500 and that of glutenin, another wheat protein, is in the millions.

There are many kinds of proteins. In plants, proteins are commonly classified on the basis of their origin. Seed proteins are the primary reserve forms. The proteins found in the endosperm of wheat (commonly referred to as gluten) are responsible for the characteristic elastic property of dough made from wheat flour. Proteins in plant tissue occur primarily in the cytoplasm and chloroplasts. Nucleoproteins, proteins that complex with nucleic acids, are a major constituent of chromosomes. Enzymes, the major regulatory materials of living organisms, are proteins.

Organic Acids and Alcohols

Plants manufacture a great variety of organic acids. A number participate in metabolic cycles, such as those involved in the biological combustion of sugar to carbon dioxide. Organic acids are found in the vascular sap and commonly accumulate in certain plant organs, particularly the fruit (citric acid in lemons, malic acid in apples).

Alcohols occur only in very low concentrations in the uncombined state, and they are generally found combined with organic acids as esters. Fats are esters of an alcohol and fatty acids. The characteristic odors and flavors of fruit are due to a combination of the volatile organic acids, esters, and other compounds such as ketones and aldehydes.

Aromatic Compounds

Aromatic compounds, in contrast to aliphatic, or "straight-chain," carbon compounds, have structures containing at least one benzene ring. Many, but not all, of these compounds have very characteristic odors, hence the name *aromatic*. The simple phenolic compounds (those containing one benzene ring), such as vanillin and methyl salicylate, are responsible for the odors of vanilla and wintergreen.

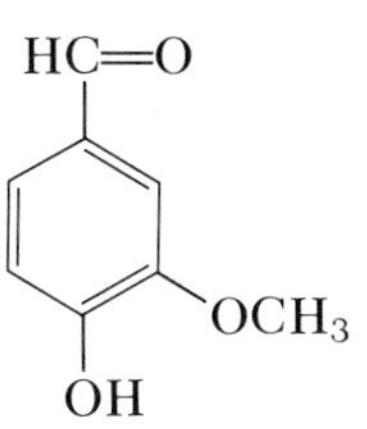

Vanillin

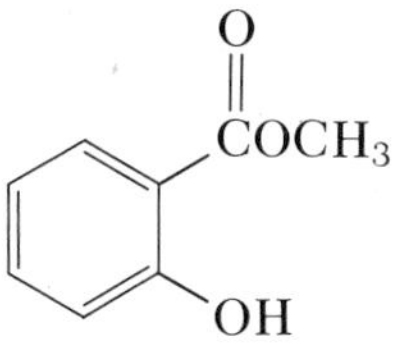

Methyl salicylate

The volatile phenolic compounds (specifically, the phenylpropane derivatives) are responsible for the characteristic flavors and odors of cinnamon, clover, and parsley.

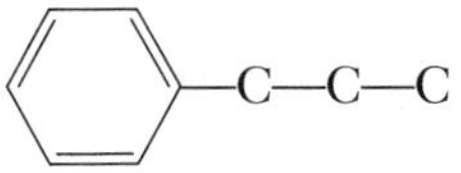

Phenylpropane skeleton

Two commercially important plant constituents, lignins and tannins, are mixtures of complex aromatic and carbohydrate materials. Lignins (polymers of phenolic acid) harden the cellulose cell wall into an inelastic and enduring material resistant to microbial decomposition. Tannins have the property of precipitating proteins, and they are used to transform animal hides into leather. Tannins are extracted from a number of plants—the word *tan* originally referred to the bark of the oak tree. Unripe persimmons are high in tannin, as any person bold enough to sample one will discover: The extremely bitter, astringent taste is due to protein precipitation on the tongue.

Flavonoids are a group of aromatic compounds characterized by two substituted benzene rings connected by a 3-carbon chain and an oxygen bridge.

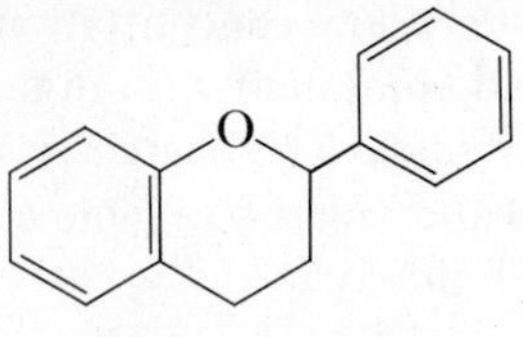

Flavonoid skeleton

The flavonoids are commonly attached to sugars to form glycosides. Among the flavonoids are many common pigments, including the anthocyanins and flavones. Anthocyanins are responsible for the reds and blues of many fruits and flowers; flavones, for the yellow of lemon. The insecticide rotenone is related to the flavonoids.

Terpenoids and Steroids

Such compounds as essential oils, steroids, alkaloids, and various pigments are related in the chemical sense to fused units of the 5-carbon substance isoprene. Two units of isoprene may form a terpene unit, the basic structure of a number of important plant constituents called terpenoids.

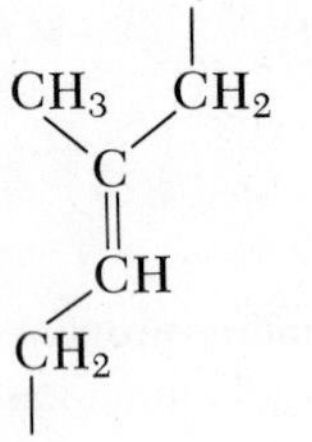

Isoprene skeleton

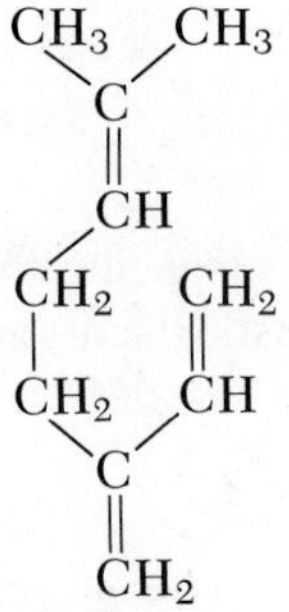

Myrcene,
an open-chain terpene

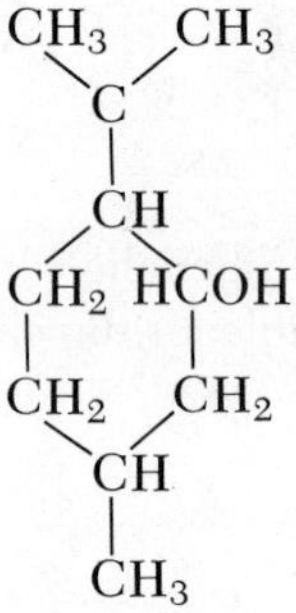

Menthol,
a cyclic terpene

The essential oils are mixtures of volatile, highly aromatic substances that have distinctive odors and flavors. They are formed in specialized glands, ducts, or cells in various plant parts. Although called "oils," these compounds are not lipids (nor are they necessarily "essential" to the plant—the term refers to the notion of *essence*, the belief that these aromatic substances possessed the virtues of the plant in concentrated form). An example is menthol, which is the major constituent of both peppermint oil and turpentine. The odors, flavors, and other properties of essential oils make them economically important for a variety of uses.

Resins—gummy exudates of many plants—are terpenoids consisting of three to six isoprene units. Resins are often associated with essential oils (and have the ability to harden as the oils evaporate). Resin synthesis occurs near wounded tissue, and the resins serve to retard water loss and entry of microorganisms. The ability of plants to produce resins is lowered if the general vigor of the plant is poor.

Steroids are complex cyclic terpene compounds composed of eight units of isoprene (tetraterpenoids). Although their function in plants is not well understood, it is known that many of them have important metabolic effects in animals. Cortisone, sex hormones, and vitamin D are steroids.

The carotenoids are complex tetraterpenes—yellow to red pigments that occur in many different kinds of tissue. Common members of this group include carotene (composed exclusively of carbon and hydrogen) and xanthophyll (composed of oxygen in addition to carbon and hydrogen). The most widespread carotenoid is the orange pigment β-carotene, which, when split in the digestive tract of animals, gives rise to two molecules of vitamin A.

Rubber is a high-molecular-weight terpenoid containing 3000 to 6000 isoprene units. Although it is produced in many dicotyledonous plants, only a few produce enough for commercial purposes. The most important is the rubber tree, *Hevea brasiliensis*. The rubber is extracted from latex, a milky, sticky liquid exuded by glandular structures associated with the phloem.

Nonprotein Nitrogen Compounds

Nitrogen is found in a wide variety of compounds in addition to amino acids and proteins. These include the nitrogen bases and a heterogeneous group of compounds called alkaloids. Heterocyclic rings containing both carbon and nitrogen are a common structural form (Figure 4-8).

The nucleoproteins, the source of the genetic material, are composed of proteins and nucleic acids. Nucleic acids are high-molecular-weight polymers of nucleotides, each formed from three constituents: a sugar (specifically ribose or deoxyribose), phosphoric acid, and a nitrogenous base having the structure of either a purine or pyrimidine ring. In deoxyribonucleic acid (DNA), the genetic material found in chromosomes, there are two purines (adenine and guanine) and two pyrimidines (cytosine and thymine). Thus there are four nucleotides, depending on the specific base involved.

The sequence of each of the four possible nucleotides spells out a message that is decipherable by the cell. This genetic code specifies the sequence of amino acids in protein. Enzyme specificity is determined by its sequence of amino acids (as well as by its structural configuration); thus, DNA controls the destiny of the cell. The gene, in essence a functional portion of the DNA molecule, is the genetic information passed from generation to generation through the gametes.

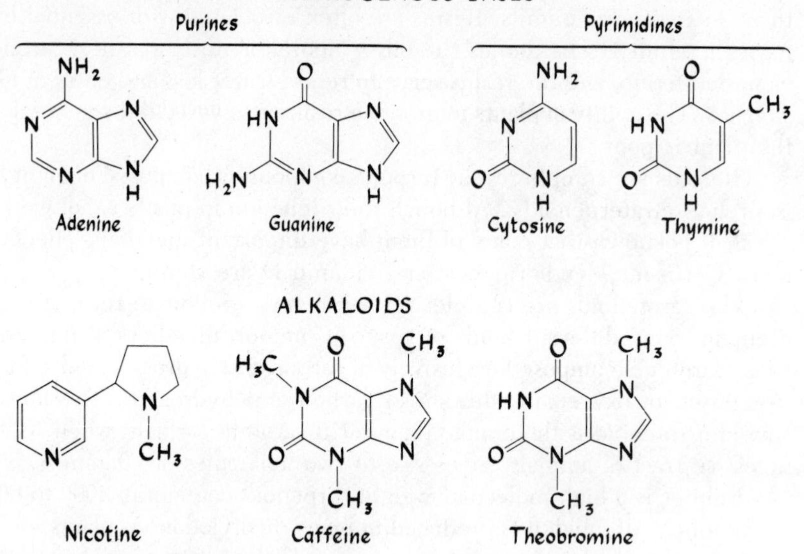

FIGURE 4-8. Cyclic nitrogen compounds.

Alkaloids, as their name indicates, are organic bases. (Although the nitrogen bases discussed above fit into the broad definition of alkaloids, they are usually treated separately.) The function of many alkaloids in plants is obscure. Nevertheless, the many effects that these substances have on human physiology render them of extraordinary pharmacological interest. A list of alkaloids derived from plant sources, along with their medicinal properties or other uses, is shown in Table 4-5. It is of interest to note that the most popular beverages—coffee, tea, maté, and cocoa—all contain stimulating alkaloids: The first three contain caffeine; the last, theobromine.

A number of extremely important pigments are complex alkaloids called porphyrins. Chlorophyll is a magnesium-containing porphyrin. The cytochromes are iron-containing porphyrins that function in energy transfer. Hemoglobin, the oxygen carrier in animal blood, is also a porphyrin.

Vitamins

The relationship between diet and disease is very old. That eating liver could cure night blindness was recorded by Hippocrates 24 centuries ago. The consumption of lime juice by British sailors to prevent scurvy (hence the term "limies") has been traced to a suggestion by James Lind in 1757 that eating fresh fruits and vegetables might help prevent the disease. Early nutritionists found that animals fed on rations of purified forms of known foods sickened and died, yet thrived when small amounts of certain substances, such as water-soluble extracts from yeast or wheat germ, were added.

TABLE 4-5 Origin and uses of some plant alkaloids.

Alkaloid	Common plant source	Medicinal and other uses
Atropine	*Datura stramonium, Atropa belladonna, Duboisia* spp.	Relaxes gastrointestinal tract, affects parasympathetic nervous system
Caffeine	*Thea sinensis, Coffea arabica, Ilex paraguayensis, Cola acuminata*	Stimulates central nervous system
Cocaine	*Erythroxylon coca*	Surface anesthetic, stimulates central nervous system
Colchicine	*Colchicum autumnale*	Induces chromosome doubling, relieves symptoms of gout
Emetine	*Cephaelis ipecacuanha*	Emetic effect, an amoebicide
Ephedrine	*Ephedra gerardiana*	Stimulates central nervous system, relieves nasal congestion
Hydrastine	*Hydrastis canadensis*	Antihemorrhagic effect
Morphine	*Papaver somniferum*	Analgesic effect
Nicotine	*Nicotiana tabacum*	Insecticide
Pelletierine	*Punica granatum*	Vermifuge (tapeworm)
Pilocarpine	*Pilocarpus microphyllus*	Diaphoretic (sweat inducing) effect
Quinine	*Cinchona* spp.	Antimalarial medicine, a cardiac depressant
Reserpine	*Rauwolfia serpentina*	Tranquilizer, sedative effect
Strychnine	*Strychnos nux-vomica*	Poison, stimulates central nervous system
Tubocurarine	*Chondodendron tomentosum*	Relaxes muscles
Theobromine	*Theobroma cacao*	Diuretic effect, a stimulant
Yohimbine	*Pausinystalia yohimba*	Aphrodisiac effect

In 1912, Frederick G. Hopkins and Casimir Funk suggested that specific human diseases, such as beriberi, rickets, and scurvy, were caused by the absence of certain nutritional substances in the diet. These were termed vitamines ("vital amines"), because the first such substance isolated, thiamin (vitamin B_1), was an amine (a compound containing an amino group ($-NH_2$). When other such essential substances were isolated and analyzed, they proved not to be amines, but the term **vitamin** was retained to refer to any essential growth factor (other than a mineral) required in very small amounts. Many essential vitamins have been discovered since that time. Although the functions of some vitamins in plants are unknown, many have been shown to be coenzymes, which are required for the function of particular enzymes. The water-soluble B vitamins tend to play a part in energy transfer.

Respiration

Energy is required to run the machinery of the cell. The energy incorporated in the chemical bonds of the sugars formed from photosynthesis cannot, of course, be harnessed by the cell from high-temperature combustion but must be provided at low and constant temperatures in delicately controlled reactions. Respiration, the process of obtaining energy from organic material, is accomplished with great efficiency in the cell. It is, in a superficial sense, the reverse process of photosynthesis:

$$C_6H_{12}O_6 + 6O_2 \longrightarrow 6H_2 + 6CO_2 + \text{energy}$$

The captured energy of light is released by the low-temperature oxidation (removal of hydrogen) of sugars. Although a small part is lost as heat, the useful energy is channeled into chemical work, initially as high-energy phosphates, and later in the synthesis of organic materials required in growth and development (Figure 4-9).

The biologic combustion of sugar is accomplished through an extremely complicated series of reactions involving many specific enzyme systems and

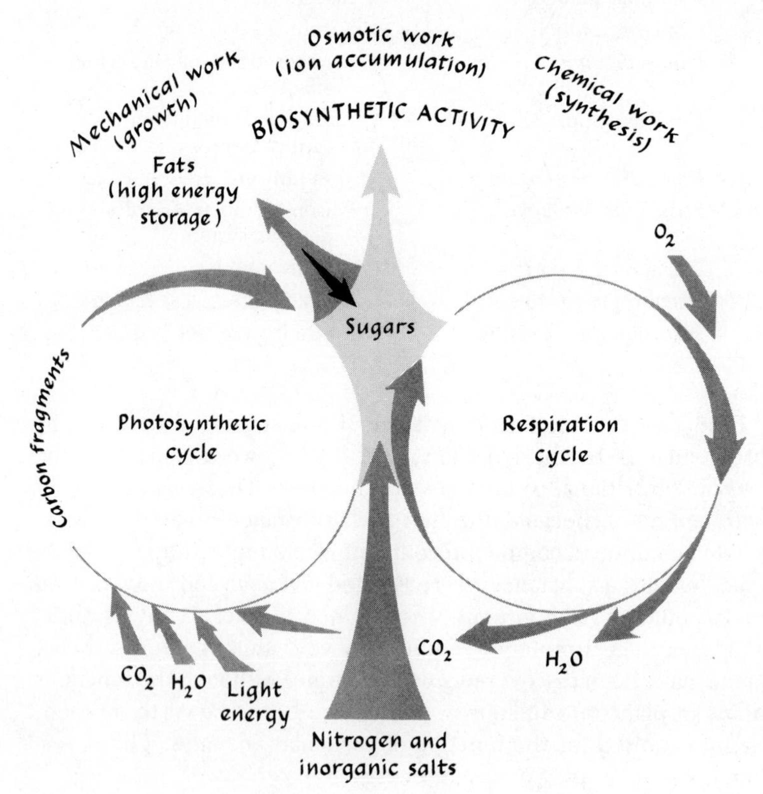

FIGURE 4-9. The respiration cycle in green plants.

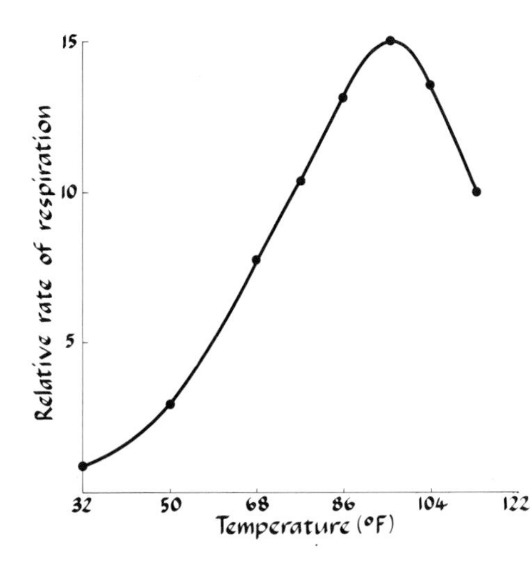

FIGURE 4-10. The relationship between temperature and the respiration rate of germinating pea seedlings. [Adapted from J. Bonner and A. W. Galston, *Principles of Plant Physiology*, W. H. Freeman and Company, copyright © 1952; after data of Fernandes.]

energy carriers. The steps from 6-carbon sugar to carbon dioxide involve the transformation of phosphorylated sugars to 3-carbon pyruvic acid ($CH_3COCOOH$). This step, known as **glycolysis,** is common to many organisms. Organisms that do not require oxygen (**anaerobes**) transform pyruvic acid to alcohol or lactic acid. Organisms that do require oxygen (**aerobes**) transform pyruvic acid to water and carbon dioxide. This involves the participation of a number of plant acids in the cyclic series of steps known as the **citric acid cycle,** or (in honor of its discoverer) the **Krebs cycle.**

The rate of respiration depends on many factors: It is highest in rapidly growing tissues and lowest in dormant tissues. The rate of respiration is greatly influenced by temperature: As Figure 4-10 shows, it approximately doubles for each 18°F (10°C) rise over a range of 40 to 97°F (4 to 36°C). Other factors influence the respiration rate, including the availability of oxygen and carbohydrates and the age and condition of cells and tissues. Respiration is a common feature of all living material, and consequently the detection of evolved carbon dioxide is often utilized as a test for life.

Selected References

Bonner, J., and J. E. Varner (eds.), 1976. *Plant Biochemistry.* New York: Academic Press. (An advanced treatise.)

Goodwin, T. W., and E. I. Mercer, 1983. *Introduction to Plant Biochemistry,* 2d ed. Oxford: Pergamon. (The biochemistry of plant metabolism.)

Noggle, G. R., and G. J. Fritz, 1976. *Introductory Plant Physiology.* Englewood Cliffs, N.J.: Prentice-Hall. (A solid introductory text.)

Salisbury, F. B., and C. Ross, 1978. *Plant Physiology,* 2d ed. Belmont, Calif.: Wadsworth. (A modern text in plant physiology.)

5 *Differentiation and Development*

After considering the plant's physiological processes in Chapter 4, we can now examine the organism as an integrated mechanism capable of an irrevocable increase in size and complexity. This chapter will begin with the problem of differentiation and regulation. The plant, however, is more than the sum of its physiological and developmental processes. The orderly cycle of development that the whole plant undergoes involves complex but orderly patterns of change in cells, tissues, and organs. This cycle begins with seed germination and progresses through juvenility, maturity, and flowering. Upon fruiting the essential cycle of plant development is completed. In perennials the plant is ready to begin a new cycle after a period of quiescence. In annuals and biennials fruiting is a signal to the organism to enter the final phases of plant growth—senescence and death (Figures 5-1 and 5-2).

DIFFERENTIATION

Explaining differentiation is one of the great problems of biology. There are two levels of complexity to the problem. The first concerns development of individual cells. Unicellular organisms are capable of undergoing complex transforma-

---

Sections of this chapter are based on a chapter in the first edition written by C. E. Hess.

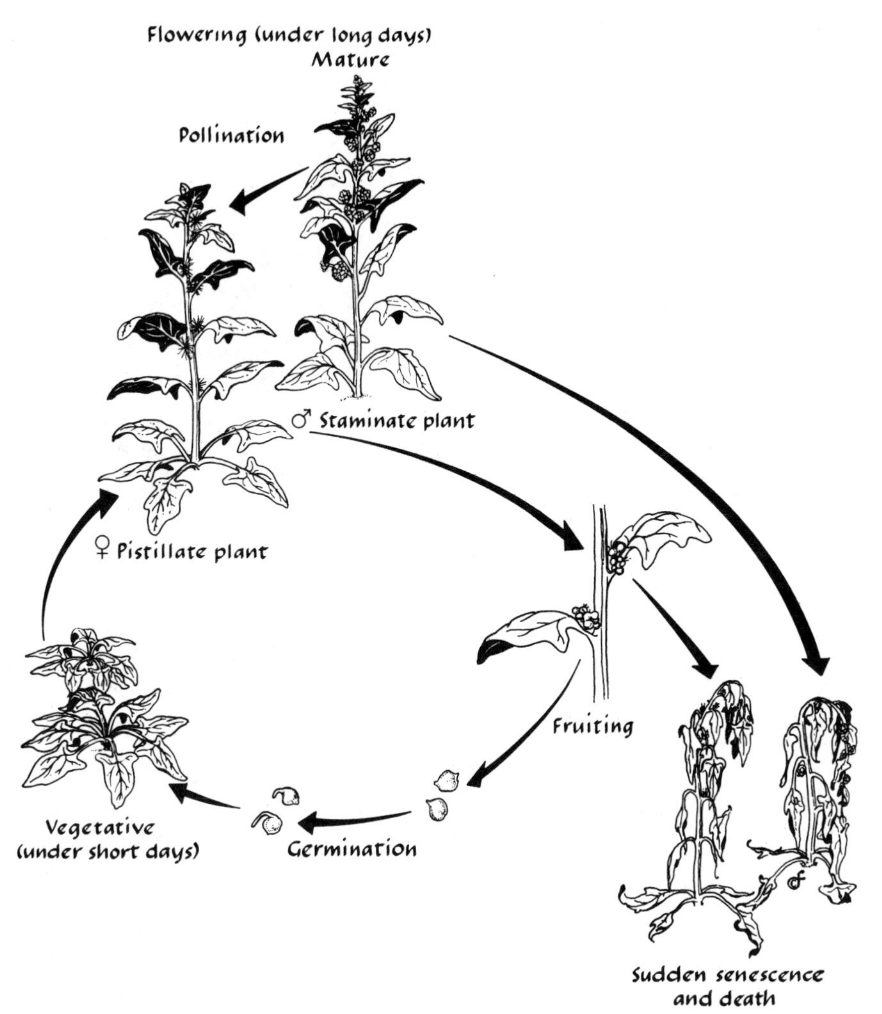

FIGURE 5-1. The developmental history of a herbaceous annual (spinach): The seed of spinach quickly germinates under proper environmental conditions for growth. If the seedling is grown under short-day conditions, it forms a distinct vegetative stage (rosette). Under long-day conditions, the stem elongates to form a seedstalk and initiates flowers. Since spinach is normally dioecious, staminate and pistillate flowers form on separate plants. Soon after the staminate plants flower, and soon after the pistillate plants fruit, they undergo rapid senescence and die.

tions with no apparent internal stimulus other than their own genetic makeup. The differentiation of individual cells must involve some systematic turning on and off of the multiplicity of complex instructions potentially receivable from the genetic material. Differentiation in multicellular organisms comes about as a result of differential growth within and among cells. This is accomplished in an orderly and systematic way, with the mitotic process in cell division ensuring genetic continuity of all cells. The differentiation of genetically identical

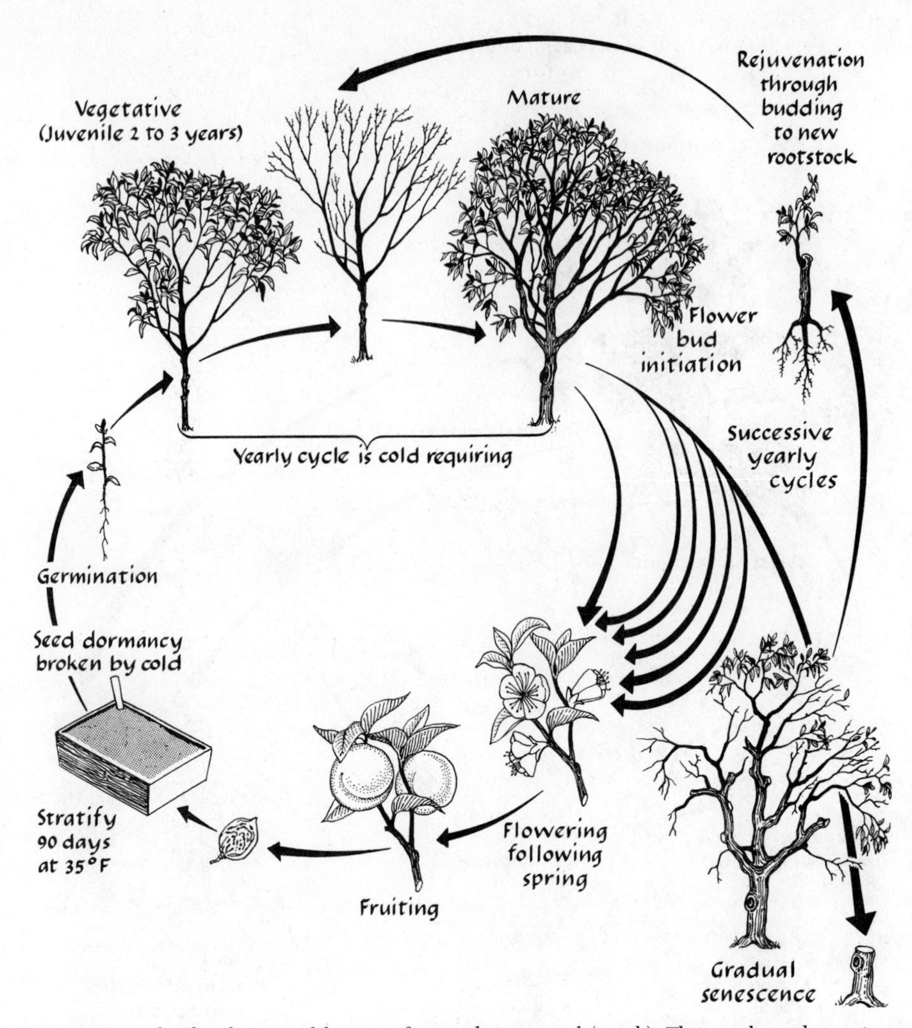

FIGURE 5-2. The developmental history of a woody perennial (peach): The peach seed germinates only after a period of cold treatment. The juvenile stage lasts about 2 to 3 years. The yearly cycles must be interrupted by a period of cold. Usually flower buds are first initiated in the third or fourth year and open in the following spring. The plant may live for 50 to 60 years, and it undergoes a very gradual deterioration. However, the plant may be rejuvenated from buds close to the base of the tree. If vegetative growth is continually budded onto new rootstocks, the "original" plant may remain, in effect, immortal.

cells in multicellular organisms is indeed a mystifying process. The way in which this is accomplished appears to depend on the interaction between the cell's genetically controlled processes and their external environment. In a multicellular organism the environment of one part of a cell may be quite different from that of another part. Investigations of tissue and organ differentiation have shown that many of these differences involve the interaction of substances produced by different parts of the organs. For example, to culture tomato roots artificially, not only must root meristem be present, but also thiamin and pyridoxine (familiar as vitamins in the B complex), which are normally provided by the leaves.

In higher organisms particular cells take over the control of differentiation through the action of "chemical messengers." Naturally occurring organic substances that affect growth and development in very low concentrations and whose action may be involved in sites far removed from their origin are known as **hormones** (Figure 5-3). A great many such substances occur in plants. Mi-

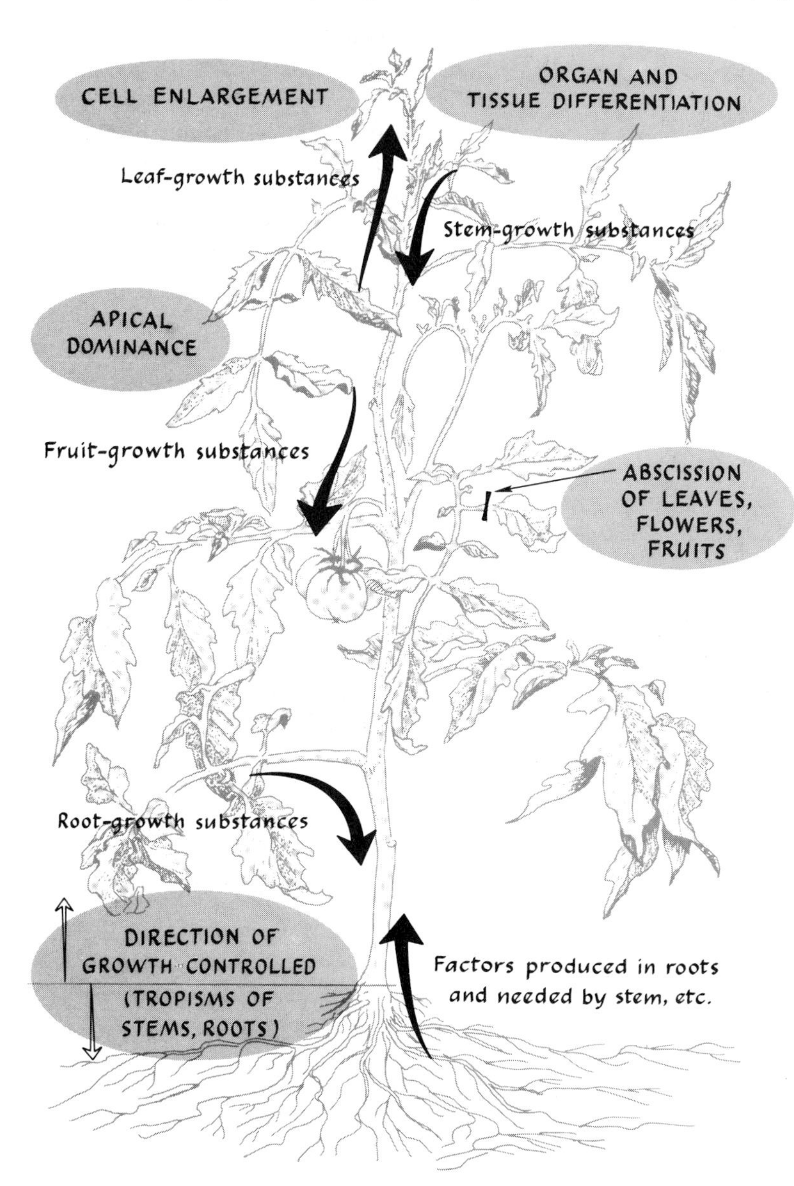

FIGURE 5-3. Plant growth and development is directed by organic substances produced in various parts of the plant and translocated to others. One group of such substances, called auxins, are produced in the plant extremities. Auxins are associated with various growth functions, as indicated by the shaded portions of the diagram. [Adapted from J. Bonner and A. W. Galston, *Principles of Plant Physiology*, W. H. Freeman and Company, copyright © 1952; and A. C. Leopold, *Auxins and Plant Growth*, University of California Press, 1955.]

nute amounts (as low as one part per billion) exert measurable physiological effects. The term **growth regulators** has been coined to include all naturally occurring and synthetically copied, or created, substances that affect growth and development. They may be either inhibitors or promoters of growth; and a single such substance may be variously an inhibitor *and* a promoter, depending upon its concentration.

Growth Regulators: Regulators of Growth and Development

From continuing research in the chemical control systems affecting differentiation in plants, a number of important groups of hormonelike substances have been defined: **auxins, gibberellins, cytokinins, ethylene,** and **abscisic acid.** These substances are so grouped on the basis of structural and physiological similarities, but their functions overlap and interact. There are also undoubtedly other groups. The basis for the chemical control of many distinct growth patterns, such as flowering, remains unknown. In addition, many naturally occurring compounds known to inhibit growth and developmental processes fit no well-defined group. The chemical nature of some of these regulators will be discussed here, and later, in Chapter 15, we will discuss how application of them can be used to direct the growth of plants.

Auxins

The auxins, the first class of growth regulators to be discovered, have received considerable attention from plant physiologists. Auxins are growth-promoting plant hormones. Cell elongation, the simplest example of anatomical differentiation, is directly affected by auxin concentration. Their mode of action appears to involve alterations in the plasticity of the cell wall. This fundamental property of auxins has been used in assaying their activity. The basic bioassay for auxin activity consists of measuring the elongation of dark-grown oat coleoptile sections. Similar tests assay the rate of curvature of longitudinally halved stems in response to auxin application, as in the split-pea test (Figure 5-4). The most common natural auxin is indoleacetic acid (IAA).

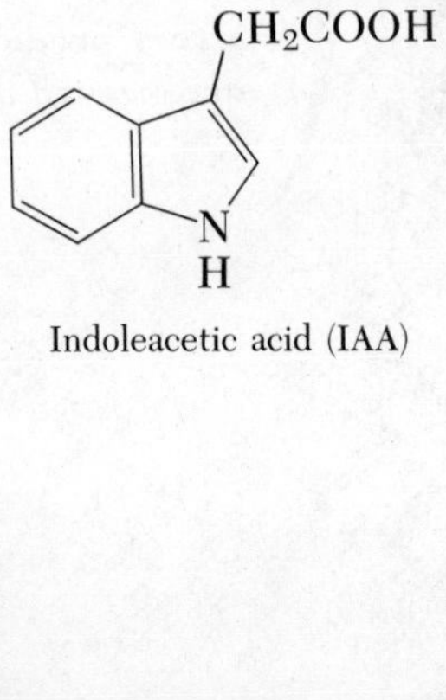

CH_2COOH

N
H

Indoleacetic acid (IAA)

FIGURE 5-4. The split-pea test measures the biological activity of auxins. Sections of stems are split with a razor blade and placed in the solution to be tested. The amount of inward curvature is proportional to the auxin concentration. The petri dish on the left contains auxin; the dish on the right is the control and shows no activity. [Photograph courtesy Purdue University.]

Some very fundamental growth responses have been shown to be controlled by auxins. Phototropism, the bending toward light of the stem apex (growing point) can be explained as the result of differential cell elongation caused by the accumulation of auxin on the darkened side of the meristem (Figure 5-5). The cells on that side elongate and cause the stem to bend toward

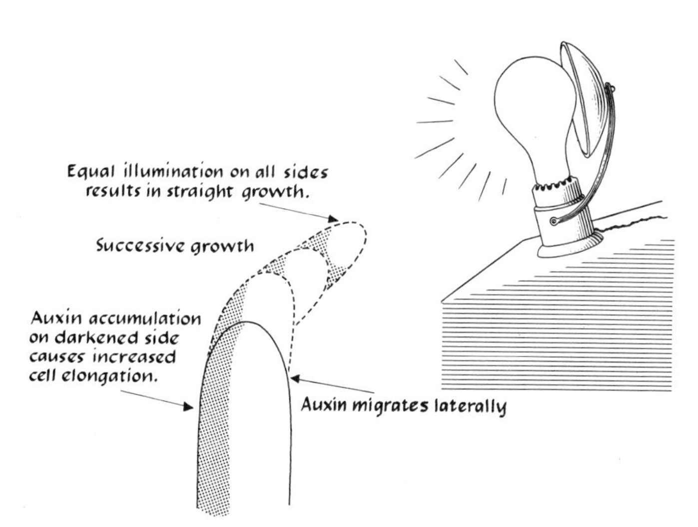

FIGURE 5-5. Phototropism results from the redistribution and inhibition of auxin in the growing point by light. The subsequent accumulation of auxin on the darkened portion elongates cells on that side and bends the seedling toward the light.

the light. Similarly, auxins greatly affect growth patterns in the plant. For example, **apical dominance**—the inhibition by the growing point of the growth of axillary buds below it—appears to be a function of auxin distribution. Auxins are produced in greatest abundance in a vigorously growing stem apex, and high concentrations emanating therefrom have been shown to inhibit bud break. Removal of the auxin-producing stem tip increases lateral bud break and subsequent branching, usually directly below the cut. Thus the form of plants can be changed by the manipulation of apical dominance through pruning.

Auxins are at present the best understood of the many substances affecting plant development. They are formed in the stem and possibly in the root apices, whence they move to the rest of the plant. Their distribution, however, is not uniform. The resultant concentration of auxins in various parts of the plant has been correlated with inhibition and stimulation of growth (Figure 5-6) as well as with differentiation of organs and tissues. Such processes as cell enlargement, leaf and organ abscission, apical dominance, and fruit set and growth have been shown to be influenced by auxins. Auxins have been associated with flowering and sex expression in some plants. Research on auxins has had a deep impact on agriculture. Chemical substances that are chemically similar to IAA and that possess growth-stimulating activities are called synthetic auxins. They have had tremendous agricultural and horticultural applications and are used as herbicides, to promote rooting and fruit set, to control fruit drop, and to thin fruits with a minimum of labor.

The role of auxins has been associated with so many diverse growth systems in plants that it has at times been tempting to suggest that they must behave as "master" hormones affecting growth and differentiation. However, the importance ascribed to any particular developmental material may be influenced by

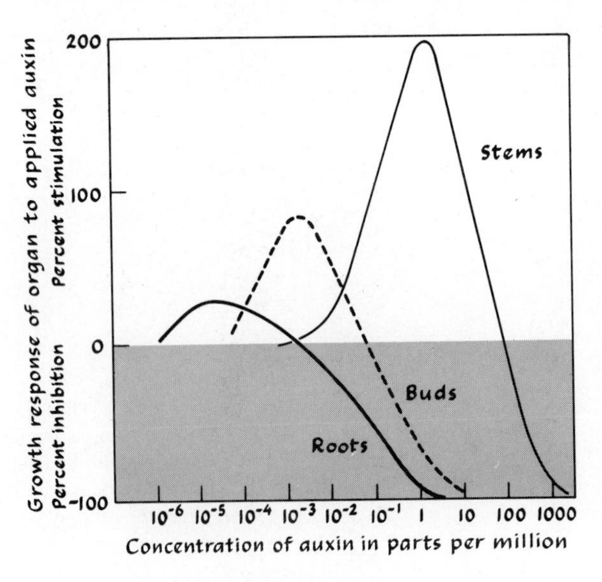

FIGURE 5-6. The effects of auxin concentration on the growth of roots, buds, and stems. [Adapted from L. Machlis and J. G. Torrey, *Plants in Action*, W. H. Freeman and Company, copyright © 1956; after Thimann.]

the intensity of study that it receives. Growth and development appear to depend on interactions of many factors. The conclusion is now inescapable that auxins, though important substances, constitute only one class of the many significant materials involved in differentiation and growth.

Gibberellins

The gibberellins are a group of at least 45 closely related, naturally occurring terpenoid compounds. They were discovered through studies of excessive growth of rice occurring in response to a fungal disease. The gibberellins affect cell enlargement and cell division in subapical meristems. Their most startling effect is the stimulation of growth in many compact plant types. Minute applications transform bush beans to pole beans or dwarf maize to normal maize (Figure 5-7). This effect is utilized as a biological assay. In addition to the dwarf-reversing response, gibberellins have a wide array of effects in many developmental processes, particularly those controlled by temperature and

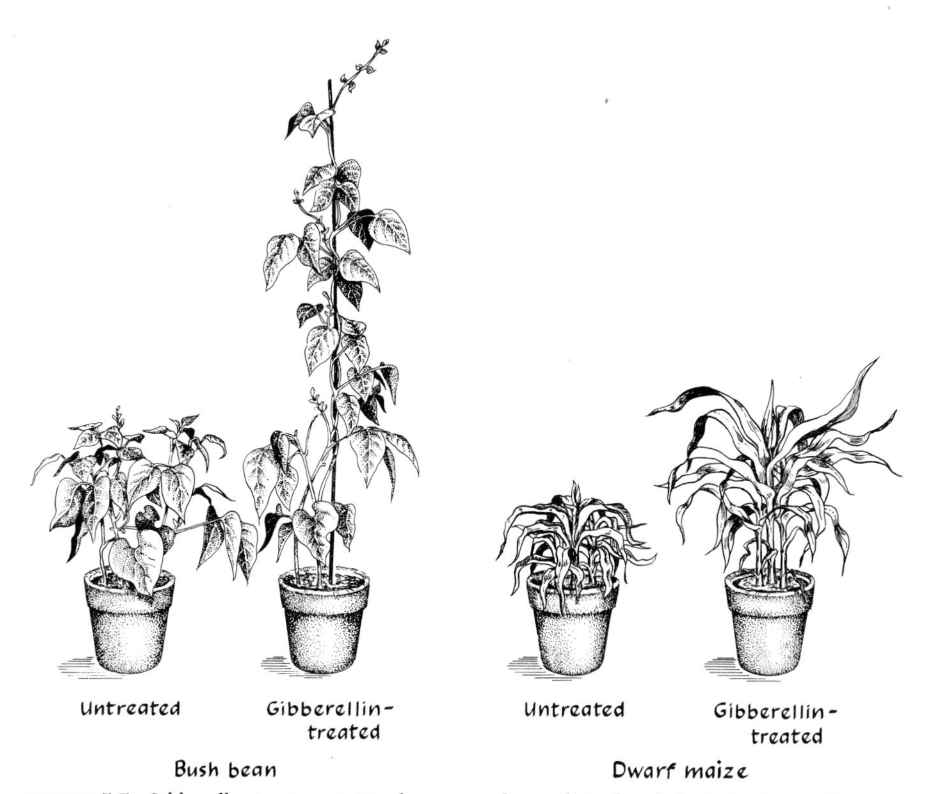

FIGURE 5-7. Gibberellin treatment stimulates normal growth in dwarf plants. In beans, 20 μg changes the bush to a vine habit; in maize, 60 μg causes a change from dwarf to normal growth.

light (photoperiod), including seed and plant dormancy, germination, and seed-stalk and fruit development. In brewing, barley is treated with gibberellic acid in the malting stages to increase the enzyme content of the malt. The use of gibberellins has had impressive agricultural applications, particularly in grape production. In Japan, seedlessness is induced in grapes, and in California the 'Thompson Seedless' cultivar ('Sultanina') shows a remarkable increase in berry size after gibberellin treatment.

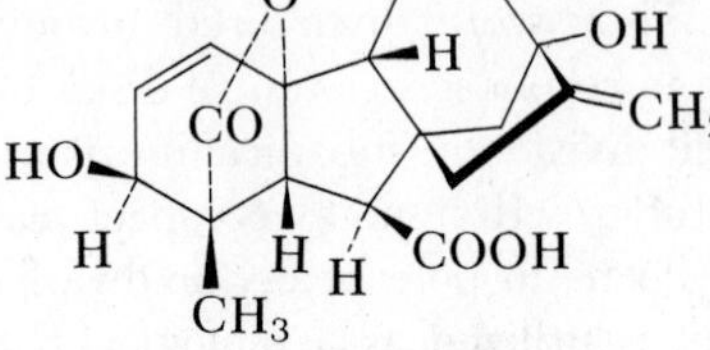

Gibberellic acid (GA_3)

Cytokinins

The term *cytokinins* (or simply *kinins*) has been applied to a group of chemical substances that have a decisive influence in the stimulation of cell division. Many cytokinins are purines. Kinetin, the original cell-division stimulant isolated from yeast, has been identified as 6-furfurylamino purine. The synthetic cytokinin 6-benzylamino purine (also known as benzyladenine or BA) stimulates shoot initiation in woody plants, including conifers.

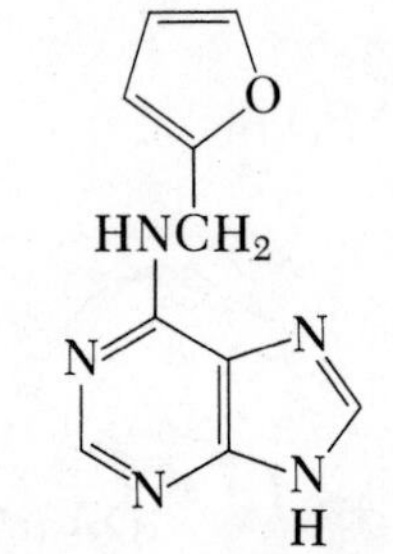

6-Furfurylamino purine (Kinetin)

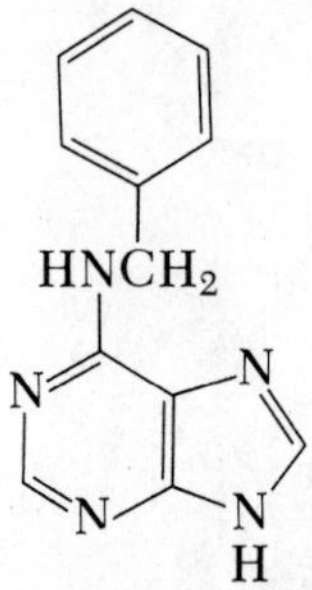

6-Benzylamino purine (BA)

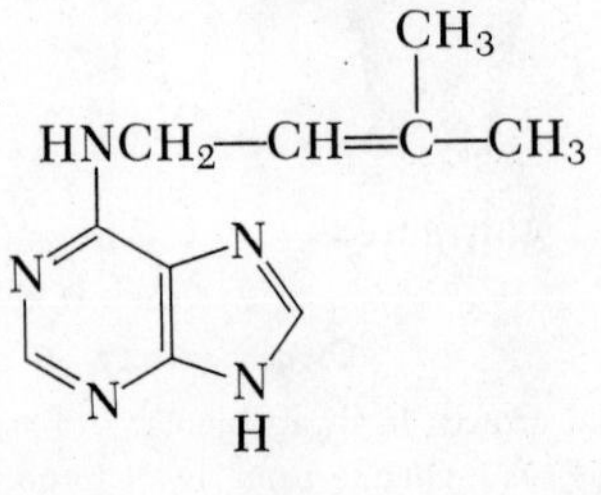

6-(γ,γ-Dimethylallylamino)purine (2iP)

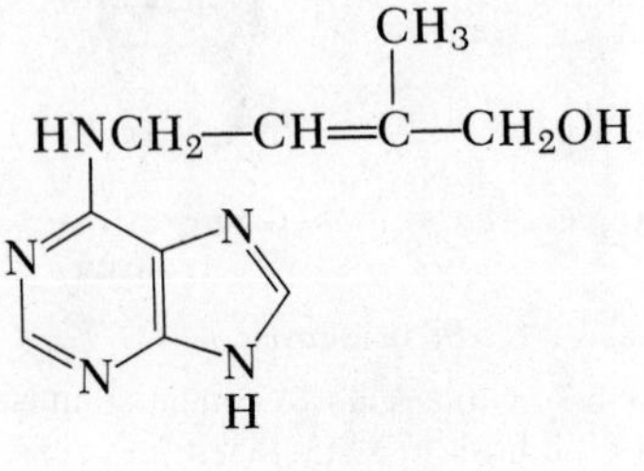

Zeatin

6-(Benzylamino)-9-(2-tetrahydropyranyl)-9*H*-purine (PBA)

Many cytokinins have been detected in research involving tissue culture. Vascular strands laid on top of cultured pith tissue stimulate differentiation in otherwise nondividing cells, an effect also induced by adenine and kinetin. Cytokinins and auxins interact to affect differentiation (Figure 5-8): High auxin and low cytokinin gives rise to root development; low auxin and high cytokinin gives rise to bud development; equal amounts result in undifferentiated growth. Cytokinins are found in abundance in fruit and seeds (for example, maize and coconut endosperm) and are probably important in promoting growth and differentiation of the embryo. Tumor cells are also rich in cytokinins.

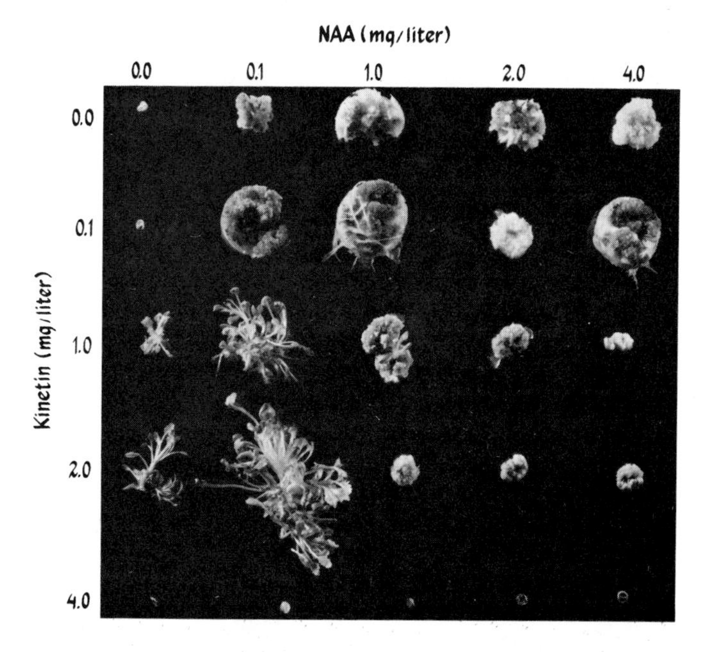

FIGURE 5-8. Effect of various levels of a cytokinin (kinetin) and an auxin (naphthalene-acetic acid, or NAA) on organogenesis in tissue culture. Leaf discs of salpiglossis were cultured for 40 days.

Cytokinins affect such diverse physiological mechanisms as leaf growth, light response, and aging. Recently, cytokinins and other plant hormones have been found to be active in regulating protein synthesis, possibly by turning gene transcription on and off or acting through transfer RNA to control translation of the gene product.

Ethylene

Although it has been known for many years that ethylene has a number of striking effects on plant growth and development, only recently has it been considered a regulatory hormone. Thus, it has been a standard practice to ripen bananas through the introduction of ethylene, and ethylene (as well as such closely related compounds as acetylene and calcium carbide) has been used to obtain uniform flowering in pineapple. Ethylene has enormous effects on dark-grown seedlings, causing swelling and disorientation; such effects were used as an assay method before the days of gas chromatography. Ethylene also influences cell division; thus, tomatoes grown in high concentrations of ethylene show extensive rooting up and down the stem. It is wise not to store budwood or hardwood cuttings in places where apples are stored: Apples produce ethylene as they ripen, and ethylene will cause splitting and bark-peeling. In the past decade it has been clearly shown that the natural or "endogenous" ethylene produced by plants influences (1) the natural course of development in etiolated seedlings, (2) abscission, (3) floral initiation (in some plants), and (4) fruit ripening. Ethephon, or (2-chloroethyl)phosphonic acid, a material that is transformed into ethylene in the plant, has a wide array of interesting effects, including the induction of uniform ripening in pineapple and tomato, sex conversion, induction of pistillate flowers in androecious (staminate) plants of cucumbers, and the stimulation of latex flow in rubber trees.

$$H_2C{=}CH_2$$

Ethylene

Abscisic Acid

Abscisic acid (ABA), referred to at one time as "dormin" and "abscisin II," is a natural inhibitory compound that affects bud and seed dormancy and leaf abscission. Similar synthetic substances have been reported. It has a wide range of effects: In addition to promoting dormancy in buds and seeds and accelerating leaf abscission, it also promotes flowering in some short-day plants. The dormancy response may be through an effect of RNA and protein synthesis. Some effects of ABA seem to be reversed by gibberellins (seed dormancy) or by cytokinins (stomatal closure).

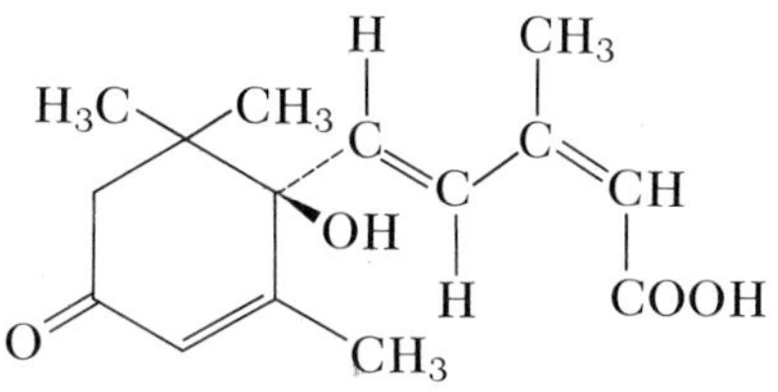

Abscisic acid

Inhibitors

Natural as well as synthetic inhibitory substances are often placed together as a diverse class of growth regulatory materials (Figure 5-9). Natural inhibitors help control such processes as seed germination, shoot growth, and dormancy. Several synthetic inhibitors have found important agricultural applications. Maleic hydrazide has been effective in preventing the sprouting of onions and potatoes. A number of materials that inhibit the natural formation of gibberellins by the plant act to dwarf plants. Chlorphonium chloride (phosphon) and chlormequat are used to reduce the height of many ornamental flowering plants without unduly interfering with flowering time or flower size. A gibberellin suppressor, daminozide, in addition to dwarfing, affects fruit maturity, hardiness, and many other plant processes.

VEGETATIVE PHYSIOLOGY

Germination

Germination includes all the sequential steps from the time the seed imbibes water until the seedling is self-sustaining (Figure 5-10). In the simplest terms, germination involves the enzymatic conversion of complex reserve substances to simple soluble substances that are readily translocated to the embryonic plant. Here some substances are oxidized through respiration to release energy, and others are utilized in synthesis. The seed must be provided with ample supplies of oxygen (to satisfy the respiratory requirements) and water (to provide a medium for enzymatic activity and synthesis). The temperature of the

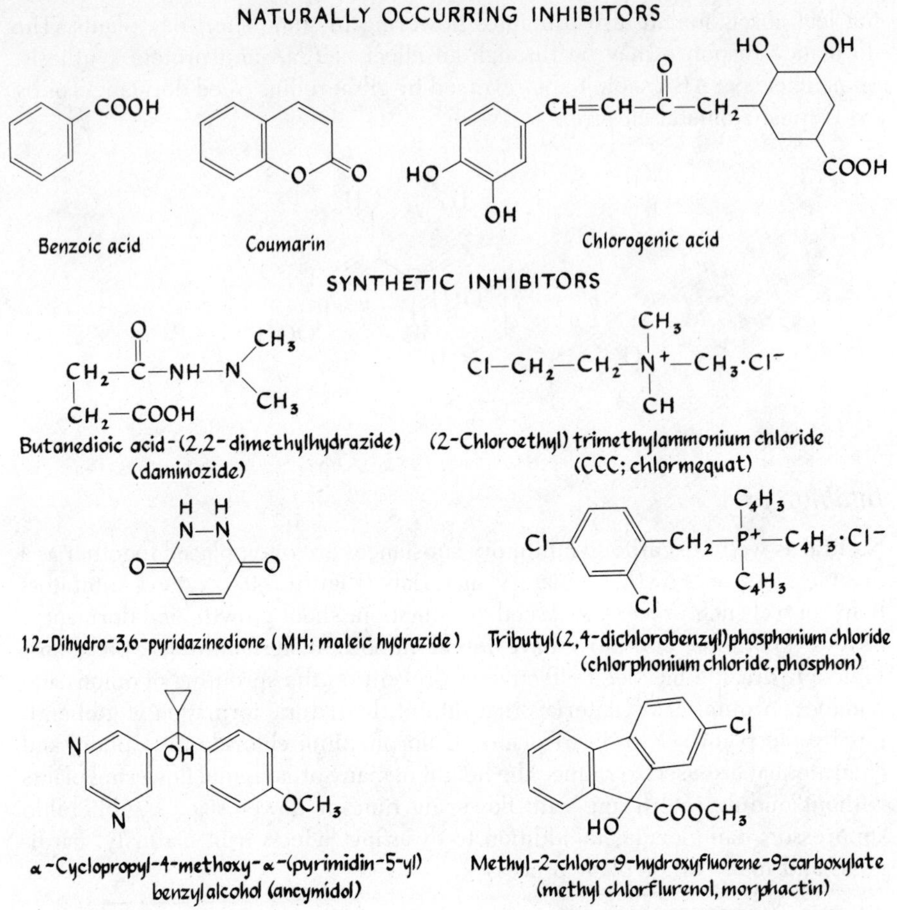

NATURALLY OCCURRING INHIBITORS

Benzoic acid

Coumarin

Chlorogenic acid

SYNTHETIC INHIBITORS

Butanedioic acid-(2,2-dimethylhydrazide) (daminozide)

(2-Chloroethyl) trimethylammonium chloride (CCC; chlormequat)

1,2-Dihydro-3,6-pyridazinedione (MH; maleic hydrazide)

Tributyl(2,4-dichlorobenzyl)phosphonium chloride (chlorphonium chloride, phosphon)

α-Cyclopropyl-4-methoxy-α-(pyrimidin-5-yl) benzyl alcohol (ancymidol)

Methyl-2-chloro-9-hydroxyfluorene-9-carboxylate (methyl chlorflurenol, morphactin)

FIGURE 5-9. Structural formulas of some naturally occurring and synthetic inhibitors. [After R. J. Weaver, *Plant Growth Substances in Agriculture*, W. H. Freeman and Company, copyright © 1972.]

environment must be such that the biochemical processes of degradation and synthesis can operate. Although light is not usually required, in some plants (for example, in certain cultivars of lettuce) it can either trigger or inhibit the germination process.

Seed Dormancy

Seed that is viable and yet fails to germinate in the presence of favorable environmental conditions is said to be **dormant.** The cause of dormancy may be physical or physiological.

Physical dormancy may be caused by hard, impervious seed coats, which

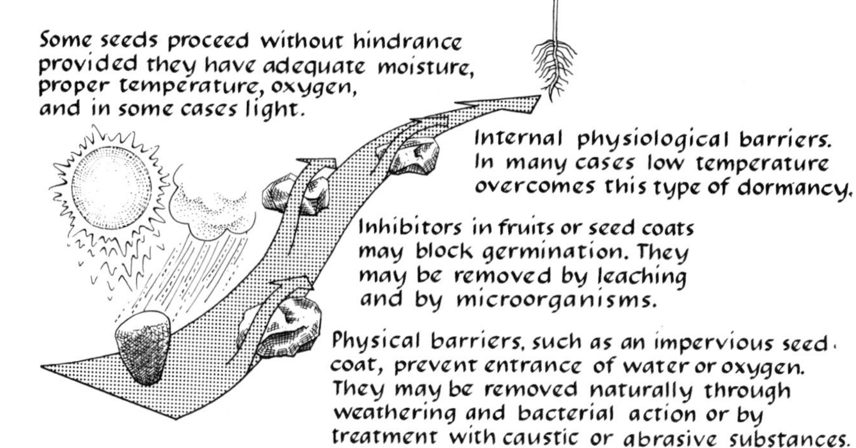

FIGURE 5-10. Germination is the route from seed to seedling. Barriers to germination (dormancy) may prevent its occurrence even though environmental conditions favorable to growth are present.

provide a barrier to the entrance of water and, in some plants, oxygen. The legume family provides the greatest number of examples of this type of dormancy. Some hard seeds resist water imbibition by means of a "one-way valve" in the hilum, the scar formed from the attachment of the seed to the pod. Although the seed may lose moisture in a dry environment, a moist environment effectively closes the valve. Under natural conditions these seeds do not germinate until soil microorganisms or weathering have sufficiently weakened the seed coat to permit the entrance of water. Thus, setting the stage for germination may require a number of years rather than only a small part of one season. Seeds having an impervious coat may be artificially worn or weakened in order that germination can take place uniformly and without delay (see Chapter 12).

Physiological dormancy may be due to a number of mechanisms. These commonly involve growth-regulating systems of inhibitors or promotors. Inhibiting substances that block the germination process may be present in the flesh of the fruit, in the seed coat, or even in the endosperm of the seed. Germination of seed within a tomato is a rare occurrence, yet as the flesh of the tomato is removed and the seeds are rinsed, germination takes place without delay. This inhibition of germination is due not to the low pH or high osmotic values of the tomato flesh but to a specific chemical entity. Although the inhibitor in tomato fruit has not been isolated and characterized, there is evidence that it belongs to a group of compounds known as the unsaturated lactones. A naturally occurring member of this group is coumarin, a substance responsible for the aroma of freshly cut hay.

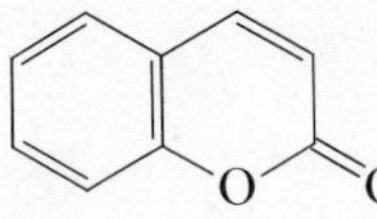

Coumarin

The sesquiterpenoid compounds also contain inhibitors. The best known is abscisic acid (ABA), previously described. Dormancy in apple seeds and in buds of sycamore maple (*Acer pseudoplatanus*) has been correlated with the presence of ABA. The mode of action of such substances as coumarin consists of blocking or inactivating enzymes essential to the germination process. For example, it has been shown that the activity of α- and β-amylases is blocked in the presence of a germination inhibitor. The amylases are essential for the hydrolysis of starch, that is, the conversion of complex insoluble carbohydrates to simple soluble forms. The inhibitory effects of abscisic acid are attributed to its inactivating effect on substances that promote germination, such as gibberellic acid. As with dormancy imposed by the impervious seed coat, inhibitors delay and extend the period during which germination takes place. Leaching by rain and degradation by microorganisms are examples of how germination inhibitors are removed from the fruit and seed coats.

Physiological dormancy may also be due to such internal factors as an immature embryo. In some seeds, at the time the fruit is ripe, the embryo consists of only a few undifferentiated cells. The seeds of the American holly (*Ilex opaca*) and ginsing (*Panax* spp.) are examples. Therefore, an **afterripening** process must occur during which the embryo differentiates at the expense of the endosperm. In other seeds, particularly in woody species of the temperate zone, the causes of internal dormancy are more complex. In addition to the inhibitors that block germination, a germination stimulator is also required, which is produced as a prerequisite for, or concurrently with, germination. Both the removal of the blocking action of the inhibitor and the production of the stimulatory substance occur during a condition known as **cold stratification.** The prerequisites for cold stratification are the presence of moisture and a temperature above freezing but below 50°F (10°C). A temperature of approximately 41°F (5°C) appears to be optimum. The fate of the inhibitors during the cold stratification period is not completely clear, but the production of a germination stimulator has been demonstrated.

The morphological site of internal dormancy is primarily in the plumule. In some seeds, however, it is found in the radicle; in others, both the plumule and the radicle are dormant. Under such conditions it is often necessary to expose the seed to alternating cold and warm temperatures. For example, with *Viburnum* it is necessary to provide a warm temperature (70 to 80°F, or 21 to 27°C) for the radicle to develop. Exposure to low-temperature stratification is then necessary. As a result, the inhibitor content decreases, and growth-promoting substances increase. Germination then takes place when the seeds are reexposed to warmth.

More complex are the germination requirements of the tree peony (*Paeonia suffruticosa*). The seed must first be exposed to low-temperature stratification to break the dormancy of the radicle. Then a warm temperature is required for the radicle to develop. As soon as it emerges, the seed must be returned to low-temperature stratification to break the dormancy of the plumule. After the second stratification period, the seed germinates upon return to warm temperatures.

The fact that the germination of a seed is blocked by one form of dormancy does not exclude the possibility that another form may also be present. Such a condition is called **double dormancy** and is characteristic of several members of the legume family. A well-known example is the redbud (*Cercis canadensis*). Both physical and physiological blocks to germination are present, the former being a seed coat that is impervious to water and the latter being an internal dormancy that can be broken by exposure to cold. The sequence of events to which seeds with double dormancy must be exposed is very precise. The physical barrier to germination must be removed before any attempt is made to remove the physiological barrier. If a seed coat is impervious, water essential for biochemical reactions cannot enter, and cold treatments to remove inhibitors or to promote the synthesis of a stimulator will not be effective. In nature, seeds with double dormancy often require two years for germination. The first year is required for soil microorganisms and weathering to remove the physical barrier by rendering the seed coat permeable to water and oxygen. The seed is not yet ready to germinate, however, because the internal dormancy has not been broken. During the second winter internal dormancy is broken, and germination occurs the following spring.

Dormancy of seeds is a biological mechanism that provides protection against premature germination when environmental conditions may not be favorable for seedling growth. Thus, in nature, seeds that have a cold-requiring internal dormancy need an exposure to the low temperatures of winter. Seeds from woody plants of the temperate zone will not germinate during the late fall to face the unfavorable growing conditions of winter but are delayed by internal dormancy until the following spring.

The germination of seeds of most of our common vegetable crops are often not blocked by either physical or physiological forms of dormancy. In contrast, such plants as the woody ornamentals and many weed species possess, almost without exception, one or more of the major types of dormancy. Thus, it would appear that dormancy mechanisms may be eliminated through an intensive breeding program.

Juvenility

A seed is considered germinated when it has produced a plant that, under proper environmental conditions, is potentially capable of continuous and uninterrupted growth. From the time this stage is reached until the first flower primordium is initiated, the plant is considered to be in the vegetative phase of growth. If during the vegetative phase the plant cannot be made to flower, regardless of the environmental conditions imposed, it is said to be **juvenile.** Juvenility and maturity, however, are relative terms. In many species these growth phases blend into each other. The end of the juvenile period is indicated when the plant responds to flower-inducing stimuli.

The juvenile phase is characterized by the most rapid rate of growth in the plant's lifetime and, in some plants, by distinct morphological and physiological features. The juvenile phase varies in length from one to two months for annuals to a period of a few years for most fruit trees, but some plants, such as bamboos, require scores of years to reach maturity. Among the morphological features that are associated with juvenility, and which are lost or altered at maturity, are the presence of thorns (*Citrus*), the presence of leaf lobes (ivy), the lack of leaf lobes (*Philodendron*), and the angle of the branches with respect to the main axis of the plant (spruce). The geotropic response varies with developmental stage in some plants. In its juvenile phase of growth, English ivy (*Hedera helix*) is a trailing vine, climbing only with support; in the mature stage of development, the plant grows upright (Figure 5-11). The ability of juvenile plants to readily initiate adventitious roots is a common feature of many species. This ability decreases or is lost entirely in mature forms.

One physiological concept of juvenility postulates that the apical meristem, although constantly laying down new cells, actually goes through an aging process. This assumes that no correlative growth substances are involved. A more likely hypothesis is that the cause of juvenility is the presence of substances

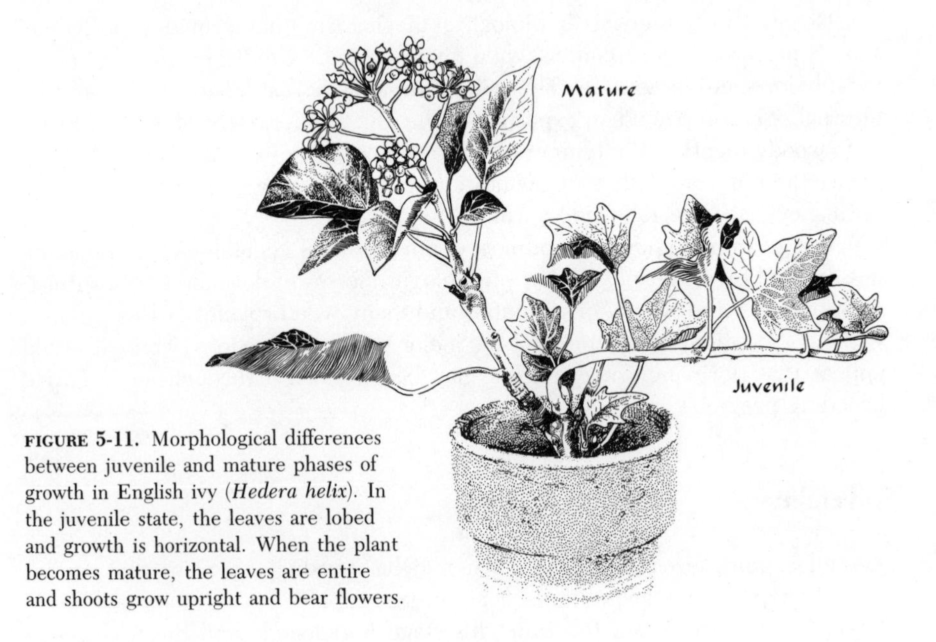

FIGURE 5-11. Morphological differences between juvenile and mature phases of growth in English ivy (*Hedera helix*). In the juvenile state, the leaves are lobed and growth is horizontal. When the plant becomes mature, the leaves are entire and shoots grow upright and bear flowers.

emanating from the seed or from the juvenile root system. An explanation for the transition from the juvenile to the mature state is that the "juvenile factor" gradually becomes exhausted as the plant grows or is rendered ineffective as the distance from the apex to the root system increases. In apple, a certain height (about 6 feet) must be achieved before seedlings flower. The most effective way to reduce the time to first flower is to grow the seedlings in such a way that they attain a large size as rapidly as possible. However, reducing the time to the appearance of first flower may not necessarily mean that the juvenile phase has been reduced.

Some support for the notion of a disappearing juvenile factor is found in grafting experiments involving mature shoots and a juvenile stock of *Hedera helix*. If the mature branch is grafted on a juvenile stock, juvenile shoots develop at first on the mature branch. After a few weeks' growth the juvenility gradually disappears, and the shoot again becomes mature (Figure 5-12). This particular experiment lends support to the "exhaustion" concept, which assumes that the new shoots on the mature scion first utilize the juvenile factors in the stock. As growth proceeds the juvenile factors become exhausted, and the mature phase is then resumed.

Additional support for the idea of hormonal regulation of juvenility is found in the fact that the application of gibberellic acid to the mature phase of *Hedera helix* will cause rejuvenation. In apples, growth retardants such as daminozide have been used to reduce the juvenile phase and induce flowering 1 or 2 years earlier than normal. The mode of action of the growth retardants may be the blocking of gibberellin synthesis.

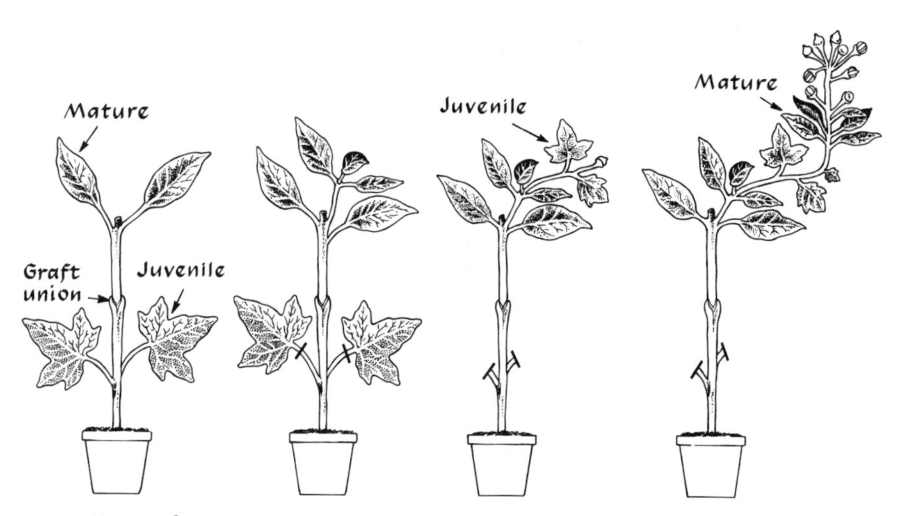

FIGURE 5-12. Induction of juvenility in English ivy, and reversal to the mature state, by grafting.

Environmental factors also can influence the duration of the juvenile phase. Growing birch (*Betula verrucosa*) in uninterrupted periods of long daylight, or crab apples (*Malus hupehensis*) in a CO_2-enriched atmosphere, accelerated vegetative growth and thereby shortened the time required to reach the mature phase of development. Ultimately, however, the duration of the juvenile phase is under genetic control. In a breeding program, therefore, it is possible to reduce the juvenile phase by using parents that transmit precocious flowering and by selecting for this character.

Bud Dormancy

Vegetative growth is not a continuous process but is associated with periods of arrested development. One type of arrested growth is brought about by unfavorable environmental conditions. For example, the growth of bluegrass, a common turf species, ceases under moisture stress, and the above-ground parts may die under continued drought. When conditions are more favorable, growth resumes from underground rhizomes. Many plants similarly survive periods of temperature extremes by undergoing a period of quiescence.

Bud dormancy, as is true of seed dormancy, implies that growth is temporarily suspended even though all the external conditions normally required for growth are provided. For example, woody plants of the temperate climates develop vegetative buds at each node throughout the growing season. The lack of growth of these buds after formation is initially an expression of apical dominance controlled by auxin distribution (see Chapter 14). At this time, these buds can be induced to grow by pruning, which removes the growth-inhibiting apical meristem. With the onset of autumn, however, the buds develop a true dormant condition and will not grow even if the plant is moved to a warm greenhouse. The degree of dormancy varies not only between species but also between buds on the same plant. The flower buds of many trees such as peach, cherry, and apple have a lower chilling requirement than do the vegetative buds, and therefore flower before the leaves emerge. In *Forsythia* the cold requirement is so minimal that it may be seen to flower in the fall after a brief period of cold weather.

In woody plants of the temperate climates, the onset of dormancy is conditioned by short day length and low temperature. Bud dormancy in these plants is probably internally regulated by the formation of growth-inhibiting substances. It is broken naturally by cold temperatures. This cold reaction is localized, for if an isolated stem of a dormant plant is cold-treated while the remainder of the plant is kept at a warm temperature, only the dormancy of the treated stem buds is broken.

The survival value of dormancy is clear. A physiological mechanism that prevents growth is a biological check on the perversity of weather. If woody plants lacked internally imposed dormancy they might initiate growth under favorable periods in the late fall only to have their succulent new growth succumb to succeeding severe weather. After there has been enough cold weather to break dormancy, growth is limited only by favorable temperature.

The period of cold treatment required to break dormancy not only varies with the species but is sensitive to selection within species. For example, peaches have been selected for cold requirements that vary from 350 to 1200 hours below 45°F (7°C). Low-chill-requiring cultivars are selected for southern areas where the periods of cold weather may be brief. If cultivars having high chill requirements are grown too far south, they will leaf out poorly or not at all in the spring (Figure 5-13). It is the cold-requiring dormancy brought on by short day length that prevents the production of temperate fruit crops in subtropical regions.

A

B

FIGURE 5-13. The effects of insufficient chill in the peach. (*A*) A peach seedling photographed at the South Coast Field Station in California on May 31, 1961, when only 300 hours below 45°F (7°C) had accumulated. The tree, which has a relatively long chilling requirement, shows severe injury as a result of insufficient chill. The tree has flowered and fruited because flower buds have a lower chill requirement than leaf buds. (*B*) The same clone on June 6 at Yucaipa, where the minimum temperatures are significantly lower as a result of the elevation (2700 feet), has leafed out normally. [Photographs courtesy J. W. Lesley, University of California, Division of Agricultural Sciences.]

Maturity

When a plant becomes potentially capable of reproduction it is said to be **mature.** The maturity of a plant is unquestionably attested to by the development of flowers. In many plants, however, physiological and morphological changes characteristic of maturity take place before the macroscopic expression of flowering becomes apparent. In English ivy, for example, leaf shape is greatly modified as the mature state is reached. Leaves lose the lobed shape characteristic of the juvenile state and become entire (i.e., nonlobed) as maturity is attained. The last leaf formed prior to the flower bud is almost reduced to a bract.

When a plant reaches maturity, it is capable of flowering, but it will not necessarily do so. The environment to which the plant is exposed at the time of maturity determines whether the plant will exhibit the ultimate expression of the mature state—the flowering response.

REPRODUCTIVE PHYSIOLOGY

Flowering

Flowering is a term that refers to a wide spectrum of physiological and morphological events. The first event, the most critical and perhaps the least understood, is the transformation of the vegetative stem primordia into floral primordia. At this time subtle biochemical changes take place that dramatically alter the pattern of differentiation from leaf, bud, and stem tissue to the tissues that make up the reproductive organs—pistil and stamens—and the accessory flower parts—petals and sepals. Among the events that follow initiation are development of the individual floral parts, floral maturation, and anthesis.

Once a meristem has been biochemically signaled to change from the vegetative to the reproductive state, microscopic changes in its configuration become apparent (Figures 5-14 and 5-15). Growth of the central portion is re-

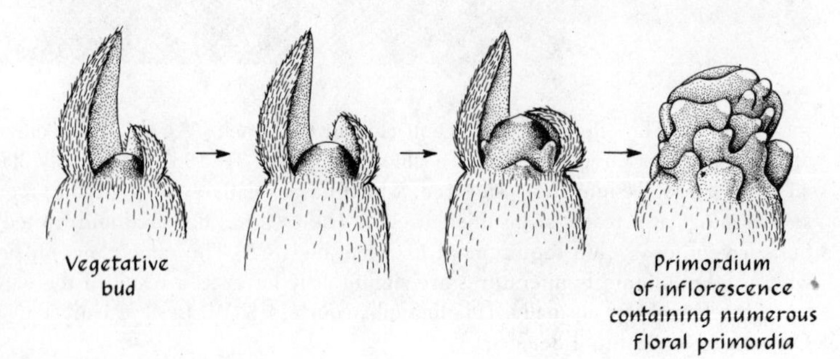

FIGURE 5-14. The transition of a vegetative bud to a floral primodium in the cocklebur. [From J. Bonner and A. W. Galston, *Principles of Plant Physiology*, W. H. Freeman and Company, copyright © 1952; after J. Bonner and J. Thurlow.]

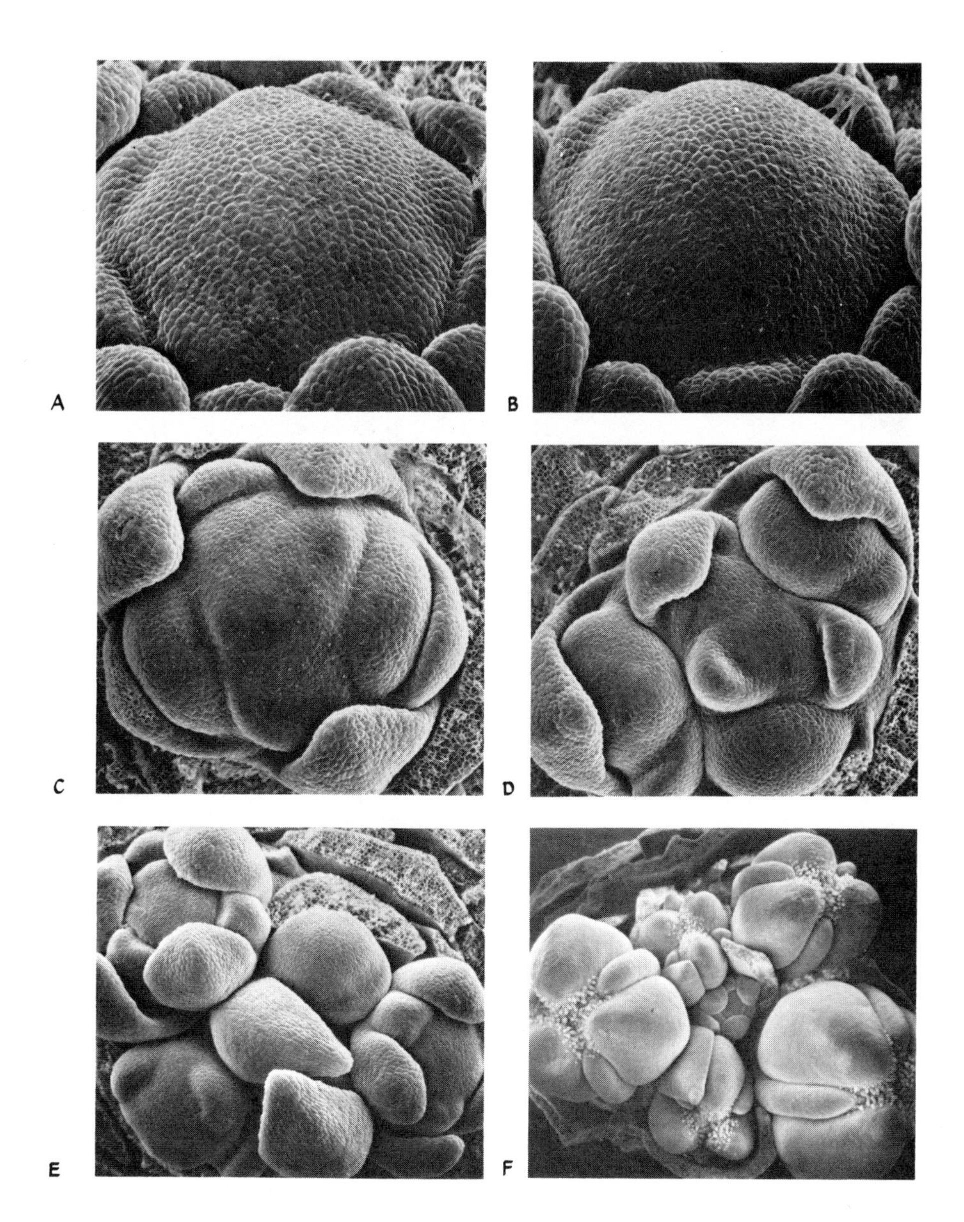

FIGURE 5-15. Morphological changes associated with the transition from vegetative to reproductive buds in Easter lily (*A* to *F*) and kalanchoe (*G* to *M*).
Easter lily: (*A*) Vegetative meristem surrounded by leaf primordia in late October (× 95). (*B*) The prefloral meristem becomes domelike in early January (× 95). (*C*) Early reproductive stage in mid-January showing floral primordium regions. Each primordium differentiates into a bract and a floral bud (× 77). (*D*) Reproductive meristem showing five floral buds, late January (× 77). (*E*,*F*) Reproductive meristem showing continued development of floral buds (× 34, × 10). [Continued on following page.]

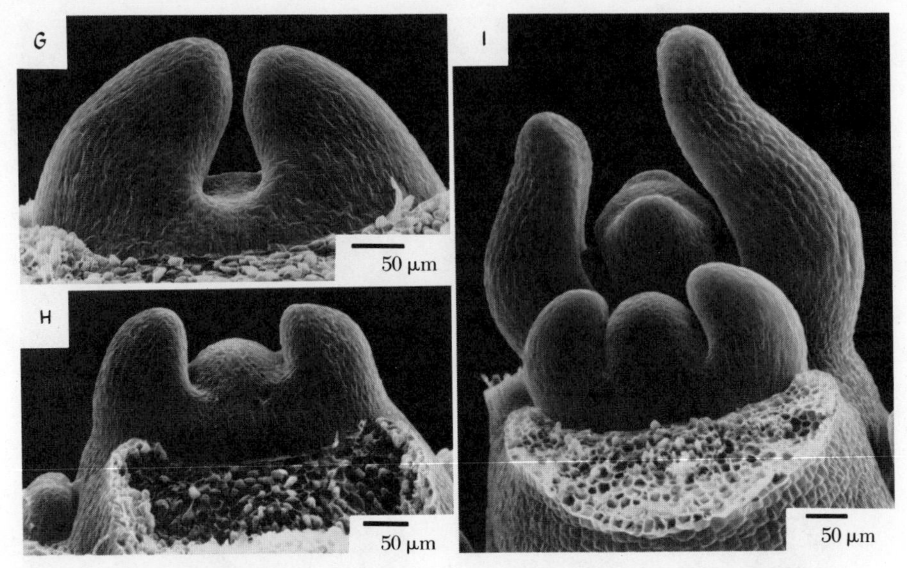

Kalanchoe: (*G* to *I*) Vegetative and early flower development. (*G*) Vegetative meristem. (*H*) Early reproductive meristem. (*I*) Sepal primordia present in primary flower, axillary flower present. (*J* to *M*) Late floral development. (*J*) Petal primordia present. (*K*) Stamen primordia present. (*L*) Petals and stamen elongated, stigma primordia present. (*M*) Stamen elongated, stigma developed, sepals and petals removed. [Parts *A* to *F* from A. A. De Hertogh, H. P. Rasmussen, and N. Blakely, *J. Am. Soc. Hort. Sci.* 101:463–467, 1976; parts *G* to *M* from T. A. Nell, J. M. Fischer, T. J. Sheeham, and J. E. Barrett, *J. Am. Soc. Hort.* 107:900–904, 1982.)

duced or inhibited, and the meristem becomes flattened in contrast to the conical shape characteristic of the vegetative condition. Next, small protuberances develop in a spiral or whorl arrangement around the meristem. Although this phase of reproductive differentiation is quite similar to vegetative differentiation, a basic difference does exist in that there is no elongation of the axis between the successive floral primordia as there is between the leaf primordia. In most plants, once the transformation from the vegetative to the reproductive state has been made, the process is irreversible and the floral parts will continue development until **anthesis**—the point at which the flower is fully open—even though the environmental conditions that existed during initiation are changed. By the time anthesis takes place, meiosis has already occurred, and pollen and embryo-sac development are complete. At this stage the plant is prepared for the next major step in its development—fruiting.

Carbohydrate–Nitrogen Relationship

Floral initiation has been studied primarily through manipulations of elements of the plant's environment, particularly nutrition, light, and temperature (see Chapters 7, 9, and 10). The rapid advances in the knowledge of the purely nutritional aspects of plant physiology in the latter part of the nineteenth cen-

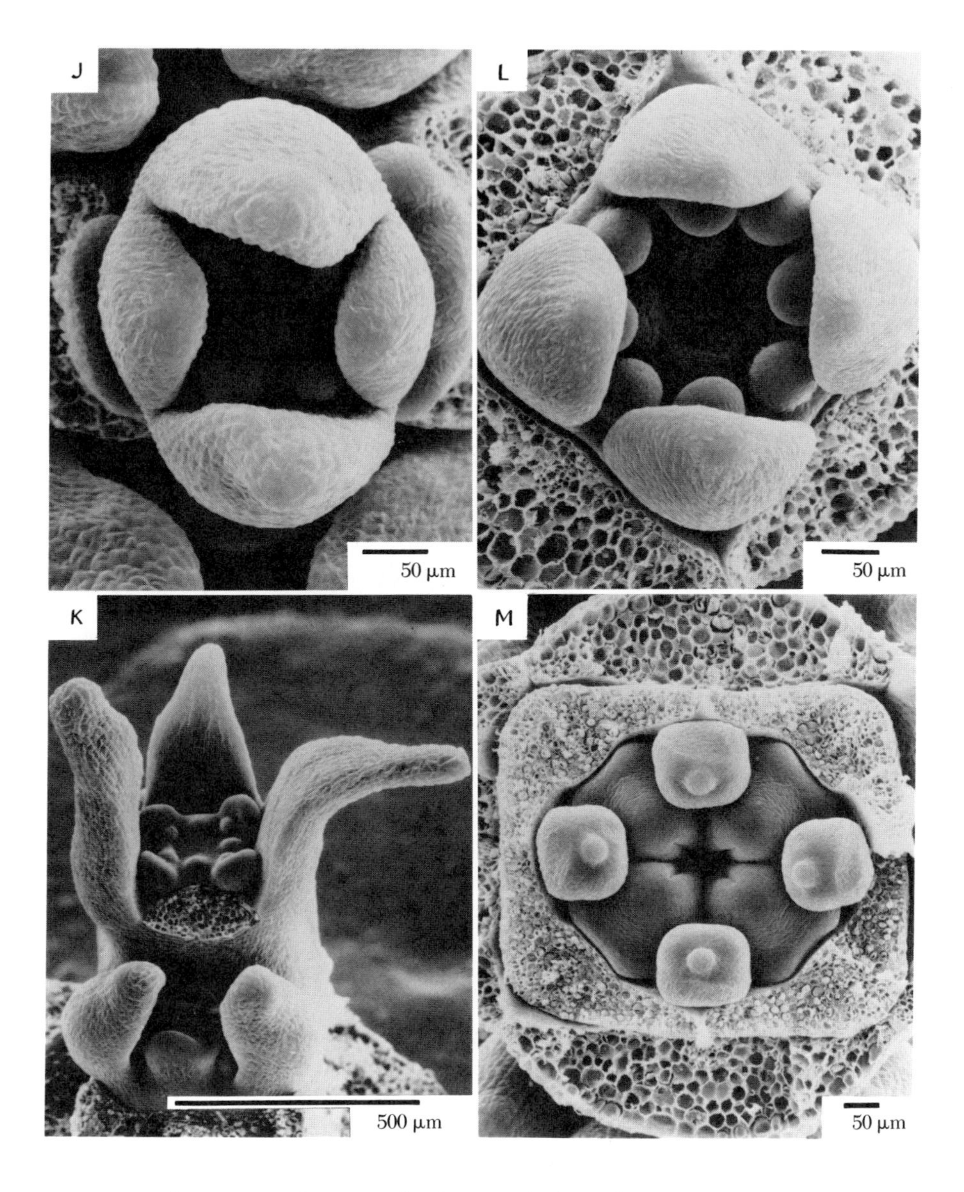

tury created an atmosphere in which it was believed that perhaps all aspects of plant growth could be explained or regulated by an alteration or adjustment of a plant's nutrition. This approach culminated, with respect to flower initiation, with the concepts of E. J. Kraus and H. R. Kraybill, who proposed, in 1918, that the initiation of flowering was regulated by the **carbohydrate–nitrogen relationship** of the plant. When tomato plants were grown under conditions favoring photosynthesis, and at the same time were supplied with an abundance of nitrogen fertilizers, vegetative growth was lush and flowering was reduced. But when the nitrogen supply was reduced while photosynthesis was maintained at a high level, vegetative growth was reduced and flowering was abundant. With the combination of low nitrogen and low photosynthesis, both vegetative growth and flowering were reduced. This concept of nutritional control of flowering was readily accepted, and it stimulated investigations on the effect of the carbon-nitrogen relationship in other plants. The results of such studies, however, indicated that plants will flower over an extremely wide range of carbon-nitrogen ratios. In view of our present appreciation of the tremendous physiological effects of minute quantities of growth-regulating substances, it is not difficult to understand why the gross ratios of total carbon compounds to total nitrogen compounds does not provide a consistent indication of the physiological condition of the plant. This is not to say that the nutritional status of a plant lacks importance in regard to flower initiation. The nutritional status can directly influence the degree or quantity of flowering but can only indirectly affect the qualifying event of initiation. As an example, it is possible, through the continued removal of new growth, to cause a tomato plant to initiate a flower when only the cotyledons are present (Figure 5-16). It can be demonstrated that the presence of new growth has an inhibitory effect upon floral initiation. From these results it is tempting to suggest that the reason nitrogen reduces flowering in the tomato is only incidentally associated with the carbohydrate-nitrogen relationship. Perhaps the stimulation of new growth by nitrogen inhibits flower initiation. In support of this concept is the observa-

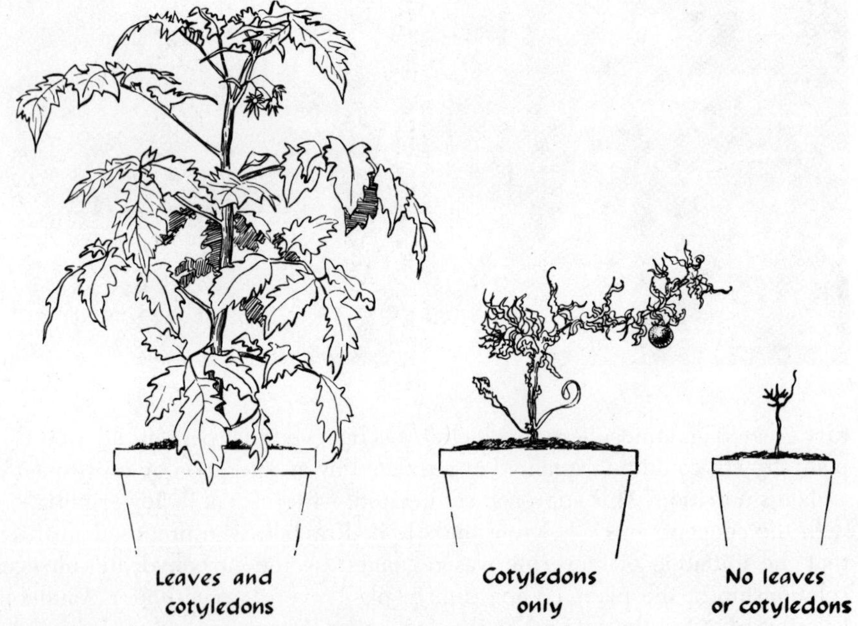

FIGURE 5-16. The presence of young leaves on tomato plants inhibits flower initiation. [Adapted from D. DeZeeuw, *Meded. Landbouwhogesch. Wageningen*, 54(1):1–44, 1954.]

tion that almost any means by which growth can be reduced, such as bending a branch from an upright to a downward position, results in increased floral initiation.

The concept that flower initiation is triggered by minute chemical changes was suggested by J. von Sachs in 1865. However, because of the popularity of the nutritional concepts of growth control, Sachs's theories received little support. In 1920, 2 years after the Kraus and Kraybill hypothesis was published, a discovery was made that caused a revolution in the concepts of flower initiation and provided direct support to the postulations of Sachs.

Photoperiodic Effect

The discovery of photoperiodism by W. W. Garner and H. A. Allard, scientists working for the United States Department of Agriculture, was made in conjunction with a breeding experiment. A new cultivar of tobacco was developed that flowered only in the greenhouse during the fall. Since most cultivars of tobacco flower in the summer, Garner and Allard attempted to provide the new cultivar with environmental conditions that would cause it to initiate flowers in the summer in order to make additional crosses. But no matter what method was used in an attempt to reduce vegetative growth and initiate flowering—altering the nutrition of the plant, allowing it to become pot-bound, withholding water—all attempts were failures. An attempt was then made to vary an environmental factor that had not been previously considered—namely, day length. The result was the discovery that, by artificially shortening the daily exposure to light during the summer, the new cultivar of tobacco could be made to flower as profusely as it did in the fall. Garner and Allard's research stimulated a great amount of investigation in the field now called **photoperiodism**—the growth response of a plant to the length of day, or, more precisely, to the length of the light and dark periods. It was soon found that a great number of plants responded to variations in day length. Some plants responded exactly as did the new cultivar of tobacco, but others responded in exactly the opposite way; that is, they flowered only when the days were long or were artificially lengthened. As the results of these many investigations accumulated, it became apparent that a majority of plants fell into one of three categories: short-day, long-day, and day-neutral plants. **Short-day** plants initiate flowers only when the day length is below about 12 hours. These include many of the spring- and fall-flowering plants, such as chrysanthemum, salvia, cosmos, and poinsettia. **Long-day** plants initiate flowers only in day lengths exceeding 12 hours. They include almost all the summer-flowering plants of the temperate zones, such as beet, radish, lettuce, spinach, and potato. **Day-neutral** plants apparently can initiate flowers during days of any length. They include the dandelion, buckwheat, and many tropical plants that either flower on a year-round basis or, if they do not, can be shown to be affected by other environmental conditions. The tomato is a typical example of a day-neutral plant, but it has been demonstrated that, with proper control of other environmental factors and under certain temperatures, the tomato will initiate a greater number of flowers under short day length. As the study of plant response to day length continues other categories have been added. For example, there are nonobligate long- or short-day plants. These are plants that will flower regardless of the day length but that will flower earlier or more profusely when the day is either long or short. A petunia will flower either during long or short days, but it will flower better during long days. It is therefore classified as a nonobligate long-day plant. In still another category are the plants that flower only after an alteration of day lengths and are known as "long-day, short-day plants." Such plants require first an exposure to long days and then to a period of short days.

The discovery of photoperiodism is particularly significant in that it clearly demonstrates the hormonal control of flower initiation. The mature or newly expanded leaf is the perceiver of changes in day length. In some plants the leaves need only to be exposed to one light–dark cycle of the proper day length to cause flower initiation. In the majority of plants, however, several to many cycles are required. Once the leaves have received the photoperiodic message, it has been postulated, they produce a substance, or a precursor of a substance, called **florigen.** Unfortunately, however, florigen remains only a name, for it has proven to be one of the most elusive of all plant-growth substances. Its transport from the leaves to the growing point can be demonstrated up and down stems, across graft unions, and from one plant to another, but the substance has not been isolated. Present thinking is that florigen *per se* may not exist, but that it may be represented by a particular balance of already known growth factors.

A major advance that may lead to an understanding of the flowering process is the isolation of the pigment system in the leaf called **phytochrome,** which specifically receives the photoperiodic message. Phytochrome, a blue-green pigment present in small amounts in all plants, was discovered by H. A. Borthwick and S. B. Hendricks, scientists with the United States Department of Agriculture, and first reported in 1959. Three important characteristics of photoperiodism aided in the discovery of phytochrome. First, interruption of the dark period with a small amount of light prevents flowering in short-day plants and permits flowering of long-day plants (Figure 5-17). Interruption of the light period with brief intervals of darkness has no effect on flowering. The night-interruption reaction is extremely critical in some plants. It can be shown that a very weak light (3 lux) is enough to interfere with the dark period in cocklebur. Second, red light is the most effective portion of the light spectrum for producing the night-interruption effect. Third, light from the far-red portion of the spectrum can completely reverse the effects of the red light. Furthermore, the effect of the far-red exposure is reversed again by exposure to red light. The direction of the photoconversion of phytochrome is determined by the light quality (that is, red or far-red) of the final exposure.

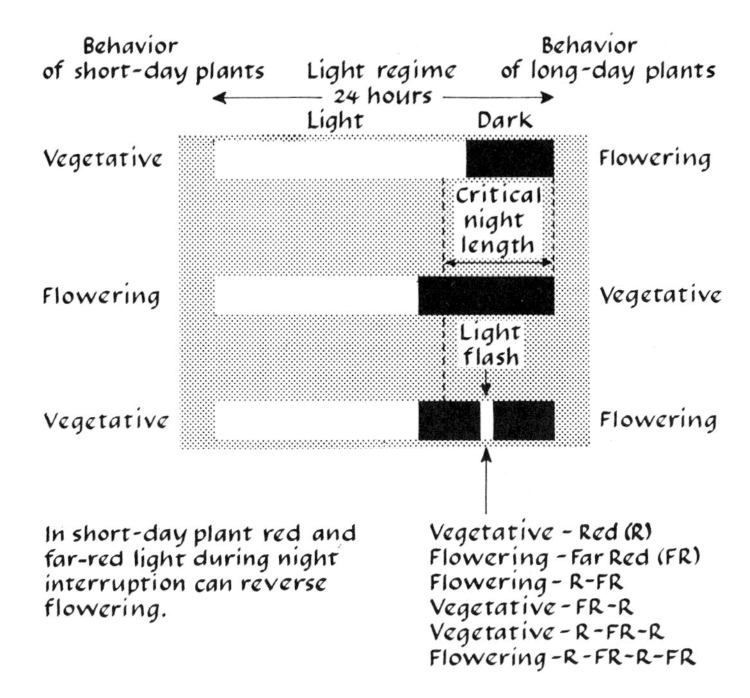

FIGURE 5-17. Photoperiodism is the control of flowering by the length of the light and dark periods to which the plant is exposed. The length of the dark period rather than the length of the light period is the critical factor. Short-day plants flower under periods of long night. Long-day plants flower under periods of short night. The flowering effect can be reversed by substituting red and far-red light during night interruption. [Adapted from A. W. Galston, *The Life of the Green Plant*, Prentice-Hall, 1961.]

The reaction is completely reversible over a considerable period of time, as long as the time lapse between exposures does not exceed a specific amount. Under short-day conditions an interruption of the dark period with red light causes a molecular shift in phytochrome that brings about flowering in long-day plants and prevents flowering in short-day plants. The reason for this differential response is not known. The action of far-red light can reverse the red effect. In addition, this shift is also influenced by the length of the dark period and by temperature. This may explain the light–temperature relationship in photoperiodism and may explain why the light break is most effective at the middle of the dark period. The characterization of phytochrome is still under way. It is known to be a protein with chromophoric (pigment) groups. The basic molecular unit of phytochrome is estimated to have a molecular weight of about 128,000 with three chromophores per molecule. The chromophore is a pigment very similar to the chromophore of c-phycocyanin, an algal chromoprotein.

Phytochrome appears to have many effects upon growth and development. It has been shown to be responsible for the promotion and inhibition of germination in seeds of some plants. In lettuce, seed germination can be inhibited or promoted by alternating red and far-red light. Although all seeds contain phytochrome, many plants do not require this "light signal" to germinate.

Although the discovery of phytochrome has made a very significant contribution toward explaining the mechanism of flower initiation, a vast amount of experimentation must be done before the substances specifically involved in the transition of the apical meristems from the vegetative to the flowering condition can be isolated and their mode of action determined.

The emphasis placed upon photoperiodism should not be interpreted as being an indication that it has exclusive control over flowering. Temperature, for example, has both an indirect and a direct effect upon flower initiation. It can influence flowering indirectly by modifying the plant response to a given photoperiod. Thus, a poinsettia will initiate and develop flowers in 65 days when grown under short-day conditions at 70°F (21°C). When grown under the same conditions at 60°F (15.5°C), however, 85 days are required before flower initiation occurs. A more striking example is that of the strawberry. At temperatures above 67°F (19.5°C) the June-bearing strawberry behaves as a short-day plant and will not initiate flowers in day lengths longer than 12 hours. At temperatures below 67°F (19.5°C) its response is that of a day-neutral plant, and the plant will initiate flowers even in continuous illumination.

Vernalization

The direct effect of temperature upon flowering was recognized more than one hundred years ago by the American J. H. Klippart. He described a process whereby it was possible to "convert" winter wheat into spring wheat. (Winter wheat is planted in the fall and produces a crop the following year; spring wheat produces a crop in the same year it is planted.) The process consisted essentially of reproducing the process that occurs in the field. Klippart partially germinated the winter wheat and then prevented further development by maintaining the seed at a temperature near freezing. The treated seed was planted in the spring, and it produced a crop in the same year. Therefore, the normal biennial habit of the winter wheat had been eliminated, or, more correctly, the cold requirement was satisfied. Klippart's research did not receive much attention until his work was repeated by the controversial Russian agronomist T. Lysenko.[1] The phenomenon was characterized as being the effect of temperature, during one or more of the developmental phases of plant growth upon future flowering behavior and was named vernalization by Lysenko. Since Lysenko's work in 1928, a considerable amount of research has been devoted to mechanisms of the cold response. A vernalized plant—that is, a plant which has been given a cold treatment—can be grafted onto a nonvernalized plant, and both will flower. The implication is that a substance has passed from the vernalized plant, across the graft union, to the nonvernalized plant. A similar experiment has been conducted with annual and biennial forms of the same plant, such as henbane. The annual form requires no cold treatment to flower and, when grafted onto the biennial form, causes the latter to flower earlier

[1]In the late 1940s Lysenko was placed in charge of Soviet genetics research. He became notorious for his autocratic methods and his controversial and now completely discredited views about genetics and plant breeding.

than it would normally. In wheat, this difference between annual (spring) and biennial (winter) forms has been shown to be controlled by a single gene, which might indicate that a single compound is involved. Although the flower-inducing substance produced during vernalization has not been isolated, a clue to its identity is found in the fact that the cold requirement of some plants can be partially or completely replaced with gibberellic acid.

Further evidence that temperature can have a profound effect upon flowering is provided in the phenomenon of **devernalization.** Plants exposed to a cold period of the proper temperature (41°F, or 5°C) for a period of time normally sufficient to induce flowering (usually at least 6 weeks) can be reverted back to their original nonflowering condition by exposure to high temperature. Onion growers take advantage of devernalization on a commercial scale. Onion sets are stored during the winter at temperatures near freezing to retard spoilage. The onion sets are vernalized at this temperature, and, if they were planted directly from cold storage in the spring, the set would quickly flower and no bulb would be formed. Therefore, the onion sets are exposed to temperatures above 80°F (27°C) for 2 to 3 weeks before planting. The set, now devernalized, will form bulbs instead of flowers.

Although photoperiodism and vernalization appear to have many similarities, and are definitely interrelated, the stimuli produced in the two responses to environment do not seem to be identical. Evidence for this supposition is taken from experiments in which it is possible to separate the effects of vernalization from those of photoperiodism. For example, most biennial plants when supplied the required exposure to low temperature still will not flower unless they are given the proper photoperiod. Similarly, where gibberellins replace the cold requirement, flower initiation will not occur unless the day length is correct.

In addition to these temperature effects, it has been shown that the alternation of warm and cool temperatures also influences flowering. This is the phenomenon of **thermal periodicity.** The classic example is the tomato, which will initiate more flowers when grown at a cycle of about 80°F (27°C) during the day and 63 to 68°F (17 to 20°C) during the night than when grown at higher or lower night temperatures. This phenomena has been utilized in greenhouse tomato culture.

Moisture Effects

Moisture is another environmental factor that may influence flower initiation. For example, if rhododendrons are subjected to a period of rainy weather during the fall, which is when they normally initiate flower buds, most of the buds will be vegetative. Many other woody plants respond similarly. It can be regularly observed that flowering in the spring is much more abundant after a dry summer and fall than after a wet summer and fall. Those who emphasize the nutritional role in flower initiation interpret this effect of moisture as being due to a favorable carbohydrate–nitrogen relationship. In rainy weather, the production of carbohydrates is reduced to a level that is too low to provide the balance necessary for flowering. As pointed out before, however, this interpretation appears to be an oversimplification of a rather complex phenomenon.

Time of Flower Initiation

The time at which the initiation of flowers takes place is of particular importance to the horticulturalist. Many perennial plants, both woody and herbaceous, initiate flower primordia from several to many months before flowering (Figure 5-18). In the apple, for example, flower buds are initiated from June to August, depending upon the cultivar. Flowering occurs during the following spring. In June-bearing strawberries the flowers are initiated in August and September, when the days become short. Attempts to obtain two crops from plants that normally produce one have been made in Holland. There, the growers subject the plants to a short photoperiod immediately after the first crop is harvested rather than waiting for it to occur naturally in late summer or fall.

Regulation of Flowering

Evidence that flowering is regulated by an endogenous chemical stimulus led many investigators to attempt to induce flowering by external application of growth regulators. In some cases the results have provided substantial practical applications. For example, auxins (see page 142) have been used extensively to induce flowering in pineapple, although they are now being replaced by the ethylene-generating compound ethephon. Apparently, the auxins caused flowering indirectly by stimulating the formation of ethylene by the pineapple tissues.

FIGURE 5-18. The crocus flowers in the early spring from buds initiated the previous summer. [Photograph by J. C. Allen & Son.]

Gibberellins also can cause flowering in some long-day plants or in some plants that require a cold treatment prior to flowering. Examples are cabbage, beet, carrot, and endive. However, gibberellins have not been used extensively for this purpose on a practical basis.

Sex Expression

The environment, besides having a profound influence upon flower initiation, can also influence the subsequent differentiation of the flower. This phenomenon is best seen in sex expression. In the normal sequence of the growth of cucurbits, for example, the first flowers produced are staminate (Figure 5-19). As growth continues there is an alternation of staminate and pistillate flowers; eventually only pistillate flowers are initiated. If the plants are grown under long-day conditions and cool temperatures there is an increase in the ratio of

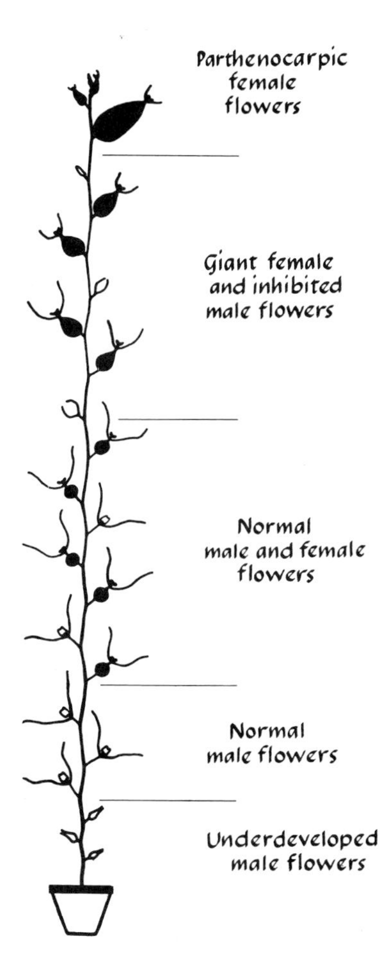

FIGURE 5-19. The normal sequential development of staminate and pistillate flowers on the acorn squash. [Adapted from J. P. Nitsch, E. B. Kurtz, J. L. Liverman, and F. W. Went, *Am. J. Bot.* 39:32–43, 1952.]

pistillate to staminate flowers. The effect of nutrition, photoperiod, and temperature has been demonstrated to affect sex expression in many plants.

Sex expression can also be modified by the application of growth regulators. Auxins and ethephon induce femaleness, whereas gibberellins induce maleness. Some of the effects of specific growth regulators on various plants are discussed in Chapter 15.

Fruit Development

Fruit development can be conveniently divided into four phases: (1) initiation of the fruit tissues; (2) prepollination development; (3) postpollination growth; and (4) ripening, maturation, and senescence. The origin of the fruit is found in the initiation of the floral primordia, which usually develops concomitantly with the flower. The increase in fruit size in the prepollination phase of development is primarily the result of cell division. After pollination, cell enlargement is responsible for the major portion of size increase. However, in some large-fruited plants (for example, watermelon and squash) cell division continues for some time after pollination, the final size being a consequence of increases both in cell number and in cell size. The stimuli and nutrients for prepollination growth are supplied primarily by the main body of the plant. In plants having perfect flowers, the stamen primordia are often differentiated before the ovary primordium, and they have been shown to be a source of growth stimulus. Surgical removal of the immature stamens in the flower bud adversely affects the growth of the ovary if the operation is performed at an early stage of development. Chemical extraction of the unripe anthers reveals the presence of large amounts of auxin.

Pollination

One of the most critical points in the growth and development of a fruit is pollination. Pollination has at least two separate and independent functions. The first is the initiation of the physiological processes which culminate in "fruit set" or, more precisely, in inhibition of fruit or flower abscission. The second function of pollination is to provide the male gamete for fertilization. That these two functions are indeed separate can be demonstrated by the use of dead pollen. In the orchid the use of dead pollen results in fruit set and some growth, but fertilization is not possible. A more precise demonstration of the multiphase function of pollen is the use of a synthetic auxin (Figure 5-20). Here again fruit set is obtained. The fact that water extracts of pollen are also effective in inducing fruit set has led to the postulation that the pollen contains an auxin. But the minute amount of auxin present in the pollen that lands on a stigma usually cannot account for the auxin response obtained. Instead it seems that the pollen contributes either an enzyme that converts auxin precursors present

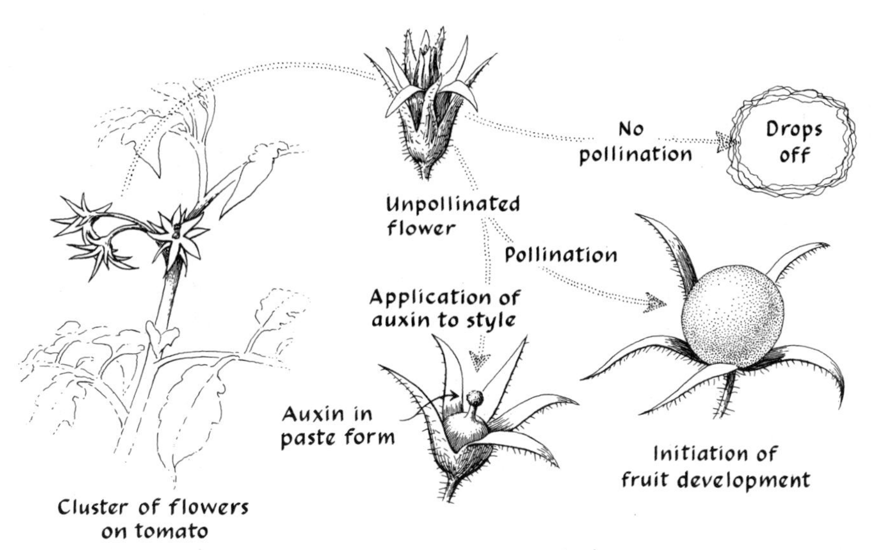

FIGURE 5-20. In addition to providing the male gametes, pollination prevents abscission of the flower. This role of the pollen may be replaced by the application of auxins. [From J. Bonner and A. W. Galston, *Principles of Plant Physiology*, W. H. Freeman and Company, copyright © 1952; after Avery and Johnson.]

in the stigma to auxin or provides a synergist that renders effective the auxin already present.

Even if pollination takes place and fruit set is obtained, fertilization is not absolutely assured. Sometimes pollen does not germinate, or if it does, the pollen tube may burst in the style. The germination of pollen is dependent upon the presence of a medium of the proper osmotic concentration (as shown by the effect of various sugar concentrations upon germination) and is stimulated by the presence of certain inorganic substances, such as manganese sulfate, calcium, and boron. In the lily, it has been shown that the highest concentration of boron occurs in the style and stigma.

In addition to the presence of organic and inorganic substances present in and on the stigma, which stimulate germination, there is also evidence for the existence of substances that chemically attract the growth of the pollen tube. If, for example, a slice of lily stigma is placed on an agar medium containing sucrose and boron, and is surrounded by a ring of pollen placed at a distance from the stigma, all the pollen tubes will grow in the direction of the stigma.

If the growth of the pollen tube is very slow, the style, or even the entire flower, may be shed. This may be artificially prevented, however, by the application of auxins such as α-naphthalene acetamide. This technique is particularly valuable to plant breeders, who through the use of auxins are now able to obtain seed from strains of petunia, cabbage, and marigold that previously had been considered self-sterile. When fruit set and growth are obtained either by pollination or by the application of an auxin, but fertilization does not take place, the fruit is said to be "parthenocarpic." Although no seeds are present, seedlike structures may develop, as in the case in some seedless cultivars of holly, orange, and grape.

The effect of pollination on fruit set and the influence of the resultant seed on fruit growth make pollination a crucial phase in the production of many fruit crops. For example, in apple and pear, in which genetic incompatibility prevents self-pollination within the same clone, more than one pollen-fertile cultivar must be present to assure pollination. Because pollen is effectively transferred by honeybees, hives are often moved to the orchard for this purpose (Figure 5-21).

Although pollination has the two major functions of fruit set and fertilization, there is still another interesting but uncommon effect called **metaxenia,** the direct influence of pollen on the maternal tissues of the fruit. In the date palm, a dioecious plant, the pollen, in addition to causing fruit set and pollination, affects both the size of the fruit and the date of ripening. This influence of the pollen variety can be demonstrated by using one strain of pollen to fertilize a number of female inflorescences of different date cultivars. This effect is in contrast to the direct effect of pollen on embryonic or endosperm tissue, which is referred to as **xenia.** The different colored kernels of Indian corn are a good example of xenia.

FIGURE 5-21. Commercial growers of apple and many other crops depend on honeybees (*Apis mellifera*) as pollination vectors (*A*). In good orchard management, hives of domestic bees (*B*) are moved into the orchard during bloom to ensure good fruit set because wild bee populations may be too small to complete the job, especially when poor weather restricts bee activity. [Photographs by J. C. Allen & Son.]

Postfertilization Development

After fertilization, the plant enters into a phase of physiological activity that is second in intensity only to germination. The developing fruit no longer depends primarily upon the parent plant for a source of growth stimuli. Instead, these stimuli are provided by the developing seed within the fruit. The role of the seed can be empirically demonstrated by observing that misshapen fruits result from uneven distribution of seeds. In many fruits, a direct correlation exists between weight (or length) and seed number. This effect of the seed on fruit development is mediated through chemical substances. For example, extracts of immature seeds can stimulate growth of unpollinated tomato fruit. Furthermore, it is possible to correlate various physiological events in the development of a fruit with the presence of growth substances. It has been demonstrated that the auxin levels reach a low at the time of flower drop, and particularly during the natural abscission of partially developed fruit that occurs when an especially heavy crop is obtained in fruit trees, known in horticultural terminology as the "June drop." The relationship between natural auxin level and fruit development is shown in Figure 5-22. The growth of the strawberry receptacle also has been prevented by removal of the achenes, which are one-seeded fruits If one or two achenes were left intact, the receptacle would grow only in the area directly under the achene (Figure 5-23). The achenes were extracted and found to contain high levels of auxin, whereas the receptacle tissue was comparatively low in auxin content. Finally, the addition of an auxin in the form of indoleacetic acid (IAA) dramatically stimulated the growth of an acheneless receptacle.

Other growth regulators are involved in fruit development. Grapes have two peaks of fruit growth; the first period has been correlated with increases in auxin content and the second with the biosynthesis of gibberellins. A sequence of growth regulators has also been suggested, with cytokinins playing a key role

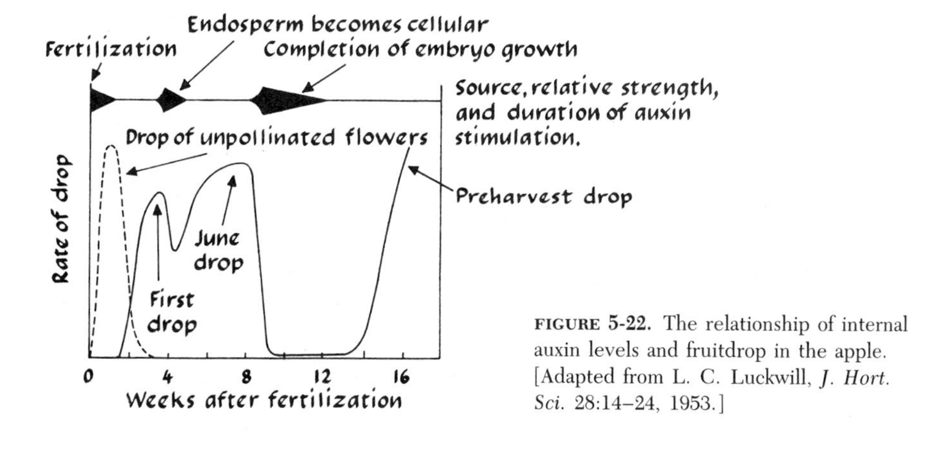

FIGURE 5-22. The relationship of internal auxin levels and fruitdrop in the apple. [Adapted from L. C. Luckwill, *J. Hort. Sci.* 28:14–24, 1953.]

FIGURE 5-23. Developing seeds are a source of stimulus for fruit growth. In the strawberry, when all the achenes except one are removed, only the receptacle tissue directly under the achene will enlarge. If additional achenes remain, additional areas of the receptacle will develop. The stimulus from the seeds can be replaced in part by the application of auxins. [From J. Bonner and A. W. Galston, *Principles of Plant Physiology*, W. H. Freeman and Company, copyright © 1952; after Nitsch.]

during cell division, followed by auxins and gibberellins during subsequent periods of growth. Practical use of the role of gibberellins in fruit development is seen in the table-grape industry. Application of a spray containing 20 ppm of gibberellic acid doubles the size of 'Thompson Seedless' grapes (Figure 15-11).

Nutrition and Fruit Growth

Although the control center of fruit growth is located in the seed, the raw materials for fruit development are supplied by other parts of the plant. Thus, the nutrition and moisture available to the plant directly affect fruit size. It has been calculated that at least 40 leaves on a mature apple tree are required to support the growth of one apple. If the 40:1 ratio is substantially reduced by an abnormally high fruit set, the quality and size of the individual fruits is greatly reduced. Therefore, it is now a common orchard practice to reduce the number of fruit artificially. This process, known as **fruit thinning,** will be discussed in Chapter 15.

Ripening

A final, dramatic physiological event marks the end of maturation and the beginning of senescence of the fruit. This is the **climacteric.** It is characterized by a marked and sudden rise in the respiration of a fruit prior to senescence, and it takes place apparently without the influence of external agents. The respiration rate then returns to a level equal to or below the level that existed prior to the climacteric. The climacteric is associated not only with the quantitative burst in carbon dioxide production but also with qualitative changes related to ripening, such as pigment changes. The transition from green to yellow in certain cultivars of apple, pear, and banana takes place during or immediately following the climacteric. The peak of acceptability, or "edible ripeness," of pears coincides with the peak of the climacteric. In apple, banana, and avocado, maximum acceptability is reached immediately after the climacteric. Finally, a marked increase in the susceptibility of fruits to fungal invasion follows the climacteric.

The occurrence and causes of the climacteric are currently under study. Almost all fruits studied exhibit a characteristic rise in respiration after the harvest, with the exception of most citrus fruits. (The lemon, when held in an atmosphere containing at least 33 percent carbon dioxide, will exhibit a climacteric.) The degree and duration of the rise varies considerably between species, from a short intense peak for the avocado to a longer, less definite peak for the apple. Explanations for the climacteric rise in respiration are varied. One hypothesis holds that, prior to the climacteric rise, the acidity of the cytoplasm decreases to a critical level that in turn increases the permeability of the cell membranes. Then, fructose, which previously had been accumulated in the vacuole, can move back into the cytoplasm and provide a substrate for increased respiration.

A second hypothesis involves adenosine diphosphate (ADP) and adenosine triphosphate (ATP), compounds that are instrumental in energy transfer in the cytoplasm. During respiration, a part of the energy released is preserved in the conversion of ADP to ATP; that is, the addition of a phosphate to ADP in the presence of energy results in the formation of a high-energy bond. This bond can later be broken to release energy for use in synthesis. The rate of respiration may be limited by the amount of ADP available for accepting a high-energy phosphate bond. It is therefore postulated that, during fruit maturation, when there is rapid cell enlargement and a high demand for protein synthesis, there is a shortage of ADP. But as the fruit matures, ADP becomes available and the respiration rate increases. Evidence in support of this hypothesis is that the addition of ADP to tissues of immature fruit causes an increase in respiration, but as maturation progresses the response decreases until it is completely lost during the climacteric.

As with any physiological event, temperature has a profound effect upon maturation and the climacteric. For example, a comparison of respiration has been made between apples held at 36°F (2°C) and apples held at 73°F (23°C). The maximum respiratory activity is 5 to 6 times as high at 73°F as at 36°F, and it takes 25 times as long to reach the climacteric at the lower temperature. However, the total amount of CO_2 liberated during the time between harvest and the end of storage life was approximately the same for both temperatures, equivalent to 16 to 20 percent of the reserve carbohydrates initially present in the fruit. In pears a pattern was established similar to that for apples at the two temperature extremes, but the rates of CO_2 evolution were much higher and the storage life much shorter.

It is becoming clear that fruit ripening, like other phases of plant development, is a DNA-controlled process. There is a rise in RNA in fruits at climacteric followed by increases in enzyme proteins such as the synthetases, hydrolases, and oxidases. Therefore, the climacteric is associated with both degradative and synthetic reactions. Ethylene and perhaps other volatiles appear to play key roles in the ripening process, and ethylene has been referred to as a fruit-ripening hormone. It is true that ethylene can accelerate fruit ripening; treatments to remove ethylene from within tissues of preclimacteric fruit can delay ripening. How ethylene exerts its effect on the initiation of ripening is not yet clear.

SENESCENCE

Senescence refers to the erosive processes that accompany aging prior to death. This process is one of the most baffling and least understood of the developmental processes.

Senescence in plants may be **partial** or **complete** (Figure 5-24). **Partial senescence** is the deterioration and death of plant organs such as leaves, stems, fruits, and flowers. Examples are the death and abscission of cotyledons in bean

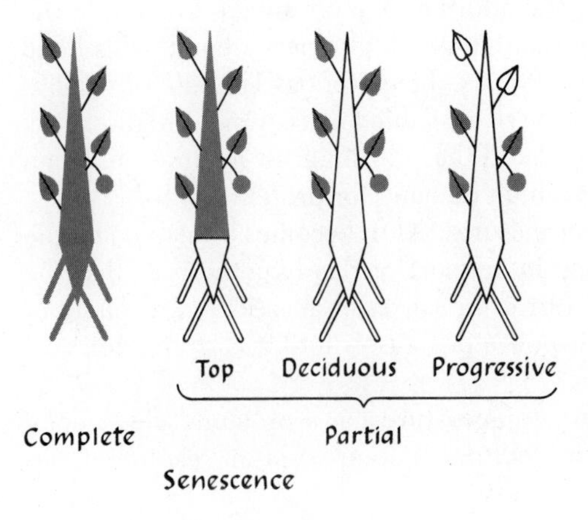

FIGURE 5-24. Patterns of plant senescence. The dead portions of the plant are indicated by shading. [After A. C. Leopold, *Plant Growth and Development*, McGraw-Hill, 1964.]

plants, the death of 2-year-old raspberry canes, or the death of the entire shoot of tulip in the early summer. **Complete senescence** is the aging and death of the entire plant except for the seeds. The termination of the life cycle of true annuals and biennials is often sudden and dramatic. After fruiting, whole fields of such crops as spinach and maize die in a synchronized pattern during the early or middle part of the growing season. In contrast, the senescence of such perennial plants as the apple or peach appears as a gradual erosion of growth and viability. In addition, perennials can be rejuvenated. Mature apple trees may be revitalized by severe pruning and fertilization or by encouraging the growth of adventitious buds at the base of the tree. The senescence of annual plants is relatively irreversible and is associated with flowering and fruiting. Spinach plants kept vegetative under short days will not become senescent. After having been induced to flower, however, they will surely die.

The older concepts explaining the senescence of annual plants are associated with a depletion hypothesis. It is suggested that during flowering and fruiting essential metabolites are drained from the main plant and accumulate in the fruit and seed. By the time the fruit is mature, the plant is depleted to the point that further growth is impossible, and death rapidly follows. It can be

FIGURE 5-25. The separation of bolting and flowering effects on senescence. Treatments from left to right: (1) short day; (2) short day + gibberellic acid; (3) long day; (4) long day + gibberellic acid. Note that the phenomenon of bolting does not lead to senescence unless it is associated with flowering induced by long days. Plants shown are spinach. [From J. Janick and A. C. Leopold, *Nature*, 192:887–888, 1961.]

shown that the removal of fruits can significantly postpone senescence and death. It has also been observed that the greater the number of flowers or fruits on an individual plant the more rapid is senescence and death.

However, the depletion concept of senescence is an oversimplification of a complex process. Unpollinated pistillate plants and staminate plants of a dioecious plant such as spinach also become senescent. In these plants senescence can be postponed by the removal of developing flowers, but death cannot be averted. Clearly, there is no depletion of the plant to account for this effect. It can further be shown that the seed-stalk formation that accompanies flowering is distinct from senescence, since spinach plants induced to bolt through the use of gibberellic acid, but kept vegetative, do not become senescent (Figure 5-25). Apparently, the biological signal that causes an annual plant to become senescent is associated only with the flowering process and is merely accentuated through fruiting. The relationship between this signal and flowering is not at all clear.

Selected References

Audus, L. J., 1972. *Plant Growth Substances*, 3d ed. New York: Barnes & Noble. (An advanced treatise on the chemistry and physiology of growth regulating materials.)

Bernier, G., J. Kinet, and R. M. Sachs, 1981. *The Physiology of Flowering*. Vol. 1, *The Initiation of Flowers*. Vol. 2, *Transition to Reproductive Growth*. Boca Raton, Fla.: CRC Press. (A physiological analysis of the flowering process.)

Khan, A. A. (ed.), 1982. *The Physiology and Biochemistry of Seed Development, Dormancy and Germination*. Amsterdam: Elsevier. (An advanced treatise dealing with all aspects of seed physiology.)

Halevy, A. H. (ed.), 1985. *Handbook of Flowering*. (5 vols.) Boca Raton, Fla.: CRC Press. (A comprehensive treatise on flowering presented species by species.)

Leopold, A. C., and P. E. Kriedemann, 1975. *Plant Growth and Development*, 2d ed. New York: McGraw-Hill. (An excellent text on plant physiology.)

Salisbury, F. B., 1982. Photoperiodism. *Horticultural Reviews* 4:66–105. (A review of plant responses to the relative lengths of day and/or night with special emphasis on flowering.)

Smith, H., 1975. *Phytochrome and Photomorphogenesis: An Introduction to the Photocontrol of Plant Development*. New York: McGraw-Hill. (Photobiology of plants.)

Thimann, K. V., 1977. *Hormone Action in the Whole Life of Plants*. Amherst, Mass.: University of Massachusetts Press. (An advanced text on phytohormones and plant growth and development.)

Wareing, P. F., and I. D. J. Phillips, 1979. *The Control of Growth and Differentiation in Plants*, 2d ed. New York: Pergamon. (An introduction to the processes underlying and controlling development in plants.)

Wilkens, M. B. (ed.), 1984. *Advanced Plant Physiology*. London: Putnam. (An up-to-date look at growth and development with each chapter contributed by a specialist.)

Reproduction 6

Reproduction is the sequence of events resulting in the replication and perpetuation of organelles, cells, organisms, and species. Thus, reproduction is a fundamental property of life. Reproduction expresses itself in growth, in the creation of new individuals, and in the perpetuation of species. All involve a self-duplication of the life-controlling mechanism. An understanding of this process is the basis of modern biology. In this chapter we shall explore two phases of reproduction: (1) the reproduction of cells and (2) the reproduction of multicellular plants.

The persistence of life in the face of the certainty of death is due to the ability of individual organisms to start anew through offspring. As we shall see, plants reproduce themselves in two fundamentally different ways—*asexually* and *sexually*. **Asexual,** or **vegetative, reproduction** occurs by means of ordinary cell division and differentiation so that the new individuals are usually identical copies called clones. (Vegetative reproduction is common in lower organisms and many plants but is not common in higher animals. In humans, identical twins are a consequence of asexual reproduction.) In **sexual reproduction,** special nuclei (called gametes) are fused, causing a recombining of genetic information *if* the fused nuclei are genetically different. Thus, sexual reproduction may create variability.

In horticulture, the utilization of the reproductive processes to perpetuate plants is called **plant propagation.** Plant propagation, the basis of agriculture and horticulture, is discussed in Chapter 12.

REPLICATION OF LIVING SYSTEMS

The principle behind the mechanism by which living material replicates itself is somewhat analogous to that of a typesetting machine (for example, a linotype). The machine arranges the type in a certain sequence to transmit information—that is, produce the printed page. New type is continually being formed from a mold, or template, of the original type. In this analogy the reproducing element is type; in living systems the reproducing elements are the nitrogen bases in DNA.

In the English language, 26 letters of the alphabet produce a dictionary of tens of thousands of words. The same results can be achieved with the two symbols of the Morse code. The four symbols of the genetic alphabet, or code, are the four nitrogen bases that, along with one sugar group and one phosphate group each, make up the **nucleotides** (or "building blocks") of DNA. These bases are adenine, guanine, thymine, and cytosine (see page 133). If two symbols can express 26 letters, thousands of words, and thus an infinite series of messages, is it not conceivable that four bases can make up the code by which the information for maintaining living matter is transmitted? The way in which this is accomplished is a subject of intense study. The general theory is that the sequence of bases in DNA directs the sequence of amino acids in the synthesis of protein, specifically enzymes. The specificity of an enzyme is related to its amino acid sequence, and it is not unreasonable to assume that enzymes are responsible for the biochemical control of living systems.

The replication of DNA appears to be accomplished by a system quite analogous to the template system of the type machine. The template model, however, is based on chemical bonding rather than physical impression. Chemical analysis of DNA shows that, although the relative quantity of the four nitrogen bases varies, the amount of adenine is always the same as that of thymine, and the amount of guanine is always the same as that of cytosine. This suggests that adenine and thymine, and guanine and cytosine, occur in pairs. The significance of this becomes apparent when the structure of DNA is considered (Figure 6-1). DNA normally occurs in double strands, and each strand is connected by linkages between the nitrogen bases. But because of the configuration of the bases, adenine is always paired with, and opposite to, thymine; and guanine is always paired with, and opposite to, cytosine. Thus if the sequence of one strand is given, the sequence of the opposite strand is fixed. Each strand, then, is the "template" for the other.

The theory of DNA replication follows nicely from its structure. The separation of the strands of DNA is followed by a realignment of complementary bases on each strand (Figure 6-2).

The structure of DNA accounts for both its ability to replicate and synthesize. The actual synthesis of protein is carried out not in the nucleus but in the cytoplasm. The information from the nucleus must be transferred to the sites of

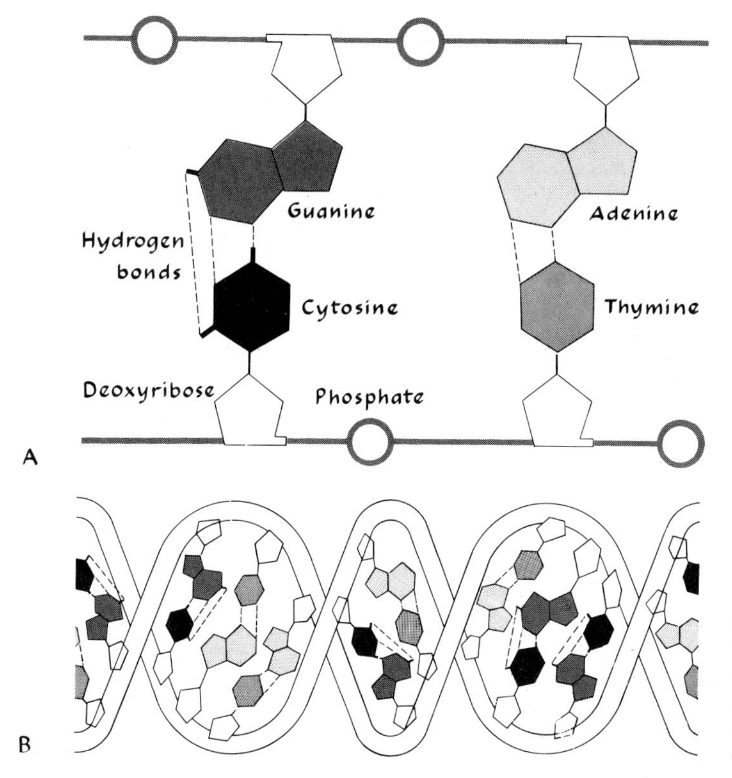

FIGURE 6-1. (*A*) A portion of the DNA molecule, shown untwisted. The molecule consists of a long double chain of nucleotides made up of phosphate-linked deoxyribose sugar groups, each of which bears a side group of one of the four nitrogenous bases. Hydrogen bonds (*broken lines*) link pairs of bases to form the double chain. The bases are always paired as shown, although the sequence varies. (*B*) The double-helix form of the DNA molecule. [After V. G. Allfrey and A. E. Mirsky, "How Cells Make Molecules," copyright © 1961 by Scientific American, Inc., all rights reserved.]

synthesis in the cytoplasm, the ribosomes. The information is carried by a substance called ribonucleic acid, or RNA. (The components of RNA are almost identical to those of DNA except that the sugar portion of the molecule is slightly different—ribose takes the place of deoxyribose, and the nitrogen base thymine is replaced by a closely related base called uracil.) The particular form of RNA that carries the genetic instructions, in the form of a complementary copy of the DNA series of bases, is called messenger RNA. Another form of RNA, transfer RNA, brings the amino acids to the ribosomes to construct protein (Figure 6-3).

Protein synthesis can be achieved outside of the living organism by placing together the component parts: ribosomes (from bacteria), an energy-generating system, amino acids, and RNA. Very elegant experiments involving the direction of protein synthesis have been possible through the use of artificial RNA. In this way, the genetic code has been deciphered; that is to say, the sequence of bases in RNA has been associated with the incorporation of a particular amino acid into protein. The message has been shown to be in the form of a three-letter code; that is, a sequence of three bases directs each amino acid. For example, RNA composed only of uracil (UUU) incorporates the amino acid phenylalanine; the sequence UUA incorporates leucine. The four-letter alpha-

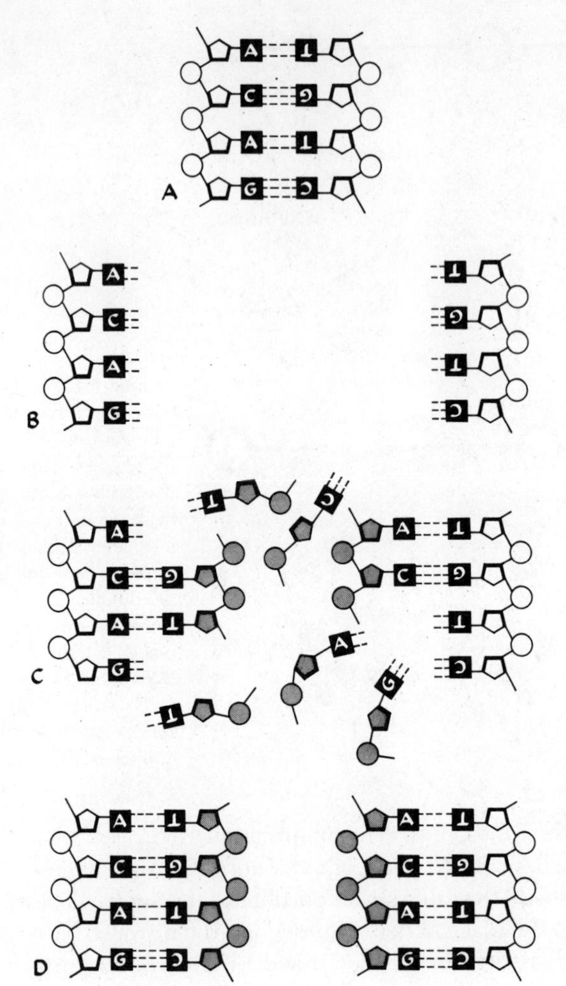

FIGURE 6-2. DNA replication. The linked DNA strands (*A*) separate in the region undergoing replication (*B*). Free nucleotides (indicated by shading) pair with their appropriate partners (*C*), forming two complete DNA molecules (*D*). [After F. Crick, "Nucleic Acids," copyright © 1957 by Scientific American, Inc., all rights reserved.]

bet and base sequence is shown in Table 6-1. The sequence of amino acids in the polypeptide (protein unit) of the enzyme is related to the three-letter code sequence of DNA. Thus the four-letter alphabet represented by the four nitrogen bases is used to make three-letter "words," each corresponding to one of the approximately twenty amino acids. Each sentence spells out the amino acid sequence for one protein. This "sentence" is the gene!

The complete set of instructions is separated into chromosomes, linear structures within the nucleus. A chromosome is in essence a very long molecule of DNA. A specific function-controlling sequence of nucleotides within that molecule of DNA is a gene. Our understanding of the molecular, or "fine," structure of the gene is derived from studies that began in the late 1940s. A

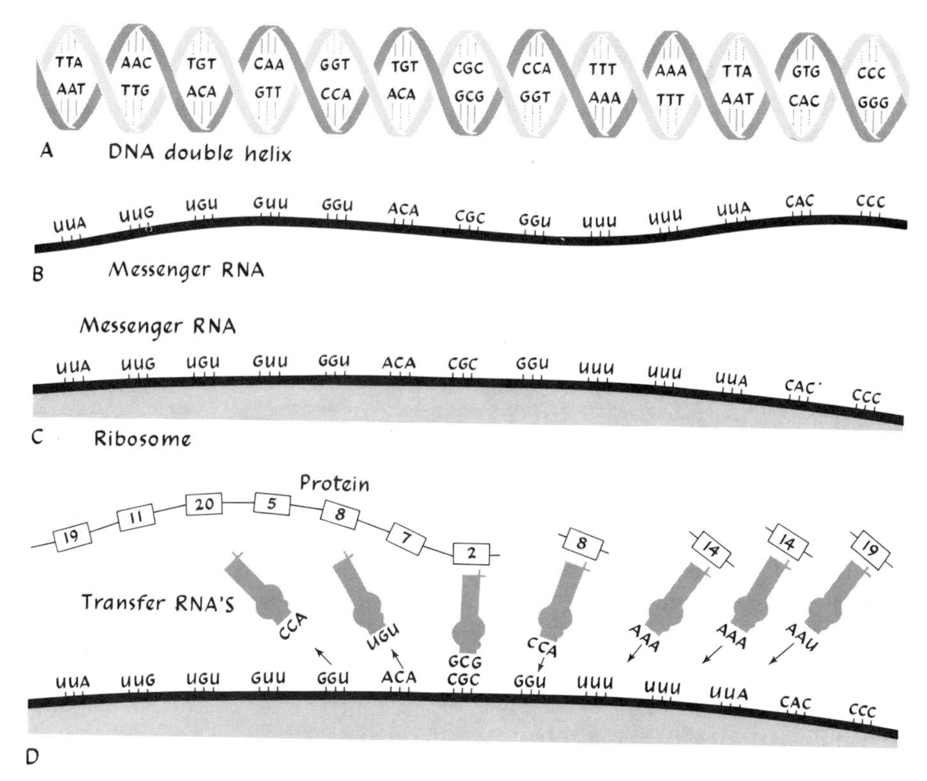

FIGURE 6-3. A schematic representation of protein synthesis as related to the genetic code. The complementary code, here derived from the dark-lettered strand of DNA (*A*), is transferred into RNA (*B*) and moves to the ribosome (*C*), the site of protein synthesis. Amino acids (*D*, numbered rectangles) are presumably carried to the proper sites on messenger RNA by transfer RNA and are linked to form proteins. [After M. Nirenberg, "The Genetic Code," copyright © 1963 by Scientific American, Inc., all rights reserved.]

tremendous amount of information on the transmission and action of genes has accumulated from work on a wide spectrum of organisms, including crop plants. These investigations trace back to the beginnings of genetics, conveniently marked by the rediscovery in 1900 of Mendel's 1865 paper on inheritance. The theory that the units of inheritance, the genes, are particulate bodies on chromosomes and that they are transmitted from cell to cell and from generation to generation is the contribution of "classical" genetics and was clearly established by 1925.

Studies of many organisms have indicated that genes function through the regulation of chemical events. Genes control each step in the biochemical pathways of metabolism. The discussion of genes and gene action will be continued in Chapter 17.

TABLE 6-1. The genetic-code dictionary.*

First base	Second base U	Second base C	Second base A	Second base G	Third base
U (Uracil)	Phenylalanine	Serine	Tyrosine	Cysteine	U
	Phenylalanine	Serine	Tyrosine	Cysteine	C
	Leucine	Serine	*Nonsense*	*Nonsense*	A
	Leucine	Serine	*Nonsense*	Tryptophan	G
C (Cytosine)	Leucine	Proline	Histidine	Argenine	U
	Leucine	Proline	Histidine	Argenine	C
	Leucine	Proline	Glutamine	Argenine	A
	Leucine	Proline	Glutamine	Argenine	G
A (Adenine)	Isoleucine	Threonine	Asparagine	Serine	U
	Isoleucine	Threonine	Asparagine	Serine	C
	Isoleucine	Threonine	Lysine	Argenine	A
	Methionine	Threonine	Lysine	Argenine	G
G (Guanine)	Valine	Alanine	Asparagine	Glycine	U
	Valine	Alanine	Asparagine	Glycine	C
	Valine	Alanine	Glutamine	Glycine	A
	Valine	Alanine	Glutamine	Glycine	G

SOURCE: This code is derived from the work of M. Nirenberg and others.

*The three bases of RNA (remember that uracil replaces the thymine of DNA) specify the amino acids indicated. The *nonsense* combination acts as "periods" to stop polypeptide synthesis.

CELL REPRODUCTION

The basic difference between sexual and asexual reproduction lies in the way the chromosomes are distributed during cell division—**mitosis** and **meiosis** (Figures 6-4 and 6-5).

Mitosis

Mitosis is the type of cell division that occurs in growth. It is a synchronized division in which both the chromosomes and the cell divide (Figure 6-4). The chromosomes split longitudinally, each half then duplicating its missing half from raw materials available in the nucleus and moving to opposite ends of the cell. The two resulting daughter cells thus receive exactly the same number

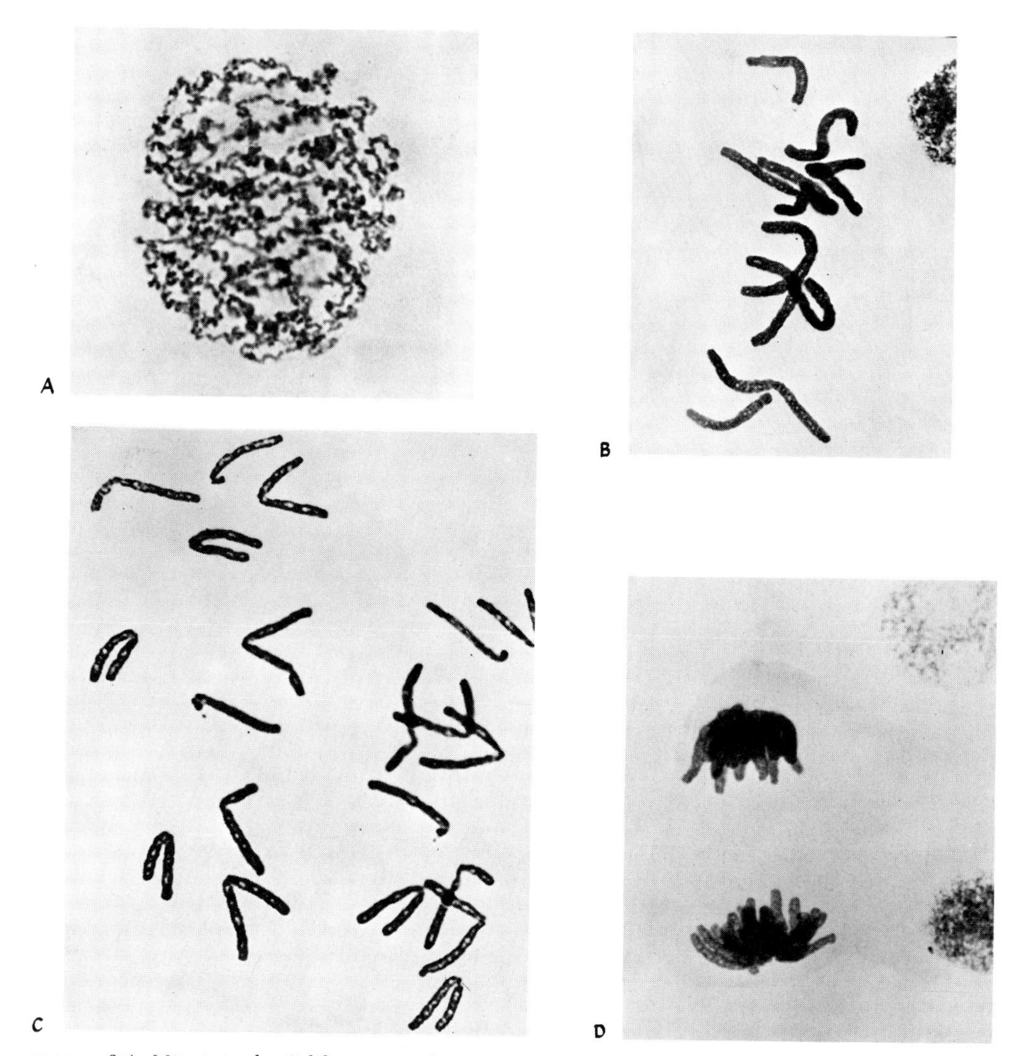

FIGURE 6-4. Mitosis in the California coastal peony. The vegetative cells of this species have 10 chromosomes ($2n = 10$). Although mitosis is continuous, the process is broken down into a number of stages for descriptive purposes. (*A*) *Prophase:* During this stage the nucleus becomes less granular and the linear structure of the chromosomes can readily be discerned. Note the chromosomes coils. (*B*) *Metaphase:* The 10 chromosomes line up in the equatorial plate, which appear as an "equatorial line" due to the smearing process. The chromosomes have reduplicated, and each appears visibly doubled. (*C*) *Anaphase:* The chromosomes have separated, and approach the poles of the cell. It is possible to pick out the pairs in each group and to match the daughter chromosomes, which have just separated from each other. (*D*) *Telophase:* During this stage the contracted chromosomes are pressed close together at each end of the cell. A wall subsequently forms across the cell, making two "daughter" cells with the same number and kind of chromosomes as exist in the original cell. [Photographs courtesy M. S. Walters and S. W. Brown.]

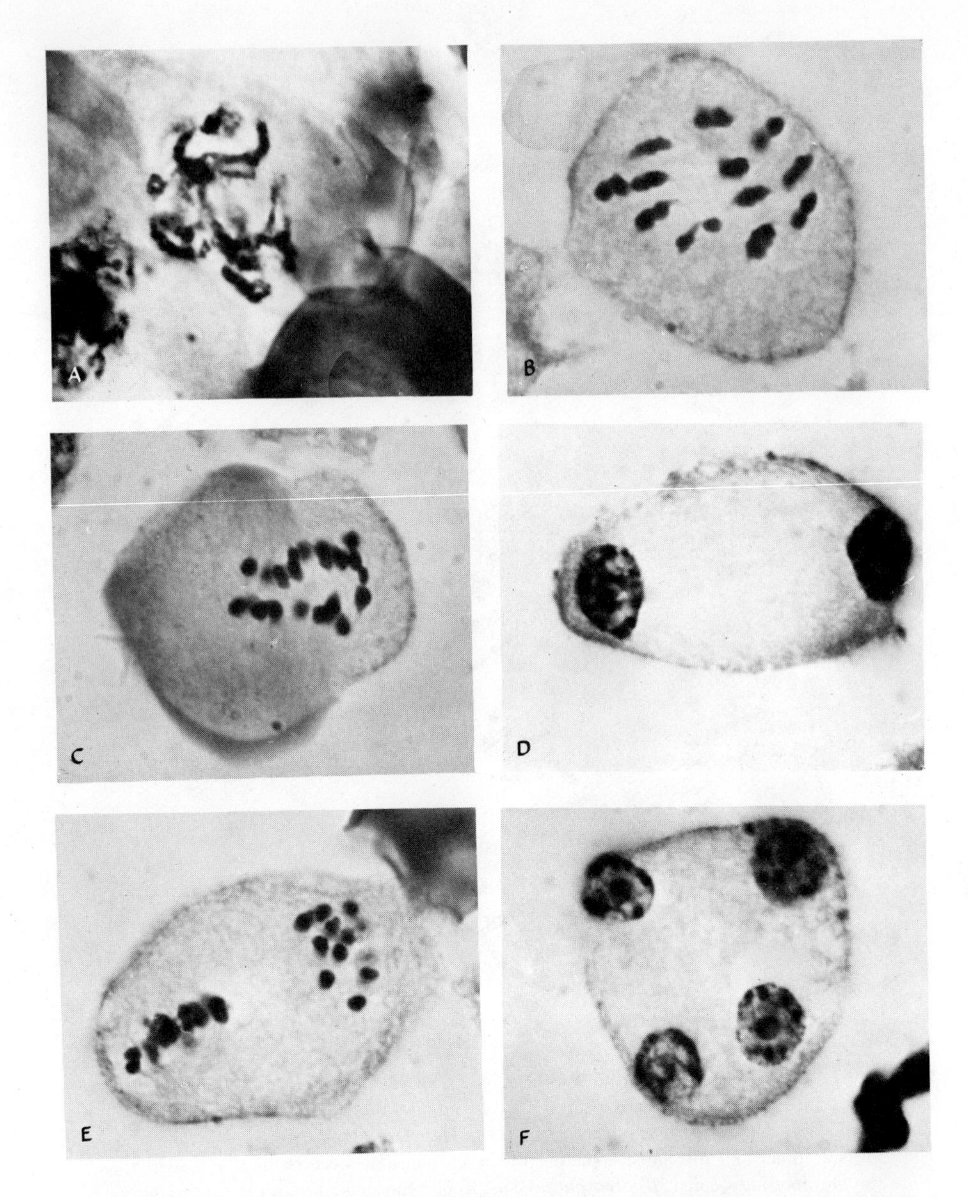

FIGURE 6-5. Meiosis in pepper (*Capsicum*). This species has 24 chromosomes in the vegetative cells ($2n = 24$). Note that there are two divisions. (*A*) In the prophase of the first division, the chromosomes reduplicate and pair. A pair of visibly reduplicated chromosomes can be seen in the bottom of the cell. (*B*) At metaphase of the first division, the 12 chromosome pairs line up on the equatorial plate (face view). Each consists of 2 doubled chromosomes (4 chromatids). (*C*) Anaphase of the first division. (*D*) Telophase of the first division. (*E*) Metaphase of the second division: One plate is a face view, the other is a side view. (*F*) Telophase of the second division. Walls will form four pollen grains, each containing 12 chromosomes, half as many as the original cell.

and kind of chromosomes. The division of the cell distributes more or less equally the other constituents of the cell.

The differentiation of cells into tissues and organs ordinarily is not related to any chromosomal difference. That each cell contains all the necessary genetic material implies that any cell holds the potential to give rise to the entire organism. The production of an entire plant from a single carrot parenchyma cell has been achieved. The formation of shoot buds from roots, or of roots from leaves, is a potential result of the genetic continuity made possible by mitosis. The differential patterns of growth would indicate that the instructions carried in the chromosome can be turned on and off. This system itself is under the control of the genes.

Meiosis

Meiosis is a specialized sequence of two cell divisions that occurs prior to the formation of gametes. Meiotic divisions produce cells containing half the number of chromosomes that were in the original cell. In somatic (ordinary vegetative) cells the chromosomes ordinarily occur in pairs (homologues), which are morphologically similar and contain the same kind of genes, although the two members of any given gene pair may not be identical because of some mutation (nonidentical members of a given gene pair are termed **alleles**). The combination of different alleles, as well as different genes, is responsible for the genetic variability between living things. The sexual process is one mechanism that provides for the reassortment and recombination of genetic factors, a feature that permits organisms to adapt through time to an ever-changing environment. The reassortment of genetic factors between and within chromosomes is accomplished by meiosis; the recombination is accomplished by fertilization.

Although in meiosis the cells divide twice, the chromosomes duplicate themselves only once. This explains why each of the four resulting cells has the haploid number of chromosomes–that is, half the number of chromosomes present in the somatic cells. Each of these four cells may give rise to a gamete. Fertilization, the fusion of two gametes, subsequently restores the diploid number of chromosomes.

The way in which meiosis differs from mitosis becomes apparent at the first division (Figure 6-5). The chromosomes divide longitudinally, as in mitosis (into half chromosomes called chromatids), but the homologous visibly split chromosomes pair (synapse). During synapsis there may be an actual exchange (crossover) of segments between chromatids of homologous chromosomes. The precise way in which this occurs is still not clearly understood. The attraction between the synapsed chromosomes changes to repulsion and each doubled chromosome moves to opposite poles of the cell without separation of the chromatids. The resultant shuffling of chromosomes occurs at random. In the second division, which immediately follows the first, the chromatids separate (Figure 6-5).

The most obvious effect of meiosis is the reduction in chromosome number from the diploid to the haploid number necessary to accommodate the fusion of gametes at fertilization and to maintain a uniform chromosome number from generation to generation. The less obvious but equally important effect in the long run is the reassortment of the genetic material in the gametes. The reassortment involves not only chromosomes but also those chromosome segments involved in crossovers.

SEXUAL REPRODUCTION

Sexual reproduction is based on a fusion of gametes. For the process to be orderly a mechanism must maintain chromosome number and kind constant. This mechanism is the sequence of alternating mitotic and meiotic cell divisions and a cell fusion, or fertilization (Figure 6-6). This sequence of events is termed the **life cycle.**

The number of chromosomes in the gametes is known as the n number (also called the gametic, or haploid, number). The number of chromosomes in the fused gametes (the zygote) is therefore the $2n$ number. In plants the $2n$ number is called by various terms: sporophytic number, somatic number, diploid number. Of these, sporophytic is the most precise because the zygote divides mitotically to form the sporophyte (literally: the plant bearing spores). Each ordinary vegetative cell of the sporophyte is $2n$. The term **somatic** is widely used but is based on the incorrect assumption that cells of an organism were predestined to be either **soma** (body) cells or **germ** (reproductive) cells. This concept is no longer held, but the terms **somatic cells** for $2n$ and **germ cells** for n cells have been retained. Finally, the term *diploid number* is only precise for plants that have two sets of chromosomes in the sporophytes. This point will be made clear in later discussions of polyploidy. At first encounter, botanical nomenclature appears to be designed to be abstruse and perplexing; however, with increased familiarity, we realize that botanical nomenclature properly used conveys precise information.

In higher animals, including humans, the body cells are $2n$ and only the gametes are n. In plants and many microorganisms the situation is more complex because the $2n$ and n phases may be separate entities. This is known as the **alternation of generations.**

Life Cycles

The series of events making up the life cycle of plants contains an alternation of $2n$ (sporophytic) and n (gameteophytic) stages. This is illustrated for sweet corn in Figure 6-6.

After the haploid stage is produced by the sporophyte through a meiotic division, the resulting haploid (n) cells undergo mitotic division to produce

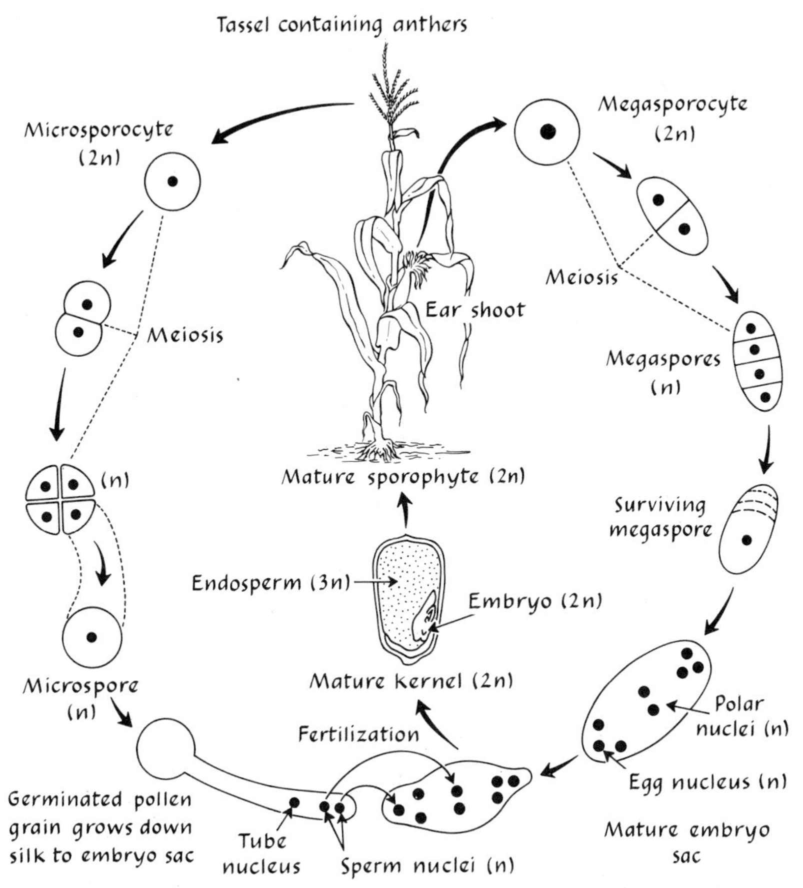

FIGURE 6-6. The life cycle of sweet corn (*Zea mays*), illustrating pollen and embryo-sac formation. The cell that undergoes meiosis in the anther is called a **microsporocyte** or **pollen mother cell.** Each microsporocyte gives rise to four functional microspores, each of which develops into a pollen grain (microgameteophyte) that produces two male gametes (sperm nuclei) by mitosis. The megasporocyte undergoes meiosis within the ovule to produce four megaspores of which only one survives. This gives rise to the embryo sac (megagameteophyte) that produces a female gamete (egg nucleus) by mitosis. [Adapted from A. M. Srb, R. D. Owen, and R. S. Edgar, *General Genetics*, 2d ed., W. H. Freeman and Company, copyright © 1965.]

multicellular entities called the gameteophytes (i.e., the plant bearing gametes) that produce gametes by mitosis. Male and female gametes are produced in different structures, which may or may not be on the same gameteophyte. Thus, sex is a function of the gameteophyte stage. The fusion of the male and female gametes (sperm and egg) to form the zygote restores the $2n$ condition. A series of mitotic divisions gives rise to the embryo, and subsequently the sporophyte, to complete the cycle.

In seed plants, the sporophytic ($2n$) generation is the dominant and independent one, and the gameteophytic (n) generation is short and dependent, or parasitic, on the sporophyte. In more primitive plants, such as the ferns, these two generations lead an independent existence. In plants that are even more primitive, such as the mosses, the gameteophytic generation is independent and of long duration while the sporophytic generation is transitory and dependent.

In higher plants the "male" gameteophyte (microgameteophyte), or pollen, is produced in anthers and usually consists of three nuclei: two sperm nuclei (the gametes) and a tube nucleus. The "female" gameteophyte (megagameteophyte) is produced in a specialized region of the ovary called the ovule. Although there are various patterns, the developed female gameteophyte commonly consists of eight nuclei and is known as the **embryo sac.** One nucleus is the female gamete—the egg. Two other nuclei (the polar nuclei) migrate to the middle of the embryo sac and eventually contribute to the formation of the endosperm, a tissue that supports the young embryo and may or may not persist in the seed. The function of the five remaining nuclei is unclear.

Pollination

Pollination is the transfer of pollen from the anther to the stigma. The transfer of pollen within the same flower (or to any flower on the same plant or clone) is known as **self-pollination;** the transfer of pollen to a flower on another plant (which is likely to have a different genetic makeup) is **cross-pollination.** Self-pollination may be accomplished by gravity or by actual contact of the shedding anther with the sticky surface of the stigma. In cross-pollination wind and insects are the important agents of pollen transfer. Most plants both self- and cross-pollinate naturally, with the proportion of each mode depending upon functional or structural features of the flower or upon genetic incompatibility. Plants may be grouped by their usual method of pollination as **self-pollinated** (cross-pollination less than about 4 to 5 percent), **cross-pollinated** (cross-pollination predominant), and **self- and cross-pollinated.** The mode of pollination has important consequences to the plant's genetic structure (see Chapter 17).

The likelihood of reproducing a particular plant exactly depends on its natural method of pollination. Seed propagation duplicates plants that are highly self-pollinated because such plants tend to be homozygous. Since cross-pollinated plants are highly heterozygous, they can be duplicated exactly only by asexual methods. Nevertheless, a high degree of uniformity in some characters may be achieved in the seed propagation of cross-pollinated plants by constant selection.

Natural self-pollination is achieved through functional and structural features of the flower. Perfect flowers—those that contain both stamens and pistils—lend themselves to self-pollination. In the violet the pollen sheds before the flower is open; in the tomato the pistil grows through a sheath of anthers.

Cross-pollination is brought about in many different ways. Stamens and pistils may occur in separate flowers on the same plant, as in maize and cucumbers (in which case the plant is termed **monoecious**), or staminate and pistillate flowers may be borne on different plants, as in spinach, asparagus, hemp, hops, and date palm (in which case the species is termed **dioecious**). Furthermore, many perfect-flowered plants cross-pollinate. This is achieved by anatomical or physiological features of the flower that prevent self-pollination (selfing). For example, the differential maturation of stamens or pistils will prevent natural selfing. The structural features of the flower that insure cross-pollination are often adapted to pollen transfer by insects. Among the special adaptations that aid in insect pollination are petal color, odor, and the presence of nectar. In some plants an intimate interdependence exists between structural features of the flower and insect pollination (Figure 6-7).

Incompatibility (self-sterility) is a physiological mechanism that prevents self-fertilization. A genetic factor (or factors) serves to prevent pollen tubes produced by a plant from growing in the style of the same plant. Incompatibility factors prevent self-pollination in such crop plants as alfalfa, cabbage, tobacco, and apple.

Fertilization

After a pollen grain lands on the stigma of the pistil, it absorbs water and such substances as sugars and forms a pollen tube that literally grows down the style to the embryo sac. The pollen tube penetrates the embryo sac, where one male gamete unites with the egg to form the **zygote.** After mitotic division the zygote becomes the **embryo** of the resultant seed. In angiosperms the other male gamete fuses with the two polar nuclei of the embryo sac and forms the **endosperm** (which is therefore $3n$), a nutritive tissue that supports the developing embryo. This complete process is referred to as **double fertilization.** Gymnosperms do not have an endosperm; the analogous nutritive tissue is formed from mitotic divisions in the gametophyte.

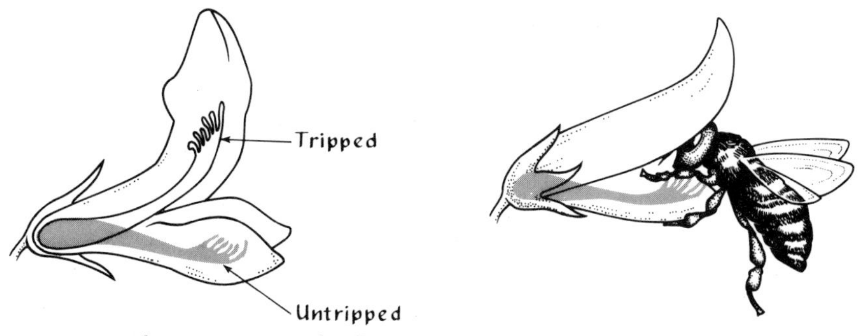

FIGURE 6-7. Flower structure and pollinating mechanism in the legume alfalfa. Pollen is deposited on the bee when the staminal column (pistil and stamens) is tripped by the bee landing on the flower. [After J. M. Poehlman, *Breeding Field Crops*, Holt, Rinehart & Winston, 1959.]

Seed Maturation

Fertilization initiates rapid growth of the ovary and subsequent development of the seed. Usually the ovary will not develop unless it contains viable and growing seed. Common exceptions (**parthenocarpy**) are seedless grapes and seedless oranges, in which the developing seed breaks down at an early stage of development.

In the developing seed the development of the embryo is preceded by endosperm growth (Figure 6-8). When rapid growth of the embryo begins, it does so at the expense of the endosperm. The amount of endosperm at maturity varies with different plants.

In some seeds the endosperm is a prominent tissue surrounding the embryo at maturity. The endosperm thus serves as a source of nourishment during and after germination. Most of the edible portions of the seeds of grasses such as wheat, maize, oat, and barley consist of endosperm. In other plants the endosperm may be completely absorbed at seed maturity, in which case stored food may be in the embryo itself, as in the cotyledons of bean seeds.

Seed development is completed with the formation of hardened integuments, or seed coats. In one-seeded fruits, these integuments include maternal ovarian tissue.

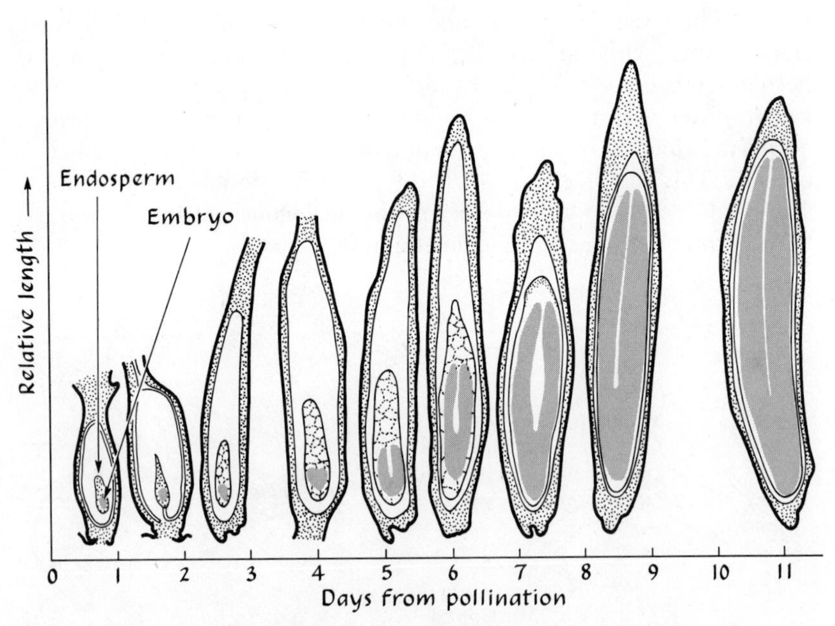

FIGURE 6-8. Growth of one-seeded fruit of lettuce. Note that early growth of the endosperm precedes that of the embryo. [Adapted from H. T. Hartmann and D. E. Kester, *Plant Propagation: Principles and Practices*, Prentice-Hall, 1959; after Jones.]

Genetic Consequences of Sexual Reproduction

The genetic consequences of sexual reproduction are more fully discussed in Chapter 17, but a brief discussion will be given here. Plants that are continually self-pollinated, such as tomato or pea, contain essentially similar pairs of genes on homologous chromosomes (that is, they are **homozygous**). Homozygous plants will reduplicate themselves exactly by sexual reproduction; they breed true because the gametes that combine are essentially identical. Plants that tend to cross-pollinate, such as petunia or cucumber, will have many dissimilar pairs of genes on homologous chromosomes (that is, they are **heterozygous**). Sexual reproduction constantly rearranges these genetic factors. Cross-pollinated plants do not breed true; they **segregate.**

The problem of reproducing a particular plant exactly thus depends on its natural method of pollination. Seed propagation will duplicate naturally any plant that is highly self-pollinated because such plants tend to be homozygous. A particular cross-pollinated plant can be duplicated exactly only by asexual methods because it is heterozygous. However, a high degree of uniformity in some character may be achieved in the seed propagation of cross-pollinated plants as a result of constant selection. For example, in well-selected petunia cultivars each plant may have the same flower color, although it can be demonstrated that the plants are not identical for all characters.

Selected References

Curtis, H., 1983. *Biology*, 4th ed. New York: Worth. (Chapters 11 through 18 have an outstanding review of modern genetics.)

Drlica, K., 1984. *Understanding DNA and Gene Cloning: A Guide for the Curious.* New York: Wiley. (A popular introduction to genetic engineering technology.)

[Photograph courtesy USDA.]

HORTICULTURAL ENVIRONMENT

II

Soil 7

Soil means different things to different people. To most people, soil is dirt—something to be washed from clothes or swept up in a dustpan. To the geologist, soil is the unconsolidated weathered part of the earth's surface, an insignificant part of our planet. To the civil engineer, soil is part of the earth's solid surface that supports structures—the trivial upper 30 cm is considered merely a nuisance. To the soil scientist and agriculturist, soil is that part of the earth's crust in which the roots of crop plants grow. Finally, to many horticulturists, soil is any medium, regardless of source, used to grow plants.

In this chapter we shall consider soil as the *naturally* weathered part of the earth, including organic matter, that supports the growth of plants. Soil is a vital component of the plant's environment and can be manipulated to affect plant performance. As we shall see, horticulturists also are interested in the creation of "synthetic soils" using nonsoil ingredients: typically materials synthesized from mineral materials, such as perlite or vermiculite, and such organic components as pine bark and decomposed sewage. Synthetic soils sometimes contain true soil material.

Soil, both natural and synthetic, has three primary functions in sustaining plant life. It supplies (1) mineral elements, (2) water, and (3) a medium of anchorage for many terrestrial and aquatic plants.

Despite its importance, soil is not required by plants, and many horticulturists grow plants without any natural soil whatsoever. Nevertheless, field crop culture relies on soil and mass culture of most plants would be inconceivable without it.

SOIL SYSTEMS

Soils consist of a number of interacting systems: (1) inorganic minerals, (2) organic matter, (3) soil organisms, (4) soil atmosphere, and (5) soil water. Although these systems are interconnected, we shall discuss them individually.

Inorganic Minerals

Inorganic soil minerals are derived from the chemical weathering of underlying minerals or organic matter. Soil minerals may be original, relatively unchanged from the parent minerals (such as quartz in the form of sand), or secondary, formed by the weathering of less resistant minerals (clays are examples of secondary minerals). The amount of inorganic material varies from more than 99 percent of the weight of sandy and clay soils to as little as 1 percent in some organic soils. The inorganic component consists of a mixture of particles that differ in size, composition, and physical and chemical properties.

Soil Texture

The mineral particles of the soil can be arranged according to size from very fine to very coarse. The term **soil texture** refers to the size of the individual mineral particles (Table 7-1). The designations used to describe soils by texture (Figure 7-1) are sand, silt, clay, and loam. The nontechnical terms *lightness* and *heaviness* also refer to soil texture. Heavy soils are high in clay and other fine particles; light soils are low in clay and high in sand and other coarse particles.

Silt particles, which are smaller than sand particles, are more or less unweathered, but their surfaces are coated with a clayey matter. Thus, the properties of silt are intermediate between sand and clay.

The clays, the smallest of the soil particles, show distinct chemical and physical properties. Clays are colloidal, composed of small, insoluble, nondiffusable particles larger than molecules but small enough to remain suspended in a fluid medium without settling. Most soils contain organic colloidal particles as well as the inorganic colloidal particles of clays.

Clays are viscous and gelatinous when moist but hard and cohesive when dry. They are composed of particles called **micelles,** which are formed from the parent materials by a crystallization process; they are not merely finely divided

TABLE 7-1. Classification of soil particles.*

Particle	Diameter (mm)
Coarse sand	0.5–1
Medium sand	0.25–0.5
Fine sand	0.1–0.25
Very fine sand	0.05–0.1
Silt	0.002–0.05
Clay	<0.002

SOURCE: *Soil Taxonomy,* 1975, USDA Handbook 436.

*The International Classification (Atterberg) System refers only to soil particles under 2 mm.

Coarse sand	0.2–2 mm
Fine sand	0.02–0.2
Salt	0.002–0.02
Clay	<0.002

rock. The micelles are sheetlike (laminar), with internal as well as external surfaces, and tend to be held together by chemical linkages or ions between the plates. Their tremendous surface area relative to their volume is one of their most significant features.

The structure of clay micelles may be complex (Figure 7-2). Clay particles are negatively charged and therefore attract, retain, and exchange positively

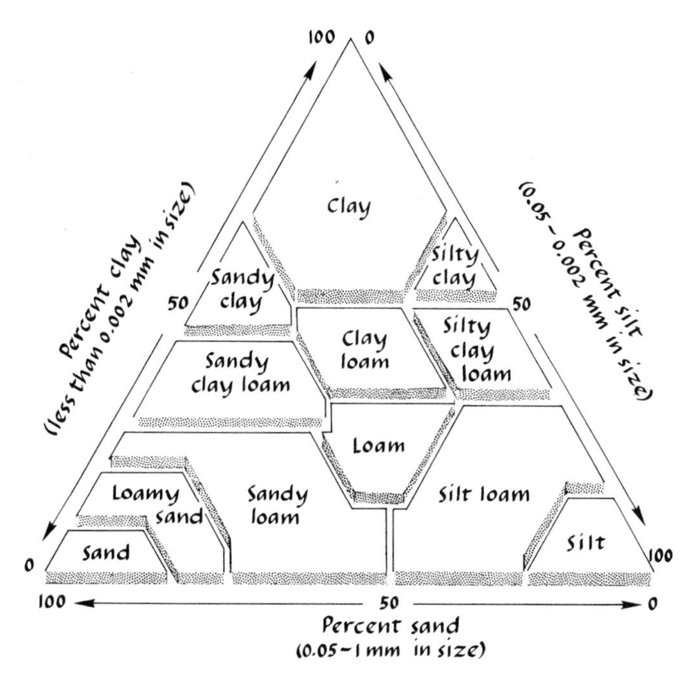

FIGURE 7-1. Soil texture triangle. To find the textural name of a soil, locate the points corresponding to the percentage of clay and silt on each side of the triangle. The silt line is projected inward parallel to the clay side; the clay line is projected inward parallel to the sand side. The lines will meet in the class name of the soil. [Adapted, with permission of the publisher, from T. L. Lyon, H. O. Buckman, and N. G. Brady, *The Nature and Properties of Soils,* 5th ed., copyright 1952 by the Macmillan Company.]

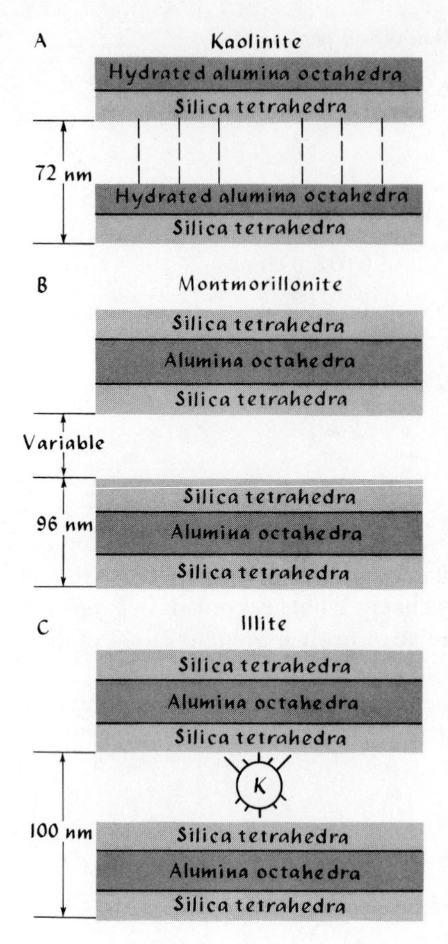

FIGURE 7-2. The crystalline structure of clay. (*A*) The simplest clay mineral is called kaolinite. The micelles of kaolinite consist of two different layers, one of silica and the other of alumina. Kaolinitic clay micelles are relatively large and are bound together tightly. The distance between the alumina and silica layers is rigidly fixed and does not increase when water is adsorbed on the surfaces of the individual micelles. The internal space is not available for surface reactions. Such clays do not shrink greatly when dry or expand much when hydrated. (*B*) Montmorillonite is more complex, and each micelle consists of an alumina layer sandwiched between two layers of silica. These clays are not bound together tightly, and they swell when wet because the hygroscopic surfaces between layers absorb water and force the layers apart. All surfaces can adsorb water and minerals. (*C*) The micelles of illite consist of an alumina layer between silica layers, and adjacent layers are held together by potassium atoms. Surface adsorption can take place on the surfaces between crystals to a limited degree. Illite is less abundant than kaolinite or montmorillonite.

charged ions. On the particles collects adsorbed water, which acts both as a lubricant and as a binding agent. Clay platelets behave like a stack of wet poker chips. To a large degree, this characteristic explains the plasticity of clay. Clay soils are low in both organic matter and weakly hydrated cations such as calcium, and they become sticky or puddled when wet. The unique chemical and physical properties of clays are responsible for most of the important properties of soils.

Soil texture affects the retention of water and the rate of water infiltration (Table 7-2). Coarse soils permit the rapid infiltration and percolation of water, so there is no surface runoff even after a heavy rain. In contrast, clay soils are so finely textured that very little water penetrates to lower levels, especially after the surface clays become wet and expand. Coarse soils, however, are incapable of retaining large quantities of water.

TABLE 7-2. Relationship of soil texture to soil water retention.

Soil type	Water capacity (millimeters per meter of soil)
Very coarse texture—very coarse sands	33–63
Coarse texture—coarse sands, fine sands, and loamy sands	63–104
Moderately coarse texture—sandy loams and fine sandy loams	104–146
Medium texture—very fine sandy loams, loams, and silt loams	125–192
Moderately fine texture—clay loams, silty clay loams, and sandy clay loams	146–208
Fine texture—sandy clays, silty clays, and clays	133–208
Peats and mucks	167–250

Soil Structure

The **structure** of soil refers to the arrangement of the soil particles into aggregates. Sands and gravels are examples of soils with a simple structure. These soils have very little cohesion, plasticity, and consistency (the resistance of the particles in the soil to separation). Simple-structured soils are usually composed of materials that are relatively resistant to weathering, such as quartz sand. Most agricultural soils have a compound structure; their particles aggregate, or stick together.

Soil structure develops when small colloidal soil particles clump together or **flocculate** into granules. **Granulation** is promoted by freezing and thawing, the disruptive action of plant roots, the mixing effects of soil fauna, the expansion and contraction of water films, and the presence of a network of fungal hyphae. However, unless the granules are stabilized by coatings of organic matter or by their own electrochemical properties, they will coalesce into clods.

Soil structure is very important for agricultural soils to facilitate aeration, water movement, and root penetration. Highly granulated soils are well aerated and have a high water-holding capacity because of the increased volume of the soil pore space. This pore space is occupied by water and air in varying proportions, the soil acting as a huge sponge. The total pore space of soil, which is typically about 50 percent of the total volume, is not so important as the size of the individual pores. Clay soils have more total pore space than sandy soils, but because their pores are small, air and water move through them slowly.

When the small pores of clay soils are filled with water, aeration is greatly limited. Large pore spaces become filled and are drained by gravity, whereas small pores absorb and retain water by capillary action. Capillary water is of the utmost significance to plants: It is the soil solution most used by them.

The crumbly nature of good agricultural soils depends on soil texture and on the percentage of humus (decomposed, stable organic matter). Clay soils low in organic matter typically have poor structure. In order to maintain good compound structure, they must be carefully managed. If worked when too wet, the structure may be damaged. When clods are exposed, they become dry, hard, and difficult to work back into the soil. In heavy soils, organic matter must be added to maintain good structure. In sandy soils, where structure is not so critical, it is necessary to add organic matter to increase their water- and nutrient-holding capacities.

Exchange Capacity

The capacity of a soil to retain and exchange such cations as H^+, Ca^{2+}, Mg^{2+}, and K^+ is called its **exchange capacity.** Exchange capacity, a measure of the chemical reactivity of the soil, varies inversely with particle size (Figure 7-3). Because a given volume of small soil particles has much more surface area than an equal volume of large particles, fine soils accumulate and retain many times more cations than do coarse soils. These cations are attracted by the negative charge of the soil's colloidal particles, clay and humus. When these particles in a soil are saturated with hydrogen ions, the soil has a strong acid reaction. Soil reaction and pH are discussed more fully in the next section.

The exchange capacity of the soil's colloidal particle is tremendously important. Nutrients that would otherwise be lost by leaching are held in reserve, and when exchanged, become available to the plant.

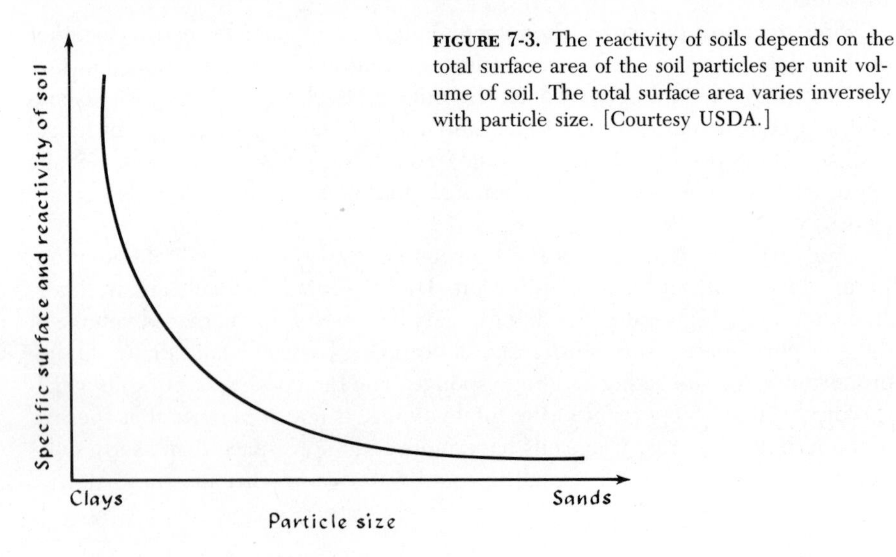

FIGURE 7-3. The reactivity of soils depends on the total surface area of the soil particles per unit volume of soil. The total surface area varies inversely with particle size. [Courtesy USDA.]

The process of cation exchange is not a random one. Cations differ in their ability to replace one another; if present in equal amounts,

$$H^+ \text{ replaces } Ca^{2+} \text{ replaces } Mg^{2+} \text{ replaces } K^+ \text{ replaces } Na^+$$

If one cation is added in large amounts it may replace another by sheer force of number (mass action). This is essentially what occurs with the addition of fertilizer.

The release of hydrogen ions in soils tends to promote the exchange of cations, making them available to plants. Hydrogen ions are made available by the dissociation of carbonic acid formed from the carbon dioxide released by roots and the even larger quantity of carbon dioxide released by the enormous population of microorganisms in respiration and in the decomposition of organic matter:

$$CO_2 + H_2O \rightleftharpoons H_2CO_3 \rightleftharpoons H^+ + HCO_3^-$$

The cations are replenished by the decomposition of the inorganic fraction of the soil, by the degradation of organic materials, and by the application of fertilizer. The cations in a productive soil are in equilibrium among the soil particle, the soil solution, the soil microorganisms, and the plant.

The cation exchange capacity of a soil is expressed in milliequivalents (meq) of H^+, which give the number of milligrams of H^+ that will combine with 100 g of dry soil. The exchange capacity of a soil depends on the percentage of humus it contains and on the percentage and composition of its clays. Clays differ markedly in their ability to exchange cations. Kaolinite has an exchange capacity of 10 meq/100 g, whereas this capacity in montmorillonite clay is about 100 meq/100 g. The exchange capacity of humus ranges from 150 to 300 meq/100 g. The ranges of exchange capacity for various soils are given in Table 7-3.

TABLE 7-3. Cation-exchange capacity ranges for various soil types.

Soil type	Cation-exchange capacity (meq/100 g)
Sands	2–4
Sandy loams	2–27
Loams	7–16
Silt loams	9–30
Clay and clay loams	4–60
Organic soils	50–300

SOURCE: Adapted from T. L. Lyon, H. O. Buckman, and N. C. Brady, 1952, *Nature and Properties of Soils* (Macmillan).

SOIL REACTION. **Soil reaction** refers to the acidity or alkalinity of the soil. It is expressed in terms of pH, the logarithm of the reciprocal of the hydrogen ion concentration, and is usually expressed in units from 0 to 14:

$$\text{pH} = \log \frac{1}{[\text{H}^+]}.$$

Table 7-4 gives the concentration of moles of H^+ and OH^- for pH values of 0 to 14. Note that the molar concentration of H^+ times the molar concentration of OH^- equals a constant of 10^{-14}. The pH of the soil is regulated by the extent of the colloid fraction charged with hydrogen ions. For example, a clay particle charged with abundant H^+ acts as a weak acid and imparts an acid reaction or low pH. Similarly, a clay particle charged with mineral cations imparts an alkaline reaction, or high pH (Figure 7-4).

The proper soil pH (6 to 7) is vitally important in plant growth. Abnormally high soil pH (above 9) and low pH (below 4) are, in themselves, toxic to plant roots. Within the normal range the pH determines the behavior of certain nutrients, precipitating them or making them unavailable (Figure 7-5). For example, the chlorosis found in some plants grown in the soil with a high pH is a result of iron deficiency caused by the precipitation of iron compounds. Soil organisms, especially bacteria, are also affected by pH. Vigorous nitrification and nitrogen fixation require a pH above 5.5.

TABLE 7-4. The concentration of H^+ and OH^- with varying pH.

pH	Soil reaction			H^+ Concentration (moles per liter)*	OH^- Concentration (moles per liter)†	Reaction of common substances
0				10^{-0}	10^{-14}	
1				10^{-1}	10^{-13}	
2				10^{-2}	10^{-12}	
3	Acidity	very strong	Soil range	10^{-3}	10^{-11}	Lemon juice
4		strong		10^{-4}	10^{-10}	Orange juice
5		moderate		10^{-5}	10^{-9}	
6		slight		10^{-6}	10^{-8}	Milk
7		neutral		10^{-7}	10^{-7}	Pure water
8	Alkalinity	slight		10^{-8}	10^{-6}	Sea water
9		moderate		10^{-9}	10^{-5}	Soap solution
10		strong		10^{-10}	10^{-4}	
11		very strong		10^{-11}	10^{-3}	
12				10^{-12}	10^{-2}	
13				10^{-13}	10^{-1}	
14				10^{-14}	10^{-0}	

*1 mole of H = 1 g †1 mole of OH = 17 g

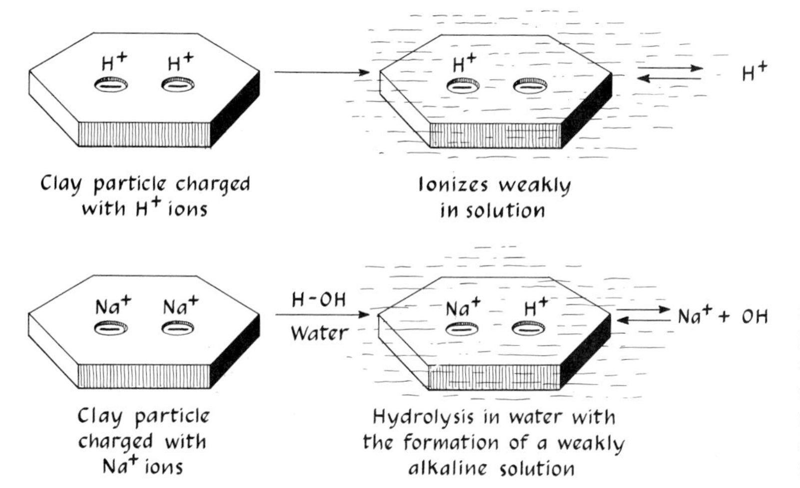

FIGURE 7-4. The soil reaction depends on whether the clay particles are charged with hydrogen ions or mineral cations. [From J. Bonner and A. W. Galston, *Principles of Plant Physiology*, W. H. Freeman and Company, copyright © 1952.]

Soil Organisms

A fertile soil is literally alive. Although insects and earthworms are the most obvious of its living inhabitants, bacteria, fungi, and other microorganisms constitute the great bulk of soil organisms (Table 7-5). The total weight of soil organisms (excluding higher plants) in the upper 30 cm (about 1 foot) of fertile agricultural soils is impressive, accounting for as much as 7000 kg/ha (about 6250 pounds per acre), which equals the weight of 20 to 30 marketable hogs.

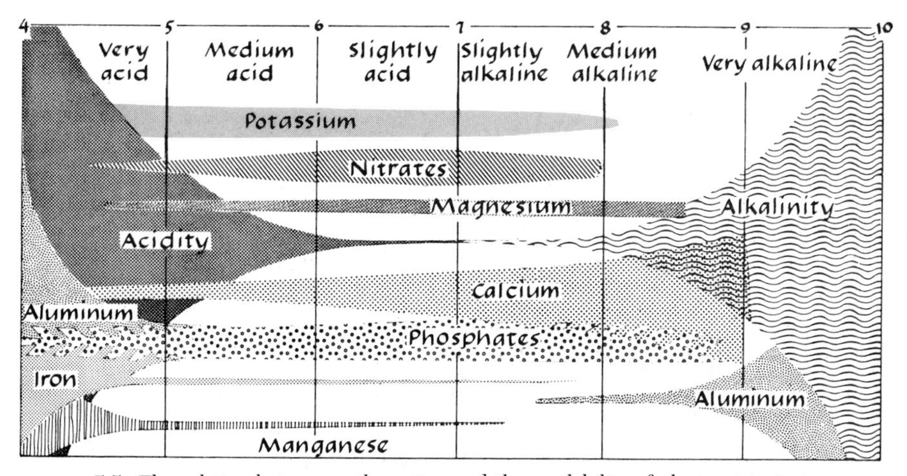

FIGURE 7-5. The relation between soil reaction and the availability of plant nutrients to crops. The thickness of the various bands indicates the relative availability of the nutrients they represent. [Courtesy Virginia Poltechnic Institute and State University.]

TABLE 7-5. Average weight of organisms in the upper 30 cm of soil (in kg/ha).

Organism	Low	High
Bacteria	560	1120
Fungi	1680	2240
Actinomyces	895	1680
Protozoa	225	450
Algae	225	335
Nematodes	28	55
Other worms and insects	895	1120
Total	4508	7000

SOURCE: O. N. Allen, 1957, *Experiments in Soil Bacteriology (Burgess).*

This is, however, only about 0.1 percent of the weight of the soil that the organisms occupy. The organic matter of the soil is derived not only from decomposed plant and animal tissue but from the microorganisms themselves.

Soil organisms play an important part in the development of soil from inorganic minerals. In fact, a soil does not usually develop until the inorganic material is "invaded" by various kinds of organisms.

Higher plants must be considered the principal soil organisms. Roots of trees and other higher plants penetrate and expand crevices in rock, which may be split by the tremendous forces roots exert when growing. Thus roots perform an important function in the continuing process of soil formation. Roots exude many kinds of organic acids and other substances that hasten the solution of soil minerals and make them available to plants. Living roots give off carbon dioxide, which raises the carbonic acid content of the soil solution and increases the rate at which soil minerals dissolve. When leaves fall or when plants die, their organic matter becomes incorporated into the soil, contributing to its fertility. Channels remaining after dead roots have decayed serve as pathways for the movement of soil water.

Bacteria are present in all soils. Estimates of the weight of living and dead bacteria in forest soils have been as great as 6300 kg/ha (about 5600 pounds per acre), although fertile agricultural soils typically contain about 20 percent of this amount.

Bacteria in the soil decompose organic matter produced by plants and other organisms, releasing the minerals they contain, which then become available for another cycle of plant growth. Some kinds of bacteria, such as those in the genus *Azotobacter,* convert molecular nitrogen from the atmosphere into nitrogenous compounds that can be used by plants. Other species live in symbiotic association with the roots of certain kinds of plants and perform the same function. **Nitrogen-fixing bacteria** supply nitrogen to the plant, which in turn supplies carbohydrates to the bacteria. Some bacteria play roles in soil formation and degradation, but these processes are poorly understood.

Not all soil bacteria are helpful; many species are **pathogenic,** causing crop diseases and large economic losses. Other harmful types oxidize ferrous iron to the slightly soluble ferric form, thereby contributing to the formation of an impermeable soil horizon called a hardpan. Hardpans may prevent the drainage of excess water and inhibit root penetration.

Fungi perform many of the same functions as bacteria in decomposing organic matter and in cycling nutrients. **Saprophytic fungi** are nonharmful fungi that decompose the dead organic matter of soils. Some species of fungi live in close symbiotic association with roots (particularly of woody plants) and exchange organic and inorganic materials. Such an association is known as a **mycorrhiza.** Water and minerals move from the fungi to the roots, and carbohydrates and other organic materials move in the reverse direction. Some mycorrhizal fungi actually penetrate the interior of the roots; others remain on the outside. The importance of these fungi is becoming increasingly apparent. Most species of pines cannot survive without them, and there is ever-stronger evidence that they may be associated with the roots of practically all plant species.

Most types of algae are aquatic, but several kinds, mainly the blue-green algae (division Cyanophyta) and the green algae (division Chlorophyta), are also found in the soil, where they hasten the weathering of soil minerals. Algae normally occur near the soil surface, since they require light for photosynthesis. **Lichens** are made up of an alga and a fungus living in a symbiotic relationship. They occupy a special niche in the process of soil formation. They slowly dissolve the rocks they become established on, and the dust they trap accumulates to form soil.

Arthropods (phylum Anthropoda) are among the more obvious soil fauna. Into this large phylum fall the crayfish (important in poorly drained areas of the southeastern United States), mites, ants, centipedes, millipedes, sow bugs, insects, and numerous other species that physically invade and cultivate the soil and contribute to its organic matter when they die.

The segmented worms (phylum Annelida), commonly called earthworms, and round, nonsegmented eelworms, or nematodes (class Nematoda of the phylum Aschelminthes), are both important in agricultural soils. Earthworms recycle subsurface soil to the surface in the form of castings. Nematodes are economically significant because some of them are pathogenic to crop plants and others distribute parasitic fungi and increase the extent of root-rot diseases.

A number of vertebrate animals, including many species of burrowing mammals, must be considered part of the soil fauna. Badgers, gophers, moles, voles, foxes, shrews, mice, ground squirrels, and woodchucks frequent the soil and, when present in great numbers, may become serious pests.

Soil Organic Matter

Soil organic matter is the fraction that is derived from living organisms. Most apparent is the surface **litter** composed of undecayed leaves, branches, reproductive parts, and other plant residues. **Duff,** or partially decayed litter, is frequently matted together with fungal mycelia, and in this condition it is called **leaf mold.** Duff forms when soil is moist enough to supply the water essential for microbial activity and when litter is thick enough to prevent evaporative water losses. Leaf mold is an important component of forest soils, but it is seldom found in cultivated agricultural soils. Plant roots and the excreta, sheddings, and bodies of soil organisms, although not so apparent, also contribute to soil organic matter.

The plant and animal material is decomposed by enzymatic digestion carried out by soil microorganisms. The decomposition of simple carbohydrates (starches and sugars) occurs fairly rapidly and releases carbon dioxide in the soil. Water-soluble proteins are decomposed readily to amino acids and then to ammonium compounds. Through the action of certain "nitrifying" bacteria ammonium compounds are transformed to nitrates, in which form they are again available to plants. The decomposition of organic materials, however, is not complete. Certain substances, such as lignins, waxes, fats, and some proteinaceous materials resist decomposition, but, through complex biochemical processes, they form the dark, noncrystalline, colloidal substance called **humus.** Humus can absorb even higher amounts of nutrients and moisture than does clay. Yet, unlike clay, its plasticity and cohesiveness is extremely low. Thus, small amounts of humus greatly affect the structural and nutritive properties of soil.

One of the most essential contributions of organic matter to soil is that it increases water-holding capacity. Organic matter acts like a sponge: It can absorb large amounts of water relative to its weight. Because it is porous, it also readily permits the infiltration of water. In addition, organic matter is a source of mineral elements, which are made available when it decomposes. The term **mineralization** designates the decomposition of organic matter by bacteria, fungi, and other organisms to form water and carbon dioxide with the release of minerals. This important type of **chemical cycling** consists of (1) the absorption of minerals through roots and their incorporation into chemical compounds of various kinds in plants; (2) the death of plants and plant parts; and (3) the decomposition of the plant material and the release of its minerals into the soil. The minerals are then reabsorbed and recycled. The high adsorptive capacity of organic matter also contributes to the retention and exchange of mineral cations. Organic matter may retain large quantities of minerals and thus prevent their loss from the soil by leaching. It may account for as much as 90 percent of the absorptive and adsorptive capacities of sandy soils.

Organic matter helps maintain the structure of cultivated soils. Finely divided organic matter covers mineral particles and keeps them from sticking together. Clay soils with an appropriate quantity of organic matter are less sticky and more readily cultivated. Crop growers say that soils that are friable and easy to cultivate have good **tilth.**

Soils are frequently classified according to their content of organic matter. One such classification is as follows:

Soil name	Percent organic matter
Mineral	0–10
Muck	10–40
Peat	40–100

Although most agricultural soils are mineral soils, mucks and peats may also be rendered extremely productive under proper management. These two soil types develop when organic matter from living plants is covered by water and does not decompose. Once these soils are exposed to air by cultivation and drainage, bacteria and fungi begin to decompose them, with the result that the soil surface is actually lowered. In the Florida Everglades, drained and cultivated peat soils subsided, in fact, nearly 2 m (about 6 feet) during 40 years of cultivation.

Soil Atmosphere

The soil atmosphere exists in the pore spaces that are not filled with water. These pores contain the same gases as the atmosphere above the ground, but in different proportions. The soil atmosphere is not necessarily a continuous system, for it may contain isolated, unconnected pore spaces. Also, soil atmosphere is very variable depending on soil type, organic matter content, and season.

The humidity of the soil atmosphere is nearly 100 percent much of the time. The carbon dioxide content of the soil is greater than that of the air above it because of the decomposition of organic matter, and it increases with depth because carbon dioxide diffuses very slowly into the air above the soil. Conversely, the oxygen content of the soil atmosphere is less than that of the air above the soil, and it decreases with depth. Oxygen in the soil is used in respiration carried on by roots and microorganisms, and it is slowly replaced by diffusion from the atmosphere.

As water is added to a soil, the air in the soil is squeezed out. Consequently, plant roots may be deprived of oxygen in flooded or very wet soils. For some plants, even a few days of flooding may be disastrous, especially during the growing season. Soil atmosphere will be discussed in Chapter 11.

Soil Moisture

Soil moisture includes free water as well as capillary water, hygroscopic water, and water vapor. Soil moisture is discussed in Chapter 8.

SOIL ORIGINS

In nature, land plants are intimately associated with soil, the reservoir of nutrients and moisture. The soil is far more than an inorganic mass of debris—it is a biological system in a state of dynamic equilibrium. The genesis of soil from the earth's crust begins with a disintegration process that finely subdivides the parental rock. Leaching and the subsequent action of leached materials on the original mineral substances form entirely new substances. It is the biological action of plant and microorganism, however, that transforms the subdivided minerals into the complex material known as soil.

Soil genesis is a continuing process. The process may be viewed from a vertical slice though shallow soil, where bedrock is just slightly beneath the soil surface. The three distinct gradations from bedrock to "topsoil" are known as horizons (Figure 7-6). The morphology of these horizons makes it possible to classify soil into types in order that its structure and fertility may be predicted.

The three major horizons are designated (from the top down) as A, B, and C, and may be further subdivided. The underlying bedrock is called the R horizon. The A horizon is the zone of leaching. Roots, bacteria, fungi, and small animals (for example, earthworms and nematodes) are most abundant here. This horizon is poor in soluble substances and has lost some clay and some iron and aluminum oxides.

The B horizon is the zone of accumulation. It is less abundant in living matter, but higher in clay and in iron and aluminum oxides. It is thus stickier when wet and harder when dry. The C horizon consists of weathered rock material, often true parent material. When hardpans are present, they are a part of this horizon. The R horizon is the underlying consolidated bedrock, such as limestone, granite, or sandstone, from which the overlying horizons were formed.

Soil is formed as a result of all the interacting forces that affect the parent rock materials, the movement by air and water of particles in and out of soils, and the composition and fate of living organisms that inhabit the soil.

Soil Parent Materials

Rocks of unconsolidated materials are the soil's parental material, which determine its physical and chemical properties. The wide variety of soils found within relatively small areas frequently reflects a diversity of parent materials. Residual materials are those formed by the weathering of rocks still in their

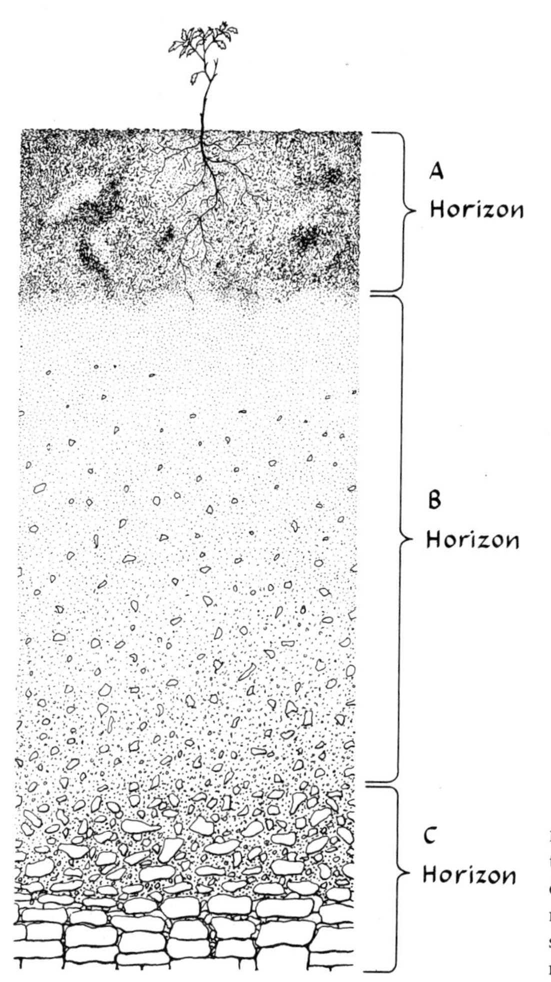

FIGURE 7-6. A soil profile. Each soil type has a distinctive profile that is determined by the particular combination of factors that produced the soil. Distinct zones, called horizons, make up the profile.

place of origin. Transported materials are those that have been moved to the place of soil formation by wind, water, gravity, ice, or a combination of these forces.

Unconsolidated materials that are deposited downslope are the **colluvial** (gravity-deposited) parent materials of soils. Soils deposited by water in the bottomlands along rivers are called **alluvial.**

The three main kinds of rocks are igneous, sedimentary, and metamorphic. **Igneous rocks** are formed when molten lava or magma solidifies. **Sedimentary rocks** are formed when sediments that have accumulated in water or have been deposited from the air are compressed and cemented into rocks. **Metamorphic rocks** are formed when igneous, sedimentary, or previously metamorphosed rocks are altered and recrystallized by heat or pressure, or occasionally when infiltrated by solutions of minerals that subsequently precipitate.

Weathering

The combination of physical and chemical forces that transform the parental materials into soil is known as **weathering.** Many chemical changes take place in the parental materials, both on the surfaces and within rocks. They include hydrolysis, carbonation, oxidation, and hydration.

Hydrolysis is a decomposition process in which water is one of the reacting agents. The hydrolysis of potassium aluminum silicate by carbonic acid and water produce kaolinite, an important clay mineral:

$$2KAlSi_3O_8 + H_2CO_3 + H_2O \longrightarrow Al_2Si_2O_5(OH)_4 + K_2CO_3 + 4SiO_2$$

Carbonation may be illustrated by the decomposition of calcite (calcium carbonate) to calcium bicarbonate:

$$CaCO_3 + CO_2 + H_2O \longrightarrow Ca(HCO_3)_2$$

Calcium bicarbonate is highly soluble and is readily leached from soils.

Oxidation involves the loss of electrons to a receptor, which is frequently oxygen. Iron is a familiar substance that is easily oxidized. Iron pyrite is an important soil mineral that is oxidized to yield a hydrated iron oxide and sulfuric acid:

$$4FeS_2 + 10H_2O + 15H_2O \longrightarrow 4FeO(OH) + 8H_2SO_4$$

Hydration is the combination of molecular water with a compound. In the hydrated form, the compound may be more subject to weathering. The mineral anhydrite (calcium sulfate) is hydrated to gypsum (hydrous calcium sulfate) by the addition of water:

$$CaSO_4 + 2H_2O \longrightarrow CaSO_4 \cdot 2H_2O$$

Temperature is a major factor in the weathering of parent materials because changes in temperature cause differential expansion and contraction of minerals and rocks. In rocks exposed to solar radiation, a very large temperature gradient is established within the first few millimeters beneath the surface and the hot outer surfaces may expand and flake off.

Low temperatures cause water to freeze in the cracks of rocks and split them apart (as noted earlier). Freezing is probably the main weathering agent at high latitudes and altitudes. The rubble formed by freezing has characteristic sharp edges.

The dissociation of water molecules into hydrogen (H^+) and hydroxyl

(OH^-) ions is much greater at high temperatures than at low temperatures. Since H^+ ions are particularly effective in dissolving minerals, the rate of weathering is faster when temperatures are higher. If soil is frozen or very cold throughout the year, bedrock may be weathered only a few centimeters deep. On the other hand, weathering in the tropics may extend to a depth of 50 m (more than 160 feet).

Rainfall is frequently the most influential climatic factor in soil development. Soil development is greater in regions of moderate to high rainfall than where rainfall is sparse, as in deserts. The change is achieved by the pounding of raindrops, the erosion of streams and rivers, and leaching.

Topography frequently plays a substantial part in soil development. If the land is flat, low, and poorly drained, water may accumulate and drain so slowly that biological activity is arrested for long periods. Under such conditions, soil material, both inorganic and organic, may remain unchanged until the land is drained. When topography is very steep, parent materials may be so subject to erosion by water, wind, and ice that soil material is removed from the site as soon as it is weathered from the parent material. The most favorable topography for soil development is a slope steep enough to carry off excess water, but not so steep as to permit the removal of weathered materials that lack the protection of a vegetative cover.

The significance of the events in soil formation cannot be truly appreciated unless their duration is recognized. A mature soil is not formed in a decade or even 10 decades, but over thousands of years. Soil development and vegetational succession are related processes. When vegetational development reaches a climax, in which the vegetation is in balance with its environment and is relatively unchanging, the soil will also have reached full development and will be in balance with its environment. The soil is then said to be mature. The absolute length of time required for complete maturation varies widely. In the tundra of the far north, cold weather precludes rapid biological and physical weathering; consequently, 5000 to 10,000 years might be required to form a soil only very few centimeters deep. In the tropics, however, the much faster rates of reaction can lead to the development of 2-to-3-m-deep soils in a much shorter time.

Soil age is determined by the degree to which it has developed toward maturity. Soil may be chronologically young (in terms of absolute years) and yet may be mature in its zonal structure—the soil profile.

Major Soil Groups

Differences in soil-forming processes lead to different zonal soils, soils that have developed profiles. The major soil groups are based on climatic differences. In the humid regions, three broad groups have been recognized: tundra, podzolic, and lateritic soils. In the arid and semiarid regions, specific soils that have developed in response to low rainfall include the chernozems, or black soils, and various desert soils.

The **tundra soils** are formed in very cold climates, where biological activity is minimal. Organic matter accumulates on the surface, and the deeper horizons remain frozen throughout the year (permafrost). In winter, tundra soils are completely frozen. Their horizons are only poorly differentiated. These soils are of little agricultural significance.

In cold, humid climates a particular combination of soil-forming processes, collectively called **podzolization,** produces podzolic soils. Podzols, the most extreme type, have a deep accumulation of litter and humus, a strongly acid reaction, and an upper horizon from which iron and aluminum oxides (the so-called sesquioxides) have been leached. (The name *podzol* is derived from the Russian word meaning "ash beneath," referring to the color in the lower part of the A horizon). Fungi are the main soil-forming organisms. True podzol soils are found at the higher latitudes and coincide with the boreal forest. The mild summers and long, cold winters in these latitudes permit the accumulation of the litter and humus necessary for their formation.

Podzolization is a factor even in temperate climates, and sometimes under special conditions in the subtropics. The widespread forests of the temperate zone grow predominantly on podzolic soils. They are potentially good agricultural soils, topography permitting, but they are easily degraded by poor cropping practices.

In the humid tropics and subtropics, the combination of soil-forming processes is collectively called **laterization;** these processes form lateritic soils. **Laterites,** the extreme type, have a very shallow accumulation of litter and humus, a neutral soil reaction, and an accumulation of iron and aluminum oxides in the upper horizons. Bacteria are the primary soil-forming organism. Laterites are typical of tropical rain forests and are characterized by extremely intensive chemical weathering. Even the most resistant compounds, such as silica, are broken down and leached by the heavy precipitation in the tropical rain forests. These soils are so deeply weathered that some geologists regard laterization as a geologic rather than a soil-forming process. The name *laterite* is derived from the Latin word *later,* which means brick; these soils are widely used as building materials.

In climates with low precipitation, some soils accumulate carbonates on and near the surface. This process, called calcification, tends to produce very alkaline soils, which are characteristic of deserts and steppes.

Chernozems (meaning "black soils") are formed in areas that receive 40 to 60 cm (15 to 25 inches) of rain per year. Grass is the natural vegetation. The upper horizon is dark, owing to the accumulation of organic matter. In the temperate zones, these soils are among the most fertile agriculturally and produce most of the world's grain.

Desert soils are only slightly weathered and leached because of low rainfall. Nutrients other than nitrogen are present in moderate or large quantities. Desert soils may become very productive under irrigation and proper management.

Soil Classification

Soils may be classified on the basis of origin—that is, parent materials—or of morphology. The National Cooperative Soil Survey has initiated a system, the United States Comprehensive System of Soil Classification, which contains 8000 soil series and about 80,000 soil types and phases. Emphasizing soil properties as they exist today, this system is much more utilitarian than the older classification. It permits a higher degree of accuracy in classifying soils and greater efficiency in mapping, and it allows the classification of soils that for various reasons could not be included in other systems.

The classification system used previously was developed and published in 1936 by C. F. Marbut, chief of the U.S. Soil Survey at that time. A drastic modification, outlined in the 1938 Yearbook of Agriculture *(Soils and Men)*, was used until 1965. But these classifications were based on assumptions of what the soil was when the country was settled, even though cultivation and erosion might have changed them greatly. Since agreement couldn't be achieved about the genesis of many soils, it was difficult to classify them.

The Comprehensive System differs from all other systems in two important respects: (1) The nomenclature of the higher categories is entirely new, and (2) definitions of classes are much more quantitative and specific.

A new language was created for the higher categories of classification to eliminate confusion with names or words formerly in use. Soil scientists selected the diagnostic properties for each class, and classicists from the University of Illinois and the University of Ghent proposed Greek and Latin names for them.

In the new system, there are 10 **orders** or soils, each defined by a few distinguishing characteristics (Table 7-6). As an example, one of the orders is designated "Oxisol," the *oxi* referring to oxide and *sol* to soil. These are the soils of tropical regions; they contain large amounts of iron and aluminum oxides.

Subcategories include suborders, great groups, subgroups, families, series, and soil types:

Suborders are named according to the distinguishing features of the horizons. Thus, an Aquox is a wet soil, gray or bluish in color, with large amounts of oxides.

Great groups define the soils still more specifically, whereas **subgroups** may be based on minor properties that are also found in great groups. **Families** are based on properties important to the growth of plants.

TABLE 7-6. Soil orders according to the comprehensive system of classification.

Order	Formative syllable	Derivation	Meaning	Diagnostic features	Older equivalents
1. Entisol	ent	Meaningless syllable	Recent soil	Very weak or no profile development	Regosols, lithosols, alluvial, some low humic gley
2. Vertisol	ert	Latin: *verto*, turn	Inverted soil	Self-mulching; expanding lattice clays; subhumid to arid climates	Grummusol, regur, black cotton, tropical black clays, smonitza, some alluvial
3. Inceptisol	ept	Latin: *inceptum*, beginning	Young soil	Weak profile development but no strong illuvial horizon, cambic horizon present	Brown forest, subarctic brown forest, tundra, ando, and some lithosols, regosols, and humic gley
4. Aridisol	id	Latin: *aridus*, dry	Arid soil	Soils of arid regions; often have natric, calcic, gypsic, or salic horizons	Desert, red desert, sierozem, reddish brown, solonchak, some regosols, and lithosols
5. Mollisol	oll	Latin: *mollis*, soft	Soft soil	Thick, dark A, stratum; usually develops under grassy vegetation	Chernozem, brunizem (prairie), chestnut, red dish prairie, some humic gley, rendzinas, brown, reddish chestnut, and brown forest soils
6. Spodosol	od	Greek: *spodos*, wood ash	Ashy soil (podzol)	Illuvial horizon shows accumulation of iron and organic colloids; weak to strongly cemented hardpan	Podzols, brown podzolic, groundwater podzols
7. Alfisol	alf	Meaningless syllable	Aluminum-iron soil (pedalfer)	Argillic horizon of relatively high base saturation (>35%); usually under boreal or deciduous broad-leaf forest	Noncalcic brown, gray-wooded; many planosols, some half-bog soils
8. Ultisol	ult	Latin: *ultimus*, last	Ultimate (of leaching) soil	Argillic horizon of low base saturation (<35%); plinthite often present; humid climate; usually forest or savanna vegetation	Red-yellow podzolic, reddish-brown lateritic, rubrozem, some gley and groundwater laterites
9. Oxisol	ox	French: *oxide*, oxide	Oxide soil	Argillic horizon very high in iron and aluminum oxides	Latosols, and most groundwater laterites
10. Histosol	ist	Greek: *histos*, tissue	Tissue (organic) soil	Organic surface horizon (>30% organic matter) more than 6 inches thick	Bog and some half-bog soils

SOURCE: *Soil Taxonomy*, 1975, USDA Soil Conservation Service, Agriculture Handbook 436.

The **soil series** is a collection of individual soils that have essentially the same differentiating characteristics, including color, zonal development, and the depth limit of each horizon. The **series** is classified according to **soil type,** the lowest category of classification, which is based on texture.

Land-capability Classification

The capability classification of land has been widely used in the United States. In this system, soils are grouped according to their potential productivity and their limitations on crop production without special treatment. Soils are grouped into eight classes according to their susceptibility to damage and their potential for sustained production. Details of these classes are provided in the following paragraphs. The first four classes are suitable for cultivation if good management practices are followed; Classes V through VII are not suited for cultivation, but can be used for grazing and forest management; Class VIII land is useful only for recreation, wildlife, and watersheds.

Class I soils are potentially the most productive of all soils. They are nearly level and not particularly susceptible to erosion. They are deep and well drained, with a good nutrient supply and water-holding capacity, and have all the other physical and chemical characteristics required for intensive crop production. They can be used for crops, pasture, range, forests, and wildlife. Irrigated soils may be placed in Class I if water is supplied by a permanent irrigation system.

Class II soils have some limitations. They may have gentle slopes, moderate susceptibility to erosion, a soil depth that is less than ideal, slightly unfavorable soil structure, wetness that can be corrected by drainage, slight climatic limitations, moderate salinity or alkalinity, or other slight limitations. Farm operators have a smaller choice of crops to produce on these soils than they do on Class I soils. Class II soils may be used for crops, pasture, range, forests, and wildlife.

Class III soils have severe limitations that reduce the choice of plants, frequently requiring the use of special conservation practices. They may have moderately steep slopes, high susceptibility to wind or water erosion, overflows that may result in crop damage, only slightly permeable subsoil, a shallow depth to rock or hardpan that restricts rooting, poor fertility, and moderate salinity or alkalinity, either all in combination or singly. They may be used for crops, pasture, forests, range, or wildlife. Management of these soils must be very good if they are to remain productive.

Class IV soils have very severe limitations. They may be on steep, erodible sites, have a history of erosion, be too shallow or excessively wet, be severely alkaline, have poor water-holding capacity, or be subject to damaging overflows. Although these soils may be used for crops, pasture, forests, range, or wildlife, extremely careful soil conservation practices must be applied. They should not be used for crops that require extensive cultivation.

Class V soils have little or no susceptibility to erosion, but other limitations restrict their use to pasture, range, forests, or wildlife. They may be bottom-land soils subject to frequent overflow, nearly level soils in areas where a short growing season limits the production of cultivated crops, level soils that are rocky, or ponded areas in which drainage is not feasible.

Class VI soils have severe limitations. They may be steep, highly erodible or eroded, stony, shallow, wet, subject to overflow, saline, or alkaline, or they may have a poor water-holding capacity. They can be used for pasture, range, forest, or wildlife, but special practices, such as water control by contour ditches, water spreading, or drainage, are usually necessary. Some of these soils are adapted to such special crops as sodded orchards and blueberries.

Class VII soils have such severe limitations that their use is restricted to grazing, forest, or wildlife. They are more severely restricted than Class VI soils because one or more of their continuing limitations cannot possibly be compensated for or corrected. They may be too steep, stony, erodible, wet, alkaline, saline, or deficient in some other manner.

Class VIII soils are useful only for recreation, wildlife habitats, watersheds, or esthetic purposes. They cannot be managed well enough to grow crops, grasses, or trees. Uncorrectable limitations may include erosion or erosion hazard, wetness, stoniness, salinity or alkalinity, or poor water-holding capacity.

The capability classification of land has been used in the past solely for agricultural purposes. However, land-use planners and municipal officials are now beginning to realize that this system can be applied in many other fields as well, such as in planning the location of new subdivisions. For example, a shallow soil with an impermeable layer close to the surface would preclude the use of septic tanks and drainage fields.

SOIL CONSERVATION

Agriculture is conducted in a finite and decreasing land area throughout the world's temperate climates. Its basis is a layer of topsoil that averages only 7 inches over the earth's surface. This delicate mantle of topsoil cannot be exploited indefinitely; it must be conserved and renewed.

Although the reduction of the productive capacity of soils through the loss of fertility and structure is considerable, the most serious problem is erosion. Nutrients can be added by fertilization, but the loss of topsoil cannot be so easily or quickly remedied. The loss of soil due to wind and water is a national problem. Erosion clogs rivers with silt, intensifies droughts and floods, and compounds poverty. Yet soil conservation need not be practiced for purely altruistic reasons. Conservation practices yield immediate rewards in terms of plant growth, and they must be considered the basis of sound soil management.

Erosion of the soil is a natural process influenced by climate, topography, and the nature of the soil itself. Where permanent and undisturbed plant cover exists, erosion is more or less gradual and in equilibrium with soil-formation processes. Erosion is accelerated in the absence of plant cover. Areas that, because of climate or topography, are unable to support a permanent plant cover undergo a "geologic" erosion, such as is found in the Grand Canyon. The accelerated soil erosion resulting from cultivation or overgrazing comes about principally through the action of water in humid climates (Figure 7-7) and the action of wind in arid climates.

The maintenance of vegetative cover is basic to soil management. Vegetative cover retards erosion by breaking and cushioning the beating force of the rain (Figure 7-8) by increasing the absorptive capacity of the soil, and by holding the soil against both water and wind. The soil cover increases the infiltration of water through the soil by preventing the clogging of the soil pores by fine surface particles. The techniques used for increasing soil cover include sod culture (as in orchards), crop rotation, cover cropping, and mulching.

Water erodes the soil by literally carrying it away. The carrying power of moving water increases with its speed and volume. The volume of water depends upon the amount of rainfall and the rate at which it is absorbed by the soil. The speed with which this water moves is directly related to the slope of the land and the amount of cover. Any technique that either increases absorption or reduces the speed of the runoff will help to prevent soil erosion.

FIGURE 7-7. Mountain runoff has severely eroded this vineyard in Santa Fe Springs, California. [Photograph courtesy USDA.]

FIGURE 7-8. The beating force of a raindrop striking wet soil. Soil particles and globules of mud are hurled in all directions. [Photograph courtesy U.S. Soil Conservation Service.]

The absorptive capacity of the soil may be increased by deep plowing, by raising the amount of organic matter in the soil, or by increasing drainage. Thus, the burning or removal of organic matter is a poor conservation practice. Where natural drainage is poor, the installation of drainage tiles beneath the surface may be necessary to remove water and introduce air.

Controlling erosion by reducing the speed of runoff may be accomplished in a number of ways. Most basic is **contour tillage,** in which plowing, cultivation, and the direction of the "row" follow the contour of the land rather than its slope. This slows runoff and therefore increases the ability of the tilled soil to absorb water (Figure 7-9). The use of **intertillage,** or **strip cropping,** which

FIGURE 7-9. Contour planting of a field of pineapple in Hawaii. [Photograph courtesy Dole Corp.]

alternates strips of sod and row crops planted along the contour, helps to slow runoff by interposing barriers with high absorptive capacity, and this alternation also provides the benefits of rotation. On steeper slopes, where greater amounts of surface water must be accommodated, the use of **waterways** (permanently sodded areas) facilitates water removal and minimizes erosion.

Where contour cultivation and strip cropping are not sufficient to check erosion, **terraces** constructed on the contour must be used. Terracing, an ancient practice, consists of cutting up a slope into a number of level areas. Terraces appear as steps on the hillside. Although the boundaries of ancient terraces were made of stone, modern terraces are created by building low, rounded ridges of earth across the sloping hillside (Figure 7-10). Terracing slows down the speed of surface runoff, and, although it is designed primarily to prevent erosion, it also facilitates the storing of available water. Thus, terracing is an important practice in areas of low rainfall.

Some of the erosion caused by wind, especially in open prairies or plains, may be checked by planting **windbreaks**—one or more rows of trees or shrubs planted at right angles to the prevailing winds (Figure 7-11). The effectiveness of a windbreak is local and is related to the thickness and the height of the trees. The maintenance of a permanent plant cover in conjunction with windbreaks effectively reduces wind erosion. On organic soils, rows of small grain crops such as oats are used as temporary windbreaks to protect vegetable seedlings.

FIGURE 7-10. A terraced peach planting on a fine sandy loam. The terrace ridges have been left in the rye cover in order to protect the land against wind erosion. [Photograph courtesy U.S. Soil Conservation Service.]

FIGURE 7-11. Tree plantings act as windbreaks to reduce wind velocity and thus protect crops as well as homes and livestock from cold and hot winds. Shelterbelts consisting of 5- to 10-row strips of trees help to control wind erosion, contribute substantially to the saving of moisture, and protect crops from damage by high winds. [Photograph courtesy USDA.]

SYNTHETIC SOILS[1]

Horticultural production of container-grown plants under controlled environments, such as in greenhouses, requires a standardized growing medium. The ideal substrate for such plants should be light, inexpensive, easily managed, well drained, free from weeds and pests; it should also provide adequate nutrition. Various types of growing media have been developed to meet these needs.

Growing media are divided into two types: **soil mixes** that contain natural soil and **soilless mixes** that do not. All soil mixes that contain a substantial amount of nonsoil ingredients may be considered as **synthetic soils.** The nonsoil ingredients of these soils are listed in the following subsection.

Inorganic Components

Sand: Silica sand with particle size between 0.5 mm and 2 mm is preferred for mixes. Sand that is too fine may cause cementing if soil particles are included. Sand is inexpensive but very heavy, weighing about 1.6 metric tons per cubic meter (100 pounds per cubic foot). High pH limestone-derived sands are unsatisfactory for soil mixes.

[1]Various terms have been used in connection with the generic term *synthetic soils.* They include (in alphabetical order) amended soils, artificial media, greenhouse soils, growing media, plant substrate, potting composts, potting mixes, potting soils, and soil mixtures.

Vermiculite: A lightweight mica heated to about 1090°C (2000°F) is free of disease organisms and weed seed. It contains considerable magnesium and potassium, has high water-holding capacity, high cation-exchange capacity, and good buffering characteristics.

Perlite: This white material is derived from lava heated to 760°C (1400°F). It holds three to four times its weight in water and is more resistant to physical destruction than vermiculite, but it has no cation exchange or buffering capacity. It is most useful for increasing aeration in mixes.

Calcined clay: This is baked montmorillonite clay.

Pumice: This natural glass is of vulcanic origin.

Cinders: These coal residues must be leached to remove sulfates.

Synthetic inorganic aggregates include expanded polystyrene flakes and urea formaldehyde foams.

Organic Components

Peat: This substance is the remains of wetlands vegetation preserved under water in a partially decomposed state. Compaction varies with the type of vegetation, state of decomposition, mineral content, and acidity. Peat may be classified as moss peat, reed sedge peat, or peat humus.

1. *Moss peat (peatmoss)* originates from sphagnum mosses and is the least decomposed of the peats. It is usually purchased in compressed bales. Moss peat has high water-holding capacity (15 times its dry weight) and is highly acidic (pH 3.2–4.5). A moss peat derived from hypnum moss is sold in uncompressed form and is not as satisfactory for mixes.
2. *Reed sedge peat* originates from grasses, seeds, sedges, and other swamp vegetation. Its rapid rate of decomposition and low fiber content make it unsatisfactory for synthetic soil mixes.
3. *Peat humus (muck)* consists of undefined peat in an advanced state of decomposition. Its color is dark brown-black. Although often sold with high moisture, it has a low moisture-holding capacity and is unsatisfactory as an ingredient for growing media because it does not improve drainage or aeration.

Sphagnum moss: This material is the dehydrated residue of various species of *Sphagnum*. It must be shredded before use. Sphagnum moss is acid (pH 3.5–4.0) and does not contain many plant nutrients. However, it is fungistatic and inhibits the growth of organisms that cause a disease of germinating seeds and seedlings called damping off (see page 321).

Wood residues: These materials include tree barks, wood chips, shavings, and sawdust. They are used to replace peat in soil mixes because of cost. Wood residues must be weathered to degrade or leach toxic material. Nitrogen must be added to compensate for the tie-up of nitrogen by microorganisms during decomposition.

Plant residues: Various plant by-products, such as corn cobs, bagasse (the sugarcane residue that remains after the stems are crushed), straw, and peanut and rice hulls have been used in soil mixes, depending on price and availability, as total or partial substitutes for peat.

Manures: Animal manure, as well as spent mushroom compost, is not recommended in synthetic soil mixes because pasteurization results in the accumulation of ammonia to toxic levels. In most areas, manures have been replaced by peat because of expense and handling difficulties.

Synthetic Soil Mixes

Various synthetic soil mixes have been developed (Table 7-7). There are no set recipes because the source and type of soil vary. In general, the amount of clay in soil rises as the amount of nonsoil ingredients in the mix increases. These mixtures must be pasteurized (71°C or 160°F, for 3 minutes) or sterilized with chloropicrin or methylbromide–chloropicrin mixtures.

The most widely known soil mixes are the John Innes Composts developed in England. They are mixtures of composted loamy soil, peat, and sand to which are added fertilizers (Table 7-8).

Soilless mixes

Soil variability and weed and disease problems led to the development of soilless mixes. At the present, many are commercially available. Most are based on a combination of peat mixed with other organic amendments, and inorganic aggregates to improve aeration. Peat is difficult to wet (hydrophobic) so that proper watering is problematic. In the nursery industry where tremendous amounts of soilless mixes are used, barks are replacing peat. These media dry

TABLE 7-7. Common ratios of ingredients in synthetic soil mixes.

	Recommended ratio		
Soil type	Soil	Peat	Perlite or calcined clay
Sand	1	1	0
	1	1	1
	2	3	3
Clay	3	5	5

TABLE 7-8. Composition of the John Innes Composts.

Type of mix	Ratio			Fertilizer amendment	Concentration	
	Loam	Peat	Sand		Kilograms per cubic meter	Pounds per cubic yard
Seed	2	1	1	Superphosphate	1.2	2.0
				Limestone	0.6	1.0
Potting	7	3	2	Fertilizer*	3.2	5.4
				Limestone	0.6	1.0
Potting (acid-loving plants)	2	1	1	Superphosphate	1.2	2.0
				Flour of sulfur	0.6	1.0

*Base fertilizer: 2 parts hoof horn meal (or blood meal)
2 parts superphosphate (8.6% P)
1 part sulfate of potash (40% K)

out early and require about 1.3 to 1.9 cm (½ to ¾ inch) of water a day. Plants in these light media do not transplant well to heavy clay soils.

The University of California mixes were developed to provide pathogen-free substrates for the nursery industry. These mixes are based on various combinations of fine sand and finely shredded peat. The 1:1 mixture (48 percent moisture-holding capacity) requires fertilizer additions as given in Table 7-9.

TABLE 7-9. Fertilizers required for University of California mixes.

Ingredient	Concentration: Kilograms per cubic meter	Concentration: Pounds per cubic yard
For immediate use:		
Hoof and horn or blood meal (13% N)	1.47	2.49
KNO_3	0.15	0.25
KSO_4	0.15	0.25
Single superphosphate	1.47	2.49
Dolomite lime	4.42	7.45
Calcium carbonate lime	1.47	2.49
For stored mixes:		
KNO_3	0.15	0.25
KSO_4	0.15	0.25
Single superphosphate	1.47	2.49
Dolomitic lime	4.42	7.45
Calcium carbonate	1.47	2.49

TABLE 7-10. The Cornell Peat-lite mixes.

Type of mix	Per cubic meter	Per cubic yard
Peat-lite, mix A:		
Sphagnum peat	0.5 m^3	0.5 yd
Vermiculite	0.5 m^3	0.5 yd
Ground limestone	2.96 kg	5 lb
Superphosphate	0.59–1.19 kg	1–2 lb
Calcium or potassium nitrate	0.59 kg	1 lb
Fritted trace elements	0.73 g	2 oz
Wetting agent	1.23 g	3 oz
Peat-lite, mix B:		
Sphagnum peat	0.5 m^3	0.5 yd^3
Horticultural grade perlite	0.5 m^3	0.5 yd^3
Ground limestone	2.96 kg	5 lb
Superphosphate	1.19 kg	2 lb
Potassium nitrate	0.89 kg	1.5 lb
Fritted trace element	0.73 g	2 oz
Wetting agent	1.23 g	3 oz

The University of California mixes contain sand and are heavy. The lighter Cornell Peat-lite mixes were developed for container growing of bedding plants (Table 7-10). Being soilless, they do not require sterilization although moss peat should be pasteurized to eliminate or decrease inoculum. There are two mixes: Mix A uses vermiculite and mix B uses perlite.

Selected References

Black, C. A., 1968. *Soil-Plant Relationships,* 2d ed. New York: Wiley. (An in-depth discussion of soil fertility.)

Brady, N. C., 1974. *The Nature and Properties of Soils,* 8th ed. New York: Macmillan. (A basic and widely used text on agricultural soils.)

Russell, E. W., 1973. *Soil Conditions and Plant Growth,* 10th ed. New York: Longman. (The most famous work on agricultural soils. The first seven editions were written by Sir John Russell.)

U.S. Department of Agriculture, 1957. *Soil.* (USDA Yearbook, 1957.) (A broad, nontechnical treatment.)

U.S. Department of Agriculture, Soil Survey Staff, 1975. *Soil Taxonomy.* (USDA Handbook 436.) (The most recent approach to the comprehensive classification of world soils.)

Water 8

Water is a pervasive yet unique substance in the environment. We are familiar with it in all three of its phases—as a liquid, a gas, and a solid. In a sense, plants as well as animals are largely columns of water with various amounts of other dissolved substances. Plants especially use large quantities of water. Thus, water has a tremendous influence on plant performance and productivity.

A study of water in relation to plants may be conducted on many levels. In this chapter, water in the soil will be discussed because plants obtain most of their water from this source. Water's place in plant physiology has been briefly covered in Chapter 4, which clarified the uptake (absorption), distribution (translocation), chemical use (metabolism), and loss (transpiration) of water in plants. Water as moisture in the atmosphere will be discussed in Chapter 9, "Air." Finally, the earth's hydrologic cycle will be described in Chapter 19, "Horticultural Geography."

Water is a constituent of all cells, the amount varying with the tissue in question. It may be as low as 3 percent in shelled peanut seed, about 40 percent in dormant wood, and up to 95 percent in such succulent fruits as the watermelon. Water is the solvent system of the cell and it provides a medium for transfer within the plant. It maintains the turgor necessary for the intricacies of transpiration and plant growth. In addition, water itself is required as a nutrient for the production of new compounds. One-third of the weight of carbohydrates and proteins is derived from chemically combined water.

The water in a plant is in a continual state of flux. A net loss of water causes growth to stop, and a continued water deficiency eventually produces irreversible alterations of the plant that result in death. Under hot, dry conditions, this may occur quite rapidly in plants that are not structurally adapted to prevent water loss. However, in some plants, seed or pollen may be stored for very long periods in the dry state if the temperature is controlled.

The high percentage of water in plants and its capacity as a nutrient carrier and solvent do not alone explain the high rate of water utilization by plants. The water requirements of plants, expressed as the number of units of water absorbed per unit of dry matter produced, varies from about 50 in conifers to 2500 in leafy vegetables! Most crop plants range from 300 to 1000. While growing, the plant continuously absorbs water from the soil and gives it off in transpiration. This loss of water is a by-product of carbon fixation. Carbon dioxide, which provides the carbon necessary for growth, enters the plant through the water films surrounding the spongy mesophyll of the leaf. As this film evaporates, it is replenished from the tissues of the plant, which in turn draw water from the vascular system.

The transpirational loss of water by the plant may be considered as an exchange of water for carbon, and in this sense transpiration is necessary for plant growth. Rapidly growing plants thus require large quantities of water, greatly in excess of the amount found in the plant itself. The rate of water loss depends largely on the temperature, the relative humidity, and the amount of air movement. Radiation from the sun provides the energy needed to change the state of the water from film to vapor. This "boiling off" of water is responsible for the dissipation of a large part of the total energy received by the plant from the sun.

SOIL MOISTURE

The amount of soil moisture that is of benefit to plants has definite limits. Too much water may be as troublesome as not enough. Excess water is not toxic; rather, it is the lack of aeration in waterlogged soils that causes damage. Plants can be grown satisfactorily in water solutions when aeration is provided (Figure 8-1).

The amount of water in a soil may be measured in a number of ways. The expression of soil moisture in inches of water per foot of soil is useful for some purposes (1 acre-inch is equivalent to approximately 27,000 gallons). Expressing soil moisture in terms of the **field capacity** of a soil takes into account the physical condition of the soil, and thus has greater agricultural significance. The field capacity of a soil is the maximum amount of moisture that is retained after the surface water is drained and after the water that passes out of the soil by gravity (free water) is removed.

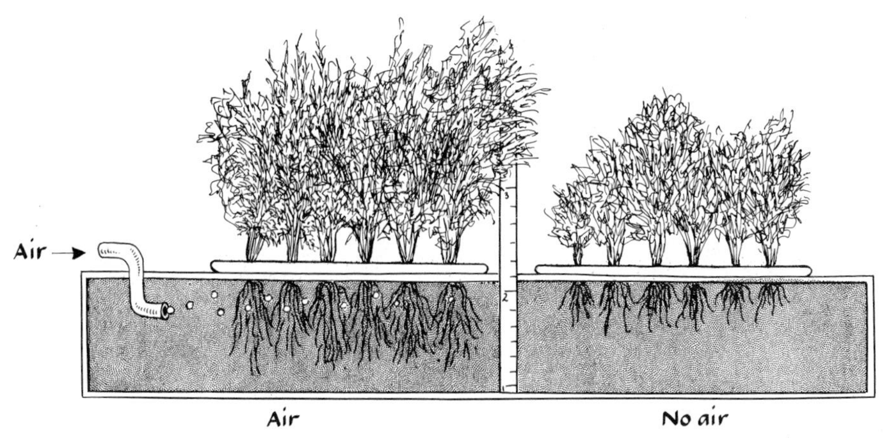

FIGURE 8-1. Effect of aeration on asparagus plants grown in a nutrient solution containing all the essential elements. [Adapted from D. R. Hoagland and D. I. Arnon, California Agricultural Experiment Station Circular 347, 1950.]

The water content of a soil may also be expressed in terms of the availability of water to the plants that grow in it. The moisture content at which irreversible wilting occurs is known as the **permanent wilting point.** The percentage of available water present depends upon the soil, but it is relatively independent of the test plant. The moisture left in the soil but unavailable to the plant is known as **hygroscopic water** and **chemically combined water.** The hygroscopic water is held tenaciously by the soil in "atomically" thin films. The amount of hygroscopic water varies with the surface area of the soil particles, and is therefore highest in clay and organic soils.

The total amount of soil moisture present is not so important as is its availability. The **available moisture** is the difference between the permanent wilting point and the field capacity. This water is often referred to as **capillary water.** It is retained in the smaller soil pores, where the capillary forces prevent water drainage, and as films around the soil particles (Figure 8-2). Soils differ in their ability to hold moisture, and this ability depends upon their texture (Table 8-1). Although sandy soils afford better drainage and aeration, they have a lower water-holding capacity than clay soils. The total amount of capillary water can be increased in a sandy soil by raising its content of organic matter. The total quantity available to a crop will depend on many factors, among which are the type and depth of soil, the depth of rooting of the crop, the rate of water loss by evaporation and transpiration, the temperature, and the rate at which supplemental water is added. In addition, the amount of available water itself is a factor: The smaller the proportion of water in a soil, the greater is the tenacity with which the water is held.

Because the rate of water extraction from soil is a function of root concentration, it decreases with the depth of the root zone. About 40 percent of the

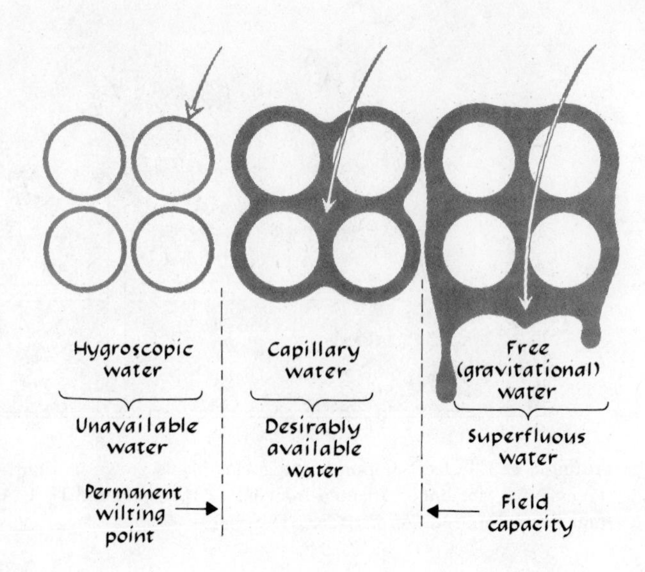

FIGURE 8-2. The classes of soil moisture. All the capillary water is not equally available to plants. As the capillary water is depleted, the tension by which this water is held in the soil increases from 1 atmosphere of pressure at field capacity to about 15 atmospheres at the permanent wilting point. The amount of capillary water present increases with the fineness of the soil pore space. [Adapted from J. Bonner and A. W. Galston, *Principles of Plant Physiology*, W. H. Freeman and Company, copyright © 1952.]

total water is extracted from the upper quarter of the root zone, 30 percent from the second quarter, 20 percent from the third quarter, and 10 percent from the bottom quarter. Under maximum transpiration, sufficient water for maximum growth cannot be obtained when the upper quarter of the root zone is depleted.

TABLE 8-1. Feel chart for the determination of moisture in medium- to fine-textured soils. With sandy soil, the balls are more friable and fragile throughout the whole range.

Degree of moisture	Feel	Percent of field capacity
Dry	Powder dry	0
Low (critical)	Crumbly, will not form a ball	Less than 25
Fair (usual time to irrigate)	Forms a ball, but will crumble upon being tossed several times	25–50
Good	Forms a ball that will remain intact after being tossed five times, will stick slightly with pressure	50–75
Excellent	Forms a durable ball and is pliable; sticks readily; a sizable chunk will stick to the thumb after soil is squeezed firmly	75–100
Too wet	With firm pressure, some water can be squeezed from the ball	In excess of field capacity

SOURCE: W. Strong, 1956, in *Sprinkler Irrigation Manual* (Wright Rain, Ringwood, England).

Measuring Soil Moisture

The amount of water in soil may be expressed in several ways. Experienced persons can approximate moisture content rather closely by feel—rubbing a small amount of soil from the root zone between the fingers to determine its consistency (Table 8-1).

The simplest way to describe soil moisture quantitatively is as a percentage of the dry weight of the soil from which it has been extracted. To determine moisture content gravimetrically, soil samples from the field are usually weighed and then oven-dried until their weight remains constant. The weight of water lost is used to compute the percentage of soil moisture.

Another useful method to state soil moisture gives the tenacity or tension with which it is bound to the soil. **Free water** is held with very low tension. **Capillary water** is bound more tightly, but not extremely so. **Hygroscopic water** is bound very tightly, and the tension required to remove it is high. Tension may be expressed in atmospheres of pressure or in one of several other units (Table 8-2).

The present terminology used by soil scientists expresses the tension or retentive force that restricts water behavior in soils in thermodynamic terms. Thus, water that moves freely in soil possesses high **potential energy** (or **high potential**); water that is restricted (i.e., under tension) in the soil has **low potential energy** (or **low potential**). A unit to express water potential is a bar, a metric unit, or a kilopasqual, an SI unit (See Table 8-2). Note that 1 bar is roughly equivalent to 1 atmosphere.

TABLE 8-2. Units for measuring water tension.

Soil water in relation to plant growth	Water tension units			
	Bars*	Kilopasquals	h = cm water	Binding force (Pf)†
Free water	0	0		
Field capacity	0.33	33		−0.48
Available water‡	1	100	1000	3
	10	1000	10,000	4
Permanent wilting point§	15	1500	15,000	4.2
Unavailable water	100	10,000	100,000	5
	1000	100,000	1,000,000	6

*A bar is roughly equivalent to an atmosphere of pressure. 1 bar = 1 atmosphere = 76 cm of mercury or 1013 cm of water. Water tension infers negative pressure so bars are not given as negatives in this context.

†Pf = $\log_{10}$ h, where h = the tension measured in centimeters of water.

‡Note that the available water is not equally available to plants.

§Some soils, as sands, may run out of water before 15 bars can be reached.

The total soil water potential (Ψ soil) is considered in relative terms. The value is set at 0 bars for free water. The energy available in soil water retained by the soil is always less than zero (i.e., negative bars), reflecting the fact that energy is required to extract it.

Various forces bind the water to the soil (often described in such terms as "tension," "suction," "capillarity," or "chemical forces"). A study of these and other forces belongs under the domain of soil physics. Suffice to say that soil water potential decreases continuously from 0 bars in the soil's wettest condition to about −10,000 bars in its driest state and that the force required to extract water from soil increases as the soil dries out.

The greatest advantage to using water potential or tension to express soil moisture is that the critical values regarding plant growth, wilting coefficient and field capacity, are the same for all soils, even though the total amount of extractable water may be vastly different. The reason is that the critical factor for plants is not the amount of water in the soil, but rather the force or work required to extract it.

The portion of water plants use, available water, is between −0.33 (−1/3) bars (*field capacity*) and −15 bars (*permanent wilting point*). Plant growth where water potential is more than −1/3 bars (i.e., closer to zero) is usually restricted because of oxygen deficiency, while water held at a potential less than −15 bars is unavailable in quantities needed to support plant growth. Thus, plants will wilt even when water remains in the soil. The actual amount of water in soils varies with soil texture, as shown in Figure 8-3. Some sandy soils can run out of water before 15 bars are reached.

The tension with which water is held in the soil can be measured with an apparatus called a **tensiometer** (Figure 8-4). When the tensiometer is filled with water and placed in moist soil, water is extracted from a porous ceramic cup until it comes into equilibrium with the soil water. The tension in the ceramic cup is read from a manometer or a vacuum gauge.

Moisture content can be estimated by measuring the electrical conductivity of a volume of soil. Conductivity is not reliable, however, because duplicating measurements is difficult and because other factors besides moisture affect it, such as great differences in texture from one spot to another and uneven distribution of soil minerals.

A rapid technique uses the ability of sorption blocks (called Bouyoucos blocks after their inventor) made of gypsum or other porous material. They are buried in the soil and absorb water in proportion to the amount present in the soil. The percentage of soil moisture is determined by weight or by direct measurement of the electrical conductance or resistance between electrodes inserted in the block.

A modern technique for measuring the moisture content of soils is to use the thermalization and scattering of fast neutrons. Fast neutrons have an energy of about 2 million electron volts, but after colliding elastically with the

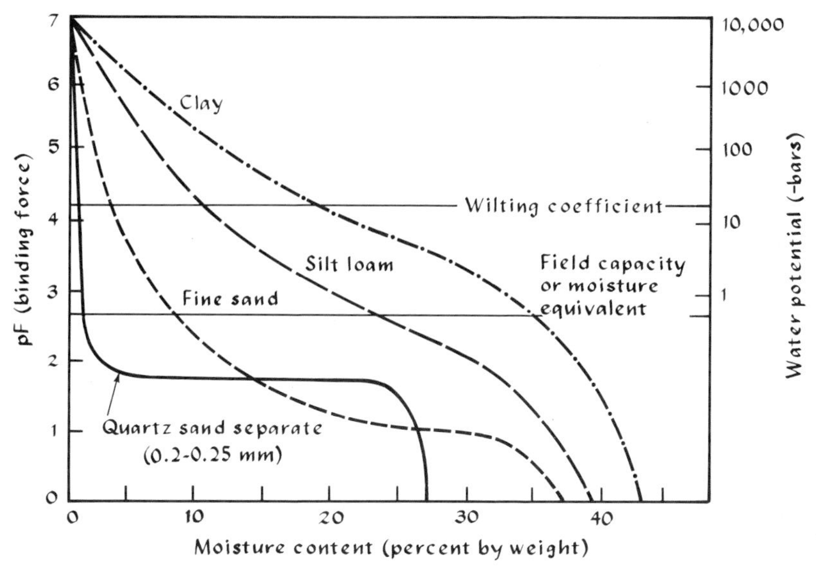

FIGURE 8-3. As the moisture content of soil increases, the soil water is less strongly bound. Note that at the wilting point and at field capacity a clay soil may contain four times as much water as a sand. [After L. D. Baver, *Soil Physics*, 3d ed., Wiley, 1956.]

hydrogen in water, the speed of the neutrons comes into thermoequilibrium with the speed of the molecules composing the material in which they are, and they become slow, or thermal, neutrons. Thermal neutrons have energies of about 0.025 electron volt. If a source of fast neutrons is placed in the soil, their number is a function of the amount of water in the soil, and is detected by a

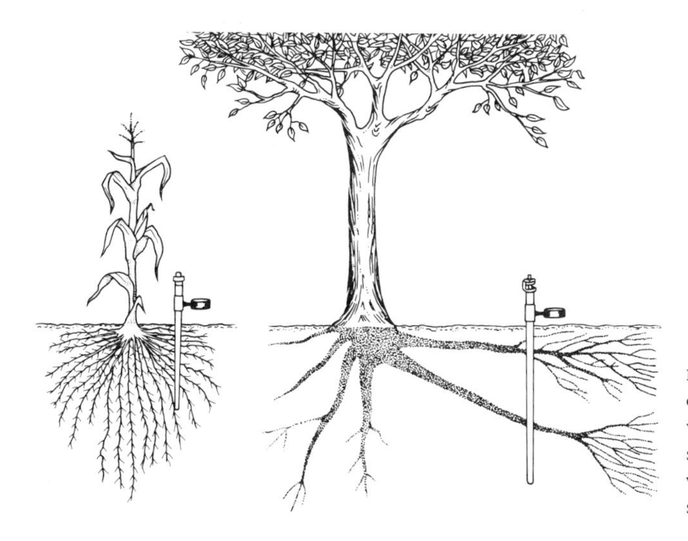

FIGURE 8-4. Tensiometers are placed in the vicinity of roots to measure the tension with which water is held in soil.

thermal neutron counter. This technique has been found useful and various devices have become commercially available. However, it does not distinguish between water in the soil, water in the roots, and hydrogen in soil organic matter.

Water Movement

The movement of water through a soil is related to the amount of water present. Water applied to a soil moves through the soil only as fast as field capacity is attained (Figure 8-5). The rate of water movement depends to a great extent on soil texture. Because the pore size is smaller in heavy clay soils, water moves much more slowly through these than through loamy or sandy soils.

The upward movement of water through a soil is caused by capillary action. Since this is a surface-tension phenomenon, the height the water will reach is inversely related to the diameter of the openings in the soil. Thus, the finer the soil spaces, the greater the distance of capillary movement. The upward rise of capillary water from the water table (the depth at which all the soil is at field capacity) is a factor in the replenishment of water lost to the plant and evaporated from the soil. This evaporative loss of water is restricted to the upper portion of the soil, since more and more pressure is needed to pull the water, depending on the height of the water column. During a period of extended drought, the shallow-rooted plants can easily be recognized.

WATER MANAGEMENT

Water management in agriculture is principally concerned with the regulation and use of water to obtain plant growth. In addition to the conversation of water, it involves controlling the effects of water excesses and water shortages. Because of the close association between water and soil, the control of moisture must be an integral part of soil management.

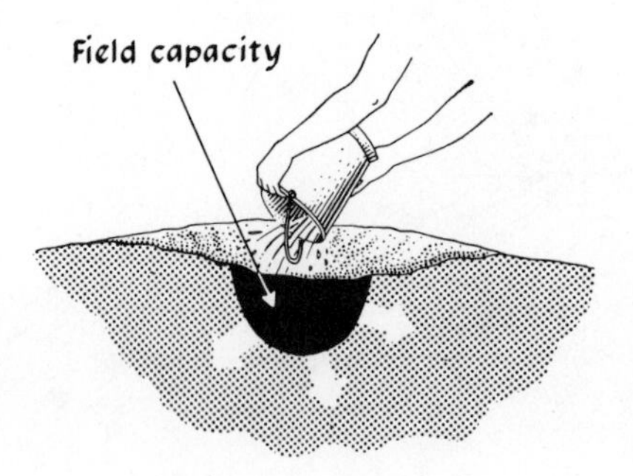

FIGURE 8-5. The rate of movement of water added to the soil is related to the speed at which field capacity is attained.

Irrigation

And a river went out of Eden to water the garden; and from thence it was parted, and became into four heads.
GENESIS 2:10

The great ancient civilization of Egypt, Babylon, China, and Peru were dependent upon irrigation for the abundant agriculture necessary to support their large populations. They developed complex systems of water distribution that involved ditches, canals, and waterways. Mechanical devices, such as the dragon wheel and Archimedes' screw, were used to move water uphill. Although irrigation has long been a vital part of the agriculture of the semiarid, subtropical climates, it has not been widely practiced in the humid temperate climates until recently. At present, supplemental irrigation is increasing as a component technology of horticulture. This development has been brought about by improvements in irrigation technology, such as increased pump efficiency, lightweight tubing, and sprinkler systems.

Types of Irrigation

There are four general methods of soil irrigation: surface irrigation, sprinkler irrigation, subirrigation, and trickle irrigation. In **surface irrigation,** the water is conveyed directly over the field, and the soil acts as the reservoir for moisture. In **sprinkler irrigation,** water is conveyed through pipes and is distributed under pressure as simulated rain. In **subirrigation,** the water flows underground as a controlled water table over an impervious substratum and provides moisture to the crop by upward capillary movement. In **trickle irrigation,** water is applied slowly to individual plants through small tubes.

SURFACE IRRIGATION. The application of water by surface irrigation is utilized in arid and semiarid regions where the topography is level (Figure 8-6). The water is conveyed to the fields in open ditches at a slow, nonerosive velocity. Where water is especially scarce, pipelines may be used, since they eliminate losses due to seepage and evaporation. The distribution of water is accomplished by various control structures or by siphons. Its flow must be carefully controlled, for water is usually fairly expensive in dry regions, and excessive leaching of water-soluble nutrients and erosion of soil may occur with too rapid a flow. Drainage canals must be provided to remove waste water and eliminate ponding. The advantages of surface irrigation over sprinkler irrigation are that it requires less power and that less water is lost to evaporation. But, since the system depends on gravity flow, it is inefficient in distribution because more water is supplied to those areas closest to the source. Another serious objection to surface irrigation is its deleterious effect on soil structure. Heavy soils become puddled under the heavy load of water, resulting in a loss of soil aeration and in subsequent baking and cracking when the soil dries out.

FIGURE 8-6. Furrow irrigation in a lettuce field in the Salinas Valley, California. [Photograph by Ansel Adams, courtesy Wells Fargo Bank, San Francisco.]

The distribution of water over the field in surface irrigation may be accomplished either by flooding the entire field in a continuous sheet (**flood irrigation**) or by restricting the water to some type of furrow (**furrow irrigation**). The field must be almost level, or low spots will get too much water and high spots will get none. Flood irrigation is common in horticultural crops that are tolerant of excessive water, as are cranberries. Furrow irrigation is the most widely used method of applying water to row crops. Although it is fairly efficient in water utilization, it requires high labor costs. Constant supervision is necessary to prevent furrow streams from uniting and forming large channels. On rolling land and with closely growing crops, water may be applied to small furrows that guide, rather than carry, the water. This approach is called **corrugating irrigation. Soakers** (perforated hoses that allow water to seep into the soil at a low, uniform rate) provide a form of surface irrigation that can be employed in greenhouses and in homes and gardens.

SPRINKLER IRRIGATION. Sprinkler irrigation, although not new, has come into widespread use since 1945, owing to the introduction of light aluminum pipe, quick couplers, and improved nozzles. Until then, sprinkler systems consisted of permanent "overhead" installations that were confined to use in small market gardens. Portable, lightweight pipe has put sprinkler irrigation into a prominent position in agricultural technology. It has proved practical as a means of providing supplemental water in the so-called humid climates. Its advantages lie in its even and controlled application rate. Although in arid climates, evaporation may be higher with sprinkler irrigation than with surface irrigation, the controlled application rate results in a more efficient use of water. The slower rate of application reduces runoff, erosion, and compaction of the soil. In addi-

FIGURE 8-7. Three large diesel engines of this type provide power for sprinkler irrigation on a large Midwestern sod farm. [Photograph courtesy International Harvester Corp.]

tion, sprinkler irrigation can be used on land that is too steep to permit the proper use of other methods, and makes full use of land by eliminating the need to devote some space to a permanent distribution system. The limitations of sprinkler irrigation lie in the high initial cost, although operational labor costs are reduced. The power requirements are high, for water pressures of 15 to 100 pounds per square inch (psi) must be maintained (Figure 8-7). The chief operational difficulty is high wind, which disturbs the sprinkler patterns and results in uneven water distribution.

Sprinkler-irrigation equipment operates with nozzles or perforated pipe. Rotating sprinkler heads, the most widely used type, apply water to circular areas at rates from 0.2 inch per hour to more than 1.0 inch per hour (Figure 8-8). In order to obtain proper coverage, each wetted circle should overlap adjacent circles from one-fourth to one-half their area.

SUBSURFACE IRRIGATION. Subsurface irrigation consists of creating and maintaining an artificial water table. In order for such a system to function properly, the ground must be level, and subsurface soil must be permeable enough to permit the rapid movement of water laterally and vertically. A bottom barrier, such as an impervious layer of soil or rock, must be present to prevent the loss of water through deep percolation. A distribution system of ditches and laterals permits the artificial water table to be raised or lowered by pumping water into or out of the system. There are, however, relatively few places in the United States where these specialized conditions exist. A good example is the San Luis Valley of Colorado, an important potato-producing area. Subirrigation is also fairly common in organic soils.

The exact water-table level in organic soils can be very important. For most plants, the most favorable growth is achieved with a water table 24 to 36 inches from the surface of the soil, but such crops as mint and celery require a water

FIGURE 8-8. Sprinkler irrigation is practical as a result of portable, lightweight, aluminum pipe. The sprinkler pattern must be overlapped by about 40 percent in order to achieve uniform application of water. [Photograph courtesy USDA.]

table about 18 inches from the surface. The water table also affects organic soil subsidence, which is the "loss" of soil due to drying, settling, compaction, slow oxidation, and wind erosion. The rate of subsidence increases as the water table drops. One of the problems associated with a high water table, however, is caused by excessive, unexpected rainfall. Unless the water table can be quickly lowered, uncontrolled flooding may result and cause extensive crop damage.

TRICKLE IRRIGATION. Trickle irrigation (also called drip, high-frequency, or daily-flow irrigation) employs a ramifying system of water-conducting plastic tubes that deliver water, by means of "emitters" or leads, to individual plants. Water is applied frequently, but very slowly and in small amounts so that little is lost by evaporation or by flowing to portions of the field not occupied by plant roots. Water is generally applied at a rate of 1 to 2 gallons per hour under low pressure (15 psi or less). Since moisture stress tends to equalize throughout the plant, as little as 25 percent of the root system need be wetted.

Trickle irrigation is an extremely versatile method. It has proved useful in the floriculture industry to automate the watering of plants in individual pots (Figure 8-9). In the greenhouse, each terminal tube of the system ends in a valve that permits individual control and is weighted to keep it in place in the pot. The entire system can be automated for watering and fertilizing with a timing or sensing device (such as a scale to weight the watered pot).

Trickle irrigation has proved to be of greatest value to orchard systems, although it is also used extensively for strawberries and row vegetable crops. The system is flexible because fast and simple coupling and uncoupling are accommodated by either piercing or plugging the plastic tubes that carry the water. Various types of emitters precisely regulate the flow and reduce the pressure from the "leader" line. A popular type of emitter is a thin tube or pathway through a nozzle with an inside diameter of about 0.9 mm (0.036 inch). The length of the tube regulates the pressure—the longer the tube, the slower the rate of delivery. In this way, the delivery of water can be equalized to plants located at different distances from the water source.

Trickle irrigation is attractive economically in many horticultural situations. The principal benefit is that plants can be supplied with very nearly the precise amount of water they require. In arid regions, trickle irrigation may have dramatic consequences because evaporation from traditional irrigation systems causes salts to accumulate on the soil surface. Additional water supplied by trickle irrigation in excess of the amounts required for plant needs will leach these potentially damaging salts from the soil zone close to the roots. Providing this water by trickle irrigation is far more efficient than either wetting the entire field or supplying water through furrows. The waste of water between rows and through deep percolation is eliminated, which eliminates fertilizer contamination of deep underground water supplies.

When saline water is used for irrigation, salt buildup can be controlled by continuous leaching. Salts are pushed out to the periphery of the root profile by an advancing front of water discharged from the emitters. The roots are able to take up water freely from the middle of the wet zone, where soil-moisture tension is low and the salt level remains nearly the same as it is in the irrigation water.

The delivery system of trickle irrigation is very light compared with other systems, and can easily be moved by one person. Fertilization is efficiently combined with irrigation by the addition of soluble fertilizers to the system. Trickle irrigation has been used on slopes as high as 50 percent without erosion problems because delivery is in a closed system under pressure, only small amount of water are applied, and there is no runoff.

Because most of the soil surface stays dry, especially between plants, the crop field is essentially free of weeds. These can be a problem in wet areas near emitters, but they can be controlled by selective herbicides.

Some of the problems of drip irrigation include the high costs of installation and difficulties in keeping equipment operational. Bacterial slime may build up in the plastic tubes, and small soil particles may block the water outlets, because the low water pressure used in the system is frequently insufficient to unclog them. In some areas, ants seeking water cut the tubes open, and rodents have been known to use the plastic tubing to sharpen their teeth. Additives in water to decrease bacterial growth, and in the plastic tubing to repel insects and mammals, should solve these problems.

Trickle irrigation, first developed in Israel, has gained wide acceptance worldwide, especially in arid areas with high labor and water costs. It is now the fastest growing method of irrigation, with extensive acreage in the United States (particularly California), Australia, Israel, South Africa, and Mexico. In the United States, trickle irrigation increased from barely 100 acres in 1960 to more than 800,000 acres in 1985.

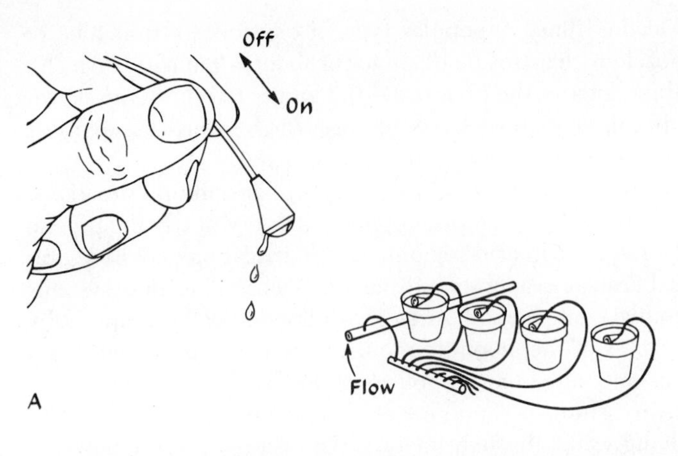

FIGURE 8-9. Trickle irrigation: (*A*) The Chapin System of trickle irrigation for greenhouse watering uses weighted valves (*left*) to deliver water to individual pots (*right*). (*B*) Trickle irrigation installation for tomatoes in a plastic greenhouse. (*C*) Trickle irrigation systems used in the field. Delivery systems may be modified for different crops. In an orchard densely planted with small trees, pipes can be laid parallel to the rows of trees and fitted with numerous emitters. With large trees, the pipes can be looped around each tree and fitted with several emitters per loop. For row crops, the pattern may be one pipe per row or one pipe in every other row. (*D*) Nozzlelike emitters reduce water pressure by causing the water to move along an elongated helical path. The emission holes are made larger when pressures are low so that clogging is reduced. The emitters are preset during construction to emit from 1 to 2 gallons of water per hour. The hook at the right attaches to the delivery pipe at a right angle; water enters the conical tip from the center of the pipe. [Courtesy Merle Jensen; Parts (*C*) and (*D*) after Shoji, "Drip Irrigation," copyright © 1977 by Scientific American, Inc., all rights reserved.]

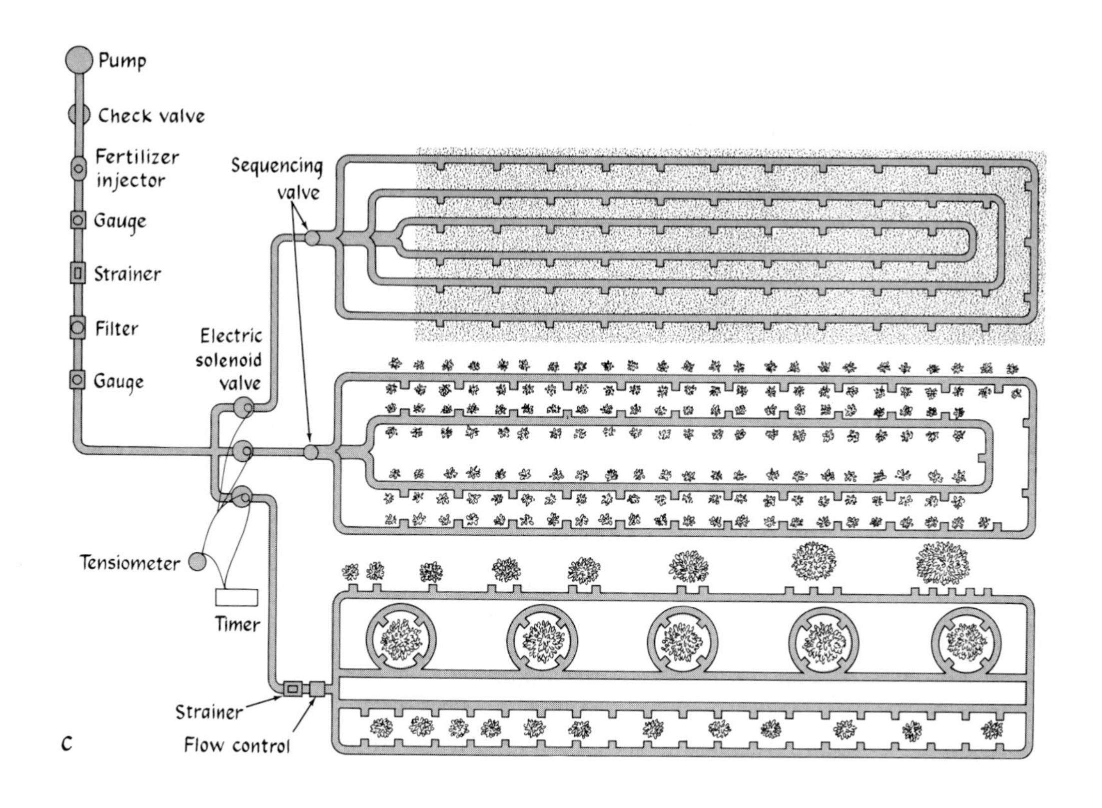
Pump
Check valve
Fertilizer injector
Gauge
Strainer
Filter
Gauge
Sequencing valve
Electric solenoid valve
Tensiometer
Timer
Strainer
Flow control
C

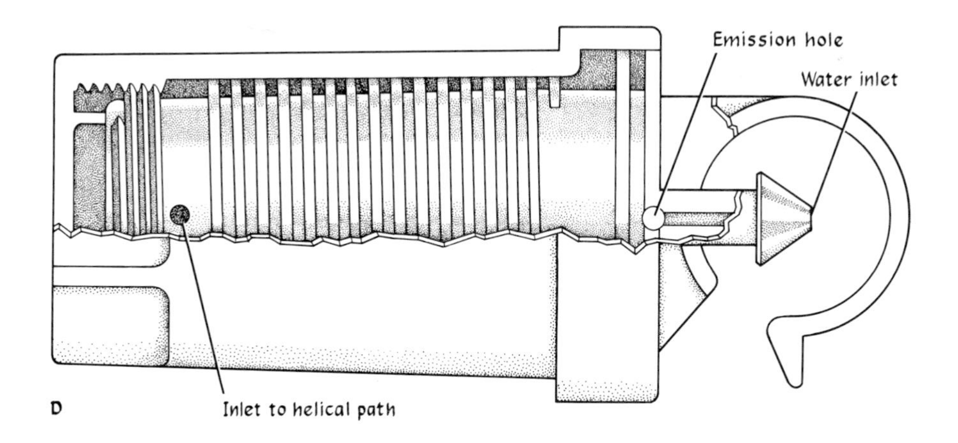
Emission hole
Water inlet
Inlet to helical path
D

Determining Irrigation Requirements

Irrigation, though capable often of yielding enormous benefits, can also be a wasteful and harmful practice if applied incorrectly. Determining when to water and how much water to supply is one of the main problems in irrigation. It requires the establishment of an accounting system of sorts that will indicate whether or not the available water will meet crop needs in spite of losses due to evaporation, transpiration, runoff, and percolation. The net deficiency not compensated by natural precipitation may be made up by irrigation. Timing can be extremely critical In snap beans, for example, moisture stress during flowering and pod formation seriously depresses the yield by promoting flower abscission and ovule abortion.

There are two approaches to the determination of irrigation requirements. One is the measurement of soil moisture, from which the moisture available to the plant is determined. The other is calculation of water availability from meteorological data.

Measuring soil moisture accurately is essential for determining irrigation requirements. At present, the gravimetric method, although slow and laborious, must be considered the most accurate. The tensiometer and other rapid moisture-measuring devices have not proved altogether successful, owing in part to the random variation of soil moisture, the difficulty of achieving intimate contact of a sorption block with the soil, and the problem of determining the best place in the root zone for making the measurement.

Meteorological and climatological data offer a powerful tool for estimating the amount of available water. The procedure involves the calculation of **consumptive use**—the water lost by evaporation and transpiration—which is probably the best index of irrigation requirements. Consumptive use varies with a great number of factors: temperature, hours of sunshine, humidity, wind movement, amount of plant cover, the stage of plant development, and available moisture.

High rates of water consumption are associated with a high percentage of plant cover and with hot, dry, windy conditions. Because optimum plant health is associated with an adequate, uninterrupted supply of water, the peak requirements must be considered. Crops have the highest water requirements in the fruiting or seed-forming periods (Figure 8-10). An approximate relationship between peak moisture use by horticultural crops and climate is given in

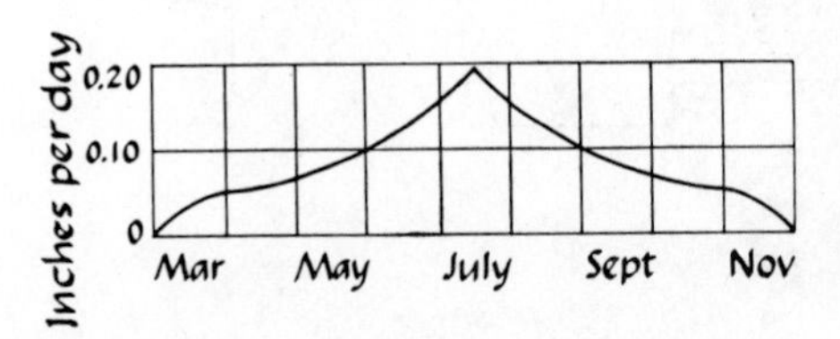

FIGURE 8-10. Rate of water use by peaches in a cool, dry climate. [Adapted from W. Strong, in *Sprinkler Irrigation Manual*, Wright Rain, Ringwood, England, 1956.]

TABLE 8-3. Average peak moisture use (in inches of water per day) for commonly irrigated horticultural crops.

Crop	Cool climate		Moderate climate		Hot climate	
	Humid	Dry	Humid	Dry	Humid	Dry
Potato	0.10	0.16	0.12	0.20	0.14	0.24
Tomato	0.14	0.17	0.17	0.22	0.23	0.27
Bean	0.12	0.16	0.16	0.20	0.20	0.25
Vegetables	0.12	0.15	0.15	0.19	0.20	0.23
Deciduous orchard	0.15	0.20	0.20	0.25	0.25	0.30
Deciduous orchard with cover	0.20	0.25	0.25	0.30	0.30	0.35
Citrus orchard	0.10	0.15	0.13	0.19	0.18	0.23

SOURCE: W. Strong, 1956, in *Sprinkler Irrigation Manual* (Wright Rain, Ringwood, England).

Table 8-3. Plants differ in their water requirements largely in relation to their ground-covering ability and their depth of rooting (Table 8-4). The relationship between water requirements and depth of rooting may be somewhat less than expected because deep-rooted plants obtain most of their moisture from the upper part of the root zone.

The amount of available moisture provided by rainfall may be very much less than the total rainfall. Owing to evaporation from the soil and the slow rate of infiltration, showers of less than ¼ inch during hot summer days may contribute very little to available soil moisture. On the other hand, a high proportion of water from heavy precipitation may be lost by runoff. The effectiveness of precipitation, therefore, depends upon the intensity of rainfall, as well as upon the amount, in relation to temperature and the absorbing capacity of the soil.

It has been possible to determine a satisfactory consumptive-use index for a particular area by using monthly averages of mean temperature and hours of sunshine. Empirically derived constants are available for adjusting these values for different crops. Any water deficit can be calculated by subtracting the consumptive-use requirements from the available water present. Potential evapotranspiration (the combination of evaporation and transpiration when the surface is completely covered with vegetation and there is an abundance of moisture) can be calculated by using the Bellani black-plate atmometer, a relatively simple instrument used to measure evaporation (Figure 8-11), and by estimating the percentage of ground cover.

Because not all of the irrigation water applied is available to the crop, the amount applied must be based on **irrigation efficiency**, the percentage of applied irrigation water that actually becomes available for consumptive use.

TABLE 8-4. Normal root-zone depths of mature irrigated crops grown in a deep, permeable, well-drained soil.

Crop	Root depth (feet)	Crop	Root depth (feet)
Alfalfa	5–10	Grass pasture	3–4
Artichoke	4	Ladino clover	2
Asparagus	6–10	Lettuce	½
Bean	3–4	Onions	1
Beet (sugar)	4–6	Parsnip	3
Beet (table)	2–3	Pea	3–4
Broccoli	2	Potato	3–4
Cabbage	2	Pumpkin	6
Cantaloupe	4–6	Radish	1
Carrot	2–3	Spinach	2
Cauliflower	2	Squash	3
Citrus	4–6	Sweetpotato	4–6
Corn (sweet)	3	Tomato	6–10
Corn (field)	4–5	Turnip	3
Cotton	4–6	Strawberry	3–4
Deciduous orchard	6–8	Walnut	12+
Grain	4	Watermelon	6

SOURCE: D. G. Shockley, 1956, in *Sprinkler Irrigation Manual* (Wright Rain, Ringwood, England).

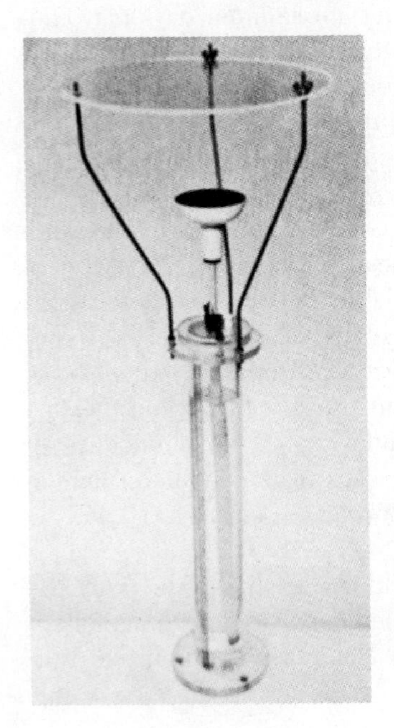

FIGURE 8-11. The Bellani black-plate atmometer measures evaporation. Potential evapotranspiration can be calculated by taking readings with this instrument and estimating total plant cover. [Photograph courtesy W. H. Gabelman.]

Water should be applied to bring the soil up to field capacity at a depth commensurate with the bulk of the feeder root system. The rate must be consistent with the absorptive properties of the soil. Irrigation is best applied when the water tension in the zone of rapid water removal goes above four atmospheres or when 60 percent of the available water in the root zone is depleted. The amount of water that can be efficiently utilized is primarily related to the level of soil fertility. The maximum benefits of irrigation are dependent upon the existence of a readily available nitrogen supply.

Drainage

Drainage is the removal of excess gravitational water from the soil. Under conditions of good natural drainage, surplus surface and soil water is rapidly removed to streams and rivers. The poor natural drainage of some areas is a result of several factors. Such areas may have a high natural water table caused by an impervious layer that prevents downward percolation, resulting in **waterlogged soils.** Others may simply be low-lying in relation to surrounding drainage. Some areas are subject to flooding brought about by the overflow of streams and rivers. Flooding can be averted either by building protective levees or by controlling the rate of water movement. This can be accomplished, as part of a program of upstream watershed management, by controlling excess runoff or by constructing dams and reservoirs to restrict the flow in times of excess water movement.

The facilitation of natural drainage is both a land-reclamation practice and a cultural practice. The permanent drainage of wetlands has been a significant factor in the expansion of agriculture in the eastern United States, where some of the most productive cropland was formerly "worthless" marsh and swamp. Not all wetlands are suitable for drainage, but they still remain valuable for wildlife, forest, and recreational use. As a cultural practice, drainage consists of removing the excess water that interferes with plant development and with the performance of such operations as tillage and harvesting. It is necessary when the natural removal of water by runoff, percolation, and evapotranspiration is too slow. Drainage extends the potential growing season by permitting earlier tillage in lands that are otherwise too wet in the spring.

Excess water can be removed by surface or subsurface drainage. Surface drainage refers to the removal of surface water by developing the slope of the land. Subsurface drainage is accomplished by the construction of open ditches and tile fields to intercept ground water and carry it off. The water enters the tiling through the joints, and drainage is achieved by the effect of gravity on the water tiles. Drainage design, which requires determining the depth, size, and number of drains to be installed, is an application of the physics of groundwater movement.

Water Conservation

Water conservation is of national concern. For a nation to prosper, an abundant source of high-quality water must be available for agricultural and industrial use as well as for human consumption and sanitation. The misuse of water resources leads to alternate flood and drought, problems that affect all of us.

Water conservation implies the proper stewardship of our water resources as a whole. It may involve large programs to control flooding, to develop hydroelectric power, and to facilitate navigation. These are projects that require national effort. Water conservation also involves the control of water resources on a smaller scale. It must therefore be a part of the water management of every individual enterprise.

Soil-management practices developed for the efficient utilization of water involve the control of soil erosion and the conservation of soil moisture through the control of runoff and the implementation of methods to increase the water-absorbing capacity of the soil. It may include such practices as mulching or the close mowing of sod in orchards, both of which are designed to reduce the removal of water from the soil. Because horticultural crops are great users of water, the extreme practices of dryland farming, such as fallowing to conserve moisture without resorting to irrigation, are usually not practical. In fallowing, the ground is left unplanted for a whole year and is cultivated only to eliminate weeds, in an effort to build up soil moisture. For horticultural crops to be grown where water is insufficient, irrigation is essential. But irrigation depends on large sources of water; a source of water supply must be developed if organized irrigation facilities are not available and if lakes or streams with sufficient flow do not adjoin the property. Irrigation wells offer one possibility. These are large-volume wells capable of supplying great quantities of water. Storage ponds or reservoirs are becoming increasingly important as sources of irrigation water. They are usually made by constructing an earthen dam across a gulley or an intermittent or spring-fed stream (Figure 8-12).

The problem of water rights has social as well as economic implications, which are reflected in our laws. The **Riparian law**, the common law involving water rights with respect to rivers and streams, has established a legal framework for disputes concerning water diversion and distribution. In this common law, property rights do not involve complete water rights except for personal use. Neither the landowner nor anyone else owns the water or may divert it from its normal flow. However, because of differences in water availability from one region to another, the common law has been modified throughout the United States. The right to use water from streams for irrigation is variable and depends on state law. Similarly, the right to pump underground water for irrigation differs widely from state to state. The legal codes must be clearly understood in situations concerning irrigation and drainage procedure.

FIGURE 8-12. Artifically created storage ponds are an important source of irrigation water. [Photograph courtesy U.S. Soil Conservation Service.]

Selected References

Elfing, D. C., 1982. Crop response to trickle irrigation. *Horticultural Reviews* 4:1–48. (A current review of trickle irrigation with special reference to horticulture.)

Hagan, R. M., H. R. Haise, and T. W. Edminster (eds.), 1967. *Irrigation of Agricultural Lands.* Madison, Wis.: American Society of Agronomy. (A broad compilation of irrigation practices.)

Hansen, V. E., O. W. Israelsen, and G. E. Stringham, 1980. *Irrigation Principles and Practices,* 4th ed. New York: Wiley. (A good reference on irrigation practices.)

Teare, I. D., and M. M. Peet (eds.), 1983. *Crop-Water Relations.* New York: Wiley. (Physical and physiological aspects of water transfer and the water relations of 12 specific crops including bean and potato.)

U.S. Department of Agriculture, 1955. *Water.* (USDA Yearbook, 1955.) (An excellent discussion of water in relation to agriculture.)

Wright Rain Ltd., 1956. *Sprinkler Irrigation Manual.* Ringwood, Hants., England: Wright Rain. (A compilation by English workers.)

9 *Light*

SOLAR RADIATION

The sun is the primary source of energy available to earth. This energy is transformed across 150 million km (93 million miles) of space in the form of radiation. **Solar radiation** has two phases, one electric and one magnetic, and unless deflected, travels in a straight line at 300,000 km (about 186,000 miles) per second. Note that the radiant energy occurring as visible light is but a small fraction of the frequency range of a continuous electromagnetic spectrum (Figure 9-1).

Radiation of the electromagnetic spectrum is described in two seemingly contradictory models—a wave model and a particle model. In the wave model, **radiant energy** is described in terms of its ***wavelength***, the distance between successive peaks or troughs, and its ***frequency***, the number of cycles per unit time. Wavelength is inversely proportional to frequency. Radiant energy can also be defined as bundles or packets of energy called **photons** or **quanta.** Radiant energy varies inversely with frequency; the product of energy and frequency is a constant equal to the velocity of light. Thus, energy is highest in shortwave, high-frequency cosmic rays and lowest in longwave, low-frequency radio waves.

Solar radiation received at the edge of the earth's atmosphere has a frequency range from about one-millionth of a nanometer (nm = 10^{-9} meters) to

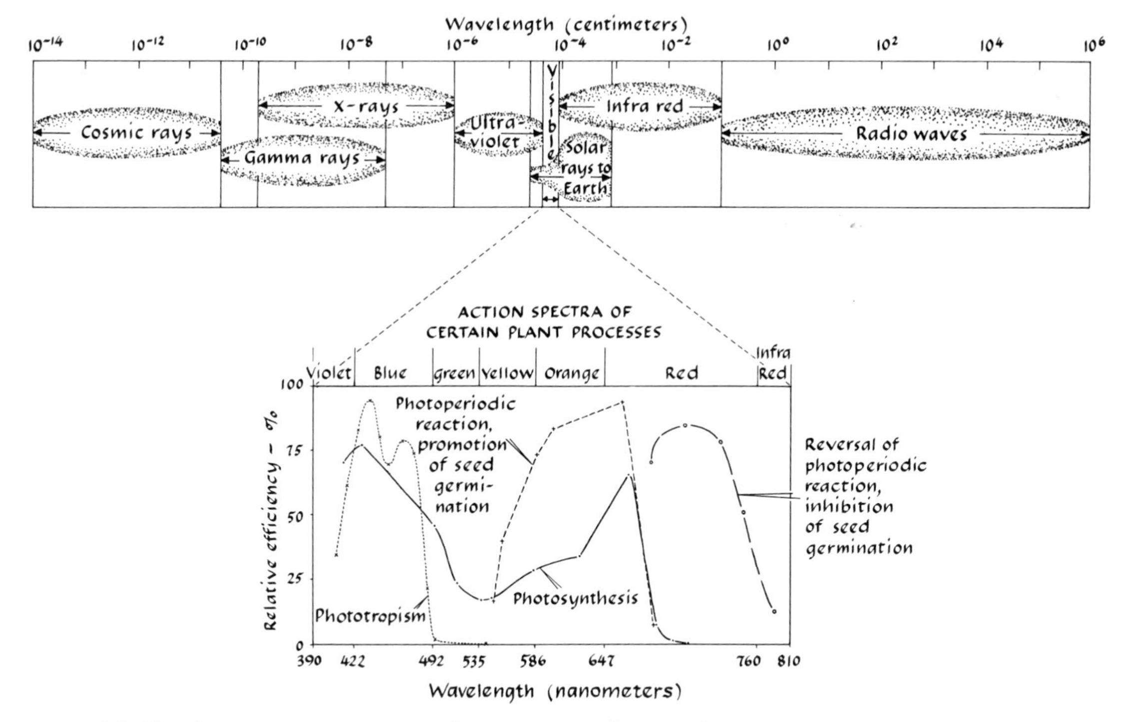

FIGURE 9-1. The electromagnetic spectrum and action spectra of certain plant processes. [Adapted from L. Machlis and J. G. Torrey, *Plants in Action*, W. H. Freeman and Company, copyright © 1959.]

over 3 billion nm. As the energy encounters the earth's atmosphere, various gases screen out the incoming radiation (Figure 9-2). For example, ozone absorbs and thus filters out much of the damaging shortwave, high-energy ultraviolet, and water vapor filters out much of the far red and infrared by diffusing, reflecting, and absorbing this radiant energy. The radiation that pierces the atmospheric window to be intercepted by the earth's surface is most abundant in the 400 to 700 nm range—the range that the human eye perceives as light and that plants use in photosynthesis.

The amount and quality of solar energy that any portion of the earth's surface receives depend on the energy's duration and intensity. The duration of intercepted radiation changes with the seasons because of the variation in day length and cloud cover. The intensity of the intercepted radiation is related to the angle at which the solar rays penetrate the earth's atmosphere. Because the earth is spherical, the rays of the sun falling on the poles are oblique as compared to those falling on the equator (Figure 9-3). These rays are spread over a larger surface of the earth and pass through a thicker layer of atmosphere, which filters out a larger quantity of rays. When the sun is directly over the equator, its rays must penetrate an air mass at the poles equivalent to 45 air masses at the equator.

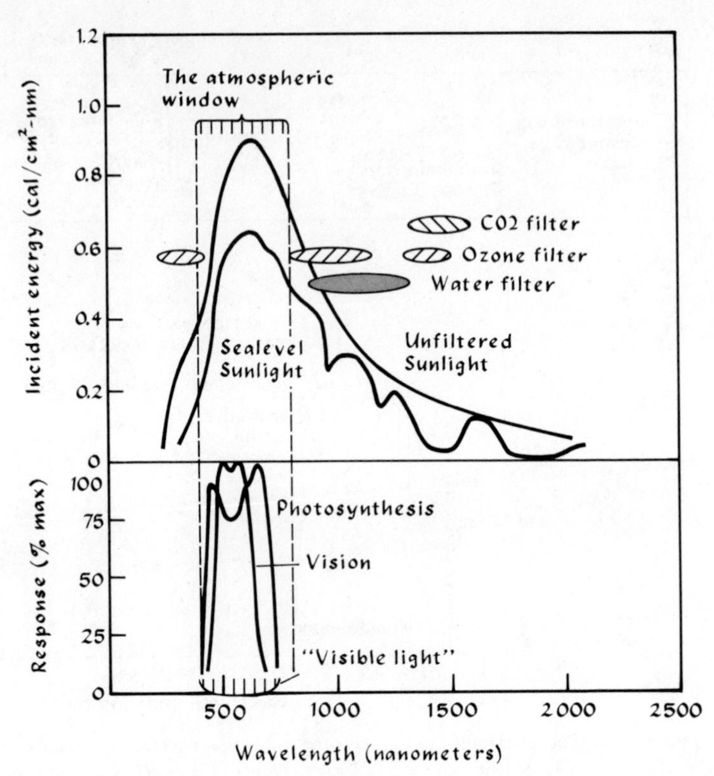

FIGURE 9-2. The incident radiation from the sun is most abundant in the wavelengths used for such biological reactions as photosynthesis and vision. [From: A. C. Leopold and P. E. Kriedemann, *Plant Growth and Development*, 1975, McGraw-Hill.]

Absorption of Solar Radiation

The quantity and quality of radiation depend on the temperature of the radiating body. The higher the temperature, the greater the rate of radiation and the richer the proportion of short wavelength (high-frequency) radiation. Thus, the high-temperature solar radiation consists mostly of shortwave radiation in the visible or near-visible portions of the spectrum. This shortwave solar energy is absorbed at the earth's surface, where it is transformed into heat. The earth

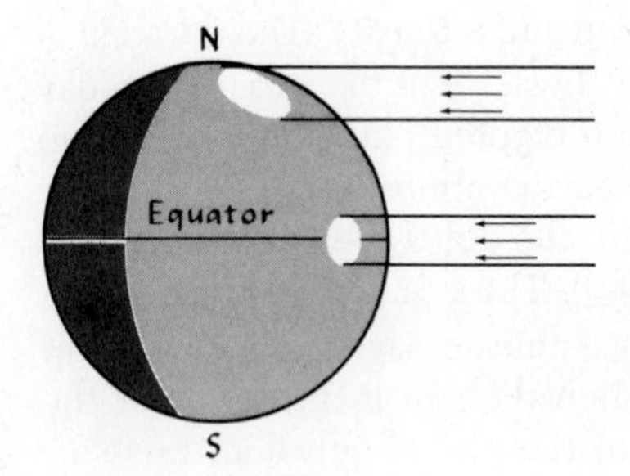

FIGURE 9-3. The effect of latitude on the dispersion of solar energy over the earth's surface during the equinox. [After W. E. Reifsnyder and H. W. Lull, "Radiant Energy in Relation to Forests," in *USDA Tech. Bull.* 1344, 1965.]

then becomes a radiating body at a lower temperature (average, 57°F, or 14°C). The earth's radiation is in the form of long waves (low-frequency radiation).

Water vapor, the most significant of the atmosphere's absorbing gases, absorbs only about 14 percent of the incoming shortwave radiation but 85 percent of the earth's longwave radiation. This phenomenon tends to keep surface temperatures much higher than they otherwise would be. The atmosphere thus acts as a pane of glass, transparent to the sun's short waves, but opaque to the earth's long waves; hence, the name *greenhouse effect* for this phenomenon (Figure 9-4). The earth therefore receives most of its heat only indirectly from the sun.

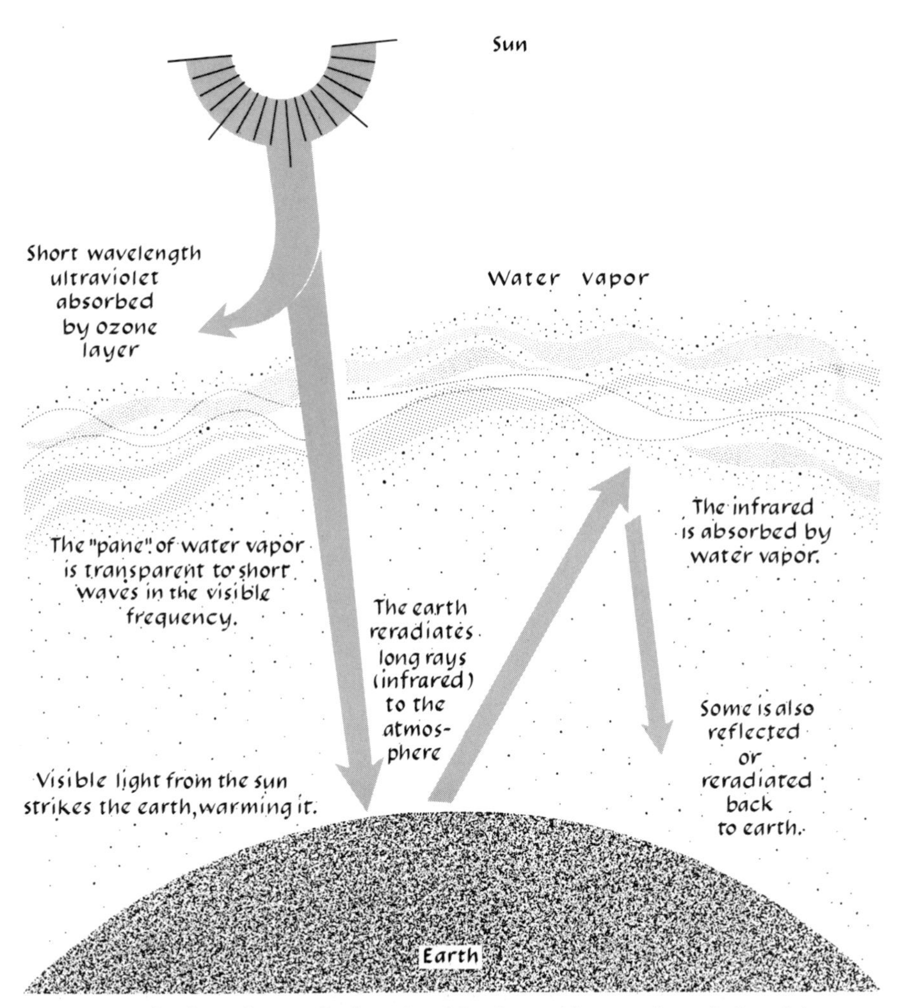

FIGURE 9-4. The "greenhouse effect" produced by the earth's atmosphere. [Adapted from A. W. Galston, *Life of the Green Plant*, Prentice-Hall, 1961.]

MEASURING RADIANT ENERGY

Solar radiation may be measured in various units. Unfortunately, the many choices lead to confusion because some units are widely used by engineers, others are used principally by plant scientists, and many are relatively new. In horticulture, radiant energy may be discussed in terms of heat energy, light energy, photosynthetic energy, electrical energy, or chemical energy.

For engineering purposes, radiant energy is often expressed as heat energy. The familiar Btu (British Thermal Unit) is the energy required to raise the temperature of 1 pound of water 1 degree Fahrenheit. Its metrical equivalent, the gram calorie (calorie with a lowercase *c*) is the amount of heat energy necessary to raise the temperature of 1 gram of water 1 degree Celsius (specifically, from 14.5 to 15.5°C): 1 Btu equals 252 gram calories. Note that 1 kilocalorie (sometimes written with a capital *C*) equals 1000 gram calories. Energy may also be expressed as the amount of heat energy per unit area: 1 langley equals 1 gram calorie per square centimeter.

In the new International System (SI) of units, the universal unit for energy is the joule (J) whether it be heat energy, radiant energy, or food energy.

$$1 \text{ Joule} = 1/4.1868 \text{ calories} = 0.2388 \text{ calories}$$

$$1 \text{ calorie} = 4.1868 \text{ joules}$$

The electrical term watt (W) corresponds to joules per second:

$$1 \text{ watt} = \text{Js}^{-1}$$

Note the joule is a measure of the amount of energy, whereas the watt measures the rate of energy, that is, energy per unit time.

Light and Vision

Light quantity or intensity refers to the concentration of light waves, that portion of the electromagnetic spectrum perceived by the human eye. The wavelength most sensitive to the human eye is 555 nm, although the eye can detect ranges from 380 nm to 780 nm. Various photometric units may be used to describe light intensity.

In the English system, the footcandle (*abbr.*, ft-c) originally designated the amount of illumination (luminosity) shed by a standard candle and intercepted on 1 square foot of surface with a curvature radius of 12 inches.

The lux (*abbr.*, lx) is the metric counterpart of the footcandle (1 lx =

10.8 ft-c). Terminology for light has undergone transformation since the adoption of SI units. "Light intensity" (which is measured in lumens, *abbr.*, lm) is defined as the radiant flux emitted from a light source, while "illuminance" (which is measured in luxes, or lm/m^2) is defined as the radiant flux intercepted per unit area. When reporting illuminance, the measured range of wavelength should be given (400 to 700 nm is the photosynthesis range).

Light and Photosynthesis

Although the wavelength stimulating vision straddles the same bands (about 400 to 700 nm) that are effective in photosynthesis, the action spectra are not identical (Figure 9-2). That is, how the human eye "sees" light differs from how chlorophyll responds to light. The part of the radiant energy spectrum effective in photosynthesis is called photosynthetically active radiation (PAR). Instruments that measure light for the human eye (as photographic light meters do) cannot be precise instruments for plant research. New terms have been developed to accommodate the energy associated with the twin peak wavelengths involved in photosynthesis.

The energy associated with photosynthetic radiation per unit time and per area is identified as photosynthetic photon flux density (PPFD). The energy is described in quantum units as einsteins per second per square meter ($E = s^{-1}m^{-2}$) where an einstein (E) is the energy in 1 mole or Avogadros number of photons (6.02×10^{23}). Because the einstein is not an SI unit, the mole has been substituted: 1 einstein equals 1 mole.

The PPFD is thus the energy associated with the precise wavelengths responsive in photosynthesis and is obtained by adding (i.e., integrating) all the energy at each wavelength in the action spectra for photosynthesis.

Many kinds of instruments have been devised for measuring solar radiation. Most are useful for measuring a particular narrow range of energies. Instruments useful for measuring all wavelengths of solar radiation are radiometers. They are generally sensitive in the range from 300 to 100,000 nm. Net radiometers contain two sensitive elements, one facing the sun and the other facing the earth. Longwave radiation from the earth is subtracted from incoming radiation to give net values.

One of the standard instruments used for measuring direct incoming radiation from the sun is the Eppley normal incidence pyrheliometer. It is designed so that a blackened surface containing 15 thermocouples intercepts a narrow beam of solar radiation and transforms it into heat energy. This heat energy causes the flow of a small current that is measured by a meter.

It is frequently desirable to measure the radiation impinging on a hemispherical surface, for this is the geometry to which a plant is exposed. An Eppley pyrheliometer is one of the instruments used for this purpose. It has a

16-junction thermopile that is exposed to all radiation passing through a hemispherical glass bulb 7.6 cm (3 inches) in diameter. Other instruments that measure radiation are based on either the differential bending of metals or the evaporation of liquids.

The intensity of visible light may be measured by various techniques and devices, such as photocells, exposure meters, photographic paper, and extinction meters. Instruments designed to measure visible light are called photometers or light meters. They are standardized to respond to 555 nm, as does the human eye. For plant research, a spectral radiometer, which measures radiant energy received at various wavelengths, can be used to describe the quality and quantity of radiation. In most horticultural research, photosynthetic radiation is measured by a quantum sensor and expressed as μmol sec^{-1}m^{-2} (micromoles per second per square meter) in the wavelength between 400 and 700 nm. Even this measurement is not precise because the radiation effective in photosynthesis is between 400 and 700 nm with peaks at 400 to 510 nm and 610 to 700 nm. Put another way, the green leaf "sees" two wavelength peaks best in respect to photosynthesis. The green color of plants is due to the fact that chlorophyll absorbs light in the violet and blue wavelengths (400 to 510 nm) and in the red wavelength (610 to 700 nm) and transmits and reflects the wavelength between 510 and 610 nm, which we perceive as green. Furthermore, the chlorophylls of higher plants absorb higher in the red than in the blue wavelengths. A meter that senses precisely the intensity of wavelength used in photosynthesis has yet to be developed. For practical purposes, light meters and photosynthetic meters are roughly comparable in ordinary outdoor environments. However, they diverge when determining energy from lamps because many lamps do not have continuous spectra but contain energy at very narrow bands called line spectra.

Most lamps used for human vision can be used for regulating plant growth. This energy is usually expressed in electrical energy, watts per square meter (Wm^2 = Js^{-1}m^{-2}). To compare lamp efficiency for plant growth, however, only energy in the region of 400 to 700 nm must be considered.

Converting the various energy units for light is not straightforward. Many horticulturists are familiar with the footcandle even though this unit is inappropriate for measuring anything but visual light. Furthermore, the footcandle is not a metric unit. It is useful to remember that full sun is about 10,000 ft-c, overcast weather is about 1000 ft-c, and indoor light near a window is about 100 ft-c (Table 9-1).

To convert footcandles to lux, multiply by 10.8. Remember 1000 lux equals 1 klx.

$$10{,}000 \text{ ft-c} = 10{,}800 \text{ lux} = 10.8 \text{ klx}$$

One cannot convert the photometric unit lux to μmol s^{-1}m^{-2}, the quantum unit applicable to photosynthesis, without specifying the wavelength or

TABLE 9-1. Approximate intensity of starlight, moonlight, and sunlight in photometric units (footcandles and lux) for visual light and quantum units (photon flux density) for photosynthetic light.

Terrestrial light	Photo reaction	Photometric unit		Quantum unit
		ft-c	lux	μmol sec^{-1}m^{-2}
Starlight		0.0001	0.001	
Moonlight		0.02	0.2	
Sunlight	Photoperiod induction	0.3	3.2	
	Indoors near window	100	1080	20
	Overcast weather	1000	10,800	200
	Maximum photosynthesis (individual leaf)	1200	12,960	240
	Direct sunlight	10,000	108,000	2000

source of light. However, for sunlight, divide lux by 54 (for blue sky, divide by 52):

$$\text{lux(daylight)}/54 = \mu\text{mol s}^{-1}\text{m}^{-2} \text{ photosynthetic active radiation (PAR)}$$

$$108{,}000 \text{ lux}/54 = 2000\ \mu\text{mol s}^{-1}\text{m}^{-2} \text{ PAR}$$

For using different sources of light (as different lamps), different constants must be used, as seen in the following table:

Lamp source	Divisor (lux to μmol s^{-1}m^{-2})
Incandescent	50
Cool-white fluorescent	74
High-pressure sodium	82
Mercury delux	84

THE PLANT IN RELATION TO LIGHT

Plants grown in the absence of light but supplied with a source of food from storage organs (for example, seed, tuber, or bulb) are yellow and have greatly elongated, spindly stems (Figure 9-5). The same plants, when provided with light, develop the green color associated with chlorophyll and photosynthesis, and assume the normal stem structure. The morphological expression of light

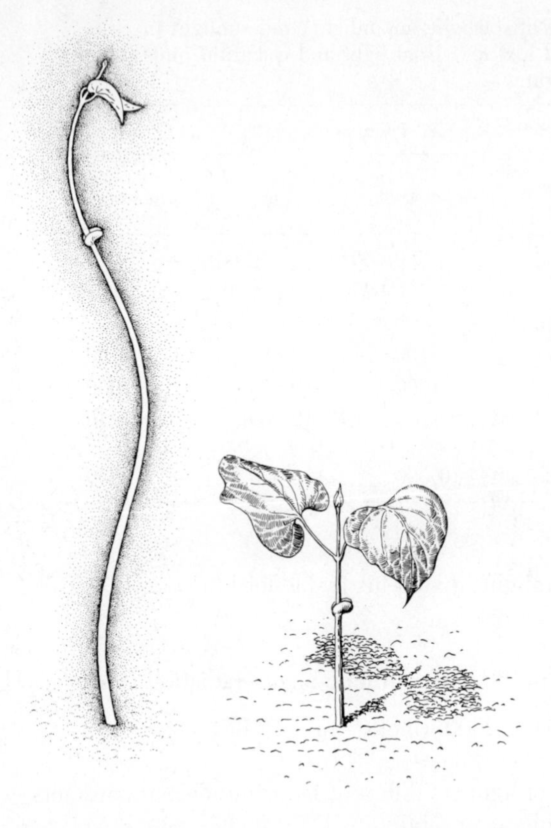

FIGURE 9-5. Etiolated (*left*) and light-grown (*right*) bean seedlings. [Adapted from J. F. Bonner and A. W. Galston, *Principles of Plant Physiology*, W. H. Freeman and Company, copyright © 1952.]

deficiency is called **etiolation,** and is related to the effects of light on auxin distribution and synthesis. The dependence of chlorophyll development on light is utilized in the production of **blanched,** or white, asparagus and celery that are preferred in Europe. Asparagus is dug instead of cut to obtain blanched spears. Blanched celery is produced by mounding the base of the growing plant with some opaque material, such as soil or paper. In the United States, self-blanching types are grown that are naturally somewhat lighter and produce a thicker stalk that shades the inner portion of the plant.

Some anthocyanin pigments also require light for their development. The 'Sinkuro' cultivar of eggplant develops purple pigment only in the presence of light, and the fruit is white under the calyx. Similarly, apples produced on the inner branches of the tree do not develop as intense a pigmentation as do those on outer limbs. Apple fruits can be imprinted with slogans or designs by means of tapes that have transparent letters or lines on an opaque background (Figure 9-6).

Light influences a great many other plant responses. These include germination, tuber and bulb formation, flowering, and sex expression. This effect of light on plant development is often related to the length of the light and dark periods or photoperiod, as discussed in Chapter 5.

FIGURE 9-6. Light influences the formation of some anthocyanin pigments. These labeled apples were produced by affixing the tapes shown below to the fruits on the tree before the natural formation of pigment. [Photographs courtesy Purdue University.]

Quality and Quantity of Light

The radiant energy required by plants is confined almost entirely to the visible spectrum. Growth is optimum when the entire range of that spectrum (that is, white light, or sunlight) is provided. As previously explained, light energy, described in terms of particles called photons (quanta), is inversely proportional to the wavelength. Thus, visible light of different wavelengths, which we see as different colors, provides different energy requirements. The light reactions of the plant (photosynthesis, phototropism, photoperiodism) are based on photochemical reactions carried on by specific pigment systems that respond to various wavelengths (Figure 9-1). For example, those portions of the visible spectrum that result in phototropism are the violet, blue, and green regions. The red portion, most effective in photosynthesis, is ineffective in phototropism. The pigments that absorb wavelengths effective in phototropism are yellow, perhaps carotenoids or flavonoids. It is intriguing that the pigments manufactured by plants are transformed by animals to compounds active in their photoreceptive reactions—vision.

The range in light intensity over the earth is enormous. Light intensity at full sun is a billion times brighter than starlight. The intensity of various light conditions is shown in Table 9-1. The different light reactions of the plant vary in the intensity required for their initiation and in the effect of the intensity on the rate of the reaction.

The intensity and quality of light reaching the plant vary with the season, the latitude, and the weather conditions affecting the water vapor in the atmosphere. Thus, during the winter, light often becomes a limiting factor in greenhouses, although heat may be provided artificially. The northern areas, which are in almost continuous light for a part of the year, provide abundant photosynthesis where temperature is not limiting. The huge size of the potatoes and cabbages produced in Alaska is due to the abundant light energy available in this region during the summer.

Light and Photosynthesis

The rate of photosynthesis is related to the availability of the raw materials water and carbon dioxide and to the energy provided in the form of light and heat. These simple requirements are abundantly provided in the temperate and tropical areas of the earth and sea.

The photosynthetic rate is proportional to the intensity of light up to about 13 klx, or 240 μmol $s^{-1}m^{-2}$. Chlorophyll thus is able to use efficiently only a portion of the incident light energy on a sunny day, which may be more than 108 klx, or 2000 μmol $s^{-1}m^{-2}$. Because of shading effects, however, a maximum amount of light intensity is required to provide all a plant's leaves with optimum quantities of energy. The rate of photosynthesis is sharply curtailed during low light intensity in cloudy weather. Not all plants, however, respond to high light intensity. Some require as little as one-tenth of full sunlight. These differences in light intensity requirements enable the classification of plants as sun plants or shade plants.

Only about 1 percent of the light received by the leaf during sunny days is utilized in photosynthesis. The remainder is reflected, reradiated, transformed into heat, or utilized for transpiration (Figure 9-7).

LIGHT MODIFICATION

The control of light is a significant part of the technology of horticulture. The quality, intensity, and duration of light has manifold effects on the many physiological processes of the plant.

Satisfying Photosynthetic Requirements

Plant growth depends on the fixation of carbon during photosynthesis. Although most plants grow best in the high light intensities of full sun (108 klx = 108,000 lux = 10,000 ft-c = 2000 μmol $s^{-1}m^{-2}$), a single leaf is light-saturated at about 13 klx. The higher intensities are needed, however, to compensate for leaf shading. Growth is much reduced at lower light intensities. Most plants cannot grow when exposed to less than 1 to 2 klx, the level of light in an average room. The **compensation point** is that light intensity at which plants will maintain themselves but will not grow. Foliage plants grown for decor are selected for their ability to maintain themselves at this level. For optimum appearance they must usually be replaced within the year unless more light is provided for growth. During the winter the light intensity available above plants in a greenhouse is often between 3 and 11 klx. This low light intensity and the short day length often severely limit plant growth.

Because of the high energy requirements of photosynthesis and the present cost of power, it is not economically feasible even in greenhouses to use supple-

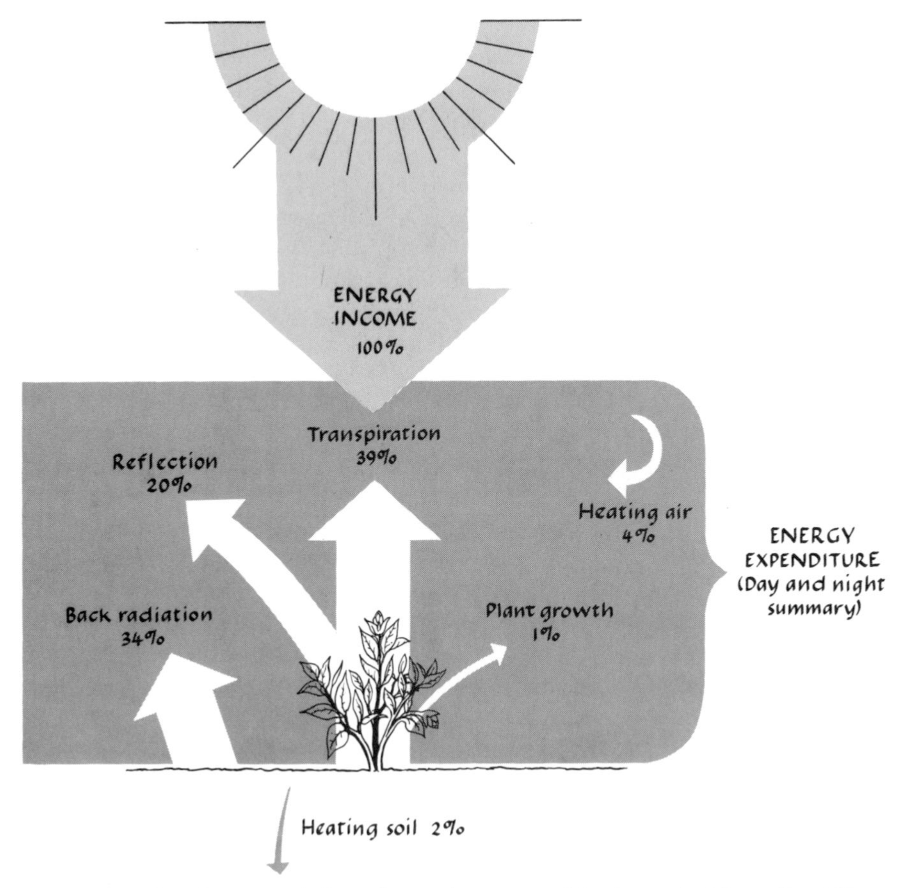

FIGURE 9-7. The solar energy balance during the summer (May-September) in southeast England. The energy expenditure is expressed as the percentage of the total that is sufficient to evaporate 36 inches of water. [Adapted from H. L. Penman, in *Sprinkler Irrigation Manual*, Wright Rain, Ringwood, England, 1956.]

mental light to increase photosynthesis, except for high-value florist crops. Supplemental light is used widely to increase growth in experimental studies or for indoor decorative plantings when cost is not a limiting factor. In outdoor cultivation the efficiency of light utilization may be increased by various cultural practices such as spacing, training, and pruning.

Control of Day Length

The control of day length by either supplemental illumination or shading has become a standard practice in florist-crop production. The artificial lengthening of the day or an interruption of the dark period makes it possible to promote flowering in long-day plants or to prevent or delay flowering of short-day plants. Similarly, under naturally long days, shading with black opaque cloth prevents flowering of long-day plants and promotes flowering in short-day plants. In this way, plants can be induced to flower out of season. In the culture of chrysanthemums, the most important florist crop in the United States, it is standard practice to control flowering by manipulating the photoperiod. The extension of the photoperiod by illumination is economical because of the low light intensity required for the process. The commercial control of flowering makes it possible to produce a continuous supply of many florist crops. It allows flower production to be synchronized more closely with market demand, which, in the United States, is governed by the season and by proximity to particular holidays. The alteration of photoperiod is a valuable tool for the breeder who may wish to cross plants that do not normally flower simultaneously.

An increase in photoperiod for controlling flowering can be achieved by extending the day length to about 17 or 18 hours. The same effect can be achieved, however, by interrupting the middle of the dark period for about 3 hours. Thus, in terms of power, alteration of the dark period is more efficient than extension of the day length. This effect can be made even more efficient by the use of brief light flashes (4 seconds per minute).

The reduction of photoperiod is achieved by screening the plants with black cloth, as noted above. Because of the low intensity of light required to stimulate the plants, care must be taken to darken them completely. Both indoor and outdoor culture commonly make use of black curtains that can be moved along fixed tracks.

Light Utilization and Modification

A number of strategems are used to obtain optimal utilization of light in horticultural crop production. They include choice of location and crop species, various practices to maximize light interception by altering the spatial distribution of plants, shading, and the use of supplemental illumination.

Geographic Location

The duration and intensity of light depend on climate and geography. Day length and the amount of light markedly change from the tropics to polar regions; consequently, location has a profound effect on the choice of species and the horticultural industry (see Chapter 19). The total amount of light received is also affected by local topographical features and atmospheric conditions that determine the number of sunny and cloudy days during the growing season. Light greatly influences quality and appearance in many crops. The bright clear summer days in the state of Washington favor the production of high-quality, attractive apples. Similarly, grape production is intensive on sunny slopes that receive abundant reflected light from nearby rivers.

Increasing Light Interception by Plants

Within a given location various techniques are used to maximize the amount of light captured by crop plants. Light that falls on bare ground or on competing weeds is wasted in a horticultural sense. Thus, the efficiency of the light utilized may be raised by any system that increases crop cover. These systems include increasing plant density, improving early growth of plants, or removing weed competition. For row crops, spacing within and between rows determines density (see also Chapter 15). In the past, distance between rows was determined by the nature of cultivation equipment. The trend toward narrow rows and high plant populations is an attempt to make maximum use of available light early in development. Similarly, any technique that encourages rapid growth of seedlings increases the efficiency of light utilization.

The distribution of plants is also a factor in efficient light utilization. In general, equidistant plant spacing is more efficient than any other spacing because it delays the onset of light competition (Figure 9-8).

The orientation of rows affects light use efficiency. Crops planted in rows oriented east and west utilize light more effectively than those planted north and south. The reason is that north-south rows are self-shading as the sun sweeps the sky, while the shade from east-west rows falls on the row middles. Row orientation, however, is usually governed by prevailing slope or by convenience.

Weed control may be considered in part a practice to eliminate competition for light between crop and weed. This is discussed further in Chapter 16.

Competition for light between parts of tree and vine crops can be modified by training and pruning to use the light most efficiently. Training means the orientation of the plant in space, while pruning refers to the judicious removal of plant parts. Training and pruning may affect the light-intercepting capacity of trees by exposing their productive portions to light and by adjusting fruit load to the light-absorbing capacity of foliage, which increases quality. Pruning and training as horticultural practices are discussed more fully in Chapter 14.

Shading to Reduce Light

In some cases it may be necessary to reduce the amount of light that plants receive. Shading is important in many nursery operations to reduce light and temperature and to lower moisture requirements. In permanent lath houses, shade is determined by the spacing between the laths. In recent years, plastic screen has been manufactured with various meshes that allow for the passage of different amounts of light. It is thus possible to change the light intensity throughout the course of a growing season by changing the screen.

Some crops are adapted to lower light levels and must be shaded. Examples include various ornamentals grown for interior use, tea, ginseng, and tobacco grown for cigar wrappers where the shade promotes the growth of large, thin, undamaged leaves.

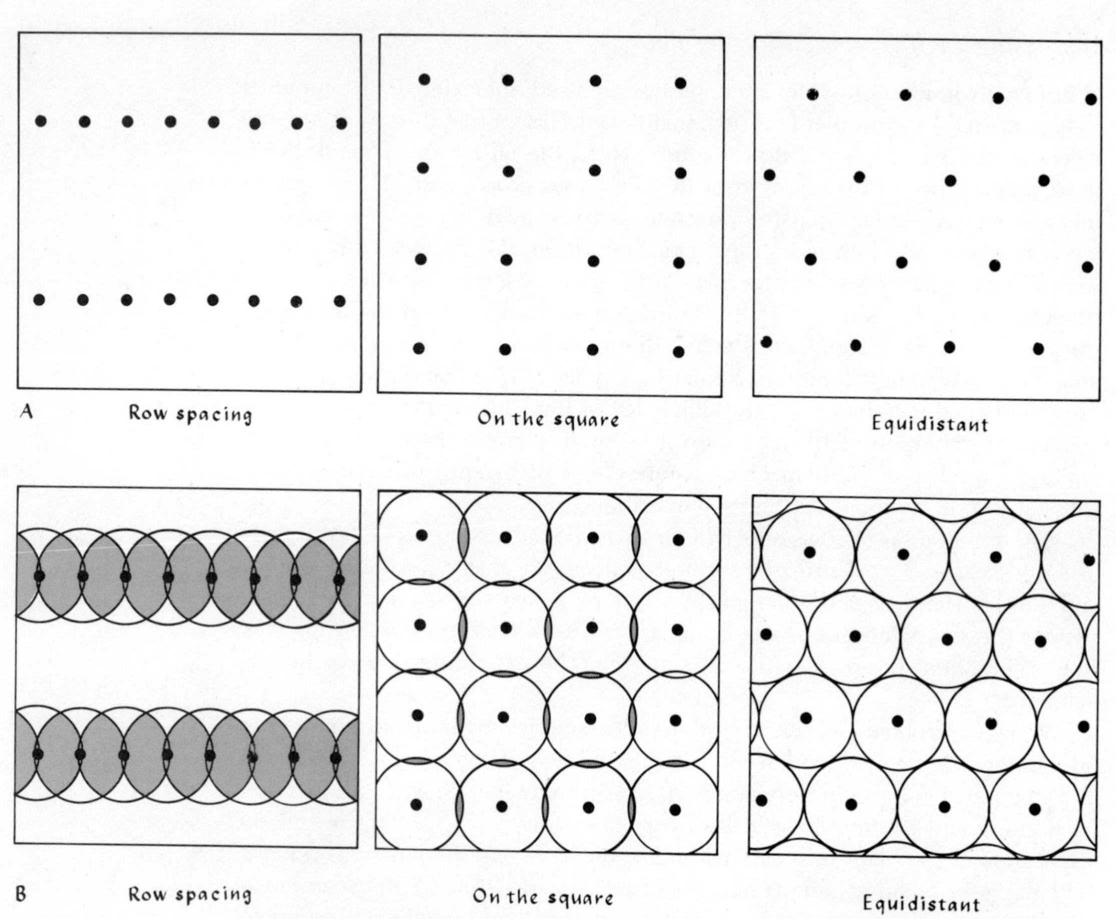

FIGURE 9-8. (*A*) Plant-distribution patterns. (*B*) Note that as plants grow, the canopies overlap. The most efficient distribution for light use is the equidistant pattern because the overlap is delayed and more light is captured.

Shading is used to inhibit pigment development in crops in which lack of color is desired. In England the desired blanching of celery by mounding the base of the crop with soil was mentioned earlier. Similarly, the Japanese grow some apples in bags to produce the translucent appearance prized for gifts.

Supplemental Illumination

Supplemental illumination may be used either to satisfy a crop's photosynthetic requirement or to extend day length in photoperiod control. Artificial light sources differ greatly in their spectral distribution (Figures 9-9 and 9-10).

Incandescent lamps emit light from a tungsten filament heated to extremely high temperatures (about 4670°F, or roughly 2850°C), and produce a

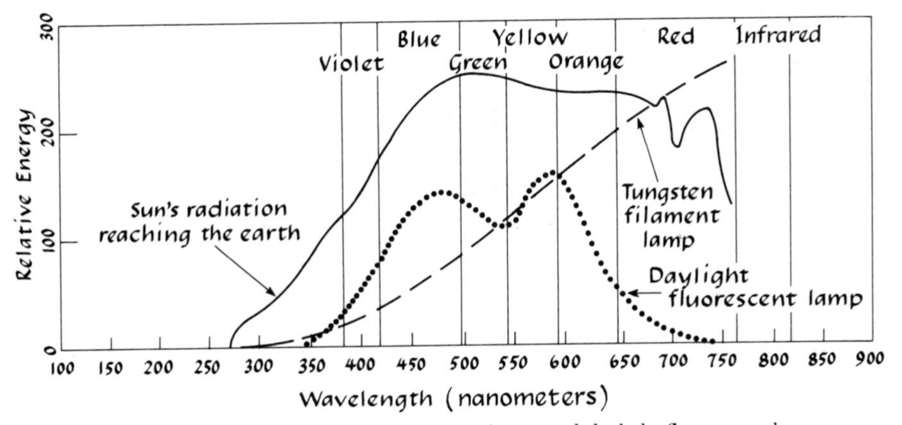

FIGURE 9-9. Spectral emission of tungsten filament lamps and daylight fluorescent lamps compared with the spectrum of the sunlight reaching the earth. The weak line spectra of the mercury discharge from the fluorescent lamps is not shown. The chart offers no quantitative comparison with respect to the energy output of the sources. [Based on data from General Electric Co.]

continuous spectrum from blue to infrared. The radiation within the visible spectrum lies mainly in the red and far red, although the greater part of the overall emission is in the invisible infrared.

Fluorescent lamps emit light from both low-pressure mercury vapor and fluorescent powder. Their emission spectrum contains both the continuous spectrum from the fluorescent material and the line spectrum of the mercury vapor. Light from ordinary fluorescent lamps is low in red and deficient in far red. This is why fluorescent bulbs are cool. The spectral distribution of different types of lamps and light sources will vary. For example, special fluorescent lights are available that will produce light richer in red.

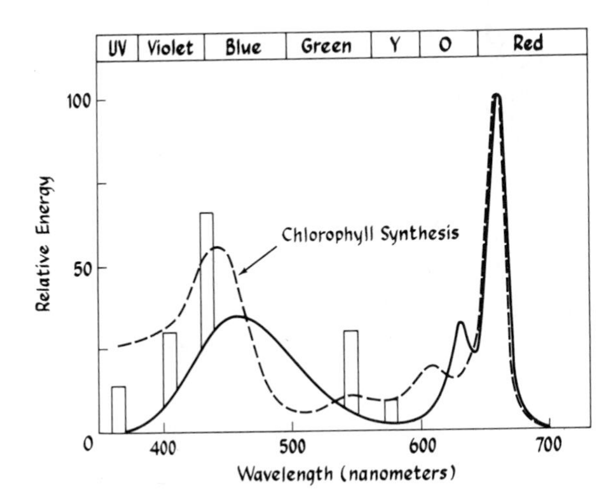

FIGURE 9-10. Spectral energy distribution for fluorescent lamps (Gro-Lux) designed to enhance vegetative plant growth compared with the spectral requirement for chlorophyll synthesis. [Courtesy Sylvania Electric Products.]

A

B

FIGURE 9-11. (*A*) A movable high intensity lighting system designed for winter production of lilies in the Netherlands. (*B*) Close-up of high-intensity lamp used to increase photosynthesis and plant growth in greenhouses. [Photographs courtesy T. C. Weiler and P. A. Hammer.]

High intensity discharge (HID) lamps are a third class of lamp, some of which are appropriate for plant lighting (Figure 9-11). In these lamps electrical discharge occurs in a quartz tube filled with mercury or sodium vapors under pressure and surrounded by a UV absorbent glass envelope coated with phosfors. The lamp must be connected to electrical supplies through special ballasts as with fluorescent lamps.

Because of the energy lost in heat (the infrared radiation), incandescent tungsten lamps are inefficient, since only about 5 to 8 percent of the energy input is transformed into the light range required by plants (400 to 700 nm), as compared to more than 15 percent for fluorescent lights and over 20 percent for HID lamps. Consequently, fluorescent and HID lights are more efficient in providing the energy required for the acceleration of photosynthesis. Their use is increasing in greenhouses, despite the relatively high cost of fixtures and installation.

Tungsten lamps have proven very efficient for controlling flowering and for promoting vegetative growth in many woody plants because they provide large amounts of red light so important in the photoperiodic effect. Fluorescent light that is poor in red is preferable for supplemental illumination of those plants that become etiolated and spindly in far-red light. Tungsten light is used where elongation is desirable, as in the culture of asters or hyacinths.

The most satisfactory growth in completely artificial light is obtained with a combination of light sources that complement each other to produce a spectrum closer to that of sunlight than any one of them does separately. In experimental growth chambers, where high light intensities are desirable, combinations of tungsten, fluorescent, and HID lamps may be used. Special care must be taken, however, to dispose of the infrared radiation (heat).

Selected References

Bickford, E. D., and S. Dunn, 1972. *Lighting for Plant Growth.* Kent, Ohio: Kent State University Press. (A handbook on light and plant growth, including many applied uses.)

Cathey, H. M., and L. E. Campbell, 1980. Light and Lighting Systems for Horticultural Plants. *Horticultural Reviews* 2:491–537. (A review of lighting technology in horticulture.)

Langhans, R. W. (ed.), 1978. *A Growth Chamber Manual: Environmental Control for Plants.* Ithaca, N.Y.: Cornell University Press. (A comprehensive guide to plant growth chambers.)

10 *Temperature*

The temperature of a substance is a measure of the relative speed with which its atoms are vibrating. All vibrations cease at absolute zero (−273°C). This chapter will include a review of the effect of temperature on biological and horticultural systems and the control of temperature in horticultural production. The control of temperature for the storage of horticultural products will be discussed in Chapter 18 and the relation of temperature to climate will be covered in Chapter 19.

TEMPERATURE MEASUREMENT

The temperature of a substance may be expressed in terms of several different scales. The Fahrenheit scale (°F) is in common use in the United States, but the Celsius scale (°C), based on a system in which 0 is the freezing point and 100 the boiling point, is used in most other parts of the world and in the scientific community. A table that connects Celsius and Fahrenheit scales is provided in the appendix of this book. For some scientific purposes, the Kelvin (K) scale, whose zero point is absolute zero, is used.

Temperatures can be measured by various devices. The most common are based on the expansion or contraction of liquids, solids, or gases when temperatures change. As temperature increases, molecules move faster and require more space causing substances to expand. Mercury and alcohol are the liquids most commonly used in inexpensive household and fever thermometers. Liquid-in-glass thermometers, although relatively simple, can be made very precise.

When extremely precise temperature measurement is required, thermocouples are frequently the first choice. A **thermocouple,** or **thermoelectric couple,** consists of two pieces of dissimilar metals in contact. As the point of contact is heated or cooled, the electrical potential between the two metals changes. Since the change in potential is proportional to the change in temperature, a thermocouple can be calibrated to measure temperature.

Thermisters measure temperature by guaging the change in electrical conductivity that metal undergoes when its temperature changes. These devices make use of certain metal alloys that demonstrate very great change in conductivity for very small change in temperature. Various other devices that measure temperature rely on the melting point of solids, longwave radiation (an infrared radiometer measures surface temperature), sonic frequencies, and the speed of chemical reactions.

HEAT TRANSFER

The transfer of heat energy is accomplished by radiation, conduction, convection, and reflection. **Radiation** is an organized flow of energy through space. It does not travel in the form of heat, which involves molecular motion, but in the form of electromagnetic waves. When radiation is absorbed on a surface, it usually produces a rise in temperature. In this case, radiation is transformed into heat energy. In **conduction,** the energy flows through a conducting medium from the warmer point to the cooler one. The transfer of heat through the soil takes place by conduction. The ability of a substance to conduct heat (**conductivity**) varies with the material, as shown in Table 10-1. Heat may move by

TABLE 10-1. Heat conductivity of various substances.

Substance	Value (cal/cm-sec-°C)
Silver	1.0
Iron	0.1
Water	0.0013
Dry soil	0.0003
Sawdust	0.0001
Air	0.00005

convection, the circulation of warmed air or water, whose density has changed as a result of heat. Air near a stone radiating heat warms and becomes less dense than the cooler air farther away, and the warm air is pushed upward. Similarly, cool water sinks. Heat, as well as light, may be **reflected** from a surface. A sheet of polished metal will reflect both heat and light in the same way. The persistence of snow in mild weather is a result of its high reflective property.

BIOLOGICAL EFFECTS OF TEMPERATURE

Physical and chemical processes are controlled by temperature, and these processes in turn control the biological reactions that take place in plants. For example, temperature determines the diffusion rates of gases and liquids in plants.

The solubility of various substances is temperature-dependent. Carbon dioxide is perhaps twice as soluble in cold water as it is in warm water. The inverse is true of most solids; sugar is much more soluble in warm water than in cold water.

The rate of reaction is affected by temperature; usually, the higher the temperature, the faster the reaction. Thus, temperature has a very marked and important effect on respiration. However, the relationship of temperature to the biochemical reactions that occur in plants is seldom directly proportional because of certain complicating factors. For example, the end products produced, such as sugars, may accumulate and block further reactions. In some reactions the availability of raw materials may be a limiting factor.

It is useful to think in terms of the Q_{10} of reaction velocities. The symbol Q_{10} refers to the rate of change in reaction activity that results from a change in temperature. The rates of uncatalyzed chemical reactions increase approximately 2.4 times for each 10°C (18°F) rise in temperature. Of course, this implies a physical system in which limiting processes are minimal. For the overall growth process, the Q_{10} of chemical reactions is more likely to be only 1.2 to 1.3 because many factors slow the rates of chemical reactions in cells and living systems.

Temperature affects the stability of enzyme systems. At optimal temperatures, these systems function well and remain stable for long periods of time. At colder temperatures, they remain stable but are nonfunctional, while at high temperatures they completely break down. An enzyme system that is stable at 20°C (68°F) may be active for only a half-hour at 30°C (86°F) and for only a few seconds at 38°C (100°F).

The equilibrium of various kinds of systems and compounds is a function of temperature. For example, the balance between sugars, starches, and fats is altered when temperature changes. During the fall season, sugars in some species decrease in quantity, whereas starches and fats increase (some plants store food mainly as fats, others as starches). When spring arrives, however, there is a change from both starch and fats to sugars, which are translocated to actively growing parts of plants.

Since temperature has strong effects on physiological and biochemical reactions in plants, it also determines the rates of various plant functions, such as the absorption of mineral elements and water. Not only is the viscosity of water greater at low temperatures, but cytoplasmic membranes through which water must pass seem to be less permeable. Photosynthesis is notably slower at low temperatures, and consequently the growth rate is slower. Temperature also affects the rate of cytoplasmic streaming within individual cells.

In view of its profound influences on the biochemical reactions of the plant, the critical effects of temperature on growth and many developmental processes are not unexpected. Among these processes are flowering, sex expression, seed-stalk formation, dormancy of seed and plant, as well as various maturation processes (see Chapters 4 and 5).

The Plant in Relation to Temperature

The minimum and maximum temperatures to support plant growth generally lie between 40 and 97°F (4.5 and 36°C). The temperature at which optimum growth occurs varies with the plant and differs with the stage of development (Figure 10-1). In addition, different parts of the same plant will withstand varying minimum temperatures. Roots of cold-acclimated plants are more sensitive to low temperature than stems, and flower buds are more tender than leaf buds.

A number of growth processes show a quantitative relationship to temperature. Among them are respiration, part of the photosynthetic reaction, and

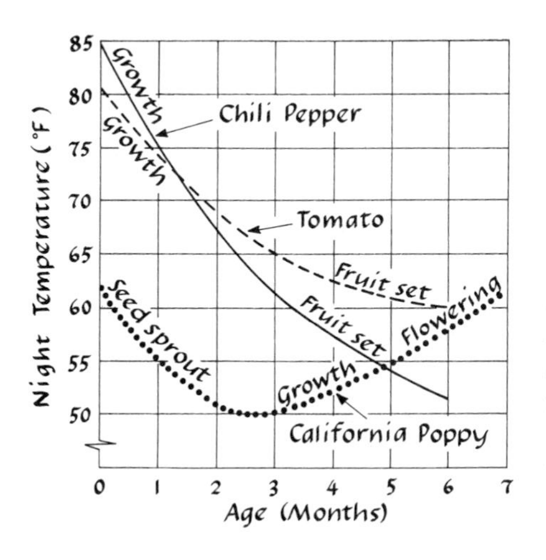

FIGURE 10-1. Optimum night temperature varies with the species and with the stage of development in the life cycle. [Adapted from "Climate and Agriculture," by F. W. Went. Copyright © 1957 by Scientific American, Inc. All rights reserved.]

various maturation and ripening phenomena. In addition, such plant processes as dormancy, flowering, and fruit set are temperature-critical. The optimum temperature for plant growth depends, then, on the species and cultivar, and on the particular physiological stage of the growth process. Plants grown under uniform, constant temperature do not grow or produce fruit as rapidly as do those grown under alternating night and day temperatures. Most plants require a lower night than day temperature. Some plants need a period of cold temperature to complete their annual or life cycle. Finally, there are various types of injury associated with extremes in temperature.

Low-Temperature Injury

Low-temperature injury is one of the main factors that determine the distribution of plant species. Damage caused by low temperature differs with season, tissue, and species. Some plants or plant tissues are damaged by temperatures slightly above freezing. The **chilling injury** is noted in many crops of tropical origin, such as peanut, velvet bean, sweetpotato, and many of the cucurbits. Chilling injury is related to membrane damage in sensitive species.

Temperatures below the freezing point of water often cause permanent damage resulting in death. This is dramatically seen after the first fall freeze, when most herbaceous plants are killed. Yet some woody plants when dormant survive temperatures below −40°C (−40°F) and many tissues can be shown to survive the temperature of liquid nitrogen (−196°C, or −320°F). This is indeed a remarkable adaptation.

Cold injury in many crops by temperatures below the freezing point of water is due to the formation of ice that causes mechanical injury to the cell. When plant tissue freezes, ice forms either outside cell walls (extracellular freezing) or inside them (intracellular freezing). Plants that tolerate freezing tolerate extracellular freezing without ill effects, but intracellular freezing disrupts the integrity of the cell and is invariably lethal. The transition of water to ice and the ability to avoid intracellular freezing are the critical factors in cold-temperature injury. Many tissues are not injured by temperatures below freezing if ice crystals are not formed. Thus, any mechanism that avoids or delays ice formation confers frost hardiness in tissues. There are two mechanisms to avoid injury from ice in critical tissue: lowering the temperature at which freezing occurs (freezing-point depression) or maintaining water in a liquid or at least in a noncrystalline state below the freezing point (supercooling).

Although pure water freezes at 0°C (32°F), the freezing point of water with dissolved solutes is lower. This is the principle of antifreeze. Thus, the freezing point of most succulent herbaceous tissue ranges from −1 to −4°C (30 to 24°F). Ice avoidance by freezing-point depression can be accomplished by increasing solute accumulation or by decreasing the amount of water in tissues.

Woody tissues that are properly acclimatized in the fall can survive slow rates of freezing because ice forms only extracellularly, that is outside cell protoplasts. The protoplasts of hardy woody plant species, when properly hardened, have the remarkable ability to maintain supercooled water.

Tissues differ in their ability to withstand low temperatures. Citrus fruit peel is impaired by temperatures above 0°C (32°F), while leaves are not damaged by temperatures below freezing if ice is not formed. Flowers of hardy species of woody plants have little ability to survive low temperatures or to be hardened. Fruit buds of the peach will withstand less cold than vegetative buds even if acclimated. Thus, temperatures of −26°C (−15°F) in midwinter will destroy peach crops the following year but the trees will survive. Roots are less hardy than above-ground stems.

Plants that are resistant to cold injury (**hardy** plants) as compared with those susceptible to cold injury (**tender** plants) appear to have a higher proportion of unfreezable water (**bound water**) in the cells, an accumulation of soluble carbohydrates, and a lower water content. The free water associated with succulent tissue freezes at 32°F (0°C). The **osmotically held water**, caused by an increase in sugarlike substances, has a lowered freezing point depending on the concentration and acts as an antifreeze. The bound, or colloidally held, water freezes at a still lower temperature. The winter injury that occurs in grapes after a season of unusually heavy production (often due to inadequate pruning) is associated with a low sugar content of the tissues, which renders them susceptible to cold injury. Thus differences in cold resistance of particular plants may be induced by factors that tend to increase sugar accumulation.

The variation in cold resistance among plant species is probably related to their ability to supercool, that is, to bind water in nonfreezable forms. The higher the proportion of bound water, the hardier the plant. The increase in the ability of plant tissues to withstand the stress of cold temperatures is known as **hardening.** This capability is discussed later in this chapter.

There are other types of cold injury. During winter it is also associated with tissue desiccation brought about by decreased water absorption by roots. This type of winter kill is a common injury to evergreens, in which transpiration during the winter is a contributing factor. Evergreens may be protected from winter kill by covering them with black plastic which cuts down or eliminates transpiration losses. (Clear plastic will cause overheating on sunny days.) Other types of winter injury include winter sunscald of thin-barked species, frost splitting of tree trunks (Figure 10-2), and mechanical injury due to frost heaving.

The "heaving" of soil caused by alternate freezing and thawing injures the plant by the mechanical ripping of the root system. This problem may be overcome by procedures such as mulching that tend to prevent premature thawing. A substantial portion of freezing injury is associated with unseasonably high temperatures in the winter. In temperate regions, unseasonably high temperatures in late winter often initiate growth prematurely. This effect renders the

FIGURE 10-2. Bark splitting as a result of low-temperature injury. [Courtesy Fenton E. Larson.]

plant extremely susceptible to subsequent cold weather. This is often noticed on the southern side of trees (in the Northern Hemisphere) where insolation is greatest. Similarly, the early blooming of fruit trees induced by unseasonably warm weather is feared because of the increased danger of frost injury to flower buds.

High-Temperature Injury

High-temperature injury is often related to desiccation. The "burning up" of plants during unusually hot weather is generally a result of excessive water loss in transpiration unbalanced by water uptake. Excessive water loss is very noticeable when unusually warm, dry, windy weather occurs after transplanting. Soil surface temperatures under these conditions may be high enough to interfere with root growth. Young transplants often "burn off" at the soil line. Extremely high air temperatures (46 to 54°C, or 115 to 130°F) may be lethal to the plant as a result of the coagulation of protein. However, some heat-resistant plants appear to increase their chances of survival at high temperatures by the presence or production of special heat-resistant proteins.

The cessation of growth in hot weather is a reflection of an altered metabolic balance. When the respiration rate rises more rapidly than the rate of photosynthesis, a depletion of food reserves will result. As in all biologic processes, the critical temperatures vary with the material.

Hardening

Hardening, in the broad sense, refers to processes that increase the ability of plant to survive the impact of unfavorable environmental stress. In its more restricted meaning, hardening refers to processes that enable plants to withstand cold injury, just as the term *hardiness* usually refers specifically to cold-hardiness.

Cold-hardiness is a variable characteristic that differs greatly with species and with seasonal change. For example, many plants that survive extreme winter freezing may be severely injured by spring frost. The natural change in hardiness in woody plants is related to temperature, water availability, and day length. The onset of cooler weather and shorter days in the fall brings about dormancy and a physiological toughening in some plants.

Cold resistance of the whole plant develops in a number of steps. As the end of the growing season approaches, the rate of growth slows and carbohydrates begin to accumulate. Reserve foods accumulate during this period partially because they are not being utilized in growth and respiration. They are translocated to other plant parts, particularly roots, where they remain over the winter as insoluble fats and starches, and are used at least to some extent in respiration. Also during this period, water in the individual cells moves from the protoplasm to the vacuoles. These changes in water relations and in the availability of certain carbohydrates promote stability of the protoplasmic proteins, which have a higher proportion of bound water during the winter than during the summer. There seems to be a surface effect on protoplasmic colloids, so that they do not clump or precipitate so readily. During the onset of winter, the concentrations of colloidal and dissolved materials become much greater than during the summer, which may contribute to cold resistance. Sections of the hardened stems of woody plants can survive immersion in liquid nitrogen (−196°C, or −320°F).

There are several ways to condition plants for cold weather. One of the simplest techniques is to withhold water and excessive fertilizer, especially nitrogen, toward the end of the growing season. Nitrogen stimulates vegetative growth, and if excessive nitrogen is applied late in the season, plants may continue to grow until late in their growing period rather than to harden off prior to winter. Succulent green shoots, stimulated by the addition of nitrogen, are not hardy, but are quickly killed as temperatures approach freezing.

Promoting plant vigor during the growing season increases the supply of carbohydrates. Plants with high levels of reserve carbohydrates are better able to withstand cold weather than those in poor condition. Respiration continues to some extent throughout the winter, and there must be a supply of carbohydrates to maintain life processes during this period. Some grape cultivars overbear, and if the excess fruit is not removed before it matures, the plants are subject to winter injury.

As a general rule, resistance to cold injury increases with age until senescence. When exotic ornamental shrubs and other plants are introduced into a region, seeds are not planted directly in the field. Rather, they are planted in nurseries and carefully sheltered and protected for several years to get them past the juvenile stage. After they have ceased to be tender young plants and are hardened to some extent, they survive much better than if they are planted directly in the field. Some species of ornamentals must be 6 to 8 years old before they will survive field planting.

The hardening of transplants may be achieved by any treatment that materially checks new growth. This goal may be accomplished by gradually exposing plants to cold, by withholding moisture, or by a combination of these two treatments. In general, a period of about 10 days is sufficient to harden plants. This treatment is designed to produce a stocky, toughened plant in contrast to a soft, tender, "leggy" plant. Hardened plants are often darker green in color, and hardened crucifers have more waxy covering on the leaves. The induced cold resistance brought about by the hardening treatment in such cool-season crops as cabbage or celery may be considerable. Unhardened cabbage plants show injury at 28°F (about 2°C), whereas hardened plants can withstand temperatures as low as 22°F (5.5°C). In such warm-season crops as tomato, the degree of cold-hardiness imposed is slight, but hardening inures the plant to transplanting shock and hastens plant establishment under adverse conditions. The reduced growth rate enables the plant to withstand desiccation until the root system becomes established. Tender plants, which respire rapidly, have little chance of survival if warm, windy, dry conditions follow transplanting. The cessation of growth under the hardening treatment, however, may severely interfere with subsequent performance; care must therefore be taken to avoid overhardening. Under ideal transplanting conditions, hardening may not be necessary.

FROST

Frost is a thin layer of ice crystals deposited on soil and plant surfaces as a result of freezing temperatures. Two types of weather conditions produce freezing temperatures: (1) rapid radiational cooling, which results in a **radiational frost,** and (2) the introduction of a cold air mass with a temperature below 32°F (0°C), which produces an **air-mass freeze.** Frost is common under conditions of clear, calm skies; a freeze, on the other hand, may occur with an overcast sky and, usually, when there is considerable wind. Both frost and freeze involve temperatures at or below the freezing point. Because of its local nature, frost may occur when the so-called official temperature (taken relatively nearby, and usually at a height of 6 feet) is above freezing. Glaze damage (Figure 10-3) can

FIGURE 10-3. Glaze damage: limbs have been broken off by the weight of their ice coating. [Photograph courtesy USDA.]

occur when relative humidity is high and temperatures drop below freezing. The weight of the ice coating on tree limbs causes them to break.

The earth radiates energy, as do all bodies. At night the earth receives no solar radiation, resulting in a net loss of radiation. Frost occurs when the loss of heat from the ground permits the temperature to drop to or below the freezing point.

Frost is of vital horticultural concern. It defines the season for annual crops in the middle latitudes and is potentially extremely destructive to perennial fruit crops that flower in the early spring. The danger of frost is usually not due to actual tree damage but to injury of temperature-sensitive flower parts (Figure 10-4). The year's crop is thus at the mercy of spring frost.

Frost Conditions

Conditions favorable for radiational frost are those that are conducive to rapid and prolonged surface cooling. For example, the introduction of low-temperature polar air followed by clear, dry, calm nights will facilitate upward radiation. The presence of cloudy, humid air causes reradiation back to the earth and thus prevents frost conditions. The main effect of fogging in the control of frost is the creation of an artificial reradiating surface, although there is some release of heat as the fog droplets cool. Strong evaporation after rainfall, especially on plant cover with a large surface area, will increase cooling. The absence of wind leaves the coldest air undisturbed and next to the ground.

FIGURE 10-4. Frost is a serious hazard to the orchard in bloom. [Photograph by J. C. Allen & Son.]

Frost conditions are usually quite variable, since local conditions modify radiation. Site is an important factor in frost. For example, slopes are protected by "air drainage." Since cold air is denser than warm air, it moves downhill. The circulation pattern of the displaced warm air produces a relatively warm area, or **thermal belt,** on the slope itself. It is this microclimatic factor that makes slopes good sites for fruit planting. Low areas, on the other hand, collect cold air and become "frost pockets." This condition results in a phenomenon referred to as **temperature inversion,** in which air temperatures increase with altitude (Figure 10-5). Artificial wind machines prevent frost damage by mixing in the upper warmer air (Figure 10-6).

The frost protection afforded by large bodies of water is due to the high specific heat of water as compared to that of the land; water absorbs and gives

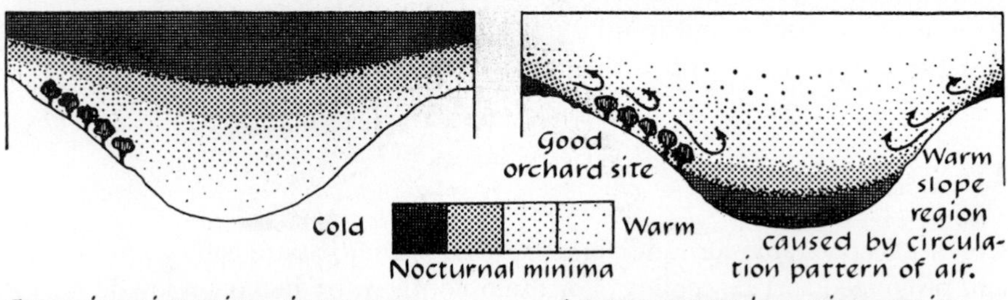

FIGURE 10-5. Schematic representation of the origin of a thermal belt as a result of a temperature inversion on a hillside. [Adapted from R. Geiger, *Climate Near the Ground,* Harvard University Press, 1957.]

FIGURE 10-6. Wind machines protect citrus from frost during temperature inversions by mixing the warmer upper air into the colder air below. [Photograph courtesy Food Machinery and Chemical Corp.]

up heat slowly. Solar radiation penetrates water more deeply than land, and the continuous internal movement of water results in heat distribution throughout the water mass. Thus, large bodies of water become heat reservoirs in the fall and cold reservoirs in the spring. Because of their great heat capacity, large expanses of water moderate temperatures. In the winter and early spring, the influence of water keeps temperatures moderately low and prevents premature plant growth. In the late spring, the water may provide enough heat to prevent frost. Similarly, in the fall there is a warming influence that tends to delay the advent of frost conditions. This temperature lag is felt mainly along the windward side of large bodies of water.

Many factors affect frost. Anything that prevents the accumulation of radiation during the day will increase frost. For example, vegetation that shades the soil will reduce the amount of heat stored in the day. Thus, a sodded or mulched area is more liable to become frosted than one under "clean cultivation." The necessity for controlling frost is one of the main reasons why peaches are not grown under permanent sod. The exposure of the slope also affects the quantity of radiation received. In the Northern Hemisphere, southerly slopes receive considerably more radiation than northerly slopes do.

Heat from the lower layers of the earth moves up by conduction. Consequently, the conductivity of the soil will affect frost at the surface. Frost on

FIGURE 10-7. Frost is frozen dew. Hoarfrost results when sublimation occurs over several hours, leaving heavy deposits of ice crystals in the form of scales, needles, feathers, and fans. [Photograph by J. C. Allen & Son.]

muck is a serious hazard because organic soils tend to be poor conductors of heat as compared with mineral soils. More important than soil type in the occurrence of frost is the amount of soil moisture. By replacing air (a poor conductor) with water (a better conductor), the danger of frost can be reduced. Thus, frost damage may be prevented in muck areas by flooding.

The **white frost** commonly seen in the morning results from frozen dew (Figure 10-7). Its occurrence depends on the **dewpoint**—the temperature at which relative humidity reaches 100 percent. When the air temperature is below the dewpoint but above 32°F (0°C), water vapor condenses in the form of dew. White frost occurs when the air temperature is below both the dewpoint and 32°F. If the humidity is low, frost damage may occur when the air temperature is below 32°F but above the dewpoint. This is known as a **black frost** because the only visible indication of it comes when the vegetation turns dark as a result of cold injury.

The change in state from water to ice results in the release of energy, the **heat of fusion**. Consequently, if temperatures do not get too low, the freezing of water or the occurrence of a white frost actually protects vegetation from lower temperatures. This phenomenon is exploited in the use of sprinkler irrigation as a method of frost protection. The ice forming on the plant releases heat and acts as a protective buffer against cold injury.

TEMPERATURE CONTROL

Plant growth shows a marked response to small changes in temperature, and if extremes in temperature persist for even short periods, they will lead to irreversible changes of state, resulting in the death of the plant or parts of it. Methods for the control of temperature in the culture of horticultural crops vary greatly. For the great majority of crops there is no active control, but rather, an adaptation through selection of location, site, and choice of plant (discussed further in Chapter 19). For some field-grown horticultural plants, an active attempt may be made to modify and ameliorate extremes in temperature through cultural practices such as mulching and various techniques of frost control. The regulation of temperature in greenhouse culture can be complete, including artificial heating and cooling.

Cultural Practices

Mulching

Mulches are insulating substances spread over the surface of the soil (Figure 10-8). Although one of their chief purposes is the regulation of soil temperature, they serve many other functions. They conserve soil moisture because they reduce evaporation by lowering the soil temperature and by increasing the absorptive capacity of the upper layer of the soil. Erosion is reduced as a result of decreased surface runoff and the shielding effect of the mulch to driving rain. Mulch is commonly applied for this reason to newly planted lawns and seed beds, especially on sloping areas. Mulches may control weeds and eliminate the need for cultivation by smothering weed growth and cutting off light from the soil surface. They offer protection to flowers and fruit from mud-splattering rain, an especially important safeguard in such low-growing crops as strawberries. In addition, mulches may be a source of organic matter and nutrients for the soil. Mulching is often desirable for its own sake, since its pleasing appearance provides an attractive background for flowers and other plant materials.

Mulches may be applied during the period of active growth (**summer mulch**) or be restricted to late fall to provide cold-weather protection (**winter mulch**). Although the benefit of the summer mulch is attributed to a number of

FIGURE 10-8. Corncobs make an inexpensive mulch for apple orchards in the Corn Belt states. The mulch is spread around the drip line of the tree.

factors (for example, moisture conservation and weed control), the principal benefit of winter protective mulch is its influence on the temperature of the soil.

The temperature-stabilizing effect of summer mulches is due to insulation, heat absorption, and shading. The surface of bare, dark-colored soils on a sunny midsummer day may be higher by 17°C (almost 30°F) than the air temperature. The reduction in soil temperature attained as a result of mulching appears to increase nutrient availability. It also improves root growth and, ultimately, the performance of many plants.

The practice of using plastic sheeting as a summer mulch has shown a tremendous increase throughout the world in the production of fresh market vegetables. This system is a standard practice for strawberry production in California (Figure 10-9). Plastic mulch has a great influence on soil temperature, soil moisture, and weeds. Clear plastic raises soil temperatures, encouraging early production; but it also stimulates weed growth unless it is used in conjunction with chemical weed-control measures. Opaque black plastic shades the soil and controls weeds.

The use of foams to insulate plants temporarily to control frost damage is a recent innovation that incorporates the principles of a mulch (Figure 10-10). The foam is a combination of surfactant, stabilizer, and protein material (such as gelatin). Application is made the day before frost is expected; the foam dissipates a few hours after sunrise.

The temperature-regulating purpose of a winter protective mulch is two-

FIGURE 10-9. Strawberries grown on a plastic mulch in California. [Photograph courtesy Victor Voth.]

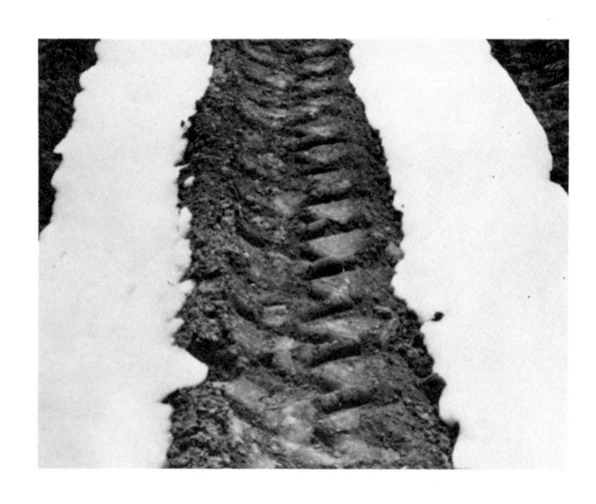

FIGURE 10-10. Foam applied to young vegetable seedlings for frost control. [Photograph courtesy USDA.]

fold. One effect is to temper extremely low winter temperatures. This is achieved through the insulation effect provided by the mulch, which conserves ground heat. The other effect is to stabilize and buffer soil temperature and prevent recurring freezing and thawing that rip and injure plant roots through soil heaving. During winter warm spells in cold climates, a mulch tends to keep the ground frozen by providing insulation and shading. Thus, not only does a mulch "warm" the plant under extreme winter weather, but it also keeps it cold during unseasonable warm spells. By keeping them under a winter mulch, spring-flowering plants may be delayed from early blooming to avoid the damaging effect of spring frost.

The application of winter mulch is usually made after a light freeze so as not to delay dormancy. Tender plants, such as roses, may be protected by mounding the crown with soil. After the mound is frozen, it is covered with an insulating organic mulch. Winter mulching is a standard practice in strawberry culture. After the plants have become dormant in the fall but before heavy injurious freezes, the entire planting is usually covered with 5 to 7 cm (2 to 3 inches) of straw (Figure 10-11). The plants are uncovered in spring, when growth can no longer be prevented. The excess straw is then moved to the middle of the rows. If frost is expected during flowering, re-covering the plant provides a measure of protection.

Most mulching materials consist of plant refuse or by-products: leaves, straw, sawdust, corn cobs, peat, tobacco stems, pine needles, wood chips, or paper. The main virtues of summer mulches are relatively independent of the material. Inorganic substances, such as rockwool or gravel, are also effective. A good mulch must be economical, available, and easy to handle. It must also be stable so that it will not easily wash or blow away. Mulches used around the home must be unobjectionable in odor and appearance.

Some of the problems associated with mulching materials arise from their tendency to act as sources and harborers of plant pests—weeds, disease-pro-

FIGURE 10-11. Mechanized mulching of strawberries. [Photograph courtesy Friday Tractor Co., Hartford, Mich.]

ducing microorganisms, and rodents. Because of the disease problem, the refuse of the plant being protected—such as its own fallen leaves—should not be used as mulch. Straw that has been improperly handled may contain weed or grain seed, which may contribute to the weed population the following spring. Fresh straw should be prespread and moistened during warm weather to induce germination of any seeds it contains before it is used as a mulch. The use of mulch in orchards must be accompanied by vigorous measures for rodent control, lest rodent populations build up to damaging levels in the favorable environment that a deep mulch affords.

Many mulching materials are highly inflammable and present a fire hazard. Straw mulch in particular should not be placed very close to buildings. Unless partially decomposed, fresh leaves make unsatisfactory mulch because they tend to pack closely and may smother plants. Although organic mulches decompose and will eventually contribute plant nutrients, the high carbon content of many of these materials may contribute to nitrogen deficiency. This is especially true if the mulch is later plowed under, but this risk can be avoided by applying extra nitrogen.

Frost Control

A number of techniques may be used to avoid the destructive consequences of spring frost. The prevention of frost injury involves three strategies: (1) escape, (2) reduction of heat loss, and (3) addition of heat.

Spring frost may be escaped by late planting. However, although the probable effect of frost conditions may be predicted, it is not always practical to plant at what is calculated to be the last frost-free date. Even this date is only a statistic. Escaping spring frost by late planting is possible with annual crops, but in locations with short growing seasons this strategy sometimes backfires because the crops may be exposed to early fall frost.

The spring culture of seedlings in protective structures with field transplanting after frost-free dates circumvents the dangers of frost, but transplanting is usually much more expensive than direct seeding. Frost control for perennial plantings, however, must depend on more substantial procedures.

In fruit plantings, spring frost could be escaped if bloom could be delayed. Recently such postponement has been accomplished experimentally by overhead misting: Evaporative cooling delays the onset of bud break by as much as two weeks. However, some problems have been encountered with this method, including reduced fruit set. Escaping spring frost by choosing the right location and site, coupled with late-blooming, frost-resistant cultivars, remains the preferred method in the fruit industry.

A new method to escape freeze damage has been proposed that involves interfering with the transition of liquid water to ice crystals. The transition of water to ice requires nucleating particles that include certain bacteria, especially species of *Pseudomonas* and *Erwinia*. This suggests that the presence of **ice-nucleating active (INA) bacteria** are indirectly responsible for freeze injury; thus, eliminating them from the surface of plants should prevent freeze injury or at least reduce temperatures at which freeze injury could be expected. Reducing INA bacteria might be accomplished by bacteriocides or use of nonnucleating bacterial antagonists or inhibitors. Covering foliage of crop plants with genetically engineered bacteria that will not nucleate has been proposed, but controversy about releasing these strains to the environment has prevented full testing.

Various cultural practices used in the control of frost involve techniques that either encourage the conservation of heat or add heat directly to the immediate environment of the plant. The conservation of heat is achieved by any method that will increase daytime absorption of heat by the soil or prevent its loss at night. The addition of heat may be accomplished in a number of ways—for example, by using heaters, flooding, spray irrigation, or artificial air movement.

The reduction of heat loss is accomplished through the use of such devices as hot caps (Figure 10-12), plastic tunnels, and cold frames. A recent practice is the use of stable foams to insulate plants against frost. Such foams (Figure 10-10) are nontoxic combinations of surfactants, stabilizers, and protein materials (gelatin) selected for durability, low cost, and insulating properties.

Heat may be obtained from the earth by improving the conduction characteristics of the soil, or it may be obtained from the air by disturbing the temperature inversion. Heat may be added indirectly, by sprinkler irrigation or by techniques that increase the daytime absorption of insolation such as by cultivating the soil. Frost control by sprinklers is the preferred technique for straw-

FIGURE 10-12. Hot caps protect early tomatoes in California's San Luis Rey district. The hot cap is made of a translucent paper and acts as a miniature greenhouse. [Photograph courtesy USDA.]

berry production. Heat may be added directly using various kinds of heaters and heating systems (Figure 10-13) to ward off cold temperatures, especially in high-value fruit crops. Solid petroleum wax candles 20 cm (8 inches) in diameter burn with a low flame for 8 hours. Two such heaters beneath a grapefruit tree will raise the average air temperature within the canopy of the tree by about 4°C (7°F).

In the past, heaters were commonly used in regions where citrus crops were subject to occasional frost. Growers, warned by the United States Weather Bureau of anticipated frost, filled their heaters with enough oil to burn throughout the danger period and placed them in their orchards in strategic locations to be lit just before the cold front arrived. The radiant fraction of the thermal output of heaters provides the most protection, although part of the convection heat is also useful, particularly under conditions of good inversion where wind machines can redirect some of the warm layer overhead back into the orchard. Although the smoke of smoky heaters ("smudge pots") may be of some value as a radiation "blanket" (if atmospheric conditions permit its accumulation in an inversion ceiling during the night), when more than one night of protection against radiation frost is required, this blanket may be costly, for the smoke will reduce the incoming solar radiation during the succeeding day.

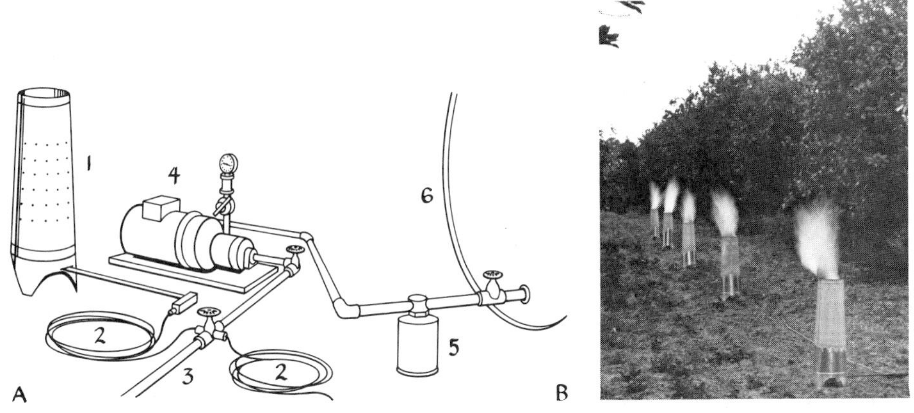

FIGURE 10-13. A popular central-distribution system for heating orchards. Fuel oil, liquefied petroleum gas, or natural gas is supplied to the burners from a central fuel source via an underground truckline network. (*A*) Lightweight portable burners (*1*) are attached to flexible feeder lines (*2*) which, in turn, are attached to a trunk line (*3*). Trunk lines are attached to an oil pump (*4*). A filter (*5*) traps impurities from the central storage tank (*6*). Valves control oil flow and burning rate. (*B*) Heaters in action in a Midwestern apple orchard. [Courtesy Spot Heaters, Inc.]

"Smudging" now is illegal in California citrus-growing areas because of the air pollution that it causes.

The escalation in the price of oil and the increase of antipollution ordinances have reduced orchard heating as a horticultural practice. Many Florida citrus growers have switched to sprinklers to control winter frost damage. The exploitation of protected sites and the development of increased cold-hardiness or frost-avoidance mechanisms by breeding is probably the best long-term strategy to reduce frost injury.

Plant-Growing Structures

Cold Frames

An inexpensive form of temperature control for seedlings and transplants during the early spring can be achieved with the cold frame. A **cold frame** is an enclosed ground bed, usually sunken, with a removable sash. Heat is provided through the trapping of solar energy. Temperatures inside the cold frame increase relative to the air during the day when the sash is in place, owing to the "greenhouse effect" discussed in Chapter 9. Heat is stored in the soil during the night, and plants can be protected even though outside air temperatures dip below freezing. With especially low temperatures, insulating material such as straw is sometimes placed over the sash. Temperatures are maintained during the day by raising or removing the sash. Cold frames are commonly used for starting early transplants from seed or for hardening off greenhouse-grown transplants.

Hotbeds

Hotbeds are essentially cold frames provided with a supplemental source of heat. Additional heat may be obtained by fermentation, hot water, steam, or electricity. Fermentation heat is provided from decaying organic matter, most commonly strawy manure, placed under the plants. Hot air, steam, or hot-water systems are arranged to heat the soil by conduction. Electrical heating also supplies ground heat through the use of a soil-heating cable. Thermostatically regulated electrical heating facilitates precise temperature control. Such systems can be easily installed; the operating cost, of course, depends upon local electrical rates.

Greenhouses

Greenhouses (in England they are known as glass houses) are usually elaborate, permanent structures equipped not only to regulate temperatures but to provide increased environmental control of plant growth (Figure 10-14). Because of the great amount of control that must be achieved in greenhouses, this type of culture becomes an extremely specialized operation.

In ordinary greenhouses, temperature is regulated through a heating and ventilation system similar to that of the hotbed. In cold climates, a central coal or oil furnace supplies the heat. In Europe, portable "steam plants" are available for this purpose. Peripheral steam heating is the most commonly used

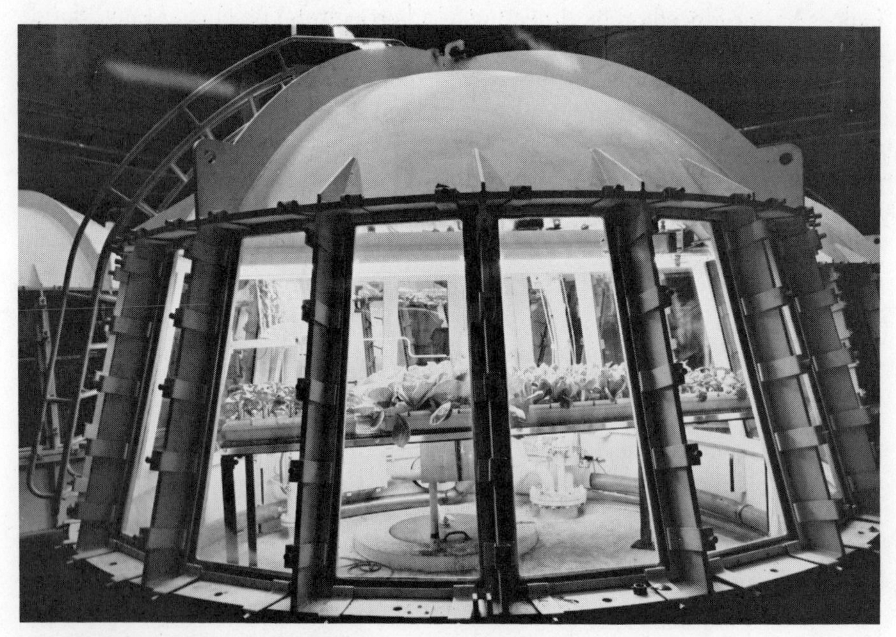

FIGURE 10-14. Experimental greenhouse used by the National Aeronautics and Space Administration (NASA) to investigate growth of plants in space stations.

distribution system, although heating pipes may be placed under benches in large greenhouses. Ventilation is provided at the sides and top of the structure. Automatic controls are available for both heating and ventilation.

Temperature control during cold weather is a matter of adjusting heating and ventilation to take maximum advantage of solar heat. In warmer weather, however, it becomes increasingly difficult to maintain reasonable temperatures for plant growth with an ordinary ventilation system. Some greenhouse cooling is achieved by shading the glass with a whitewash spray. The whitewash is made in such a way that it will weather off naturally by fall. Fan-and-pad cooling provides an economical system for lowering summer greenhouse temperatures. In this method, cooling is achieved by the evaporation of circulated water through a pad of excelsior or some other coarsely porous material with a high ratio of surface area to volume (Figure 10-15). Fans opposite the cooling pads draw the cooled air across the greenhouse. The efficiency of the system increases as the humidity goes down. Even in the hot, humid midwestern United States, temperatures can be kept at least on a par with the outdoor shade.

The use of refrigeration equipment is not economical for commercial greenhouse cooling, although it is used to obtain uniform temperatures for experimental conditions. Refrigeration equipment is widely used in greenhouses for storage purposes.

Plastic films have proved to be a convenient and inexpensive substitute for glass and have found a ready market in construction of cold frames, hotbeds, small sash houses, and greenhouses (Figure 10-16). However, the high initial

FIGURE 10-15. Fan-and-pad installation for cooling greenhouses. Cooling pads are at right, fans are in roof at left. [Photograph courtesy Acme Engineering and Manufacturing Corp., Muskogee, Okla.]

FIGURE 10-16. Plastic greenhouses are convenient, inexpensive structures. The polyethylene plastic is removed in the spring, when temperatures get too high, and is replaced in the fall. The frame can then be covered with shade cloth and the structure converted to a shade house. (*A*) A scissor-type, truss-rafter plastic greenhouse in the process of being covered. (*B*) A gothic-rafter plastic greenhouse. The insulated pipe carries steam from the greenhouse-range. (*C*) A gothic-type plastic greenhouse constructed from aluminum pipe used to overwinter strawberries. The strawberries are grown in plastic tunnels with plastic mulch. [Photographs courtesy P. H. Massey, Jr., and Merle H. Jensen.]

construction cost of glass greenhouses still compares favorably with the cost of plastic greenhouses on a long-term depreciation basis. The light-absorbing qualities of plastic are similar to those of glass. At present, a number of different types of plastic coverings are available. These vary from polyethylene films to the more rigid plastics (Figure 10-17). Since some polyethylene films disintegrate under the influence of ultraviolet light during the summer, they must be replaced each fall. This requirement provides a unique advantage in that, with the plastic removed, the summer cooling problems are eliminated entirely. Ultraviolet-resistant polyethylene is now available. Shade cloth may be substituted for plastic on the frame during the summer. This means of temperature control is, of course, not possible with the more permanent plastic coverings.

Air-supported plastic greenhouses are being developed to produce crops in inhospitable desert climates using an integrated power, water, and food system (Figure 10-18).

Cloches and Plastic Tunnels

The use of a portable, tentlike glass sash (**cloche**) over individual plants has long been used in European market gardens to facilitate early vegetable production. The use of this technique declined because of the tremendous labor inputs required. However, the principle has been revived on a large scale for winter vegetable production, with the introduction of **plastic tunnels** (Figures 10-19

FIGURE 10-17. Rigid plastic (fiberglass) covered greenhouses. [Photographs courtesy University of California.]

A

B

C

FIGURE 10-18. Controlled environment in an integrated system providing power, water, and food for desert coast areas: a concept pioneered by the University of Arizona, the University of Sonora, and the Rockefeller Foundation. Fresh water is produced from sea water by desalting facilities that harness heat from small engine-driven generators. Heating and cooling is achieved with sea water, which is always about 76 to 78°F (24 to 26°C). Humidity is very high and conserves moisture; diseases have not been a problem since spores are removed by seawater spray. Yield per acre is often higher than field production. (*A*) An aerial view of greenhouses built by the University of Arizona's Environmental Research Laboratory and the Arabian principality of Abu Dhabi on the Persian Gulf. Half of the greenhouses are air inflated (*center*) and the remainder are supported structures. (*B*) Close-up of the supported greenhouses. (*C*) Close-up of the air-inflated greenhouses. (*D*) Diagrams of greenhouse system. Water is supplied by a desalting plant. Arrows indicate the air-flow pattern. (*E, F*) Vegetable production in the greenhouses. [Courtesy Carl O. Hodges.]

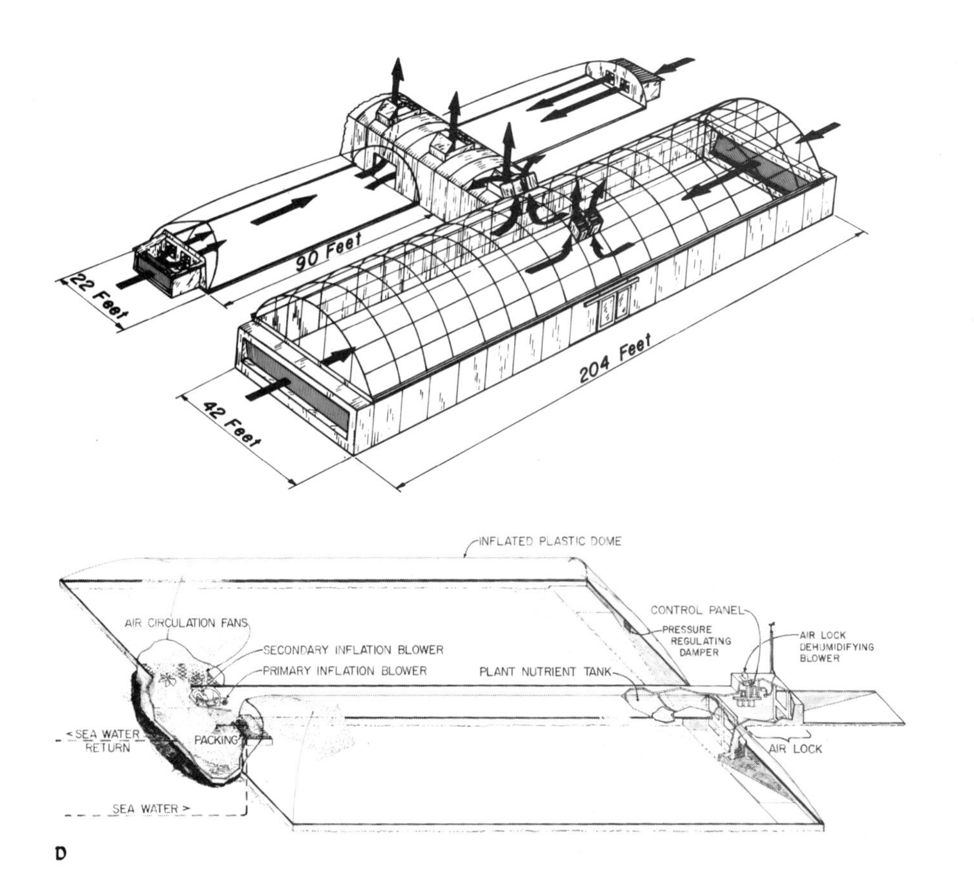
90 Feet
22 Feet
204 Feet
42 Feet
INFLATED PLASTIC DOME
CONTROL PANEL
AIR CIRCULATION FANS
PRESSURE
REGULATING
DAMPER
AIR LOCK
DEHUMIDIFYING
BLOWER
SECONDARY INFLATION BLOWER
PRIMARY INFLATION BLOWER
PLANT NUTRIENT TANK
< SEA WATER
RETURN
PACKING
AIR LOCK
SEA WATER >

D

E

F

FIGURE 10-19. Production of melons under plastic tunnels in Israel. (*A*) Early muskmelon production. Polyethylene is slit as temperatures increase. (*B*) Broad tunnels held down by elastic bands for watermelons. (*C*) Covering wire hoops with polyethylene to create narrow tunnels. (*D*) Combination of plastic mulch and plastic tunnels. [Photographs courtesy J. Rudich.]

and 10-20). Such tunnels are made from sheets of polyethylene laid down mechanically, usually over wire hoops. Soil may be used to seal the sides of the plastic. The increase in temperature within the tunnel encourages early production. A number of systems ameliorate built-up heat as temperatures rise with the approach of summer. In one system the plastic may be temporarily slipped off the hoops during periods of warm weather. In others, the plastic is first perforated for ventilation and then slit at spatial intervals as the temperature rises. Plastic tunnel and plastic mulch may be combined; in some cases the plastic tunnel is converted to a plastic mulch.

Shade Houses

Shade houses may either be large walk-in structures (Figure 10-21) or low, covered frames. Although shading is commonly used to reduce temperature, it

A

B

FIGURE 10-20. Use of plastics in California. (*A*) Plastic tunnels. (*B*) Plastic row covers to protect staked tomatoes from rain and wind from November to January in San Diego County. [Courtesy B. J. Hall.]

is also useful to protect such shade-loving plants as chrysanthemum, hydrangea, azalea, and various foliage plants from leaf damage caused by high light intensity. This protection is accomplished through the use of such materials as lath or screening. In addition, various types of "shading cloth" are available that can be used to cut down light intensity by different amounts.

Propagation beds are often located in shade houses in order to reduce excessive transpiration. Owing to the inadequacy of their root systems, excessive heat is especially injurious to newly rooted cuttings and transplants. Less watering is required under shade, since transpiration and soil evaporation are reduced.

FIGURE 10-21. Shade house in Florida used for research on ornamentals. The plants shown are leatherleaf fern (*Polystichum adiantiforme*). [Photograph courtesy Charles A. Conover.]

Selected References

Downs, R. J., and H. Hellmers, 1975. *Environment and the Experimental Control of Plant Growth.* New York: Academic Press. (A discussion of controlled environments, such as air-conditioned greenhouses, plant-growth chambers, and phytotrons, especially for research.)

Evans, L. T. (ed.), 1963. *Environment Control of Plant Growth.* New York: Academic Press. (A collection of papers dealing with the effects of many environmental factors on plant growth.)

Levitt, J., 1980. *Response of Plants to Environmental Stresses,* vol. 1, *Chilling, Freezing and High Temperature Stress.* New York: Academic Press. (A scholarly work on temperature stress.)

Li, P. H., and A. Sakai, 1978, 1982. *Plant Cold Hardiness and Freezing Stress,* vols. 1 and 2. New York: Academic Press. (A series of technical papers based on two international seminars.)

Went, F. W., 1957. *The Experimental Control of Plant Growth.* Waltham, Mass.: Chronica Botanica (Classic studies of controlled plant growth carried out at the Earhart Laboratory. Particularly valuable for data on climatic responses of individual crop plants.)

Air 11

Air is composed of about 78 percent nitrogen, 21 percent oxygen, 0.9 percent argon, 0.03 percent carbon dioxide, and about 1 to 3 percent water vapor. In addition, there may be trace amounts of inorganic and organic compounds, many of which result from photochemical reactions between sunlight and various combustion products. These "pollutants" may interfere with the normal growth of plants and people, and possibly inflict dangerous consequences on our planet.

Air was not seriously treated as an environmental factor in plant growth until recently because, under normal conditions, it does not limit growth. In the past, the atmosphere varied relatively little and normal movement of the air kept its constituents in equilibria. This is no longer the case, since proximity to urban and industrial areas can considerably degrade air quality. The soil atmosphere has always varied greatly as has the atmosphere of confined spaces, such as greenhouses. Finally, storage atmosphere has tremendous effect on product viability and quality (see Chapter 18).

AIR CONSTITUENTS

Oxygen

Oxygen is abundant in the air and only during flooding are plants deprived of the amount they need. However, its availability in the soil is often a critical factor in plant growth. Poorly aerated soils have low oxygen and high carbon dioxide content. This impedes respiration in roots and stunts root growth, thus reducing water and nutrient uptake. High carbon dioxide levels also have a toxic effect on roots. The death of large trees when their roots are covered over with extra soil or pavement is a dramatic example of the effects of an altered ratio of oxygen to carbon dioxide in the soil. Good drainage in soils is important because excess moisture in poorly drained soils reduces oxygen levels. Aeration is essential in synthetic soils used in the rooting of cuttings. This is why coarse sand, which increases pore size, is a popular ingredient for these mixes. Oxygen is required for germination of many seeds. Thus, overwatering can prevent seed germination.

Aquatic plants, or plants adapted to marshy or boggy conditions, may be structurally altered so that the oxygen requirement of the roots is satisfied by oxygen absorbed through the leaves. Other water-loving plants apparently have adapted in some manner to low oxygen concentrations.

Oxygen and carbon dioxide levels have a great effect on fruit and plant storage. This topic will be discussed in Chapter 18.

Carbon dioxide

Although carbon dioxide is a very small part of the atmosphere (0.03 percent equals 300 ppm), it is critical to plants as a source of carbon. Carbon dioxide is probably never limiting in most field cultures, but it may drop below normal level in greenhouses because photosynthesis may deplete CO_2 in the air when ventilation is restricted. In a tightly closed greenhouse, CO_2 levels may be reduced to 150 to 200 ppm. Thus, CO_2 enrichment can increase crop yields and improve plant quality and so has been a standard practice in intensive greenhouse production.

Measurement of atmospheric carbon dioxide indicates that CO_2 levels have been increasing at an ever-rising rate (Table 11-1). Preindustrial levels prior to 1860 were estimated at 274 ppm and early industrial levels at 290 ppm. In 1958, the mean CO_2 level was 316 ppm and in 1982 it was 340 ppm! Thus, the increase represents 66 ppm in 122 years. The present rate of increase is 1.5 ppm per year.

Rising atmospheric CO_2 levels are due to the increased combustion of fossil fuels in the world. The consequences could be enormous because, although CO_2 is used by plants to fix carbon, CO_2 strongly absorbs and emits radiation in the infrared. The earth's surface and atmosphere remain at temperature equilibrium by absorbing incoming sunlight and radiating an equal amount of infrared energy to outer space. By being transparent to visible light and partially opaque to the infrared, CO_2 prevents normal reradiation of infrared energy to space (the **greenhouse effect**), thus interfering with the earth's energy budget. Anything interfering with the flow of energy from sun to earth and earth to outer space could change atmospheric temperature and climate.

TABLE 11-1. Increasing atmospheric CO_2 from 1860 to 1982.

Year	Atmospheric CO_2 (ppm)	Estimated rate of increase per year
1860	274	
1900	290	0.5
1958	314	1.0
1982	340	1.5

SOURCE: David M. Gates, 1982, in E. R. Lemon (ed.), 1983, *CO_2 and Plants* (AAAS Selected Symposium 84).

A rise in the earth's mean temperature would change precipitation patterns (possibly decreasing precipitation in the central United States) and could raise ocean levels owing to the melting of the polar icecap, which would result in the dislocation of millions of people living in coastal areas. However, although all scientists agree that CO_2 levels in the environment are increasing, the evidence for higher earth temperatures in response to recent CO_2 increases is controversial.

If increasing CO_2 will indeed raise mean temperatures, efforts to decrease CO_2 must be sought. An increase in nuclear energy with a corresponding decrease in fossil fuel use could reduce net CO_2 emission. Another concept is to substitute energy from renewable carbon sources, such as alcohol, produced from plant residues or other plant hydrocarbons, rather than to obtain energy from carbon fixed in the past and stored in the form of coal and oil. If all energy were obtained from carbon fixed annually, there would be no net change in the amount of CO_2 in the atmosphere.

Nitrogen

Nitrogen is an inert gas in the atmosphere that is unavailable to plants unless transformed into nitrate (NO_3^-) or ammonia ions (NH_4^+). This process is known as **nitrogen fixation.** Some nitrates are formed in the air from the heat of lightning and are incorporated into the soil by rainfall. However, this is only a small fraction (5 to 7 kilograms per hectare, or 5 to 6 pounds per acre, per year). Most of the nitrogen is fixed by microorganisms. These include free-living bacteria such as *Agrobacter* and *Clostridium,* some blue-green algae (very important in rice culture), and certain bacteria, usually species of *Rhizobium,* that are found in a symbiotic association (that is, a mutually beneficial relationship) in the roots of many plants—typically, legumes. The nodular deformation of the roots are sites of these bacteria. In return for the nitrogen, the plant system provides the energy for these bacteria. Attempts to incorporate the nitrogen-fixation ability by gene transformation may not be a good idea, even if it could be in fact accomplished. The plant would have to divert some of its energy from growing to nitrogen fixing, causing a yield reduction.

Traditionally, crop growers have assisted the fixation of atmospheric nitrogen through the use of legume rotations: After harvest, the growers plowed down the legumes as a green manure and planted a different crop in the field. This technique has been replaced in intensive agricultural and horticultural systems by the expanded use of nitrogen-containing fertilizers. In the past, most nitrogen fertilizers were organic by-products, but at present most are produced by the **chemical fixation** of atmospheric nitrogen. Air is essentially distilled to obtain nitrogen, which is combined with hydrogen obtained from natural gas (or by-products of petroleum) to produce ammonia (NH_3). Ammonia can be stored under pressure and injected into the soil, or it can be combined with carbon dioxide to produce urea

$$CO_2 + 2NH_3 \longrightarrow CO(NH_2)_2 + H_2O$$

or with nitric acid to produce ammonium nitrate

$$NH_3 + HNO_3 \longrightarrow NH_4NO_3$$

Thus the price of nitrogen fertilizer is tied to the cost of energy and natural gas. Nitrogen is the critical element in plant nutrition. The most efficient way to fix nitrogen, biological or chemical, will be the most appropriate system. In horticultural intensive production, it is difficult to replace the use of nitrogen fertilizer.

Water Vapor

In Chapter 10 we were concerned with water as a liquid in the soil. Here, we shall consider water as a gas in the earth's atmosphere and soil.

Atmospheric humidity may be described in various ways. **Absolute humidity** is the amount of water vapor per unit volume of air, and it is expressed as grams per cubic meter or as grains ($1/7000$ of a pound) per cubic foot. **Specific humidity** is the weight of water vapor per unit weight of air (including the weight of the water vapor). Thus, absolute humidity is a function of volume and specific humidity is a function of weight. A common unit of water vapor is **relative humidity,** the quantity of water vapor in the atmosphere expressed as a percentage of the maximum quantity the air is capable of holding at a given temperature. This quantity increases with temperature. Thus, the **drying power** of air, which is proportional to the water-vapor deficit below saturation, is related to both relative humidity and temperature. As the temperature goes down, the relative humidity increases and drying power decreases. At high temperatures, small differences in relative humidity represent large differences in drying power; at low temperatures, differences in relative humidity represent smaller differences in drying power.

Atmospheric water may also be expressed as **vapor pressure,** that part of the total atmospheric pressure due to water vapor. Vapor pressure, like atmospheric pressure, is expressed as millimeters of mercury (mm Hg). (Remember the atmospheric pressure at sea level is usually equivalent to 76 cm of mercury, or 1000 cm^2 of water. In the metric system, 1 mm Hg equals 1.32 millibars.) The **vapor pressure gradient** is the difference between the actual vapor pressure and the vapor pressure needed to saturate the air at the same temperature. This value is useful for calculating rates of evaporation and transpiration.

Forms of Atmospheric Moisture

There are several forms of atmospheric moisture, including vapor, rain, snow, sleet, hail, and dew. Atmospheric moisture is important to plants in two different ways. As precipitation, it supplies moisture to soil; as water vapor, it decreases the loss of water from plants (and soil) by transpiration (and evaporation).

Rain forms when atmospheric moisture condenses at temperatures above the freezing point. When the condensation occurs at temperatures below the freezing point, **snow** results. **Sleet** is formed by rain falling through a layer of air with temperatures below the freezing point. If air currents carry the frozen particles of sleet back up into the frigid layer of air so that they accumulate a thicker covering of ice before falling again, **hail** is produced.

Dew forms at night when atmospheric moisture condenses on cold surfaces. It can do so only when a clear sky allows the soil and plants to lose enough heat so that their temperature falls below the dew point. Clouds, haze, and dust prevent the loss of heat at night, when the energy balance of the earth is normally negative.

A number of plant species are able to use the moisture from fog and mist. The epiphyte Spanish moss (*Tillandsia usneoides*) grows on, but receives no sustenance from, the branches of several tree species. It is never found in areas that are not subject to mist and high relative humidity.

Fogs and mists are low clouds, and moisture from them is not measured by conventional rain gauges although it affects plant growth. The vegetation that develops in areas with frequent fogs may be far richer and more abundant than might be expected from precipitation records.

Measuring Atmospheric Moisture

The most common instrument for measuring humidity is the **psychrometer,** an instrument consisting of two thermometers, one of which has a wet wick around the bulb. The two thermometers, fastened side by side, are spun so that a stream of air passes over their bulbs. The thermometer without a wick will indicate the true air temperature, while the other will show a lower temperature because the water evaporating from the wick causes a heat loss. The relative humidity can be obtained from the two temperatures by using the appropriate calculations or tables.

The **hygrometer** contains substances that are affected directly by the moisture content of the environment. This instrument is made from hairs and thin animal membranes that stretch when they become moist. Both the hairs and the membranes are connected through a series of levers to a chart or a needle calibrated to indicate relative humidity.

The amount of water vapor may be determined chemically by exposing the chemical cobalt chloride to the air. This substance changes color as humidity changes.

Humidity may be measured electronically by using **resistance psychrometers.** A nonconducting glass or plastic element is coated with lithium chloride or some other hygroscopic salt, and the electrical resistance of the material when placed between two electrodes is measured with a meter. The salt will adsorb more water at higher humidities; therefore, the electrical resistance between electrodes decreases as atmospheric moisture increases.

Liquid precipitation may be measured in various ways, but the **rain gauge** is the most common. Basically, a bucket is calibrated to read accumulated water in millimeters or inches. If the gauge is to be read infrequently, the presence of mineral oil will prevent evaporation. **Snow gauges** are important for measuring winter precipitation, but the amount of snow must be converted to its equivalent in water, usually about one-seventh of the volume.

The simplest device to measure evaporation is an open pan of water placed in an exposed location. Evaporation is measured merely by noting changes in the water level. The **atmometer,** a more sophisticated device, consists of a porous clay bulb connected by a water-filled tube to a reservoir of water. As solar radiation strikes and heats the outside of the clay bulb, the water in the tube evaporates and is replaced by water that moves up from the reservoir. Another device, the **evaporimeter,** consists of a wick wetted by water from a reservoir. As water evaporates from the wick, it is replaced by water from the reservoir. In both the atmometer and evaporimeter, the amount of water lost from the reservoir indicates the relative rate of evaporation.

AIR POLLUTANTS

Urbanization and industrialization have introduced various harmful substances into the atmosphere. **Air pollution** is esthetically offensive and can be a genuine health hazard to human beings as well as to vegetation and many horticultural plants.

The quality of the air is becoming a significant factor in plant growth, especially in horticultural operations that are close to urban and industrial areas. About 125 million tons of air pollutants are released annually in the United States. They include carbon monoxide, 52 percent; oxides of sulfur, 18 percent; hydrocarbons, 12 percent; particulate matter, 10 percent; and oxides of nitrogen, 6 percent. Air pollution can be traced to its principal sources: 60 percent to transportation vehicles, largely the automobile; 19 percent to industry; 12 percent to power-generating plants; and 9 percent to space heating and refuse burning. Particulate materials can damage plants, but the gaseous contaminants are even more injurious. The most harmful to plants are ozone and peroxyacetyl nitrate (the main phototoxicants in photochemical smog) and sulfur dioxide (released by combustion of sulfur-containing fuels and the smelting of certain ores). Other compounds injurious to crops include fluorides, ethylene, nitrogen dioxide, chlorine, and hydrogen chloride.

The principal air pollutants that are toxic to plants will be discussed in turn here.

Ozone

Ozone is most abundant in urban environments; automobile engines are the chief source. Nitric oxide (NO) formed in the engine passes into the atmosphere and oxidizes quickly to form nitrogen dioxide (NO_2). This chemical reacts with other atmospheric components and releases oxygen (O), which combines with molecular oxygen (O_2) in the atmosphere to form ozone (O_3).

Whereas 100 to 200 ppm is a normal level of ozone in the atmosphere, the daily maximum in the Los Angeles basin, for example, has reached 10,000 ppm. Daily peaks frequently reach 3000 ppm in the summer and fall. Ozone is very reactive once in the atmosphere, however, so photochemically produced ozone disappears soon after it ceases to be produced.

Ozone has many effects on plants (Figure 11-1), some that are visible and some that are not. Photosynthesis may be decreased and growth impaired. Stomata may close when ozone is in the atmosphere, limiting gas exchange. Respiration may be increased, resulting in a reduction of carbohydrates.

Ozone injury first presents a shiny, oily appearance on upper leaf surfaces. If the exposure is sublethal and soon halted, the leaf will return to normal. In tissues that have been killed, the shiny areas appear dull, gray, and water-soaked. The groups of cells subsequently turn white, giving the leaves a stippled look. In monocotyledons, there is a chlorotic stipple between the veins. Bronzing and early senescence of the leaves occur in some woody ornamentals. Very sensitive species, such as white pine (*Pinus strobus*,) are affected even when far removed from urban environments. White pine needle blight (emergence tip burn) and stipple in grapes are ozone-induced disorders. Stipple begins with small brown or black lesions on young leaves, ultimately leading to bronzing and defoliation.

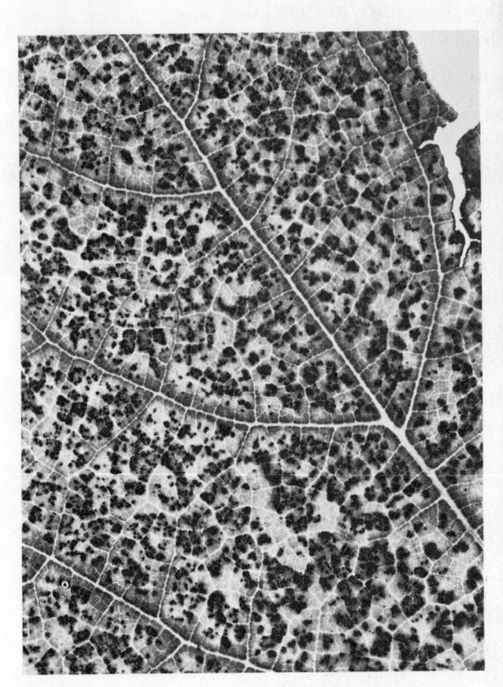

A

B

FIGURE 11-1. Ozone (O_3) injury to grape leaves (*A*) results in lesions (called stipple) that are distributed interveinally over the entire leaf surface. Excessive exposure to ozone can reduce photosynthesis and cause leaf injury and death. (*B*) On tomatoes, ozone injury is concentrated on the terminal leaves, which become yellow near the tip and die gradually as the damage extends toward the base. [Courtesy Walter Kender, and USDA.]

Sulfur Dioxide

Sulfur dioxide (SO_2) is released into the atmosphere from burning coal. Power plants, refineries, and home furnaces are major contributors, and additional SO_2 is cast into the air when certain ores of nickel, copper, and iron are smelted.

SO_2 may be toxic to vegetation and threaten human and animal health. It enters leaves through stomata, passing into the intercellular spaces of the mesophyll where it is absorbed on the wet cell walls. The mechanism by which SO_2 damages plants is controversial, but protein synthesis interference has been implicated. Injury may also result if sulfur accumulations are large enough to be directly toxic to cells: It causes them to shrink and collapse, and chloroplasts to disintegrate.

The external symptoms of SO_2 damage are varied (Figure 11-2). Red clover leaves develop white spots, whereas white pine tips lose color. Both the spots

FIGURE 11-2. Sulfur dioxide fumes have damaged the white birch leaf on the left. The leaf on the right is from a tree grown in filtered air. [Courtesy USDA.]

and the pine tips will green again in a few days if the source of SO_2 is removed; if not, the affected tissue dies (becomes necrotic). Studies with various crops indicate that crop yield is directly proportional to the degree of leaf necrosis.

Fluoride

Fluorides are produced by processes in the aluminum, glass, ceramic, and phosphate industries. They escape when ores containing such minerals as mica, hornblende, and cryolite are smelted, or when fluoride compounds are manufactured or used for catalysts and fluxes. Either gases or particulate forms may be released.

Fluoride enters plants by diffusing through stomata into intercellular spaces, where it is absorbed into the cells and transported through vascular tissues to the tips and margins of leaves. Flouride compounds can accumulate in chloroplasts, cell walls, nuclei, and cytoplasm. Although plants may normally have a fluoride content of from 1 to 20 ppm, those in industrial areas may contain 100 to 1000 ppm. A few plants can accumulate large amounts of fluoride. For example, tea plants may contain as much as 1300 ppm without apparent damage. Such accumulation can occur even in environments that have no industrial pollution.

Exactly how fluorides injure plants is not known, but they seem to interfere with enzymes that affect many metabolic processes. Fluoride may precipitate the calcium within the plant, causing a calcium deficiency.

Fluoride causes mottling and necrosis at the tips and margins of leaves of sensitive broad-leaved species. On grasses, tips are killed, and stippling or mottling is found along the margins. On conifers, necrosis starts at the tips of the current year's needles. The fruits of some species are extremely sensitive, even more so than the leaves. In peaches, fluoride causes the soft-suture disease, a premature ripening along one or both sides of the suture line. This area overripens or rots before the rest of the fruit matures.

Fluoride pollution does not always damage plants, however. Conifers, aspen, grapes, and orange trees may actually increase in growth upon exposure to small doses of atmospheric fluoride. But if the amounts are large, crop production is decreased and the composition of natural vegetation may be changed. Fluoride damage may also occur in sensitive species where the fluoride content of water is high or as a result of fluoride contamination of some fertilizers including superphosphate.

Smog

The term **smog** was coined in London in 1905 by combining the words *smoke* and *fog*. Research has shown that it contains dusts, oxides of nitrogen, hydrocarbons, ozones, sulfur dioxide, aldehydes, and numerous other compounds. The components of smog most damaging to plants (Figure 11-3) seem to be oxidants, which include ozone, nitrogen oxides, and peroxides.

Hydrocarbons and nitrogen oxides from automobile exhausts react in a normal atmosphere in the presence of sunlight to yield toxic gases, which cause plant damage and eye irritation. The end products from this series of reactions are a group of compounds collectively called peroxyacyl nitrate (PAN). PAN and ozone (photochemical oxidants) cause great damage to plants (Figure 11-4). Even if exposure is as slight as 5 ppm for 10 minutes, some sensitive species are damaged. PAN inflicts several types of injury on plants. For example, glazing and bronzing occur on spinach, lettuce, endive, and sugar beets.

FIGURE 11-3. Petunia plants react very strongly to smog, which has become a common component of the atmospheric environment of most cities. [Courtesy California Statewide Air Pollution Research Center.]

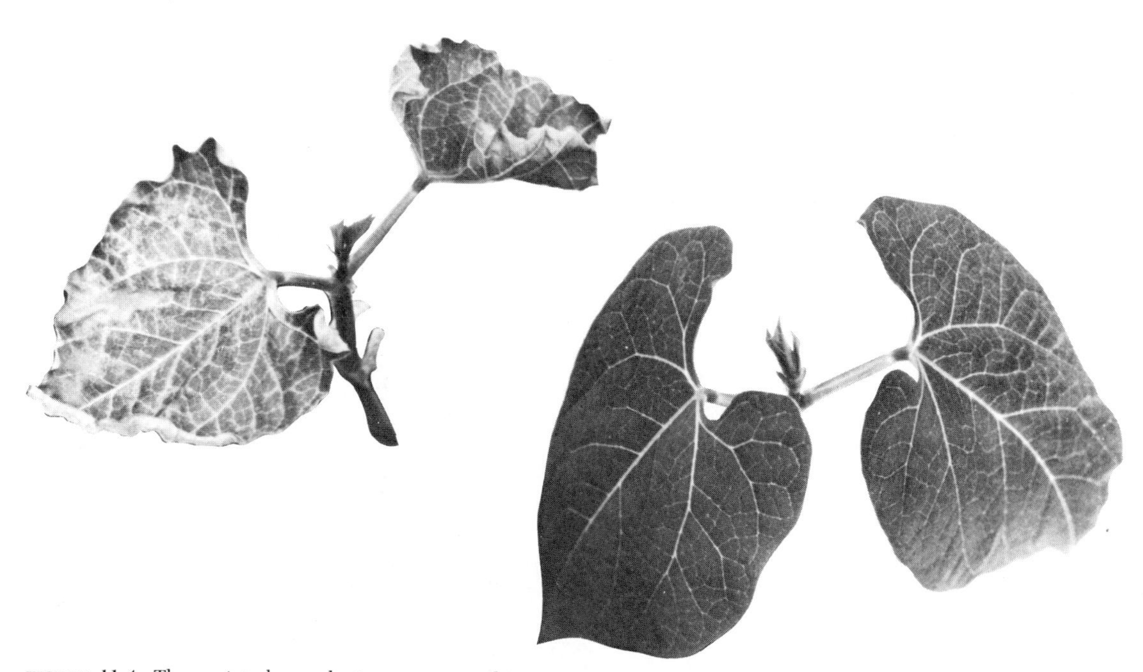

FIGURE 11-4. These pinto-bean plants were exposed to air containing PAN (at a concentration of 5 ppm) for ten minutes. The plant on the left was in the light during the exposure while the plant on the right was in the dark. The variation between the two illustrates the photochemical nature of the phytotoxic effects of PAN. [Courtesy EPA.]

Other Gaseous Pollutants

Additional harmful gases include ammonia, chlorine, and other gases that have been accidentally spilled into the atmosphere, causing local problems. Ethylene, a natural plant product, will injure plants if it is present in too great a quantity (Figure 11-5). Ethylene gas is a hazard in postharvest shipment of flowers. Budwood stored with apple fruit will split, form callus, and deteriorate as a result of exposure to ethylene.

Suspended Particles

Suspended particles or dust (often referred to as "particulate matter") may be injurious to plants. These can be introduced into the air by smelters, smokestacks, mining operations, cement plants, quarries, strip mines, mineral-processing operations, and other industries, or by various agricultural practices that increase soil erosion.

Dust, when it is most damaging, can sandblast seedlings. Wind-blown particles can erode the epidermis of young seedlings at the soil surface. Such damage occurs frequently in the coastal plain of the southeastern United States.

When particulate matter coats leaves, photosynthesis can be impaired. Finer particles may obstruct stomatal openings so that gas exchange is inhibited. Airborne particles settling on crops such as lettuce and celery may impair quality; many fruit and ornamental crops are similarly damaged.

Perhaps the most subtle of all dust-caused damage may be indirect: Dust particles can carry pesticides. Even particles that have been airborne for as much as 1600 km (1000 miles) may have a coating of pesticides. This means that the spraying practices of a grain farmer in the Midwest, for example, could have a direct effect on grape production in the East.

FIGURE 11-5. Ethylene can injure greenhouse crops, such as the orchids shown here. [Courtesy California Statewide Air Pollution Research Center.]

Potentially damaging particles come from the smoke produced by controlled and uncontrolled burnings. Dusts from various sources, aerosols, and other kinds of particulate matter, including smoke, all cause injury to shade trees.

Acid Rain

Acid rain or acid precipitation refers to the presence of acid-causing compounds in rain that have been associated with extensive ecological damage. The pH of unpolluted rain is about 5.6 due to natural carbonic acid formed from atmospheric carbon dioxide. Rain at pH 4.0 to 4.3 appears to be typical in the northeastern United States, where acid rain is reported to be gravest and rain episodes as low as pH 2.2 to 3.0 have been reported. The acid components in the rain have been implicated in plant injury and death, the destruction of aquatic life, as well as in the literal dissolution of building materials such as stone and bronze.

Acid rain can be produced naturally—by volcanic eruptions, for example—but the main contributors are human activities in industrial societies, such as the creation of sulfur dioxide and nitrogen oxide by burning fossil fuels. Coal-burning utilities are responsible for over 80 percent of human-made SO_2 in the atmosphere. The SO_2 and NO_2 emission, when catalyzed by sunlight, becomes transformed into dilute sulfuric and nitric acid—acid rain. Other acid-forming compounds have also been implicated. Injury to plant and aquatic life from acid rain may be due in part to alterations in soil chemistry produced by acidified conditions that cause the release of metals such as aluminum.

Although the precise cause and effect of acid rain and biological health hazards remain elusive, it seems clear that the pollution of our industrial society is the culprit. The only control must be the reduction of sulfates from emissions by a combination of various techniques. These include the substitution of low sulfur coals—especially in the summer during peak electrical usage for air conditioning, the use of coal washing to reduce soluble sulfates, and various chemical techniques to remove sulfates from gases at industrial sites (sulfur scrubbers). The control is expensive but it is a necessary cost of continued use of coal energy.

Selected References

Diagnosing Vegetation Injury Caused by Air Pollution. 1978. Applied Science Associates; U.S. Environmental Protection Agency (EPA-450/3-78-005). (A basic manual on air pollution injury to plants and a pictorial guide to symptoms.)

Grace, J., E. D. Ford, and P. G. Jarvis (eds.), 1981. *Plants and Their Atmospheric Environment*. New York: Wiley. (Proceedings of a conference on the relation between vegetation and the atmosphere.)

Lemon, E. R. (ed.), 1983. *CO_2 and Plants: The Response of Plants to Rising Levels of Atmospheric Carbon Dioxide*. (AAAS Selected Symposium.) Boulder, Colo.: Westview Press. (An assessment of the biological effects of CO_2 in the atmosphere.)

Mansfield, T. A. (ed.), 1976. *Effect of Air Pollutants on Plants*. Cambridge, Eng.: Cambridge University Press. (A collection of essays on experimental studies on air pollution and plant metabolism.)

Ormrod, D. P. (ed.), 1978. *Pollution in Horticulture*. New York: Elsevier. (A presentation relating the nature of pollution problems in horticulture to factors that must be considered for their solution.)

Unsworth, M. H., and D. P. Ormrod, 1982. *Effects of Gaseous Air Pollution in Agriculture and Horticulture*. Woburn, Mass.: Butterworth, Scientific. (Proceedings of an international conference devoted to effects of air pollution on agriculture with special emphasis on horticultural crops.)

Climatron, St. Louis, Missouri. [Photograph by Piaget, Courtesy Frits Went.]

THE TECHNOLOGY OF HORTICULTURE III

Propagation 12

Propagation refers to the controlled perpetuation of plants. The basic objective of plant propagation is twofold: to achieve an increase in number and to preserve the essential characteristics of the plant. There are two essentially different types of propagation: sexual and asexual. **Sexual propagation** is the increase of plants through seeds formed from the union of gametes. **Asexual propagation** is the increase of plants through ordinary cell division and differentiation. The essential feature of asexual propagation is the regeneration of missing plant parts. Thus, a stem cutting initiates roots, a root cutting develops shoot buds, and a leaf cutting initiates both roots and shoots.

SEED PROPAGATION

Seed is the most common means of propagation for self-pollinated plants, and is extensively used for many cross-pollinated plants. It is often the only possible or practical propagation method. Propagating from seed has many advantages. It is usually the cheapest method of plant propagation. Seeds also offer a convenient method for storing plants. When kept dry and cool, they remain viable from harvest to the following planting season. Some seeds remain productive for very long periods: Those of Indian lotus *(Nelumbo nucifera)* remain viable for as long as 1000 years; and seeds of Arctic tundra lupine, uncovered after being frozen for an estimated 10,000 years, germinated and produced normal plants.

Another advantage is that seed propagation provides a method for starting "disease-free" plants. This is especially important with respect to virus diseases, since it is almost impossible to free an infected plant of viruses. Most virus diseases are usually not seed-transmitted. The major disadvantages to seed propagation, besides genetic segregation in heterozygous plants (see Chapter 6), is the long time required by some plants to reach maturity from seed. For example, 8 years is usually needed for pears to fruit from seed. Potatoes do not produce large tubers the first year when grown from seed. Thus, asexual propagation not only provides trueness to type but, in many plants, saves several years.

Seed Production and Handling

Although seeds of some vegetables (for example, tomato and watermelon) are grown in the same locations used for crop production, most seed crops are grown in specialized locations. (The principal horticultural seed-producing areas of the United States are in the western states, principally Idaho and California.) The limitation may be imposed by specific flowering requirements, such as cold induction or photoperiod. In addition, commercial seed production entails specialized requirements: low moisture at harvest to permit proper maturation of the seed and to reduce the incidence of fungal and bacterial disease, and isolation to prevent contamination in cross-pollinating species. For example, sweet-corn seed cannot be produced in the Midwest because the abundance of pollen from field corn or popcorn would result in undesirable hybrids. Very little tree and shrub seed is grown commercially. This seed is collected from natural stands, nurseries, and arboreta. Seed from fruit trees, the seedlings of which are often used as rootstocks, is obtained as a by-product of the fruit-processing industry.

A considerable amount of hand labor is required in the harvesting of flower and vegetable seed. This is particularly true of species in which the seed head, or pod, shatters easily, or of plants on which the seed matures gradually, such that at any one time there may be mature seed pods, flowers, and flower buds. Examples are aquilegia, delphinium, salvia, petunia, and pansy. One great advantage of hand picking is that the cleaning operations are greatly simplified. With other crops the entire plant is cut and placed on sheets of canvas or in windrows to dry and is then threshed. This procedure is used with carnation, centaurea, daisy, gypsophila, hollyhock, larkspur, French marigold, lychnis, phlox, scabiosa, snapdragon, verbena, sweetpea, nasturtium, and morning-glory.

In many crops the seed must be extracted or milled from the fruits and then cleaned. The separation of the seed from fleshy fruits, such as the tomato, is accomplished through fermentation of the macerated pulp. Seed is removed from dried pods or seed heads by milling. Cleaning is facilitated by differences in size, density, and shape of the seed in comparison with the plant debris or other seeds that may have been harvested incidentally. Screens may be used to separate large particles from the seed. Small, light fragments are blown out by passing an air stream through the seed as it moves from one screen to another or is passed across a porous bench or against an inclined plane. The heavier seed remains at the base while lighter material moves up the plane. Seeds or particles of the same density as the crop seed but of a different shape can be removed on an "indent machine." A wheel covered with indentations is passed through a mass of seed, and each "indent" picks up a seed. The size and shape of the indentation are determined by the crop being cleaned. Some seeds, particularly beans and peas, can be separated on the basis of color. Single seeds are picked up by suction through perforations on a hollow wheel and then are passed through a photoelectric cell. If the cell detects a seed of the wrong color, a device releases the vacuum and ejects the seed. Throughout the milling and cleaning operations, extreme care must be taken in the adjustment of the machinery. Seeds that are chipped or damaged may be less viable or may produce abnormal seedlings.

Seed Storage

The storage life of seeds varies greatly with the species. With any species the longevity is greatly affected by storage conditions. Most seeds retain the highest viability in a relative humidity of 4 to 6 percent, although those of some plants (for example, silver maple and citrus) lose their viability under low moisture. The best temperatures for storing many seeds have been shown to range between 0 to 32°F (−18 to 0°C). Actual storage conditions required for seed depends ultimately on the species and on the length of storage time desired. For most seeds, temperatures of 32 to 50°F (0 to 10°C) and a relative humidity of 50 to 65 percent are adequate to maintain full viability for at least one year.

Germination

Germination, the series of events from dormant seed to growing seedling, depends upon seed viability, the breaking of dormancy, and suitable environmental conditions. Because germinating seeds and young seedlings are vulnerable to certain diseases (for example, damping-off disease), protection must be provided.

Viability

Seed viability refers to the percentage of seed that will complete germination, the speed of germination, and the resulting vigor of the seedlings. The viability of seed lots can be determined by standardized testing procedures. Probably the most significant measure of viability is the **germination percentage,** the percentage of seed of the species tested that produces normal seedlings under optimum germinating conditions (Figure 12-1). Germinating tests are usually performed on moistened absorbent paper under rigidly controlled environmental conditions (Figure 12-2). The length of the test varies, for some species are notoriously slow to germinate. Perhaps the greatest problem is in distinguishing between dormant and nonviable seed. Seed dormancy must be overcome to obtain a reliable test. A rapid chemical test involving tetrazolium (2,3,5-triphenyltetrazolium chloride) makes it possible to evaluate viability in nongerminating dormant seed. Living cells treated with the chemical turn red, whereas nonliving cells show no color.

Breaking Dormancy

Breaking dormancy and creating a suitable growing environment are necessary to initiate the germination process. Many treatments can be used to break dormancy, the appropriate ones being determined by the particular type of dormancy (see Chapter 5). They include scarification, dry storage, stratifica-

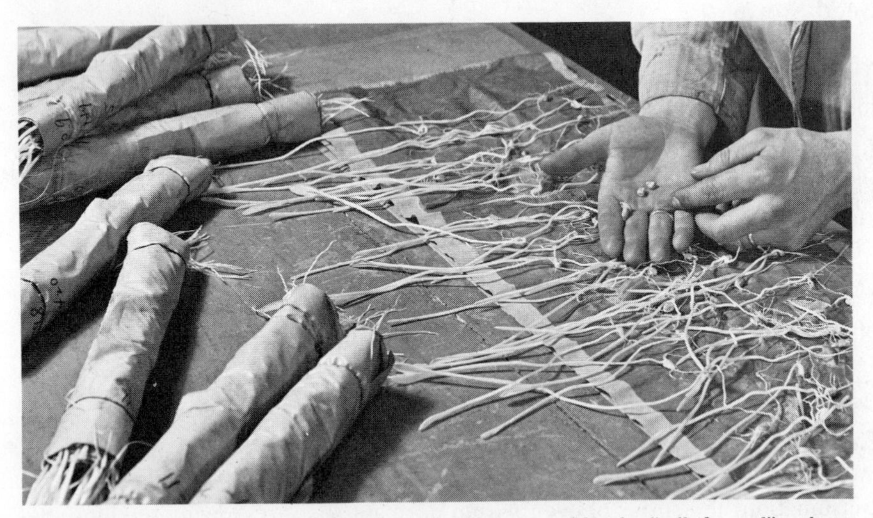

FIGURE 12-1. Germination of sweet corn seed is being tested by the "rolled towel" technique. One hundred seeds are placed on moist paper toweling, which is sealed in wax paper, rolled, and stored at a standardized temperature. After 7 days, the percentage of germinated seeds is determined. [Photograph by J. C. Allen & Son.]

FIGURE 12-2. Seed germinator with complete environmental control, including control of light. This germinator was designed for the Indiana State Seed Laboratory. [Photograph by J. C. Allen & Son.]

tion, embryo culture, or various combinations of these treatments, along with suitable environment control.

SCARIFICATION. The germination of seed that contains an impervious seed coat may be promoted by **scarification**—the alteration of the seed coat to render it permeable to gases and water. This is accomplished by a number of techniques, mechanical methods involving abrasive action being the most common. The action of hot water (170 to 212°F, or 77 to 100°C) is effective for the seed of honey locust. Some seeds are best scarified by the corrosive chemical action of sulfuric acid.

DRY STORAGE AND STRATIFICATION. Seed that will not germinate immediately after harvest requires **dry storage** for a period of days or months. The physiological basis of this type of dormancy is not clear but it has been associated with the evolution of volatile germination inhibitors. The afterripening of some seeds requires a period of moist storage known as **stratification.**

Cold-stratification—the afterripening of dormant embryos by storing them at high moisture and low temperature—is a prerequisite for the uniform germination of many temperate-zone species, such as apple, pear, and redbud. The cold-stratification of apple and pear seed involves storing the moist aerated seed at around 32°F (0°C). The germination percentage increases with time until the third month of treatment. The stratification medium consists of moist soil, sand, and peat, or such synthetic substances as vermiculite. An effective means of preventing the loss of moisture and of providing an adequate exchange of oxygen and carbon dioxide consists in sealing the seed in polyethylene bags containing moist blotting paper.

Warm-stratification—moist storage above approximately 45°F (7°C)—promotes germination in some species as a result of microbial decomposition of the seed covering. Seed of such plants as viburnum, which exhibits two different types of dormancy (**double dormancy**), are first **warm-** and then **cold-stratified.** For redbud seed, a combination of scarification and cold-stratification is used.

EMBRYO CULTURE. The aseptic growth of excised embryos (often with associated parts, as placenta tissue) in artificial media is known as **embryo culture.** This specialized technique is used to facilitate seed germination in some species. For example, the embryos of many early-ripening peaches (such as 'Mayflower') are not sufficiently mature to germinate when the fruit is ripe. A serious impediment to the breeding of early-ripening peaches, this problem can be overcome, however, by excising the embryo from the pit and culturing it under aseptic conditions in media that provide certain nutrients. Tukey's Solution is one such medium. This technique, called **embryo rescue,** is also used to produce viable seedlings from interspecific crosses that have defective endosperms. A number of new types of crucifers from interspecific crosses have been developed with the aid of embryo culture. Embryo culture is also used to circumvent dormancy caused by inhibitory substances associated with the seed coat. For this reason the technique is used with viburnum.

Tukey's Solution for culturing mature and relatively immature embryos.

Stock chemical	Relative amounts
KCl	5
$CaSO_4$	1.25
$MgSO_4$	1.25
$Ca_3(PO_4)_2$	1.25
$Fe_3(PO_4)_2$	1.25
KNO_3	1

Note: Use 1.5 g of mixture per liter of water.

The routine germination of orchid seed involves culture in artificial media. This seed, which is almost microscopic, consists of a very simple, undifferentiated embryo and contains no reserve food. The stage between germination and the developing seedling is called a **protocorm.**

Knudson's Solution B for growing orchid seedlings.

Chemical	mg/liter water
$Ca(NO_3)_2 \cdot 4H_2O$	1000
$(NH_4)_2SO_4$	500
KH_2PO_4	250
$MgSO_4 \cdot 7H_2O$	250
$FePO_4 \cdot 4H_2O$	50
Agar	17,500
Sucrose	20,000

Environmental Factors Affecting Germination

The germination of seed that does not require afterripening, or of seed that has had this requirement satisfied, depends upon external environmental factors—namely, water, favorable temperature, oxygen, and, sometimes, light.

The amount of water required for germination varies somewhat from species to species. For example, celery seed requires that soil moisture be near field capacity, whereas tomato seed will germinate with soil moisture just above the permanent wilting point. For most seed, excessive wetness is harmful, since it prevents aeration and promotes disease. Moisture must be maintained during germination, however, lest the germinating seedling dry out. Shading to conserve moisture is recommended until germination is complete. The use of glass over seed flats conserves moisture, but care should be taken to prevent the seeds from getting too hot.

The effect of temperature upon germination varies by species and is related somewhat to the temperature requirement for optimum growth of the mature plant (Table 12-1). In general, the germination rate increases as temperature rises, although the highest germination percentage may be at a relatively low temperature. An alternating temperature is usually more favorable than a constant temperature. Because of its critical role in respiration, oxygen is necessary for seed germination in all plants except some water-loving species (for example, rice and cattails). The maintenance of proper drainage and tilth in seed beds promotes rapid germination, largely as a result of good aeration (Figure 12-3).

TABLE 12-1. Soil temperatures for vegetable seed germination.

Minimum					
32°F (0°C)	40°F (4°C)		50°F (10°C)	60°F (16°C)	
Endive	Beet	Parsley	Asparagus	Lima bean	Okra
Lettuce	Broccoli	Pea	Sweet corn	Snap bean	Pepper
Onion	Cabbage	Radish	Tomato	Cucumber	Pumpkin
Parsnip	Carrot	Swiss chard		Eggplant	Squash
Spinach	Cauliflower	Turnip		Muskmelon	Watermelon
	Celery				
Optimum					
70°F (21°C)	75°F (24°C)	80°F (27°C)	85°F (29°C)		95°F (35°C)
Celery	Asparagus	Lima bean	Snap bean	Radish	Cucumber
Parsnip	Endive	Carrot	Beet	Sweet corn	Muskmelon
Spinach	Lettuce	Cauliflower	Broccoli	Swiss chard	Okra
	Pea	Onion	Cabbage	Tomato	Pumpkin
		Parsley	Eggplant	Turnip	Squash
			Pepper		Watermelon
Maximum					
75°F (24°C)	85°F (29°C)	95°F (35°C)		105°F (41°C)	
Celery	Lima bean	Asparagus	Eggplant	Cucumber	Squash
Endive	Parsnip	Snap bean	Onion	Muskmelon	Sweet corn
Lettuce	Pea	Beet	Parsley	Okra	Turnip
Spinach		Broccoli	Pepper	Pumpkin	Watermelon
		Cabbage	Radish		
		Carrot	Swiss chard		
		Cauliflower	Tomato		

SOURCE: H. T. Hartmann and D. E. Kester, 1959, *Plant Propagation: Principles and Practices* (Prentice-Hall).

The effect of light in stimulating or inhibiting the germination of some seed (discussed in Chapter 5) is a reversible red–far red phenomenon. To produce good stocky plants, ample light must be supplied during early seedling growth.

The action of certain salts has been shown to influence germination. At concentrations of 0.1 to 0.2 percent, potassium nitrate will increase germination in a number of plants, and for this reason it is used in seed testing. Seed treatment with salts has recently been used to achieve rapid field germination. In general, however, high concentrations of salts caused by overfertilization inhibit germination.

FIGURE 12-3. Good seedbed preparation promotes rapid seed germination. [Photograph courtesy Ford Motor Co., Tractor and Implement Division.]

Disease Control

Disease is a critical factor in the germination process. This is especially true for seed that must be stratified or that requires considerable time for germination. The control of these diseases is an integral part of the technology of seed propagation.

The major diseases of germinating seeds are grouped under a single name: **damping off.** These diseases are caused by several separate fungi, mainly species of *Pythium*, *Rhizoctonia*, and *Phytophthora*. Damping off is expressed either by the failure of the seedling to emerge or by its death shortly after emergence. A common symptom is the girdling of young seedling stems at the soil surface. Damping off usually occurs only in young, succulent seedlings during or shortly after germination, but older plants may be affected in severe cases. Damping off can be severe in both greenhouse and field soils, and it is often a limiting factor in the success of seed propagation. Protection of seedlings from damping off and other diseases involves both the direct control of the pathogens and the regulation of environmental conditions such that they favor the rapid growth of the plant rather than the growth of the pathogens.

SEED AND SOIL TREATMENT. A number of seed treatments are available either to eliminate the pathogens from the seed or to provide protection to the seedling when planted in infested soil. These consist in coating the seed with a suitable fungicide, such as cuprous oxide or calcium hypochlorite. A common seed treatment is a 5-minute dip in a 10 percent solution of Clorox (which is a 5.25 percent solution of sodium hypochlorite). Several compounds intended for seed treatment are available commercially. The treatment of seeds in hot water—122°F (50°C) for 15 to 30 minutes—has been used for seed-borne diseases of vegetables (for example, alternaria of onion). Such treatment must be precise, however, or the seed may be seriously injured.

Soil may be treated by applying fungicides to the upper surface or by applying heat. Raising the soil temperature to 180°F (82°C) for 30 minutes ("pasteurization") is always recommended for potting soils to control weeds and nematodes, as well as damping-off organisms. Complete soil sterilization interferes with the availability of nutrients and should be avoided. Sphagnum moss has proved satisfactory as a germination and stratification medium for some seeds because inhibitors and low pH prevent the growth of many of the damping-off organisms. The use of such sterile media as sand, vermiculite, or perlite may be desirable for seed germination. Care must be taken, however, to avoid recontamination of sterilized soil. The absence of natural predators (bacteria and other fungi) may result in great damage if such soil becomes infested with a pathogen.

CONTROL OF THE ENVIRONMENT. Any environmental effect that encourages rapid plant growth more than it does the buildup of pathogens is effective in the control of such seedling diseases as damping off, because older seedlings appear to resist attack. The temperatures most favorable to damping-off fungi are approximately 70 to 85°F (21 to 29°C). Thus, damping off tends to be severe when cool-season crops are germinated at temperatures that are too high, and vice versa. For best control, germinating temperatures should be optimum for the crop. This principle can be utilized in the control of damping off in the field by regulating planting dates. Good viable seed and rapid seedling growth are important. Many of the fungi responsible for damping off are water-loving,[1] and are encouraged in wet soils. Cloudy weather and periods of poor drying encourage the damping-off complex; consequently, frequent and shallow watering should be avoided after planting.

Sanitation to reduce the buildup of organisms responsible for damping off should be practiced in the greenhouse, where this trouble is a perpetual problem. This special care involves eliminating plant refuse, disinfecting the walks and the potting area, keeping unsterilized soil out of the potting area, and general cleanliness.

Planting

Seed may be sown in a permanent location (**direct seeding**) or it may be planted first in some container from which the young plants can be transplanted once or many times before permanent planting. The growing of **transplants** makes it possible to provide precise environmental control during the critical stages of germination and early seedling growth. Many ornamentals are grown from transplants, as are vegetables for early production.

Direct Seeding

Plants that are difficult to transplant, or those for which the individual value of the plant does not justify the trouble and expense that transplanting entails, are grown by direct seeding. Many of the common vegetables (for example, beans, sweet corn, and radishes) are always grown by direct seeding (Figure 12-4). Although direct seeding requires much less labor and trouble than transplanting, one of its limitations is weed control. However, the recent advances in chemical weed control have made the direct seeding of some crops, such as tomato, economically feasible.

Precision spacing is important in direct seeding to prevent the need for extensive thinning or replanting. Because such spacing is difficult to accomplish with small seeds, attempts have been made to "**pelletize**" such seeds by coating them with some suitable material, usually clay with additives. The increased size of the seed facilitates planting, and the coating material may be treated with fertilizer to encourage rapid seedling growth. Pelleting has been somewhat successful with lettuce, but with present materials it is of doubtful value for most crops, since the pelletizing materials can reduce or retard germination. But this is an area of seed technology that can be expected to change dramatically.

[1]The fungi of class Phycomycetes, to which both *Pythium* and *Phytophthora* belong, are often referred to as water molds.

FIGURE 12-4. Direct seeding with an eight-row planter. [Photograph by J. C. Allen & Son.]

The planting of pregerminated seed—either suspended in a protective gel or delivered to the soil with a small amount of water—is a new technology called **fluid drilling.** The potential advantages of this method over conventional drilling systems using dry seed are that growers can proceed with germination under ideal conditions prior to sowing, eliminating the variable effects of uncontrolled seed-bed environment, and they can exploit the gel as a carrier for nutrients, plant growth regulators, and pesticides to encourage early seedling growth. Experimental studies with fluid drilling on vegetable crops have indicated higher and more uniform germination rates that produce higher yields and greater crop uniformity.

Transplants

The growth of transplants is a specialized part of seed propagation. Seed may be germinated first in seedling flats containing specially prepared media, and the seedlings transplanted later to suitable containers. If the seedlings transplant with difficulty (as do those of the cucurbits), the seed may be planted directly into individual containers, such as small plastic pots or 3-inch veneer plant bands. The germination medium in the seedling flats is usually sand or a sand-soil mixture. Sand has the advantage of being well drained and relatively easy to keep free of disease-producing fungi. Germination media are lacking in nutrients, but this is not essential as long as the seeds are transplanted soon after emergence. Supplemental feedings with nutrient solutions can be provided. The depth of planting in seedling flats depends on the size of the seed. As a rough approximation, the planting depth should be one to two times the largest seed diameter. Very fine seed may simply be sprinkled over the surface of the soil.

Seedlings grown in flats should be transplanted as soon as they are large enough to handle. The transplanting operation must be done carefully to prevent injury. (In many plants, however, the destruction of the tap root results in a more fibrous root system, which may be advantageous.) The transplanting operation is best made with the soil medium just wet enough to be impressionable, but not wet enough to be sticky. A "dibbler" is useful for making the planting hole.

A number of containers, made of various materials, are available for transplanting—for example, flats, pots, and bands. Containers made of a decomposable organic material such as peat are proving of value, especially for retail flower transplants. In their manufacture, peat pots are treated with a fungicide to prevent decomposition by mold and with nitrogenous fertilizer to overcome nitrogen deficiencies commonly associated with the use of organic materials. Wooden plant bands are best soaked in nitrogenous fertilizer for the same reason.

Field transplanting is a part of both seed and vegetative propagation. Transplants may be planted along with the soil in which they were grown, or they may be "bare-rooted." The transplanting of bare-rooted plants of many crops (tomato, strawberry, and many kinds of nursery stock) is well adapted to mechanization (Figure 12-5). Transplanting machines are often equipped to apply water and starter solution. Bare-root nursery plants are usually covered with mud before transplanting to prevent drying.

FIGURE 12-5. A tomato transplanting operation. Starter solution is being added to the transplanter. [Photograph by J. C. Allen & Son.]

VEGETATIVE PROPAGATION

Vegetative propagation involves nonsexual reproduction through the regeneration of tissues or plant parts. In many cases, this process is a completely natural one; in others, it is more or less artificial, depending on the degree of human interference and regulation. Among the many methods of vegetative propagation, the best one to use depends on the plant and the objectives of the propagator. The advantages of vegetative propagation are readily apparent. Heterozygous material may be perpetuated without alteration. In addition, vegetative propagation may be easier and faster than seed propagation, as seed-dormancy problems may be completely eliminated and the juvenile stage reduced. Vegetative propagation also makes it possible to perpetuate clones that do not produce viable seed or any seed at all—for example, 'Washington Navel' orange, 'Gros Michel' banana, and 'Thompson Seedless' grape.

The various methods of vegetative propagation are summarized in the following list.

1. Utilization of apomictic seed (as with citrus)
2. Utilization of specialized vegetative structures
   - Runners (strawberry)
   - Bulbs (tulip)
   - Corms (gladiolus)
   - Rhizomes (iris)
   - Offshoots (daylily)
   - Stem tubers (potato)
   - Tuberous roots (sweetpotato)
3. Induction of adventitious roots or shoots
   - Layerage (regeneration from vegetative part while still attached to the plant)
   - Cutting propagation (regeneration from vegetative part detached from the plant)
4. Graftage (the joining of plant parts by means of tissue regeneration)
5. Tissue culture

Utilization of Apomictic Seed

Apomixis refers to the development of seeds without the completion of the sexual process. It is therefore a form of nonsexual or vegetative reproduction. The most significant type of apomixis is that in which the complete meiotic cycle is eliminated. The seed is formed directly from a diploid cell, which may either be the nonreduced megaspore mother cell or some cell from the maternal ovular tissue. As a result, a heterozygous cross-pollinating plant will appear to **breed true.**

Although apomixis is widespread within the plant kingdom, it is not commonly exploited as a means of asexual propagation. It is utilized in the propagation of 'Kentucky' bluegrass, citrus rootstocks, and mango. These species, however, are only partially apomictic, and the seed they produce is derived from both the sexual process and apomixis.

Utilization of Specialized Vegetative Structures

The natural increase of many plants is achieved through specialized vegetative structures. These modified roots or stems are often also food-storage organs (bulbs, corms, and tubers), although in some plants they function primarily as natural vegetative extensions, as do runners. These organs enable the plant to survive adverse conditions, such as the cold period in temperate climates or the dry period in tropical climates, and provide the plant with a means of spreading. These specialized structures renew both the plant and themselves through adventitious roots and shoots, and they are commonly utilized by people as a means of propagation. When these structures subdivide naturally, the process is called **separation;** when they must be cut, the process is called **division.**

Stem Modifications

BULBS. **Bulbs** are shortened stems with thick, fleshy leaf scales (Figure 12-6). In addition to their development at the central growing point, bulbs produce buds at the axils of their leaf scales that form miniature bulbs, or **bulblets.** When grown to full size, bulblets are known as **offsets.**

The development of bulbs from initiation to flowering size takes a single season in the onion, but most bulbs, such as those of the daffodil and hyacinth, continue to grow from the center, becoming larger each year while continually producing new offsets. The asexual propagation of bulb-forming plants is commonly achieved through the development of scale buds. Various stages of development may be utilized from the individual scales, offsets, or the enlarging mature bulb itself. In hyacinth propagation the bulb is commonly wounded to encourage the formation of adventitious bulblets. The bulblets develop into usable size in 2 to 4 years.

CORMS. Although they resemble bulbs, **corms** do not contain fleshy leaves, but are solid structures consisting of stem tissues, complete with nodes and internodes. The gladiolus, crocus, and water chestnut *(Eleocharis tuberosa)* are examples of corm-forming plants. In large, mature corms, one or more of the upper buds develop into a flowering shoot. The corm is expended in flower production, and the base of the shoot forms a new corm above the old. By season's end, one or more new corms may have developed in this manner. **Cormels,** or miniature corms, are fleshy buds that develop between the old and new corms. They do not increase in size when planted, but produce larger

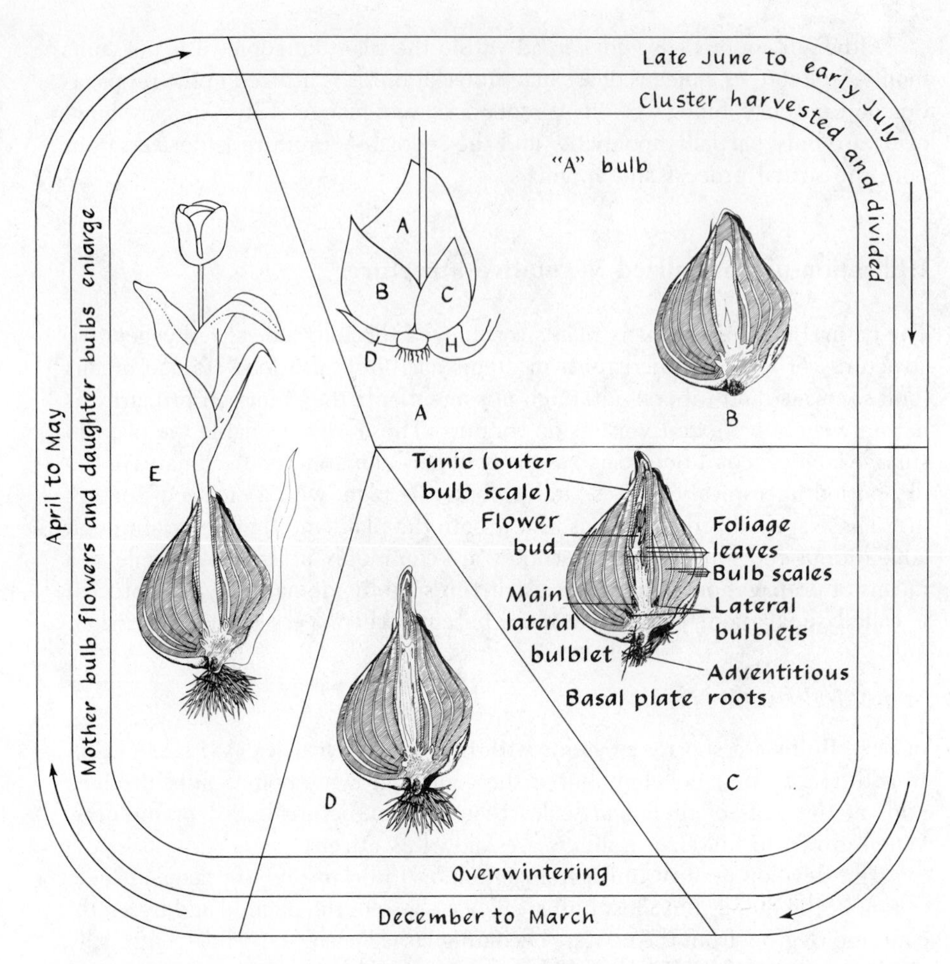

FIGURE 12-6. Annual growth and development cycle of a flowering tulip. (*A*) Harvested cluster. (*B*) Separated "A" bulb. (*C*) Mother bulb with developed root primordia and shoot prior to planting. (*D*) Rooted bulb in overwintering environment. (E) Mother bulb at anthesis, small shoot is "H" bulb. [From A. A. De Hertogh, L. H. Aung, and M. Benschop, *Horticultural Reviews* 5:77, 1983.]

corms from the base of the new stem axis. These require 1 to 2 years to reach flowering size. This is the usual method of propagating corm-producing plants. Corms may also be increased by division, but this is not commonly practiced because of disease problems.

RUNNERS. **Runners** are specialized aerial stems that develop from the leaf axils at the base or crown of a plant having rosette stems (Figure 12-7). They provide a means of natural increase and spread. Among the plants propagated by runners are the strawberry, strawberry geranium (*Saxifraga sarmentosa*), and

FIGURE 12-7. Runnering in the strawberry geranium (*Saxifraga sarmentosa*). [Photograph courtesy E. R. Honeywell.]

bugle weed (*Ajuga*). The commercial propagation of strawberries is done through runner-plant production. Leaf clusters, which root easily, are formed at the second node of the runner. These rooted plants may in turn produce new runners. Runnering is photoperiod sensitive, being commonly initiated under a day length of 12 hours or longer. Dormant plants are dug by machines in the fall or in the spring. The yield in plants per mother plant varies with the variety, but under optimum conditions it may be as high as 200:1. A field increase of 20:1 or 30:1 is common. Some species of strawberries are nonrunnering, and many of the ever-bearing cultivars usually form relatively few runners. These plants may be vegetatively propagated by crown divisions, but the increase in plants is much lower than in those with runners.

RHIZOMES. Horizontal cylindrical stems growing underground are called **rhizomes** (Figure 12-8). They contain nodes and internodes of various lengths, and they readily produce adventitious roots. Rhizomes may be thick and fleshy (iris) or slender and elongated ('Kentucky' bluegrass). Growth proceeds from the terminal bud or through lateral shoots. In many plants the older portion of the rhizome dies out. If new growth proceeds from branching, the new plants eventually become separated. Rhizomatous plants are easily propagated by cutting the rhizome into several pieces, as long as each piece contains a vegetative bud.

TUBERS. Fleshy portions of underground rhizomes (of which the potato is probably the best known example) are known as **tubers.** The potato is propagated by planting either the whole tuber or pieces containing at least one "eye," or bud. If the entire tuber is planted, the terminal eye commonly inhib-

FIGURE 12-8. Root system and rhizomes of the tawny day-lily, *Hemerocallis fulva*. [Photograph courtesy G. M. Fosler.]

its the other buds, but this apical dominance is destroyed when the tuber is cut. Commonly, the "seed pieces" are kept at 1 to 2 ounces (25 to 50 g) to provide sufficient food for the young plant. The seed pieces may be cured for awhile to heal the cut surface (see Chapter 18). Chemical treatments are used to prevent disease.

OFFSHOOTS. In many plants, lateral shoots develop from the stem, which, when rooted, serve to duplicate the plant (Figure 12-9). These have been called, in horticultural terminology, **offsets, suckers, crown divisions, ratoons,** or **slips,** depending on the species. Lateral shoots may be referred to collectively as **offshoots.** The increase of bulbs and corms by offsets is a similar phenomenon. Propagation of plants that produce offshoots is easily made by division. In temperate climates, rooted offshoots of outdoor perennials may be divided either in fall or spring.

Root Modifications

TUBEROUS ROOTS. Roots as well as stems may be structurally modified into propagative and food-storage organs. Fleshy, swollen roots that store food materials are known as **tuberous roots** (Figure 12-10). Shoot buds are readily formed adventitiously. Tuberous roots of some species may contain shoot buds at the stem end as part of their structure.

The sweetpotato is commonly propagated from the formation of rooted adventitious shoots called **slips.** In the dahlia, the roots are divided, but each tuberous root must incorporate a bud from the crown. In the tuberous begonia, the primary tap root develops into an enlarged tuberous root with buds at the stem end. This root can be propagated by division, but each section must contain a bud.

FIGURE 12-9. Pineapple may be propagated from slips—leafy shoots originating from axillary buds borne on the base of the fruit stalk. They may also be grown from the crown that issues from the top of the pineapple or from suckers that grow lower on the stem. [Photograph courtesy Dole Corp.]

SUCKERS. Shoots that arise adventitiously from roots are called **suckers**, although the term has been commonly used (less precisely, perhaps) to refer to shoots originating from stem tissue. The red raspberry, for example, is propagated by suckers abundantly produced from horizontal roots, and suckering may be stimulated by extensive pruning. The rooted suckers are usually dug during the period of plant dormancy.

Induction of Adventitious Roots and Shoots

The regeneration of structural parts in the propagation of many plants is accomplished by the artificial induction of adventitious roots and shoots. When the regenerated vegetative part is attached to the plant, the process is called **layerage;** when the regenerated vegetative part is detached from the plant of origin, the process is called **cutting propagation.** These two processes, although technically different, are part of the same phenomenon—namely, the ability of vegetative plant parts to develop into a complete plant.

Layerage is often a natural process. In the black raspberry, the drooping stem tips tend to root when in contact with the soil; in strawberries, the runners form natural layers. Because the regenerated stem is still attached and nourished by the parent plant, the timing and techniques of layerage are not as critical as in cutting propagation, in which the vegetative part to be regenerated

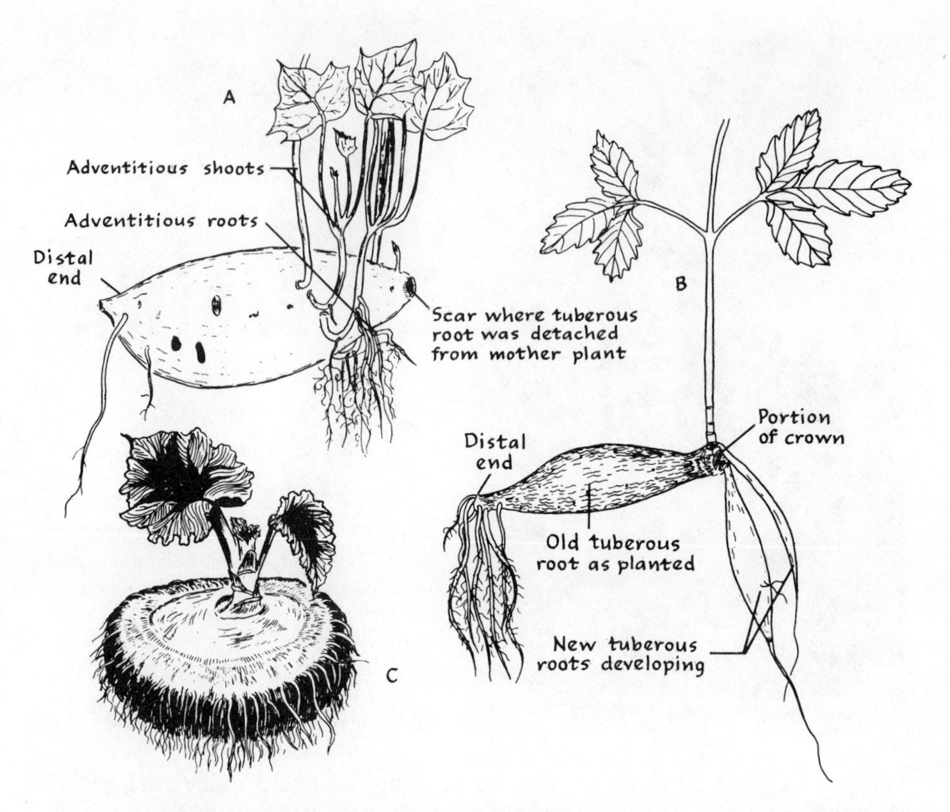

FIGURE 12-10. Tuberous roots of sweetpotato (*A*), dahlia (*B*), and tuberous begonia (*C*). The tuberous root of the sweetpotato and dahlia disintegrate in the production of the new plant. The tuberous root of the begonia enlarges each year. [From H. D. Hartmann and D. E. Kester, *Plant Propagation: Principles and Practices*, Prentice-Hall, 1959.]

is severed from the parent plant. Rooting may be facilitated by such practices as wounding, girdling, etiolation, and disorientation of the stem, which affects the movement and accumulation of the carbohydrates and auxin needed to stimulate root initiation.

Layerage is a simple and effective means of propagation that can be practiced in the field. It is particularly suited to the amateur because of the high degree of success possible with only a minimum of specialized facilities. However, it is relatively slow, offers less flexibility than techniques of cutting propagation, and requires much hand labor. For these reasons, layerage is not adaptable to large-scale nursery practices, and it is normally used only for plants that are most naturally adapted to this method of propagation, or in which propagation by cuttings is difficult. Different types of layerage are illustrated in Figure 12-11.

Cutting propagation is one of the most important means of vegetative propagation. The term "cutting" refers to any detached vegetative plant part that

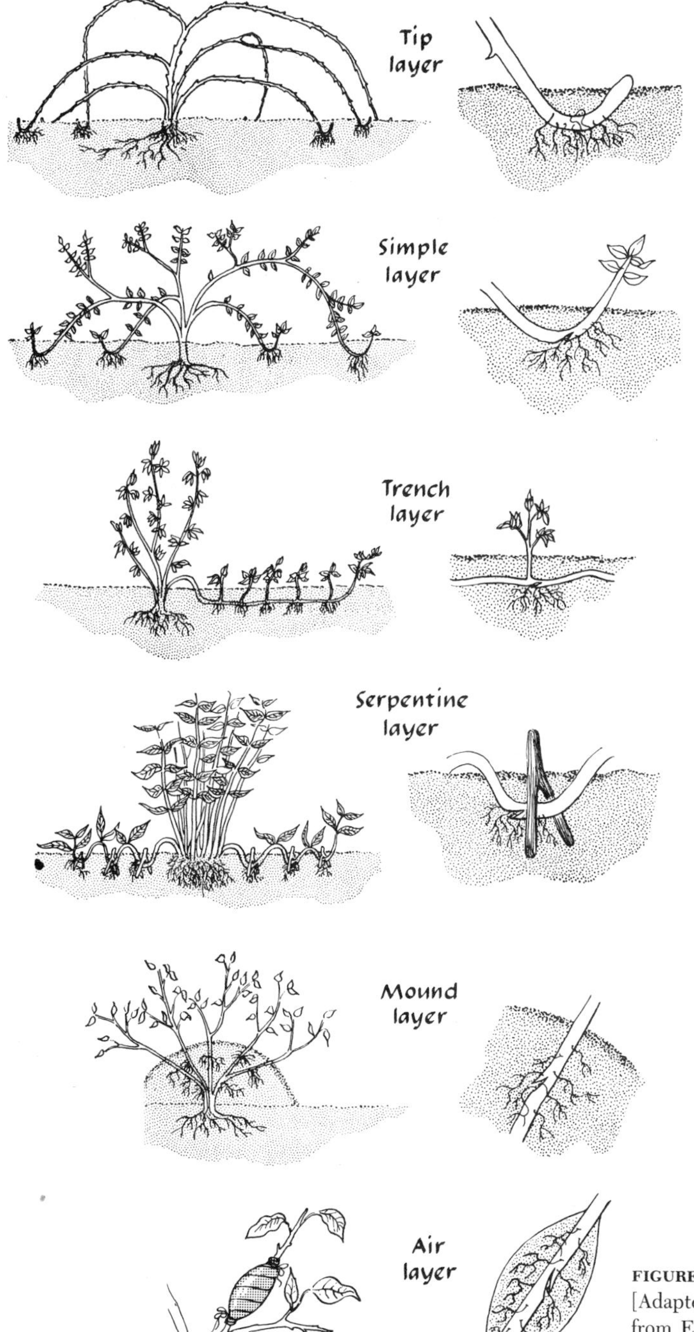

FIGURE 12-11. Methods of layering. [Adapted, with permission of the publisher, from E. L. Denisen, *Principles of Horticulture*, copyright 1958 by the Macmillan Company.]

can be expected to regenerate the missing part (or parts) to form a complete plant. Cuttings are commonly classified by plant part (root, stem, leaf, leaf bud). In **stem cuttings** or **leaf-bud cuttings** a new root system must be initiated; in **root cuttings** a new shoot must be initiated; and in **leaf cuttings** both roots and shoots must be initiated.

Anatomical Basis of Adventitious Roots and Shoots

The formation of adventitious roots may be divided into two phases. One is initiation, which is characterized by cell division and the differentiation of certain cells into a root initial. The second phase is growth, in which the root initial expands by a combination of cell division and elongation. Although the two processes usually occur in sequence, in some plants, such as the willow, the time between initiation and development is well separated.

Root initials are formed adjacent to vascular tissue. In herbaceous plants that lack a cambium, the root initials are formed near the vascular bundles close to the phloem. Thus, roots will appear in rows along the stem, corresponding to the major vascular bundles. In woody plants, initiation commonly occurs in the phloem tissue, usually at a point corresponding to the entrance of a vascular ray.

The production of both adventitious roots and shoots from leaf cuttings commonly originates in secondary meristematic tissues—cells that have differentiated but later resume meristematic activity. In *Kalanchoë* (*Bryophyllum*) new plantlets form during leaf development from meristematic regions on the leaf edges (Figure 12-12).

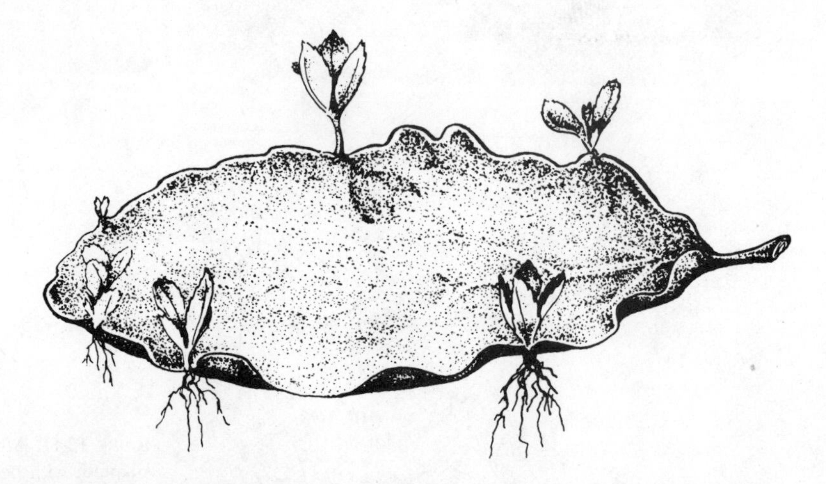

FIGURE 12-12. In *Kalanchoë*, new plants arise from meristem located in notches at the leaf edge. Shoot and root primordia are present in adult leaves. [From J. Mahlstede and E. S. Haber, *Plant Propagation*, Wiley, 1957.]

Adventitious roots and shoots may be derived from different kinds of tissue; for example, in leaf cuttings of African violet the roots are initiated from cells between the vascular bundles, whereas shoots are initiated from cells of the epidermis or cortex. In the sweetpotato, on the other hand, roots and shoots may be derived from callus tissue formed on the cut surface.

The formation of complete plants from pieces of root depends upon both the initiation of adventitious shoots and the extension of new root growth. Adventitious shoot buds develop from cells of the phloem parenchyma and from rays. New roots originate from older tissues through latent root initials, although new root initials may arise adventitiously from the vascular cambial region.

Physiological Basis of Rooting

The variables that enable a stem to root depend upon the plant and its treatment. Some insight into the physiological basis of rooting has been developed from studies on **easy-** and **difficult-to-root** plants. The capacity for a stem to root has been shown to be due to an interaction of inherent factors present in the stem cells as well as to transportable substances produced in leaves and buds. Among these transportable substances are auxins, carbohydrates, nitrogenous compounds, vitamins, and various unidentified compounds. Substances that interact with auxins to affect rooting may be referred to as **rooting cofactors.** In addition, such environmental factors as light, temperature, humidity, and the availability of oxygen play an important role in the process. The physiological factors involved in rooting are only beginning to be understood; it is still not possible to effect rooting in many plants, including blue spruce, rubber tree, and oak.

Auxin levels are closely associated with adventitious rooting of stem cuttings, although the precise relationship is not clear. The normal rooting of stems appears to be triggered by the accumulation of auxins at the base of a cutting. The increase in rooting produced by the application of indoleacetic acid or auxin derivatives supports this concept. However, it is certain that auxins are only a part of the stimulus, for rooting of many difficult-to-root cuttings is not improved by auxins alone. Other specific factors that either stimulate rooting (as does catechol) or inhibit it have been isolated. It is expected that more such factors will be found.

The presence of leaves and buds strongly influences the rooting of stem cuttings. In many plants the effect of buds is due primarily to their role as a source of auxins, whereas the rooting stimulation provided by leaves is related in part to carbohydrate production. But in many plants the effect of leaves and buds can be shown to be due to additional transportable cofactors that complement both carbohydrate and auxin application (Figure 12-13).

An important component of the capacity for a stem to root is the nutritional status of the plant. In general, high carbohydrate levels are associated with

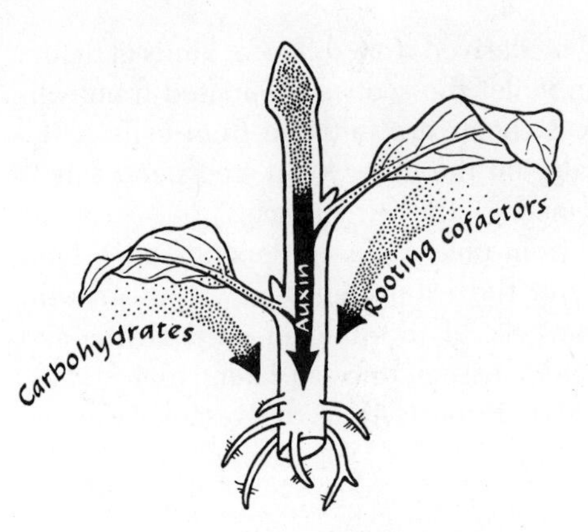

FIGURE 12-13. The rooting of a cutting depends on auxin, carbohydrates, and the presence of rooting cofactors. The cofactors interact with auxin to trigger rooting.

vigorous root growth. The nitrogen levels of the plant will affect the number of roots produced. Although a low nitrogen level will increase this number, an outright deficiency will inhibit rooting.

The accumulation of auxins as well as of carbohydrates explains in part the effectiveness of ringing and wounding in stimulating rooting. In addition, wounding stimulates root initiation by some unknown process. Callusing of the wounded surface also raises the efficiency of water absorption. This wounding effect is utilized to increase the absorption of applied auxins.

The effectiveness of stem rooting varies with the stage of development and the age of the plant, the type and location of stem, and the time of year. Owing to the great variation among species, precise conclusions concerning the relationship of these factors to rooting cannot be made. In general, rooting ability is associated with the juvenile stage of growth. Such plants as English ivy, apple, and many conifers become very difficult to root when they reach the mature stage. Mature, difficult-to-root plants may be made easy to root by a reversion to the juvenile stage. Generally, adventitious shoots from the base of mature plants tend to assume juvenile characteristics. In mature plants that become more difficult to root with increasing age, these adventitious shoots may be induced by severe pruning. A form of layerage called **stooling** maintains the juvenile stage of growth by continued pruning at the base of the plant. The stem bases are mounded with soil to facilitate rooting.

The ability of a stem to root is also affected by its position on the plant; lateral shoots tend to root better than terminal shoots. Vegetative shoots also are likely to root better than flowering shoots. These differences may be related in part to auxin levels and the amount of stored food.

Cuttings vary in their ability to root, depending upon the type of stem tissue from which they are derived. Cuttings may be made from succulent nonlignified growth (**softwood cuttings**)[2] or from wood up to several years old

---

[2]Softwood cuttings of plants that are normally nonwoody are sometimes referred to as **herbaceous cuttings**, whereas those of woody plants prior to lignification are known as **greenwood cuttings**.

(hardwood cuttings). Although almost all types of cuttings of easy-to-root plants root readily, softwood cuttings of deciduous, woody plants taken in the spring or summer generally root more easily than hardwood cuttings obtained in the winter. Dormant hardwood cuttings are used when possible, however, because of the ease with which they can be shipped and handled. Dormant cuttings must be stored until the shoot's rest period is broken, although rooting is less affected by dormancy. The time that softwood cuttings are taken varies greatly with different plant species. Such cuttings of azalea root best in the early spring; for other broadleafed evergreens, the optimum time for rooting softwood cuttings may be from spring to late fall.

Environmental Factors Affecting Rooting

HUMIDITY. The death of the stem as a result of desiccation before rooting is achieved is one of the primary causes of failures in propagation by cuttings. The lack of roots prevents sufficient water intake, although the intact leaves and new shoot growth continue to lose water by transpiration. Leaves or portions of the leaf may be removed to prevent excessive transpiration. However, this practice is not desirable because the presence of leaves encourages rooting. The use of mist (Figure 12-14) maintains high humidity and also reduced leaf temperature by maintaining a water film on the leaf. This makes it possible to expose the cuttings to light of greater intensity, so that photosynthesis can proceed at a high rate. The use of automatic controls to produce an intermittent mist is desirable because it eliminates the accumulation of excess water, which may be harmful to many plants, and because it enables the grower to maintain higher temperatures in the rooting medium.

FIGURE 12-14. An in-bench mist installation. The supply pipe runs along the bottom of the bench. The polyethylene wind barrier eliminates the problem of drift. The deflection-type nozzle produces a mist by directing a fine stream of water against a flat surface. An intermittent mist, commonly 4 seconds on and 56 seconds off, is controlled by a timer. [Photograph courtesy Purdue University.]

TEMPERATURE. The use of bottom heat to maintain the temperature of the rooting medium at about 75°F (24°C) facilitates rooting by stimulating cell division in the rooting area. The aerial portion may be kept cool to reduce transpiration and respiration. Daytime air temperatures of 70 to 80°F (21 to 27°C) and night temperatures of 60 to 70°F (16 to 21°C) are optimum for the rooting of most species.

LIGHT. Light in itself appears to inhibit root initiation (or, conversely, the lack of light encourages it). Softwood and herbaceous cuttings indirectly respond to light because of its role in the synthesis of carbohydrates. However, deciduous hardwood cuttings that contain sufficient stored food, and to which artificial auxins can be supplied, root best in the dark. The role of light in inducing rooting thus varies with the plant and with the method of propagation. The reason that the absence of light favors root initiation in stem tissues is not clear. Root promotion may be achieved by the use of opaque coverings that etiolate the stem. Etiolation probably affects the accumulation of auxins and other substances that are unstable in light.

ROOTING MEDIA. Rooting media must provide sufficient moisture and oxygen and must be relatively disease free. It is not necessary that a rooting medium be a source of nutrients until a root system is established. The medium may have an effect on the percentage of cuttings rooted and on the type of roots formed. Various mixes containing soil, sand, peat, and artificial inorganic substances such as vermiculite (expanded mica) and perlite (expanded volcanic lava) have been widely used. Perlite used alone or in combination with peat moss has proven especially effective because of its good water-holding properties, drain-

FIGURE 12-15. Rooted cuttings of woody ornamentals planted in a mixture of perlite and peat. Fertilizer must be added. They will be marketed as container-grown stock. [Photograph courtesy Perlite Institute.]

FIGURE 12-16. (*Left*) A home propagator made up of a large pot filled with sand. The hole of the smaller inner pot is plugged with cork and kept filled with water. Uniform moisture is maintained by seepage from the small pot's porous sides. (*Right*) Cuttings that are easy to root, such as coleus, may be propagated in water. [Photograph courtesy E. R. Honeywell.]

age, and freedom from root-rotting diseases (Figure 12-15). Sand or water alone may be satisfactory for some easy-to-root cuttings (Figure 12-16). When water is used alone, improved results are achieved with aeration.

Graftage

Graftage involves the joining together of plant parts by means of tissue regeneration, in which the resulting combination of parts achieves physical union and grows as a single plant. The part of the combination that provides the root is called the **stock;** the added piece is called the **scion.** The stock may be a piece of root or an entire plant. When the scion consists of a single bud only, the process is referred to as **budding.** (Budding and propagation by cuttings represent the most important commercial methods of asexual propagation.) When the graft combination consists of more than two parts, the middle piece is referred to as an **interstock,** body stock, or interpiece.

There are two basic kinds of grafts: **approach** and **detached-scion.** In the approach graft the scion and stock are each connected to a growing root system. In the detached-scion graft, the kind most commonly used, only the stock is rooted, and the scion is severed from any root connection. The approach graft is used when it is difficult to obtain a union by ordinary procedures. The roots of the scion act as a "nurse" until union is achieved, at which time the scion is severed from its own roots.

The Graft Union

The **graft union** is the basis of graftage. It is formed from the intermingling and interlocking of callus tissue produced by the stock and scion cambium in response to wounding. The cambium, the meristematic tissue between the xylem (wood) and phloem (bark), is continuous in perennial woody dicots. Monocotyledonous plants with diffuse cambium cannot be grafted. Callus tissue is composed of parenchymatous cells. Under the influence of the existing cambium, the callus tissue differentiates new cambium tissue. This new cambium redifferentiates xylem and phloem to form a living, growing connection between stock and scion (Figure 12-17). The basic technique of grafting consists in placing the cambial tissue of stock and scion in intimate association such that the resulting callus tissue produced from stock and scion interlocks to form a continuous connection. A snug fit is often obtained by utilizing the natural tension of the stock and/or scion. Tape, rubber bands, or nails may also be used to facilitate contact. Various types of budding and grafting are shown in Figure 12-18. Natural grafts may be formed as a result of the close intertwining of roots or stems.

Although there is usually no actual interchange of cells through the graft union, the connection is such that many viruses and hormones pass through unhindered. This principle is utilized in virus identification. Plants containing a suspected virus, but which may not show obvious morphological symptoms, are grafted to a plant that is sensitive and will show the symptoms of the virus. This process is known as **indexing** (Figure 12-19).

Graft Incompatibility

Owing to the lack of any antibody mechanism, plants tolerate grafting better than do animals. The ability of two plants to form a successful graft combination is related in large part to their natural relationships. The inherent failure of two plants to form a successful union—their **graft incompatibility**—may be structural as well as physiological. Graft incompatibility may be the cause of a high percentage of grafting failures; poor, weak, or abnormal growth of the scion; overgrowths at the graft union; or poor mechanical strength of the union, which in extreme cases results in a clean break at the graft. Incompatibility may be manifested immediately or it may be delayed for several years. In some cases, incompatibility has been traced to a virus contributed by one of the graft components, which was itself virus tolerant. In such cases, incompatibility is due to the virus sensitivity of the other component. If the sensitive component is the root stock, the entire tree may be adversely affected.

In general, grafting within a species results in compatible unions. Grafts made between species of the same genus or species of closely related genera vary in their degree of compatibility. For example, many, but not all, pears (*Pyrus* species) may be successfully grafted on quince stock (*Cydonia oblonga*), but the reverse combination, quince scions on pear stock, is not successful. Incompatibility may be bridged by an interpiece composed of a variety compatible with both components. A few exceptional graft combinations involving species of different families have been experimentally produced in herbaceous plants.

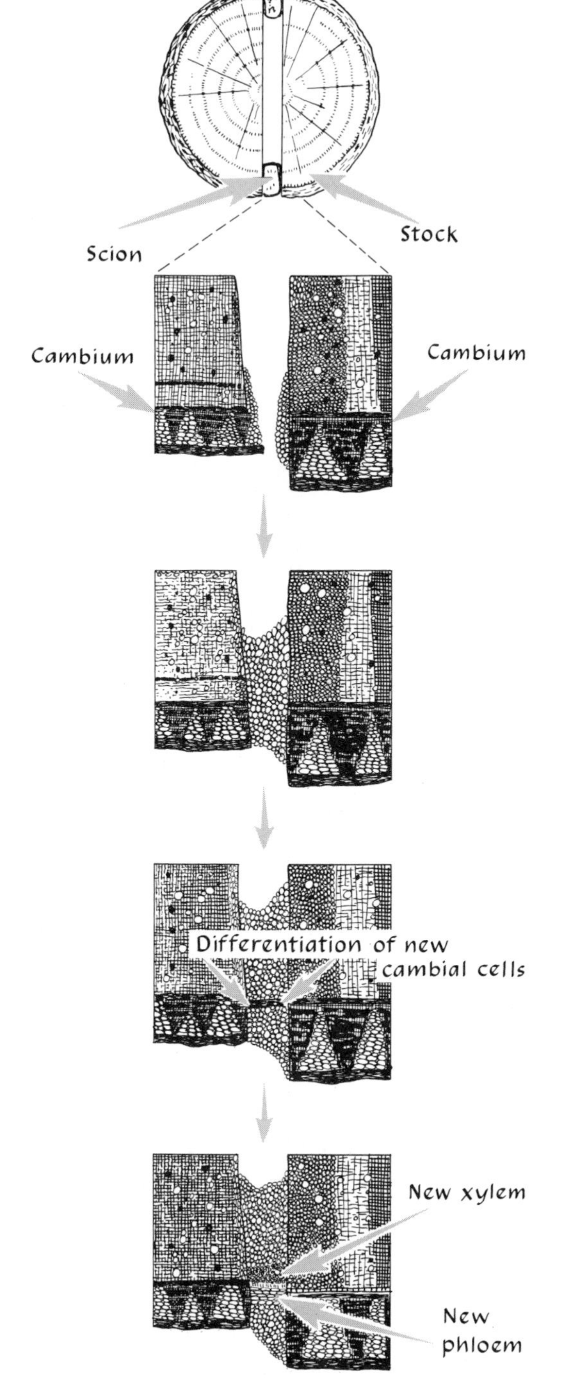

FIGURE 12-17. Developmental sequence during the healing of a cleft-graft union. The graft union is formed from the redifferentiation of the callus tissue under the influence of the stock cambium. [From H. D. Hartmann and D. E. Kester, *Plant Propagation: Principles and Practices*, Prentice-Hall, 1959.]

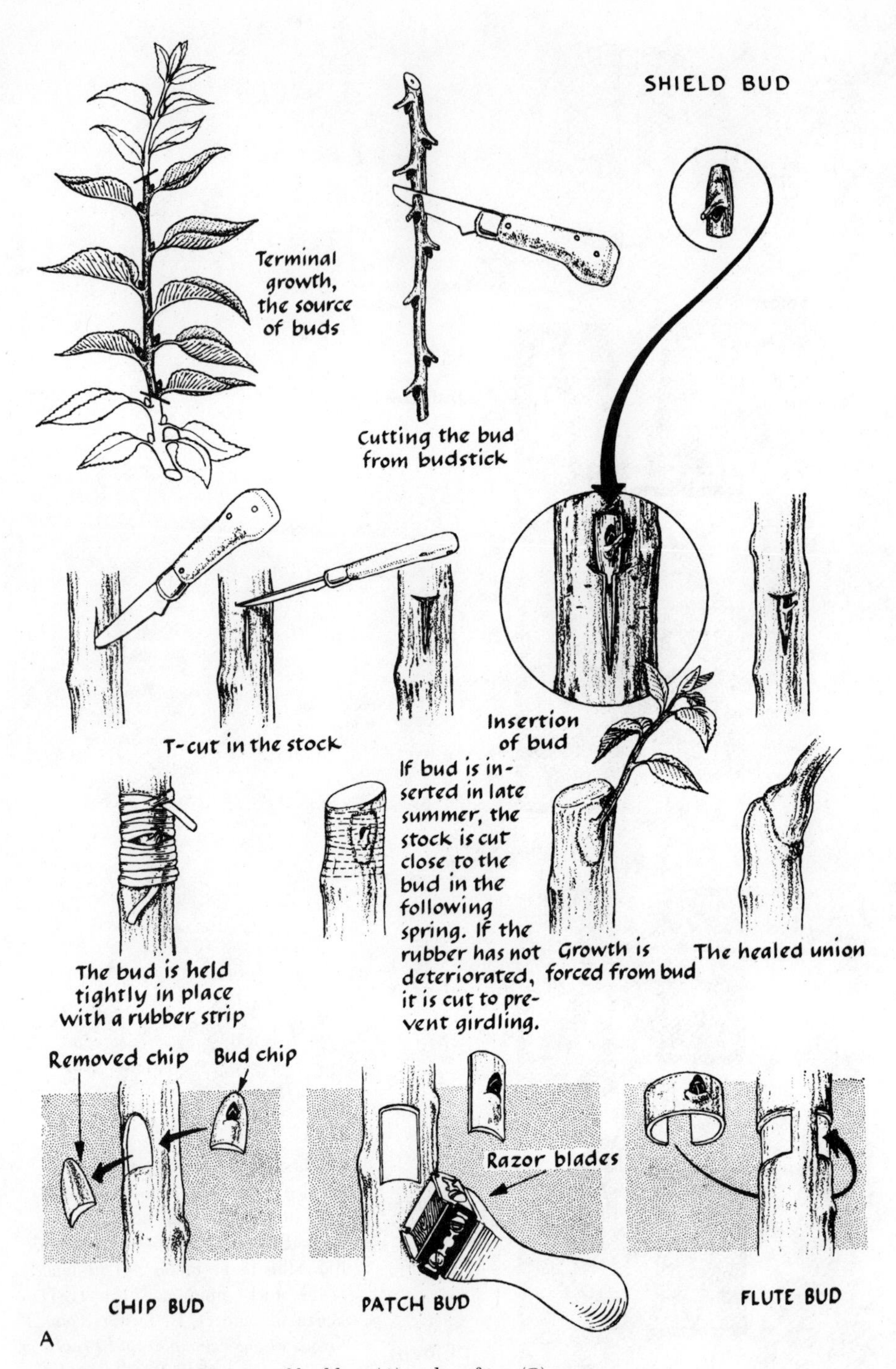

FIGURE 12-18. Techniques of budding (*A*) and grafting (*B*).

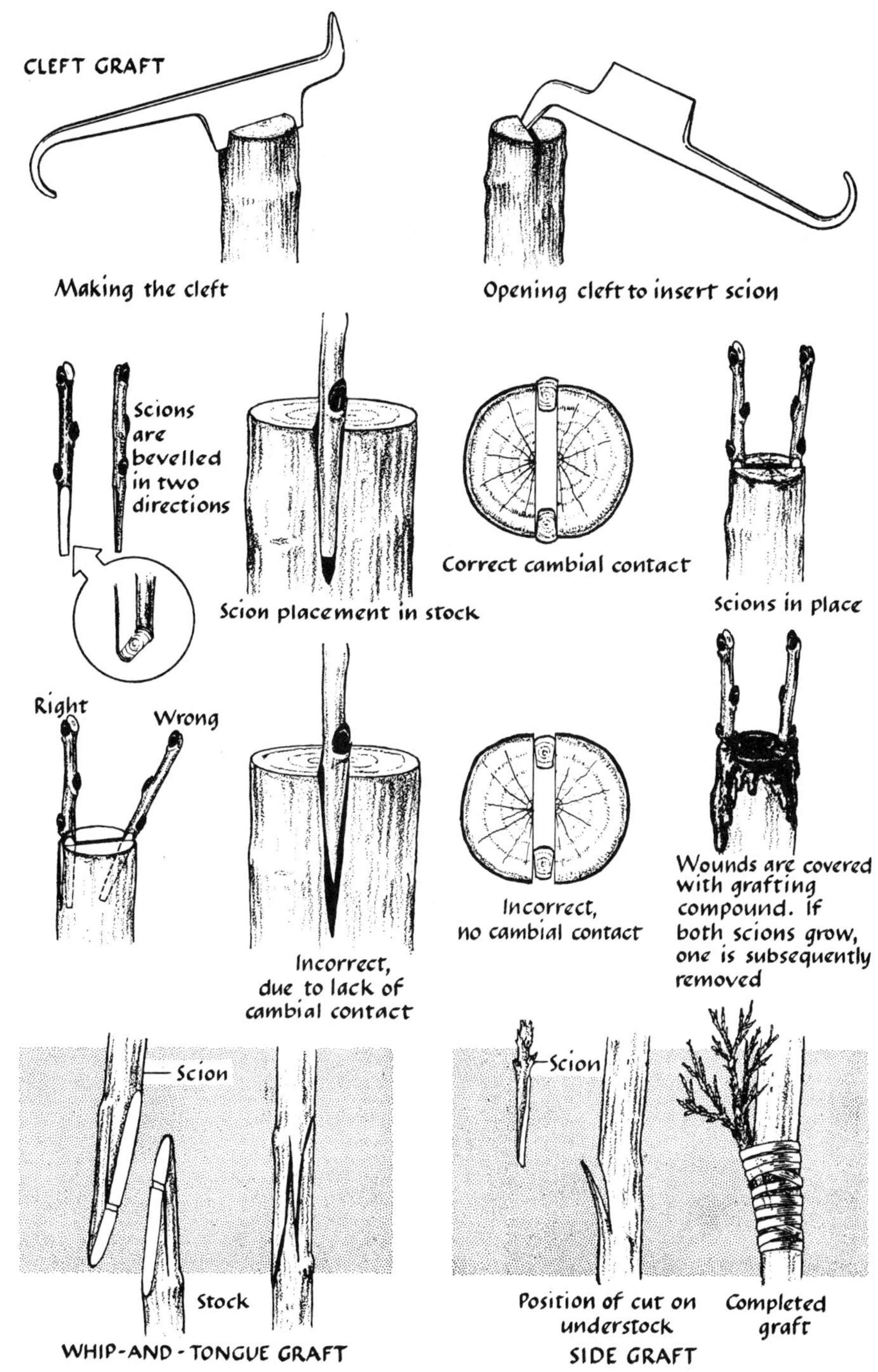
CLEFT GRAFT
Making the cleft
Opening cleft to insert scion
Scions are bevelled in two directions
Scion placement in stock
Correct cambial contact
Scions in place
Right
Wrong
Incorrect, due to lack of cambial contact
Incorrect, no cambial contact
Wounds are covered with grafting compound. If both scions grow, one is subsequently removed
Scion
Stock
WHIP-AND-TONGUE GRAFT
Scion
Position of cut on understock
Completed graft
SIDE GRAFT

B

FIGURE 12-19. Indexing strawberries by means of the leaf graft. (*A*) The middle blade of the trifoliate leaf of a strawberry is cut to a wedge and inserted into the split petiole in place of the removed blade of the sensitive indicator, *Fragaria vesca*. The blade is usually held in place with a self-sticking latex tape. (*B*) After 2 months the grafted leaf from each of two different plants is still alive. The older leaves have been removed to show the new growth. The normal regrowth on the left indicates that the inserted leaf was free of any viruses to which the indicator plant is sensitive. The stunted, deformed growth on the plant at the right indicates transmission of a virus from the excised leaf to the indicator plant. [Photograph courtesy Purdue University.]

Factors Influencing Grafting Success

In addition to the inherent compatibility of the plants, there are a number of factors that affect successful "take" in grafting. Skillful grafting or budding techniques are, of course, necessary. Success then depends upon environmental factors that promote callus formation. Callus formation is usually optimum at about 80 to 85°F (26.5 to 29.5°C). After the grafting of dormant material (bench grafting), the completed graft is best stored under warm and moist conditions for a week or two to stimulate callus formation.

It is very important that high moisture conditions be maintained to prevent the scion from drying out. Waxing the tissue after grafting serves to prevent desiccation of the delicate callus tissue. Special waxes are available that consist of various formulations of resins, beeswax, paraffin, and linseed oil. Bench grafts should be plunged in moist peat to prevent desiccation during the period of healing at warm temperature. The use of plastic films has proved successful in conserving moisture. Oxygen has been shown to be required for callus formation in some plants (for example, the grape). Waxing should not be used with such plants because of its effect in limiting the oxygen supply.

The percentage of "take" in grafting is often improved if the stock is in a vigorous state of growth. The scion, however, should be dormant to prevent premature growth and subsequent desiccation. Growing trees are often grafted with dormant scions from refrigerated storage. In summer budding, irrigation should be supplied before budding in order to invigorate the stock. The leaf buds of most temperate woody plants, when inserted in late summer, remain naturally dormant until the following spring.

Grafting of Established Trees

In addition to its role as a method of propagation, grafting is useful in many instances for cultivar change, repair, or invigoration of older established trees. Grafting to affect growth is discussed in Chapter 15.

TOPWORKING. Owing to the long time required for growth of many fruit or nut crops, it is often desirable to utilize the existing framework and root system of a larger, established tree. It becomes possible to rework a tree and rapidly bring an improved cultivar into production. The regrafting of scions onto a large growing tree to utilize an existing framework is called **topworking** (Figure 12-20). A cleft graft is usually used.

FIGURE 12-20. One-year-old olive scions growing on a 30-year-old stump. [Photograph courtesy USDA.]

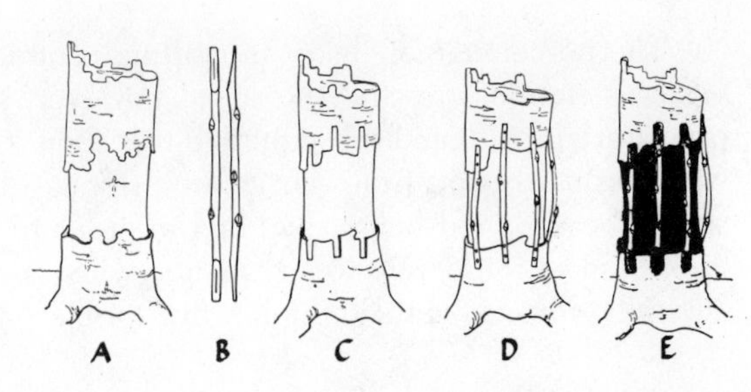

FIGURE 12-21. Five steps in the development of a bridge graft. [From Ontario Department of Agriculture Publication 439, 1961.]

In topworking, the length of the scion inserted may affect subsequent growth. The few buds from short scions all tend to produce vigorous vegetative growth. Bud break on long scions is confined to the terminal buds; the basal buds often grow into fruiting spurs. This technique is utilized to bring new cultivars into early bearing for observation.

INARCHING, BRIDGING, AND BRACING. **Inarching** is a form of "repair" grafting that involves reinforcing the existing root system of an already growing tree. It is used when the existing root system is weak and must be replaced to save the tree. A number of seedlings or rooted cuttings are planted around the tree and are then approach-grafted to the stem. The grafts are usually held in place by nailing.

Bridge grafting, as the name implies, is a means of "bridging" an intact root and stem when the connection between them has been damaged. This treatment can save a tree from death when the stem has been girdled by mice (Figure 12-21).

Bracing the framework of a tree can be accomplished by means of grafting. The technique involves the twisting of young branches around one another to encourage the formation of a natural graft.

Tissue Culture

Tissue culture (or ***in vitro*** **culture**) is the aseptic growth of cells, tissues, or organs in artificial media. Although the culture of plant cells and tissues has long been a tool of the physiologist, these techniques are now increasingly used as a means of rapid plant propagation. Plant propagation by tissue culture (also known as **micropropagation**) involves a sequence of steps, each of which may necessitate a specific set of conditions. Three distinct steps are usually necessary: (1) establishment of an aseptic culture (**explantation**), (2) multiplication of the propagule (**proliferation**), and (3) preparation of the propagule for an independent existence outside of culture by hardening and acclimation (**acclimatization** and **establishment**).

The explantation step requires aseptic surgical removal of a suitable tissue or organ—the **explant.** Almost every plant organ has been successfully cultured, but the most suitable material differs from species to species. Shoot tips are extensively used; other explants in use include sections of stems, leaves, petals, cotyledons, roots, fruits, embryos, shoot apical meristems (the actual growing point), and nucellar tissues.

Explants taken from the true shoot meristem (the apical meristematic dome), which is less than 0.1 mm in height, are often virus free even when the plant is infected. The true meristem is usually too small for use in *in vitro* propagation, but it is used whenever the establishment of an entirely virus-free tissue culture is especially important. Shoot-tip pieces as long as 1 mm have been used successfully in the establishment of pathogen-free cultures, however. Larger shoot tips establish more quickly and growth is faster.

Although environmental factors such as light may be critical to successful tissue culture, growth and differentiation of the explant depend primarily on the culture medium. Various formulas have been developed that make use of combinations of inorganic salts (Table 12-2), sugars, vitamins, amino acids, purine bases, such organic complexes as coconut "water" (the liquid endosperm) or yeast extract, and growth regulators. The key factor controlling morphogenesis is the auxin-cytokinin balance. A high cytokinin-to-auxin ratio promotes shoot formation while a high auxin-to-cytokinin ratio fosters root formation (see Figure 5-8). Species and tissues differ in the critical balance required as well as in their reactions to different auxins and cytokinins. A critical mass of tissue is usually needed before differentiation can proceed.

TABLE 12-2. Inorganic constituents of various media for culture of tissues, organs, and embryos (in mg/liter).

Compound	Knop medium	White medium	Heller medium	Murashige and Skoog medium*
KNO_3	200	80		1900
$Ca(NO_3)_2 \cdot 4H_2O$	800	200		
$NaNO_3$			600	
NH_4NO_3				1650
KH_2PO_4	200			170
$NaH_2PO_4 \cdot 4H_2O$		17	125	
Na_2SO_4		200		
$MgSO_4 \cdot 7H_2O$	200	360	250	370
KCl		65	750	
$CaCl_2 \cdot 2H_2O$			75	440

SOURCE: H. T. Hartmann and D. E. Kester, 1968, *Plant Propagation: Principles and Practices*, 2d ed (Prentice-Hall).

*Also contains $MnSO_4 \cdot 4H_2O$, 22.3; H_3BO_3, 6.2; $ZnSO_4 \cdot 4H_2O$, 8.6; KI, 0.83; $Na_2MoO_4 \cdot 2H_2O$, 0.25; $CuSO_4 \cdot 5H_2O$, 0.025; and $CoCl_2 \cdot 6H_2O$, 0.025; all in mg/liter.

Also add Na_2EDTA, 37.3, plus $FeSO_4 \cdot 7H_2O$, 27.8 (as 5 ml/liter of stock solution of $FeSO_4 \cdot 7H_2O$, 5.57 g, plus 7.45 g Na_2EDTA per liter of water).

Actual propagation after the explant is established may be achieved by one of three pathways: (1) asexual formation of embryos (**asexual embryogenesis**); (2) proliferation of axillary shoots (**axillary branching**); and (3) formation of adventitious shoots from callus (**organogenesis**).

Asexual Embryogenesis

Asexual embryogenesis cannot yet be induced in all species, and its applicability to commercial propagation is confined to relatively few plant families. This process may proceed directly from the cultured explant or indirectly from a callus intermediary, undifferentiated growth induced from the cut surface of plant tissues. In certain plants, microspores in anthers proliferate embryos directly or indirectly from callus, yielding haploid plants (see Figure 6-6).

During asexual embryogenesis, embryo development proceeds through globular, heart-shaped, and torpedo-shaped stages, thus mimicking normal embryo development from zygotes. The asexually formed embryos may be produced in very large numbers: Literally hundreds of embryos may be formed from a relatively small volume of tissue in a culture tube. Each embryo is potentially a separate plant. Asexual embryogenesis was first observed in carrot, but it can occur in many other plants, such as *Kalanchoe,* asparagus, poppy, cacao, and citrus (Figure 12-22). Synthetic carrot seeds have been produced experimentally by proliferating asexual embryos *in vitro* and encapsulating them with a water-soluble resinous film (Figure 12–23).

Very efficient clonal increase has occurred within some orchid genera by means of tissue-culture techniques utilizing a system that appears to be related to embryogenesis. These techniques are potentially very valuable because the usual method of vegetative propagation of orchids (by offshoot) is extremely slow. In some genera (*Cymbidium, Cattleya,* and *Dendrobium*), the culture of shoot tips results in the development of bulblets that resemble protocorms (orchid plants in the "embryonic" stage between germination and developing seedlings). These protocormlike bulblets may be multiplied by cutting them into quarters. Vegetative propagules of orchid produced in this manner are called **mericlones.** Vegetative increase may be extremely rapid, with the number of propagules quadrupling every 10 days, but such rapid propagation is not possible with all orchids. Species of *Vanda* and *Phaleonopsis* do not form such protocormlike bodies.

Rapid Axillary Branching

Enhancement of the development of axillary buds has wide applicability to tissue-culture propagation. Although the proliferation rate is the slowest of the three tissue-culture proliferation procedures, the tissues are under the normal control of the apical meristem and are not likely to produce genetically deviant plants—especially not those of the sort caused by such chromosome irregulari-

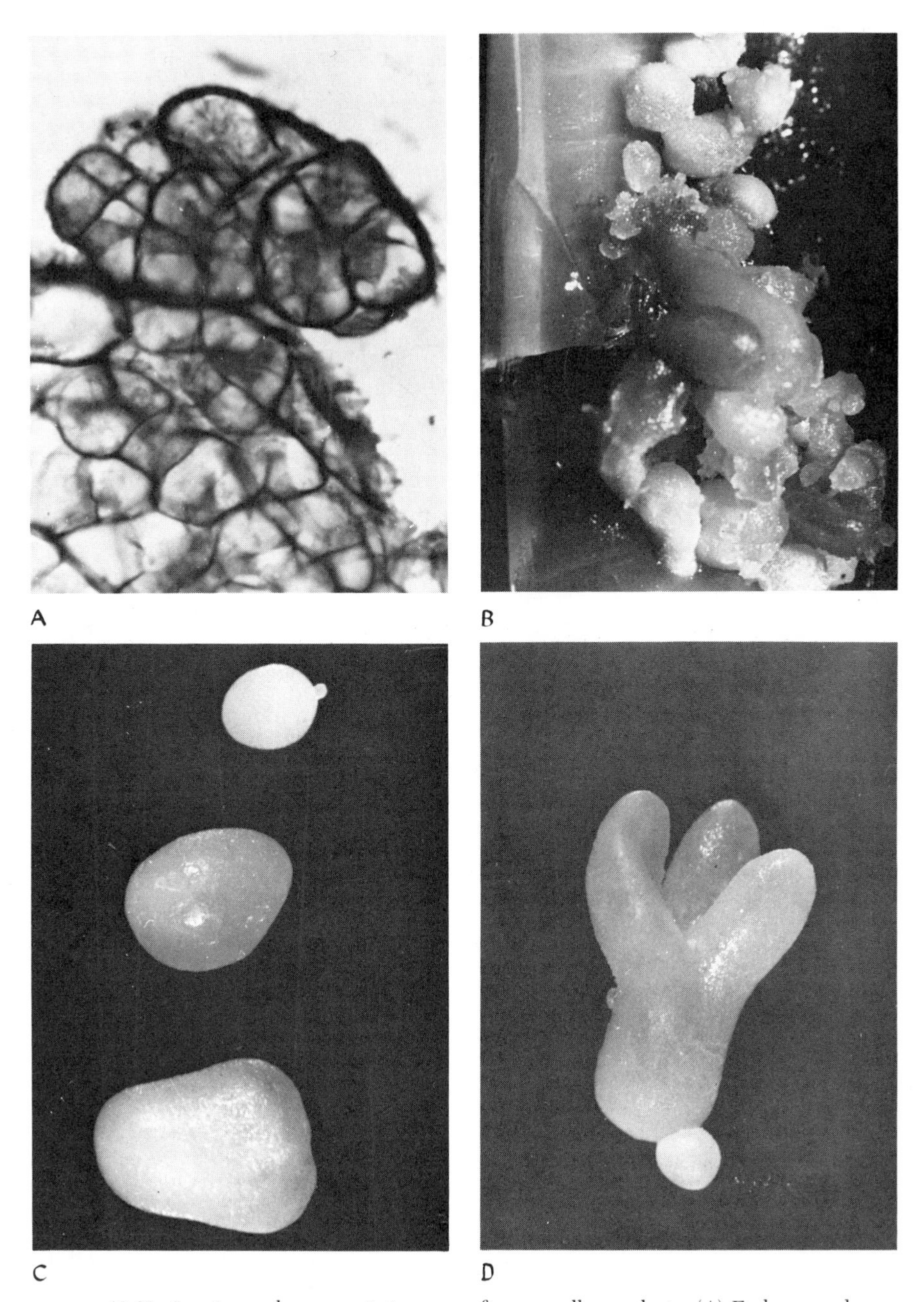

FIGURE 12-22. *In vitro* embryogenesis in orange from nucellus explants. (*A*) Embryos and embryogenic callus. (*B*) Cross section of callus showing cluster of developing proembryos. States of embryo development: (*C*) globular, elongated, heart-shaped, and (*D*) cotyledonary. Note a globular embryo bud at the base of the most advanced stage. [From J. Kochba and P. Spiegel-Roy, *HortScience* 12:110–114, 1977.]

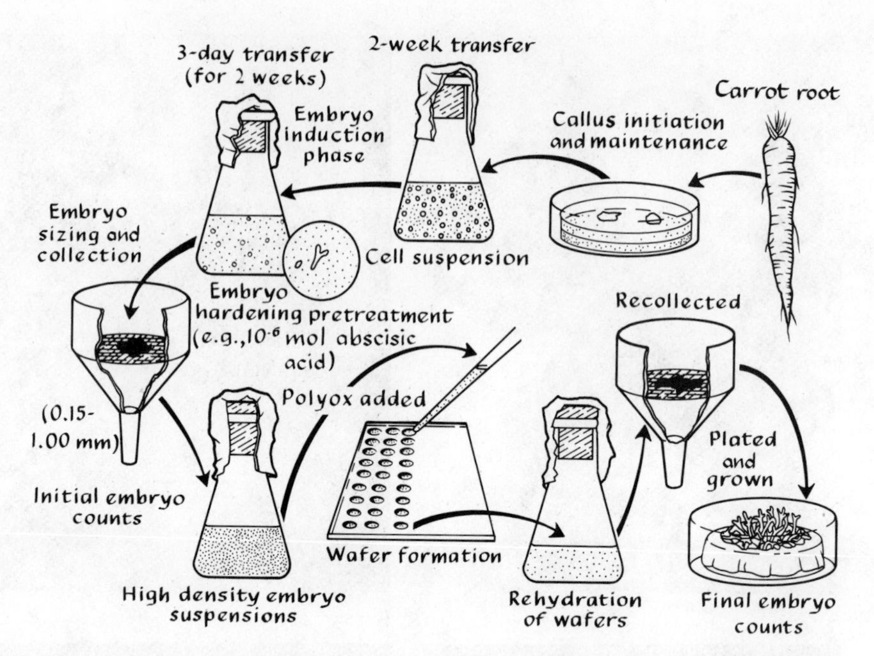

FIGURE 12-23. An experimental system to produce synthetic seed from the encapsulation of asexual embryos using a water-soluble resin (Polyox). (From S. Kitto and J. Janick, *J. Am. Soc. Hort. Sci.*, 110:277–282, 1985.)

ties as chromosome doubling. Shoot tips are usually proliferated with high cytokinin-to-auxin ratios; benzylamino purine (BA) and kinetin are the cytokinins commonly used. The proliferated shoot must be separated and induced to form roots for propagation to be complete. Carnations (Figure 12-24) and many tropical foliage plants are typically propagated by axillary-bud proliferation.

Organogenesis

The formation of shoots from callus (or other tissues, such as leaf discs) is a very rapid way to propagate certain plants. However, the process of forming adventitious shoots via organogenesis is associated with a higher incidence of genetically aberrant plants than is the proliferation of axillary buds. Although the process is controlled by the cytokinin-auxin balance, organogenesis has yet to be demonstrated in many plants, especially many woody species. Intense effort is now being applied to the problem of developing organogenetic propagation methods for all horticulturally important species. The induction of roots from callus without the induction of shoots is common; root initiation appears to be a physiologically simpler process than shoot initiation. Of course, when shoots

A

B

FIGURE 12-24. (*A*) Proliferation of axillary shoots in carnation after 5 weeks of growth in a liquid medium containing 2.0 mg/liter kinetin and 0.02 mg/liter NAA with a single replacement of the medium after 3 weeks. (*B*) Close-up showing single shoots and clumps. [From E. D. Earle and R. W. Langhans, *HortScience* 10:608, 1975.]

are induced they must be rooted to complete the propagation cycle. Boston fern and geraniums (Figure 12-25) are among the species propagated by organogenesis.

Plant tissue culture has had a significant impact on clonal multiplication of floricultural crops, including foliage plants and potted plants. The potential rate of increase per year may be a millionfold greater than with conventional methods. Other uses of tissue culture include the production, isolation, and maintenance of disease-free plants (Figure 12-26) and the storage of germplasm. The use of tissue culture for plant improvement will be discussed in Chapter 17.

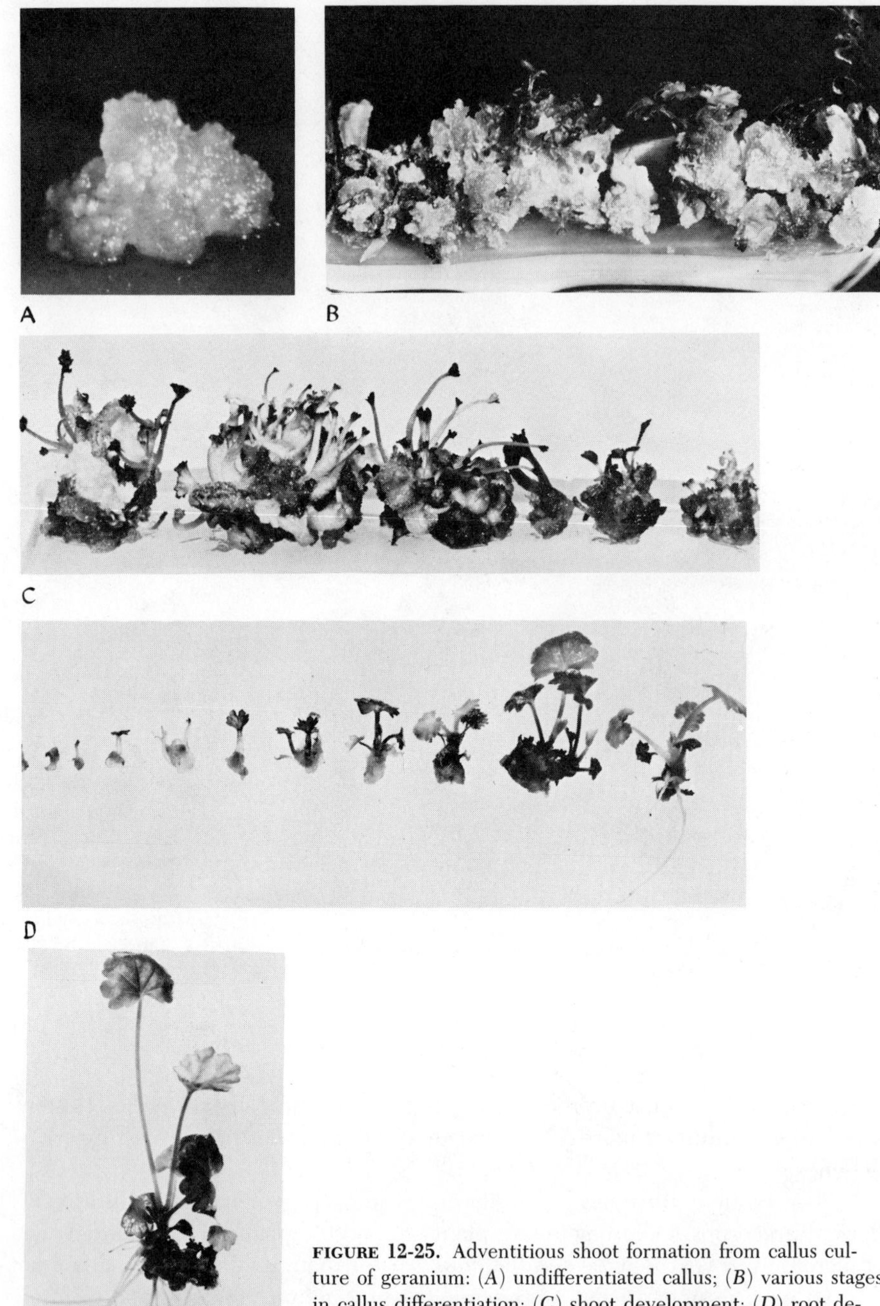

FIGURE 12-25. Adventitious shoot formation from callus culture of geranium: (*A*) undifferentiated callus; (*B*) various stages in callus differentiation; (*C*) shoot development; (*D*) root development; (*E*) a plantlet 2 inches long derived from callus culture. [Photographs courtesy A. C. Hildebrandt.]

A

FIGURE 12-26. The long-term maintenance of "pathogen-free" plants is one of the promising uses of tissue culture technology. Disease-free strawberry plantlets (*A*) have been stored in the dark at 4°C for as long as 6 years. (*B*) The same plantlets after transplanting. [From R. H. Mullin and D. E. Schlegel, *HortScience* 11:100–101, 1976.]

B

Selected References

Evans, D. A., W. R. Sharp, P. V. Ammirato, and Y. Yamada, 1983. *Handbook of Plant Cell Culture*. Vol. 1: *Techniques for Propagation and Breeding*. New York: Macmillan. (An up-to-date treatise on plant cell and tissue culture that is part of a projected five-volume work, of which Volumes 2 and 3 were issued in 1984.)

Garner, R. J., 1979. *The Grafter's Handbook*. New York: Oxford University Press. (A guide to grafting.)

George, E. F., and P. D. Sherrington, 1984. *Plant Propagation by Tissue Culture: Handbook and Directory of Commercial Laboratories*. Eversley, Eng.: Exegetics Ltd. (A useful laboratory manual for micropropagation.)

Hartmann, H. T., and D. E. Kester, 1983. *Plant Propagation: Principles and Practices*, 4th ed. Englewood Cliffs, N.J.: Prentice-Hall. (The most valuable and authoritative book on the subject.)

International Plant Propagators Society. *Proceedings.* P.O. Box 3131, Boulder, CO 80307. (The society's proceedings, published since 1951 by academic and industry horticulturists, include a wealth of materials on individual species with particular emphasis on woody plants. A cumulative index is available for Volumes 1 through 30.)

Mayer, A. M., and A. Poljakoff-Mayber, 1982. *The Germination of Seeds,* 3d ed. New York: Pergamon Press. (A monograph on the physiology of seed germination.)

Murashige, T., 1977. Plant propagation through tissue cultures. *Annual Review of Plant Physiology* 25:136–166. (A review of micropropagation.)

Tomes, D. T., B. E. Ellis, P. M. Harvey, K. J. Kash, and R. L. Peterson (eds.), 1982. *Applications of Plant Cell and Tissue Culture of Agriculture and Industry.* Guelph, Ontario, Canada: The University of Guelph. (An assessment of the agricultural potential of cell and tissue culture.)

U.S. Department of Agriculture, 1961. *Seeds.* (USDA Yearbook, 1961). (The story of seeds.)

U.S. Forest Service, 1974. *Seeds of Woody Plants in the United States.* (USDA Agricultural Handbook 450). (A manual on all aspects of the seeds of woody plants and their proper handling.)

Mineral Nutrition 13

The great bulk (92 to 95 percent) of plants is composed of carbon, hydrogen, and oxygen. Carbon is supplied from carbon dioxide; hydrogen, from water; and oxygen, from water and air. The remaining 5 to 8 percent is made up of 13 essential mineral elements and some nonessential elements, such as silicon or aluminum, that may be swept up in the plant along with essential elements. Of these 13 elements, 6 are required in relatively large quantities and are called **macroelements.** They include nitrogen, phosphorus, sulfur, calcium, potassium, and magnesium. The remaining 7, called **microelements,** are needed in trace amounts. They include iron, zinc, manganese, copper, boron, molybdenum, and chlorine. Because all these 13 elements are normally taken up by plant roots from the soil, plant nutrition is usually entwined with the study of soil fertility—the ability of the soil to provide the essential elements. Soil by itself, however, is not needed to provide mineral nutrients. This fact will be demonstrated in the discussion of hydroponics, the soilless culture of plants, at the end of this chapter. The control of mineral nutrition in either a soil or a soilless medium is one of the basic technologies of horticulture.

ESSENTIAL ELEMENTS

The essential mineral elements will be discussed in terms of soil relations, plant functions, deficiency and toxicity symptoms, and fertilizer practices.

Macroelements (N, P, K, Ca, S, Mg)

Nitrogen (N)

SOIL RELATIONS. Nitrogen, an inert, odorless, tasteless gas, is unavailable to plants unless it is "fixed" in the soil in soluble inorganic forms. Nitrogen is also found tied up in organic matter in the soil. Although the level of nitrogen in the typical horticultural plant ranges from 1.5 to 2.5 percent, nitrogen-containing

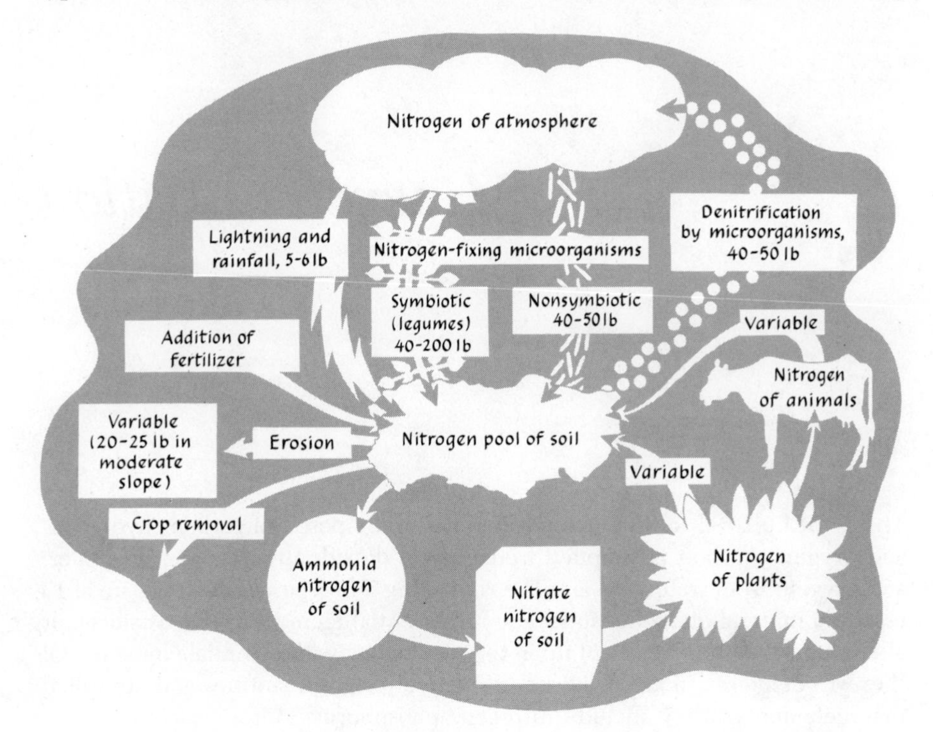

Nitrogen added		Nitrogen removed	
Method	lb/acre	Method	lb/acre
Plant and animal residues	Variable	Crop harvested	Variable
Nitrogen fixation		Leaching	
Symbiotic	40–200	Crop rotation	5–10
Nonsymbiotic	40–50	Bare soil	60–70
Lightning and rainfall	5–6	Erosion (moderate slope)	20–25
		Denitrification	40–50

FIGURE 13-1. The nitrogen cycle. Nitrogen removal by crops must be compensated for by nitrogen addition. [Adapted from J. Bonner and A. W. Galston, *Principles of Plant Physiology*, W. H. Freeman and Company, copyright © 1952.]

compounds represent about 25 percent of the total. In fully expanded leaves, nitrogen should be 2.5 to 4.5 percent dry weight.

Nitrogen is the nutrient most limiting to plant growth. Its main available forms in the soil are nitrate (NO_3^-) and ammonium ions (NH_4^+). The nitrite ion (NO_2^-) can be utilized by the plant, but it tends to be unstable and toxic in high amounts. The transformation of nitrogen-containing compounds to available forms is called the **nitrogen cycle.** This circuitous route of nitrogen from element to protein and back is largely biological, as shown in Figure 13-1.

The transformation of atmospheric nitrogen into forms available to plants, **nitrogen fixation,** is accomplished by certain species of bacteria (Table 13-1). The most efficient of these bacteria are symbiotic; that is, they convert atmospheric nitrogen to combined forms only in association with the roots of legumes. The breakdown of the complex proteins of organic material into amino acids is also accomplished largely by bacterial action. But the nitrogen from this process is available only after the death and disintegration of the bacteria involved in this decaying process. Soil microorganisms have the first call on nutrients. This is especially true for material with a carbon to nitrogen ratio by weight greater than 10:1. The breakdown of amino acids to forms of nitrogen available to plants takes place by transformations known as **ammonification** and **nitrification.**

Ammonification

$$\text{Amino acids from degradation of protein} \xrightarrow[\text{Many bacteria}]{} NH_4^+ \text{ (ammonium ion)}$$

Nitrification

$$NH_4^+ \xrightarrow[\textit{Nitrosococcus, Nitrosomonas}]{} NO_2^- \text{ (nitrite ion)} \xrightarrow[\textit{Nitrobacter}]{} NO_3^- \text{ (nitrate ion)}$$

TABLE 13-1. Nitrogen-fixing bacteria.

Type	Genus	Requirements
Symbiotic	*Rhizobium*	Carbohydrates, inhibited by nitrates and ammonium
Nonsymbiotic		
Anaerobic	*Clostridium*	Carbohydrates, inhibited by nitrates
Aerobic	*Azotobacter*	Calcium, traces of molybdenum

The bacteria involved in nitrification are autotrophic and aerobic; that is, they do not require organic nutrition, but they do need oxygen. Thus, they are greatly affected by soil aeration, temperature, and moisture.

The removal of nitrogen from the soil is partly biological. In addition to its removal by plants (which is permanent when a crop is harvested), certain bacteria convert nitrates back to atmospheric nitrogen. This process of denitrification is an anaerobic one. Thus, loss of proper aeration results in the loss of available nitrogen. Furthermore, nitrates are readily soluble in water, and if they are not utilized by microorganisms or higher plants, they may be lost by leaching. In summary, the level of available nitrogen is dependent upon the content of organic matter and the microbiological activity of the soil. Consequently, the amount of nitrogen available is related to cropping practice. The available soil nitrogen is, of course, greatly affected by the application of fertilizer. Quickly available forms of inorganic nitrogen probably account for most of the nitrogen in intensively cropped soils today.

NITROGEN METABOLISM. Except for the nitrogen-fixing legumes and a few other plants that depend on their bacterial partners, plants rely on the nitrate and ammonium ions in the soil solution as a source of nitrogen. Since microorganisms in the soil rapidly oxidize the reduced forms of nitrogen, the most common source of the element taken up by plants is the nitrate ion (NO_3^-). Before combining with organic acids to form amino acids, nitrate is reduced through a series of enzymatic reactions. The reduced nitrogen groups NH_2 and NH_3 combine with the carbon frameworks formed during the oxidation of sugars to form amino acids. The ammonia (NH_3) initially combines with the organic acid α-ketoglutaric acid to form the amino acid glutamic acid, an energy-requiring process. Other amino acids are constructed via the transfer of the amino group (NH_2) from this amino acid to other organic acids from the metabolic pool (a process known as **transamination**).

Amino acids join to form proteins through a complex series of reactions regulated by the nucleic acids present in the cell. The primary sites of protein synthesis in plants are the tissues in which new cells are being formed, such as tips of stems and roots, buds, cambium, and developing storage organs. The green leaf is also an important site of protein synthesis, since both the carbon frameworks and the inorganic nitrogen necessary for the formation of amino acids are readily available there. Leaf protein is in a continuous state of turnover. When leaves are placed in the dark, are excised from the plant, or reach the state of senescence, protein breaks down and soluble nitrogen compounds are formed that can be reincorporated into amino acids in other leaves or organs of the plant.

Plants are able to survive in an inorganic world because they can convert inorganic carbon (carbon dioxide from the air) and nitrogen (nitrates and ammonium ions) to organic forms. Because the nitrogen metabolism and the carbohydrate metabolism of the plant are closely linked, both processes are markedly affected by gross changes in the environment of the plant, such as heavy nitrogen fertilization, reduced light intensity, or drought. These effects are reflected in altered patterns of growth and development.

PLANT FUNCTIONS. Nitrogen is an integral component of the many cell constituents, including amino acids, nucleic acids, and the chlorophyll molecule. Each chlorophyll molecule contains four nitrogen atoms for every magnesium atom. Because most of the nitrogen in a plant is in the form of proteins, protein may be estimated by multiplying the nitrogen content by 16.5.

The relation between nitrogen and stored carbohydrates (known as the carbohydrate-to-nitrogen ratio) affects the type of plant growth. High nitrogen encourages succulent vegetative growth and can delay flowering and reproductive growth; reproductive processes dominate with abundant carbohydrates and moderately available nitrogen. Plant growth is suppressed with deficiencies of nitrogen. (The carbohydrate-to-nitrogen ratio is discussed further on pages 160–164).

Of the three elements (N-P-K) commonly supplied by commercial fertilizers, nitrogen has the quickest and most pronounced effect. For example, the addition of nitrogen quickly imparts a dark green color to leaves. Nitrogen is not abundant in soils and is readily leached; it is therefore likely to be the limiting factor in plant growth.

DEFICIENCY AND TOXICITY SYMPTOMS. The first symptoms of nitrogen deficiency are reduced growth and pale green to yellowish leaves that are often smaller than normal. Older leaves are most affected because nitrogen is very mobile in the plant. As it becomes scarce, proteins in the older leaves degrade to amino acids, which are translocated to young leaves to build protein. With insufficient nitrogen, plants become stunted, their symptoms increasing as the season progresses. Fruits of nitrogen-deficient perennials are usually smaller and early maturing. Fall coloring is often more intense and older leaves may abscise sooner.

Excessive nitrogen results in shoot elongation, weak succulent tissues, and abnormally dark leaves. In perennial fruit crops, excessive nitrogen reduces fruit color and maturity is delayed. Excessive nitrogen may diminish fruit flavor and storage life.

FERTILIZERS. Plants absorb nitrogen as NO_3^- and/or NH_4^+ regardless of source. Evidence has accumulated that both forms of nitrogen are required, and many plants perform best with 50 to 60 percent in the NO_3^- form and 40 to 50 percent in the NH_4^+ form. In fields where NH_4^+ is converted quickly to NO_3^- by nitrifying bacteria, the nitrogen form (NO_3^- or NH_4^+) is not too critical. However, in greenhouses or nurseries where a constant supply is required, it is best to supply most nitrogen directly as NO_3^- because excessive ammonia can be detrimental. Acid-loving plants (ericaceous plants such as rhododendrons and blueberries), however, require the ammoniacal form.

Nitrogen may be supplied in the form of organic (natural) fertilizers or by inorganic forms. In organic forms (animal manures, seed press cake, fish meals, sewage sludge, garbage tankage, organic matter composts), the nitrogen is tied up in proteins, which must be degraded to ammonia and/or nitrates before plants can absorb them. Organic nitrogen has the virtue of being released slowly and steadily, but organic materials are usually expensive when calculated on a nutrient basis because the amount of nitrogen is low. For example, cattle manure is only 1 to 2 percent nitrogen. In horticulture field crops, most nitrogen is now supplied by inorganic fertilizers: principally urea (46 percent nitrogen), ammonium nitrate (38 percent nitrogen), and mixtures of the two called nitrogen solutions (28 percent nitrogen). In greenhouses where NO_3^- is desirable, the common sources are ammonium nitrate, calcium nitrate (15.5 percent nitrogen), and potassium nitrate (14 percent nitrogen). Ammonium sulfate (20.6 percent) is used in special situations where the ammonium ion is required, as in growing acid-loving plants.

Producers of container-grown plants commonly use slow-release inorganic complete fertilizers (particles coated with materials such as waxes or resins that dissolve slowly) containing nitrogen as well as phosphorus and potassium.

Nitrogen may also be supplied by foliar applications of urea. Foliar-applied nitrogen permits precise control of nitrogen levels in perennial fruit crops.

Phosphorus (P)

SOIL RELATIONS. Phosphorus is never found free in nature. It is released into soils by weathering of rocks that contain the mineral apatite, $Ca_5F(PO_4)_3$, and is also found in organic forms (Figure 13-2). However, a soil may have abundant

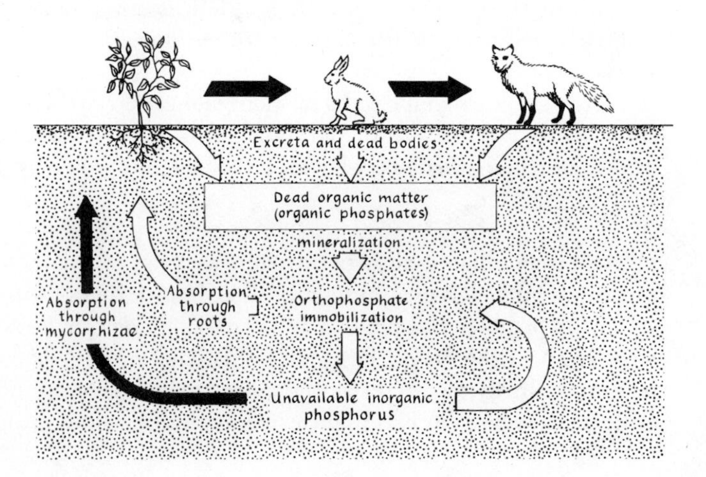

FIGURE 13-2.
The phosphorus cycle.

phosphorus, yet little of it may be available for plant growth if conditions are unfavorable.

Phosphorus, unlike nitrogen, is relatively stable in the soil (Figure 13-3). Phosphorus is "tied up," or fixed, in compounds containing calcium, magnesium, iron, or aluminum. The availability of phosphorus to the plant is low and is primarily related to soil pH. At very low pH (2 to 5), applied phosphorus is precipitated out of the soil solution as complex aluminum and iron compounds. At high pH (7 to 10), phosphorus becomes fixed in complex calcium compounds. At pH 5 to 7, it is in the form of mono- or dicalcium phosphates and is most available for plant use. The concentration of phosphorus in the soil solution is low at best. In fertile agricultural soils, only 0.5 to 1.0 ppm of phosphorus is in solution, as compared with 25 ppm of nitrogen. However, because the mobility of phosphorus in the soil is low, leaching is slight.

When phosphorus is applied to soil, it seldom moves far because of complex reactions with clay and organic matter. Consequently, it is usually placed close to the crop row rather than broadcast. Even when applied in a water-soluble form, it usually diffuses no more than a few centimeters (an inch or so) from the application point before it is converted within 2 or 3 days to an insoluble compound.

Losses of phosphorus from the soil are usually due to crop removal and to its becoming tied up in unavailable forms in the soil. The phosphorus cycle is illustrated in Figure 13-2.

PLANT FUNCTIONS. Phosphorus is a component of nucleoproteins, phospholipids, and compounds with high-energy phosphate bonds as ADP and ATP and thus it is critical to the cell's energy-transfer system. Plants require less phosphorus than they do nitrogen and potassium, with optimum leaf concentrations ranging from 0.2 to 0.3 percent dry weight. Seeds are particularly high in phosphorus.

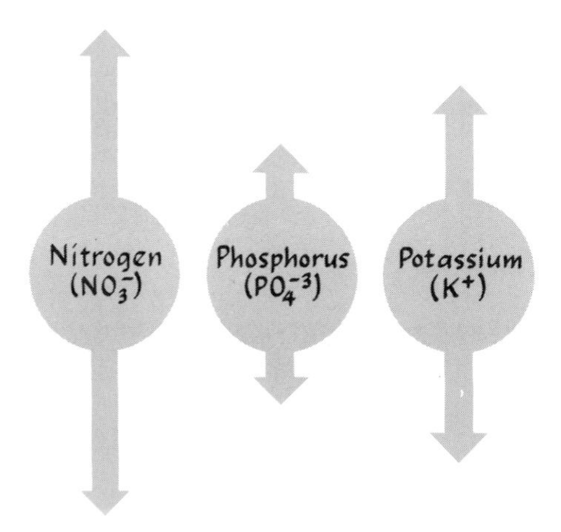

FIGURE 13-3. The relative mobility of nitrogen, phosphorus, and potassium in the soil. The high movement of nitrogen is due to the complete solubility of nitrates in the soil solution. The movement of phosphorus is regulated by the low solubility of the phosphorus compounds that are formed in the soil. Although the potassium compounds in the soil are soluble, the movement of potassium is controlled by its exchange properties with the colloidal fraction.

DEFICIENCY AND TOXICITY SYMPTOMS. Phosphorus is extremely mobile, so deficiency symptoms are first found in older tissues. Older leaves lose their sheen and appear dull and dark; red, blue, or yellow pigment shows through the green—usually along main veins on the undersides of the leaves—and necrosis begins at leaf tips. Young leaves will be very small.

Symptoms of excess phosphorus are hard to diagnose because they are associated with deficiencies of other essential heavy metals, such as zinc, copper, iron, or manganese.

FERTILIZERS. The most common **phosphorus fertilizer** is single or triple superphosphate. Single superphosphate is produced by treating phosphate rock (apatite) with sulfuric acid; it contains about 8.6 percent available phosphorus.

$$2Ca_5F(PO_4)_3 + 7H_2SO_4 + 3H_2O \longrightarrow 3CaH_4(PO_4)_2H_2O + 7CaSO_4 + 2HF$$

Triple superphosphate, 20 percent elemental phosphorus, is produced by treating phosphate rock with phosphoric acid (H_3PO_4). Single and triple superphosphates have fluorine as a contaminant and may produce toxicity symptoms in susceptible species such as tuber and foliage plants.

Other phosphate fertilizers include ammonium phosphate, calcium metaphosphate, potassium phosphate, and finely ground phosphate rock. Phosphate rock contains a high percentage of phosphorus but is not readily available to plants. Ammonium and potassium phosphate and phosphoric acid (H_3PO_4) are used in liquid fertilizer applications.

Seedlings have high phosphorus requirements and adequate quantities must be made available close to the seed or transplant. Therefore, phosphorus applications are often made as bands of fertilizer under the seed. The use of a starter solution, water containing a liberal amount of phosphorus (about 1500 parts per million), is recommended for many transplants. The phosphorus is supplied in the form of soluble phosphate salts, such as monoammonium or diammonium phosphate and monopotassium phosphate. A popular starter solution contains 1 kg of a mixture of diammonium phosphate and monopotassium phosphate (10–52–17 analysis = 10N–22P–14K) per 140 liters of water (about 6 pounds per 100 gallons). When an extensive root system is established, plant requirements are met by lower concentrations of phosphorus.

Potassium (K)

SOIL RELATIONS. Potassium in the elemental form is a white metal, but it is so reactive it must be stored under oil to prevent its reacting with oxygen.

The total potassium in mineral soils is usually higher than phosphorus or nitrogen. However, typically only 1 or 2 percent of total potassium is readily available to plants; it is located as exchangeable ions on soil colloids or in the soil solution. The unavailable potassium is in nonexchangeable positions between layers of clay crystals or in unweathered minerals such as micas and feldspars. Although potassium is plentiful in mineral soils, the low solubility of the primary potassium-rich minerals results in low availability from this source. The primary minerals do, however, continually renew the supply of exchangeable ions. Potassium tends to be low in organic soil. The leaching of potassium varies greatly, depending upon the type of clay and the amount of organic matter in the soil.

PLANT FUNCTIONS. Although potassium has several functions in plants, it is not found in organic form. It is involved in the inactivation of enzymes that regulate photosynthesis, respiration rates, carbohydrate metabolism, and translocation. Potassium is the most abundant cellular cation and serves in regulating water movement in stomata. Plants contain about 1 percent of potassium by dry weight. Leaf tissues should contain about 3.5 to 4.5 percent of this element.

DEFICIENCY AND TOXICITY SYMPTOMS. Potassium is mobile in the plant and therefore deficiencies first appear in older leaves. The early symptoms include chlorosis, but scorching in the margins of older leaves is the most prominent. Plants deficient in potassium are low-yielding with poorly developed root symptoms.

Specific symptoms of toxicity seldom occur, but high potassium concentrations may reduce uptake of magnesium, manganese, and zinc.

FERTILIZERS. Potassium chloride (muriate of potash) containing 50 percent potassium is the most common fertilizer source; others include potassium sulphate (sulphate of potash) and potassium nitrate (38 percent potassium, 13 percent nitrogen). A few plants (beets, turnip, cabbage) can replace their requirements for potassium with sodium.

Calcium (Ca)

SOIL RELATIONS. Calcium, a silvery white metal, occurs in an abundant element making up 3.6 percent of the earth's surface and about 1 percent of United States soils. Although there is less calcium than potassium in the soil, a higher proportion is readily available; most is held on the soil colloid as an exchangeable cation.

Calcium, besides being an essential element, has a strong influence on ion absorption by soil particles and on the availability of other elements. Thus, calcium-containing compounds as lime (calcium oxide, CaO, or limestone) are frequently applied to acid soils in order to reduce soil acidity, which reduces phosphorus fixation and increases the availability of magnesium and molybdenum. Excess lining, however, can reduce manganese and iron availability, which are less soluble at high pH. Calcium also promotes good soil structure by increasing aggregation and it encourages nitrifying and nitrogen-fixation bacteria.

PLANT FUNCTIONS. Calcium is a constituent of cell walls and is found as calcium pectate in the middle lamella, the cementing layers between cells.

DEFICIENCY AND TOXICITY SYMPTOMS. Calcium pectate is immobile. Thus, calcium is not retranslocated and deficiency symptoms occur in younger portions of the plant. Terminal leaves become small without a patterned chlorosis and other leaves get thick and brittle. Stunted growth and curled, distorted leaves with necrotic spots are the classic deficiency symptoms (Figure 13-4A, B). Stem tips may die. Roots in calcium-deficient plants become stubby and brown.

Because calcium is so immobile, rapidly growing fruits may show severe calcium problems even when enough calcium exists in the rest of the plant for normal growth. Fruit symptoms include cracking, pits in the skin, lenticel breakdown, and various internal disorders. Blossom-end rot of tomato fruits (Figure 13-4C) and bitter pit, cork spot, water core (a glassy appearance of the flesh unrelated to water), and internal breakdown of apple fruits are disorders associated with low fruit calcium. Unfortunately, it is difficult to elevate calcium levels in the fruit.

FERTILIZERS. Calcium in the form of limestone is usually added to the soil to regulate soil pH rather than to achieve the ostensible purpose of correcting soil deficiencies. Various forms include **agricultural meal** (finely ground calcium and magnesium carbonates); **marl** (a sedimentary rock composed of precipitated calcium carbonates); **oyster shells** and **shell rock; basic slag** (a by-product of iron melting containing calcium silicate and calcium oxide); **gypsum** (calcium sulfate); or **agricultural hydrate, slaked lime, caustic lime,** or **hydrated lime,** all names for $Ca(OH_2)$, produced from reacting CaO (burnt lime) with water.

Calcium is also provided in various fertilizers applied to supply other ele-

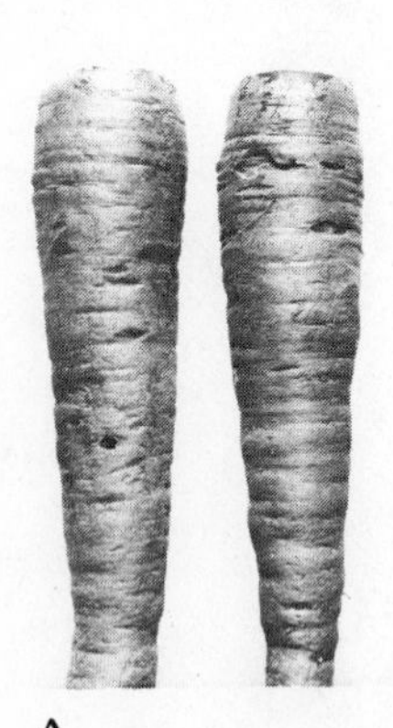

A

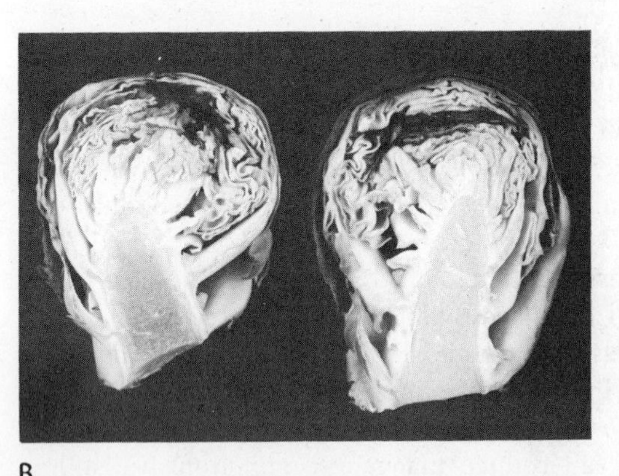

B

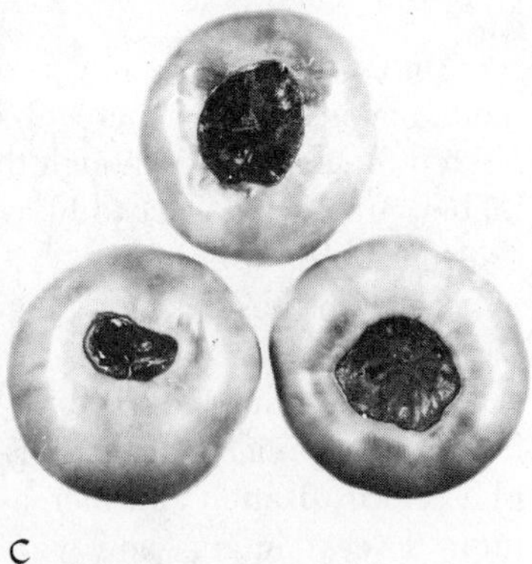

C

FIGURE 13-4. Calcium-deficiency symptoms: (*A*) cavity spot in carrot; (*B*) internal browning in Brussels sprouts; (*C*) blossom-end rot in tomato fruits. [Courtesy D. N. Maynard.]

ments. Superphosphate contains more calcium (19 to 22 percent) than phosphorus (8.6 percent); triple superphosphate contains 14 percent calcium and 21 percent phosphorous.

Because actual shortages were rare in most agricultural soils, little attention was placed on calcium under the assumption that this element would be self-regulating if soil pH were maintained at a near-neutral range. However, the myriad of fruit disorders in horticultural crops that are now known to be associated with localized calcium deficiencies has strengthened efforts to increase calcium uptake and distribution within the plant.

Sulfur (S)

SOIL RELATIONS. The earth's crust contains about 0.06 percent sulfur as sulfides (combinations of sulfur with other elements except oxygen); sulfates (combinates involving the SO_4^{2-} ion); and elemental sulfur. Sulfur occurs in rocks mainly as the mineral pyrite (FeS_2, "fool's gold"), but 70 to 90 percent of the total sulfur of many surface soils is in organic matter.

Sulfur undergoes a continuous turnover in the soil. It is added by rainfall near industrial regions where rain absorbs sulfur dioxide from the air. SO_2 in very low frequencies can be absorbed by plants to supplement sulfur requirements, but even at concentrations as low as 1 ppm, it is very toxic. Thus, SO_2 is considered a soil pollutant (see page 307). Most soil sulfur derives from the weathering of iron sulfide to soluble sulfates which become incorporated into organic form (Figure 13-5). The soil's sulfur cycle is mediated by microorganisms as is the nitrogen cycle. Sulfur is lost from the soil by leaching, crop removal, and erosion. A chief source is organic matter; consequently, deficiencies occur in soil that is low in organic matter and distant from industrial areas.

PLANT FUNCTIONS. Sulfur is a component of several amino acids including methionine and cysteine, both essential to human nutrition. It also is a part of coenzymes and various vitamins including thiamine and biotin. The pungent odor and strong flavor of onions, garlic, mustard, horseradish, and cabbage are due to sulfur-containing compounds.

DEFICIENCY AND TOXICITY SYMPTOMS. Deficiency symptoms of sulfur are similar to those of nitrogen. Foliage tends to be light green because sulfur, although not an essential part of the chlorophyll molecule, is necessary for its synthesis. However, this element is immobile so that chlorosis from sulfur deficiency is more severe on new growth.

FERTILIZERS. In the past, sulfur deficiencies in horticultural crops were infrequent, as sulfur was commonly added to soils via superphosphate, which contains $CaSO_4$, via ammonium sulfate, or via organic fertilizer as manures. But

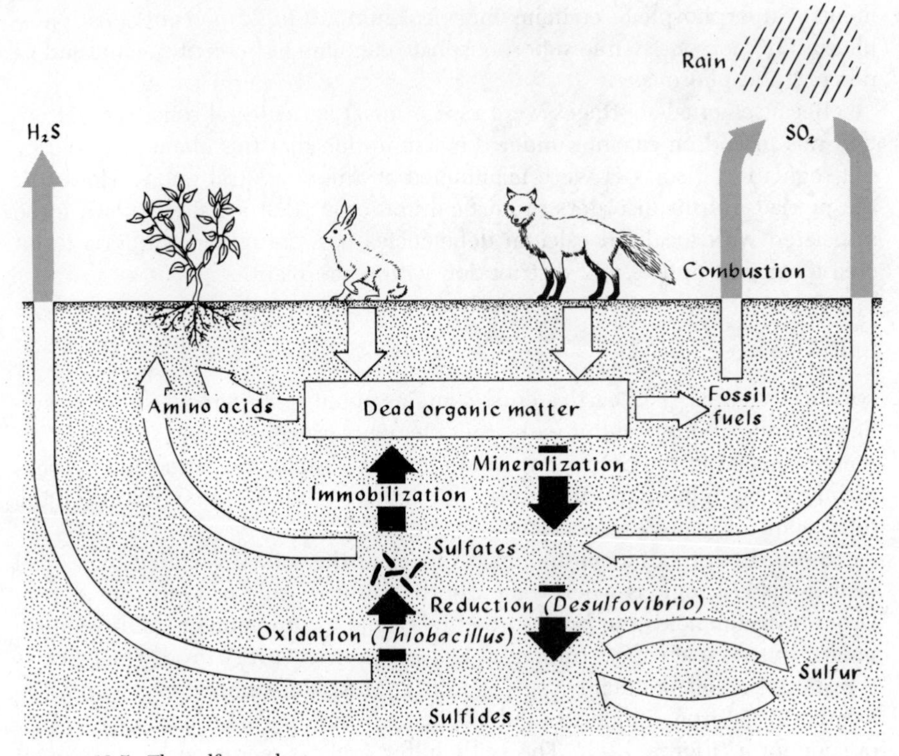

FIGURE 13-5. The sulfur cycle.

sulfur deficiencies are increasing as a result of more pollution controls and a switch to high-analysis fertilizers (replacing superphosphate, 10 to 12 percent sulfur, with triple superphosphate, 0 percent sulfur; and replacing ammonium sulfate with ammonium phosphate, potassium nitrate, or anhydrous ammonia.)

Elemental sulfur and noncalcium sulfates, when added to the soil, lower pH. This capability has been utilized as a means of lowering pH in alkaline soils (refer to page 388).

Magnesium (Mg)

SOIL RELATIONS. Magnesium is a shiny silvery metal closely related to calcium. Like calcium, it is absorbed as an ion in the soil. Magnesium occurs in the soil solution in soluble form and as an exchangeable cation. Like calcium, magnesium is sometimes deficient in acid, sandy soils in humid areas that have low cation-exchange capacity. Magnesium and calcium are similar in their soil relations, but the quantity of exchangeable calcium greatly exceeds exchangeable magnesium.

PLANT FUNCTIONS. Magnesium is the metallic portion of the chlorophyll molecule, and it has a role in the formation of amino acids and vitamins. Appearing to function as an enzyme activation, it is involved in energy transfer and phosphorus metabolism. Seeds are especially high in magnesium. Plants contain about 0.2 percent magnesium by dry weight; leaf tissue, about 0.35 to 0.55 percent.

DEFICIENCY AND TOXICITY SYMPTOMS. Symptoms of magnesium deficiency appear as interveinal chlorosis that is related to its role as a component of chlorophyll. Unlike calcium, it is relatively mobile in the plant. Thus, deficiencies appear in older leaves and move upward.

Symptoms of excess are not specific but usually appear as deficiencies of potassium or calcium. Magnesium is often a problem in floricultural crops because it is not commonly provided as fertilizer, is readily leached, and the concentration is low in sands and organic material that are widely used in synthetic media.

FERTILIZERS. Magnesium is usually added to soils by using dolomitic limestone (dolomite) $CaCO_3 \cdot MgCO_3$ rather than calcite, $Ca(CO_3)_2$, during soil liming. When magnesium deficiency is acute, foliar sprays of Epsom salts ($MgSO_4$), magnesium nitrate, or magnesium chloride are effective. Repeated sprays or soil applications are necessary to alleviate severe deficiencies.

Microelements (Fe, Mn, Zn, Cu, B, Mo, Cl)

The trace elements are required by plants in small amounts usually presented in parts per million (mg/kg) of plant dry weight.

Iron (Fe)

SOIL RELATIONS. Total iron is always high in soils but it is often unavailable; it is especially deficient in alkaline soils. Its availability in the soil is dependent on pH. Solubility and mobility are the keys to iron deficiencies. When availability is high, most iron precipitates in compounds containing iron and hydroxyl and phosphate ions. Thus, iron deficiency from high soil pH is called lime-induced chlorosis, while in acid soils iron availability sometimes increases to toxic levels.

PLANT FUNCTIONS. Iron functions in the formation of many compounds even though it may not be present in their structure. This microelement is a component of various enzymes and functions as a catalyst in the synthesis of chlorophyll; much of the iron in a plant is localized in the chloroplast.

DEFICIENCY AND TOXICITY SYMPTOMS. Symptoms of iron deficiency, termed **iron chlorosis,** is characterized by degreening of leaves resulting in a yellowish or even whitish appearance in interveinal areas (Figure 13-6). Iron is very immobile, so deficiency symptoms first appear in the very youngest leaves. New leaves may unfold completely devoid of color but veins usually turn green later. Shoot debark may occur in acute cases.

Iron toxicity usually results in manganese deficiency. Tissue levels of iron should range between 75 and 125 ppm.

FERTILIZERS. Most soils contain sufficient iron for plant growth; the main problem is to make it available by proper soil management. If soil reaction is too high, acidity may be increased by applications of sulfuric acid or ammonium sulfate.

Iron may be applied directly to the soil in an inorganic form, usually as iron sulfate, but in this form it often becomes bound in heavily limed or alkaline soils.

Perhaps the most effective way to remedy iron chlorosis is to use iron **chelates.** Chelates, from the Greek word *claw*, are complex organic molecules that bind metal ions and form a stable water-soluble complex that is relatively chemically inert. Ethylene diamine tetracetic acid (EDTA) is a widely used synthetic chelating agent; when combined with iron, it forms the iron chelate known as FeEDTA. Iron chelates bind iron at two or more positions within their structures; it is held in such a way that it remains in solution and cannot precipitate unless ion exchange occurs or unless the chelate is precipitated by phosphate. Although chelates are expensive to produce, they are widely used as foliar sprays for ornamental shrubs and other high-value crops. Often, more

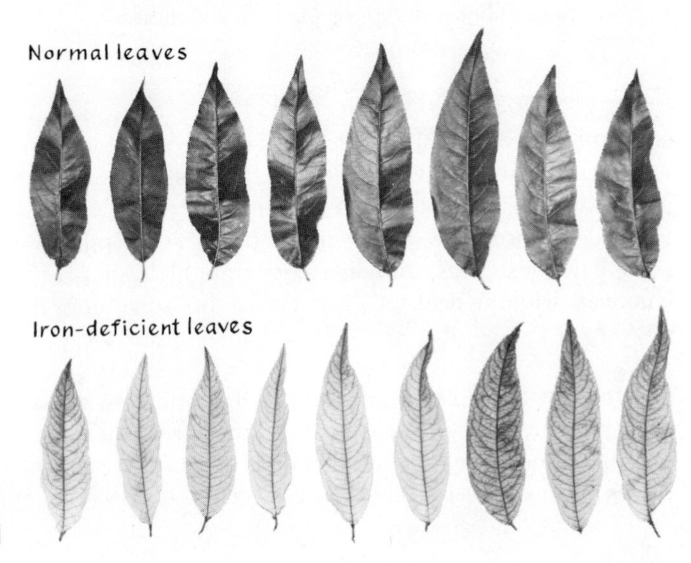

FIGURE 13-6. Iron deficiency in peach. [Courtesy National Plant Food Institute.]

than a single spray is required per season. For nursery crops, chelates are frequently applied directly to soils. They are not toxic to plants, applications of 1120 kg/ha (1000 pounds per acre) having been tried without ill effects. Chelates are decomposed by microorganisms, and hence the iron they make available may with time again become bound.

Boron (B)

SOIL RELATIONS. Total boron in most soils is low. Boron exists as borate (H_3BO_4) in most mineral soils. It is present in the soil solution or adsorbed onto soil particles, and it is the only leachable micronutrient. Boron becomes fixed and availability declines as pH increases. Its uptake may be decreased by very high or very low moisture levels. Boron deficiencies may be widespread, especially in sandy soils or muck and peat soils.

PLANT FUNCTIONS. Although the precise role of boron is obscure, it functions in diverse roles: processes such as flowering and fruiting, pollen germination, cell division, nitrogen metabolism, carbohydrate metabolism and translocation, salt absorption, water relations, and hormone transplant. The critical levels of boron in tissue range from 25 to 100 ppm and are higher in dicots than monocots.

DEFICIENCY AND TOXICITY SYMPTOMS. Short internodes and thickened stems that become tough and brittle are early symptoms of boron deficiency, followed by stunting, discoloration, and death of shoot tips (Figure 13-7A). The death of shoot tips may force lateral bud break leading to a "rosette." Leaves become thick, curled and malformed, brittle, dark and discolored. Flowering and fruit set are depressed. Roots, tubers, or fruits may become discolored and cracked (Figure 13-7B). In perennial fruit crops, symptoms often appear on the fruit before the vegetative parts are affected. Fruits may not develop normally, but show a gnarled appearance caused by depression underlain with hard corky tissue. Internal cork may be confused with calcium deficiency.

Excess boron is toxic. In stone fruits, toxicity symptoms include twig dieback, enlarged nodes, and gumming; leaf yellowing along the midrib and large lateral veins, with premature abscision; fruit splitting, corking, and premature dropping. In apple, toxicity results in early fruit maturity and shortened storage life.

FERTILIZERS. The principal boron fertilizers are borax ($Na_2B_4O_7 \cdot 10H_2O$), sodium tetraborate ($Na_2B_4O_7 \cdot 5H_2O$), and solubar or fertilizer borate ($NaB_4O_7 \cdot 4H_2O$). Other fertilizers include boron trioxide and colemanite, a form of calcium borate ($Ca_2B_6O_{11} \cdot 5H_2O$) used in sandy soils because it does not leach as rapidly as borax. Boron may be applied effectively as a foliar spray to deciduous fruit trees or to transplants.

FIGURE 13-7. Boron deficiency in pear fruits (*A*) and broccoli plants (*B*). [Courtesy National Plant Food Institute and D. N. Maynard, University of Massachusetts.]

Manganese (Mn)

SOIL RELATIONS. Manganese is found in many minerals containing iron and is available in the soil in ionic form. In alkaline soils high in organic matter and under aerobic conditions, however, it is oxidized from the manganous to the manganic form ($MnO \rightarrow MnO_2$; that is, from Mn^{2+} to Mn^{4+}), rendering it unavailable. Thus, manganese deficiency is a problem in alkaline soils. The solubility of soil manganese increases as acidity rises and at pH 5.5 or below, manganese toxicity may be a problem. Excess manganese is treated by liming.

PLANT FUNCTIONS. Manganese is involved in chlorophyll synthesis, respiration, nitrogen assimilation, and photosynthesis. It has a role as an enzyme activator for oxidation-reduction reactions. Satisfactory tissue levels vary from 50 to 100 ppm.

DEFICIENCY AND TOXICITY SYMPTOMS. Manganese is immobile, so deficiency symptoms appear in young leaves first. Characteristic symptoms consist of a mottled appearance with chlorosis between the main veins that start near the leaf margin (Figure 13-8). These symptoms are similar to iron deficiency; necrotic spots often appear as the deficiencies increase in severity.

FIGURE 13-8. Five-month-old tung seedling showing symptoms of manganese deficiency. The lower leaves are normal because they developed before the manganese in the seed was exhausted. [Courtesy USDA.]

Excess manganese produces leaf crinkling and cupping. A disorder of apple trees known as "measles," because pimples appear on the bark of 2-year-old wood, has been associated with excesses of manganese and low calcium in the bark.

FERTILIZERS. Manganese-containing fertilizers include manganese sulfate ($MnSO_4$), manganese oxide (MnO), manganese carbonate ($MnCO_3$), and manganese-chelate (MnEDTA). Soil or foliar application of $MnSO_4$ corrects manganese deficiencies except in very alkaline soils. If pH is high, chelated manganese is preferred.

Zinc (Zn)

SOIL RELATIONS. Zinc is a trace constituent of soils, but its availability is more important than its absolute amount. Zinc availability causes difficulty in alkaline soils because the insoluble calcium zincale is favored. Its deficiency has been a problem in the calcarious soils of the western United States. Zinc is readily leached from sandy soils.

Zinc toxicity may occur when soils are acidified and zinc has been applied for long periods as a fertilizer.

PLANT FUNCTIONS. Zinc has a role in protein synthesis. It is also involved in the synthesis of indoleacetic acid (IAA), the natural auxin via the synthesis of the amino acid, tryptophan, a precursor of IAA, or the conversion of tryptophan to IAA.

DEFICIENCY AND TOXICITY SYMPTOMS. The association of zinc and auxin may account for deficiency symptoms resembling auxin limitation, such as growth suppression and reduced internode length. Zinc is nonmobile, and so deficiency symptoms appear in young leaves. The initial mild symptom is an interveinal chlorosis similar to that resulting from manganese deficiency. Reduced internode elongation brings leaves that are reduced in size and closer together; hence, the terms "little leaf" or "rosette" are used to describe zinc deficiency. Nuts are often poorly filled and low in oil content where zinc deficiency is a problem.

Molybdenum (Mo)

SOIL RELATIONS. Molybdenum occurs in very low amounts in soils and, although most soils contain molybdenum, deficiencies do occur. It is only slightly soluble and is fixed more strongly in acid than alkaline soils. Therefore, liming of acid soils is the most beneficial corrective measure to eliminate molybdenum deficiency.

PLANT FUNCTIONS. Molybdenum is a component of the enzyme system that reduces nitrates to ammonia. When it is absent, the synthesis of proteins is blocked and plant growth ceases. It is also necessary for vitamin synthesis. Root-nodule bacteria of legumes require molybdenum. Seeds frequently fill poorly when molybdenum is lacking. Plants generally contain about 1 ppm or less of molybdenum. A sufficiency range is 0.15 to 1.0 ppm.

DEFICIENCY AND TOXICITY SYMPTOMS. Molybdenum is not mobile within the plant and deficiency symptoms appear on young leaves first. Common early symptoms are pale green leaves with rolled or cupped margins. Older leaves may be interveinally chlorotic and mottled. Yellow spots appear on leaves of citrus. Affected leaves are often rough, thick, irregularly wrinkled, buckled, and twisted. Molybdenum deficiency in most broad-leaved plants is called strap leaf, since affected leaves are reduced in width more than length. Leaves of crucifer become narrow, hence the name "whiptail." Potato tubers fail to develop in molybdenum-deficient plants.

FERTILIZERS. Because so little molybdenum is required, seed dusting with a molybdenum compound such as sodium molybdate may be effective. Foliar sprays of sodium or ammonium molybdate are often used to correct deficiencies; only about 80 g/ha (30 g/acre) are needed.

Copper (Cu)

SOIL RELATIONS. Most soil copper is derived from the weathering of soil minerals containing copper. Clay and loam soils usually have a copper content of 10 to 200 ppm, which is generally sufficient for most plants. Exchangeable copper occurs as Cu^{2+} or Cu^{+} with solubility higher in acid soils and declining as alkalinity increases. Sandy soils may be deficient because of leaching and low replacement. However, copper is tightly bound in organic matter complexes. Thus, organic soils are likely to be copper deficient.

PLANT FUNCTIONS. Copper is a component of several enzymes and it affects chlorophyll synthesis, and carbohydrate and protein metabolism. About 70 percent of copper is found in chloroplast. Its critical level in plant tissues ranges from 5 to 15 ppm.

DEFICIENCY AND TOXICITY SYMPTOMS. Copper deficiency symptoms vary with species and are not clear-cut. Leaf size may be reduced with chlorosis and mottling. In severe cases, shoot tips may show dieback with terminal leaves developing dead spots and brown areas. Bud break below dead shoot tips may give the plant a rosette or "witches' broom" appearance. Copper and zinc deficiencies are often associated and they produce a confusing symptom picture.

FERTILIZERS. Copper sulfate ($CuSO_4 \cdot 5H_2O$) is the most commonly applied copper fertilizer, but copper from this source is immobile and nonleachable. It becomes increasingly unavailable to plants as pH increases. It can be applied as foliar sprays of copper sulfate or chelated copper. Copper in combination with lime was once widely used as a component of a fungicide, Bordeaux mixture.

Chlorine (Cl)

SOIL RELATIONS. Chlorine was proven essential in 1954. Not only are chlorine deficiencies rare but it is extremely difficult to maintain a chlorine-free environment because of chlorine's high solubility and its ubiquitous distribution in soil, water, and atmosphere. Chlorine exists as the Cl^- ion and moves freely in the soil.

PLANT FUNCTIONS. Chlorine is present in all plants in small quantities (about 0.1 percent dry weight) and is involved in photosynthesis.

DEFICIENCY AND TOXICITY SYMPTOMS. Experimentally produced deficiencies produce wilting, stubby roots, excessive branching chlorosis, and bronzing.

Chlorine toxicity produced by extra chlorides, especially in saline soils, includes leaf chlorosis, necrosis, marginal scorching, growth suppression, and reduced yields.

FERTILIZERS. Chlorine is added to the soil by rainfall in amounts sufficient to satisfy plant requirements, and it is also supplied by many fertilizers.

SOIL MANAGEMENT

The soil provides support for the plant and is the storehouse of plant nutrients, water, and oxygen for root growth. The ability of the soil to support plant growth is often identified as its **productive capacity.** This productive capacity must be considered in terms of its fertility, the nutrient-supplying capacity of the soil, and its physical condition. It is not enough that the nutrients necessary for growth be contained in the soil; they must be released in a form readily available to the plant. Furthermore, the soil must be conserved; it is a renewable, but not easily replaceable, natural resource. Soil management is concerned with keeping land productive under sustained use and with ensuring economic crop production.

Maintenance of Soil Fertility

The fertility of the soil is only indirectly related to the chemical composition of the primary inorganic minerals. The most important factor in soil fertility is the amount and availability of the forms of the nutrients available to the plant. Such levels are related to many factors, among which are the solubility of the nutrients, the soil pH, the cation-exchange capacity of the soil, the soil texture, and the quantity of organic matter present.

The nutrient status of the soil is the net effect of many interacting factors that include the status of soil-nutrient microorganism activity, plant and organic matter, and pH. The interactions can be very complicated. This can be illustrated by mineral and pH interactions.

The term **mineral interactions** refers to the effect the supply of one nutrient can have on the uptake or utilization of another. One element may counter or depress the effect of another (antagonism) or increase the impact of another (synergism). For example, raising levels of phosphorus may depress nitrogen uptake; increasing nitrogen can depress boron uptake.

In a similar way, pH has a tremendous effect on nutrient availability. For example, the adverse effect of soil acidity is usually *not* the hydrogen ion content that affects plant growth; rather, it is the increasing concentration to toxic levels of aluminum and manganese, as well as a decrease in magnesium and phosphorus. Similarly, as the soil becomes more alkaline, a decreased solubility of copper, iron, manganese, and zinc leads to deficiency.

The recognition of soil fertility as a factor in crop response is recorded in ancient Greek writings, as is the supplemental use of manure on soils to improve plant growth. The use of cover crops, the mixing of soils, and the addition of lime and salts to increase the productivity of soils are mentioned in Roman agricultural treatises. Nevertheless, it was not until well into the nineteenth century that the role of inorganic nutrients in soil fertility was understood. The study of soil fertility today is intimately involved with microbiology, chemistry, and physics.

The maintenance of soil fertility is concerned with adjusting the current supply of available nutrients to optimum levels for economic crop production. The inherent fertility of the soil is related to the factors that contributed to its formation—namely, the parent minerals from which the inorganic part of soil was formed, the topography, the climate, the natural vegetation, and time. The fertility of "virgin" soil—soil that has not been disturbed by cultivation—reaches an equilibrium such that the nutrients released into the soil equal those that are lost from it. The inherent fertility of virgin soils, which varies greatly, may become depleted when the soil is brought under cultivation (Figure 13-9). This is brought about by crop removal (Table 13-2) and mineralization of organic matter. In addition, the removal of a permanent plant cover, row cropping, and the loss of soil structure by cultivation hasten erosion by water and

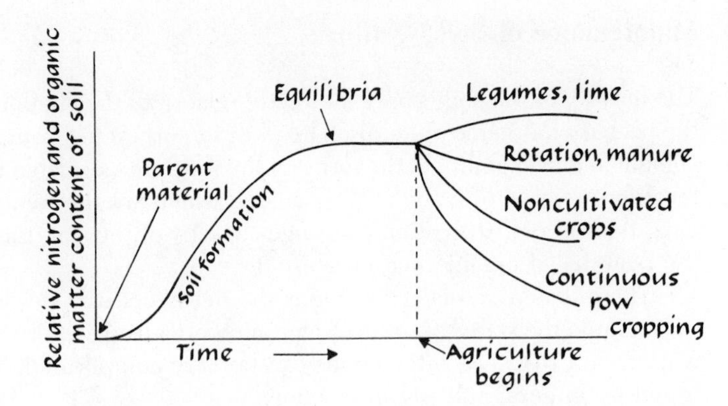

FIGURE 13-9. Influence of human activity on the fertility and organic matter of the soil. [Adapted from H. Jenny, *Factors in Soil Formation*, McGraw-Hill, 1941.]

wind. The nutrients lost may be replaced by supplemental additives in the form of fertilizers and manures. Nitrogen also may be added through the process of nitrogen fixation by bacteria on the roots of a leguminous crop. Chemical fertilizers have been the most economical source of nitrogen.

Fertilization as a Horticultural Practice

Fertilization refers to the addition of nutrients to the plant. The primary objective of crop fertilization is to achieve an optimum plant response. This may not necessarily be the greatest response; in commercial crop production, it is that point at which the value of the increased response is equal to the cost of the

TABLE 13-2. Approximate amounts of nutrients removed from soil by fruit and vegetable crops. (Note 1 lb/acre = 1.121 kg/ha.)

Crop	Yield per acre	Nutrient (in elemental form) removed (lb/acre)					
		N	P	K	Ca	S	Mg
Apple (fruit)	600 bu	34	3	32	3	12	5
Bean (seed)	40 bu	86	7	28	5	5	4
Cabbage (heads)	20 tons	120	14	133	16	32	5
Onion (bulbs)	1000 bu	154	26	121	18	48	12
Peach (fruit)	500 bu	30	3	73	3	4	5
Potato (tubers)	600 bu	126	19	150	7	14	14
Spinach (tops)	1000 bu	88	10	35	17	8	10
Sweetpotato (roots)	400 bu	53	7	93	7	9	13
Tomato (fruit)	20 tons	80	9	117	16	5	9
Turnip (roots)	500 bu	64	13	72	14	20	5

SOURCE: Data from American Plant Food Council.

additive (Figure 13-10). Fertilization beyond this level must be considered a wasteful practice. Not only is an excess subject to loss by leaching and volatilization, but it may actually be toxic to crops. Overfertilization of greenhouse soils has become a serious problem.

Materials that supply nutrient elements to plants are known as **fertilizers.** Those that supply nitrogen, phosphorus, and potassium, the major plant nutrients, are called **complete fertilizers.**

The **grade** of **analysis** of these fertilizers is a three-part number that shows the percentages of nitrogen (expressed as elemental N), phosphorus (expressed as P_2O_5), and potassium (expressed as K_2O), *in that order.* The reason phosphorus and potassium are not expressed in their elemental form is historical. A movement is currently in progress to have this changed, especially in scientific writing. To change oxide analyses to elements and *vice versa,* multiply by the following conversion factors.

Element	Oxide to element	Element to oxide
Phosphorus	0.43	2.33
Potassium	0.83	1.20

The oxide analysis is still used commercially. Thus a 10–10–10 fertilizer is 10 percent elemental nitrogen by weight, but only 4.3 percent elemental phosphorus and 8.3 percent elemental potassium. Similarly, an 80-pound bag of a 5–15–30 fertilizer will contain 4.0 pounds of N, the amount of phosphorus in 12 pounds of P_2O_5 (5.2 pounds of P), and the amount of potassium in 24 pounds of K_2O (19.9 pounds of K).

The fertilizer **ratio** is simply the analysis expressed in terms of the lowest common denominator. A 6–10–4 analysis has a 3:5:2 ratio, and a 10–10–10

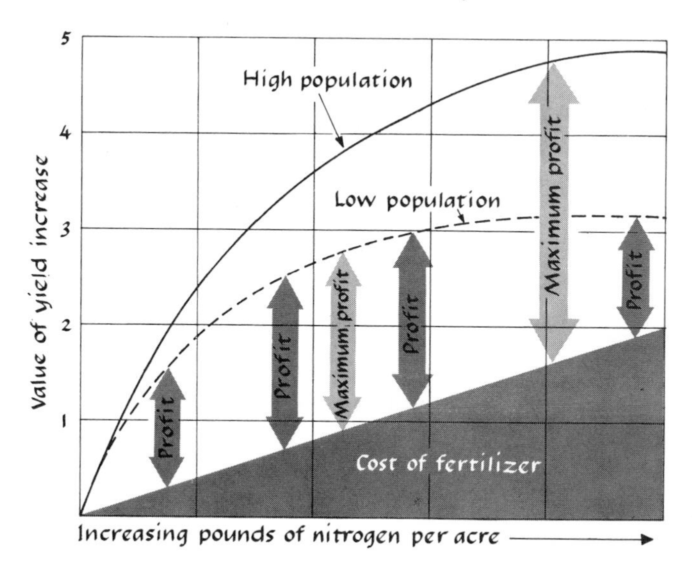

FIGURE 13-10. The yield response diminishes with increasing quantities of fertilizer. The optimum rate is obtained when the value of the yield response is just equal to the last increment of fertilizer added. At this point the profit is maximized. But the response to fertilizer may depend on other factors. As is shown above, increasing the plant population changed the response curve and made a higher rate of nitrogen profitable. [Adapted from *Soil* (USDA Yearbook, 1957).]

analysis has a 1:1:1 ratio. Fertilizers are referred to as **low-analysis** fertilizers when the amount of available nutrients is below 30 percent by weight and as **high-analysis** fertilizers when the amount of available nutrients is 30 percent or more. Because of the greater weight and the extra handling required, low-analysis fertilizers tend to be more expensive per nutrient unit than high-analysis fertilizers. High-analysis fertilizers, however, require special accuracy and care in their application. Although by law the analysis of nitrogen, phosphorus, and potassium must be stated on the fertilizer bag, many fertilizers contain, or are restricted to, other plant nutrients.

Fertilizers may be classified as natural organics or chemicals. **Natural organics** (for example, manure, blood, fish scraps, and cottonseed meal) are compounds derived from living organisms. **Chemical fertilizers,** such as ammonium nitrate and superphosphate, are synthesized from inorganic minerals. A number of forms of nitrogen-containing compounds are now synthesized, including urea and cyanamid. Urea can be synthesized directly from the air, a process that requires only a source of electrical power. Although urea and cyanamid are organic compounds in the chemical sense, they are not necessarily derived from living systems.

At present, most nitrogen fertilizer is synthesized by the Haber process, in which atmospheric nitrogen is reacted with hydrogen to form ammonia. The ammonia may be used directly as a fertilizer, or it can be used as a raw material for the manufacture of urea, nitrates, or other nitrogenous compounds. The hydrogen required by the Haber process is generally extracted from natural gas, the cost of which makes up most of the cost of the manufactured nitrogen fertilizer. The synthesis of a ton of anhydrous ammonia requires 850 m^3 (30,000 cubic feet) of natural gas.

The nutrients in natural organic fertilizers and some synthetic organics undergo gradual chemical transformations to available forms. "Urea-forms" are combinations of urea and formaldehyde—the greater the proportion of formaldehyde, the lower the solubility of urea and hence the slower the "release." Other types of slow-release materials are composed according to specifications for particle size, chemical composition, and solubility. A special slow-release system has been developed by encapsulating fertilizer pellets with a plastic coat. The coat acts as a membrane to release the plant nutrients gradually. These materials are comparatively expensive per pound of nutrient, but they have special uses in horticulture, such as for container-grown nursery stock, because they allow application costs to be reduced.

Modern fertilizers can be compounded to satisfy the different needs of the user. Thus, they may not only be constituted of different nutrients but may be a mixture of organic and inorganic forms. In this way some, or a portion, of the applied nutrients can be made available immediately, whereas the rest can be released slowly, commensurate with the needs of the crop. Decisions about whether a slow-releasing fertilizer or a number of applications of a fast-releasing fertilizer should be used are generally based on economic considerations.

The physical characteristics of fertilizer materials vary greatly. Although many fertilizers are solids, they may be applied in dissolved form as a liquid, as in irrigation water. Nitrogen may be applied in the ammonia form as a gas. In addition to soil application, nutrients may be applied directly through the foliage. Nitrogen can be efficiently applied through the leaves by spraying them with urea. The application to the foliage of such trace elements as manganese and boron has also proved practical.

LEVEL OF FERTILIZATION. The kind and level of fertilization are based on crop need in relation to current fertility levels and the alternative sources of plant nutrients. The prediction of plant-nutrient needs has been one of the main goals of plant-fertility studies. The techniques that are now used consist of correlating plant responses with chemical tests on the soil or with tests made on the plant tissues themselves. However, the total nutrient content of the soil does not give a true picture of nutrient availability. The available nutrients are related to the exchangeable cations, the soil reaction or pH, and the organic cycles. Some biological assays of the soil have been made by utilizing the responses of particularly sensitive plants or microorganisms. The relationship between these tests, soil type, and climate have been correlated for many crop plants. Quick tests have been developed, although these are often not too accurate. In many plants severe deficiencies of certain nutrients produce characteristic responses, called **deficiency symptoms,** which often can be used to diagnose the trouble (Table 13-3). The good grower will not permit nutrient shortages to become so severe that deficiency symptoms appear.

TABLE 13-3. A key to plant nutrient-deficiency symptoms based on vegetable crop responses.

Symptoms	Deficiency
A. Symptoms on leaves, stems, or petioles.	See B
Flowering or fruiting affected.	See M
Storage organs affected.	See N
Variable plant growth throughout the field. Some plants appear normal, some show severe marginal leaf necrosis, while others are stunted. Determine soil pH.	Acid or alkaline soil complex
B. Youngest leaves affected first.	See C
Entire plant affected or oldest leaves affected first.	See I
C. Chlorosis appears on youngest leaves.	See D
Chlorosis is not a dominant symptom. Growing points eventually die and storage organs are affected.	See H
D. Leaves uniformly light green, followed by yellowing and poor, spindly growth. Most common in areas with acidic, highly leached, sandy soils low in organic matter.	Sulfur
Uniform chlorosis does not occur.	See E

TABLE 13-3. *(Continued)*

	Symptoms	Deficiency
E.	Leaves wilt, become chlorotic, then necrotic. Onion bulbs are undersize and outer scales are thin and lightly colored. May occur on acidic soils, on soils high in organic matter, or on alkaline soils.	Copper
	Wilting and necrosis are not dominant symptoms.	See F
F.	Distinct yellow or white areas appear between veins, and veins eventually become chlorotic. Symptoms rare on mature leaves. Necrosis usually absent. Most common on calcareous soils ("lime induced chlorosis").	Iron
	Yellow/white areas are not so distinct, and veins remain green.	See G
G.	Chlorosis is less marked near veins. Some mottling occurs in interveinal areas. Chlorotic areas eventually become brown, transparent, or necrotic. Symptoms may appear later on older leaves. In peas and beans, the radical and central tissue of cotyledons of ungerminated seeds become brown ("marsh spot"). Most common on soils with pH over 6.8.	Manganese
	Leaves may be abnormally small and necrotic. Internodes are shortened. Beans, sweet corn ("white bud" of maize), and lima beans most affected; potatoes, tomatoes, and onions somewhat affected; uncommon with peas, asparagus, and carrots. Reduced availability in acidic, highly leached, sandy soils, in alkaline soils, and in organic soils.	Zinc
H.	Brittle tissues. Young, expanding leaves may be necrotic or distorted followed by death of growing points. Internodes may be short, especially at shoot terminals. Stems may be rough, cracked, or split along the vascular bundles (hollow stem of crucifers, cracked stem of celery). Most likely on highly leached, acidic soils and on organic soils with free lime.	Boron
	Brittle tissues not a dominant symptom. Growing points usually damaged or dead ("dieback"). Margins of leaves developing from the growing point are first to turn brown or necrotic, expanding corn leaf margins are gelatinous and necrotic, expanding cruciferous seedling leaves are cupped and have necrotic margins; old leaves remain green. Common on acidic, highly leached, sandy soils. May result from excess Na, K, or Mg from irrigation waters, fertilizers or dolomitic limestone. (Celery blackheart, brown heart of escarole, lettuce tipburn, internal tipburn of cabbage, internal browning of Brussels sprouts, hypocotyl necrosis of snap beans.)	Calcium
I.	Plant exhibits chlorosis.	See J
	Chlorosis is not a dominant symptom.	See L
J.	Interveinal or marginal chlorosis	See K
	General chlorosis. Chlorosis progresses from light green to yellow. Entire plant becomes yellow under prolonged stress. Growth is immediately restricted and plants soon become spindly and drop older leaves. Most common on highly leached soils or with high organic matter soils at low temperatures. Soil applications of N show dramatic improvements.	Nitrogen

TABLE 13-3. *(Continued)*

	Symptoms	Deficiency
K.	Marginal chlorosis or chlorotic blotches which later merge. Leaves show yellow chlorotic interveinal tissue on some species, reddish purple progressing to necrosis on others. Younger leaves affected with continued stress. Chlorotic areas may become necrotic, brittle, and curl upward. Symptoms usually occur late in growing season. Most common on acidic, highly leached, sandy soils or on soils with high K or high Ca.	Magnesium
	Interveinal chlorosis, with early symptoms resembling N deficiency (Mo is required for nitrate reduction); older leaves chlorotic or blotched with veins remaining pale green. Leaf margins become necrotic and may roll or curl. Symptoms appear on younger leaves as deficiency progresses. In brassicas, leaf margins become necrotic and disintegrate, leaving behind a thin strip of leaf ("whiptail," especially of cauliflower). Common on acidic soils or highly leached alkaline soils.	Molybdenum
L.	Leaf margins tanned, scorched, or have necrotic spots (may be small black dots which later coalesce). Margins become brown and cup downward. Growth is restricted and dieback may occur. Mild symptoms appear first on recently matured leaves, then become pronounced on older leaves, and finally on young leaves. Symptoms may be more common late in the growing season due to the translocation of K to developing storage organs. Most common on highly leached, acidic soils and on organic soils due to fixation.	Potassium
	Leaves appear dull, dark green, blue-green, or red-purple, especially on the underside, and especially at the midrib and veins. Petioles may also exhibit purpling. Restriction in growth may be noticed. Availability reduced in acidic and alkaline soils, and in cold, dry, or organic soils.	Phosphorus
	Terminal leaflets wilt with slight water stress. Wilted areas later become bronzed, and finally necrotic. Very infrequently observed.	Chlorine
M.	Fruit appears rough, cracked or spotted. Flowering is greatly reduced. Tomato fruits show open locule, internal browning, blotchy ripening or stem-end russeting. Occurs on acidic soils, on organic soils with free lime, and on highly leached soils.	Boron
	Cracking and roughness are not dominant symptoms. Fruits exhibit water-soaked lesions at blossom end, later becoming sunken, dark, or leathery (blossom-end rot of tomato, pepper, and watermelon). Common on acidic, highly leached soils.	Calcium
N.	Internal or external necrotic or water soaked areas of irregular shape (hollow stem of crucifers, internal browning of turnip and rutabaga, canker or blackheart of beet, water core of turnip). May occur on acidic soils, on alkaline soils with free lime, or on highly leached soils.	Boron
	Cavities develop in the root phloem, followed by collapse of the epidermis, causing pitted lesions. (Cavity spot of carrots or parsnips.) Common on acidic, highly leached soils.	Calcium

SOURCE: J. E. English and D. N. Maynard, 1978, *HortScience* 13:28–29.

The determination of fertility status is best made by a combination of soil testing and plant (leaf) analysis. Soil tests involve a combination of analytical procedures to determine elemental fertility status and other factors as follows:

Extractable major elements (P, Ca, Mg, K, Na)
Extractable microelements (B, Cu, Fe, Mn, Zn, Mo)
Soil water pH (1 soil:1 water)
Organic matter
Soluble salt content
Extractable nitrates (NO_3) and sulfates (SO_4)

Unfortunately, the interaction of soil and plant makes the determination of the optimum corrective measures difficult.

As a result, plant analysis is widely used. The objective is to determine the elemental content of plant tissue under the assumption that the nutrient status of the plant best integrates all the interactive forces. This approach has led to the development of "critical levels," the range of concentration at which plant growth is restricted when compared with optimal levels (Table 13-4).

In order to be useful, soil testing and tissue analysis must take into account proper sampling procedures. In addition, plant analysis requires proper tissues for analysis and must take account of seasonal fluctuations, crop load, climate, and even rootstock.

The relationship between fertility level and plant performance varies with

TABLE 13-4. Optimal levels of foliar tissue elements.

Element	Greenhouse herbaceous plants	Apple (summer)
N	2.5–4.5%	1.7–2.5%
P	0.2–0.3%	0.15–0.3%
K	2.5–5.0%	1.2–1.9%
Ca	0.5–1.5%	1.5–2.0% (summer–July)
Mg	0.35–0.55%	0.25–0.35%
S	NE*	NE*
Fe	75–125 ppm	40–500 ppm
Mn	50–100 ppm	25–150 ppm
Zn	25–100 ppm	15–200 ppm
Cu	5–15 ppm	5–12 ppm
B	25–100 ppm	20–60 ppm
Mo	0.15–1.0 ppm	0.10–0.20 ppm
Cl	100–1000 ppm	NE*

*Not established

the species and the nutrient. For example, 112 to 168 kg/ha (100 to 150 pounds per acre) of nitrogen applied prior to planting will promote optimum production of tomatoes on mineral soils, whereas this level of nitrogen will reduce muskmelon yields as a result of decreased production of perfect flowers. However, the plant response to fertility level can be discussed in general. At one end of the scale are **deficiency levels,** at which plants show definite symptoms of deficiency. At somewhat higher levels, although they may not show obvious deficiency symptoms, crop plants may show reduced yield. This has been termed **hidden hunger.** At levels above which no response to fertilizers may be demonstrated, the plant may continue to show an increasing level of nutrient absorption. This is termed **luxury consumption.** At abnormally high levels, growth is reduced and even death may occur. Maximum production presumably comes in that state of soil fertility in which a slight luxury consumption exists.

The level of crop response to fertilization is related in part to the productive capacity of the soil. Crops on soils of low productive capacity show a maximum response at a lower level of fertility than on soils of high productive capacity. Productive capacity is based on long-term nutrient availability and soil condition. Owing to the nature of forces in the soil that establish an equilibrium between the soil and the soil solution, optimum fertility cannot be achieved in one quick step. When larger amounts of fertilizer are placed on soils of a low productive capacity, much of it is wasted. These excess nutrients may be leached, may be tied up in forms unavailable to the plant, or may be poorly distributed throughout the soil with respect to the needs of the plant. However, continued applications of fertilizer at the level of optimum plant response tend to increase the productive capacity of the soil, ultimately raising its yield potential.

PLACEMENT AND TIMING. Among the important factors in the use of fertilizer are proper placement and timing. They are related to the efficiency of the plant's utilization of the fertilizer, the prevention of injury, and convenience and economy of application. To be effective, fertilizer must be applied where and when the plant needs it. Single yearly applications may not be sufficient for some nutrients, such as nitrogen, and they may not be necessary for others. Large amounts of highly soluble fertilizers should not be applied to growing plants, especially when young, because of salt injury (Figure 13-11). In perennials or in long-season annuals, it may be more efficient to control carefully the availability of nutrients throughout the season, and for this reason repeated applications are made. This is especially important with nitrogen fertilization, because excess amounts of nitrogen fertilizers are often irretrievably lost to the plant through leaching or other means.

There are various methods of fertilizer placement. A fertilizer may be applied prior to planting by scattering it uniformly over the surface of the ground (**broadcast application**). It may be dropped behind the plow at the bottom of

FIGURE 13-11. An excess of soluble fertilizers applied to growing plants such as turf grasses may burn the foliage or may kill the entire plant. [Photograph courtesy W. H. Daniel.]

the furrow (**plow-sole placement**), or placed in a band under the seed or to one or both sides (**band placement**) during planting (Figure 13-12). Another mode of application consists of applying fertilizer directly over the crop after emergence (**top dressing**) or beside the row (**side dressing**). Side-dressed applications of fertilizer are often made along with a cultivation, and are thus mixed into the soil. Fertilizer may also be applied along with mechanical transplanting, either added as a band under the plant or dissolved in the supplemental transplanting water (**starter solution**).

Nutrients may be applied in solution to the leaves (**foliar application**). Absorption of nutrients applied in this fashion is very rapid, and the plant's response may be evident within a day or two; but applications of this kind must

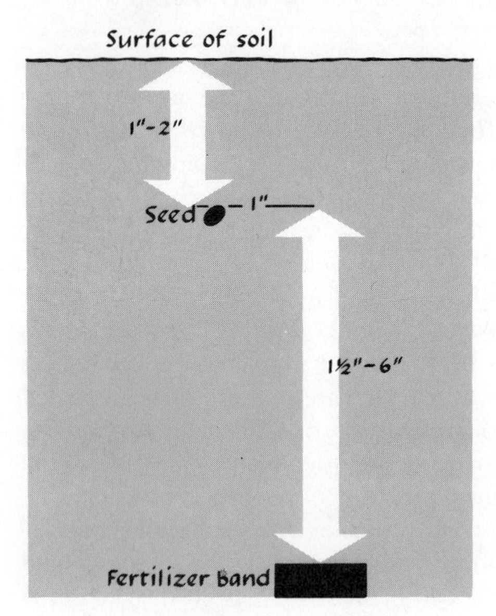

FIGURE 13-12. The proper placement of fertilizer band for many vegetable crops. Large amounts of fertilizer banded too close to the seed may inhibit growth because of salt injury. Fertilizer placed too far away may be insufficient to stimulate the early seedling growth that is so important to optimum plant performance. [Adapted from *Soil* (USDA Yearbook, 1957).]

be more frequent than applications to the soil because there is little residual effect.

The timing of fertilizer applications depends upon the nutrient and the crop as well as on the soil type and the climate. Nitrogen fertilizers must be supplied as close to the plant as possible to be of any use. The nitrate forms of nitrogen are water soluble and move rapidly within the soil. Although the ammoniacal forms of nitrogen are held on the soil colloids, they become "mobile" as soon as they are converted to the nitrate form. Therefore, nitrogen is often in short supply, especially in sandy soils. Soils high in applied organic matter may be temporarily short of nitrogen as the result of a buildup of microorganisms that feed on the organic matter. These microorganisms pre-empt available soil nitrogen for their own use. In addition, the nitrogen concentration in the soil solution may be relatively low in the spring, especially in cold, wet, poorly drained soils, owing to a lack of nitrification by aerobic bacteria and to excess denitrification by anaerobic bacteria. Consequently, many crops show a good response to spring applications of nitrogen. It is unwise to apply all the nitrogen at planting time because of possible injury, so it is often also added as a side dressing during the growing season.

In such perennial fruit crops as apple and peach, excess nitrogen is associated with poor fruit color and soft fruit, as well as with undesirable vegetative growth that occurs late in the season and leaves the plant vulnerable to winter injury. Consequently, nitrogen is usually applied only once, early in the spring, in order that any excess will have been used up by summer. In small-fruit crops, such as the strawberry, that ripen early in the summer, nitrogen is not applied to bearing patches in the spring because of the undesirable effect of fruit softening.

In contrast to nitrogen, phosphorus moves very little in the soil. Consequently, the total quantity needed during the season may be applied at one time. Because of the high phosphorus requirement of seedlings, it is important that adequate levels be made available close to the seed or transplant. The use of starter solution, transplanting water supplied with liberal phosphorus (about 1500 ppm), is recommended for many transplants. The phosphorus is supplied by soluble phosphate salts, such as monoammonium phosphate, diammonium phosphate, and monopotassium phosphate. A popular starter solution uses 3 pounds of a mixture of diammonium phosphate and monopotassium phosphate (10–52–17 analysis) in 50 gallons of water (7.2 g per liter). When an extensive root system is established, the plant requirements are satisfied by lower levels of phosphorus. Phosphorus applications are often banded under the seed to achieve the same effect (Figure 13-13). Perennial plants usually do not respond to phosphorus application because the root systems are extensive and active throughout most of the growing season.

Because of the low mobility of phosphorus and the low efficiency of phosphorus uptake by plants (less than 25 percent), it has been found profitable to build up soil phosphorus levels prior to planting horticultural crops. Once

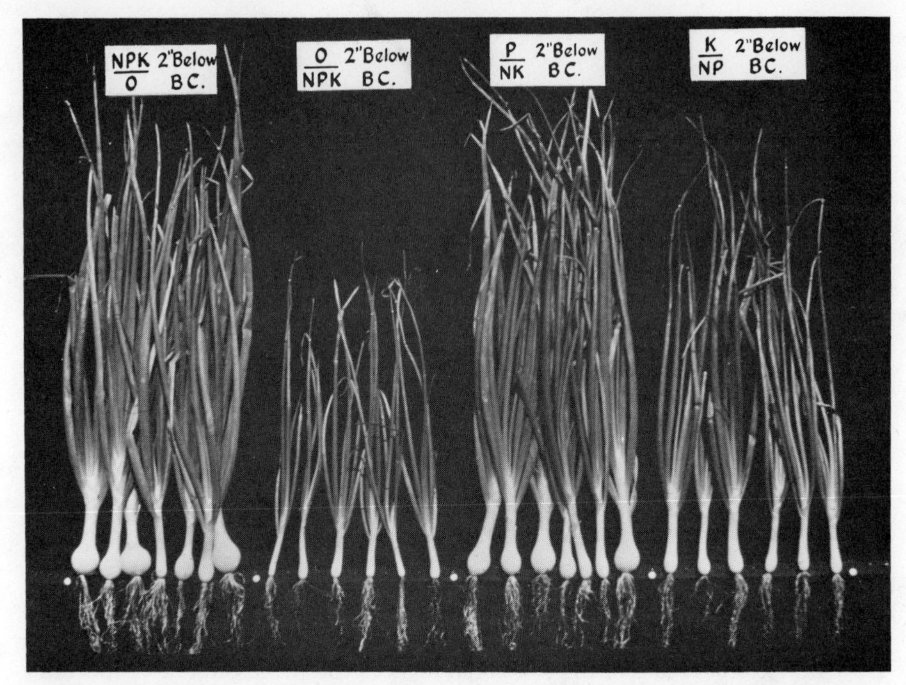

FIGURE 13-13. Response of onions to fertilizer placement. In all cases, equal amounts of fertilizer were added, either banded 2 inches below the seed or broadcast over the surface of the soil. The greatest response was obtained when the fertilizer was placed below the seed, where it was available to the young plant. The treatment series indicate that phosphorus is the critical nutrient with respect to placement. [Photograph courtesy J. F. Davis.]

phosphorus levels are brought up, supplemental additions of phosphorus need not be frequent. Turf, for example, shows practically no net loss of phosphorus.

Potassium salts are intermediate in mobility to phosphorus and nitrogen. Owing to their solubility, they cannot be placed close to seeds or plants in any great amounts. Since potassium is not as critical for seedling growth as nitrogen or phosphorus, broadcast applications are usually made before planting.

Regulation of Soil Reaction

Soil reaction, so essential to nutrient availability and root growth, is an important concern of soil management. Although plants vary in their response to pH (Figure 13-14), most horticultural crops do well with a pH between 6.0 and 6.5. There is a group of "acid-loving" plants that require conditions of low pH (4.5 to 6.0). These include gardenias and camellias in addition to most members of the family Ericaceae (which includes the azaleas, rhododendrons, cranberries, and blueberries). A low soil reaction may be used to control soil diseases in crops

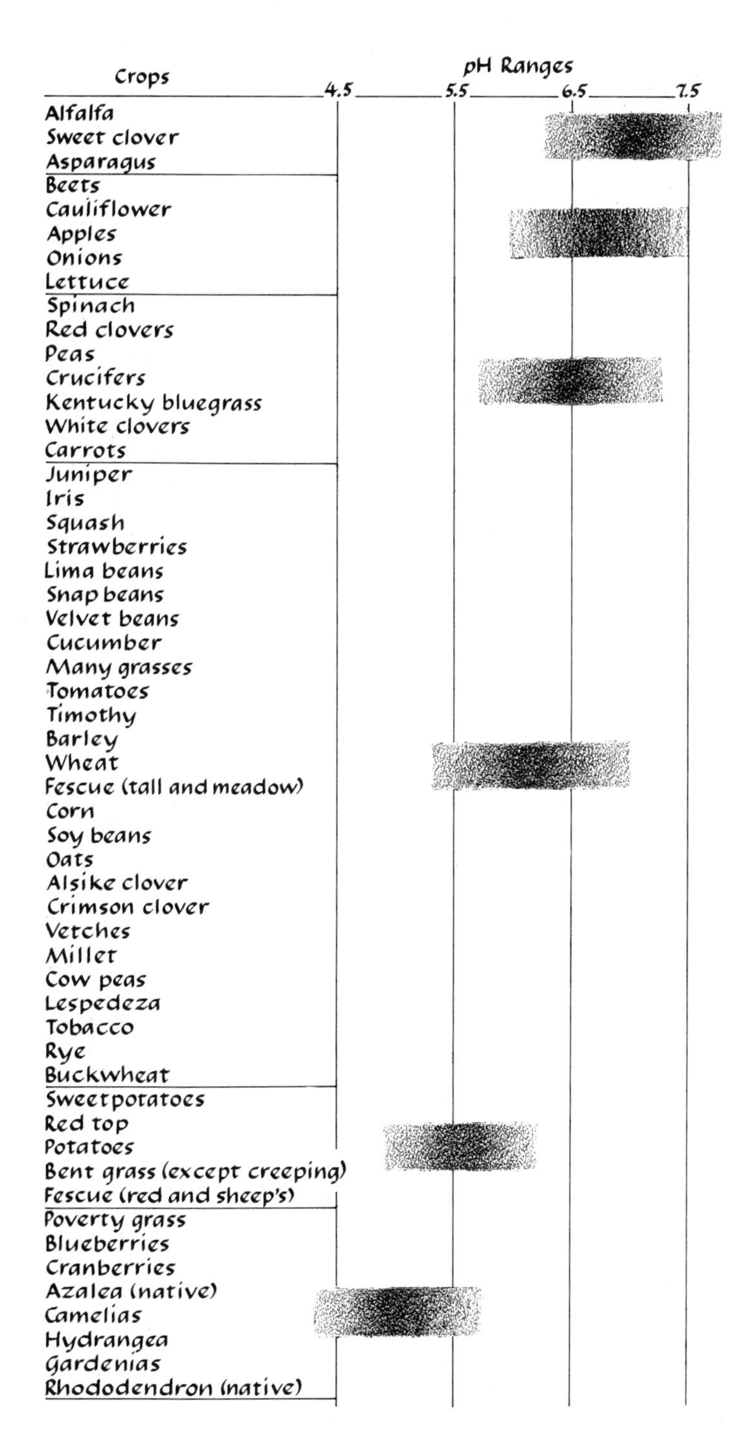

FIGURE 13-14. Suitable pH ranges from various crops and ornamental plants. [Adapted, with permission of the publisher, from T. L. Lyon, H. D. Buckman, and N. C. Brady, *The Nature and Properties of Soils*, 5th ed., copyright 1952 by the Macmillan Company.]

that prove less sensitive to low pH than the corresponding disease-producing organism. Thus, potatoes may be grown at pH 5.2 to control scab, a disease caused by a fungus that is not adaptive to acid soils. Potatoes, however, grow equally well at a higher pH in disease-free soil.

The natural reaction of soils is due to the interaction of climate with the parent materials of the soil. In general, acid soils are common where the precipitation is high enough to leach appreciable amounts of exchangeable bases from the surface layers. Thus, in the humid climates, the areas of most intense horticulture, soil acidity is often a problem. Soil alkalinity occurs generally in the more arid regions, where there is a comparatively high degree of cation accumulation. Alkaline soils may result in a high salt accumulation (salinity). This can be a problem in the irrigated soils of the American Southwest.

Soils become acid because the basic cations on the soil colloid are replaced by hydrogen ions. This process can be reversed, and the soil pH increased, by adding basic cations—for example, calcium, magnesium, sodium, and potassium. Calcium is the most economic cation for increasing soil pH. In addition, calcium has other beneficial effects. It is an essential element in plant nutrition; it is thought to promote good soil structure by promoting granulation; and it encourages certain soil microorganisms, especially the nitrifying and nitrogen-fixing bacteria. The addition of calcium or calcium and magnesium compounds to reduce the acidity of the soil is known as **liming.** Although the term *lime* refers correctly to CaO, it is used in the agricultural sense to include oxides, hydroxides, carbonates, and silicates of calcium or calcium and magnesium. Soil liming has resulted in significant increases in plant growth. The amount of liming required depends upon the degree of pH change desired, the base exchange capacity of the soil, the amount of precipitation, and the liming material and its physical form in relation to particle size.

Soil may be made more acid by placing hydrogen ions on the soil colloid. This is done by adding substances that tend to produce strong acids in the soil. Some nitrogen fertilizers will increase soil acidity, but elemental sulfur is by far the most effective substance to use for this purpose. In warm, moist, well-aerated soils, bacterial action converts sulfur to sulfuric acid.

Maintenance of Soil Condition

Tilth

Maintaining the physical condition of the soil (**tilth**) is a significant part of soil management. Soil structure and texture have a direct effect on the water-holding capacity and aeration of the soil. They influence root growth as well as the soil microorganisms that play a major part in making available the nutrients in organic matter. Crusting and puddling of the soil are indications of poor tilth (Figure 13-15).

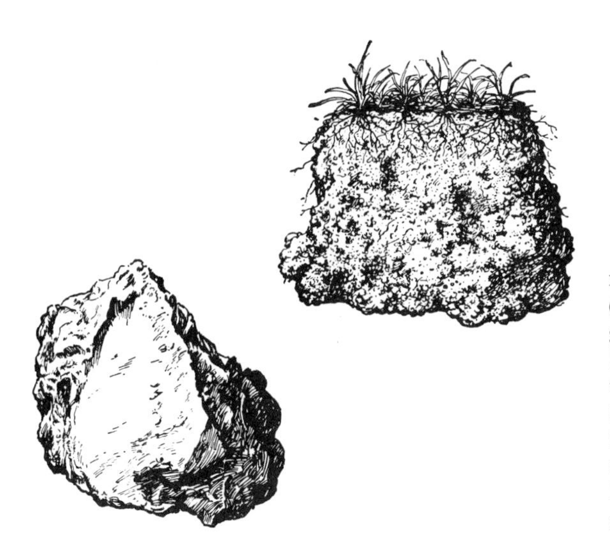

FIGURE 13-15. A puddled soil (*left*) compared with a well-granulated soil (*above*). [Adapted, with permission of the publisher, from T. L. Lyon, H. D. Buckman, and N. C. Brady, *The Nature and Properties of Soils*, 5th ed., copyright 1952 by the Macmillan Company.]

The physical condition of the soil is largely a result of the amount of organic matter it contains. Organic matter may be maintained (and in some cases, increased) in field soils by altering crop rotation and by adding supplemental organic matter in the form of manure. In potting soils, organic matter is often added in the form of peat or ground bark. The use of chemical soil conditioners such as Krilium have been proposed, although this has not proven practical. The problem of maintaining soil condition is complicated because most cultivation practices, contrary to first impression, do not aid in improving soil structure. Even the gains made by distributing plant material through the soil may be more than offset by compaction, by the loss of organic matter as a result of oxidation, and by erosion. Clay soils in particular must be handled carefully to maintain soil structure. If they are cultivated when too wet, the soil becomes puddled. The use of heavy equipment (such as mechanical harvesters) on wet clay soils seriously impairs soil structure and often leads to compaction. This is less of a problem on sandy or peat soils, whose structure is generally good under most conditions.

Increasing Organic Matter

Organic matter affects both the fertility and the physical condition of the soil. It acts as a storehouse of nitrogen and other nutrients, and it greatly influences the exchange capacity of the soil. It improves the physical condition of the soil by raising its water-holding capacity, so important to sandy soils, and by increasing aeration, which is especially necessary in clay soils.

The accumulation of organic matter reaches an equilibrium in undisturbed soil. Because the organic material of the soil is largely under the control of biologic and climatic forces, it is advisable to differentiate between temporary and long-term increases in organic matter. Unfortunately, it is easy to reduce the organic matter of a soil, and it is relatively difficult to build it up. The major loss of organic matter comes about from increased oxidation as a result of cultivation and from crop removal.

Perhaps the best way to effect a long-term increase in organic matter is through the extended use of a legume or grass sod (Figure 13-16). The organic matter produced as a result of root disintegration is protected from excessive oxidation by the constant cover and lack of cultivation. Permanent orchard plantings are kept in good tilth by the use of sod. Mixtures of shallow- and deep-rooted grasses, such as 'Alta' fescue (shallow-rooted) and 'Ladino' clover (deep-rooted), are often used to obtain a resilient floor under heavy equipment as well as to prevent compaction and to improve drainage. Sods, of course, have certain disadvantages. They are not generally used in peach culture because of frost injury (see Chapter 12) and because the grasses tend to compete with the trees for nitrogen. In annual and perennial plantings, organic matter can be built up by rotating row crops with legume sods. A standard soil management for potatoes in the northern United States involves a 3-year rotation of potatoes, grain, and clover. The grain is used to control weeds and to act as a "nurse crop" for the establishment of the clover.

The plowing down of a growing crop as a "green manure" temporarily raises the organic content of the soil, but this practice cannot be expected to produce a long-term increase in organic matter. The buildup of high populations of microorganisms may actually reduce the net organic-matter content of the soil through an unexplained breakdown of the normally more resistant fractions of the organic materials in the soil. The rapid breakdown of a green manure does release tied-up nutrients; but such materials as millet or sudan grass, which have a high carbon content in relation to their nitrogen or protein content, may create a temporary nitrogen shortage because of the utilization of the available nitrogen by microorganisms. Nitrogen must then be added to compensate for this shortage. Sweet clover and other legumes with high nitrogen content in proportion to carbon are destroyed more rapidly by microorganisms, and they release a steady supply of nitrogen as they are broken down.

The use of grasses or legumes planted in middle or late summer provides

FIGURE 13-16. Good soil structure under an alfalfa sod. Note how the roots have penetrated the soil. [Photograph by J. C. Allen & Son.]

the advantages of a sod for part of the year. These **cover crops** are probably of most value in preventing winter erosion. They are usually plowed under in the spring as a green manure. This plowing should be done early to prevent the loss of excessive moisture from the soil.

The use of manures to supplement soil organic matter is at present an expensive procedure. Manure is probably more valuable for its nutrients than for its contribution of organic matter. Similarly, the use of mulches indirectly adds to organic matter, though only in the upper surface of the soil. Their greatest contribution, however, lies in reducing the amount of cultivation, which in effect limits the oxidation of organic matter.

Regulation of Nutrient Cycles in the Soil

Soil fertility may be maintained by intervention in the nutrient cycles that connect plants, soil, animals, and microorganisms. There is the potential for a large increase in the biological fixation of nitrogen. Alfalfa, a legume, has been grown with annual yields of 16 tons per acre without the application of nitrogen fertilizers. Alfalfa is about 3 percent nitrogen; thus, in a single acre of alfalfa, the bacteria associated with the plant roots must fix at least 1000 pounds of nitrogen per year, about five times the amount generally accepted as typical. Nitrogen fixation in the soil might be further improved by artificially selecting the most efficient strains of symbiotic bacteria and by encouraging the widespread adoption of improved legumes, particularly in the tropics. It was recently discovered that bacteria capable of nitrogen fixation live in partially symbiotic associations with certain tropical grasses, including maize. The genetic manipulation of bacteria offers hope that the capacity to fix nitrogen may eventually be conferred to all crops! A recent development that shows promise is the discovery of an inexpensive substance (nitropyrin) that retards the bacterial conversion of ammonia to nitrite in the soil. Since ammonia is a cation and is retained by soil colloids, whereas nitrite and nitrate are readily leached away, nitropryin could retard the loss of nitrogen from the soil. Finally, the absorptive capacity of some plant roots is increased by an intimate association, called a **mycorrhiza,** between the roots and a fungus. The encouragement of mycorrhizal associations might benefit certain crop plants, particularly in soils of generally low fertility or where specific nutrients may be fairly scarce.

HYDROPONICS

Plants do not require soil to obtain the essential mineral elements but can absorb them from dilute solutions of mineral salts. The use of water-culture techniques is a standard method to study inorganic plant nutrition, and it has also become a horticulture technique for crop growing, identified as hydroponics. The term **hydroponics** (also known as **nutriculture**) originally referred to the growth of plants in a liquid system but now includes all systems that use nutrient solutions with or without the addition of an inert medium (such as sand, gravel, rockwood, or vermiculite) for mechanical support. Most hydroponic systems are carried out in controlled environments, typically glass or plastic-covered greenhouses, to provide temperature control.

In all variations of hydroponics, roots must be supplied with oxygen. This provision can be accomplished in various ways, such as by bubbling oxygen in the solutions, by frequent nutrient replacement, or by washing or misting exposed roots with nutrient solution. The major feature of hydroponics—the use of a dilute nutrient solution—is common to all systems and the precise composition of the nutrient solution is similar. Thus, the alternative hydroponic systems represent variation on a common theme.

Many successful formulations of nutrient solution have been devised. A popular one, developed by Hoagland and Arnon, was patterned after the proportion of nutrients found in some fertile California soils, using tomato as the test species (Table 13-5).

The precise salts and composition of nutrient solution vary but the elemental ratios are very similar. Although plants differ in their minimum require-

TABLE 13-5. Modified Hoagland's solution, prepared from six stock solutions. Stock solutions 1–4 (molar solutions) provide macronutrients N, P, K, Ca, Mg, S.

Stock solution	Concentration (g/liter)	Component	Hoagland's solution (ml/liter)
1	136	KH_2PO_4 (potassium dihydrate phosphate)	1
2	101	KNO_3 (potassium nitrate)	5
3	164	$Ca(NO_3)_2$ (Calcium nitrate)	5
4	120	$MgSO_4$ (Magnesium sulfate)	2
5 (micronutrient)		Micronutrient stock solution	1
6* (iron)		Chelated iron stock solution	1

Micronutrient stock solution

Component	Concentration (g/liter)
H_3BO_4 (boric acid)	2.86
$MnCl_2 \cdot 4H_2O$ (manganese chloride)	1.81
$ZnSO_4 \cdot H_2O$ (zinc sulfate)	0.22
$CuSO_4 \cdot 5H_2O$ (copper sulfate)	0.08
$H_2MoO_4 \cdot H_2O$ (molybdic acid)	0.02

SOURCE: Adapted from G. R. Noggle and G. J. Fritz, 1983, *Introductory Plant Physiology* (Prentice-Hall).

*Preparation of iron stock solution to provide Fe: 0.5% iron tartrate or Fe EDTA (ferric salt of ethylene dranine tetracetic acid) containing 5 mg Fe/ml.

TABLE 13-6. Comparison of elemental composition for three nutrient solutions.

Element	Solution composition (ppm)		
	Hoagland and Armon 1950	Verwer 1976	Resh 1981
N	210	173	175
P	31	39	65
K	234	280	400
Ca	160	170	197
Mg	48	25	44
S	64	103	197
Fe	0.6	1.7	2
Mn	0.5	1.1	0.5
Cu	0.02	0.02	0.03
Zn	0.05	0.25	0.05
B	0.5	0.35	0.5
Mo	0.01	0.06	0.02

SOURCE: W. L. Collins and M. H. Jensen, University of Arizona.

ments by species and stage of growth, most growing plants are fairly tolerant in the ranges currently used (Table 13-6).

Control of nutrient solution pH must be carefully monitored. Plants absorb anions (NO_3^-, PO_4^-, and SO_4^-) more rapidly than cations (Ca^+, Mg^+, and K^+). The anions are replaced by hydroxyl ions (OH^-) released by plant cells or bicarbonate ions (HCO_3^-) produced in respiration, whereas cations are replaced by hydrogen ions (H^+) produced metabolically. This results in an increase in pH (that is, an increase in alkalinity) especially when the phosphate ions, which provide buffering, are depleted. Plants grow normally in a pH range from 5 to 7 (a tenfold difference in H^+ concentration). In order to avoid problems, pH may be adjusted by adding a strong base or a strong acid, or by replenishing the solution frequently (daily or at least weekly). The pH will be more stable if nitrogen is supplied in both the ammonium and nitrate forms.

Hydroponic Systems

A great number of hydroponic systems have been devised that vary in (1) the presence and type of aggregate and (2) the frequency of nutrient solution use, that is, used once (open system) or recycled (closed system). Four different systems (Figure 13-17) are listed below.

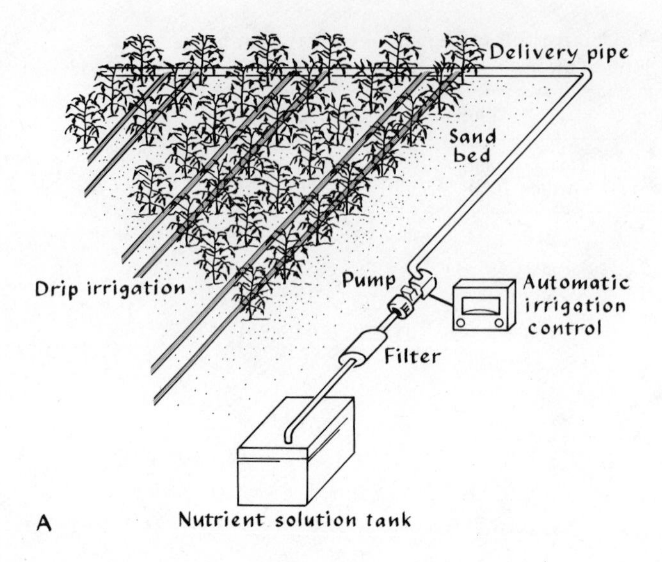

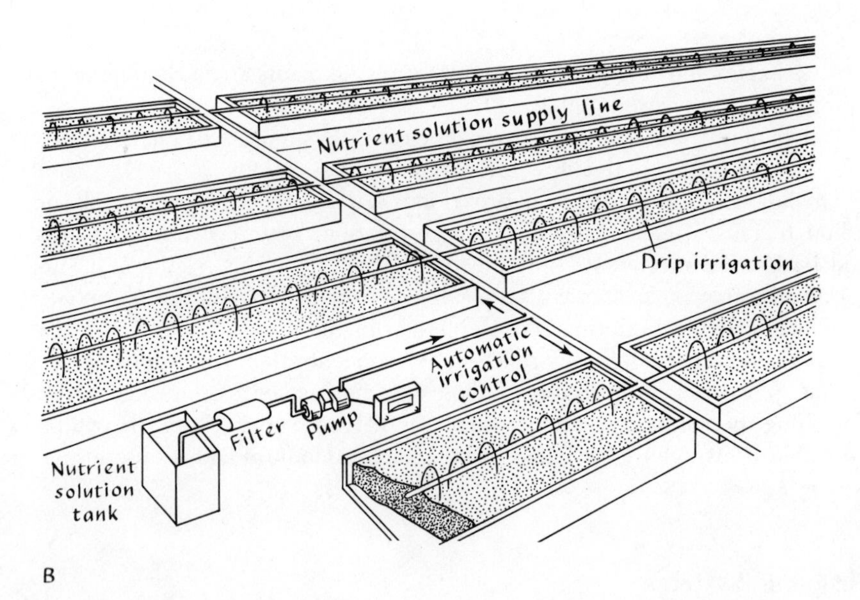

FIGURE 13-17. Hydroponic systems. (*A*) Sand culture. Nutrients are introduced in an open (nonreusable) system. (*B*, *C*) Aggregate open system. Above-grade trough systems (*B*) consist of separate waterproof beds fed with drip irrigation; or below-grade systems (trench culture) (*C*) use plastic-lined sand beds. (*D*) Nutrient film technique with tomato. (*E*) Floating hydroponic system with lettuce. [Courtesy Merle H. Jensen.]

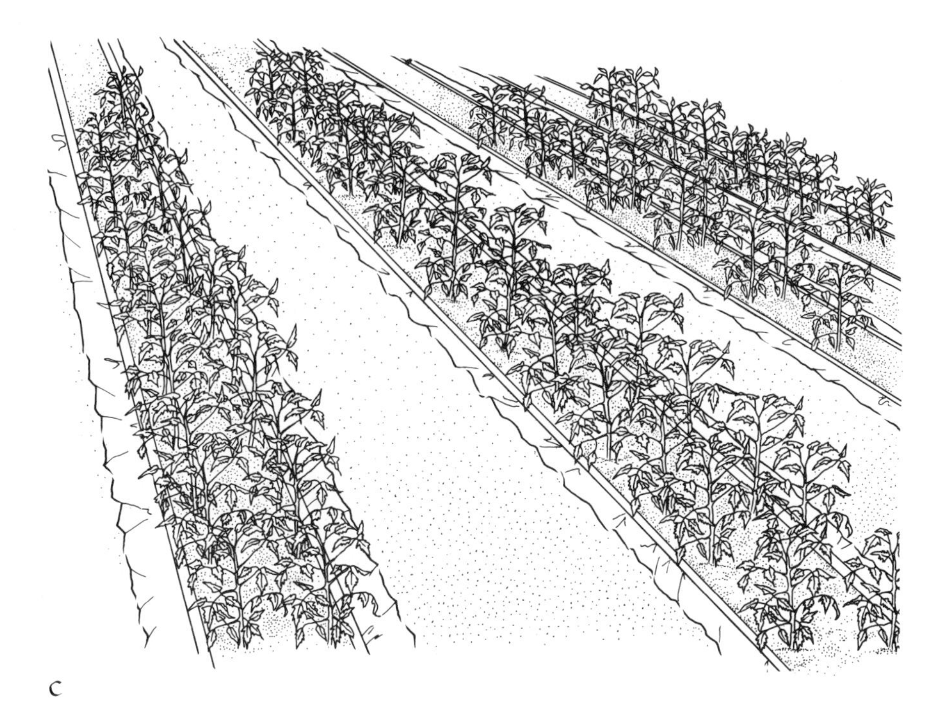

C

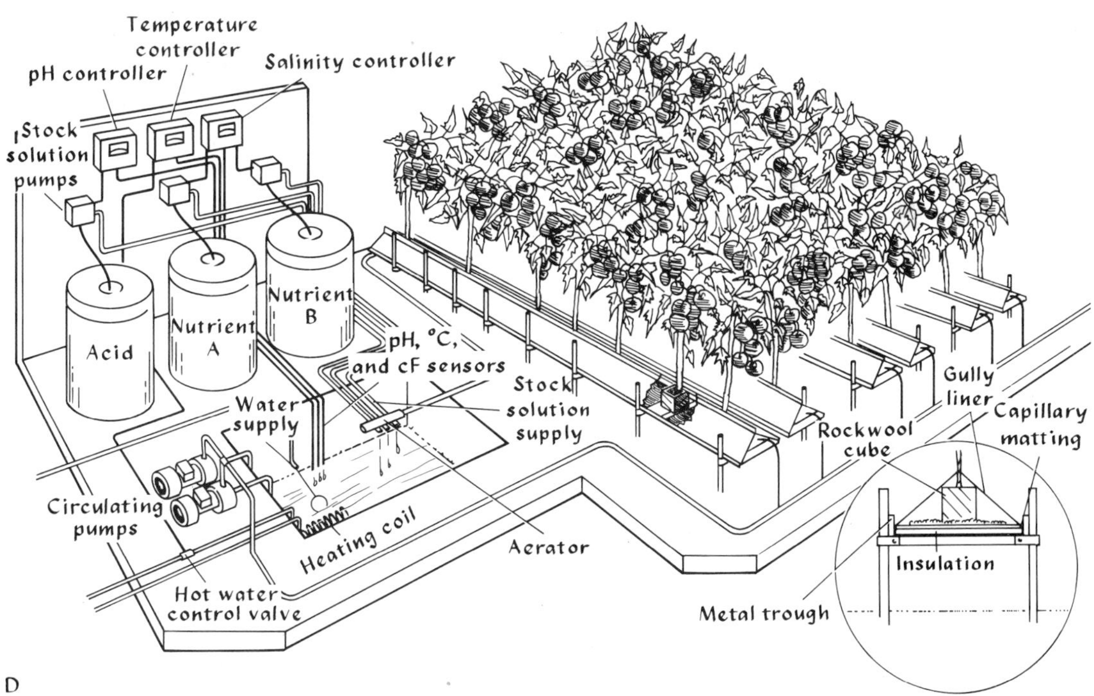

Temperature controller
pH controller
Salinity controller
Stock solution pumps
Acid
Nutrient A
Nutrient B
pH, °C, and cF sensors
Stock solution supply
Water supply
Circulating pumps
Heating coil
Aerator
Hot water control valve
Gully liner
Capillary matting
Rockwool cube
Insulation
Metal trough

D

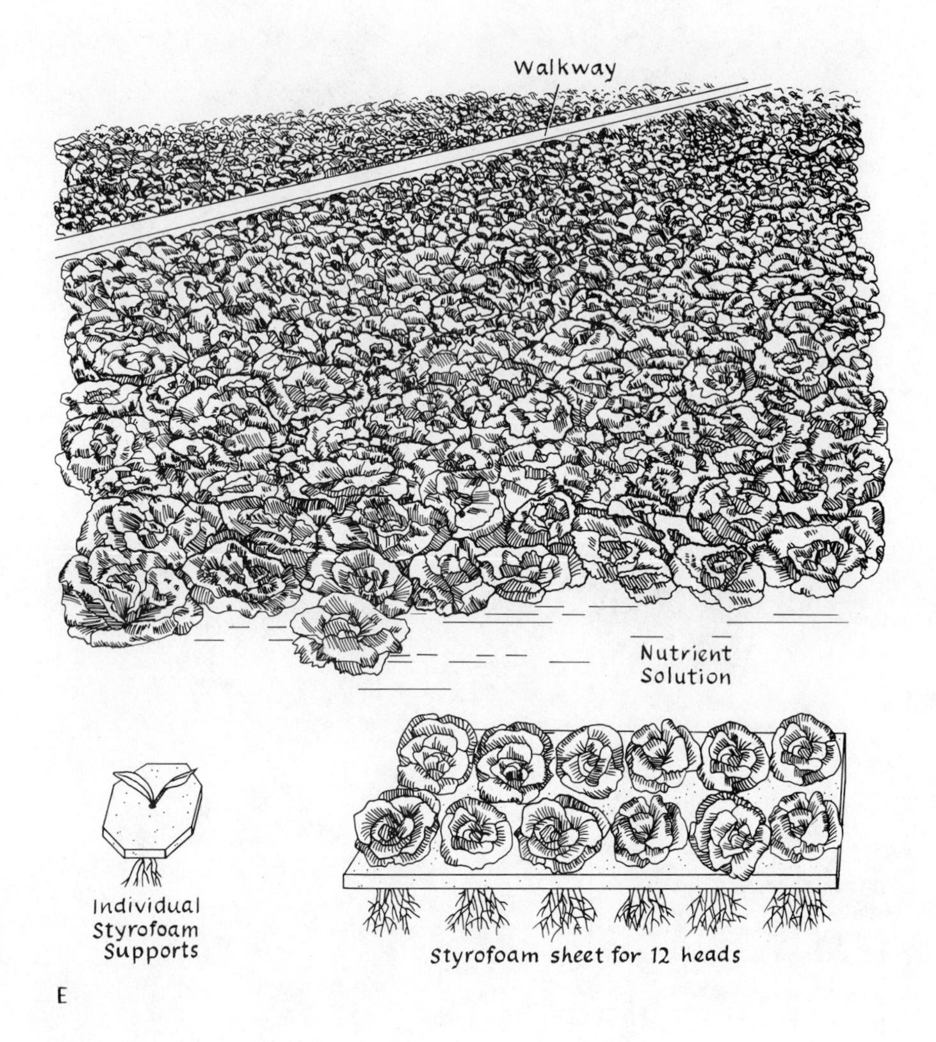

Sand Culture

Coarse sand (atop a drainage system) acts as a permanent growing medium, covering an entire greenhouse or large bed. Some large installations have been constructed in desert locations, especially in inhospitable oil-rich countries where cost is not the main consideration (see Figure 10-18).

Aggregate Open System

Seedlings are transplanted to troughs or plastic bags filled with inert or relatively inert substrates (e.g., expanded rockwood, peat) and individually irrigated with nutrient solution using a "trickle" system (see page 238). The media may be resterilized with steam and reused.

Nutrient Film Techniques

Nutrients are recirculated in a closed system. A thin film of water and nutrients is recirculated through channels containing plant roots. The closed system requires careful monitoring of nutrients, pH, and disease control.

Floating Hydroponic Systems

Leaf lettuce is planted in plastic panels that float on narrow nutrient pools called "raceways." These systems are experimental.

Selected References

Atkinson, D., J. E. Jackson, R. O. Sharples, and W. M. Waller (eds.), 1980. *Mineral Nutrition of Fruit Trees.* London: Buttersworth. (Sixty-nine essays on fruit-tree nutrition.)

Barker, A. V., and H. A. Mills, 1980. Ammonium and nitrate nutrition of horticultural crops. *Horticulture Reviews* 2:395–423. (Factors affecting nitrogen nutrition.)

Black, C. A., 1968. *Soil-Plant Relationships,* 2d ed. New York: Wiley. (An in-depth discussion of soil fertility.)

Cooke, G. W., 1982. *Fertilizing for Maximum Yield,* 3d ed. London: Granada. (On nutrition and fertilization of various types of crops for the practitioner.)

Follett, R. H., L. S. Murphy, and R. L. Donahue, 1981. *Fertilizers and Soil Amendments.* Englewood Cliffs, N.J.: Prentice-Hall. (A text covering fertilizers, plant nutrition, and economics of crop production.)

Graves, C. J., 1983. The nutrient film technique. *Horticultural Reviews* 5:1–44. (A specialized form of hydroponics.)

Jensen, M. H., and W. L. Collins, 1985. Hydroponic vegetable production. *Horticultural Reviews* 7:483–558. (A review of hydroponic technology.)

Joiner, J. N., R. T. Poole, and C. A. Conover, 1983. Nutrition and fertilization of ornamental greenhouse crops. *Horticultural Reviews* 5:317–403. (Nutrition of floricultural crops.)

Jones, J. B., Jr., 1983. *A Guide for the Hydroponic and Soilless Culture Grower.* Beaverton, Ore.: Timber Press. (A detailed description of nutrient solutions, culture, and associated diagnostic procedures.)

Maynard, D. N., and O. A. Lorenz, 1979. Controlled-relieved fertilizers for horticultural crops. *Horticultural Reviews* 1:79–140. (Current information on controlled-released fertilizer and nitrification inhibitors.)

Pearson, R. W., and F. Adams (eds.), 1967. *Soil Acidity and Liming.* (Agronomy Monograph No. 12). Madison, Wis.: American Society of Agronomy. (A collection of papers dealing with the soil chemistry of liming and the attendant crop responses.)

Russell, E. W., 1973. *Soil Conditions and Plant Growth*, 10th ed. New York: Longman. (The most famous work on agricultural soils. The first seven editions were written by Sir John Russell.)

Shear, C. B., and M. Faust, 1980. Nutritional ranges in deciduous tree fruits and nuts. *Horticultural Reviews* 2:142–161. (Deficiency and toxicity symptoms and nutrient concentrations in relation to nutritional disorders.)

Sprague, H. B. (ed.), 1964. *Hunger Signs in Crops*. Washington, D.C.: National Fertilizer Association. (A volume dealing with deficiency symptoms in selected crops.)

Tisdale, S. L., and W. L. Nelson, 1984. *Soil Fertility and Fertilizers*, 4th ed. New York: Macmillan. (A college text covering the fundamental concepts of soil fertility and fertilizer manufacture.)

Training and Pruning 14

The growth of plants can be modified to suit human needs and desires, and these modifications may be achieved by direct manipulation of the plant itself, as distinct from manipulation of the plant's environment. The direct control of growth by pruning and grafting is among the oldest of horticultural practices.

Physical techniques that control the shape, size, and direction of plant growth are known as training. **Training** is in effect the orientation of the plant in space. It may involve merely providing a support on which plants may naturally grow or, in addition, it may include the bending, twisting, or fastening of the plant to the supporting structure (Figure 14-1). Training often is combined with the judicious removal of plant parts, or **pruning.** Pruning may also be performed for other purposes—for example, to adjust fruit load, the subsequent change in form being only incidental.

The object of altering the spatial form or size of a plant is to improve its appearance or usefulness. Certain woody shrubs can be trained and pruned in a great variety of shapes, limited only by the skill of the person wielding the shears. Plant sculpture, known as **topiary,** is considered beautiful by some people and ugly by others, but it illustrates, in any case, the plasticity of the growing plant (Figure 14-2). The usefulness of a particular spatial arrangement may result from the increased efficiency of light utilization or from the facilitation of cultural operations, such as harvesting or disease control. Furthermore,

A

B

C

FIGURE 14-1. Training orients a plant in space and is an integral part of the culture of many plants. (*A*) Tomatoes are twisted around twine to maximize growing space in greenhouse production. (Tomatoes grown in the field for early market may be trained to wooden stakes.) (*B*) Cucumbers are trained to a trellis in greenhouse production. (*C*) A young pear orchard in England grown for pear cider (perry) with two types of training: The tall trees tied to the stake are permanent trees; the bushlike trees between them are temporary. See also Figures 14-15 through 14-21. [Photographs courtesy Merle H. Jensen, P. B. Lombard, and R. R. Williams.]

training and pruning may enhance the productiveness of plants and the quality of their products.

Training and pruning are well-known but by no means universal practices. Herbaceous annuals or biennials are usually grown without any attempt to alter their growth patterns. The lack of training is not so much a matter of satisfaction

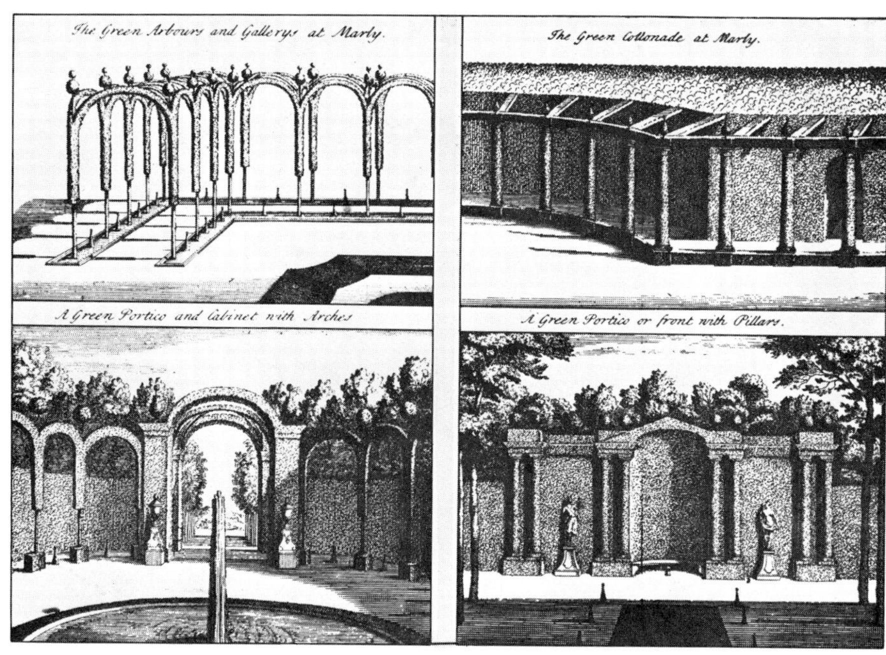

FIGURE 14-2. Examples of topiary art from *The Theory and Practice of Gardening* by Alexander Le Blond, translated by John James in 1728. Topiary is the name given to the art of training plants to resemble unnatural, ornamental shapes. This type of "bush sculpture" was very popular for ornamental plantings in the seventeenth and eighteenth centuries, but is no longer in fashion. [From R. Wright, *The Story of Gardening*, Garden City Publishing Co., 1938.]

in their performance as it is of practicality. Since there are usually many such plants in relation to the space they occupy, it is not practical to handle each one individually. Perennials, and especially woody plants, are often trained to some degree. Each individual is relatively valuable, since there are few plants per unit area, and since they are grown for extensive periods (the productive life of an apple tree may be 40 years). As plant size continually increases, the control of growth through pruning becomes necessary. The framework of a woody tree in relation to pruning is shown in Figure 14-3.

PHYSIOLOGICAL RESPONSES TO TRAINING AND PRUNING

The orientation of the plant in space has a marked physiological effect on growth and fruiting. Fruit trees planted at an angle of about 45° to the ground have been shown to become dwarfed and to flower earlier. The training of branches in a horizontal position encourages the same effect. This decrease in growth rate and increase of flowering occur naturally when the weight of a

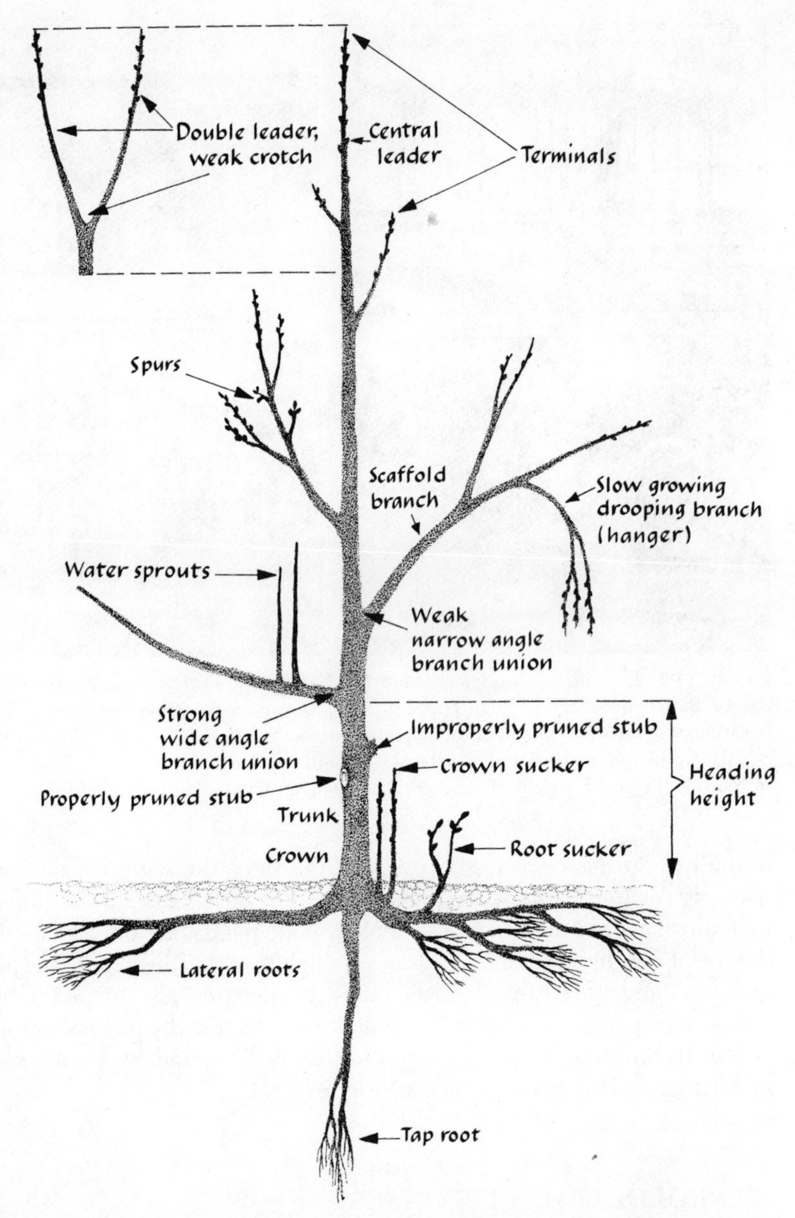

FIGURE 14-3. The plant framework in relation to pruning terminology. [Adapted, with permission of the publisher, from E. P. Christopher, *The Pruning Manual*, copyright 1954 by the Macmillan Company.]

heavy crop load bends a limb down. Thus, fruiting acts as a triggering device to keep the plant reproductive. A clear explanation of this phenomenon has not been made. It has been suggested, however, that the effect is due to a disturbance of the normal auxin movement, which in turn affects phloem transport. This explanation assumes that gravity affects the pattern of auxin distribution along the stem. The effect of the disruption of phloem transport on growth and fruiting is discussed in Chapter 15.

The response of the plant to pruning is a result of the altered relationship of the remaining plant parts and the disturbed pattern of auxin production. The effect differs to some extent, depending on whether the plant is dormant or growing when pruned.

Altered Relationship of Plant Parts

An explosion of vegetative growth normally occurs after extensive shoot pruning. The reason is that severe pruning radically alters the balance between root and shoot. The following flush of growth is caused by the diversion of water, nutrients, and stored food from an undisturbed root system into a reduced bud area (Figure 14-4). Although there is also some reduction in the amount of stored food (along with some reduction in photosynthetic area), this decrease is negligible because reserve food in the form of sugars and other carbohydrates is stored mainly in the roots and older portions of the shoot, especially during dormancy.

The increased growth that occurs after extensive pruning might indicate that the technique has a rejuvenating effect. But the additional growth does not compensate for the removed portion of the plant; the plant, pruned of vegetative growth, never quite makes up the loss. Thus, pruning is in reality a dwarfing process, although some plant parts may be selectively increased.

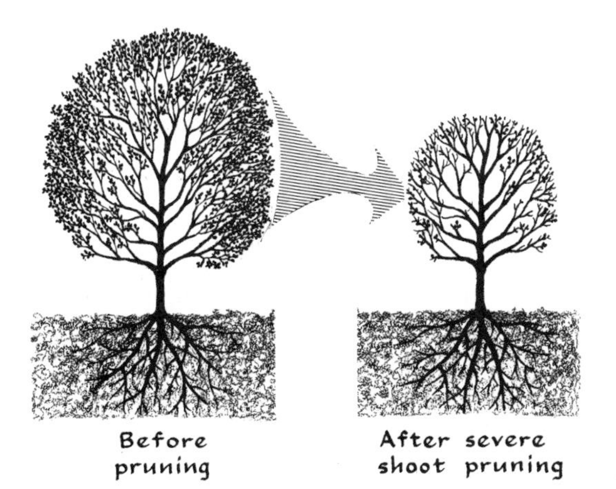

FIGURE 14-4. Shoot pruning alters the balance of root and shoot and results in increased growth of the remaining parts.

Pruning and Flowering

In general, severely shoot-pruned plants, especially if they are young, tend to remain vegetative. Conversely, root pruning encourages flowering. This can be explained in a number of ways. It has been interpreted by an extension of the carbohydrate-nitrogen "theory" of flowering (see Chapter 5). This explanation assumes that the severely shoot-pruned plant draws on its carbohydrate reserve in the promotion of growth. The resulting low carbohydrate-nitrogen balance encourages vegetative growth. Root-pruned plants reduce nitrogen accumulation, but, by slowing down vegetative growth, they conserve carbohydrates. The carbohydrate surplus supposedly promotes flowering. Another equally valid explanation may be that actively growing leaves produce substances that inhibit flowering. Thus, a rapid increase of vegetative growth would be antagonistic to flowering. The encouragement of flowering by root pruning is a direct result of slowing down vegetative growth. The precise relationship, however, is unclear; the explanation awaits a more precise elucidation of the flowering process.

Auxin Imbalance

Apical Dominance

The role of the apical meristem in inhibiting the growth of dormant buds (bud break) behind it is known as **apical dominance.** This dominance of the apical meristem differs from species to species. Thus, an actively growing bamboo is basically an unbranched stem, whereas a sprawling shrub, such as the Pfitzer juniper, grows as a many-branched structure. Both the degree of branching and the subordination of lateral growth to the main growing stem, the **central leader,** appear to be functions of apical dominance.

The branching of shoots (and roots) has been found to be influenced by auxins, which are produced in greatest abundance in a vigorously growing apex. A high auxin concentration moving down from the stem tip has been seen to inhibit lateral bud break. (Rapidly growing unbranched shoots, called watersprouts, have very high auxin levels and represent an extreme example of apical dominance.) Removal of the stem tip results in an increased amount of lateral bud break and subsequent branching, usually directly below the cut (Figure 14-5). This result is explained by the destruction of the auxin-producing meristem, although it must be admitted that the precise mechanism has not been established. Thus, pruning that merely removes the tip of the stem (heading back) can create new form changes by the destruction of apical dominance (Figure 14-6). Similarly, pruning that removes only laterals but leaves the stem tip undisturbed (thinning out) not only eliminates branching but, by increasing the vigor of its stem tip and presumably its auxin content, also limits future lateral bud break.

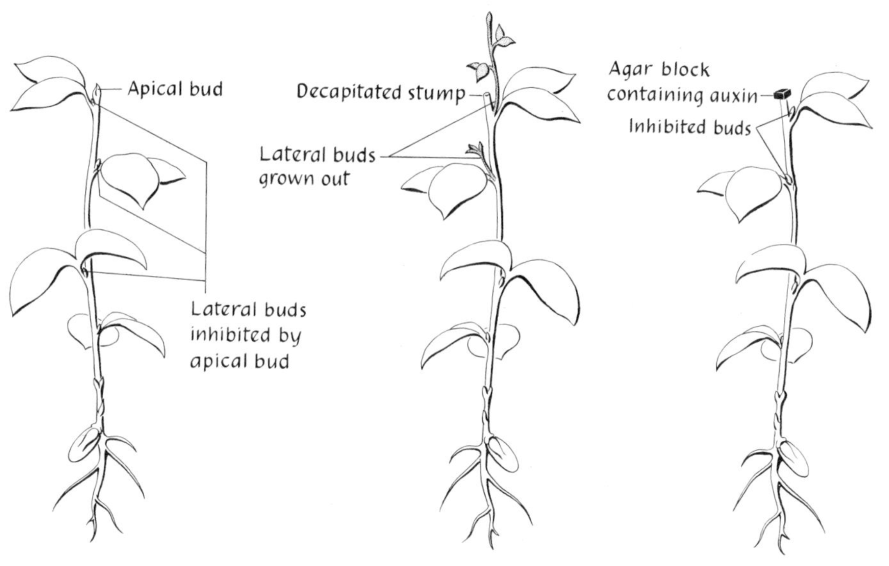

FIGURE 14-5. Apical dominance refers to the effect of the apical bud in inhibiting bud break below. Removal of the apical bud encourages lateral breaks. The growth of lateral buds can be inhibited by auxin application to the cut portion of the stem. [Adapted from J. Bonner and A. W. Galston, *Principles of Plant Physiology*, W. H. Freeman and Company, copyright © 1952.]

FIGURE 14-6. These coleus plants are of the same age and were grown under the same conditions with the exception of pruning. Removal of the apical meristem of the plant on the right has stimulated the growth of lateral shoots. [Photograph courtesy E. R. Honeywell.]

Branch Angle

There is also evidence that the angle of branching is controlled by auxins. Branches produced below an actively growing, auxin-producing apex form a wider angle with the main stem than do branches formed where the growing point has been removed. A heavily pruned young fruit tree tends to produce narrow-angled branches. In peach pruning, wide-angled scaffold branches are encouraged by permitting growth of a bud from the stub (Figure 14-7).

PRUNING TECHNIQUES

Heading Back and Thinning Out

The two basic pruning cuts, heading back and thinning out, are illustrated in Figure 14-8. **Heading back** consists of cutting back the terminal portion of a branch to a bud, whereas **thinning out** is the complete removal of a branch to a lateral or main trunk. Because the heading back of a stem destroys apical dominance, it is usually followed by the stimulation of several lateral bud breaks, depending on the species and the distance from the tip to the cut. To encourage spreading growth, the branch is usually cut back to an outward-pointing bud. Heading back tends to produce a bushy, compact plant. The shearing of a hedge is an extreme example of this type of pruning. Heading back actively growing plants is referred to as **pinching.**

A

B

FIGURE 14-7. Wide-angled scaffold branches are encouraged by proper pruning in the peach. (*A*) The central leader was summer pruned (pinched) above a bud during the first growing season to channel growth into selected scaffold branches and to encourage wide-angled attachment. (*B*) During winter pruning, the central leader was cut to a stub. The stub will be removed in a few years. [Photograph courtesy of F. H. Emerson.]

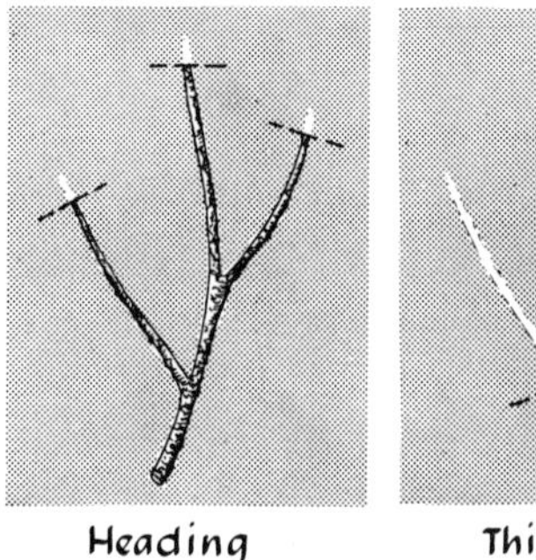

FIGURE 14-8. Heading back encourages lateral growth. Thinning out results in a more open structure.

Thinning, in contrast to heading, encourages longer growth of the remaining terminal branches. Its net result is a reduction of laterals. Thinning of weak growth tends to "open up" the tree. It usually results in producing a larger rather than a bushier plant. The rejuvenation of older trees by reducing and thereby stimulating the remaining growing points is accomplished by thinning. The thinning out of growing wood is called **deshooting.**

Timing of Pruning

Prune when the knife is sharp.

The proverbial saying might have been true long ago, but today the time to prune is influenced by a number of other factors, including convenience, the peculiarities of the species, and the effect desired. Although it is best to keep some plants more or less continually pruned, it is seldom practical to do so. Fruit trees are usually **dormant pruned.** Not only is this timing most convenient in the cycle of orcharding, but the framework of the plant can be more easily seen with the foliage off. Where winter temperatures are low, the pruning operation is usually delayed until the severest weather is past in order to reduce winter injury to fresh cuts. The pruning operation is best not carried on into the growing season because of the additional loss of translocated foods. **Summer pruning** of new growth, however, makes it possible to avoid structural faults before too much growth is wasted. This care is especially important when the tree is young. With proper heading, little photosynthetic area need be lost. However, extensive dormant pruning has been shown to be less devitalizing than summer pruning. In addition, pruning wounds made in the early spring heal better than those made at other times of the year. Extensive pruning should be avoided in the late summer since it may initiate abundant, succulent vegetative growth, which may render the plant subject to winter injury. But in any case, diseased growth or dead wood is best pruned away at once, regardless of the season. This wood, besides being unattractive or dangerous, may become a harboring place for disease-producing pests.

It is only sensible to synchronize the time of pruning so that it does not interfere with the principal functions of the plant. Thus, ornamental flowering shrubs that bloom from buds laid down the previous year should be pruned after they bloom. Similarly, it would be absurd to prune large limbs of fruit trees supporting a maturing crop.

OBJECTIVES OF PRUNING

Pruning to Control Size

Probably the most obvious effect that pruning has on perennial plants is to control their size. Since perennials grow continually, an optimum size can be maintained only by the selective removal of plant parts. Thus, the lawn is cut, the hedge is clipped, and shrubs and fruit trees are pruned in an effort to keep them within bounds (Figure 14-9).

The size of a plant may be controlled for esthetic or utilitarian reasons. In fruit production, where crops are hand-picked, large tree size makes harvesting (as well as effective spraying) extremely difficult. Certain pruning techniques can be used to reduce the height of large trees over a period of years without excessive injury.

The compensating effect of pruning on growth may increase the size of particular plant parts. This diversion of growth may be utilized to achieve an

FIGURE 14-9. The fairway is kept in bounds by constant mowing. [Photograph courtesy International Harvester Co.]

actual rise in height or spread, even though total growth is reduced. The selective removal of buds, flowers, or fruits to increase the size of the remaining parts must be considered a specialized part of pruning. This, discussed more fully in Chapter 15, is also known as **thinning** (but should not be confused with the pruning technique thinning out). The removal of buds to enlarge the flowers produced by the remaining buds is called **disbudding**.

Pruning to Control Form

The art of training and pruning to control plant form has received much attention in horticultural writings. In this context, form refers not only to the gross shape of the plant but also to its structural makeup, which involves the number, orientation, relative size, and angle of branches. The natural form characteristic of different species may be greatly modified with pruning. Plants may be trained to grow upright or to spread, and branching may be either increased or decreased.

Woody plants, especially those bearing heavy loads of fruit, must be considered as structural units because they may be torn apart in a high wind. Structural strength in fruit trees is obtained by pruning to eliminate narrow-angled branches and to achieve a well-spaced arrangement of wide-angled scaffold branches. Narrow-angled branches are weak and tend to break under pressure because of the lack of continuous cambium and the inclusion of squeezed-off bark in the crotch (Figure 14-10). For maximum strength, only one branch should develop at any point on the main stem, and the branches should be well distributed around the tree (Figure 14-11). Owing to the increase in diameter of branches, it it necessary to select branches carefully when they are young.

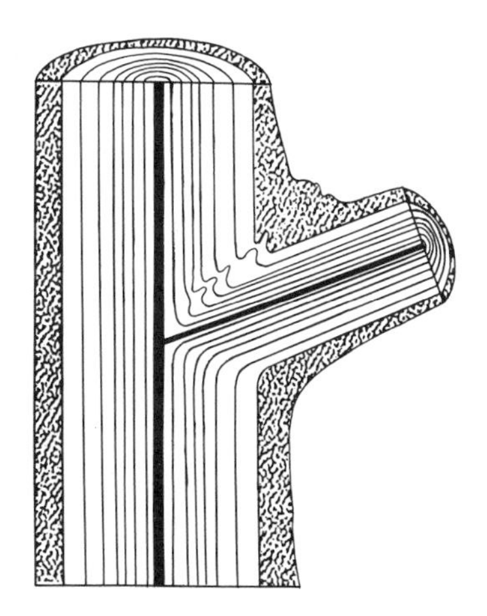

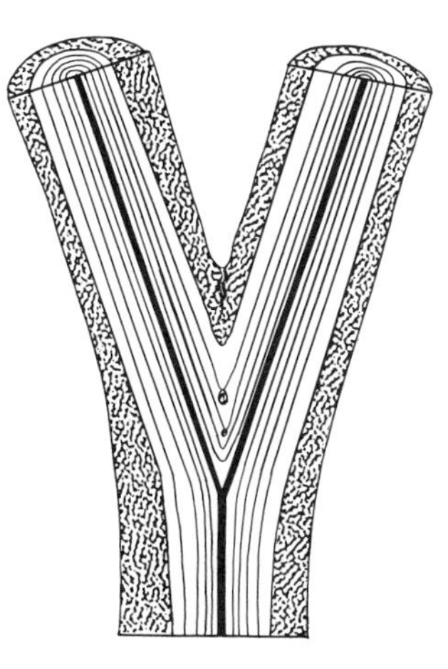

FIGURE 14-10. Narrow branch angles are weak because of the enclosure of bark and the formation of wood parenchyma in the crotch. [Adapted from A. J. Eames and L. H. MacDaniels, *An Introduction to Plant Anatomy*, McGraw-Hill, 1947.]

FIGURE 14-11. For maximum strength, scaffold branches of apple should be well spaced and evenly oriented around the tree. (*A*) In the unpruned tree the branches to be removed have been marked with paint. (*B*) The same tree after pruning. [Photographs courtesy Purdue University.]

The control of plant form may be utilized to achieve increased quality through better light distribution. The center of an unpruned apple tree is almost impervious to light and, as a result, produces few fruit, those that are produced being poor in color and quality. Opening up the tree is also useful for disease control in that it permits good spray distribution and facilitates rapid drying.

Mechanical harvesting of fruit crops requires specialized training and pruning to adapt the plant to the machine. For plants harvested by shaking, this precaution involves the development of high trunks (to allow for a single "grab" by the machine when the tree is young) and only two or three main scaffolds for use when the tree is older. Mechanical harvesting of grapes by means of a cutterbar requires special trellising and training of the vines.

Pruning for Plant Performance

Establishment of Transplants

The transplanting of large plants from natural growing sites is usually very difficult. Root pruning or repeated transplanting when the plant is young encourages a fibrous root system and allows the plant to be moved safely when large.

Proper root and shoot pruning greatly aids in reducing transplanting shock and promotes successful plant establishment. This fact is especially true in bare-rooted transplants. Light root pruning stimulates root initiation; shoot pruning conserves moisture by reducing the transpiration surface in relation to the root area.

Productivity and Quality

Pruning is often a necessary step in the control of productivity. Where vigorous bud wood is desired, as in scion orchards, heavy pruning stimulates vegetative growth. On the other hand, where flower or fruit production is the desired aim, selective pruning that eliminates weak, nonproductive wood will aid in channeling the plant's energy into flowering and fruiting. In addition, fruit and flower quality is greatly affected by the vigor of the wood that bears it as well as its location in the tree. Shoot growth (suckers) on the understock of grafted plants must be continually removed to eliminate nonproductive growth. Similarly, forest trees may be pruned of unnecessary lower branches to produce knot-free lumber.

TRAINING SYSTEMS

Training systems are carried on to control form throughout the life of the plant. Consequently, special attention must be given in the formative years. The objective is to obtain some predetermined shape in an attempt to achieve greater productivity, quality, ease of culture, or beauty.

Branch Orientation and Leader Training

The main factor that determines form is the location of the points on the main stem from which branches form and their subsequent orientation (Figures 14-12 and 14-13). The branches may be oriented around the stem to produce a "natural" shaped tree, or they may be oriented in a single plane to provide a flat shape known as an **espalier** (from the French word for shoulder). There are many variations on these two general shapes that differ principally in the height of the stem before the first branch (heading height), the angle of the branches from the main stem, and the distribution and relative length of the branches.

In the **central-leader system** of training, the trunk is encouraged to form a central axis with branches distributed laterally up, down, and around the stem. The central axis, or leader, is the dominant feature of the tree's framework, and the main direction of growth is upward. In the **open-center** or **vase system** of training, the main stem is terminated and growth is forced through a number of

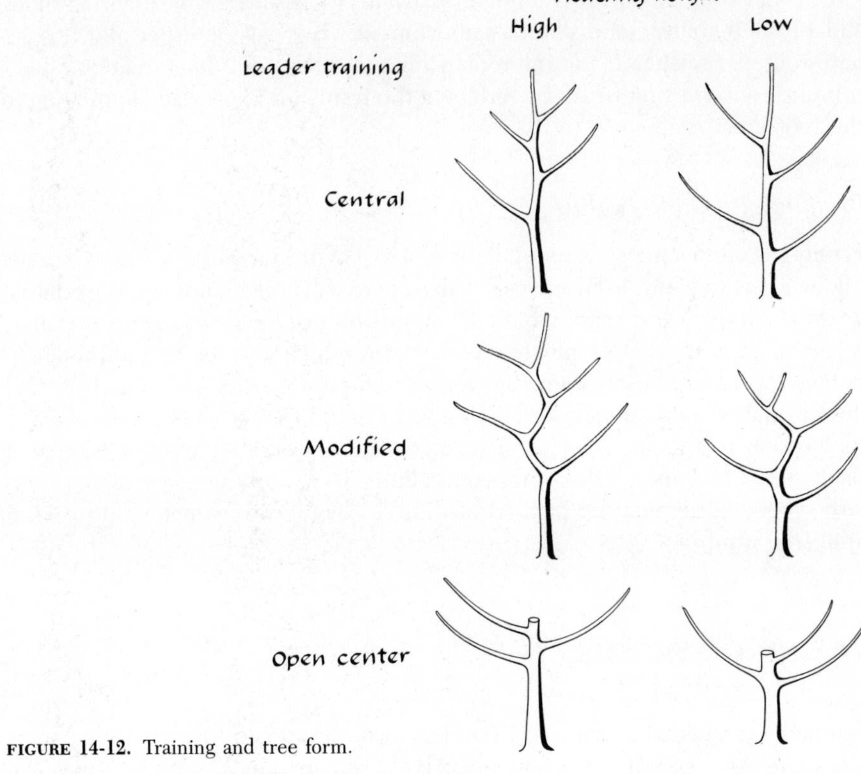

FIGURE 14-12. Training and tree form.

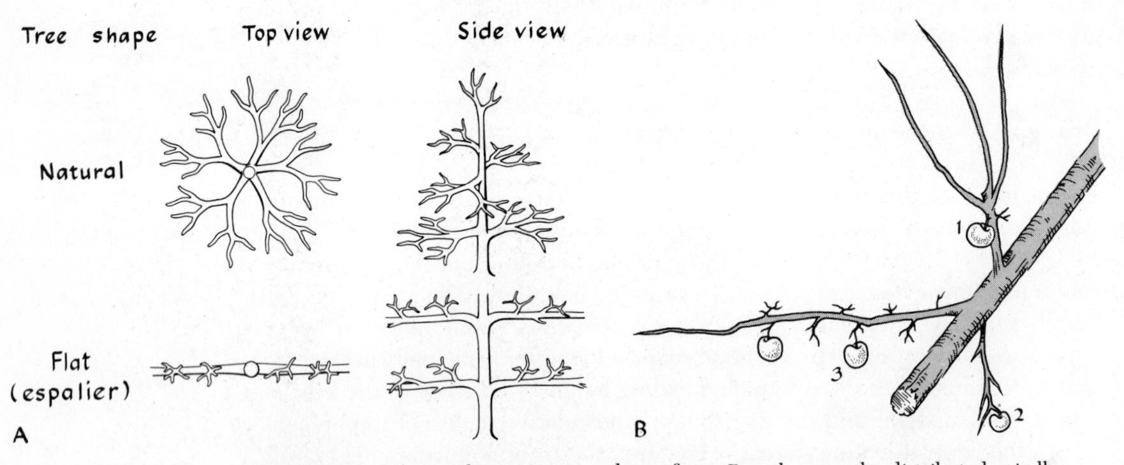

FIGURE 14-13. (*A*) Branch orientation and tree form. Branches may be distributed spirally around the stem or in a single plane. (*B*) Orientation of fruiting branches affects productivity. Upright branch (*2*) is excessively vigorous, only moderately fruitful, and produces fruits often oversize, soft, and poorly colored. Branch on underside (*1*) is heavily shaded and thus low in vigor and fruitfulness, producing small fruits of poor color. Horizontal branch (*3*) of moderate vigor and very fruitful, produces superior color fruit because of good light exposure.

branches originating rather close to the upper end of the trunk. Special pruning is required to prevent a lateral branch from becoming dominant—that is, from forming a new central leader. Although the open-center tree is lower than the central-leader tree, it has inherent mechanical weaknesses due to its narrow crotches and close branching. The **modified-leader** system is somewhat intermediate between these two types.

Originally, an espalier was a railing or trellis along which plants (usually fruit trees or vines) were trained to grow flat. Plants trained in this manner then came to be known also as espaliers. An espalier restricted to one shoot, or two shoots growing in opposite or parallel directions, is called a **cordon.** Because of the extensive pruning labor required, fruit trees have not been commercially developed as espaliers in the United States. Grapes, however, are commonly grown as espaliers, as in the widely used Kniffin system of training. Properly executed espaliers are extremely attractive as ornamentals. They are created with a combination of pruning and actual bending of the shoots while they are still succulent (Figure 14-14). Such shoots will retain the imposed shape when lignification sets in. Well-planned espaliers can be strong enough to be self-supporting eventually, although frames are required in the early years.

The particular system by which a plant is trained depends to a large extent on the species. Peaches and apricots can be pruned to an open center because of the broad angle of attachment of the branches, which produces strong crotches. In addition, the central leader in peach trees is subject to winter injury. The reason for this is not clear, but it appears to be related to the failure of the central leader to harden off (Figure 14-15). Because the narrow-angled branching of apples and pears makes the open-center tree unfeasible, they are usually trained to either the central-leader or the modified-leader system (Figure 14-16). Cherries and plums are trained to either the open-center or modified-leader system. Citrus and other evergreen fruits may be pruned lightly to establish a stronger framework, but usually little subsequent pruning is performed except to eliminate dead wood after a freeze. Mechanical hedgers are now widely used.

Tree Geometry and Planting Systems

The three-dimensional shape of the tree is ultimately based on the architecture of the framework. Various tree shapes are diagrammed in Figure 14-17. The optimum shape differs, depending on the efficiency of light interception, ease of harvest, structural strength, and pruning and training costs.

Trees may be either free-standing or supported. Free-standing trees must by necessity be based on a strong framework of structurally sound main limbs called **scaffolds.** To overcome weakness caused by a shallow root system or inherent in espalier training, trees may be supported by individual stakes (often necessary when a tree is young) or on trellises constructed of narrow strips of wood or of posts and wire. In one type of trellis, two to four wires are strung,

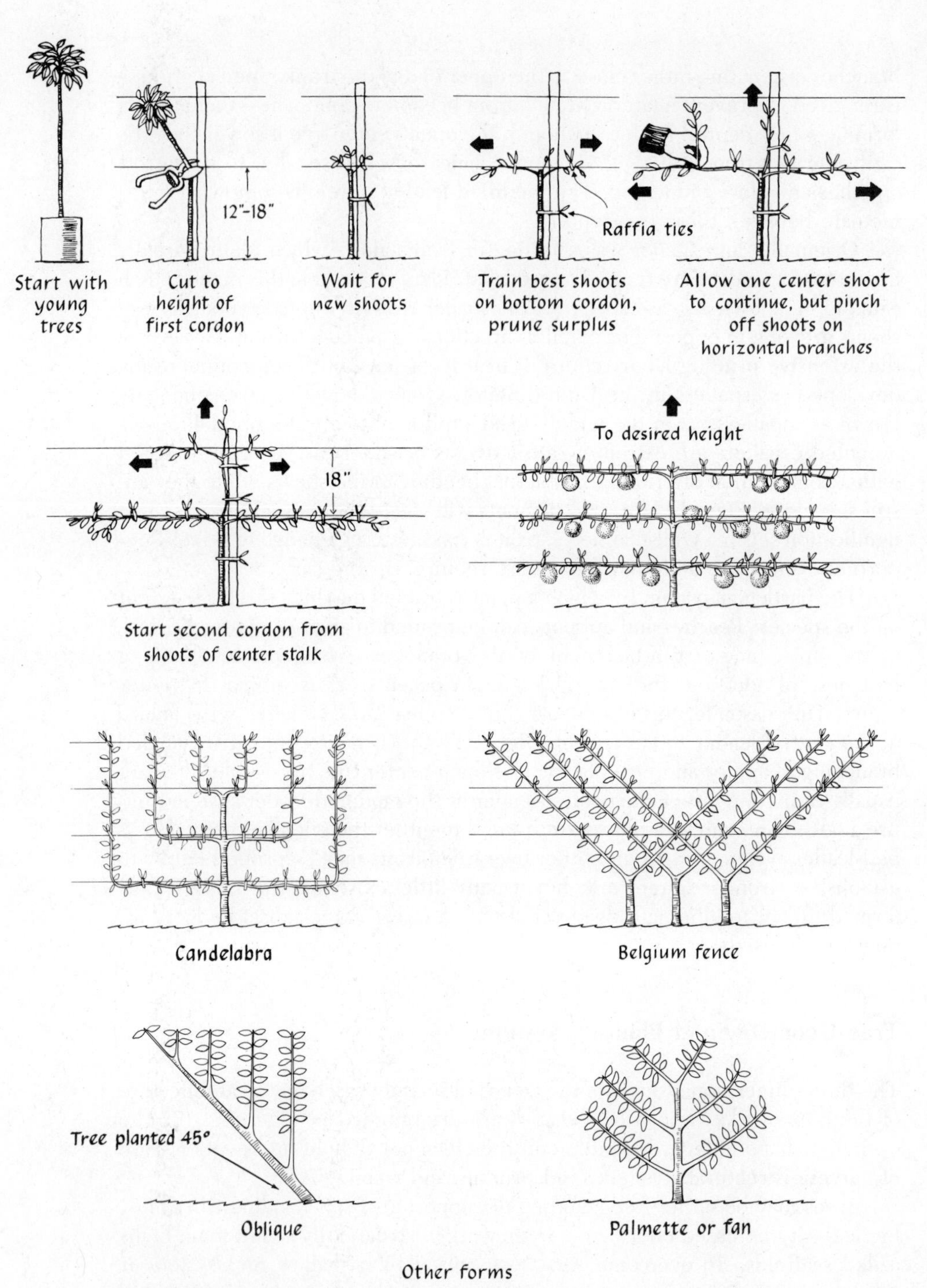

FIGURE 14-14. The creation of an espalier. [Adapted from *Sunset Pruning Handbook,* Lane Book Co., 1952.]

FIGURE 14-15. The peach is commonly pruned to an open center and is usually composed of two or three widely spaced main scaffold limbs. (*A*) A well-formed three-scaffold crotch on a 3-year-old tree. (*B*) A 6-year-old tree on which six scaffold branches were allowed to develop. All but two should have been eliminated at the start of the second year's growth. Note the weakness that has developed between the scaffold branches. (*C*) A "two-story" tree showing severe winter injury as a result of delayed maturity. The center leader should have been removed at the start of the second season's growth. (*D*) This tree was growing next to the one illustrated at the left. The two-branch wide-angle crotch shows no evidence of winter injury. [Photographs courtesy Purdue University, from Extension Circular 426, 1956.]

A

B

C

D

E

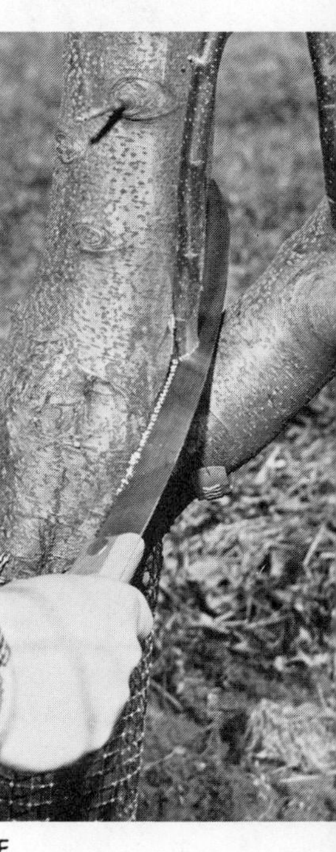

F

FIGURE 14-16. An 8-year-old apple tree trained to a modified leader. (*A*) Before corrective pruning. (*B*) After corrective pruning. The removal of the main portion of the central leader has opened up the center of the tree. Many of the pruning cuts were made to correct structural weaknesses. (*E to F*) Narrow-angled, forked branches were eliminated. Closely spaced branches, especially those growing toward the center of the tree, were removed. Intertwining branches were corrected. (Note that limb-rub injury has girdled the upright branch.) Watersprouts were removed. (The watersprout growing within the crotch would have resulted, if neglected, in extensive injury.) [Courtesy Purdue University.]

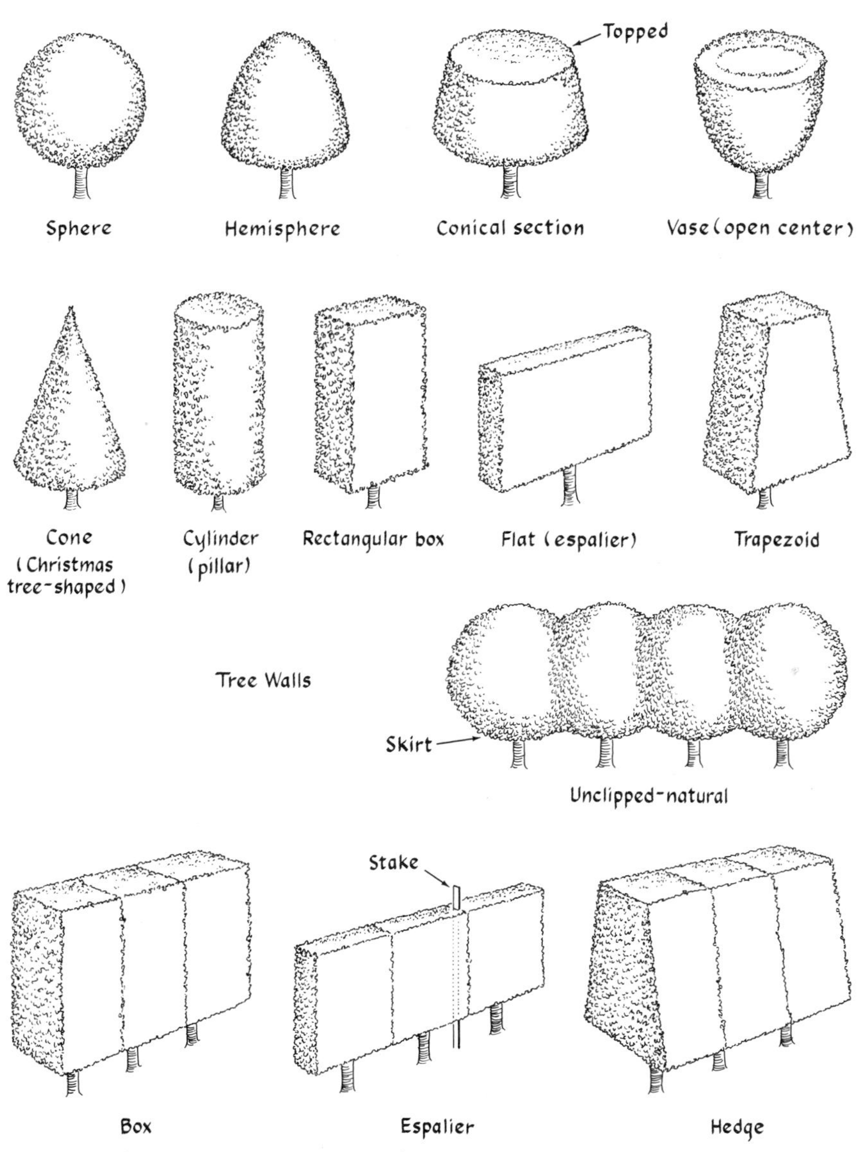

FIGURE 14-17. Various tree geometries used in apple training systems.

one above the other, across the row of supports; in another, T-shaped standards, each consisting of a post and a cross-piece, support several wires in a horizontal plane.

Planting systems involve a combination of individual plant training and patterns of plant placement in the field (Figure 14-18). In perennial fruit crops various planting systems are used to achieve maximum efficiency in the production of fruit. The variables in the systems are tree size (often modified by a combination of pruning and rootstock), tree density, and management strategies based on mechanization of harvest, pruning and training costs, irrigation, and cultivation. The fruit industry has gone through an evolutionary period as a result of changes in rootstocks, cultivars, land values, and marketing practices. The trend has been to high-density plantings of smaller trees to increase early production and maintain high-quality fruit (Figure 14-19). Planting systems based on intensive training, the use of tree supports, close spacing, and dwarfing rootstocks, which have long been practiced in Europe, have recently been modified to North American conditions, especially for the production of apples.

Renewal Pruning

In order to promote superior performance in perennial plants grown for flowering or fruiting, pruning must stimulate the most reproductive growth. It must continually renew growth to produce wood of the optimum reproductive age. Depending upon the species, this may be wood either of the current season or one or more years old.

The factors to be considered before pruning are (1) the time at which the buds are differentiated in relation to blooming and (2) the age of the wood that produces the most abundant and highest-quality buds. Flower buds may be initiated in the year of flowering, as in summer- or fall-blooming plants (for example, the rose and chrysanthemum), or on the previous year's growth, as in spring-flowering plants such as apple, lilac, peach, and brambles). Buds that differentiate the year previous to flowering may be produced on that previous year's new growth or on older wood. Many species form buds from spurs on older wood. Spurs may bear more or less irregularly for as long as 20 years. In apple, however, they are most productive for 2 to 5 years.

Plants that flower on current growth, as does the rose, are often severely dormant-pruned to encourage vigorous reproductive growth. If unpruned, the abundance of buds produces inferior individual blooms. The degree of pruning is related to the vigor of the plant: The more vigorous the plant, the greater the number of buds that are retained. During the growing season, overvigorous canes that tend to remain vegetative are headed back. Roses grown for mass effect, such as climbers, are less severely pruned, although thinning of older growth is required to stimulate vigorous new canes.

The cutting of blooms in the summer serves to invigorate the remainder of

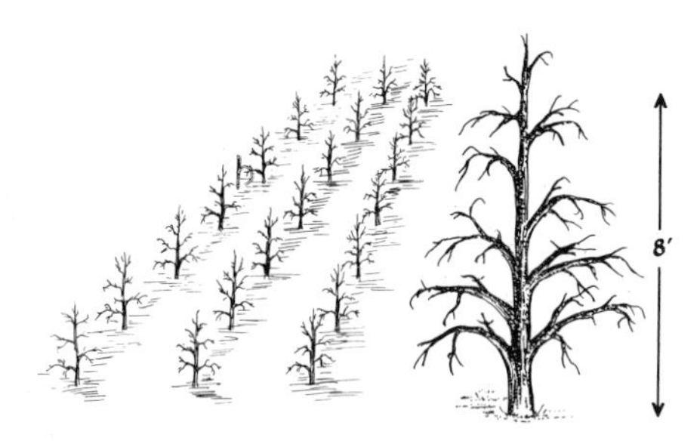

"Free-standing" tree trained to a central leader and maintained in a "Christmas tree" form.

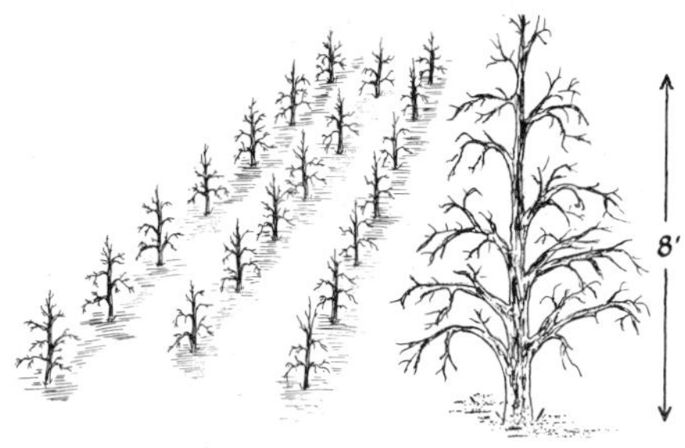

"Off-set" system using free-standing tree forming double or triple rows.

"Spindle Bush," or slender spindle bush, each tree staked for support and branch training.

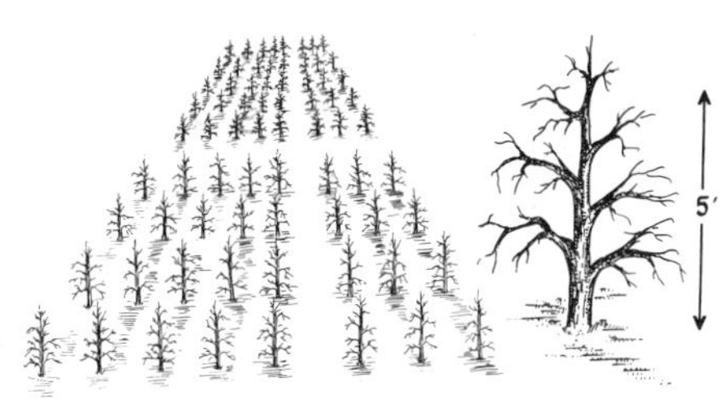

"Bed" planting system with drive rows every 5th or 6th row.

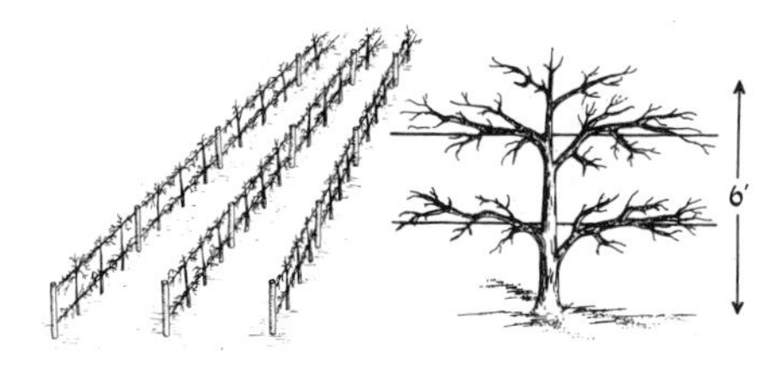

"Two Wire Trellis," each tree trained in espalier form into slender rows 10, 12, or 15 feet apart.

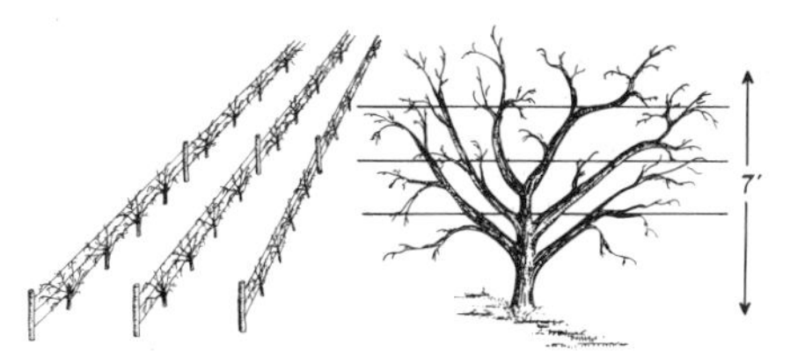

"Palmette" tree form trained on 3-wire trellis.

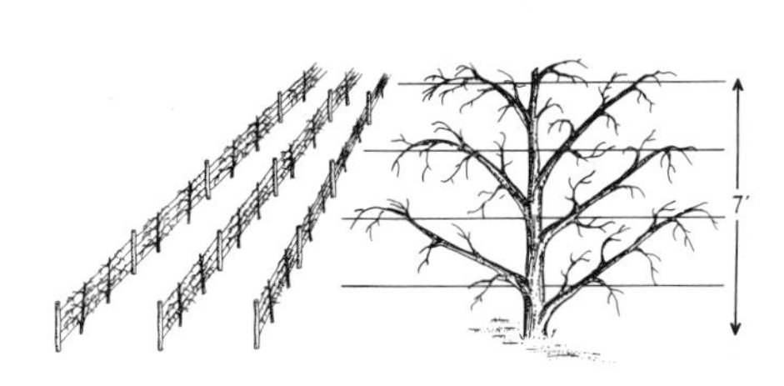

"Oblique" form trained on 4-wire trellis, each branch slanted at a 45° angle on a central leader. (The Oblique form can also be made by slanting the tree trunk at 45° and training vertical branches.)

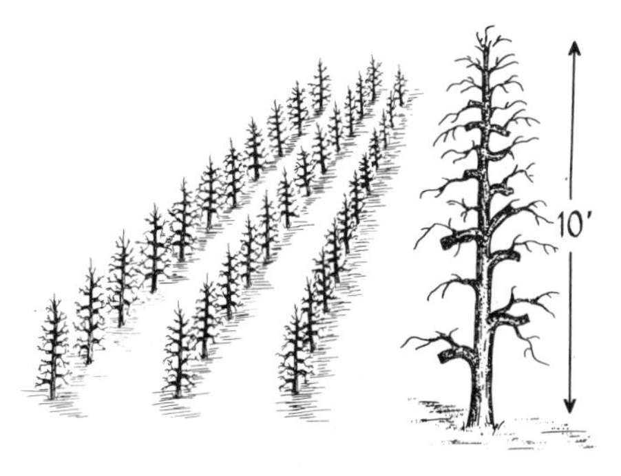

"Pillar" training system which calls for renewal of 3 and 4 year fruiting branches in alternate years to produce younger bearing branches.

FIGURE 14-18. Various tree-planting systems used in apple. [Courtesy R. F. Carlson.]

FIGURE 14-19. Evolution of high-density tree planting systems. (*A*) An old pear orchard (for pear cider or perry) planted before 1760 in Gloucestershire, England. Note the high heading, to accommodate grazing, and very wide spacing, about 50 feet apart (17 trees per acre). (*B*) A more closely spaced apple orchard in Ontario, Canada (about 1950), showing low-headed, modified-leader training. Spacing is approximately 40 × 30 feet (72 trees per acre). [Photograph courtesy Ontario Ministry of Agriculture and Food.] (*C*) Hedgerow peach in West Lafayette, Indiana, planted 10 × 14 feet apart (383 trees per acre). (*D*) A high-density planting of nectarines, trained on the free spindle system, in Italy in the 1970s (538 trees per acre). (*E*) An experimental hedgerow training system in Corvallis, Oregon, showing the shape of the tree walls after pruning. Spacing is 15 × 4 feet (726 trees per acre). [Photographs courtesy P. B. Lombard, R. R. Williams, E. W. Franklin, F. H. Emerson, R. A. Hayden, F. Loreti, and M. N. Westwood.]

the plant. Senescent flowers should be removed to prevent fruit development, since fruits drain nutrients from the plants.

Brambles (blackberry, raspberry, and related fruits) produce fruit on year-old canes. Although the roots are perennial, the canes are biennial and either weaken severely or die right after fruiting. Thus, pruning has a number of functions. Immediately after bearing, the fruiting canes are removed to encourage new shoot growth. In red raspberry (*Rubus idaeus*), new canes arise as root suckers, hence the old canes may be completely removed. Because black raspberries (*R. occidentalis*) do not produce suckers, their fruiting canes are removed above the crown. The dormant year-old canes are thinned if necessary and headed back to remove the weaker buds and to encourage branching. Black and purple raspberries are further pinched, when summer growth is 2 or 3 feet tall, to increase lateral branching.

Grapes bear fruit on the current growth from buds laid down the previous season. The greatest production is achieved from the fourth to eighth bud, and the quantity and quality of production are based on the vigor of the plant in relation to the number of remaining buds. If too many buds are left in relation to vigor, the grapes will be small and of poor quality. If too few buds are retained, yield will be reduced. Therefore, renewal pruning is an important practice in controlling yield, size, and quality in grapes. A pruning formula has been devised for the 'Concord' grape on the basis of plant vigor as determined by the weight of the prunings. Thirty buds are left for the first pound and ten buds for each additional pound of wood removed. This formula of "30 + 10" results in a "moderately" pruned vine, which gives optimum production in Ohio. In addition to fruiting canes, stubs of one or two buds, called **renewal canes,** are retained. These provide growth from which fruiting canes may be selected the following year. Thus, from a single trunk, growth is renewed each year (Figure 14-20). Other training methods differ merely with respect to the form of the plant; the renewal principle is essentially the same.

An improved system of training grapes, called the **Geneva Double Curtain,** employs a divided canopy (Figure 14-21). In essence, the system converts a wide canopy achieved by the traditional Kniffin training system into two narrow "curtains" (the leaf and shoot system formed along a single vertical plane). This is achieved by hand positioning or combing horizontal shoots so that all trend vertically downward. In a wide canopy where leaves shade one another as well as fruit, grape maturation is retarded and vines are less productive. Converting a wide canopy into narrow "curtains" exposes the leaves on the basal portion of the shoot to higher light intensities and thus to higher temperatures, which leads to more rapid maturation of the fruit and to an increase in soluble solids (sugar). In comparison to the Kniffin system, yield increases of more than 50 percent and those in soluble solids of more than 1 percent have been obtained with the Geneva Double Curtain. A further advantage is that the positioning of the shoots away from the posts of the trellis permits mechanical harvesting. This system appears to be a useful one for all

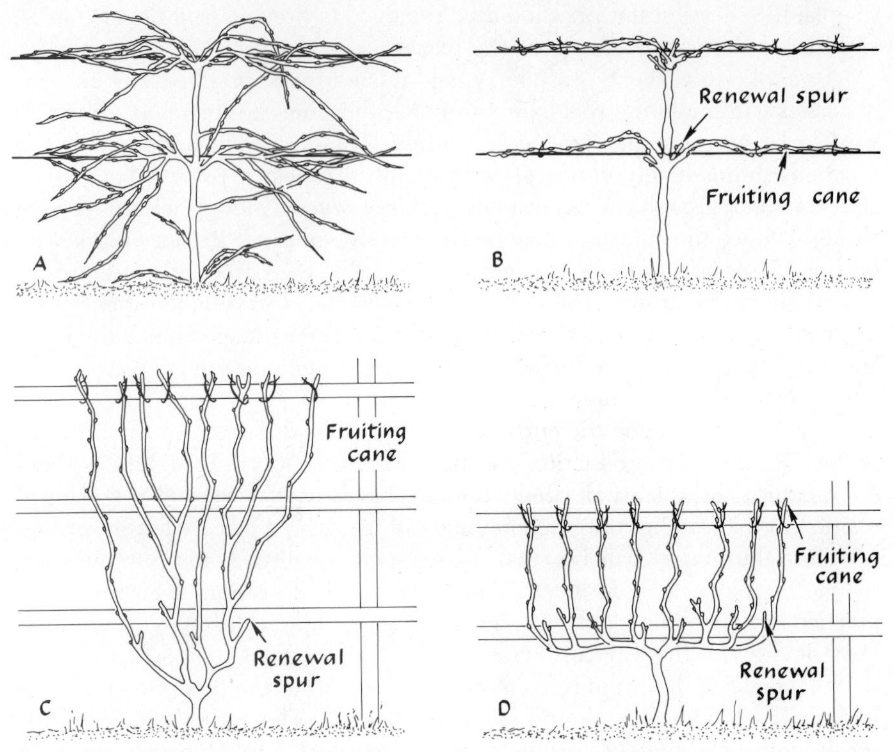

FIGURE 14-20. Some systems of pruning grapes for renewal of fruit-bearing wood. The severity of pruning is related to the vigor of the previous year's growth. (*A*) A mature dormant vine before pruning. (*B*) The same vine after pruning according to the four-cane Kniffen system. (*C*) The fan system. (*D*) The horizontal-arm spur system.

grapes, and it is especially valuable in areas of high humidity and low light intensity.

The pruning of fruit trees is done in stages. When the tree is young, pruning is principally a training operation to control form and to produce a structurally sound framework. Because of the adverse effects on early bearing, pruning must be limited at this stage. After 4 to 6 years, when the major scaffold limbs are established, renewal pruning ensures a continuing bearing surface of 2- to 4-year-old wood, from which the bulk of the crop develops.

The pillar system of training the apple—a system developed in England—is a good illustration of renewal pruning (Figure 14-22). This technique can be used with a tree structure consisting of a single leader 10 to 12 feet high. The number of bearing units maintained depends upon the vigor of the tree. Each unit consists of a 2-year fruiting limb, a 1-year-old shoot, and the current growth. Dormant pruning consists of (1) removing the spent fruiting limb (now

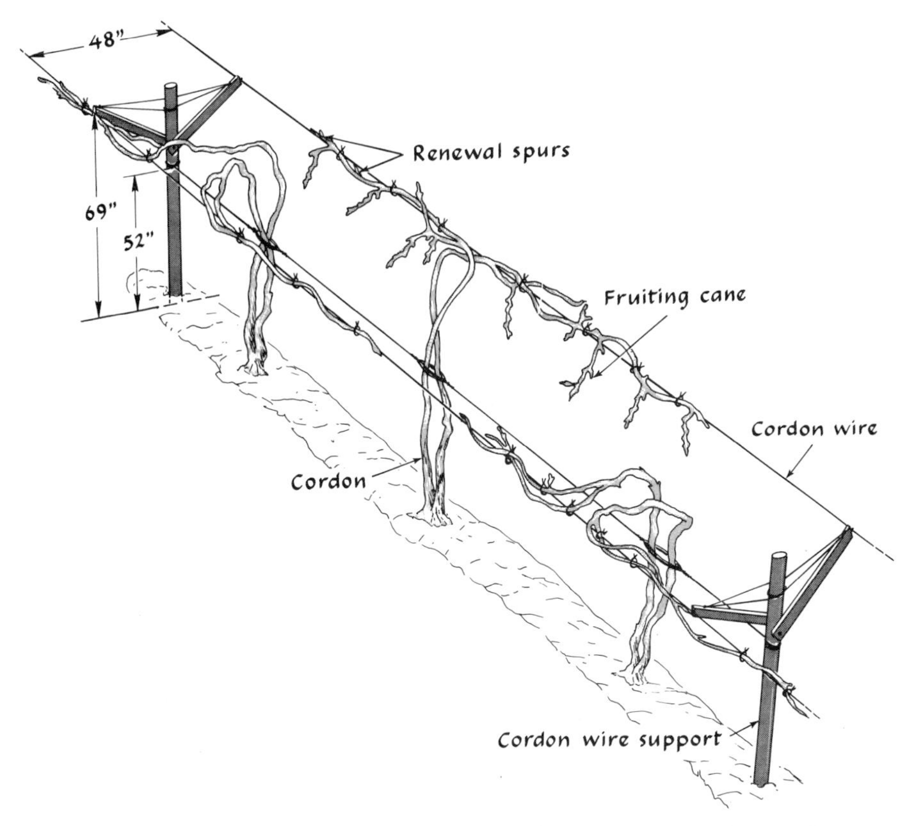

FIGURE 14-21. Diagrammatic sketch of the Geneva Double Curtain System. Spurs and canes are shown only on the middle vine. [From New York State Agricultural Experiment Station, Bulletin 811, 1967.]

3 years old), (2) heading back the 2-year terminals of the limb but leaving the fruit buds intact in anticipation of fruiting, and (3) thinning out all but 20- to 25-year-old shoots. This pruning pattern is repeated each year; thus the productive 2-year-old fruiting wood is continually renewed. High quality and annual production are achieved by controlling the bearing area. Close spacing of the trees (6 × 12 feet or 605 trees per acre) results in high production per acre. By controlling size, mechanical production practices may be facilitated and harvesting is greatly simplified. The pruning operation, although extensive, is routine; no difficult decisions need to be made. Whether the pillar system of renewal pruning will prove to be practical in the United States remains to be seen. The renewal systems presently used in this country for pruning apples differ from the pillar system in regard to the age of wood removed and the pattern of the framework.

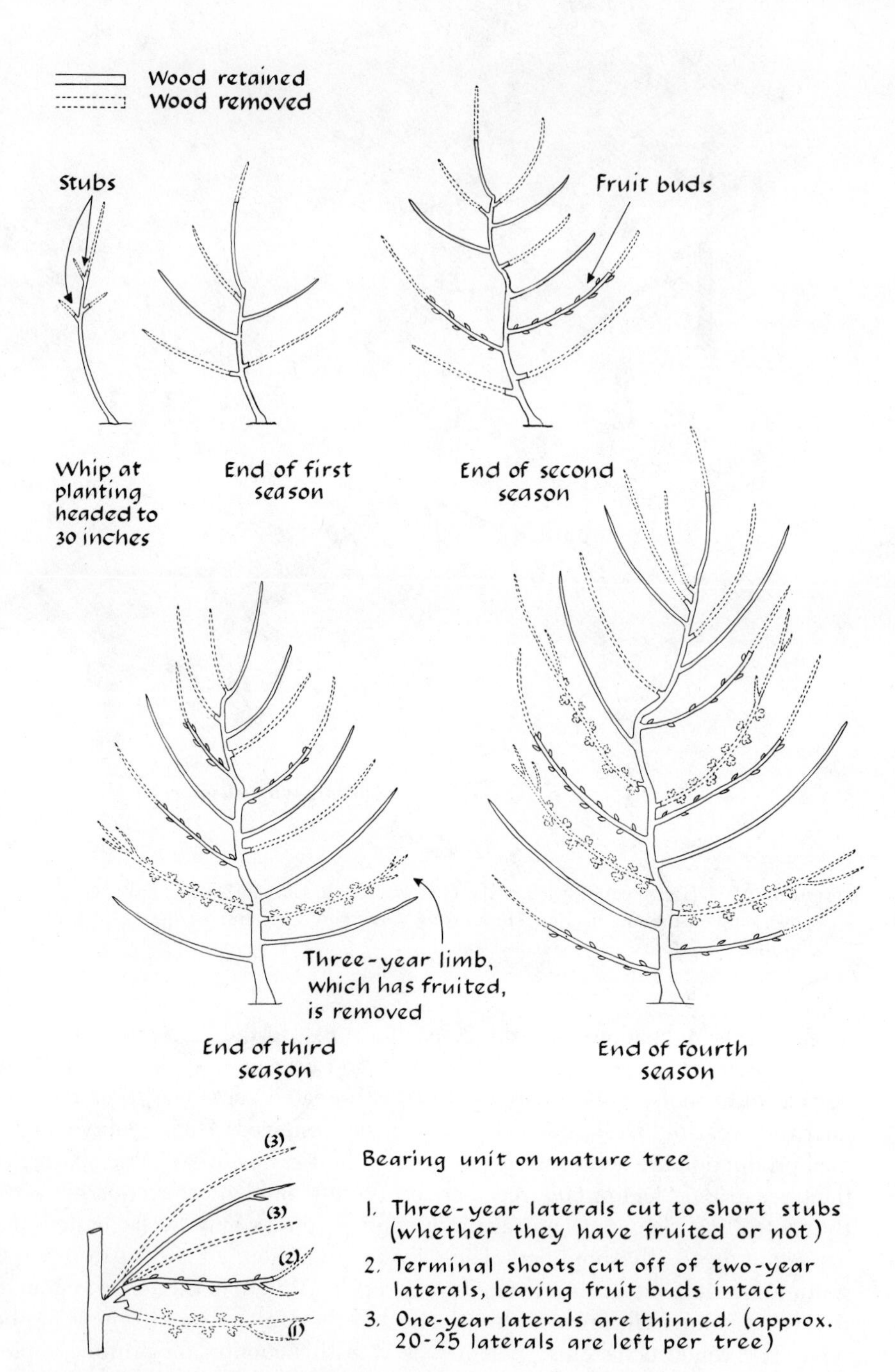

FIGURE 14-22. Pruning pillar apple trees. [Adapted from G. M. Weiss and D. V. Fisher, Canadian Department of Agriculture, 1960.]

Selected References

Christopher, E. P., 1954. *The Pruning Manual.* New York: Macmillan. (An older but still useful work on the procedures of pruning.)

Grounds, R., 1973. *The Complete Handbook of Pruning.* New York: Macmillan. (A useful work.)

Stebbins, R. L., and M. MacCaskey, 1983. *Pruning: How-to Guide for Gardeners.* Tucson, Ariz.: H P Books. (A practical, down-to-earth pruning guide for gardeners, exceptionally well illustrated.)

15 *Growth Regulation*

Horticultural science has a number of strategies for regulating growth and development. They include environmental modifications, genetic alterations, and finally, direct manipulation of the plant itself. The regulation of growth and development through environmental modifications has been covered in Part II, Horticultural Environment, and Chapter 13, "Nutrition." The genetic alterations are covered in Chapter 6, "Reproduction," and Chapter 17, "Breeding." The regulation of growth and development through direct manipulation of the plant has been introduced in Chapter 14, "Pruning and Training," which covers physical alteration of the plant to control plant shape and form, plant and organ size, and the direction of plant growth. In this chapter we shall consider the control of growth and development through biological and chemical methods.

BIOLOGICAL CONTROL

Graft Combination

One way of controlling a plant biologically is through grafting, which is practiced for modifying growth as well as for propagation. The interaction of two or more plants in a graft combination may affect both growth and productivity.

Moreover, improved disease resistance and hardiness can be achieved by the creation of a plant composed of more than one genetic component.

The practice of grafting as a means of growth control is used most extensively with fruit trees. Graft combinations of herbaceous plant material have not been fully explored, for unless the plant itself is relatively valuable, grafting is not an economical horticultural practice. In Japan, however, watermelon is grafted onto the gourd *Lagenaria* to control Verticillium wilt, and eggplant is grafted onto *Solanum integrefolium* to increase productivity.

Fruit trees are normally composed of **scion** of a particular cultivar grafted onto a **rootstock,** although more complex combinations are possible (Figure 15-1). The rootstock may either be grown from seed (**seedling rootstock**) or it may be asexually propagated (**clonal rootstock**). Some rootstocks, even though produced from seed, are in effect clonal because of apomixis, as in citrus. Seedling rootstocks are often derived from a particular clonal cultivar. Thus, the "western" pear seedlings are usually derived exclusively from the 'Bartlett' cultivar. Some of these terms were encountered in the discussion of grafting in Chapter 12, but there, the emphasis was on the techniques of grafting; here, we will discuss the effects of graft combinations.

Restriction of Growth

The use of specific rootstocks to restrict the growth of the scion variety is an ancient practice. The degree of **dwarfing** achieved varies with the species and the rootstock (Table 15-1). The physiological explanation of the dwarfing effect

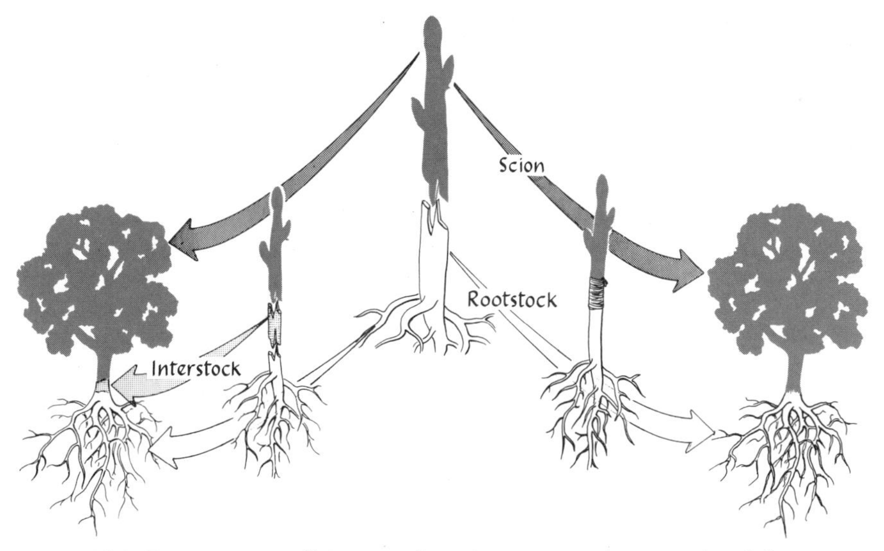

FIGURE 15-1. Fruit trees normally consist of two distinct parts, the rootstock and the scion. More complex trees may be made up of three or four components. [Adapted from J. Bonner and A. W. Galston, *Principles of Plant Physiology*, W. H. Freeman and Company, copyright © 1952.]

TABLE 15-1. Some dwarfing rootstocks.

Fruit crop	Rootstock	Effect
Apple	*Malus* clones East Malling series Malling Merton series	Complete range of dwarfing
	Malus sikkimensis "seedlings"*	Slight dwarfing
Pear	'Angers' quince clones	True dwarfing
Sweet cherry	'Stockton' Morello cherry	Height reduced to half as compared to Mazzard stocks
Peach	*Prunus domestica* clones	Slight dwarfing
	Prunus institia clones	Slight dwarfing
Plum	*Prunus besseyi* clones	Slight dwarfing
Orange	Palestinian sweet lime seedling*	Slight dwarfing
	Sour orange seedling*	Slight dwarfing

SOURCE: Adapted from K. D. Brase and R. D. Way, 1959, *New York Agricultural Experiment Station Bulletin* 783.

*Apomictic

has not been fully established. Evidence exists that it may be related to a number of causes, among which are the restriction of upward translocation of inorganic nutrients through the rootstock, the restriction of downward phloem transport, or some physiological disturbance caused by graft incompatibility.

There are abundant sources of dwarfing rootstocks for the apple (Figure 15-2). These rootstocks are all from various species of apple, and many of them derive from old European clones grown under the names "French Paradise"

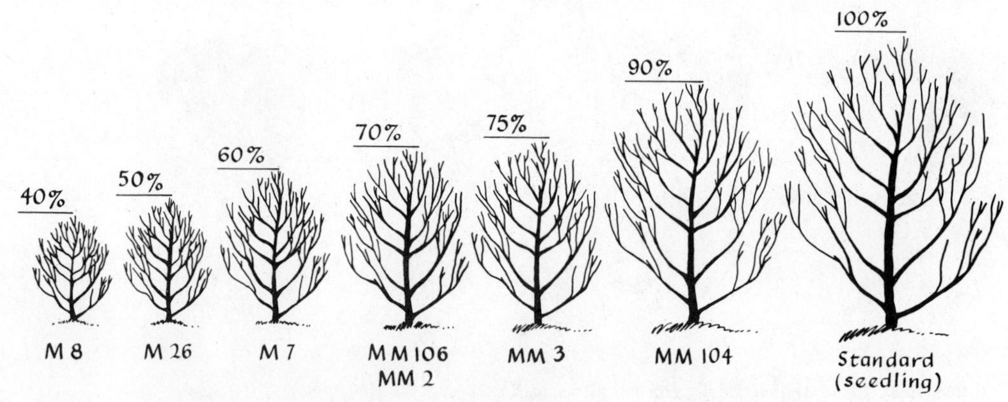

FIGURE 15-2. Comparative size of mature apple tree on various rootstocks. [Adapted from M. H. Kolbe, *Commercial Apple Production in North Carolina*, Raleigh.]

and "Doucin." These older clones were collected at the East Malling Research Station (in East Malling, Kent, England), where they were sorted out, standardized, and coded—EM 1, EM 2, and so on. Crosses of EM 9 and others with 'Northern Spy' apple have produced a number of rootstocks that show varying degrees of dwarfing and are resistant to woolly aphids. These are standardized as the Malling Merton series (MM 101, MM 102, and so on). EM 8 and EM 9 are extremely dwarfing rootstocks. They produce the dwarfing effect even if interposed between a nondwarfing rootstock and a scion variety (Figure 15-3). Thus, it is possible to avoid the shallowrooted characteristics of EM 9, which produces poorly anchored trees that are liable to tip over in wet ground after a strong wind, by using it as an **interstock** rather than a rootstock.

Pears are dwarfed by certain clones of quince, which belongs to a closely related genus *(Cydonia)*. To avoid the graft incompatibilities of certain cultivars of pear and quince, an interstock mutually compatible to both scion and stock is used as a bridge. The pear cultivar 'Hardy' can be used as such a bridge. When budding rather than grafting is employed, a shield bud of 'Hardy' inserted under the scion bud serves the same function. Dwarfing rootstocks also exist for stone fruits.

FIGURE 15-3. A comparison of fruiting and size of 7-year-old 'Delicious' apple on seedling rootstock (*left*) and on 'Clark' dwarf apple (*right*). The 'Clark' dwarf is obtained by using a stem piece of 'Clark' (a selection of 'French Paradise') as an interstock between the seedling rootstock and the scion cultivar.

Stimulation of Growth

Rootstocks may be used to compensate for poor root growth. The experimentally produced pear-apple hybrids had to be grafted on pear or apple rootstocks to survive. The upright cultivars of *Juniperus virginiana* have poor root systems, making them difficult to propagate. By grafting them onto the sturdy root system of *Juniperus chinensis* 'Hetzii,' a superior plant is created. In apple, the use of 'Virginia Crab' as a rootstock or body stock to increase the vigor of the scion was once common, but this practice is no longer recommended because of the sensitivity of 'Virginia Crab' to a virus disease known as stem pitting (Figure 15-4).

Flowering and Productivity

The induction of flowering in the nonflowering Jersey group of the sweetpotato by grafting to several species of related genera of the Convolvulaceae (morning glory family) is a striking use of grafting to achieve differentiation (Figure 15-5). Apparently, a specific flowering substance produced by the morning glory is transferred to the sweetpotato. This substance is produced in the leaves of the morning glory species in response to photoperiod and temperature effects. This technique allows the use of hybridization as a breeding method for the sweetpotato. It is of course unnecessary in the commercial production of sweetpotatoes, since they are propagated from adventitious shoots that grow from the roots.

The age at which fruit trees will begin to bear can be affected by the rootstock. The severely dwarfing rootstocks of apple and pear also encourage early bearing. Dwarfing rootstocks have been used to induce fruiting in pears, which are notoriously late-bearing, without a permanent dwarfing effect. Pears

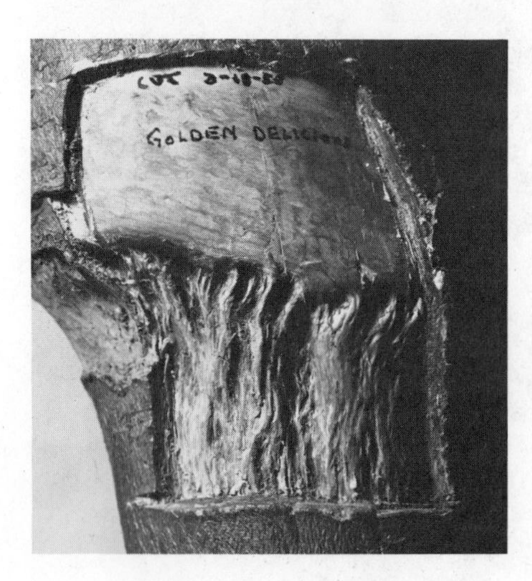

FIGURE 15-4. Symptoms of stem-pitting virus on a Virginia crab-apple bodystock. The scion cultivar, 'Golden Delicious,' does not show the symptoms, but it may carry the virus. [Photograph courtesy R. B. Tukey.]

FIGURE 15-5. Sweetpotato flowering after being grafted to a species of morning glory (*Ipomoea nil*). [Photograph courtesy S. Lam.]

grafted or budded to quince are planted with the union 6 to 8 inches below the ground. Early fruiting is stimulated by the quince rootstock. Scion rooting, however, eventually overcomes the dwarfing influence of the quince rootstock.

Although the yield of dwarfed trees is smaller per tree than that of standard trees, their reduced size allows for closer spacing. This factor, coupled with the tendency for early bearing, often results in greater per-acre yields with dwarf apples than has been achieved with standard trees on the standard spacing. Some dwarfing rootstocks actually produce more efficient fruit trees than do seedling rootstocks. In general, dwarfing rootstocks seem to have no effect on fruit size.

Some characteristics of the scion fruit may be affected by the rootstock. For example, the rough lemon rootstock lowers the sugar content of scion cultivars of orange more than other rootstocks, and rootstocks may affect such characteristics as blooming date and maturity.

Phloem Disruption

The induction of early fruiting and the control of growth by techniques that disrupt the phloem, such as girdling, scoring, or ringing, are also ancient horticultural practices. Inverting a ring of bark accomplishes the same effect (Figure

15-6). These practices, performed in early summer, are used to initiate flower-bud formation in 2- or 3-year-old clonally propagated apple trees so that they will flower and bear fruit the following year. Phloem disruption does not overcome seedling juvenility, but, like the use of dwarfing rootstocks, it is practiced to increase flowering on 4-year-old apple seedlings. The injured phloem retards the downward movement of the synthesized organic materials. The induction of flower-bud initiation is apparently caused by the accumulation of some substance above the injured phloem, for a single branch can be induced to flower on an otherwise barren tree. The effect is temporary, owing to the regeneration of phloem in the cuts and to the seam in the inverted ring. Although the inverted ring effectively blocks the downward movement through the phloem, according to Karl Sax, a proponent of the technique, this one-way movement (polarity) may eventually be reversed.

The fact that phloem disruption also produces a dwarfing effect in addition to the induction of early flowering suggests that some types of rootstock dwarfing may be related to the interference of downward phloem transport with the subsequent accumulation above the graft union of organic substances. This, however, cannot explain the dwarfing effects of all rootstocks.

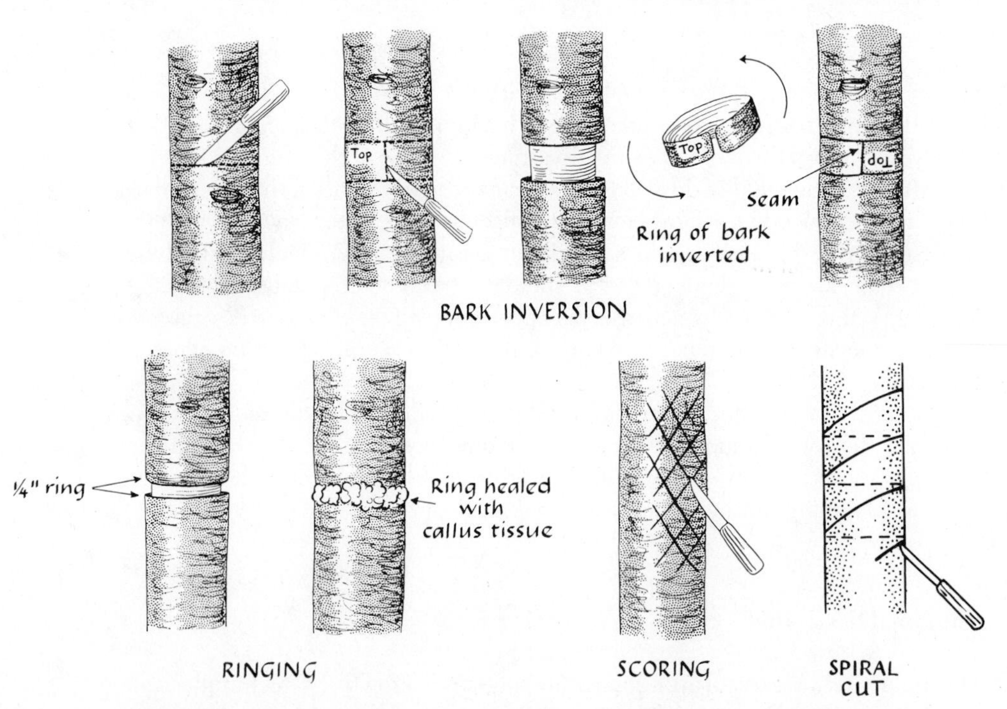

FIGURE 15-6. Dwarfing and the induction of early fruiting may be accomplished by phloem disruption.

Population Competition

Population pressures, brought about by increasing planting density, markedly affect plant performance. As plant population increases per unit area, a point is reached at which each plant begins to compete for certain essential growth factors: nutrients, sunlight, and water. The effect of increasing competition is similar to decreasing the concentration of a growth factor.

Yield, whether it be of root, shoot, flower, fruit, or seed, is usually expressed on a "per unit area" basis rather than on a "per plant" basis. One of the principal reasons is that space and the fixed costs associated with it are usually much more valuable than the costs of individual plants. Thus, the most important consideration is the effect that varying the plant population has on the yield per unit area rather than on the yield per plant. The optimum population, however, is the one that produces the greatest net return to the grower. It should be emphasized that yield must be interpreted in both quantitative and qualitative terms. The value of the total yield is not merely the total bulk, but is related to the quality of the yield (size per unit, color, appearance, culinary properties, and so on). There are two aspects to the problem of yield in horticulture. One is competition among plants as plant density changes (**interplant competition**), the other is competition among parts of the same plant (**intraplant competition**).

Interplant Competition

YIELD. The yield per unit area is equal to the yield per plant times the number of plants per unit area. When the population is below the level at which competition among plants occurs, increasing the population will have only an indirect effect on individual plant performance; the yield per unit area will rise in direct proportion to the population increase. As soon as competition among plants occurs, however, the yield per plant will decrease (Figure 15-7).

Once competition exists, the change in yield per unit area becomes a function of changes in the total weight per plant and in the proportion of the plant that is harvested as yield. A number of studies have shown that, although the yield (of dry matter) per plant decreases, the relationship is such that the total weight per unit area levels out, as seen in Figure 15-8. Thus, the yield of a particular plant part per unit area is related to its proportion of the total. If part yield is a constant proportion of the total, yield will level out with the total weight. However, if part yield decreases, yield will peak at some population, then decline, and density above or below the optimum will adversely affect yield.

For some crops—sweet corn is a good example—the yield of grain (or ears) is a decreasing percentage of total plant weight as competition increases. This means that there is an optimum population where yield is at a maximum.

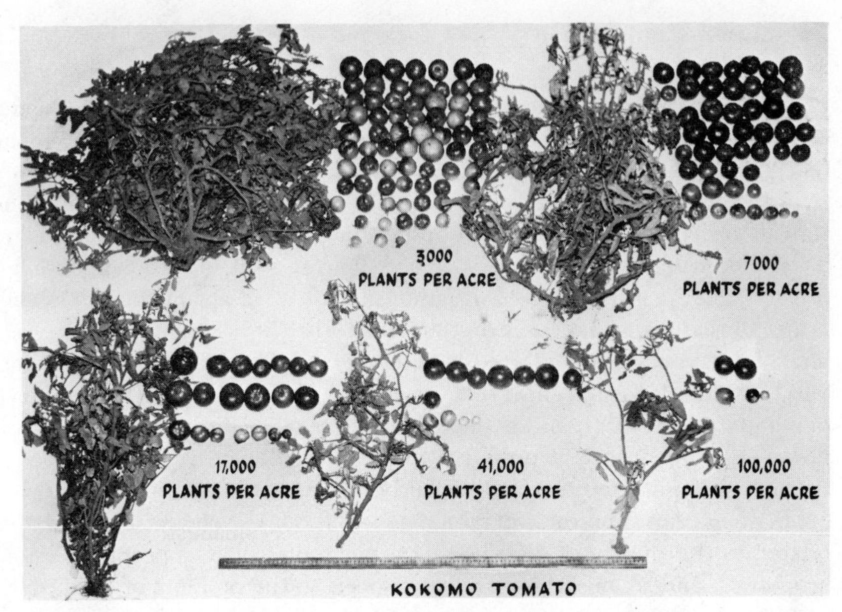

FIGURE 15-7. Increasing plant population drastically reduces plant size and yield per plant in tomato, but the yields per unit area increase with plant population and then level out. [From R. L. Fery and J. Janick, *J. Amer. Soc. Hort. Sci.*, 95:614, 1970.]

Putting restraints on yield (for example, on the ear size considered to be marketable) accentuates this effect. In many crops, part yield is a relatively stable proportion of total yield under competition. For example, the proportion of a tomato or bean plant that is fruit or seed is relatively constant over low and high populations. Fruit or seed yield then increases with population but at a de-

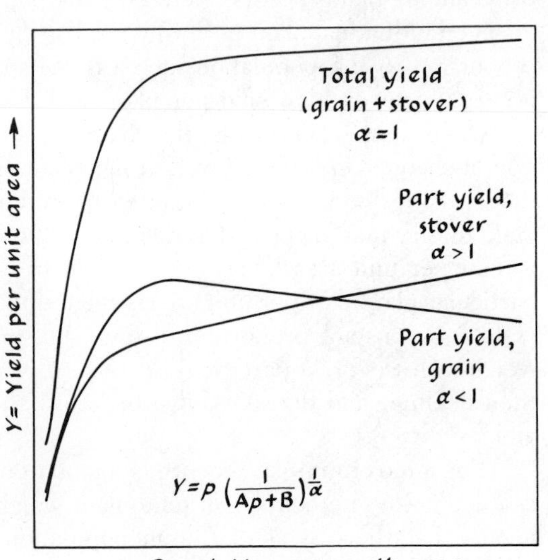

FIGURE 15-8. The relation between yield per unit area and population in maize. Note that total plant yield levels out as population increases. This appears to be true for all crops. The relation between population and the yield of some plant part (such as grain) depends on its proportion to total yield as population changes. In maize, grain yield is a decreasing proportion of total plant yield as population increases. Thus, grain yield increases up to some optimum population (usually between 12,000 and 15,000 plants per acre) and then decreases. In many crops, part yield is a fairly constant proportion of total yield as population changes, and thus part yield levels out as does total plant yield.

creasing rate, eventually leveling out (in other words, yield increases asymptotically toward a limit); and the higher the population the higher the yield, although economic considerations (for example, seed or plant costs) determine the practical limit for the population. However, if we apply restraints on yield quality to this illustration—say, by defining marketable yield in terms of fruit or seed size—yield may be maximum at a finite population.

The effect of population on plant performance thus depends on the part of the plant being measured and the restraints governing quality. Plant parts may respond to increased population pressure after competition by decreasing in size or in number or in both. The decrease in the number of plant parts (for example, the response of potato tubers) may produce a very small corresponding drop in the size of the remaining parts. These compensating responses to competition may differ with various plant parts, even on the same plant. In corn, for example, a rise in population pressure causes a decrease in ear number, ear size, and kernel number, but the kernel size remains relatively constant.

The optimum size of plant parts (fruit, tuber, bulb, flower) is not necessarily the largest size. (In pickling cucumbers, it is the smallest.) Population competition may be utilized to produce plant parts of a particular size. For example, onion seed, grown for "sets" to be replanted, is sown at the rate of 70 pounds per acre. The crowding produces a small bulb about 2 cm (3/4 inch) in diameter, an ideal size for replanting. When onion seed is planted for an edible crop and a large bulb size is desired, the planting rate is only 2 to 4 pounds per acre. Similarly, large increases in pineapple yields have been obtained by close spacing of large-fruited varieties.

Environmental factors such as nutrition or moisture levels may drastically change the level at which the response to crowding occurs. Thus, in tomatoes, the response of a given cultivar differs with the season. Of particular interest are the different types of response that may occur with different genetic types of the same plant. The dwarf maize produced with the *compact* gene appears to demonstrate a different response to competition from that of normal corn, meaning that yields may be constantly increased in proportion to population. The rearrangement of a plant population by rows increases competition in comparison to equidistant plant spacing, but this may be necessary for cultural considerations such as cultivation for weed control and access for spraying and harvesting equipment.

QUALITY OF YIELD. Population pressures affect quality factors that must be considered. For example, crowding tomatoes increases the foliage canopy, which tends to protect the fruit from being burned by the sun. On the other hand, excessive crowding of potted chrysanthemums produces undesirably spindly growth.

A high population level has an adverse effect on disease control. Plants grown exceptionally close together produce a dense cover that discourages rapid drying and produces conditions favorable for the growth of many fungi. In addition, the dense cover is impenetrable to spray application and limits chemical disease control. Sometimes dense foliage actually facilitates the spread of

For those mathematically inclined, the following formulas may be useful:

$$y_t = \frac{1}{Ap + B} \tag{1}$$

where y_t = total yield per plant, p = number of plants per unit area, and A and B are constants. The linear form of the equation

$$y_t^{-1} = Ap + B \tag{2}$$

indicates that the constants may be derived from two widely spaced populations.

Total yield per unit area (Y_t) is equal to the yield per plant times the number of plants per unit area:

$$y_t = py_t \tag{3}$$

Thus

$$Y_t = p\left(\frac{1}{Ap + B}\right) = \frac{p}{Ap + B} \tag{4}$$

The relation between total yield per plant and yield of a plant part (y_p) can be described by the following relationship:

$$y_t = Ky_p^{\alpha} \tag{5}$$

where K and α are constants. The linearized form of this equation

$$\log y_t = \log K + \alpha \log y_p \tag{6}$$

indicates it is possible to estimate α from two widely spaced populations.

By substituting equation (5) into equation (1) the yield of plant part (y_p) can be described as follows:

$$y_p = \left(\frac{1}{A'p + B'}\right)^{1/\alpha} \quad \text{or} \quad y_p^{-\alpha} = A'p + B' \tag{7}$$

where $A' = KA$ and $B' = KB$. Therefore the yield of a plant part per unit area (Y_p) is described by the equation

$$Y_p = p\left(\frac{1}{A'p + B'}\right)^{1/\alpha} \tag{8}$$

Note that when $\alpha = 1$, equation (8) is in the same form as (4).

diseases by contact, as is true of tobacco mosaic virus in tomato. High population may be utilized as a method of weed control. The increased shade produced by a dense cover of vigorous plants may permit the crop plant to outcompete weeds. This benefit is utilized in turf management.

Intraplant Competition

The relationship among parts of the same plant is a major factor in population competition. It almost invariably involves fruit and flower size. Size per unit is a significant component of value in practically all horticultural crops. For example, high yields of extremely small fruit may be economically worthless.

FRUIT SIZE. A noteworthy factor in fruit size is the leaf-to-fruit ratio. Since leaves are the carbohydrate source nourishing the fruit, fruit size will be related in any given genotype to the amount of leaf area per fruit (Figure 15-9). The seeds get first call on the carbohydrates produced by the leaves. When the seed requirements are satisfied, the extra carbohydrates become available to the fruit. Only after the requirements of the fruit are satisfied do the extra carbohydrates become available for the vegetative organs.

Leaf area may not be a constant value. Indeterminate plants (plants in which the main axis remains vegetative) have a constantly increasing leaf area as the plant grows. Under a heavy load of maturing fruit, however, many indeterminate plants stop vegetative growth. The result is a constant leaf-to-fruit ratio. In mature, bearing fruit trees, the quantity of leaves produced in any one season is relatively constant because the amount of new growth tends to be small.

There are two ways to compensate for fruit competition. One is to increase the number or efficiency of the leaves. The other is to reduce the fruit load by blossom or fruit removal—that is, by **thinning** (Figure 15-10). Increases in leaf growth and efficiency are accomplished by various cultural practices, especially by improving the plant's nutritional status. This may be self-defeating if better

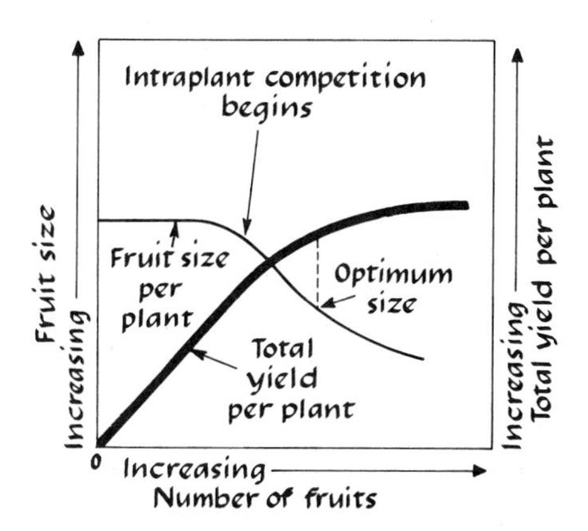

FIGURE 15-9. The relation between the number of leaves per fruit and fruit size in the 'Delicious' apple. [Adapted from data of J. R. Magness, F. L. Overly, and W. A. Luce, Washington Agricultural Experiment Station Bulletin 249, 1931.]

FIGURE 15-10. The effect of thinning on fruit size. The left side of the tree was thinned: the right side of the tree was not. [Photograph courtesy F. H. Emerson.]

nutrition also increases fruit quantity. Thinning, on the other hand, is an expensive practice confined largely to high-priced fruit crops whose value depends largely on size per unit (such as apples, pears, peaches, and plums).

YIELD AND FRUIT SIZE. Below the levels of intraplant competition, yield is directly related to the number of fruit. Under competition, however, the increase in yield levels off. The practical levels for thinning popular peach and apple cultivars occur at levels where total yield is practically unaffected. The reason is that these cultivars are potentially large-fruited. Thinning must be early to be effective. The general relationship between the number of fruit, fruit size, and yield in apples and peaches is plotted in Figure 15-11.

OTHER CONSIDERATIONS. In addition to size, there are other significant considerations in fruit competition. In apples, for instance, a relationship exists between quality and fruit load. With an unusually light crop, apples of some cultivars may get too large (more than 3½ inches in diameter), and their storage quality declines. On the other hand, with fewer leaves per fruit, there is a decrease in fruit sugars (mainly sucrose), which reduces fruit quality. A corresponding decline in anthocyanin pigments per fruit decreases red color, although this effect is relatively slight.

Competition between fruit and vegetative parts may result in severe damage to the plant; in grapes, for example, excessive fruit load renders the

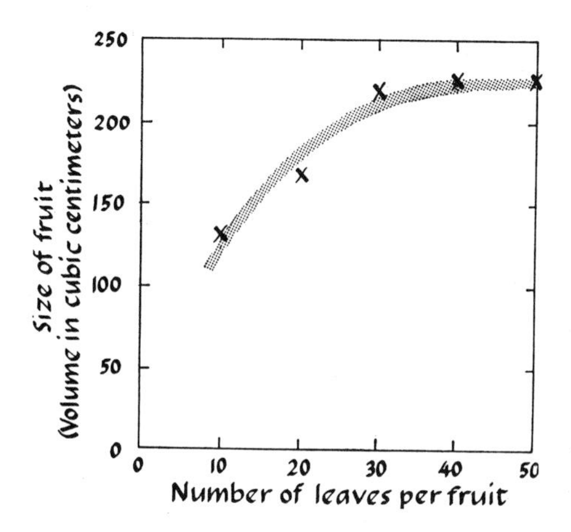

FIGURE 15-11. Idealized relation between fruit size and yield. The total yield per tree is only slightly reduced by thinning popular large-fruited varieties of apples and peaches. However, unthinned fruit may be economically worthless. Some reduction in total yield is desirable in order to keep alternate-bearing cultivars producing on an annual basis.

plant susceptible to winter injury as a consequence of low sugar accumulation by the vegetative organs. In periods of stress, such as during a drought, fruit is removed from young trees in order to prevent desiccation and subsequent winter injury. Heavy fruiting in tomatoes will increase the plant's susceptibility to foliage diseases.

In fruit trees that produce flower buds in the year previous to fruiting (especially apples and pears), fruit competition may be responsible for **biennial bearing.** The competition for nutrients among a large number of developing fruit apparently prevents the development of fruit buds for the next year's crop. Once this pattern has started, it is difficult to stop, for, in the off-bearing year, an extra abundance of flower buds will form in the absence of fruit competition. Biennial bearing of apples may be controlled by fruit thinning to prevent overbearing.

FLOWER SIZE. In plants grown for individual flowers rather than for clusters, unit size is a valuable component, just as it is in fruit. For this reason, flower size is commonly increased by reducing the number of developing flowers per shoot. This purpose is accomplished by **disbudding,** the removal of all buds but one per shoot. Disbudding is a standard practice in the culture of roses, carnations, chrysanthemums, and peonies.

Economics of Population Control

If we grant that the goal of the commercial plants producer is profit, the optimum plant population will be the one from which profits are greatest, and populations of plants or of plant parts present a number of factors that must be considered: for example, the costs of plants and planting (or thinning), as well as the costs of cultivation, disease control, and harvesting. From these figures the actual costs of production for each level of population can be determined. Similarly, the returns at each population level can be determined if the yield responses are known. The return will be the value of each unit determined from its size and quality multiplied by the total number of units of that value. The profit (or loss) at each population level may be determined by subtracting total costs from total returns.

If we could predict all plant responses, costs, and returns, it would be relatively easy to "program" the problem—that is, to substitute values into a mathematical formula and solve (with a computer, if necessary) the equation. Although this degree of control is not yet possible, it is being approached, as has been illustrated by the cling peach predictions made in California. However, empirical solutions to these problems have been obtained by trial and error through the years. The obvious advantage of the mathematical approach is its "instantaneous" utilization of changing information concerning prices, costs, and yield responses.

CHEMICAL CONTROL

The control of plant growth and differentiation through the use of chemical substances is a modern development in horticulture, although there are examples of the early use of various substances (such as salt, wine, urine, and preparations made from germinating seeds) to this end. In general, only a few isolated accounts of these ancient practices have been shown to have a real physiological basis. The advances in this area have not been made through empirical methods but are, instead, largely a by-product of investigations into growth and development. This field was given great impetus by the impact of auxin studies on horticultural technology. A number of substances are now known that have a relatively broad spectrum of effects (for example, indoleacetic acid and the gibberellins). Others merely mediate or block specific metabolic pathways. Such substances as ethyl methane sulfonate (a mutagen) and colchicine (an alkaloid from the autumn crocus, *Colchicum autumnale*, that causes chromosome doubling) have been used to achieve permanent genetic change (see Chapter 17). It can be assumed that, as knowledge expands in this area, an increasing number of growth-regulating substances will be found.

At present, the substances used to kill plants (herbicides) account for the greater part of chemical control. This fact is discussed further in Chapter 16. The following section will review chemical substances that affect physiological processes important in horticultural practice.

Rooting

The rooting of cuttings has been shown to be influenced by auxins, although auxins are by no means the only substances involved. In a cutting, the natural auxin produced in young leaves and buds moves naturally down the stem and accumulates at the cut base along with sugars and other food materials. The natural formation of roots is apparently triggered by the accumulation of an optimum auxin level in relation to these substances. In a wide variety of plants, rooting is markedly increased by the addition of synthetic auxins. Although a wide variety of such compounds has been used, the greatest degree of success has been achieved with indolebutyric acid (Figure 15-12).

$CH_2CH_2CH_2COOH$

N
H

Indolebutyric acid

Other auxins have a very narrow range of effective concentrations. Concentrations below the critical level are ineffective in root initiation, whereas those above that level not only inhibit root growth and bud development but may cause gross morphological damage. Indoleacetic acid is ineffective, probably because it is readily destroyed by the plant.

Cuttings from many plants that are naturally difficult to root, such as apple, do not respond to auxin application. An interesting facet of this problem is that the transition of cuttings of some plants from "easy-to-root" to "difficult-to-root"

Control group
no treatment

10mg added
potassium salt of
indolebutyric acid

FIGURE 15-12. The potassium salt of indolebutyric acid markedly increases the rooting of *Chaenomeles* (flowering quince). [Photographs courtesy J. S. Wells.]

is associated with the plant's change from juvenility to maturity. This relationship may be due to the formation at maturity of inhibitors that block rooting. Thus, difficult-to-root dormant grape cuttings become easy to root when leached with water to remove the inhibitor. Other plants become hard to root not because inhibitors are present but because they become deficient in, or lack, certain required substances. If these substances are applied in combination with an auxin, rooting can be promoted (see Chapter 12).

Bolting

Bolting (or seed-stalk formation) in biennial plants is induced by cold. Thus, carrots and onions do not flower unless their growth is interrupted by a cold-induced dormancy. This cold requirement may be replaced under certain conditions by treatment with gibberellins, a group of substances whose primary morphological effects are associated with stem elongation. Gibberellins apparently replace or substitute for the natural gibberellins that are produced (or that accumulate) during the cold period and are responsible for bolting. Gibberellins will also induce bolting in those plants in which the process does not require cold but that are photoperiod-sensitive, such as spinach. The bolting stimulus provided by gibberellins is independent of the flower-inducing stimulus. Thus, flowering in some chrysanthemums depends not only upon a short photoperiod but upon elongation induced by cold treatment. Gibberellins can replace the cold requirement, but the plants remain vegetative if exposed to a long photoperiod. This use of gibberellins in promoting bolting has far-reaching effects in breeding, where time is often a limiting factor.

Gibberellins have the useful effect of facilitating normal seed-stalk formation. For example, the heads of some lettuce cultivars are so tight that the seed stalk cannot push through and may break or rot unless the head is cut. The tight head is a desirable trait in crop production, but it interferes with seed production. Treatment with gibberellins may be used to encourage normal seed-stalk formation in these lettuce types.

Modification of Sex Expression

One of the most dramatic changes achieved with growth regulators has been the alteration of sex expression. Modification of sex has great potential for plant breeding. In cucumber, ethephon delays staminate flowering and thus transforms monoecious lines into all-pistillate (all-female) lines, thereby increasing yields and facilitating hybrid seed production. Similar effects have been reported with muskmelon and squash (but not watermelon). Gibberellins increase maleness in cucumbers; staminate flowers have been induced on genetically, all-pistillate (gynoecious) lines.

Although grapes have hermaphroditic flowers, certain parts of the flowers may be nonfunctional. Some clones have only functionally staminate flowers (the pistil is undeveloped). It has been possible to induce female organs (and viable seeds by selfing) in these "staminate" clones with cytokinins as well as with ethephon. Ethephon is also effective in inducing seed formation in some normally seedless clones, some of which are viable. This capability has promise for the breeding of seedless grapes.

Flower Induction

The biochemical basis of flowering still remains unknown. That the triggering mechanism is hormonal, however, is indicated by the translocation of the photoperiodic stimulation from leaf to bud and across graft unions. It has been demonstrated that the suspected flowering hormone ("florigen") is not an auxin, although an auxin is involved in the process. Thus, auxins applied to many plants after the initiation of flowering may effectively promote flowering.

Flower induction in pineapple has been achieved with ethylene as well as with auxins. Recent evidence suggests that auxins act by affecting the natural ethylene-generating mechanism of the pineapple fruit; nevertheless, the use of auxins to induce flowering in pineapple has been a standard culturing procedure. Different auxin derivatives have been used, such as sodium naphthalene acetate in Hawaii, and 2,4-D (2,4-dichlorophenoxyacetic acid) in the Caribbean area.

Biennial bearing, a serious problem with apples, can be controlled quite effectively with daminozide, the common name of butanedioic acid-(2,2-dimethylhydrazide), a growth retardant (Figure 5-1) that suppresses gibberellin formation. This effect is achieved by the reduction of shoot growth that promotes flower formation in the following season.

Fruit Set

Promotion of Fruit Set

The practice of chemically inducing **fruit set** has followed from studies of the relation of natural auxins to fruiting (see Chapter 5). The use of auxin derivatives to set fruit in the absence of pollination (parthenocarpy) has had some commercial utilization in winter production of tomatoes in the the greenhouse, where fruit set is often poor, and in fig and grape production. The auxin substances generally used for tomatoes are *p*-chlorophenoxyacetic acid and β-naphthoxyacetic acid. This practice is limited, however, since it causes fruit abnormalities, such as puffiness and premature softening. The use of auxins to set fruit in 'Calimyrna' fig eliminates the need for male trees and the practice of caprification. However, the most effective auxin for fruit set (*p*-chlorophenoxyacetic acid) produces a seedless fruit that has not proved to be acceptable, and the auxin that permits the development of the seed coat (benzothiazole-2-oxyacetic acid) is not as effective for fruit set. The use of auxin sprays to promote fruit set in some grape cultivars has eliminated the need for girdling. This has become a widely adopted practice in California for use with the 'Black Corinth' and 'Thompson Seedless' cultivars. The induction of parthenocarpy in grapes with gibberellins has made possible the commercial production of seedless grapes of the 'Delaware' cultivar in Japan. Such clones receive two gibberellin treatments, the first to induce seedlessness, and the second to increase fruit size.

Flower and Fruit Thinning

The removal of flowers and fruits to reduce crop loads has been described earlier in this chapter as thinning. The reduction in crop load by **chemical thinning** has become a standard practice in a large part of the fruit industry. It is one of the best examples of the chemical control of growth. The physiological action of these chemicals consists in preventing the completion of fertilization or in inducing embryo abortion, both of which result in natural abscission. Chemical thinning may be performed prior to fertilization (flower thinning) or after fertilization (fruit thinning).

The materials effective in thinning may be referred to either of two groups, depending on their mode of action. **Flower-thinning** compounds are composed of caustic and toxic substances (for example, phenols, cresols, and dinitro-compounds) that kill off the blossoms or render them sterile. The principal effect of an interesting substance called Mendok (sodium dichloroisobutyrate) is to induce pollen sterility. This male gametocide has been used experimentally in tomatoes (a self-pollinated crop) to induce sterility in the early clusters and thereby to concentrate fruit set. The concentrated ripening of fruits would be desirable for "once-over" mechanical harvesting. The **fruit-thinning** materials are auxin derivatives, and they bring about thinning largely through embryo abortion. It is interesting to note that auxins, which set fruit in some species, are used to remove fruit in others.

Not all auxins are necessarily effective in thinning; in fact, only naphthaleneacetic acid and its derivatives are useful in fruit thinning. This auxin is widely used in apples and is effective in peaches, pears, olives, and grapes, although *N*-1-naphthylphthalamic acid has been more widely utilized in the stone fruits. Chemical thinning with auxins is also employed to prevent fruiting in trees used as ornamentals, where only flowering is desired and fruit is considered a nuisance.

The way in which auxin derivatives cause embryo abortion is not clear. For example, the principal absorption of auxins is not through the fruit but through the foliage. The degree of thinning with auxins is greatly affected by the concentration used, the timing of the application in relation to fruit development, and the species and cultivar, as well as by such environmental factors as temperature and humidity.

The production of 'Thompson Seedless' grapes for table use has been revolutionized through the use of gibberellins, which loosen fruit clusters, greatly increase fruit size, and thin berries. In California, all table-grape acreage of 'Thompson Seedless' is treated with gibberellins (Figure 15-13). On seeded grapes, the berry-thinning effect reduces cracking and preharvest diseases. The auxin *p*-chlorophenoxyacetic acid has also been used to increase the size of pineapple fruits.

Ripening

It has long been known that ethylene and acetylene stimulate fruit ripening. Ethylene gas applied in ripening rooms is a standard practice to accelerate banana ripening. The discovery that an ethylene-generating material, ethephon, stimulates fruit ripening has made the induction of the ripening process possible in the field. The use of ethephon in pineapple encourages uniform ripening of entire fields and has very strong implications because it makes complete mechanical harvesting a possibility. A similar effect is also achieved with tomato. Ethephon appears to have tremendous commercial applications in horticulture in the control of ripening as well as other effects. It is now used to stimulate latex flow in rubber trees.

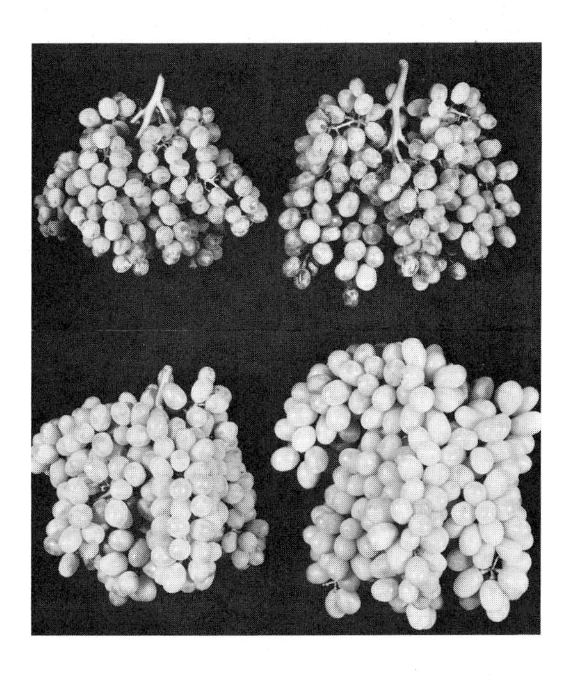

FIGURE 15-13. Effect of gibberellic acid (GA) on 'Thompson Seedless' grapes: (*top left*) control, (*top right*) 5 ppm GA, (*bottom left*) 20 ppm GA, (*bottom right*) 50 ppm GA. [Photograph courtesy R. J. Weaver.]

Preharvest Fruit Drop

The effect of auxins in inhibiting abscission has found an important horticultural application in the control of preharvest fruit drop. The natural auxin, which prevents abscission and is produced by the seed, apparently decreases with fruit maturity. A number of synthetic substances are now used to delay fruit drop, among which are naphthaleneacetic acid, 2,4,5-trichlorophenoxyacetic acid, and 2-(2,4,5-trichlorophenoxy)propionic acid. Chemical control of preharvest drop is widely used in the fruit industry for apple, apricot, pear, prune, almond, and citrus fruits, and especially in cultivars of these fruits that are prone to drop prematurely. It is an effective means of preventing fruit drop that is ordinarily accentuated after frost. The use of preharvest drop control to increase red color development after maturity cannot be recommended because overmature fruit does not store well.

Red color development can be increased by daminozide, an antigibberellin, which also delays preharvest drop in apples and appears to have other interesting beneficial effects, including the prolongation of storage and shelf life and reduction of water core and storage scald.

Dormancy

The modification of seed and plant dormancy promises to be an excellent area for chemical control, because the extension of dormancy in woody plants to avoid damage by spring frost will provide great economic benefits. With the discovery of abscisic acid, the control of dormancy has received increased attention.

In the past, much effort was expended in prolonging dormancy in horticultural products during storage. For example, serious losses of potatoes and onions result from the sprouting that occurs with the breaking of dormancy. High concentrations of auxins have been successful in prolonging dormancy, and a number of other substances that extend it without killing the tuber or bulb are being isolated. Many of these substances cannot be used on foods; however, the use of the growth inhibitor maleic hydrazide has been effective in inhibiting sprouting in onions and potatoes and can be applied to the growing plant.

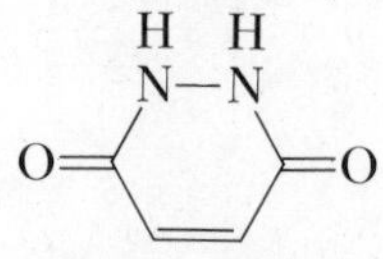

Maleic hydrazide

Although a complete biochemical basis for the chilling requirement has yet to be defined, several materials have been used to induce or terminate dormancy. Gibberellic acid has been the most effective material to substitute for low temperature in satisfying the chilling requirement, and it has been used for breaking dormancy in potatoes. It is a commercial practice in Florida to produce an early crop using freshly harvested potatoes for "seed." Gibberellins appear to overcome abscisic acid, the natural dormancy promoter.

Growth Promoters and Inhibitors

Chemical materials that promote or inhibit plant growth have promising uses in horticulture. For example, the cytokinin kinetin (6-furfurylaminopurine) promotes cell division and callus formation. The growth-stimulating effects of gibberellins are used to increase the size of celery. Combinations of the cytokinin benzylamino purine (BA) plus gibberellic acid (GA_{4+7}) applied as a foliar spray, control fruit shape in apple, specifically accentuating the elongated shape of 'Delicious' apple. This use is common to the eastern United States where this cultivar often develops a roundish shape considered undesirable in the marketplace.

Substances that inhibit or retard growth are equally desirable. The effect of maleic hydrazide in slowing the growth of turf to reduce the frequency of cutting has not proven practical, but this effort has stimulated a search for other compounds that might perform better. Several compounds have been found that dwarf plants effectively by retarding stem growth. Examples are chlorphonium chloride or phosphon (2,4-dichlorobenzyl tributyl phosphonium chloride) and chlormequat ((2-chloroethyl)trimethylammonium chloride). Such substances show promise for use in reducing the height of many ornamental flowering plants without unduly interfering with flowering time or flower size (Figure 15-14).

Daminozide, a growth retardant that acts as an antigibberellin, has a wide array of effects already discussed. Its growth-retardant properties may be of use in tomato culture to improve transplant durability and to concentrate fruit maturity by decreasing early yield.

Ancymidol (α-cyclopropyl-α-(4-methoxyphenyl)-5-pyrimidinemethanol) is an extremely powerful growth retardant. An important commercial use is to control the height of Easter lilies, but it is extensively used in other florist crops because of its effectiveness over a wide spectrum of plant species.

Pinching, Disbudding, and Sprout Control

Pinching (removal of the apical meristem) to facilitate branching is a standard practice with many florist crops (for example, carnation and chrysanthemum). It is possible to accomplish the same end through the application of such materials as methyl esters of C_8 to C_{12} fatty acids (for example, methyl decanoate) and C_8 to C_{12} alcohols. These "chemical pinching agents" destroy the growing point and facilitate bud break below. Chemical pinching of azaleas is achieved

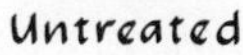

FIGURE 15-14. The growth-retardant properties of chlorphonium chloride (Phosphon). [Adapted from USDA, Agricultural Research Service 22–65, 1961.]

with dikegulac, an agent that is sprayed on foliage and is translocated to the shoot apex.

Disbudding (removal of axillary buds below a terminal) is used to enlarge the size of the terminal flower and to alter plant form in such florist crops as chrysanthemums, carnations, and roses. An active search is now going on for chemical disbudding agents; some materials show promise, but results so far have not been consistent enough for commercial use.

The control of root suckers, watersprouts, and excessive shoot development after pruning is a recurring problem in many fruit trees. The use of high (1 percent) concentrations of NAA (naphthaleneacetic acid) in paints or applied as a spray (with dormant oil and wetting agents to increase absorption) has proven effective for inhibiting these unwanted shoots (Figure 15-15).

Abscission

Substances that influence abscission of leaves and fruits have various uses in horticulture, such as the defoliation of nursery stock to improve storage, and defoliation to encourage fruiting and to facilitate mechanical harvest. Various materials defoliate plants by burning off their leaves. These include chlorates, cyanimids, borates, urea, and potassium iodide. The abscission of fruits as well as leaves is encouraged by ethylene and thus also by ethylene-generating materials and materials that increase the natural ethylene level, such as auxins and

A B

FIGURE 15-15. Control of sprouting in lemon by sprays of 1 percent NAA (naphthaleneacetic acid) plus 5 percent EHPP (ethylhydrogen 1-propylphosphonate): (*A*) control, (*B*) treated. [From R. L. Phillips and D. P. H. Tucker, *HortScience* 9:199–200, 1974.]

abscisic acid. At present, ethylene-generating materials have the most promise for loosening fruits. These include ethephon for cherries. Alsol (2-chloroethyl-tris-(2-methoxyethoxy)silan) for olives, and Release (5-chloro-3-methyl-4-nitro-1*H*-pyrazole) for citrus.

Storage Disorders

A number of physiological disorders are associated with postharvest storage of fruits. Scald, a disorder of the peel, is a common hazard in the storage of apples. Materials applied before storage to decrease this disorder (such as diphenylamine and ethoxyquin) are widely used.

Protection from Stress

The ability of growth regulators to protect plants from stress suggests their potential future use in horticulture. For example, various growth retardants, such as ancinidol and chlormequat, protect plants from such air pollutants as ozone and sulfur dioxide. Daminozide appears to accelerate recovery from drought-induced stress. Finally, substances that interfere with ethylene synthesis may prolong storage of ethylene-sensitive plants, including poinsettia and orchids. Two such substances are aminoethoxyvinyl glycine (AVG) and silver thiosulfate.

Selected References

Annual Review of Plant Physiology. Palo Alto, Calif.: Annual Reviews, Inc. (Recent advances in plant physiology brought up to date in timely, technical reviews. Published each year since 1950.)

Bleasdale, J. K. A., 1966. Plant growth and crop yield. *Annals of Applied Biology* 57:173–182. (A discussion of the effect of crop density on plant performance.)

Larson, R. A. 1985. Growth regulation in floriculture. *Horticultural Reviews* 5:399–481. (The use of growth regulators in florist crop production.)

Tukey, H. B., 1964. *Dwarfed Fruit Trees.* New York: Macmillan. (A review of dwarfing rootstocks. Reprinted 1978 by Cornell University Press, Ithaca, N.Y.)

Weaver, R. J., 1972. *Plant Growth Substances in Agriculture.* San Francisco: W. H. Freeman and Company. (A text on the applied uses of plant-growth substances.)

Wiley, R. W., and S. B. Heath, 1969. The quantitative relationships between plant population and crop yield. In A. G. Norman (ed.), *Advances in Agronomy.* Vol. 21, pp. 281–321. New York: Academic Press. (An excellent review that examines the usefulness and biological validity of the different mathematical equations that have been proposed to describe the relationships between plant population and crop yield.)

Williams, M., 1979. Chemical thinning of apples. *Horticultural Reviews* 1:270–300. (A thorough review.)

Plant Protection 16

The earth is covered with a variety of life forms in competition for food, light, and space. To a great extent, we, the ascendant species, now direct this competition to our own advantage. The efficiency of this control is commonly thought to depend on the degree of civilization or culture. The standard of living of a people is directly related to their ability to compete with other organisms.

The two great professions that deal directly with our control over our biological competitors are medicine and agriculture. Human diseases, inborn errors of metabolism not included, in large part are due to competitions by microorganisms for ascendancy in the human body. Agriculture, on the other hand, is concerned with human efforts to control and exploit plant and animal life for food and fiber. To cultivate crop plants successfully, growers must do more than assure the proper combination of environmental factors: They must interfere with the natural competition and interaction among living things. Horticulture, which infers intensive culture, requires strong human interference in the biological spectrum. The biological competitors that concern us in this chapter include pathogens, predators, and weeds. Control of these crop adversaries is called **plant protection.**

Plant pests take a heavy toll from world agriculture. In the United States the yearly loss in farm crops has been estimated at more than $8 billion. Moreover, the annual cost of pesticides and related equipment amounts to almost $1 billion for the control of insects and diseases, and over $3 billion for the control of weeds! Some species of pests could wipe out the crops in an entire section of the country, so awesome is their destructive power. Others that may not be so spectacular in the short run are, in the long run, equally destructive. These pests continually "peck away" at our abundance, but their damage tends to go unnoticed. To a particular grower or home owner, one certain pest may mean the difference between feast and famine, sun and shade. Eventually, however, all of us share the cost of this waste.

PATHOGENS AND PREDATORS

Disease

The word *disease* means, literally, "lack of ease." Broadly defined, a disease is any injurious abnormality. A distinction between these abnormalities—physiological and anatomical—is made on the basis of cause. Diseases inflicted by some biological agent are referred to as **pathogenic** diseases. **Nonpathogenic** diseases may include the adverse effects of abnormal physiological disorders; environmental factors, including extremes of heat or cold, soil fertility, water availability, and pollution (see Chapter 11); graft incompatibilities; spray injury; or disorders due to unknown causes.

The use of the word *disease* to refer to insect injury, much less to nonpathogenic disorders of plants, may be objectionable to some. Nevertheless, consideration of the term *mental disease* will indicate that the word is certainly not a narrow one. The confusion arises from the common use of the term in a restrictive sense to refer to the injurious effect caused by a pathogenic microorganism in intimate association with the host plant for extended periods of time. This use of the term would apply to the detrimental effects that viruses, bacteria, and fungi have on plants. The harmful effects of insects, mice, and birds would then be classified as injury rather than disease, the term *predator* rather than *pathogen* being applied to these pests. The distinction between a predator and a pathogen, however, is not clear-cut. Many plant-attacking nematodes as well as insects (such as the peach-tree borer) are in intimate contact with the plant for extended periods and could well be considered pathogens even in the sense of the restrictive definition. In the ensuing discussion, no special distinction will be made between pathogen and predator.

Pathogenic plant diseases may be discussed either in terms of the agent (pathogen) causing the disease or in terms of the plant's response. The specific responses of the plant are known as **symptoms,** which, together with evidence of the pathogen, called **signs,** permit the diagnosis of the disease—that is, the association of the disease with its cause. Care must be taken not to confuse the cause of the disease, the pathogen, with the disease itself.

Symptoms

Plants respond to the irritation produced by an external biological agent in a limited number of ways. The most extreme response is death, or **necrosis.** The entire plant may die, or the necrosis may be limited to specific organs, such as leaves, branches, flowers, or fruits. It may even be restricted to very small areas, resulting in spots or holes. Decline may be gradual and incomplete. For example, chloroplast breakdown, which produces yellowing or mottling, does not necessarily result in the immediate death of the plant. Another basic plant response is a reduced growth rate, which may affect the entire plant or certain of its parts, resulting in stunting, dwarfing, or incomplete differentiation. A third effect is an increased growth of an abnormal and morbid type. This reaction results in overgrowths—enlargements of organs, tissues, or cells—or tumorlike protuberances called galls. Although it is true that the basic responses of the plant are limited, the many variations involving different tissues often permit accurate diagnosis from the symptoms alone.

The Pathogen

The **pathogen** is a biological agent that produces a disease. The number of different kinds of pathogens affecting horticultural plants is truly awesome. Their effect may be as transient as an insect bite or as persistent as a virus infection.

The plant provides the pathogen with nourishment, shelter, support, or some other advantage. Competition between plant and pathogen is part of the natural order of things. Disease is not evil, malicious, or particularly unusual, nor do plants escape disease by virtue of their being "healthy." Many pathogens attack only vigorous, thrifty plants.

The association of living organisms in which one organism derives nourishment or other benefit from another, the host, is known as **parasitism.** The terms *parasite* and *pathogen* are not synonyms. A pathogen is injurious to the host at some stage of its life cycle, whereas a parasite is not necessarily injurious. Most pathogens are, however, parasitic in nature. Many disease-causing organisms may be only incidentally pathogenic; their usual mode of life may consist in living on naturally dead or decayed tissue. They are known as **saprophytes.** Some pathogens have evolved with a specific host plant to such an extent as to be **obligately parasitic;** that is, they can survive only on the living tissue of the host. The host range of some pathogens may be extremely large, or it may include only particular genotypes within a single species.

Viruses

Viruses are small infectious particles made of a core of nucleic acid surrounded by a protein sheath (Figure 16-1). Their small size (approximately one-millionth

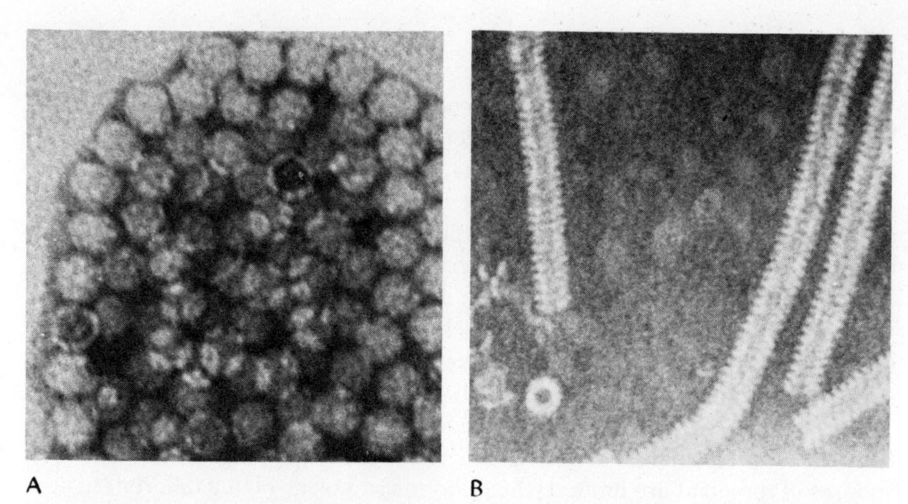

FIGURE 16-1. Plant viruses. (*A*) Portions of tobacco-mosaic virus particles. Even the spiral arrangement of the protein sheath is visible in this remarkable electron photomicrograph. The complete virus particle is about 15 × 300 nm. (*B*) The protein subunits are visible in this photomicrograph of the turnip-mosaic virus particles (30 nm in diameter). [Photographs courtesy R. W. Horne, Cambridge, England.]

of an inch) makes them visible only with the electron microscope. Particles may be short or long rigid rods, flexible threads, or "spherical" (actually 20-sided) forms. Viruses are obligate parasites; they reproduce only in living cells of the host, but they may be removed from the organism and remain capable of causing infection. Some viruses remain active in extracted plant juices for many months; the tobacco-mosaic virus remains active for years in dried material. Some can be crystallized and still cause the disease when reintroduced into the plant.

The question of whether viruses are living is a matter of terminology or philosophy. If the living system is defined as a self-duplicating entity capable of reconstructing itself from different component parts, then viruses can be considered alive. But viruses are not complete living systems, because they are unable to generate the energy required for their multiplication and must exploit the enzyme system of infected, living cells. The genes of viruses are now under intense investigation, and the complete nucleic acid structure of some viruses that infect bacteria (bacteriophage) as well as higher plants is known.

Viruses cause diseases of many organisms, including many important crop plants. Symptoms of viral diseases of plants are usually systemic, and they range from death, through mild stunting with reduced quality and yield, to no obvious effects. Symptoms may include small lesions, a spotted or mosaic pattern of green and yellowish areas, yellowing, stunting, leaf-curling and edge-crinkling, excessive branching, and color-striping or total disappearance of color. The virus is usually named by the description of the major symptom coupled with the name of the host plant where first described. Thus tobacco-mosaic virus retains its name when it infects tomatoes. Viruses are identified by a combination of symptoms, serological tests (with the use of antibodies from warm-blooded animals, usually rabbits), electron microscopy, and chemical analysis. A highly sensitive serological technique known as enzyme-linked immunosorbent assay (ELISA) is now increasingly used to detect and identify plant viruses.

The effect of viruses on the plant depends upon the sensitivity of the host cultivar and the strain of the virus. Beet curly-top, a virus disease that also affects tomatoes, results in the quick death of the plant. Some viruses produce no obvious symptoms but still cause considerable economic loss by reducing yields and performance. A combination of viruses generally complicates and increases the severity of symptoms.

Viruses are commonly insect-transmitted, many of them by aphids. A few viruses are soil-borne, and some of these are transmitted by nematodes or by fungi. Most viruses are not seed-transmitted, and even when they are, only a fraction of the seedlings become infected. Some highly infectious mosaic viruses may be transmitted by touching healthy leaves after having touched infected leaves. In laboratory assays, they are transmitted by rubbing infectious sap on healthy, susceptible leaves. Viruses are commonly transmitted through a graft union, and some pass from plant to plant through dodder, a parasitic plant.

The vegetative propagation of virus-infected plants also propagates the virus. The only satisfactory way of maintaining virus-free stocks of vegetatively propagated plants is by the perpetuation of plants that are determined to be free of the virus. Maintenance of virus-free plant stock is achieved by isolation, roguing of infected plants, and control of insect vectors (Figure 16-2). As a rule,

FIGURE 16-2. A screenhouse for growing virus-free strawberry plants. The fine screen mesh keeps out aphids, which transmit the virus. These plants are maintained as a "nuclear" source of virus-free plants for nursery growers. A screenhouse for trees is in the background. [Photograph courtesy Purdue University.]

once a plant contains a virus, little can be done to get rid of it, although heat treatment of the plant is an effective way of inactivating some viruses, and excised shoot tips (particularly if rapidly growing) may be free of virus (see Chapter 12).

Viroids

A number of viruslike diseases had long resisted many attempts at identification by conventional techniques. The infectious materials causing potato spindle-tuber disease, citrus-exocortis, and cadang-cadang (a disease of coconuts) have been shown to be low-molecular-weight RNA (about one-tenth the size of the RNA in the smallest known virus). These infectious agents represent a new class of pathogens now referred to as **viroids,** or **infectious** or **pathogenic RNA.** They may also be implicated as causal agents in some animal diseases.

Mollicutes

For many years, certain strange organisms have been known to exist that can pass through filters that trap bacteria (as do viruses) but may live on artificial media (as do bacteria). In the late 1960s, the discovery was made that these organisms cause plant diseases. Although their taxonomy is in a state of flux, some are classified as **spiroplasmas** and others are referred to as **mycoplasmalike organisms** (MLO) since they closely resemble mycoplasmas causing diseases in animals. These organisms are now implicated in several plant diseases formerly thought to be caused by viruses, particularly in a group called "yellows," which are principally spread by leaf hoppers. Examples of plant diseases known to be caused by mollicutes include aster yellows, mulberry dwarf disease, potato witches' broom, citrus little-leaf disease, and corn stunt.

The size of mollicutes varies. When grown in culture, they produce long filaments resembling fungal hyphae, hence their name, which means "fungalform." However, the filaments may break up into very small, round cells known as **elementary bodies,** ranging in size from 125 to 250 nm. These are the particles that pass through bacterial filters. From the elementary bodies mollicutes may form either branching filaments or "large cells" (500 nm) that approach the dimensions of bacteria.

Mollicutes, unlike viruses, are definitely lifelike in that they contain enzyme catalysts and energy systems, as well as genetic information in the form of DNA and RNA. Although surrounded by a membrane, they have no exterior cell wall as do most bacteria. This lack explains their apparent resistance to penicillin, which lyses (dissolves) the bacterial cell wall. However, mollicutes are inhibited by tetracycline compounds (such as chlorotetracycline); many suspected viral diseases have been proved, through the application of these chemicals, to be caused by this class of organisms.

Bacteria

Bacteria, one-celled "plants" and the smallest living organisms, are responsible for many plant diseases. Seven genera of bacteria, none of which forms spores, are plant-pathogenic. Bacteria as a group are able to enter plants only through natural openings, such as the stomata and lenticels, or through wounds. Insects are important in the transmission of bacterial diseases.

An obscure group of bacterial pathogens that are fastidious in cultural requirements, are found only in xylem or phloem, and somewhat resemble Rickettsia bacteria have recently been shown to be responsible for some well-known but previously little-understood plant diseases. They include Peirce's disease of grape and phony disease of peach.

One of the most serious bacterial diseases of plants, and the one in which bacteria were first shown to cause disease, is fireblight, a serious infection of apple and pear caused by *Erwinia amylovora* (Figure 16-3). Insects disseminate these bacteria, which penetrate the plant either through the nectar-producing glands in the flower, causing a blighting or death of the blossom (blossom blight), or through shoot terminals (shoot blight). The bacterial pathogen is also carried from infected parts to other parts of the tree by rain. The organism survives through winter in older bark lesions called cankers (Figure 16-4). The disease is extremely serious on pear and has confined commercial production in the United States to the Pacific states and the Great Lakes area. Even in these locations, however, careful control measures must be used. They involve the constant removal of blighted wood, the use of antibiotic sprays, and cultural practices that discourage rapid, succulent growth.

Symptoms of bacterial diseases include the death of tissues and the formation of galls. The soft rots common in stored fruits and vegetables are caused by pectin-dissolving enzymes produced by the bacteria. Wilts produced by some bacterial diseases (Stewart's wilt of corn, for example) are a result of vascular disturbances, specifically a "plugging" of the vascular system by masses of bacteria.

Fungi

Fungi, which cause the great majority of plant diseases, are multicelled plants. Except for some primitive types, they are characterized by a branching, threadlike (mycelial) growth. Fungi do not have chlorophyll and hence depend ultimately on green plants for their food. They may be saprophytic or parasitic, and many are both. Fungi reproduce by spores, which may be mitotically (asexually) or meiotically (sexually) produced. The life cycle of fungi is typically quite involved and comprises many different stages.

Fungi form three large, well-defined classes: Phycomycetes, Ascomycetes, and Basidiomycetes. Those whose sexual stage is not known (and that may presumably never form a sexual stage) are lumped together in a more or less artificial class (the mycologist's "trash pile") known as **Fungi Imperfecti** or **Deu-**

A

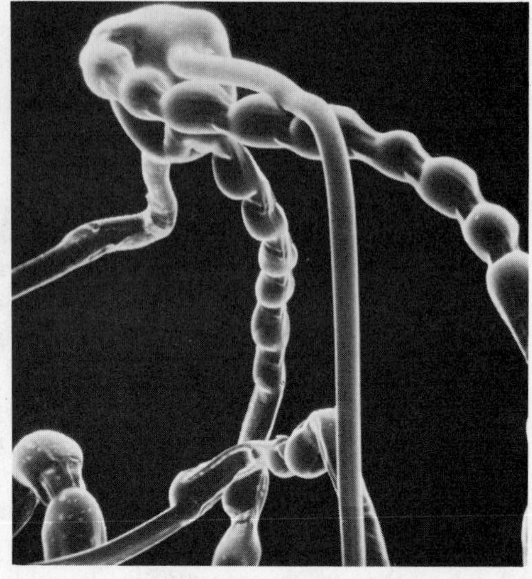
B

C

FIGURE 16-3. Fireblight, which is caused by the bacterium *Erwinia amylovora*, is one of the most devastating diseases of pear and apple. (*A*) Profuse production of bacterial strands on stem and leaf petioles of 'Bartlett' pear. The aerial strands may be smooth or beaded. (*B*) The strands contain large numbers of pathogenic bacteria embedded in a matrix. The strands are wind-disseminated, and they appear to play an important role in fireblight epidemics. (*C*) Fireblight symptoms on a 4-year-old pear tree. [Photographs (*A*) and (*B*) courtesy Harry L. Keil, U.S. Department of Agriculture; photograph (*C*) courtesy Purdue University.]

teromycetes. The **Phycomycetes** are primitive fungi whose characteristic feature is the absence of cross walls in the mycelium. Examples of well-known horticultural plant diseases caused by Phycomycetes are downy mildew of grapes, seedling damping-off (caused by a number of species), and late blight of potatoes. The **Ascomycetes,** which produce the largest number of plant diseases, are distinguished by their specialized sacs (asci), which contain the sex-

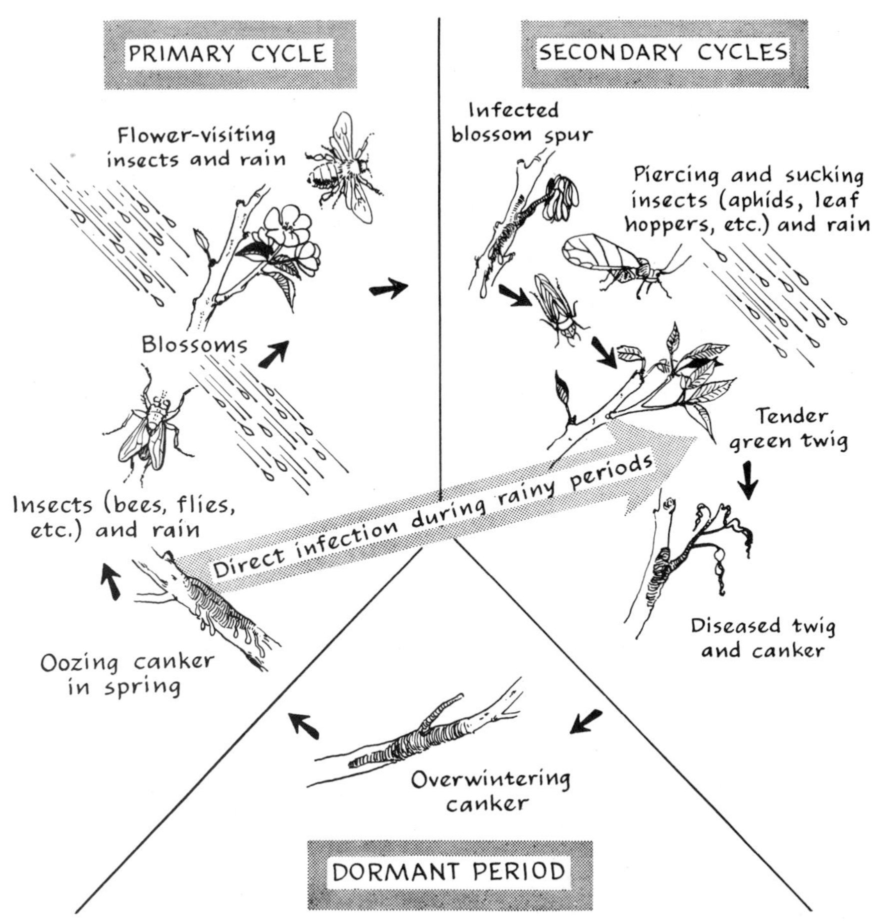

FIGURE 16-4. Life cycle of the bacterium *Erwinia amylovora,* which causes fireblight. [Adapted from E. J. Klos, Michigan State University, Extension Folder F-301, 1961.]

ual spores (Figure 16-5). Ascomycetes are responsible for the following diseases: apple scab, powdery mildews, brown rot of peaches and plums, and black spot of roses. The **Basidiomycetes,** the "higher fungi," produce a specialized sexual spore-forming structure called a **basidium.** Mushrooms, the culture of which is considered by some as part of the horticultural industry, belong to this group. Diseases caused by Basidiomycetes are among the most destructive scourges of crop plants. They are popularly called **smuts** and **rusts** because of the appearance of masses of black- or red-colored spores on the host plant, as in onion smut, corn smut, and asparagus rust. Species of rust fungi may produce many spore forms (Figure 16-6). Some rusts require two unrelated plants for the completion of their life history. Thus, different stages of the cedar–apple rust organism *(Gymnosporangium juniperi-virginianae)* infect different hosts.

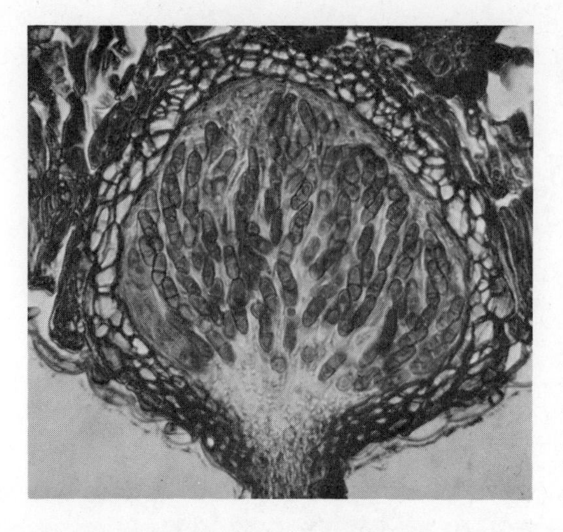

FIGURE 16-5. The fruiting body (perithecium) of *Venturia inaequalis,* which causes the apple-scab disease. Eight two-celled ascospores are contained in each ascus. [Photograph courtesy S. Maciejowska.]

Fungi are capable of entering plants by themselves, although wounds and natural openings are utilized by many. Fungal disease is spread in a great number of ways. Spores are produced in enormous numbers and are spread by water or air currents. The pathogen that causes the apple-scab disease is capable of forcibly discharging spores into the air. Mycelia growing saprophytically are a factor in the spread of soil-borne fungi. Some fungi are spread by insect

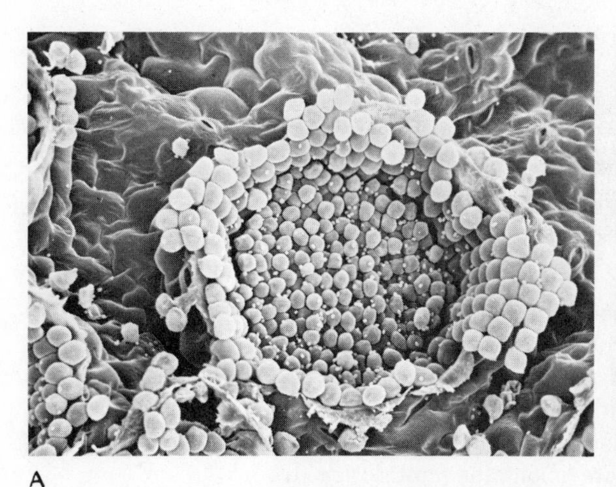

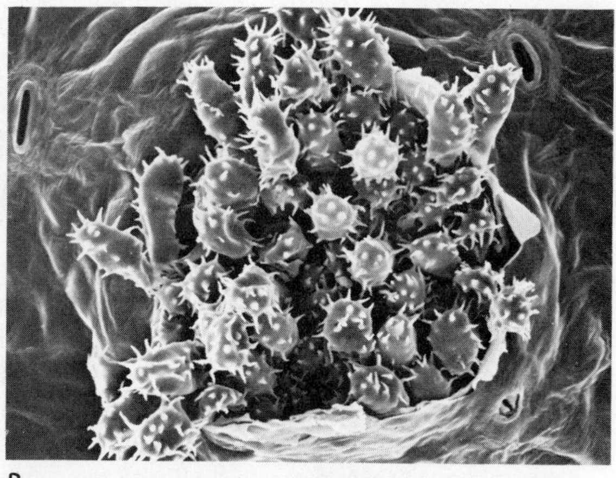

FIGURE 16-6. Fungal spores of the rust fungus *Puccinia podophylli* on May-apple: (*A*) aecial stage; (*B*) telial stage. As many as five spore forms may develop in rusts. These form in the following sequence: *pycniospores* (spermatia)—uninucleate and haploid spores of two mating types; *aeciospores*—binucleate, usually formed on side of leaf opposite the side bearing the pycniospores; *urediniospores*—binucleate, "red" spores; *teliospores*—binucleate resting spores; *basidiospores* (sporidia)—uninucleate products of meiosis. [Scanning electron micrographs courtesy H. P. Rasmussen and G. R. Hooper.]

vectors, as is the fungus that causes Dutch elm disease; others utilize wounds inflicted by insects for access, as does the fungus responsible for brown rot of plums and peaches.

Symptoms of fungal diseases are extremely varied; all parts of the plant may be affected. Plant-pathogenic fungi generally cause localized injury, although some are responsible for vascular wilts. The presence of visible (to the unaided eye) forms of the fungus, such as mycelial growth, masses of spores, or fruiting bodies, are often an integral part of symptom expression and help identify fungal diseases (Figure 16-7).

Nematodes

Nematodes are unsegmented worms (not segmented "earth worms") of the phylum Nematoda (Figure 16-8). Many species of nematodes are parasitic on plants and animals, including humans; the major portion, however, is "free-liv-

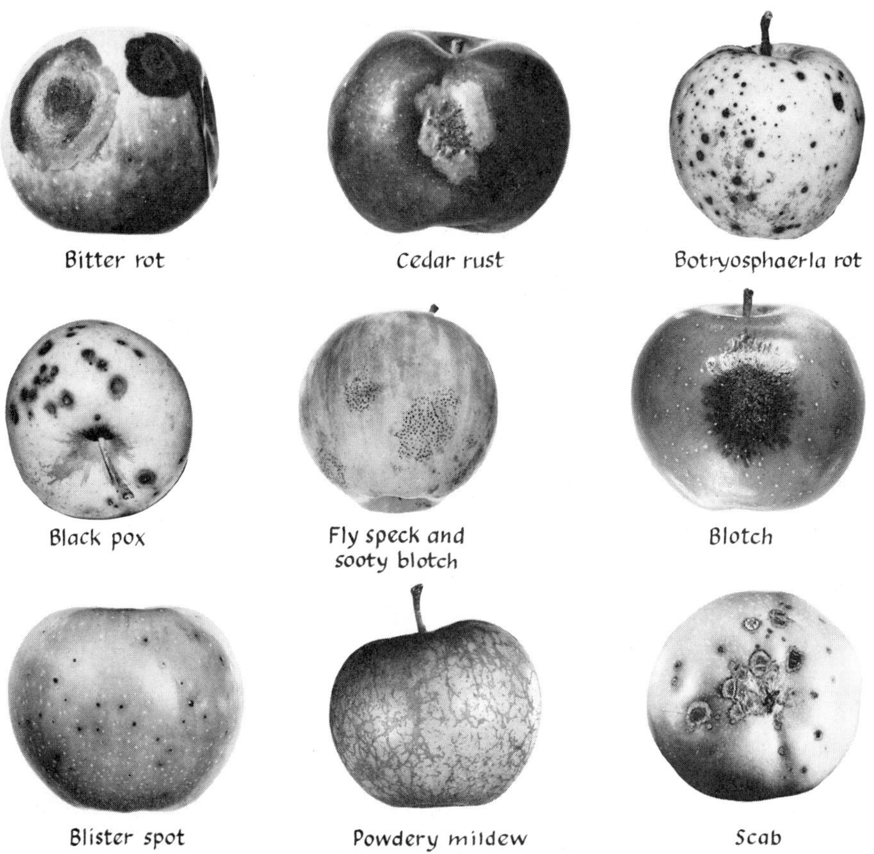

FIGURE 16-7. Fungal diseases of apple. [Photographs courtesy E. G. Sharvelle.]

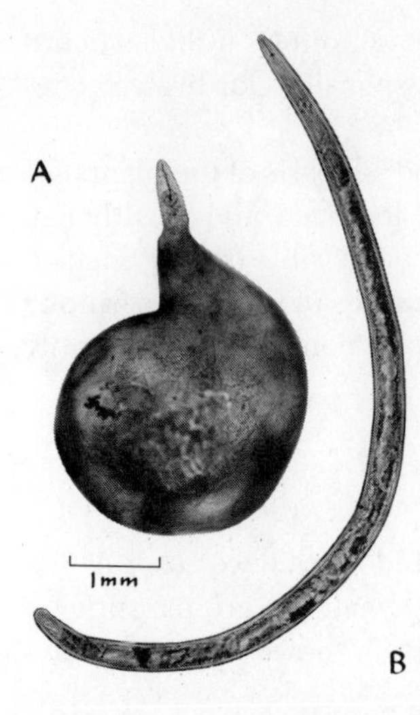

FIGURE 16-8. Nematodes of two genera, representing different morphological forms: (*A*) the swollen female of *Meloidogyne* (root-knot nematode); (*B*) the wormlike female of *Tylenchorhynchus* (stunt nematode). [Photograph by H. H. Lyon, courtesy W. F. Mai.]

ing" in the soil and represents a large group of the soil fauna. Although their size is macroscopic, they are small enough (usually less than 3 mm long) to be generally inconspicuous. The importance of nematodes in causing plant disease is becoming increasingly apparent.

The majority of nematodes that attack plants are soil-borne and generally feed on plant roots. They may feed superficially, or they may be partially or completely embedded in the root tissues. A few species of nematodes feed on the aerial portions of the plant (for example, the foliar nematode of chrysanthemum and the seed-gall nematode of bentgrass). Some are very specialized parasites and attack only a few species of plants; others have a wide host range.

Symptoms of root injury are variable. Those nematodes belonging to the genus *Meloidogyne* produce gall-like overgrowths on roots (Figure 16-9) and are known as **root-knot** nematodes. These readily observable symptoms allow positive identification of nematode injury and are used as criteria in inspection of planting stock. The infected roots eventually deteriorate, and may afford

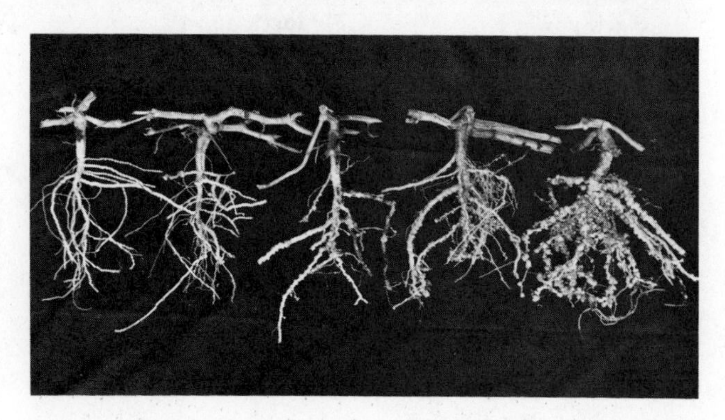

FIGURE 16-9. Increasing severity of root-knot lesion (galls) on muskmelon. [Photograph courtesy G. W. Elmstrom and D. L. Hopkins.]

access to bacterial and fungal rots. The tissues in the gall (especially the vascular tissues) become disorganized; giant cells are often formed. Thus, the above-ground symptoms appear as drought injury (that is, excessive wilting and weak, yellow growth). Other plant-parasitic nematodes, such as those of the genus *Pratylenchus* (meadow nematodes), do not form galls. Both the root damage and the above-ground symptoms resemble those of root rots. Owing to the difficulty of recognizing this pathogen, the meadow nematode may be easily transported in infected planting stock.

Arthropods

The phylum **Arthropoda** (literally, "jointed-legged") includes the invertebrate animals having an external skeleton, paired, jointed limbs, and a segmented, bilaterally symmetrical body. Arthropoda is an enormous group, and contains about 75 percent of all known animal species, of which 90 percent belong to one order, Insecta, the true insects. Almost 700,000 species of true insects are known, and the estimated number of species actually existing in the world ranges from 2 million to 10 million.

Species in two classes of arthropods, **Arachnida** (spiders and mites) and **Insecta** (true insects), are the major plant pests. Species of the class Arachnida have four pairs of legs, are wingless, and have a cephalothorax (the head and thorax are fused). Insects have three pairs of legs, almost all are winged (although some wings are rudimentary or degenerate), and all have three body regions, including a separate head. A brief classification of the arthropods showing some of the orders of true insects that attack plants is shown in Table 16-1.

The war over crop plants between people and arthropods is a continuous one. The battle lines are not clearly drawn, however. We are aided by the intense competition between insect species, and predatory insects must be regarded as beneficial. Furthermore, if we were to rid the earth of all insects, we would be much worse off than we are now, since we depend on insect species for the pollination of many of our crop plants.

Our antagonists have many built-in advantages. They include their small size, which makes them difficult to find; their power of flight; their extremely rapid rate of reproduction; and their specialized structural adaptations, which enable some species to exist in practically any location and to infest almost any plant species. Furthermore, the division of the life cycle of some insects into separate stages is an enormous advantage. This **complete metamorphosis** permits specialized structural adaptation for feeding and reproduction. The life cycle of an insect that undergoes complete metamorphosis consists of four stages: (1) **egg,** (2) **larva,** or feeding stage, (3) **pupa,** a quiescent stage in which the larva is transformed into the adult form, and (4) **adult,** or reproductive stage. Other insects, such as grasshoppers, undergo **incomplete metamorphosis,** in which the physical changes from egg to adult are gradual. The intermediate stages are known as **nymphs.**

TABLE 16-1. Partial classification of arthropods, including some of the major orders of insects that attack plants.

Class and order	Examples	Typical mouthparts	
		Larvae	Adults
Chilopoda	Centipedes		
Diplopoda	Millipedes		
Crustacea	Crayfish, lobster		
Arachnida	Scorpions, mites, ticks, spiders		
Insecta	True insects		
Gradual metamorphosis			
Orthoptera	Grasshoppers, crickets		chewing
Thysanoptera	Thrips		rasping-sucking
Hemiptera	True bugs		piercing-sucking
Complete metamorphosis			
Hymenoptera	Bees, ants, wasps	chewing	chewing-lapping
Coleoptera	Beetles	chewing	chewing
Lepidoptera	Butterflies, moths	chewing	sucking
Diptera	Two-winged flies and mosquitos	chewing	piercing-sucking-sponging

The larval stage, which often bears no obvious resemblance to the adult stage (for example, the caterpillar and the butterfly), is often the most injurious to crop plants. Thus, many of the descriptive names (tomato hornworm, apple maggot) given to insect pests refer to the larval form. The terms for the larval stages of the insect orders are not too specific. In general, the name **maggot** refers to the larval stage of species of the order Diptera (flies, mosquitos, gnats). **Miners** are dipteran larvae that tunnel within leaves. **Caterpillars** are the larval stages of species of the order Lepidoptera (butterflies and moths). The term **grub** is used for some of the soil-borne larvae of species of the order Coleoptera (beetles), although the name is also applied to any other soil-borne larvae. The larvae of "click" beetles are known as **wireworms.** The larval and adult forms of beetles that infest grains and seeds are referred to as **weevils.** The name **slug** may be applied to any slimy larva, specifically to larvae of the Hymenoptera (bees, ants, and wasps). They should not be confused with true slugs, which belong to the phylum Mollusca, along with snails and clams. **Borers** are larvae, usually of moths or beetles, that tunnel within roots or stems. The term *worm* is also used to refer to insect larvae, but this is a misnomer and should be avoided.

Insects (and mites, which we shall also consider here) injure plants in their attempts to secure food. The damage they cause directly and indirectly is enormous and varied. It includes the destruction of plants, either in whole or in part, by chewing insects; the debilitation of plants by sucking insects; the spreading of such plant pathogens as viruses, bacteria, and fungi; and the contamination of plant products by the decomposed bodies or excreta of insects.

Insects feed on plants in two distinct ways. They either tear, bite, or chew portions of the plant (**chewing insects**) or pierce or rasp the plant and suck or lap up the sap (**sucking insects**). Specialized mouth parts have been evolved for these two basic feeding patterns.

The chewing insects, adults or larvae, eat their way through the plant, riddling it with holes and tunnels (Figure 16-10). Leaf-chewing insects that do not eat the tougher vascular portions may completely skeletonize the leaves; others, less selective, may devour the entire plant. Injury caused by chewing insects that feed externally is seldom confused with anything else. Damage done by chewing insects that girdle the plant, or those that feed internally or on roots, is not as apparent. The internal feeders gain entrance to the plant from eggs deposited in the plant tissue or by eating their way in soon after being hatched on or near the surface of the plant. These internal feeders are almost impossible to control once they have entered the plant. Control consists in destroying them in their external stages. The symptoms of infestation by the

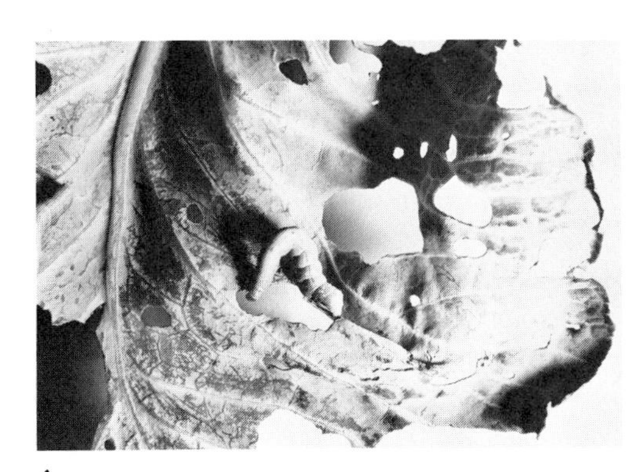

A

B

C

FIGURE 16-10. Insect damage: (*A*) cabbage looper on cabbage; (*B*) grasshopper on corn; (*C*) tomato hornworm feeding on tomato fruit. [Photographs by J. C. Allen & Son.]

peach-tree borer, which may enter the tree soon after hatching, are typical of the symptoms caused by internal insect feeders, namely, a weakened, devitalized growth of the tree and a yellowing of the foliage. A gummy exudate may be observed where the borer has wounded the trunk. Peach or plum trees infested with borers often die in a few years.

Injury caused by sucking insects (aphids, scale insects, leaf hoppers, and plant bugs) results in a distinctly different type of symptom—the curling, stunting, and deforming of plant parts, usually the stem terminals. Spotting, yellowing, and a glazed appearance of the leaves are also common symptoms. The small size of red spider mites, which have sucking mouth parts, makes them difficult to see without a magnifying glass. They are often not diagnosed until they form conspicuous webs, by which time their damage may be extremely severe.

Another symptom caused by chewing and sucking insects alike is the formation of overgrowths called galls. These are formed by an abnormal growth of plant tissue in response either to the feeding insect or merely to the presence of eggs deposited in the tissue. These galls, suggestive of cancerous growth, may be quite elaborate and structured. Some appear to do little damage to the plant; others are obviously quite injurious.

Birds, Rodents, Rabbits, and Deer

Among the vertebrates (backboned animals), birds, rodents, and deer are considered the greatest pests. Birds may become quite troublesome in such fruit crops as grapes, blueberries, and cherries. They may do severe damage to grapes; even a few pecked fruits permit the introduction of insects and rot, which may mean ruin of the whole cluster.

Rodents (including mice, voles, and moles) and rabbits are among the most serious orchard pests. They feed in winter and early spring and often completely girdle fruit trees, especially apples. Unless bridge grafting is done promptly, even large trees may be killed outright. The tunnels of moles can be a real nuisance in lawns and gardens.

In areas where they are naturally abundant or where their population is unchecked, deer may prove quite damaging to orchard and nursery stock. They are most troublesome during the winter, when their natural browse is in short supply and when the horticulturist's vigilance is at its low ebb.

The Disease Cycle

The disease cycle includes all the series of sequential changes of the pathogen and the plant in the course of the disease. It involves the life cycle of the plant as well as that of the pathogen.

Life Cycles

The **life history** of an organism includes all the diverse forms and stages through which it passes. The life history of higher organisms, including that of humans, is synonymous with the sexual cycle. Rather than the sequential steps from "womb to tomb," it is more correctly "gametes to fertilized egg to adult to gametes." The life history of lower organisms may be considerably more complex, and it is often made up of a number of continuous stages of existence called **life cycles.** These stages may involve a number of asexual cycles within the sexual cycle. In addition, many lower organisms, such as bacteria, do not ordinarily pass through the sexual stage but exist as a single continuous form or utilize asexual spores exclusively, as in the members of the fungal class Deuteromycetes (Fungi Imperfecti).

In temperate areas, the life history must adjust to the seasonal cycle. Most pathogens pass through the winter in a stage of inactivity or dormancy. This overwintering stage, although usually specific for a particular organism, may be any stage in the life history. With the advent of spring the cycle resumes. The first cycle initiated at this time is known as the **primary cycle.** Subsequent cycles within the year are all known as **secondary cycles.** In areas where the change in seasons is not distinct—that is, where there is neither a sharp temperature differential nor separate periods of wet and dry—the pathogenic cycle may be a continuous secondary cycle. Greenhouse pests often live in this cycle.

In reference to the disease, the life cycle of the pathogen can be divided into two phases: a **pathogenic phase,** in which the pathogen remains associated with the living tissues of the principal host plant; and an **independent phase** not associated with the living plant, in which the organism may be saprophytic, dormant, or pathogenic to another plant. The relative length of these two phases varies greatly. For example, the independent phase of plant viruses lasts only while they are being transported from plant to plant via insects. On the other hand, the independent stage of some pathogens may be long. Some fungi, such as *Verticillium,* live saprophytically as a natural part of the soil flora and become pathogenic only with the introduction of a suitable host plant. Similarly, some animal pests of plants may have long independent phases. Mouse damage to orchards, for instance, is done usually only in brief periods in late winter.

Pathogenic Phase

The pathogenic phase may be divided into the stages of inoculation, incubation, and infection. **Inoculation** consists in the transference of some form of the pathogen (the **inoculum**) to the plant. The pathogen may be transferred under its own power, as are insects, or by some agent of inoculation. The important vectors are wind, water, insects, and humans. In gall-forming insects, the adult form is the vector and the egg is the inoculum. **Incubation** includes all activities of the pathogen from the time it actually enters the plant until the plant reacts to it—the **infection** stage. The significant stages with respect to control are inoculation and incubation. By the time the plant reaches the infection stage, the damage has been done.

It is not possible to generalize on the relationship between the pathogenic stages and the life cycle. For example, the primary cycle of the fungus *Venturia inaequalis*, which causes apple scab, is formed when spores (ascospores) are produced from the sexual stage, which overwinters on leaves under the tree (the saprophytic stage). The secondary cycles are formed when asexual spores (conidia) are produced from primary infection. The life history of this pathogen unfolds in a single year and contains several asexual cycles (Figure 16-11).

The codling moth, which infests apple, pear, and several other fruits, demonstrates a different type of cycle. The life history may occur several times within a single year. The full-grown larval stage overwinters in silken cocoons in trees or on the ground; the pupal stage is formed in midwinter. In the spring, the grayish moths emerge and lay their eggs on the upper sides of leaves, twigs, or spurs. The eggs hatch in 6 to 20 days, and the larvae chew their way into the young fruit. This is the primary cycle. After becoming full grown in 3 to 5 weeks, the larvae crawl out of the apple, drop to the ground, and spin cocoons.

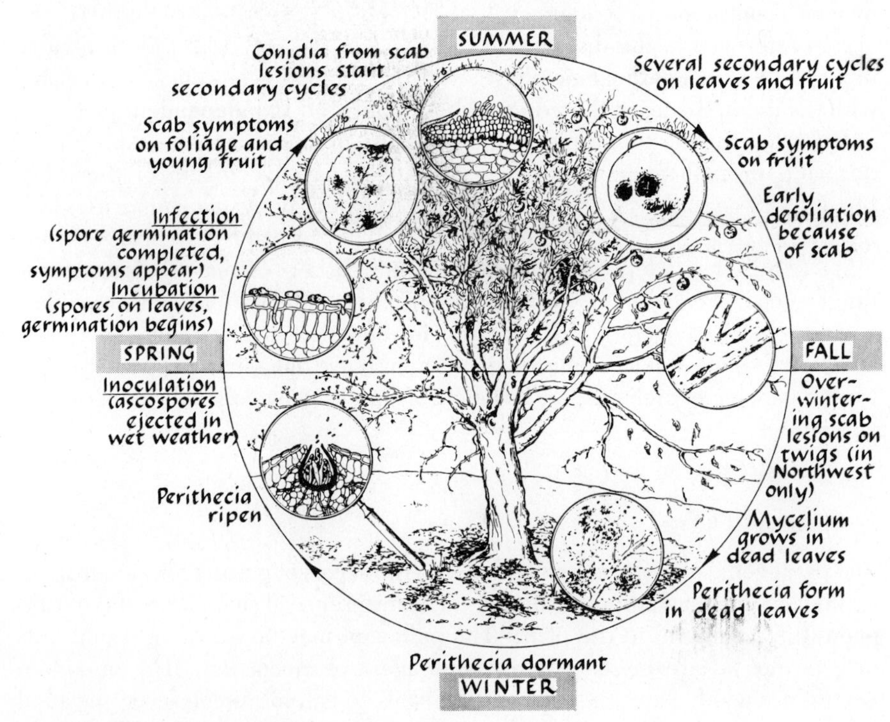

FIGURE 16-11. The life history of the apple-scab fungus (*Venturia inaequalis*). [From L. Pyenson, *Elements of Plant Protection*, Wiley, 1951.]

The moths may emerge to produce secondary cycles. There may be two or three generations, some of which may be incomplete, depending upon the latitude.

Principles of Disease Control

The study of pathogenic plant diseases in the United States is usually divided between two disciplines: **plant pathology,** which deals with diseases of viral, bacterial, and fungal origin; and **entomology,** the study of arthropods. (Nematodes may be claimed by both, but mice and other vertebrates are claimed by neither.) The applied phase of these biological sciences is directed toward the **control** of these plant pests. This approach involves both alleviating the injury and preventing the spread of the pathogen. In Europe and increasingly in the United States, the control of plant pests is considered a distinct agricultural discipline called **plant protection.**

There are two approaches to the control of pathogenic plant diseases. One is directed at the pathogen and uses techniques that prevent or restrict the pathogen's invasion of the plant. These techniques interfere at some point with the successful completion of some stage in the disease cycle. The other approach is directed toward the plant's ability to resist or at least tolerate the intrusion of the pathogen. It should be obvious that both methods of control depend on an intimate knowledge of the disease cycle. To control some diseases successfully, many different methods must be utilized; effective control requires experience, vigilance, and persistence.

The economic value of pest control is the benefit obtained by the control measure less the cost of the control. Predicting the economic benefits of any control measure is particularly hazardous because control, to be of value, must usually be applied in anticipation of the pest. Thus, many control measures can be thought of as a form of disaster insurance. In agriculture, the use of expensive methods of pest control is feasible only with high-value crops. Consequently, horticulture bears a disproportionate share of the cost of pest control.

Legal Control

The separation of the pathogen and the plant may be accomplished by legal methods. For example, **quarantines,** which prohibit the importation of certain plants, may effectively eliminate, or at least slow down, the introduction of new pests. Where outright quarantines are not feasible or practical, subjecting plant shipments to inspection helps to prevent the spread of some pests.

Cultural Control

A number of techniques may be employed to reduce the effective population level of a pathogen. They include the elimination of diseased plants or seeds (**roguing**); the removal of infected parts of the plant (**surgery**); or the removal of plant debris, which may harbor some stage of the pathogen (**sanitation**). Cultural practices that alternate plants unacceptable to the pathogen (**rotation**) may starve it out.

The effective population of the pathogen may be reduced by any method that renders the environment unfavorable to increases in its population, such as draining land to discourage water-loving fungi, pruning to reduce foliage density and to increase the rate of drying, and varying temperature and humidity conditions in the greenhouse.

Physical Control

When individual plants are valuable enough to warrant the expense, physical barriers may be employed to protect them from larger pests. (Such barriers are also effective in controlling those microscopic pathogens borne by insect vectors.) Examples are screening to keep out insects or birds, guards to protect trunks of apple trees from injury by mice, and the traditional garden fence to keep out rabbits. In Japan, developing fruits are sometimes enclosed individually in bags to protect them from insect injury.

Physical methods may be used to eliminate certain pathogens selectively or to protect the plant against the intrusion of the pathogen. Such control requires the use of techniques that will destroy or repulse the pathogen but will not damage the plant. An example of such a method is the use of heat. Immersion in hot water (110 to 122°F, or about 44 to 50°C, for 5 to 25 minutes) is a treatment for destroying pathogens that infest seeds and bulbs. Hot-water treatment is used to control some fungal diseases of grains, a bacterial disease (black rot) of crucifers, and nematodes in dormant strawberry plants. Heat-treating strawberry plants has also been effective in inactivating some viruses. Steaming of potting soils is a common control for many soil pests.

The action of a strong stream of water (**syringing**) has some value in reducing the infestation of spider mites in home or greenhouse plantings. The use of firecrackers or noisemakers to discourage birds from small fruit plantations is successful to a limited extent (although often no more effective than the old-fashioned scarecrow). In many areas, however, these are subject to antinoise ordinances.

Traps may also be used to control the pest population. Substances to lure the pest (**attractants**) are incorporated in the trap. An example is the use of "blacklight" (light of wavelengths between 340 to 380 nm) to attract insects to the trap, where they are killed by some toxic solution or by electrocution. The use of traps is generally not efficient for mice, but it has proved successful for rabbits in limited areas.

Chemical Control

The concentration of the horticultural industry has resulted in the concentration of plant pests. As a result, the entire industry is now almost completely dependent upon the use of chemical control (Figure 16-12). Commercial growers of practically all horticultural crops rely on complete "schedules" of chemical application and utilize many different compounds. The agricultural chemical industry is a vigorous one and new materials are continually being released.

BIOLOGICAL ACTION OF CHEMICALS. **Pesticide** is the generic name for all chemicals that control pests. They are usually toxic to the pest at some stage of its life cycle. (**Repellants** are compounds that may not be actively poisonous but that make the crop plant unattractive to animal predators by virtue of their odor, taste, or other physical properties. They are included in the legal definition of pesticides.) Pesticides are classed according to the organisms they control; for example, bactericide, fungicide, rodenticide. (Herbicides, chemicals that kill plants, will be discussed separately later in this chapter.)

Pesticides are generally selective in their action. Thus, chemicals that are fungicidal are usually not insecticidal. But there are exceptions; for example, Bordeaux mixture (100 gallons water; 10 pounds hydrated lime; 6 pounds copper sulfate), which is primarily fungicidal, also has some value as an insect repellant. Pesticides that are toxic to a broad spectrum of organisms, including the crops they are meant to protect, must be used in the absence of the plant, as in preplanting treatments of the soil.

It should be reemphasized that pesticides are not necessarily toxic to all stages of a particular pest, nor is it necessary that they be. Usually, a particular stage of the life cycle is especially vulnerable to chemical attack. This "weak link" may be a germinating fungal spore, the young larval stage of an insect, or the insect vector of a virus disease. For example, it is much more difficult to kill fungi after they have extensively entered the plant. However, materials that will kill or prevent spores from germinating on the plant may block further inoculation and, in effect, control the disease.

Pesticides are classified as systemics or nonsystemics. **Systemics** are actually absorbed by the plant, and may be translocated within the plant, rendering the plant itself toxic to the pathogen. Systemics must of course be restricted in their use to nonedible plants unless they break down within the plant before it is to be consumed. Much more common are the **nonsystemics**, which merely coat the plant with a toxic substance. This distinction between systemics and

FIGURE 16-12. Control of summer diseases of apple with fungicide application. Fruits on right received no spray and show a number of summer diseases, including Brooks spot, black pox, and sooty blotch. Fruits on left were grown under a full-season program of captan sprays. [Photograph courtesy J. R. Shay.]

nonsystemics is not clear-cut, however; many compounds whose main action is on the surface of the plant may be absorbed to some degree.

Insecticides are classified, according to their action, as stomach poisons or contact poisons. **Stomach poisons** are generally used against chewing insects. The poison must be ingested by the insect in order to do its work. Since sucking insects feed on plant juices, they are usually not affected by stomach poisons (unless, of course, they are also systemics). They are usually controlled by **contact poisons,** which kill by penetrating the insect body directly or through breathing or sensory pores.

A great number of compounds, both organic and inorganic, have been used to kill plant pests. The inorganic materials were usually metallic salts of such metals as copper, mercury, lead, and arsenic. A few of the organic compounds occur naturally, such as nicotine, pyrethrum, and rotenone. However, most of the present-day organic compounds are now completely synthetic, as are DDT, the organic phosphates, and the carbamates. A list of some of the important pesticides is presented in Table 16-2.

TABLE 16-2. Classification of some important pesticides (excluding herbicides) by chemical composition.

Class of material	Examples	Major type of action
Inorganic		
Arsenic	Lead arsenate	Insecticide
Copper	Copper sulfate	Fungicide, bactericide
Sulfur	Lime sulfur	Fungicide
Organic		
Naturally occurring		
Antibiotics	Streptomycin, cycloheximide	Bactericide
Nicotine	Nicotine sulfate	Insecticide
Pyrethrum		Insecticide
Rotenone		Insecticide
Oils	Diesel oil	Insecticide
Synthetic		
Benzene compounds	Chlorothalonil	Fungicide
Benzimidazoles	Benomyl, thiophanate	Fungicide
Carbamates	Maneb, zineb	Fungicide
	Carbaryl	Insecticide
Carbon tetrachloride		Fumigant
Chlorinated hydrocarbons	DDT, methoxychlor,	Insecticide
	DD	Nematocide
Heterocyclic compounds	Captan, folpet	Fungicide
Organophosphates	Malathion	Insecticide, miticide
Pyrethroids		Insecticide
Quinones	Dichlone	Fungicide

APPLICATION OF PESTICIDES. Chemicals are applied to plants in a number of formulations of the liquid, solid, or gaseous state, and consequently by a variety of methods. The aim is to obtain uniform coverage at a controlled rate. For highly active compounds, the rate of application of the active ingredient may be extremely low; therefore it is difficult to maintain adequate coverage without special equipment.

It is seldom practical to apply chemicals in their undiluted form. Consequently, their distribution is carried out by the use of an inert carrier. The carrier may be a solid, such as talc, in which case the material is applied as a **dust.** Dusts have the advantage of lightness, but there is some question as to the persistence of the material. Moreover, some materials (for example, oils) cannot be applied in this manner.

The disadvantages of dusts are obviated by the use of a liquid carrier, usually water, in which the chemical is dissolved, suspended, or emulsified. The material is applied under pressure in droplets of various sizes (Figure 16-13), but usually in the form of a **spray** (Figure 16-14). The disadvantages of

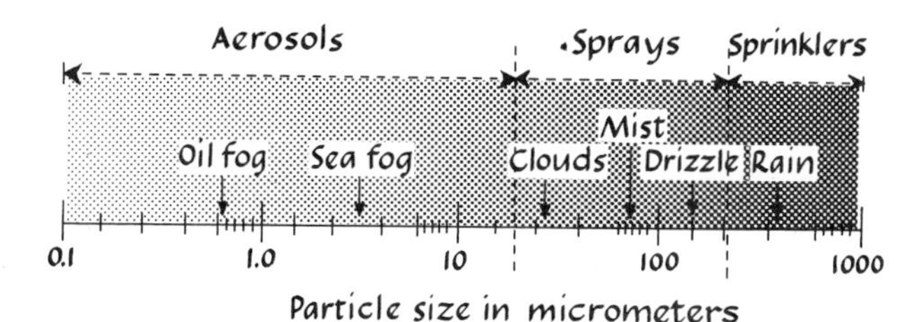

FIGURE 16-13. The drop-size spectrum.

FIGURE 16-14. Application of fungicides to tomatoes with a boom-type sprayer. [Photograph by J. C. Allen & Son.]

spraying lie in the bulk and weight of the water carrier. The trend in spraying is, therefore, to use high concentrations of the active material and achieve dispersion by a blast of air (Figure 16-15).

Another difficulty in using a water carrier is that plant surfaces, since they are heavily cutinized, are water repellant, thus causing the spray to form droplets on the surface instead of a continuous film. As a result, the deposited chemicals are irregularly dispersed. This uneven distribution may be overcome through the use of **wetting agents**—surface-active chemicals that break the surface tension. Other additives, known as **stickers,** improve the retention and persistence of the chemical application.

Volatile substances that are pesticidal in the gaseous state are known as **fumigants.** Some substances applied as sprays or dusts may also be used as fumigants; for example, nicotine sulfate may either be applied as a spray or volatilized and applied as a fumigant. Fumigants are, as a rule, extremely toxic to all life, and the problem in their use is to obtain selectivity. Thus, they have been widely used in the absence of plants to control soil pests (Figure 16-16). Some fumigants may be stored under pressure in the liquid state. Others may be solids at normal temperatures, and these are usually volatilized by heat. Such substances as paradichlorobenzene (mothballs) sublimate at normal temperatures and are applied as a solid. Soil fumigants are now being synthesized that will volatilize when wet and may be applied to the soil in granular form. Increasing in importance for enclosed areas, such as greenhouses, are **aerosol bombs,** from which the toxic chemical is dispensed by means of a gaseous carrier. In **fogging,** the active ingredient is dispersed by heat volatilization in kerosene or some other petroleum oil.

PROBLEMS OF CHEMICAL CONTROL. Although chemicals have greatly diminished the problems of pest control, they have by no means solved them. Furthermore, the use of chemical control raises many new problems.

FIGURE 16-15. Orchard speed sprayers use a blast of air as the carrier for highly concentrated sprays. [Photograph courtesy Farm Equipment Institute.]

A

B

FIGURE 16-16. (*A*) Soil fumigation is achieved in a single pass with a tractor-mounted applicator that lays a polyethylene tarp and injects a mixture of methyl bromide and chloropicrin. (*B*) The plastic strips are glued together to keep in the fumes. The polyethylene tarp is removed after 48 to 72 hours. [Photographs courtesy Bernarr J. Hall and Victor Voth.]

1. *Residues on edible products.* Severe restrictions are imposed upon the use of chemicals in order to avoid health hazards. The quantitative determination of residue and the determination of limits of safety for the pesticide operator, as well as for the consumer, must be established for each chemical by the manufacturer. The controversy over residues involves the question of the potential health hazards of minute quantities of chemicals that are toxic at higher dose levels. The solution to the problem consists in choosing pesticides that break down before consumption or are metabolized harmlessly by the body, and in timing their application such that residues may be eliminated entirely. Some chemicals leave soil residues that may interfere with subsequent plant growth.
2. *The technical problems of application.* The timing of application must be extremely precise to be effective as well as to eliminate residues. Usually the pesticide must be applied before the trouble appears; once the symptoms are obvious, it is often too late for control. Thorough application in tree

crops often requires special pruning procedures. Owing to chemical reactions that occur between certain chemicals, compatibilities must be taken into consideration when more than one pesticide is applied at one time.

3. *Spray injury to the plant.* Care must be taken to ensure that the cure is not more injurious than the disease. Spray injury can reduce yield as well as spoil the appearance or "finish" of the product. Some russetting of fruit can be traced to cuticle damage by pesticides.
4. *The development of genetic resistance by the pest.* Natural variation among pests produces types that may be resistant to the pesticide. Because of the enormous rate of reproduction of many pests, the pesticides act as a screening device for the selection of resistant types. When DDT was introduced in the 1940s, many entomologists had hopes of victory in the war against insects. Within 10 years, however, mosquitos, house flies, and many other insects had developed a resistance to DDT.
5. *The disturbance of the biological balance.* When one pesticide is used in place of another, pests that formerly appeared to cause little damage may begin to assume major importance: Pests controlled by the former pesticide may not be by the new one. For example, the older methods for controlling apple scab utilized lime-sulfur, which was also effective against powdery mildew. But, because of its adverse effect on fruit finish, lime-sulfur was replaced by organic fungicides. Although these fungicides have proven very satisfactory for scab control, their use has resulted in severe outbreaks of mildew that have necessitated additional control measures.
6. *Pollution.* Widespread attention was focused on pesticides as a source of soil and water pollution with the publication of *Silent Spring* by the biologist Rachel Carson in 1962. Carson argued that, although pesticides had been created to rid agriculture of undesirable organisms, the extent of their harm was widely spread and often unpublicized, and that the substances themselves should be considered undesirable. Residues of chemicals were found remaining in the soil long after their application, concentrating in some plants, and passing into the groundwater and ultimately into rivers and oceans. Through the 1960s, a controversy over the polluting effects of chlorinated hydrocarbons—DDT is the best-known example—and organic phosphates became public. Opponents of the use of DDT, led by ecologists and conservationists, cited disturbing evidence of its dangers and wanted its use outlawed. DDT is an extremely persistent fat-soluble molecule that is strongly concentrated as it is passed along food chains. Concentrations of DDT in the food of fish-eating birds have been responsible for the failure of those birds to produce eggs with shells strong enough to survive the incubation period.

Defenders of DDT pointed out that, over the years, a good many materials far more noxious than DDT had been released into the environment, and that it had not been proven that DDT has the deleterious effects in humans that it has in other organisms. Defenders also claimed that there were irrefutable benefits to DDT use (such as malaria control), and that its inexpensiveness and effectiveness made it unique.

The DDT story is not yet closed: Some of the nonpersistent substitutes for the persistent pesticide are extremely poisonous to applicators and handlers. Moreover, in certain areas of the world where the use of DDT has been suspended, the incidence of malaria has increased. Nevertheless, federal agencies have extended the prohibition of pesticides to include not only DDT but many other chlorinated hydrocarbons and related materials.

The problem of pesticides as pollutants is real. The necessity for nonpersistent, narrowly specific chemicals, harmful to pests and harmless to other living things, is clearly evident. Many agriculturists and horticulturists have concluded that complete reliance by pest controllers on chemicals is a dangerous concept and a wistful dream. The present feeling is that integrated control is required, with reliance on greater genetic resistance to pests and biological control, as well as judicious use of nonpersistent, narrowly specific chemical materials that are not deleterious to human health. This philosophy of pest control is known as **integrated pest management** (IPM). It strives to maximize pest control and to minimize damaging consequences of control measures by relying on a balanced, integrated strategy of pest management.

Biological Control

> So naturalists observe, a flea
> Has smaller fleas that on him prey,
> And these have smaller still to bite 'em,
> And so proceed *ad infinitum.*
>
> JONATHAN SWIFT

Biological control utilizes direct competition between organisms. Certain insects feed on other insects (Figure 16-17). In addition, there are bacterial diseases of insects, and there are virus diseases of bacteria. This competition among organisms may be directed toward the control of plant pests by the introduction of a natural parasite or predator of the pest. Thus, spores of *Bacillus thuringiensis*, a natural pathogen of caterpillars, are used as a spray material.

An example of successful biological control is the introduction of the vedalia beetle into California from Australia by Albert Koebele in 1888. This beetle feeds upon the eggs and larvae of the cottony-cushion scale, a serious insect pest of citrus. It successfully controlled the scale until the use of DDT became prevalent in the late 1940s. Then, injury to the vedalia beetle, apparently caused by the DDT, upset the biological balance and resulted in the first outbreaks of cottony-cushion scale since 1890.

A

B

FIGURE 16-17. Some insects parasitize other insects. (*A*) Cocoons of braconid wasps emerging from the body of a tomato hornworm. The cocoons contain the pupal stage of the wasp, which lays its eggs on the body of the hornworm. (*B*) The potato-beetle killer (*Perillus bioculatus*) attacking a larva of the Mexican bean beetle. [Photograph by J. C. Allen & Son.]

Biological control requires an organized attack on a pathogen. The encouraging or importing of insect predators, such as the praying mantis, by individual growers has had at best only very limited success.

The advantage of biological control is that, once put into effect, it generally works without further human interference. However, the biological balance is often more apparent than real. The upsetting of this delicate balance, as by random environmental fluctuations, is a perfectly natural phenomenon. Commercial horticulture, which by its nature disturbs the natural biological pattern, cannot afford the risk inherent in biological control and at present depends on chemical control as its main weapon against plant pests. Biological control, however, is an attractive measure, and one that is being given increasing attention.

The control of the screwworm, a severe pest of livestock, by utilizing artificially reared insects made sterile by being exposed to powerful radiation, has opened up a new approach to biologic control. Normal females are "monogamous" and mate only once. When mated to an irradiated male, they will produce only sterile eggs. The basis of the control is the relatively low population of reproductive adults during the winter. If the irradiated sterile males released continually outnumber normal males, the number of fertile eggs will continually decrease (regardless of the mating habits of the female). The greater the proportion of irradiated males over normal males, the faster will be the control. The success of this program has led to its use as a method for controlling plant-attacking insects, such as Mediterranean fruit fly.

Physiological Alteration of the Host

Plants do not have an antibody mechanism, as animals do, that can be utilized to resist disease. Thus, they cannot be made immune by vaccines. However, the physiology of the plant can be altered to affect the plant's ability either to resist invasion by the pathogen or to overcome its deleterious effects. For example, many vascular wilts caused by the fungus *Verticillium alboatrum* (for example, verticillium wilt of maple) can be compensated for by vigorous growth of the plant. Applications of fertilizer to increase vigor causes the plant literally to outgrow the pathogen. The reverse technique is utilized in fireblight of pears. Because infection and growth of the bacterium causing the disease are extremely rapid in fast-growing, succulent shoots, one method of control is achieved by slowing down rapid growth of the tree by eliminating excessive nitrogen fertilization or extensive pruning. The direct action of inorganic nutrients gives protection in some instances. For example, clubroot of cabbage appears to diminish in severity when the ratio of calcium to potassium in the soil is decreased. The effect of various levels of nutrients on disease resistance has not, however, been intensively investigated.

Genetic Alteration of the Host

The innate ability of the plant to avoid the injurious effects of the pathogen is the ideal method of control. This **genetic resistance** varies from complete absence of injury (**immunity**) through various degrees of **partial resistance.** The lack of resistance is referred to as **susceptibility. Tolerance** is a type of resistance in which the plant suffers infection and some injury, but is able to live with it without serious impairment.

Examples are known of plant resistance to viruses, bacteria, fungi (Figure 16-18), nematodes, and insects (Figures 16-19 and 16-20). The nature of plant resistance lies in the structural alterations or biochemical effects that either prevent or discourage intrusion and persistence of a particular pathogen. Some plants have resistance to whole groups of pathogens; others have resistance

FIGURE 16-18. Resistance of lima bean to downy mildew caused by *Phytophthora phaseoli:* The plants in the middle are 'Early Thorogreen' lima bean, a cultivar susceptible to the mildew. The healthy plants on either side are of the cultivar 'Thaxter.' [Photograph courtesy USDA.]

FIGURE 16-19. Corn earworm larvae cause less feeding damage and develop more slowly on the resistant sweet corn experimental hybrid 'M4399' (*left*) than on the susceptible 'Iobelle Hybrid' (*right*). [Photograph courtesy E. V. Wann, USDA.]

only to a particular species, or race, of pathogen. Where pathogen and plant are closely adapted to each other, a close relationship exists: The plant evolves a genetic resistance and the organism evolves the genetic ability for violating or overcoming this resistance. The spontaneous origin of new races of pathogens is one of the major problems the plant breeder faces in attempting to incorporate genetic resistance into an improvement program.

The combination of resistance and horticultural quality is one of the main objectives of the plant breeder (see Chapter 17). Resistance may be incorporated in the whole plant or, when the plant is composed of separate, grafted components, in part of the plant. Thus, resistance to root pests, such as the woolly aphid of apple, may be incorporated into a grafted tree through the use of an aphid-resistant rootstock. Similarly, fireblight-susceptible cultivars of pear are often grafted to a framework of the resistant 'Old Home' cultivar. This prevents an infected limb from destroying the whole tree.

FIGURE 16-20. (*A*) Muskmelon plants exposed for 8 days to mass infestation by the melon aphid, *Aphis gossypii*. The plant on the left is resistant to the aphids. Adult aphids on such resistant plants (*B*) grew slowly and were stunted as compared with aphids on susceptible plants (*C*). [Photographs courtesy G. W. Bohn, A. N. Kishaba, and H. H. Toba.]

WEEDS

Weeds as Pests

A weed may be defined as any plant that is undesirable. According to this definition, any crop plant that is out of place may be termed a weed. More typically, however, the term refers to certain naturally occurring, aggressive plants that are injurious to people or to agriculture. The degree of undesirability of weed species varies greatly. Extremely noxious weeds, if left unchecked, may completely dominate crop plants. Crop losses are usually the result of competition for light, water, and mineral nutrients. Weeds are also indirectly responsible for crop losses because they harbor other plant pests, such as viruses, fungi, and insects. In addition, weeds may lower the quality and economic value of crops. A horticultural example is the lowered quality of peppermint oil when the crop is contaminated with weeds. Because of their rank growth and unsightliness, weeds are a perpetual nuisance in turf. They represent a safety hazard along roadsides and railroad right-of-ways, and they often clog irrigation ditches and streams. Finally, poisonous weeds (such as poison ivy and poison oak) directly affect the health and comfort of people and livestock, and the pollen of such plants as ragweed is a source of misery to the millions who suffer from hay fever. The annual cost of weed control in the United States exceeds the combined losses from all other types of plant pest. The fierce competition offered by certain weeds is due to a combination of their prolific reproductive capacity and vigorous, exuberant growth.

Reproductive Capacity of Weeds

In general, the destructive power of weeds is due to their sheer number. Some weeds produce seeds in enormous quantities. Weed species differ greatly, however, in the number of seeds they produce: A single plant of the wild oat produces about 250 seeds, whereas a plant of the tumbleweed produces several million. One study of 181 perennial, biennial, and annual weed species reported an average of more than 20,000 seeds per plant.

The seeds of certain weed species have become structurally adapted for dispersal by wind, water, or animals. For example, many weeds have hard seed coats and can remain viable when passed through the digestive tract of an animal. People are among the chief disseminators of weeds through the shipment of crop seeds and plants. Most of the noxious weed species in North America are native to other parts of the world.

The large number of seeds and their efficient dispersal only partly explain the high reproductive capacity of weeds. (Many crop plants that are also prolific seed producers do not become weedlike.) The high reproductive capacity of weeds is particularly effective because of extended seed viability coupled with delayed germination brought about by dormancy. The failure of a weed seed to germinate may be due either to natural dormancy or to induced dormancy. Natural dormancy was discussed in Chapter 5. **Induced dormancy** is brought about by environmental conditions that limit germination. Seeds buried deep in the soil by tillage may lack either sufficient oxygen or the light stimulus necessary for germination. The viability of weed seed buried in the soil may be extremely long in contrast to the relatively brief viability of the seeds of most crop plants. In the eightieth year of an experiment on buried weed seeds—a study started in 1879 by Professor W. J. Beal at Michigan Agricultural College—three species—evening primrose *(Oenothera biennis)*, curly dock *(Rumex* *crispus)*, and moth mullein *(Verbascum blattaria)*—had survivors as viable seed. More than half the 20 weed species in the experiment had survivors after 25 years.

The combination of dormancy, extensive seed viability, and high seed production makes weed control exceedingly difficult. If weed seeds should all germinate at once, their control might be accomplished by a rigorous and intensive eradication program. The high weed seed population of agricultural soils makes weed control a continuing and integral part of crop culture.

Many weeds reproduce vegetatively as well as by seed. Some of the most pernicious weeds reproduce in this way: Johnson grass and quack grass, by rhizomes; wild morning glory, by roots; and wild garlic, by bulblets. Vegetative reproduction by underground stem modifications or by roots makes control by cultivation particularly difficult.

Weed Growth

The competition between weed and crop adversely affects both plants, but weeds usually win out over the crop plants. However, the exact physiological basis of the growth advantages that enable weeds to do so is not clear. Among the growth characteristics that explain the competitive ability of certain weeds are rapid germination, rapid seedling growth, and a root system that is deeply penetrating yet fibrous at the surface. Furthermore, weeds possess a natural resistance to many of the pests that plague crop plants. Their resistance to heat and cold undoubtedly gives them an added advantage.

The luxuriant growth characteristic of weeds might also have been a character of crop plants, but one that was sacrificed during "domestication" concomitant with the selection of horticultural attributes such as large fruit size. The loss of seed dormancy in most crop plants propagated by seed is an example of the loss of an adaptive trait. Another more reasonable explanation is that to survive, weeds must be uniquely adapted to their surroundings, whereas crop plants may be grown in locations far removed from the conditions to which they are best adapted. Weed species change much more dramatically than do the crops across the United States. Thus, the study of weed control must be preceded by a knowledge of the natural history and ecology of the particular species under consideration.

Weed-Control Methods

The many techniques utilized in the control of weeds may be grouped into physical, biological, and chemical methods. Weed control was given a new impetus with the discovery of 2,4-dichlorophenoxyacetic acid (2,4-D), a selective weed killer, in the middle 1940s. This chemical proved to have far-reaching effects, not only on chemical weed control, but on weed research in general. So far, however, chemicals have not proved to be the ultimate solution to the weed problem. Weeds are well endowed in their struggle for space and survival. Successful control still involves the judicious combination of many methods (Figure 16-21).

Physical Techniques

Various controls entail the physical destruction of weeds. The pulling or grubbing of weeds is the simplest and most ancient form of control. The hoe, the basic hand tool, is still widely used. Various mechanical devices have been developed to automate this process (Figure 16-22). The basic principle is to cut out, chop up, or cover the weeds and thereby destroy them. For maximum efficiency, cultivation should be carried out when weeds are very small. Weeds that are able to propagate vegetatively by means of underground parts are extremely difficult to control by cultivation and may actually be dispersed in this way.

A

B

FIGURE 16-21. Onion-seed fields occupy land for 6 to 14 months, depending on whether they are seed-planted or bulb-planted, and the onions compete poorly with weeds. Onion-seed producers must contend with three weed populations: winter, spring, and summer annuals. Control is achieved by (1) early spring cultivation, (2) midspring cultivation with an application of a preemergence herbicide, and (3) an early-summer application of herbicide when fragile seedstalks prevent cultivation. (*A*) An unweeded onion-seed field. (*B*) A field with nearly 100 percent weed control. [From P. J. Torell and D. F. Franklin, *HortScience* 8(5):419–420, 1973.]

A

B

FIGURE 16-22. Weed control by cultivation. (*A*) A 10-row onion tillivator. (*B*) A "power hoe" developed for strawberry cultivation. The operator steers with his or her feet and directs a movable hoe attachment. [Photographs courtesy Purdue University, and Friday Tractor Co., Hartford, Mich.]

Cultivation and tillage, the loosening or breaking up of the soil, are such widespread agricultural practices that many have come to believe that the loosening of the soil has beneficial functions other than the control of weeds. Yet a number of experiments have indicated that weed control is, indeed, the primary benefit of cultivation. The other advantages of cultivation, such as increased soil aeration or the conservation of soil moisture by the formation of a soil or dust mulch, may in fact be counteracted by the destructive effects of inadvertent root pruning in the surface layer, the most productive portion of the soil. In addition, extensive cultivation with heavy machinery often leads to serious compaction of the soil. Although cultivation may conserve moisture by preventing runoff, it also contributes to considerable erosion of loose soil during heavy rains. Nevertheless, cultivation is still the major form of weed control for most crops, and it should therefore be scheduled to coincide with the time most favorable for efficient weed control.

The control of weed germination by mulching has been recommended from time to time. The use of black polyethylene plastic film has been encouraging in the culture of vegetables. Mulching, however, is too expensive as a means of weed control in commercial plantings except with high-value crops, but it is a valuable practice for home gardens.

Heat may be used to control weeds: Greenhouse or cold-frame soils may be "pasteurized" by steam at 180°F (82°C) for half an hour. Weed seeds and certain "damping-off" organisms are controlled in this manner. Care must be taken to avoid sterilizing the soil, since this will destroy the bacteria involved in the nitrogen cycle. It is a good practice to avoid planting in freshly steamed soils; it is best to allow the balance of microorganisms to be restored first.

Fire has been used to destroy weeds. Flame throwers have been adapted to control them in such places as railroad beds, and have been used in weeding cotton and onions. Similarly, burning has been used to dispose of weed trash, but this practice must be considered a poor one. The burning over of muck soils to control weeds is a flagrant example of a resource waste.

Biological Techniques

The utilization of the natural competition between weeds and other organisms is the basis of biological control. The most spectacular example is the introduction of insects that feed only on certain weeds. Prickly pear (*Opuntia* species), first introduced as an ornamental into Australia prior to 1839 by the early colonists, escaped cultivation and quickly became a noxious weed. By 1925, the cactus had infested 60 million acres, and the infested area was increasing at the rate of 1 million acres annually. Finally, a natural enemy of the cactus was imported from Argentina—the moth *Cactoblastis cactorum,* whose larvae bore into the cactus and feed upon it. Ten years after the introduction of the moth, control of the cactus was almost complete. Although there were successive waves of regrowth, they were of diminishing proportions, owing to the successful establishment of the insect.

More recently, the weed *Hypericum perforatum* (Saint-John's-wort, or Klamath weed) has been successfully controlled in California by the introduction of the beetle *Chrysolina gemellata.* This beetle, native to France, had proved to be a satisfactory natural enemy of this weed in Australia.

The use of insects to control weeds can be handled only on a large scale, and, because of the problems inherent in such programs, they must be placed under the control of some national agency. Insects are usually used for controlling introduced weeds that are not attacked by native predators. In order for the introduction of insect predators to be successful, the insect must thrive in the new habitat and yet not become a pest to other agricultural crops. Thus, only those insects that are highly selective in their feeding habits can be imported. Care must also be taken to avoid introducing parasites of the imported insect.

Crop competition is an important biological method of weed control. Weed populations can be reduced by proper rotation involving well-adapted crops that can compete with weeds—for example, silage corn and alfalfa. Often it is the kind of tillage used in the rotation that brings weeds under control. For example, cultivation in corn may reduce the grass weeds that become established in small grain crops. Similarly, horticultural practices that facilitate rapid growth and good crop stands will encourage crop competition.

The use of geese to control weeds has had limited success in some horticultural crops. Geese will selectively weed strawberry patches of grass, provided there is enough grass present, but they cannot be used in fruiting patches. Fields must be fenced when geese are used, and careful management is required.

Chemical Techniques

> Buildings and walls were razed to the ground; the plough
> passed over the site, and salt was sown in the furrows made.
> A solemn curse was pronounced that neither
> house nor crops should ever rise again.
>
> A DESCRIPTION OF THE FALL OF CARTHAGE (147 B.C.E.)[1]

Substances such as common salt have been used for centuries to destroy vegetation. However, practical weed control in agriculture depends largely upon the selective destruction of weeds. In the early 1900s, a number of compounds were shown to have selective action in destroying broad-leafed weeds in grain; for example, various copper salts, sulfuric acid, iron sulfate, and sodium arsenite. But interest in these materials waned because of the unreliability of the results and the inadequacy of application equipment. The introduction of 2,4-D and other auxinlike herbicides in the 1940s rapidly transformed chemical weed control into a method of major importance. In rapid succession, many other chemicals were introduced as weed killers. Herbicides have accounted for an increasing percentage of all pesticide sales. Chemical weed killers are widely used on lawns but are seldom practical in home gardens because of the difficulty of applying them at the proper rate and the danger of injuring adjacent plants.

---

[1]B. L. Hallward, 1965, "The Siege of Carthage," in *The Cambridge Ancient History*, Vol. VIII (Cambridge, Eng.: Cambridge University Press) p. 484.

SELECTIVITY. **Nonselective herbicides** kill vegetation indiscriminately. A **selective herbicide** is one that, under certain conditions, will kill certain plants and not harm others. Selectivity in herbicides is a relative quality, and it depends to a large extent on the interaction of a number of factors: dosage, timing, method of application, chemical and physical properties of the herbicides, and the genetic and physiological state of the plants.

In order for a herbicide to cause death it must be absorbed by the plant and cause some toxic reaction. Some kill only the area of the plant actually covered (**contact herbicides**—for example, dinitro compounds, oils, and arsenates), while others are translocated within the plant (**noncontact** or **translocated herbicides**—for example, 2,4-D). Selectivity may be achieved by directing the herbicide away from the crop plant (**positional tolerance**) or by inherent morphological differences between tolerant and susceptible plants: the amount and type of waxy cuticle, which results in differential wetting and absorption (say, dinitro compounds on peas versus weeds); the angle and shape of the leaves (for example, the differences between broad-leafed plants and grasses); or the location of the growing point (for example, protected in grasses, exposed in broad-leafed plants). Diuron is an effective weed killer on grapes because it does not readily reach their deeply growing roots. Selectivity may be achieved as a function of dose. An overdose of 2,4-D will seriously affect all plants, whereas in low doses it will effectively "discriminate" between certain plants (Figure 16-23).

Physiological distinctions that result in selectivity exist between certain plants. This may be manifested in cultivars of the same species. The precise mechanism of physiological selectivity is unknown, since the exact mechanism by which many herbicides kill is still obscure. Some interfere with enzyme systems; others disturb the metabolism of the plant in some manner. Physiological selectivity may be due to differences in the plant's ability to translocate herbicides. The tolerance of carrots to Stoddard solvent is apparently due to the inherent resistance of the cell membrane to penetration.

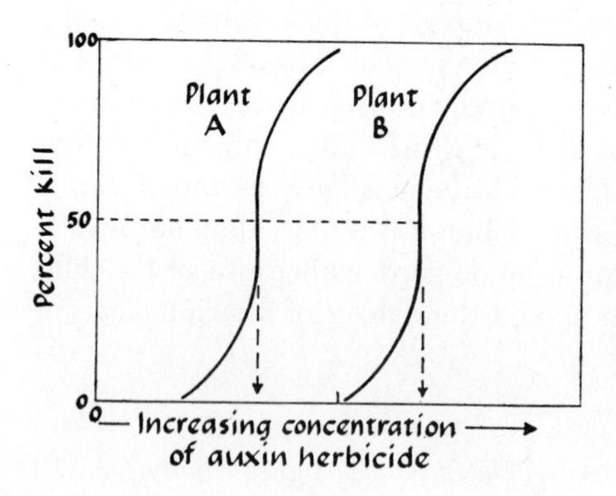

FIGURE 16-23. Selectivity of auxinlike herbicides is a function of dose. Arrows on mortality curves of two species indicate relative concentration of herbicide required to kill 50 percent of the plants. [From A. C. Leopold, *Auxin and Plant Growth*, University of California Press, 1955.]

Plants may show differences in the intensity of the toxic reaction at different stages of growth. Thus, some herbicides may be effective only during a very early stage of plant development, such as seed germination; others may be effective only at some later stage, such as flowering. Combinations of different herbicides are often required to control the many weed species usually present.

TIME AND METHOD OF APPLICATION. **Preplanting treatments** are applied before the crop is planted. Fumigants and other nonselective herbicides achieve selectivity between weed and crop by the timing of application. If preplanting herbicides are nonselective and have residual effects, sufficient time must elapse before the crop can be planted. Selective preplanting herbicides that are now available discriminate between germinating weed seeds and crop seeds or between germinating weed seeds and transplants.

Preemergence treatments are applied after the crop is planted but before it has emerged from the soil (Figure 16-24). To be effective, the herbicide must have good coverage, remain on the surface of the soil, and be relatively unleached. Timing and soil moisture are very critical. Water is often required to activate the herbicide, but too much water may leach the herbicide. Because their action is restricted to the soil surface, the herbicides are applied after the crop is planted. Physiological selectivity is often utilized so that the crop will germinate but the weed will not. In addition, selectivity may be enhanced by time. The weed seeds germinate first, since the crop seeds have a slightly delayed germination owing to the time required to imbibe water or to their depth of planting.

Postemergence treatments are applied to the growing crop. Selectivity may be physiological, or it may be due to directing the application away from the crop plant. (Care must be taken to avoid drift of herbicides.) Selectivity may be achieved as a result of plant age—for example, if the herbicide being used is

FIGURE 16-24. Preemergence application of herbicide to seed bed. Uniform and thorough coverage is essential. [Photograph courtesy Farm Equipment Institute.]

toxic only to germinating seedlings at the dose being applied. When used in this way, herbicides are usually applied immediately after a thorough cultivation, since the ground must be free of germinated weeds.

Herbicides may be applied as liquids, solids, or gases. Specialized equipment has been devised that accurately meters low dosages. This measurement is essential because of the extremely low concentrations required for some highly active substances. The increased use of granular herbicides that are absorbed by roots eliminates the need for heavy, bulky carriers for liquids.

CLASSIFICATION OF HERBICIDES. The number of chemicals known to have herbicidal activity is large and is increasing at a rapid rate. Table 16-3 presents a classification of some important families of herbicides in terms of their chemical composition. Although many compounds fit into well-defined groups on this basis, others do not appear to belong in any particular grouping. Lists of herbicides become outdated very quickly, since this technology is expanding at a rapid rate.

Oils such as diesel oil have long been used as contact, nonselective herbicides. Lighter fractions, such as stove oil and Stoddard solvent (used in dry cleaning), have proved to have selective herbicidal action. Stoddard solvent is widely used with carrots and cranberries. Heavy aromatic fractions were found to be superior to diesel oils as contact herbicides, and large numbers of these materials are used as nonselective herbicides.

TABLE 16-3. Classification of some important herbicides by chemical composition.

Class of material	Examples
Inorganic	ammonium sulphamate
	sodium arsenite
	sodium chlorite
	sulfuric acid
Organic	
Oils	diesel oils, Stoddard solvent
Phenoxyacetic acids	2,4-D, 2,4,5-T, MCPA
Chlorinated aliphatic acids	TCA, dalapon
Dinitro analines	benefin, trifluralin
Amidelike compounds	
Acetamides	Alachlor, CDAA, propachlor
Carbamates	CIPC, EPTC, IPC
Triazines	atrazine, cyanazine, simazine
Ureas	chloroxuron, diuron, binuron
Substituted phenols	Dinoseb
Diphenyl ethers	Acifluorfen, diclofop
Others	DCPA, glyphosate, paraquat, picloram

Phenoxy or auxinlike materials are those substances that have a physiological cation resembling indoleacetic acid (see Chapter 4). The most common of these are 2,4-D, 2,4,5-T, and MCPA. Note how similar their structural formulas are to one another and to that of indoleacetic acid (Figure 16-25). Various derivatives of these compounds may be achieved by different substitutions and formulations. These affect the herbicidal as well as the physical properties of the molecule. For example, the amine formulation of 2,4-D is less volatile than the ester form. Volatile esters of 2,4-D are hazardous around sensitive crops such as grapes or tomatoes. The auxin herbicides are generally highly selective with respect to dose and are effective at extremely low concentrations. They have a short residual life in the soil and low toxicity to animals. Selectivity is achieved by both differential absorption and genetic differentiation to dose. At high enough concentrations, auxin herbicides are toxic to all plants. The herbicide 2,4-D is used to control weeds in sweet corn and strawberries, and also as a broadleaf weed killer in turf.

Among the chlorinated aliphatic acids used in weed control are such compounds as TCA (trichloroacetic acid) and dalapon (2,2-dichloropropionic acid). As a group, these compounds are more selective against monocots than dicots and are thus used in the control of grassy weeds. Dalapon has the widest herbicidal use and is helpful in the control of both seedling and established perennial grasses.

Amidelike and related compounds contain the amide moiety N—C=O or the related groupings N—C=S or N—C=N. The physiological significance of their chemical structure is not known. Many important herbicides are included in this group, such as the triazines, carbamates, ureas, and acetamides. These are now some of the most useful ones in horticulture.

FATE OF HERBICIDES. Herbicides, in order to be successful agricultural tools, must dissipate so that they will not interfere with future use of the land. Their eventual disappearance may result from vaporization, chemical breakdown, biological decomposition, leaching, or adsorption on soil colloids.

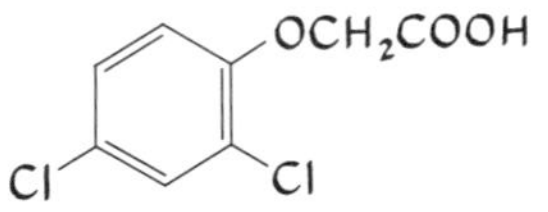

2,4-Dichlorophenoxyacetic acid (2,4-D)

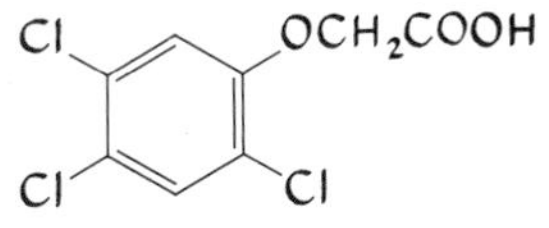

2,4,5-Trichlorophenoxyacetic acid (2,4,5-T)

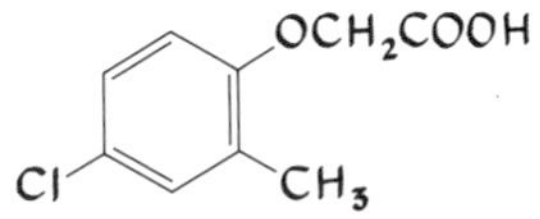

Methylchlorophenoxyacetic acid (MCPA)

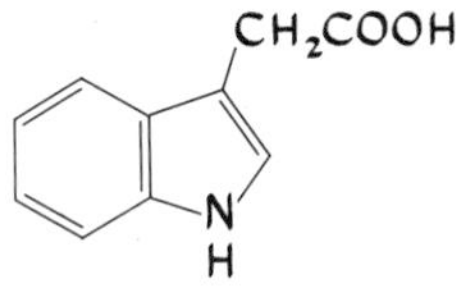

Indoleacetic acid (IAA)

FIGURE 16-25. Structures of auxinlike herbicides compared with that of indoleacetic acid.

Selected References

Agrios, G. N., 1978. *Plant Pathology*, 2d ed. New York: Academic Press. (Excellent).

Anderson, H. W., 1956. *Diseases of Fruit Crops.* New York: McGraw-Hill. (The important deciduous temperate fruits treated here, but insects are not covered.)

Anderson, W. P., 1983. *Weed Science: Principles*, 2d ed. St. Paul, Minn.: West Publishing. (A broad perspective of weeds and weed control.)

Chupp, C., and A. F. Sherf, 1960. *Vegetable Diseases and Their Control.* New York: Ronald Press. (A useful treatise on diseases of importance in the United States.)

Dixon, G. R., 1981. *Vegetable Crops Diseases.* Westport, Conn.: AVI. (An American edition of a Scottish work.)

Fry, W. E., 1982. *Principles of Plant Disease Management.* New York: Academic Press. (A modern approach emphasizing integrated disease management.)

Horsfall, J. G., and E. B. Cowlings (eds.), 1977–1980. *Plant Disease: An Advanced Treatise*, 5 vols. New York: Academic Press. (A magnum opus).

Horst, P. K., 1979. *Westcott's Plant Disease Handbook*, 4th ed. New York: Van Nostrand Reinhold. (A compendium of diseases for professional and home gardeners.)

Johnson, W. U., and H. H. Lyon, 1976. *Insects That Feed on Trees and Shrubs: An Illustrated Practical Guide.* Ithaca, N.Y.: Cornell University Press. (The best illustrated book on insect pests of ornamentals.)

Kenaga, C. B., 1985. *Principles of Phytopathology*, 3d ed. Lafayette, Ind.: Balt. (A general introductory text on plant pathology.)

Klingman, G. C., and F. M. Ashton, 1982. *Weed Science: Principles and Practices*, 2d ed. New York: Wiley. (An excellent general textbook.)

Metcalf, C. L., W. P. Flint, and R. L. Metcalf, 1962. *Destructive and Useful Insects: Their Habits and Control*, 14th ed. New York: McGraw-Hill. (The bible for economic entomology.)

Pirone, P. P., 1978. *Diseases and Pests of Ornamental Plants*, 5th ed. New York: Ronald Press. (A thorough treatment.)

Roberts, D. A., 1978. *Fundamentals of Plant-Pest Control.* San Francisco: W. H. Freeman and Company. (A comprehensive, interdisciplinary introduction).

Roberts, D. A., and C. W. Boothroyd, 1984. *Fundamentals of Plant Pathology.* New York: W. H. Freeman and Company. (A fine phytopathology text presenting specific diseases in groups according to their physiological impact upon plant welfare.)

Tattar, T. A., 1978. *Diseases of Shade Trees.* New York: Academic Press. (Pathology for arboriculturists.)

Webster, J. M. (ed.), 1972. *Economic Nematology.* New York: Academic Press. (A reference text to the important nematode pests of the world's major crops.)

Weed Science Society of America, 1983. *Weed Science Society of America Herbicide Handbook*, 5th ed. Champaign, Ill. (A list of herbicides, desiccants, plant-growth regulators, various adjuvants, and chemicals of interest to weed scientists.)

Westcott, C., 1973. *Gardener's Bug Book.* New York: Doubleday. (Despite the title, an authoritative compendium of garden pests.)

Breeding 17

Almost all the edible plants in cultivation today were domesticated before the advent of written history. Many ornamentals, such as the rose and lily, have been in cultivation for thousands of years. Plant improvement has been continuous during this time as a result of deliberate differential reproduction of certain plants by people of many cultures. This process of **selection** over the years has been extremely effective; most of our cultivated plants no longer even remotely resemble their wild ancestors. **Plant breeding**, the systematic improvement of plants, is an innovation of the last century. In the past century, the study of the mechanisms of heredity—**genetics**—has placed plant breeding on a firm theoretical basis. Plant breeding has become a specialized technology, and it is responsible for a large part of the current progress in horticulture.

Since human needs and standards change, the job of the plant breeder never ends. For example, mechanical harvesting becomes possible only with plants (or plant parts) that are ready for harvest all at once. It demands that plants be structurally adapted to the machine. The raw material for these and other changes may be found in the tremendous variation that exists in cultivated plants and their wild relatives. The incorporation of these alterations into plants adapted to specific geographical areas demands not only a knowledge of the theoretical basis of heredity but mastery of the art and skill necessary for the discovery and perpetuation of small but fundamental differences in plants.

THE GENETIC BASIS OF PLANT IMPROVEMENT

Variety is more than the spice of life; it is the very essence of it. Differences between horticultural plants range from the obvious (water-lilies versus watermelons) to the almost imperceptible variation that might exist between two apple trees of the same clonal cultivar growing side by side. Variation can be of two types, environmental and genetic. **Genetic** differences are due to the hereditary makeup of the organisms. This variation can be traced to differences in **genes,** the fundamental units that determine heredity. **Environmental** variation can be demonstrated by comparing organisms of identical genetic makeup grown in different environments. Similarly, differences between plants grown in identical environments must be due to genetic differences. The range of environmental variation is enormous, but its boundaries are determined by genetic makeup. (It is difficult to conceive of any environmental condition that will transform water-lilies to watermelons.) The genetic makeup of an organism is referred to as its **genotype.** The net outward appearance of the organism is its **phenotype.** In a stricter sense, the phenotype is the product of the interaction between a genotype and its environment.

To ask whether genetic variation or environmental variation is the more important is meaningless. The pertinent question is, *Which genotype is best suited to a particular set of environmental conditions?* And, given a particular genotype, *Which environmental conditions will permit the optimum phenotypic expression of that genotype?*

The fundamental discovery that the genotype is inherited as discrete units was first published in 1865 by Gregor Mendel, an Austrian abbot, whose discovery grew out of experimental work carried out with the garden pea. Environmental variation, however, is not transmitted, as was first firmly established by the Danish geneticist and breeder Wilhelm Johannsen in 1903, from research on seed size in the common bean. Johannsen demonstrated that inheritance of phenotypic variation is only possible when it is a result of genetic differences.

Genes and Gene Action

The chromosomes contain the code-controlling mechanism that coordinates the physiological activities of the cell and, consequently, the organism. The study of many organisms has indicated that the information provided by the chromosomes directs the formation of enzymes that affect biochemical reactions. The unit of the chromosome conferring a single enzymatic effect is called the **gene.** Structurally, the gene appears to be a portion of the deoxyribonucleic acid

(DNA) molecule, the nonprotein portion of the chromosome; the genetic code inherent in the DNA molecule affects enzyme synthesis (see Chapter 6). Considerable information has accumulated concerning gene-mediated biochemical pathways.

Gene action can be demonstrated with flower pigmentation. Two anthocyanin pigments distinguished by differing degrees of redness have been shown to be chemically differentiated by a single hydroxyl group. The synthesis of these pigments proceeds in a stepwise fashion, as shown in Figure 17-1. The addition of a single hydroxyl group changes the color of the plant's petals from bright red to bluish red. This biochemical step is gene controlled. Thus, the petals of a plant that contains a gene *(C)* controlling the transformation of pelargonidin to cyanidin will be bluish red. A plant having only an altered version of this gene *(c)* incapable of adding the hydroxyl group will have bright red petals. Alternate forms of a particular gene are referred to as **alleles** (*C* and *c* are alleles). The change in the gene's internal structure, which gives rise to new alleles, is known as **mutation.** Mutations are changes in genetic information and are ultimately responsible for the inherent variation in all living things.

Each plant species contains a characteristic number of chromosomes known as the $2n$, or somatic, number ($2n = 12$ in spinach, 14 in pea, 16 in onion, 18 in cabbage, 20 in corn, 22 in watermelon, 24 in tomato, and so on). The reproductive cells, or **gametes,** contain the haploid number (n) of chromosomes. Chromosomes occur in pairs in the vegetative cells, which therefore contain the somatic number. In meiosis one chromosome of each pair is distributed to the gametes, reducing the diploid number by one-half. Fertilization subsequently restores the diploid number in the zygote, the fertilized egg. Thus, in somatic cells of diploid plants, each gene is present twice. A particular gene (for example, *C* and its allele *c*) may be present in any one of three combinations—*CC*, *Cc*, or *cc*. A plant containing two identical genes, *CC* or *cc*, is **homozygous** for that allele. When the alleles are different, as with *Cc*, the plant is **heterozygous.**

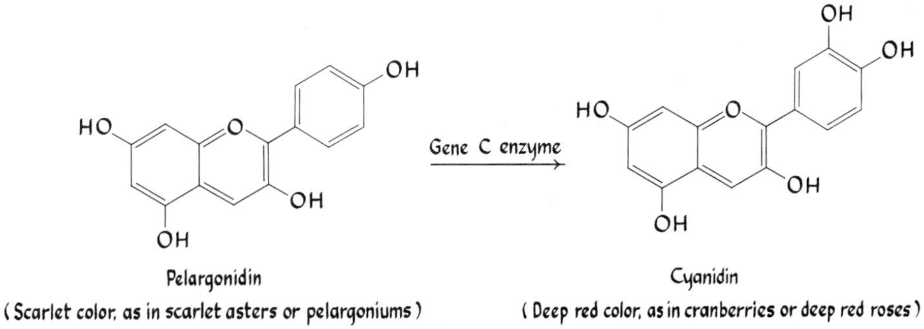

HO, O, OH, OH, OH
Gene C enzyme
HO, O, OH, OH, OH, OH

Pelargonidin
(Scarlet color, as in scarlet asters or pelargoniums)

Cyanidin
(Deep red color, as in cranberries or deep red roses)

FIGURE 17-1. The conversion of pelargonidin to cyanidin involves the addition of a single hydroxyl group and is controlled by a single gene.

What is the difference in outward expression (phenotype) when the genetic constitution (genotype) is *CC*, *Cc*, or *cc*? With reference to our example of petal color, the assumption can be made that if the single allele *c* is completely nonfunctional in regard to hydroxylation of pelargonidin, two alleles *cc* should not be any more efficient. It can be shown experimentally that flowers of *cc* plants are bright red, and chemical analysis of their petals yields no cyanidin. The difference between plants that are homozygous (*CC*) and those that are heterozygous (*Cc*) cannot be predicted. The allele may be efficient enough to produce sufficient enzyme such that plants with only one functioning allele (*Cc*) cannot be distinguished phenotypically from those having two functioning alleles (*CC*). When this is the case, it would appear that the allele *C* dominates *c*. The condition in which heterozygous plants *Cc* are indistinguishable from homozygous *CC* plants is termed **dominance.** Thus *C* is considered a **dominant,** *c* a **recessive.** If the heterozygote *Cc* is intermediate in phenotype between the two homozygous types *CC* and *cc*, dominance is said to be **incomplete.**

Assume the existence of a completely dominant gene *A* with a recessive allele *a*. The heterozygous plant *Aa* will produce two kinds of gametes (*A* and *a*) in equal proportions. A cross of these alleles, designated as *Aa* × *Aa*, will produce three kinds of progeny in a predictable ratio if all gametic and zygotic types are equally viable. It can readily be seen that the possible types of zygotes resulting from all combinations of the two kinds of gametes *A* and *a* will be *AA*, *Aa*, *aA*, and *aa* (or 1*AA*, 2*Aa*, and 1*aa*—see Figure 17-2). If we assume dominance, *AA* plants are indistinguishable from *Aa* plants. Thus the phenotypic ratio becomes 3*A*_ to 1*aa*. (*A*_ represents *AA* or *Aa*.)

The genotypes *AA* and *Aa* may be distinguished by a genetic test. The genotype *AA* will produce a single type of gamete (*A*). The genotype *Aa* will produce two kinds of gametes (*A* and *a*). By crossing plants of the phenotype *A* with themselves or with the double recessive *aa*, these two genotypes may be separated on the basis of their progeny:

AA selfed (*AA* × *AA*) ⟶ all *AA* (*AA* plants "breed true")

in contrast with

Aa selfed (*Aa* × *Aa*) ⟶ 3*A*_:1*aa* (*Aa* plants segregate)

or,

AA × *aa* ⟶ all *Aa*

in contrast with

Aa × *aa* ⟶ 1*Aa*:1*aa*

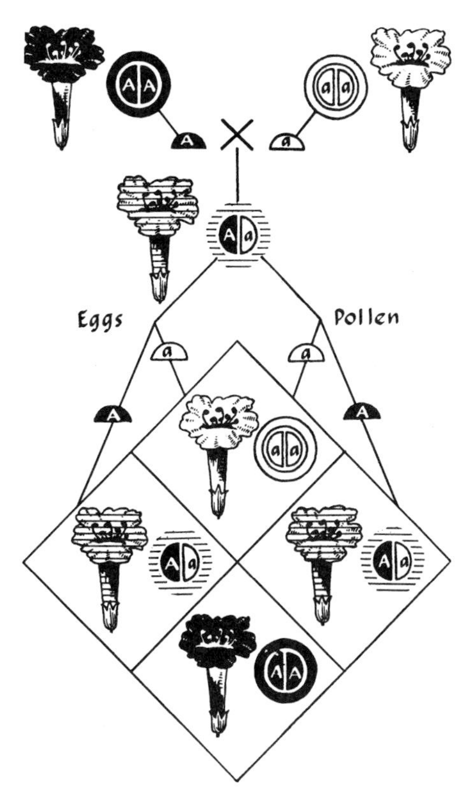

FIGURE 17-2. Mendelian segregation illustrating incomplete dominance in the four o'clock (*Mirabilis jalapa*). Note that the phenotypic ratio of 1 red:2 pink:white is also 3 colored:1 noncolored. [From "The Genetic Basis of Evolution" by T. Dobzhansky. Copyright © 1950 by Scientific American, Inc. All rights reserved.]

Generations are designated by specialized terminology. The first cross (usually referring to homozygous genotypes that differ with respect to a particular character) is the P_1, or the **parental generation.** The progeny of such a cross is the first filial generation (always identified in writing and speaking as the F_1). If the parents are homozygous for all genes concerned, the F_1 is heterozygous and nonsegregating. The F_2 (second filial generation) results from intercrossing or selfing F_1 plants:

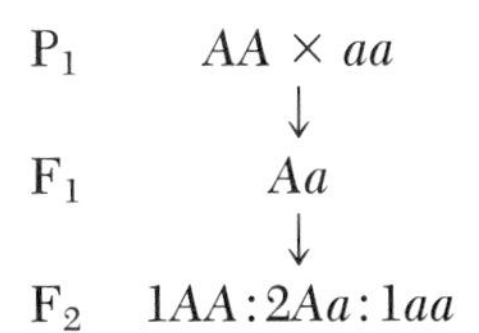

The F_2 is typically the segregating generation, which, if large enough, could theoretically include every possible genotype. Further generations are known as F_3, F_4, F_5, and so on. A cross of the F_1 with one of its parents is known as a **backcross.**

Multigenic Inheritance

Our discussion of the inheritance of a single gene difference can be expanded for examples involving two or more genes. The assortment of one pair of chromosomes at meiosis has no effect on the assortment of the other pairs; consequently, genes on separate chromosomes segregate independently of each other. Thus, a plant heterozygous for two gene pairs (*Aa* and *Bb*), each on separate chromosomes, will produce four different types of gametes (*AB*, *Ab*, *aB*, and *ab*) in equal proportions. The self progeny of the *AaBb* plant will segregate into phenotypic ratios of 3:1 for each factor, if we assume complete dominance. The combined phenotypic ratio will be 9*A_B_*:3*A_bb*:3*aaB_*:1*aabb*. This cross is diagrammed in Figure 17-3.

When two genes affect the same biochemical pathway, the phenotypic ratio gives some idea of the gene action involved. For example, two genes, both of which show complete dominance, and both of which interact to effect a single character, will transform the 9:3:3:1 ratio into a ratio of 9:7. Either gene, when homozygous recessive, will block some essential step, as shown in Figure 17-4. This phenomenon of gene interaction is called **epistasis.** Different types of gene action produce various F_2 ratios as a result of the consolidation of various genotypic classes having similar phenotypes.

Quantitative Inheritance

Characters that are continuous from one extreme to the other, such as size, yield, or quality, are of utmost significance in plant improvement. The inheritance of continuous or **quantitative** characters is an extension of the genetic principles briefly discussed for noncontinuous or **qualitative** traits. It can usually be shown that there are many genes that will each contribute a small increment of effect to modify the quantitative character. The individual effects of quantitative genes (also known as multiple factors, polygenes, and modifiers) are extremely difficult to ascertain because of the large environmental effects present in relation to the small contribution of the individual genes. As a result, it is not possible to determine the precise pattern of gene action. The special techniques used to analyze the inheritance of quantitative characters involve mathematical and statistical analyses of the variation through many generations. These techniques make it possible to predict the number of genes differing in a particular cross, as well as their average contribution and average dominance. These methods rely, however, on particular assumptions (for example, the absence of epistasis) that are difficult to prove precisely.

An example of the pattern of quantitative inheritance is shown in Figure 17-5. A cross between two homozygous lines of maize, one short-eared and the other long-eared, produced an F_1 population whose mean ear length was between those of the parental lines. The parental and F_1 variation is presumably all environmental. The distribution of the F_2 population includes the homozy-

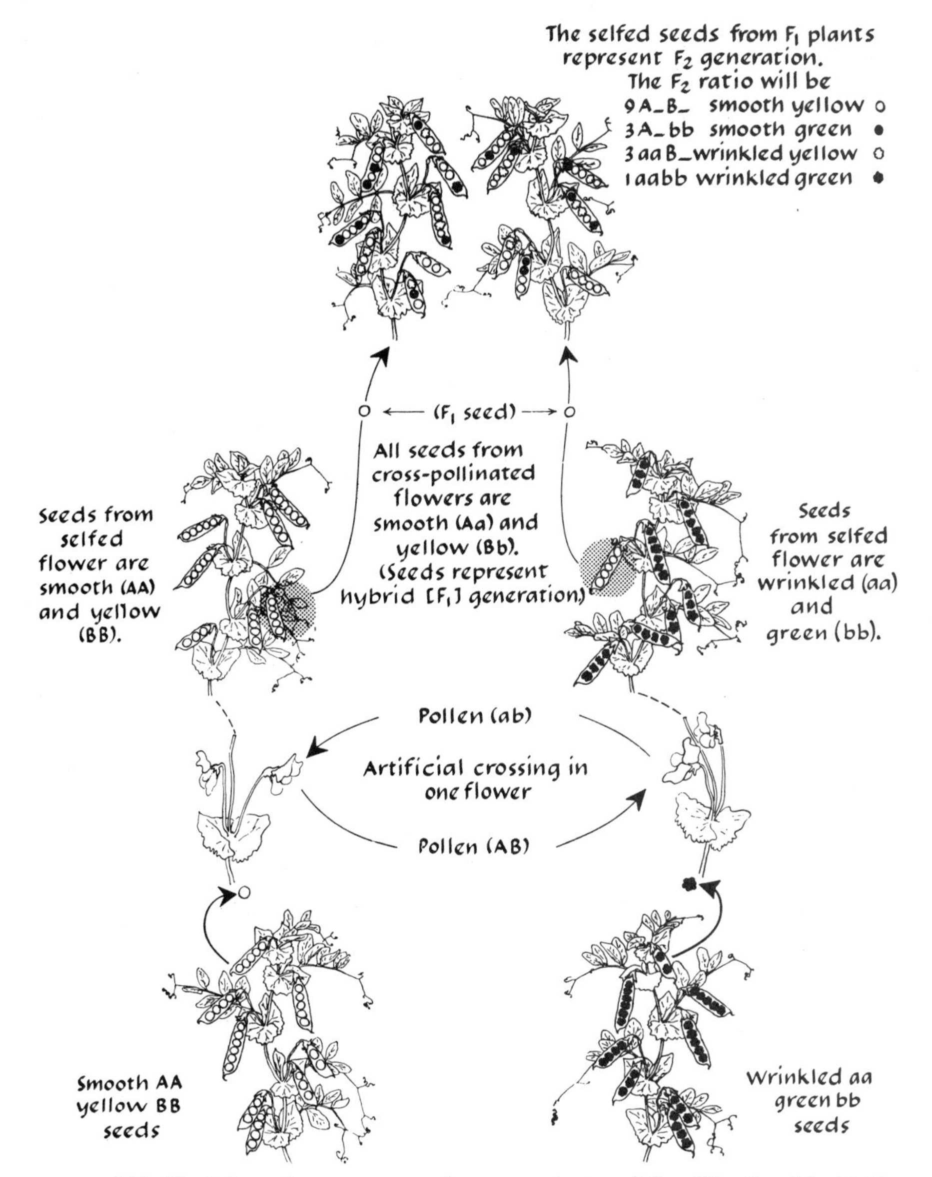

FIGURE 17-3. The independent assortment of two genes in peas. [After "The Gene" by N. H. Horowitz. Copyright © 1956 by Scientific American, Inc. All rights reserved.]

gous parental types. The F_2 can be shown mathematically to have more variation than either the F_1 or the lines of the parental generation. This increased variation in the F_2 is a result of genetic segregation.

Except for a deficiency of parental types, the data as presented are similar to the data that would be expected from a single partially dominant gene, if we

In the tomato, dominant genes R and A^t (apricot) are both required to synthesize lycopene, the red pigment in the fruit.

rr – Depletes all pigment
a^ta^t – Depletes lycopene

Genotype	
$R\_A^t\_$	Red
$R\_a^ta^t$	Yellow
$rrA^t\_$	Yellow
rra^ta^t	Yellow

(Because of differences in amount of β carotene, which produces yellow color, yellow genotypes can be differentiated by chemical analysis.)

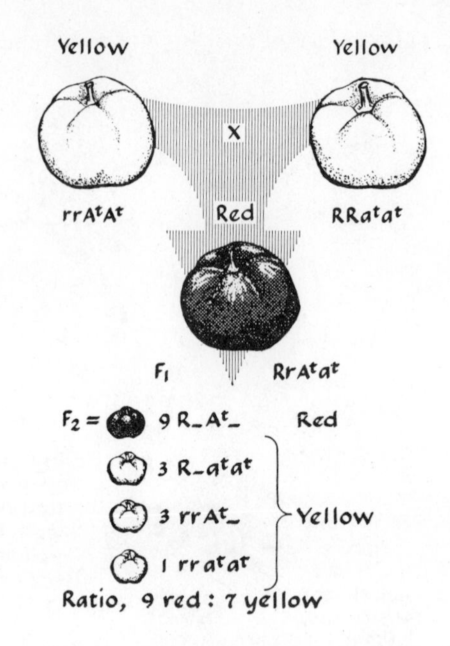

FIGURE 17-4. Epistasis in the tomato.

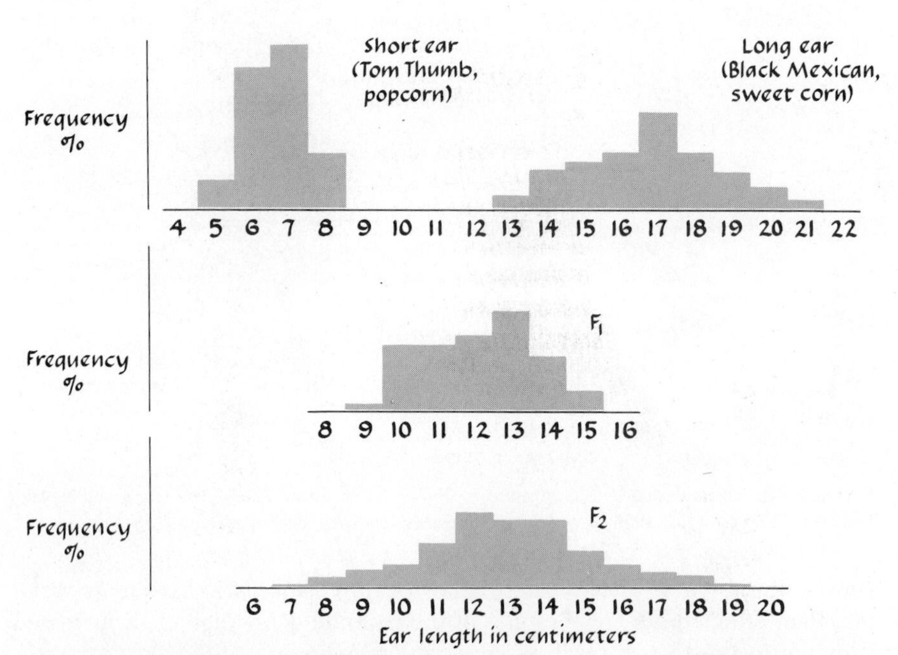

FIGURE 17-5. Distribution of ear length in a popcorn–sweet corn cross. [Adapted from A. M. Srb, R. D. Owen, and R. S. Edgar, *General Genetics*, 2d ed., W. H. Freeman and Company, copyright © 1965; after Emerson and East.]

assume enough environmental variation to obscure the differences among the three F_2 classes *(AA, Aa, aa)*. However, the major differences between a single-gene difference and a many-gene difference can be shown by selfing F_2 plants to produce F_3 lines. If we assume a single-gene difference, there will be only three genotypes in the F_2 (the genotypic ratio being 1*AA*:2*Aa*:1*aa*). Two of these would be true breeding genotypes (*AA* and *aa*), duplicating the parental types. The other genotype *(Aa)* would segregate to produce an F_3 distribution identical to the F_2. Note that each parental type would be duplicated in one-fourth of the F_2 population. It turns out that there are many more than three F_2 genotypes in this F_2!

As the number of genes differentiating the parents with respect to ear length increases from 1 to n, the number of F_2 genotypes increases from 3 to 3^n. On the other hand, the number of genotypes resembling the parental types decreases from 2(¼) of the F_2 population to $2(1/4)^n$. If we assume that only 10 genes differentiate ear length, there are 3^{10}(or 59,049) different genotypes in the F_2. To recover every genotype, a minimum population of 4^{10}(or 1,048,576) plants, would have to be grown. In this population there would be only one plant duplicating each of the parental genotypes.

The genetic value of a seed-propagated plant is the average performance of its progeny. Thus, the basic problem in the improvement of quantitatively inherited characters lies in distinguishing genetic from environmental effects. The plants that have the best appearance may not have the most desirable genotype. The great economic importance of quantitative characters demands that the breeder of horticultural plants be familiar with the basic principles governing their inheritance.

Linkage

Since plants contain thousands of genes but only a limited number of chromosomes, it is apparent that each chromosome must contain many genes. Those located on the same chromosome are not randomly assorted but tend to be inherited together. This condition is referred to as **linkage.** There is a relationship between the physical closeness of genes and the intensity or strength of their linkage. The assortment of genes on the same chromosome is related to an actual exchange of chromosome material in meiosis.

By special techniques, it is possible to locate genes on their respective chromosomes. Since the genes occur in a linear sequence on the chromosomes, their relationship to each other can be determined by their linkage strengths. In this way the topology of the chromosomes may be mapped, as shown for the tomato in Figure 17-6. The linkage map may be constructed only for genes for which mutant forms are known.

Linkage appears to occur between genes controlling qualitative and quantitative characters. In the peach, there is a particularly undesirable association

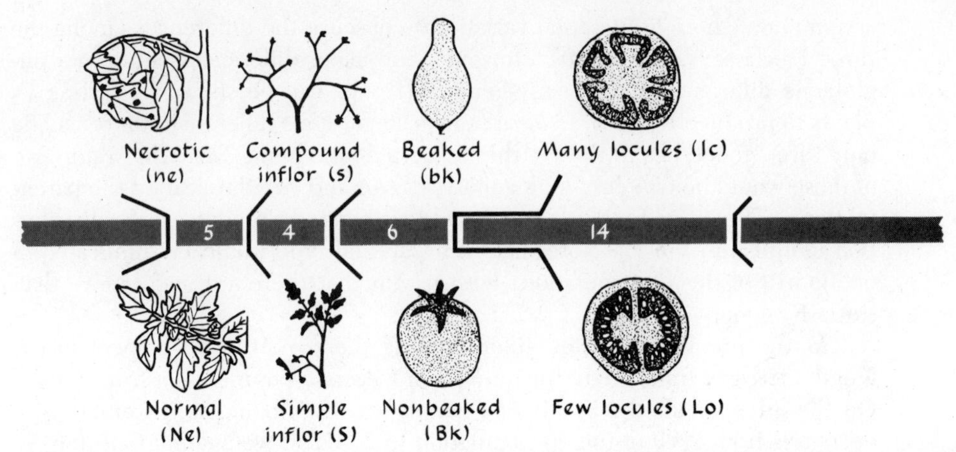

FIGURE 17-6. Map of chromosome 2 (linkage group I) of tomato. [From L. Butler, *J. Hered.* 43:25–35, 1952.]

between small fruit size (a quantitative character controlled by many genes) and the single gene for smooth skin that distinguishes the nectarine from the peach. The breaking of this apparent linkage is a prerequisite for the breeding of large nectarines. Large populations are needed to obtain the rare but desirable combination that produces large nectarines from crosses between large peaches and small nectarines. The smooth-skin gene also appears to be related to susceptibility to brown rot. Another interesting association in the peach is that between rubbery flesh (versus melting flesh) and clingstone (versus the freestone) character. Similarly, the glandless varieties of peaches and nectarines are highly susceptible to mildew. Unless recombinant types are obtained, it is difficult to prove whether such associations are actually genetic linkages or whether they are due to some physiological relationship.

Hybrid Vigor

Hybrid vigor, or **heterosis,** refers to the increase in vigor shown by certain crosses as compared to that of either parent. The classic example of this vigor is the mule, an interspecific cross of a mare and a jackass. The effect in plants has been described as similar to that caused by the addition of a balanced fertilizer. Hybrid vigor is often associated with an increased number of parts in **indeterminate** plants (whose main axis remains vegetative and in which flowers form on axillary buds—for example, cucumber) and with increased size of parts in **determinate** plants (whose main axis terminates in a floral bud such as seen in sweet corn). In perennial plants, the vigor persists year after year.

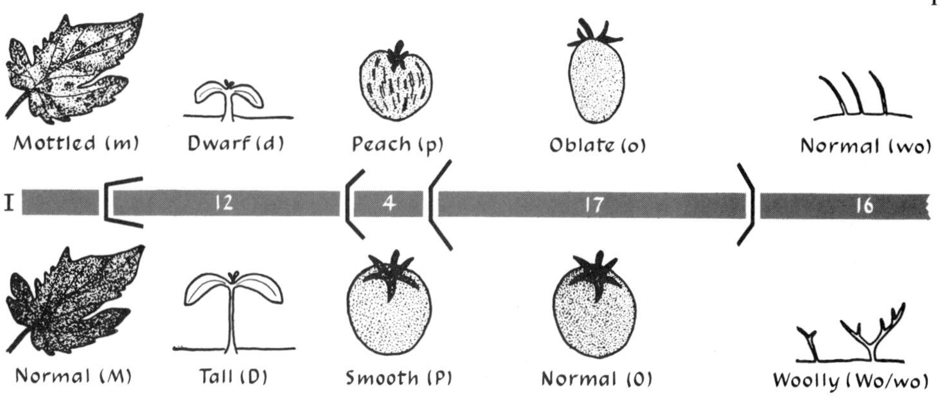

The term *hybrid* is applied loosely; any organism resulting from genetically dissimilar gametes is technically a hybrid. Thus, a plant that is heterozygous for a single-factor pair is a genetic hybrid. In horticulture and botany, the word *hybrid* is often incorrectly used to refer specifically to the result of crosses made between species. The word in the expression "hybrid corn" refers to a particular combination of inbred lines.

Genetic Consequences of Inbreeding

The phenomenon of heterosis is intimately associated with the decline in vigor brought about in some plants by continued crossing of closely related individuals (**inbreeding**). The genetic consequences of inbreeding are best explained with **selfing** (crossing a bisexual plant with itself), an extreme example of inbreeding.

Selfing has no effect on homozygous loci:

$$AA \times AA \longrightarrow AA$$
$$aa \times aa \longrightarrow aa$$

As shown previously, however, the selfing of a plant that is heterozygous for a single gene *(Aa)* produces progeny of which half are homozygous (*AA* and *aa*). This phenomenon may be generalized for any number of genes. With each generation of selfing, 50 percent of the heterozygous genes become homozygous (Figure 17-7). Continued selfing will ultimately produce plants that are homozygous for all genes. Lines of such plants are known as **inbred** or **pure**

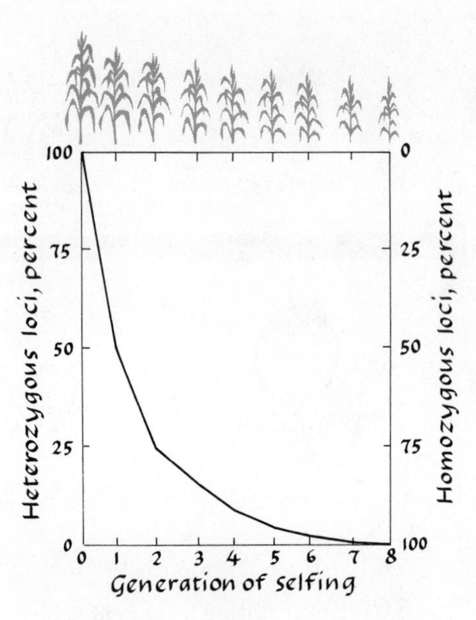

FIGURE 17-7. Selfing reduces the number of heterozygous genes by half in each generation. By the sixth generation, more than 95 percent of the genes are expected to be homozygous. Most cross-pollinated plants show a decrease in vigor as homozygosity is reached.

lines. Inbreeding can now be defined as any breeding system that increases the homozygous state.

The effect of inbreeding depends on the degree of heterozygosity and on the method of pollination of the plant. Homozygous plants (naturally cross- or self-pollinated) are unaffected by inbreeding. Nevertheless, heterozygous cross-pollinated plants tend to show a loss of vigor upon inbreeding that closely parallels the increase in homozygosity. There are, however, great differences between species and lines. Heterozygous individuals of self-pollinated plants can be obtained by artificial crosses between different homozygous lines. When these heterozygous lines are selfed, the characteristic decline in vigor is usually absent.

The crossing of unrelated inbred lines of cross-pollinated crops restores the vigor lost by inbreeding (Figure 17-8). The progenies of these crosses are referred to as F_1 hybrids. In maize some F_1 hybrids have shown increases in yield as high as 25 percent above the yield of the original open-pollinated cultivars from which the inbred lines were derived. This increased vigor is apparently due to the genotypic control, which is made possible by the selection of inbreds. A vigorous, uniform F_1 hybrid line assures an improvement of the F_1 over the segregating open-pollinated cultivars. This lack of variation in F_1 hybrids, although disadvantageous under certain conditions, is usually desirable in an age such as this that values mechanization and prepackaging.

Crossing various cultivars of self-pollinated crops (which in reality are also inbred lines) increases vigor in isolated cases, but the magnitude of the increase is very much less than it is in cross-pollinated crops.

FIGURE 17-8. Hybrid vigor in the onion from crossing two inbred lines. [Photograph courtesy K. W. Johnson.]

Genetic Explanation of Heterosis

Although the incorporation of heterosis is a most significant feature in plant improvement, its genetic basis is not completely established. The genetic explanation must reconcile both inbreeding depression and the apparent stimulation of vigor associated with heterozygosity. The two major theories proposed are now referred to as the **dominance** and **overdominance hypotheses.**

The dominance hypothesis, first proposed by D. F. Jones in 1917, is based on the assumption that cross-pollinated plants contain a large number of recessive genes concealed by their heterozygous condition. As the recessive condition is largely due to an absence of some essential gene function, these concealed recessives are largely deleterious. Upon inbreeding they become homozygous, resulting in a decline in vigor. Crossing inbred lines from diverse sources restores vigor. The recessives of each inbred are "covered up" by their dominant or partially dominant alleles carried by the other inbred, much as the wearing of two moth-eaten bathing suits might better protect the modesty of a bather than the wearing of only one. In successful combinations the inbreds mutually complement each others' deficiencies. In its simplest case this situation may be expressed by the cross

$$aaBB \times AAbb \longrightarrow AaBb$$

The hybrid *AaBb* is more vigorous than either of its homozygous parents because it contains both dominant alleles, whereas each of the parents is deficient either in gene *A* or in gene *B*.

According to the dominance hypothesis, the vigor of a hybrid is due to the large number of dominant genes it contains. The high heterozygosity is merely a concomitant and nonessential part of hybrid vigor. The theory assumes that, were it possible to obtain each heterozygous gene in the homozygous dominant condition, the vigor of this genotype would be as great or greater than that of the heterozygous type. The failure to obtain true-breeding homozygous lines that are as vigorous as F_1 hybrids is caused by the great number of genes involved and perhaps by linkages between deleterious and beneficial alleles.

The concept of inbreeding depression lies in the quantitative level of deleterious genes in the population, a direct function of the natural method of pollination and the phenomenon of dominance. The lack of a significant inbreeding depression and of heterosis in naturally self-pollinating crops is due to the absence of any deleterious "hidden" recessives. Any deleterious recessives become quickly exposed by the natural inbreeding mechanism of self-pollinating crops and are eliminated by the forces of natural selection.

The dominance hypothesis has not been an entirely satisfactory explanation to some geneticists. This dissatisfaction is due largely to the high estimates of recessive alleles in the open-pollinated population required for the phenomenon and to other mathematical considerations. The overdominance hypothesis assumes that heterozygosity *by itself* may be advantageous. The reduction of vigor upon inbreeding is a direct result of the attainment of homozygosity. This theory assumes that the heterozygous condition *Aa* is superior to either homozygous condition (*AA* or *aa*). This has been termed overdominance. There are two interpretations of overdominance at the gene level. One assumes that both alleles are physiologically active. The combined activity of both alleles in the heterozygous condition produces the heterotic effect. The alternative explanation assumes that the recessive allele is inactive but that dominance is incomplete. However, the reduced amount of essential gene product is optimum; the amount produced by the homozygous types is either deficient or inhibitory. There are experimental verifications, although rare, of both types of gene action.

It is difficult to distinguish between the dominance and overdominance hypotheses. The essential features of the hybrid effect are explained by both. The correct explanation lies in determining primary action of individual quantitative genes. In general, however, the dominance hypothesis is thought to be the most likely explanation of hybrid vigor.

Polyploidy

The condition in which an organism has more than the normal $2n$ number of chromosomes in its somatic cells is called **polyploidy.** Although many plants normally have only two complete sets of chromosomes (the diploid number), other multiples of chromosome sets may occur. Plants containing various numbers of sets are defined as follows:

Number of chromosome sets	Name of type
1	Monoploid (haploid of diploid)
2	Diploid
3	Triploid
4	Tetraploid
5	Pentaploid
6	Hexaploid
7	Septaploid
8	Octoploid

In addition to the variation of whole sets of chromosomes (**euploidy**), there is variation of the number of chromosomes within a set (**aneuploidy**). An example of aneuploid variation is shown in Figure 17-9.

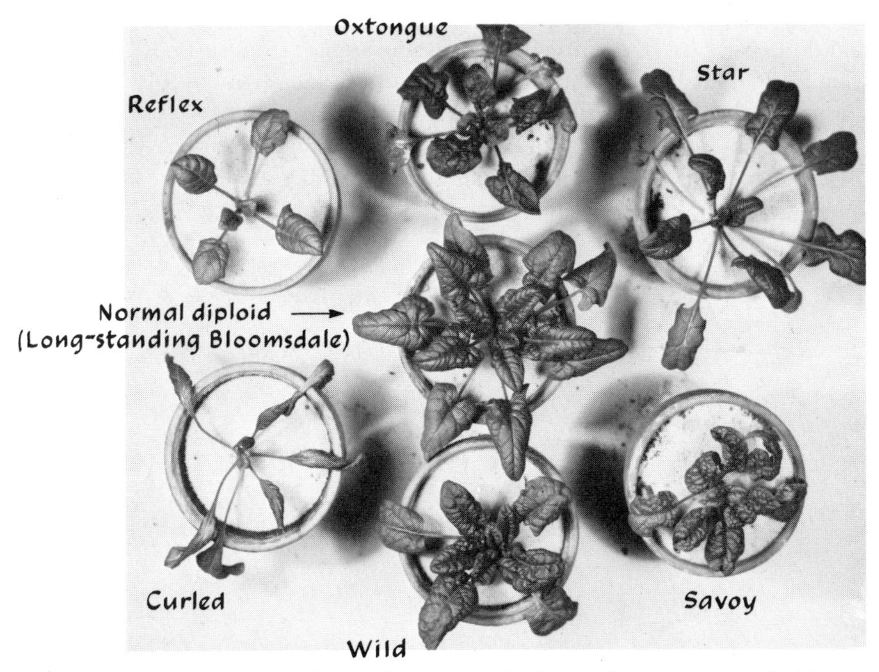

FIGURE 17-9. The six trisomics of spinach. The spinach plant in the center is a diploid having 12 chromosomes (six pairs) in the vegetative cells. The six plants surrounding it all have 13 chromosomes. Each trisomic has a different extra chromosome that confers a characteristic appearance. [From J. R. Ellis and J. Janick, *Amer. J. Bot.* 47:210–211, 1960.]

Many plants can be shown to be naturally polyploid. Some cultivars of apple and pear are triploid; tart cherries are tetraploid; cultivated strawberries are octoploid. In this regard the convention of referring to the basic number of chromosomes in a set as the x number avoids confusion. Thus, the basic number of chromosomes in the genus *Fragaria,* to which the strawberry belongs, is 7 ($x = 7$). The chromosome number in the vegetative cells of the cultivated strawberry is 56 (8 sets of 7 chromosomes), or $2n = 56 = 8x$. The chromosome numbers in the gametes, the haploid or n number, is $n = 28 = 4x$. The concept of n is halfness, whereas the concept of x is oneness.

Polyploids may occur spontaneously or may be artificially induced by special treatment. For example, the formation of callus tissue by wounding is effective in inducing tetraploidy in the tomato. An alkaloid called **colchicine,** derived from the autumn crocus, has the remarkable property of doubling the chromosome number in a wide variety of plants when applied in concentrations of 0.1 to 0.3 percent. This drug interferes with spindle formation in mitosis; the chromosomes divide, but the cell does not. Tetraploids are produced routinely with the use of colchicine. Seed treatment (usually for 24 hours) produces tetraploidy in about 5 percent of the surviving seedlings. The drug may also be applied to the growing points of seedlings or large plants by a variety of methods.

An artificially induced tetraploid may usually be distinguished from a normal diploid by its larger, thicker leaves and organs (Figure 17-10) and somewhat slower, coarser growth. Cell size is larger, and fertility is often reduced. Larger pollen size is usually a reliable indicator of tetraploids. Actual chromosome counts are necessary, however, for positive identification. The morphological changes associated with chromosome doubling vary within and between species.

FIGURE 17-10. Increased cluster size as a result of colchicine-induced polypoidy in the 'Portland' grape. (*Left*) diploid; (*right*) tetraploid. [Photograph courtesy H. Dermen.]

Chromosome Behavior of Polyploids

Tetraploids, unlike diploids, have four chromosomes of each type. The chromosome behavior of tetraploids depends on their pairing relationships at meiosis. If they pair two by two (bivalent pairing), they will separate normally. However, if all four chromosomes pair together (quadrivalent pairing), there is a possibility of a 3 to 1 distribution leading to gametes with unbalanced chromosome numbers. This type of chromosome separation is largely responsible for the reduced fertility of tetraploids.

The cross of a tetraploid with a diploid produces a triploid:

$$2x \text{ gamete} + x \text{ gamete} = 3x \text{ zygote}$$

Chromosome pairing commonly involves the three chromosomes of each type (trivalent pairing). When the chromosomes separate at the first division of meiosis, two go to one end of the cell and the third goes to the other. Since the assortment of each trivalent is independent, gametes with chromosome numbers anywhere between x and $2x$ may be formed. Consider spinach, for example, in which $x = 6$. In a triploid ($3x = 18$) the possible chromosome assortments at anaphase 1 of meiosis are 6–12, 7–11, 8–10, and 9–9. The seven types of gametes in terms of chromosome number (6,7,8,9,10,11,12) are produced in a binomial distribution and occur in the frequency of 1:6:15:20:15:6:1. Only $\frac{2}{64}$ of the gametes are x or $2x$. As the chromosome number of plants increases, the frequency of x or $2x$ gametes in triploids becomes very small. Gametophytes having unbalanced numbers of chromosomes are either nonviable or are at a great selective disadvantage. Consequently, triploids are commonly quite sterile. They may be propagated asexually (as are triploid apples and pears) or produced anew each year from tetraploid-diploid crosses (as are seedless watermelons).

Genetics of Polyploids

The single-gene ratios in diploid organisms are based on the assortment of two alleles per gene. In tetraploids, however, four alleles are present at each locus. There are two types of homozygous genotypes (*AAAA* and *aaaa*) and three kinds of heterozygous genotypes (*AAAa*, *AAaa*, and *Aaaa*). The genetic ratios from crosses involving these types differ from those of diploids. The *AAaa* heterozygote will produce gametes in a ratio of 1*AA*:4*Aa*:1*aa*. If *A* is completely dominant and produces the *A*_ phenotype even with three doses of *a* (*Aaaa*), the progeny of *AAaa* selfed will produce a phenotypic ratio of 35*A*_:1*aa*. This ratio, however, is affected by the location of the gene on the chromosome. Thus, tetraploidy tends to muddle the genetic picture. Recessive genes appear hidden, for they are expressed less frequently. As a result, genetic analysis of polyploid plants is exceedingly complex.

Use of Polyploid Breeding

Polyploidy is used in a number of important ways in breeding. Although the indiscriminate induction of tetraploidy seldom leads to anything of immediate value, manipulations of ploidy have become a valuable tool in breeding. In some plants the large size associated with higher ploidy is of value; for example, there are now a number of very attractive tetraploid snapdragon cultivars. Most lilies are bred on the tetraploid level because of the thicker petals and larger flowers associated with the increased number of chromosomes.

There has been great interest in the use of polyploidy as a means of extending variability by creating new species or by transferring genes from other species. Induced polyploidy makes it possible to overcome sterility associated with interspecific hybrids. This sterility is a result of the inability of chromosomes to pair at meiosis. Fertility is restored by doubling the chromosome number of the sterile hybrid or by crossing autotetraploids of each species. Using capital letters to designate the chromosome complement (**genome**), we can represent this as follows:

$$\begin{array}{c} AA \\ \times \\ CC \end{array} \longrightarrow \begin{array}{c} AC \\ \textit{Sterile} \\ \textit{hybrid} \end{array} \xrightarrow{\textit{chromosome doubling}} \begin{array}{c} AACC \\ \textit{Fertile} \\ \textit{amphidiploid} \end{array}$$

$$\begin{array}{c} AA \xrightarrow{\textit{chromosome doubling}} AAAA \\ \times \\ CC \xrightarrow{\textit{chromosome doubling}} CCCC \end{array} \longrightarrow \begin{array}{c} AACC \\ \textit{Fertile} \\ \textit{amphidiploid} \end{array}$$

The resulting **amphidiploid** is an allopolyploid made up of the entire somatic complement of each species (Figure 17-11). Fertility results from bivalent pairing within each genome.

In some crops the triploid condition is desirable because of increased vigor (Figure 17-12) and fruit size (pears, apples) or to take advantage of reduced fertility, such as in the banana or the seedless watermelon (Figure 17-13).

Polyploidy has not proved to be a panacea in plant breeding. It is, however, another in the arsenal of weapons breeders may use to change plants to better suit their needs. Polyploid manipulation has greatly enlarged the genetic base from which a breeder may draw. These methods undoubtedly will contribute much to future crop improvement in horticulture.

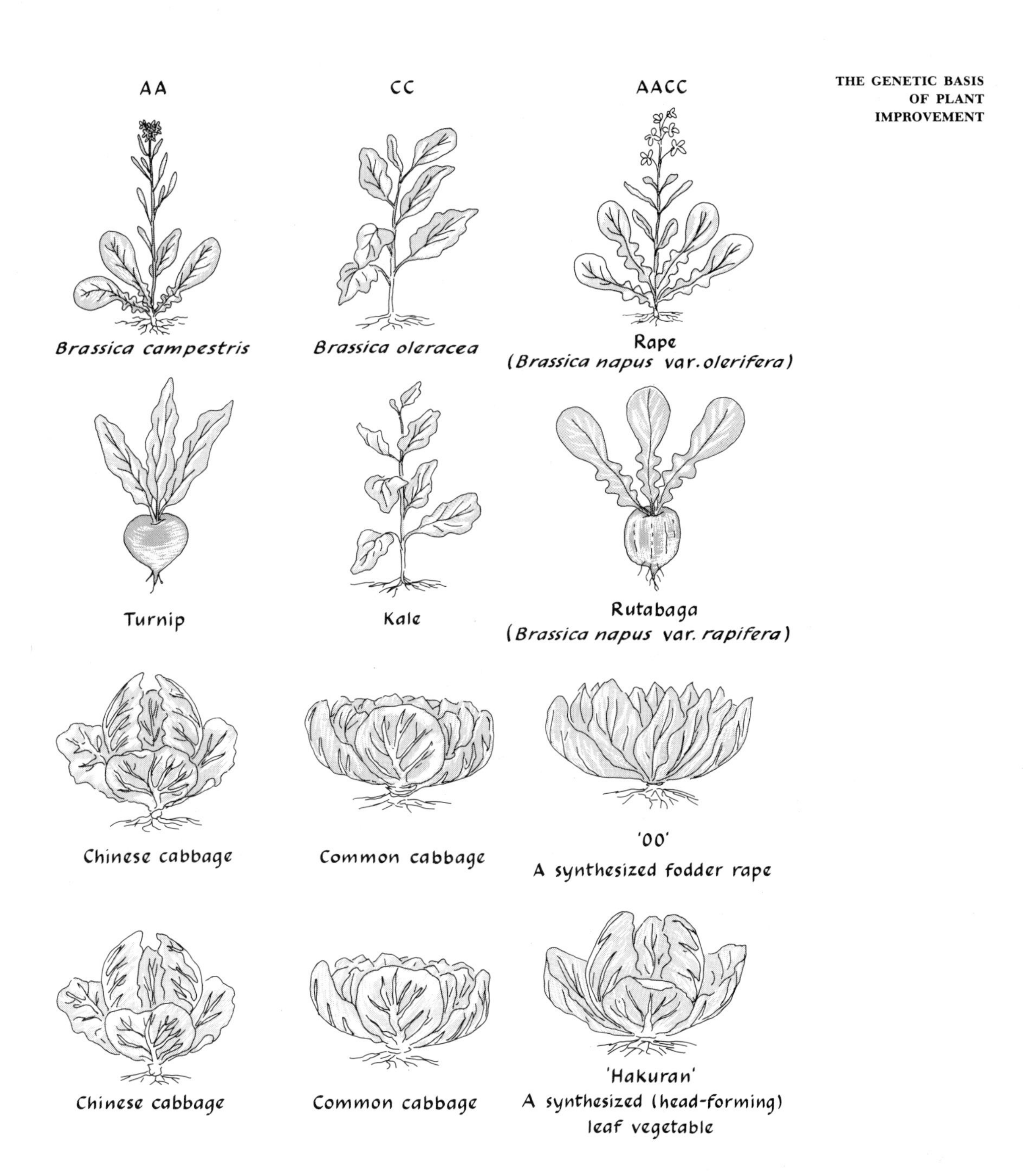

FIGURE 17-11. Natural and artificial amphidiploids in *Brassica*. [Courtesy H. Kihara, *Seiken Zihô* 20:1–14, 1968.]

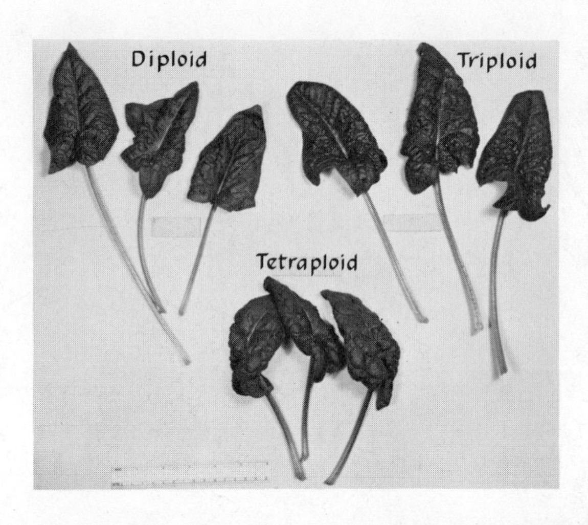

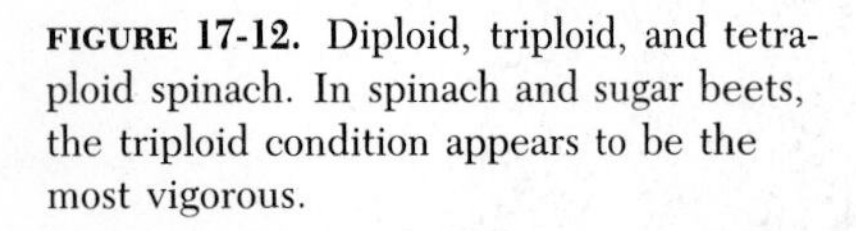
FIGURE 17-12. Diploid, triploid, and tetraploid spinach. In spinach and sugar beets, the triploid condition appears to be the most vigorous.

FIGURE 17-13. The seedless watermelon is the result of a cross between a tetraploid and a diploid. The triploid fruit matures, but, as a result of sterility, most of the seeds do not develop normally. The small underdeveloped seeds are similar in texture to seeds of cucumber in the edible stage. [Photograph courtesy Purdue University.]

BREEDING METHODS

Sources of Variation

Genetic variability is the raw material of the plant breeder. The richest source of genetic variability for a particular species has been shown to be its geographical **center of origin** (Figure 17-14). This is also a **center of diversity** where a pool of genes exists for exploitation by the plant breeder. The incorporation of this genetic variability, when in the form of closely related species, may involve such special techniques as the manipulation of chromosome numbers or the artificial culture of embryos. In the tomato, valuable genes for resistance to fusarium wilt, tomato-leaf mold, and gray leaf spot have come from a related species that grows wild in Peru, *Lycopersicon pimpinellifolium*.

For many cultivated plants, a wealth of material is available from cultivars already under cultivation. A complete collection of such widely grown plants as the apple, tomato, or rose must be worldwide in scope (Figure 17-15). The New Crops Research Branch of the Agricultural Research Service (USDA) facilitates the exploration and introduction of genetic variability for the improvement of plants and the establishment of new crops. Societies have been organized around many plants (especially ornamentals), and they often assist in the dissemination of plant material. Examples in the United States are the African Violet Society and the Tomato Breeders Cooperative. Breeders are often good sources of plant material.

Sports

Mutations of spontaneous origin, although rare, contribute to genetic variability. These changes are referred to in horticultural terminology as **sports.** Desirable mutations occurring in adapted, asexually propagated plants may result in immediate improvement such as the color sports in many apple varieties and tree types in coffee plants.

Mutations occurring somatically may involve only a sector of tissues, resulting in a **chimera** (Figure 17-16). (The original chimera, a mythical beast, was part lion, part goat, and part dragon.) Chimeras, when vegetatively propagated, tend to be unstable. The reason is that buds may be formed from tissues with or without the mutation. For example, the 'White Sim' carnation, a sport of the 'Red Sim' cultivar, is a chimera. Buds that form from internal tissues will produce red flowers. Upon close inspection, the normal 'White Sim' blooms also show small islands of red tissue. It is important to know whether a sported plant is stable before undertaking extensive propagation.

From a study of the performance of various types of natural variants under different environmental conditions, it may be possible to make an immediate improvement merely by selecting the best individuals. However, with established crops it becomes more and more difficult to make improvements in this manner. Usually, the individual desirable characters are present in many plants, no one of which contains all the desirable attributes. It then becomes necessary to attempt to recombine the characters by hybridization (sexual crossing) in the hope of obtaining plants having as many of the desirable features as possible. When these characters are spread over many cultivars and lines, the process necessitates making many crosses.

The desired characters are too often found only in plants that may be horticulturally unsatisfactory from other standpoints. Resistance to the apple scab disease is seen in species of apple with almost inedible fruit less than 1 inch in

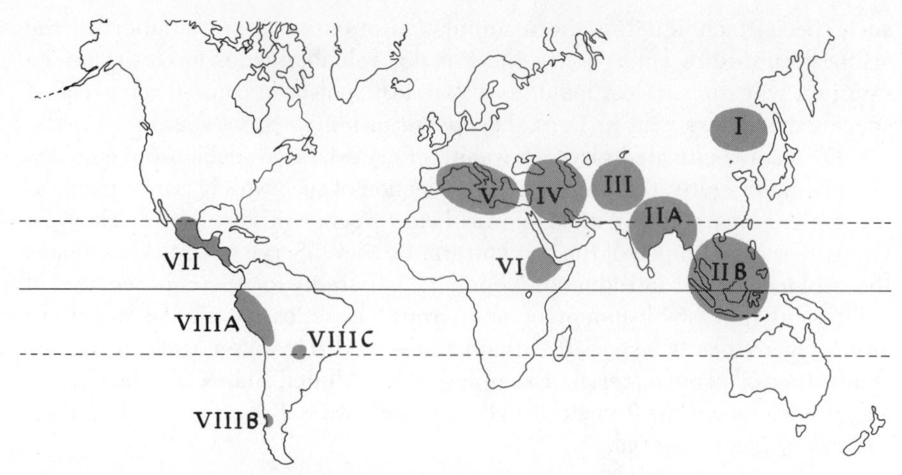

FIGURE 17-14. Centers of origin of important cultivated plants. See key on opposite page.

diameter. In addition, there are a great many characters to be considered. For example, a successful strawberry cultivar must satisfy (1) the growers, with respect to yield, season of ripening, and disease resistance; (2) the consumers, with respect to appearance and quality; (3) the nursery growers producing the plants; and (4) the shippers, handlers, and processors.

FIGURE 17-15. A collection of English cultivars of tomatoes selected for greenhouse production. [Photograph courtesy Purdue University.]

KEY TO THE CENTERS OF ORIGIN OF IMPORTANT CULTIVATED PLANTS.

Old World

I. *China*—The mountainous and adjacent lowlands of central and western China represent the earliest and largest center for the origin of cultivated plants and world agriculture. Native plants include millet, buckwheat, soybean and many other legumes, bamboo, crucifers, onion, lettuce, eggplant, various cucurbits, pear, cherry, quince, citrus, persimmon, sugar cane, cinnamon, and tea.

II. *South Asia*—This area has two subcenters. The principal one, Hindustan (IIA), is considered the center for rice, sugar cane, many legumes, and such tropical fruits as mango, orange, lemon, and tangerine. The other, Indo-Malaya (IIB), is the center for banana, coconut, sugar cane, clove, nutmeg, black pepper, and manila hemp.

III. *Central Asia*—This region is most important as the center of origin of common wheat. Native plants include pea, broad bean, lentil, hemp, cotton, carrot, radish, garlic, spinach, pistachio, apricot, pear, and apple.

IV. *Asia Minor*—At least nine species of wheat, as well as rye, are native. This area is the center for many subtropical and temperate fruits (cherry, pomegranate, walnut, quince, almond, and fig) and forages (alfalfa, Persian clover, and vetch).

V. *Mediterranean*—This is the home of the olive and many cultivated vegetables and forages. The effect of early and continuous civilizations is indicated in the improvement of crops that originated in Asian centers.

VI. *Abyssinia* (Ethiopia)—Wheats and barleys are especially rich; sesame, castor bean, coffee, and okra are indigenous.

New World

VII. *Central America*—Extremely varied native plants, including maize, bean, squash, chayote, sweetpotato, pepper, upland cotton, prickly pear, papaya, agave, and cacao.

VIII. *South America*—A number of South American centers are noted. The West-Central area (VIIIA)—now Ecuador, Peru, and Bolivia—is the center of origin of many potato species, tomato, lima bean, pumpkin, pepper, coca, Egyptian cotton, and tobacco. The island of Chiloe in southern Chile (VIIIB) is thought to be the source of the potato. The peanut and the pineapple originated in the semi-arid region of Brazil (VIIIC), and manioc and the rubber tree in the tropical Amazon region.

SOURCE: After N. V. Vavilov, 1951, *The Origin, Variation, Immunity and Breeding of Cultivated Plants*, translated by K. Starr Chester (Ronald Press).

The incorporation of these various characteristics, most of which are quantitatively determined, involves many series of crosses carried over many generations. With each generation, the amount of plant material tends to pyramid into unmanageable proportions unless rigid selection is maintained. Breeding becomes a problem in biological engineering. The main effort of the plant breeder is directed toward the recombination of desired characters in as efficient a manner as possible in terms of land, time, and labor.

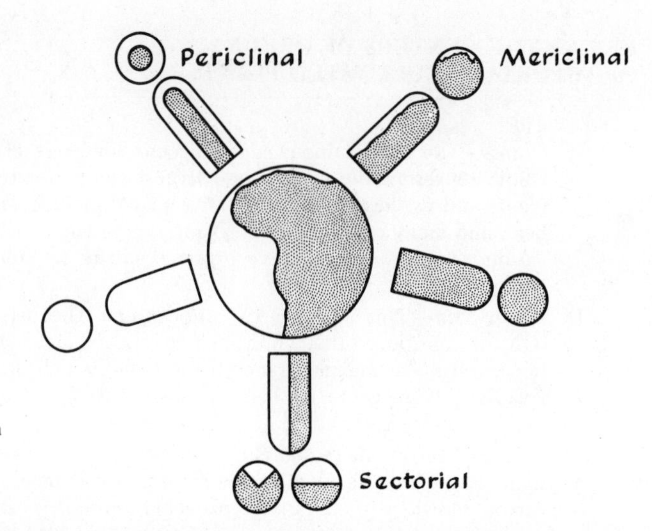

FIGURE 17-16. Plant chimeras contain genetically different tissues. Various types may be derived from buds arising from a sectored chimeral stem. [Adapted from W. N. Jones, *Bot. Rev.* 3:545–562, 1937.]

Artificial Creation of Variability

One of the limitations to plant improvement is the dependence upon naturally occurring variation. The artificial induction of mutation by radiation (X-rays, gamma rays, thermal neutrons) and by chemical means (ethyl dimethyl sulfate) provides a method for creating changes that have not occurred naturally. Since such induced changes are completely random and are largely harmful, refined techniques for screening are needed. For example, large populations of plants can easily be screened in the seedling stage for resistance to a disease if the susceptible plants died soon after inoculation. A promising application of irradiation may lie in its use in obtaining beneficial changes in asexually propagated material. In this way, otherwise adapted material may be screened in the search for a single desirable change in the hope of eliminating further generations of breeding and extensive testing.

Genetic Structure of Crop Cultivars

The method of pollination has a profound effect on the genetic constitution of the plant. The amount of cross-pollination varies from essentially none in plants such as the garden pea to 100 percent in dioecious and self-incompatible plants. Two main groups are recognized: naturally **self-pollinated** plants, in which cross-pollination is less than 4 percent, and naturally **cross-pollinated** plants, in which cross pollination exceeds 40 percent. The intermediate types are usually considered along with the cross-pollinated group.

Self-pollinated plants are ordinarily homozygous for practically all genes. The exceptions are a result of chance cross-pollination and mutation. However, any heterozygosity is quickly eliminated as a consequence of natural inbreeding. Thus, although a hybrid between two divergent types will be heterozygous for many genes, by the F_6 to F_{10} generations of natural selfing, the population will consist of many different individuals, each of which are, for all practical purposes, completely homozygous. The basic problem in the improvement of self-pollinated plants lies in selecting the best genotype from the unlimited number of different genotypes. Once a satisfactory homozygous plant is selected, the problem of genetic maintenance is small as compared to that of cross-pollinated plants.

The genes in naturally cross-pollinated, seed-propagated plants are recombined constantly from generation to generation. A cross-pollinated cultivar is typified not by any one plant but by a population of plants. The problem of improving cross-pollinated plants consists in somehow maintaining visible uniformity while avoiding the decline in vigor associated with homozygosity. In general, procedures that result in severe inbreeding cause a loss of vigor. Once a desirable population is achieved, there is still the perpetual problem of maintenance. A unique method of producing uniformity and at the same time maintaining heterozygosity has been the process of producing F_1 hybrids. At present, however, technical difficulties prevent this breeding method from being used with all cross-pollinated crops.

A large number of cross-pollinated crops (including apple, rose, and gladiolus) are normally vegetatively propagated. Here improvement depends on the selection of a single desirable genotype. The problem of genetic maintenance is solved by the elimination of the sexual process. In improving these plants by recombination, inbreeding must also be avoided. Hybridization between unrelated plants is usually made in order to obtain a vigorous population within which selection may be practiced.

Selection

The problem of the breeder confronted with a population containing many diverse genotypes is to recognize and save only the most desirable types. This process of selection differs depending on the method of reproduction of the plant.

Selection in Self-Pollinated Crops

Two fundamentally different types of populations of self-pollinated crops exist. One is a mixture of different homozygous lines as found in a collection of cultivars. Here selection consists of determining the best genotype by testing. As each cultivar is homozygous, the problem of genetic maintenance is eliminated. The best genotype can be expected to be duplicated from its selfed seed. The other type of population is a mixture of different heterozygous genotypes, as found in the F_2 population of a cross between different homozygous cultivars. As discussed previously, some completely homozygous types are theoretically expected, but their number greatly decreases as the number of genes differentiating the original cross increases. The problem of improvement is now twofold: to select the best genotype and to transform this genotype as closely as possible into a homozygous line.

The genetic value of a self-pollinated, seed-propagated plant is the average performance of its progeny. The task of selecting the best genotype from an F_2 population is difficult because of the problem of distinguishing between genetic and environmental variation. Selection of the most desirable genotypes is accomplished on the basis of progeny performance. This process, which is known as **pedigree selection,** is based on the assumption that the best homozygous genotype will be derived from the heterozygous plant that produces the most desirable progeny.

Assume that selfed seed from a number of F_2 plants are planted out to produce F_3 lines. Since each selfing brings about homozygosity in a 50 percent increase each generation, the variability within F_3 lines will be half as great as the variability between F_2 plants. That is, the plants within a particular F_3 line will resemble one another more so than will the aggregate of all F_3 lines. For example, some lines will be uniformly large and some uniformly small. By choosing between a number of lines rather than choosing a single F_2 plant, there is less chance of confusing genetic and environmental effects.

The selection process may then be repeated. What appear to be the best plants in the best F_3 lines are planted out, and selection is made again on a "pedigree" basis between F_4 lines. By the sixth to tenth generation the lines derived from single plant selections will be homozygous for more than 95 percent of their genes. Such lines are, for practical purposes, considered to be true breeding. If one of these lines is of superior type, it may be named, and it is then considered a new horticultural cultivar.

The problem of straight pedigree selection is the extreme expense it entails. Fairly extensive records must be maintained, and unless rigid selection is maintained, the program soon mushrooms to extremely unwieldy proportions.

An alternate method of selection is known as **mass selection.** In this technique, what appear to be the best plants are selected and maintained in bulk without testing the progenies separately. This may be accomplished by eliminating (**roguing out**) the undesirable plants in each generation and harvesting the remaining plants. This can be done mechanically for some crops, such as by screening for seed size or harvesting at a particular date to select for a particular season of ripening. After 6 to 10 generations, the population will consist of a heterogeneous mixture of "somewhat selected" homozygous genotypes. The progeny of any plant can be expected to form the basis of a new cultivar. The problem now is to determine the best genotype by testing. This method will be successful if the better genotypes are retained in the mass selection process either by judicious selection or merely by virtue of large numbers.

The processes of mass selection and pedigree selection can be combined. For example, F_2 plants can be pedigree selected on the basis of F_3 line performance. Thus the greatest genetic differences are exploited early, and the obviously undesirable types are eliminated. The best F_3 lines may then be bulked and carried on by mass selection until homozygosity is reached, at which time pedigree selection is resumed.

Selection in Cross-Pollinated Crops

Pedigree selection depends on genotype evaluation by inbreeding. Straight pedigree selection is undesirable as a method of improving cross-pollinated plants, in which inbreeding leads to loss of vigor, unless some procedure is set up to combine inbred lines and restore vigor. Mass selection enables a cross-breeding population to become relatively uniform for certain visible characters and to conserve enough variability to maintain vigor. Inbreeding is avoided by natural interpollination.

Mass selection often leads to a rapid improvement of cross-pollinated plants. However, it sometimes becomes difficult to increase this "genetic gain" after a certain point. To obtain more control of the genotypes that make up the cross-pollinating population, pedigree and mass selection may be combined. This goal is accomplished by selfing individual plants and pedigree-selecting them on the basis of their progeny. The selected lines are then allowed to interpollinate to restore heterozygosity. The process may be repeated for a number of cycles. There are several variations of this procedure. When the character evaluated is a fruit or a seed, selection must be made after pollination. In this case, the "female" parent is pedigree-selected, and the "male" parent is mass-selected.

In a very real sense, F_1 hybrids are an extreme form of this method. Here the inbreeding process is carried on to homozygosity, and vigor is restored by combining two inbreds (or four in a double cross). The selection of the inbred combination results in uniform, genetically controlled hybrids.

Selection of Asexually Propagated Crops

In asexually propagated plants, selection is straightforward, since any genotype may be perpetuated intact. The problem is one of testing—that is, of determining the best genotype. If the most desirable selection is still unsatisfactory and further improvement is necessary, the selection of the best genotype to be used for further crossing is more difficult because the best-performing selection does not always make the best parent. A sample of genotypes must be selected and tested as parents on the basis of their progeny performance.

Utilization of Hybrid Vigor

The production of F_1 hybrids as a source of seed-propagated, cross-pollinated plants depends upon a number of factors. The first requirement is that it must be possible to obtain and maintain inbred lines. Some inbred lines become so weak that they are hard to maintain. Inbreeding itself may be very difficult with some plants.

A number of crops resist self-fertilization because of incompatibility mechanisms. Self-incompatibility is usually controlled by alleles at a single gene locus. When pollen contains the same incompatibility allele that is present in the style, tube growth is arrested. In a few crops, such as cabbage, incompatibility may be circumvented by pollinating before the flower opens (bud pollination).

Homozygous lines may be produced in a single step, avoiding the long process of inbreeding. Monoploids are produced in certain crops, such as maize, in a very low but predictable frequency. They apparently result from a disruption in the double fertilization process. One male gamete combines with the two polar nuclei to form the triploid endosperm, but the fertilization of the egg nucleus does not occur. Monoploids may be recognized in the seedling stage by the use of special marker genes. Many of these monoploid lines double spontaneously to produce "instant" homozygous diploid types.

The second requirement for producing hybrids is that an efficient cross between the inbred lines must be possible. This cross may be exceedingly difficult in perfect-flowered plants, unless each hand pollination yields a large number of seeds. One method used in overcoming the difficulty is the incorporation of **male sterility,** which transforms one inbred into a "female" line (Figure 17-17). Another technique is the use of the self-incompatibility mechanism. If this method is used, special breeding techniques are required to perpetuate the male-sterile line.

The third requirement is an economic one. The improvement must be great enough to warrant the extra expense of hybrid seed.

The process of producing F_1 hybrids by the inbreeding process has not proved practical in cross-pollinated plants that are vegetatively propagated, although it is often suggested as a breeding method. Each clonal cultivar is essentially a hybrid, which, if duplicated asexually, need not be duplicated by crossing two unique inbred lines. The great advantage of F_1 hybrids in cross-pollinated plants is their *uniformly* high vigor. However, the process of vegetative propagation assures absolute genetic uniformity.

Hybrid seed is now commercially available for the following vegetables: broccoli, Brussels sprouts, cabbage, carrot, cauliflower, Chinese cabbage, cucumber, eggplant, muskmelon, onion, pepper, spinach, squash, sweet corn, radish, tomato, and turnip. F_1 hybrid flowers include African violet, *Ageratum,* begonia, *Calceolaria,* cyclamen, geranium *(Pelargonium),* gloxinia, Iceland poppy, *Impatiens,* marigold *(Tagetes),* pansy, petunia, *Salpiglossis,* snap-

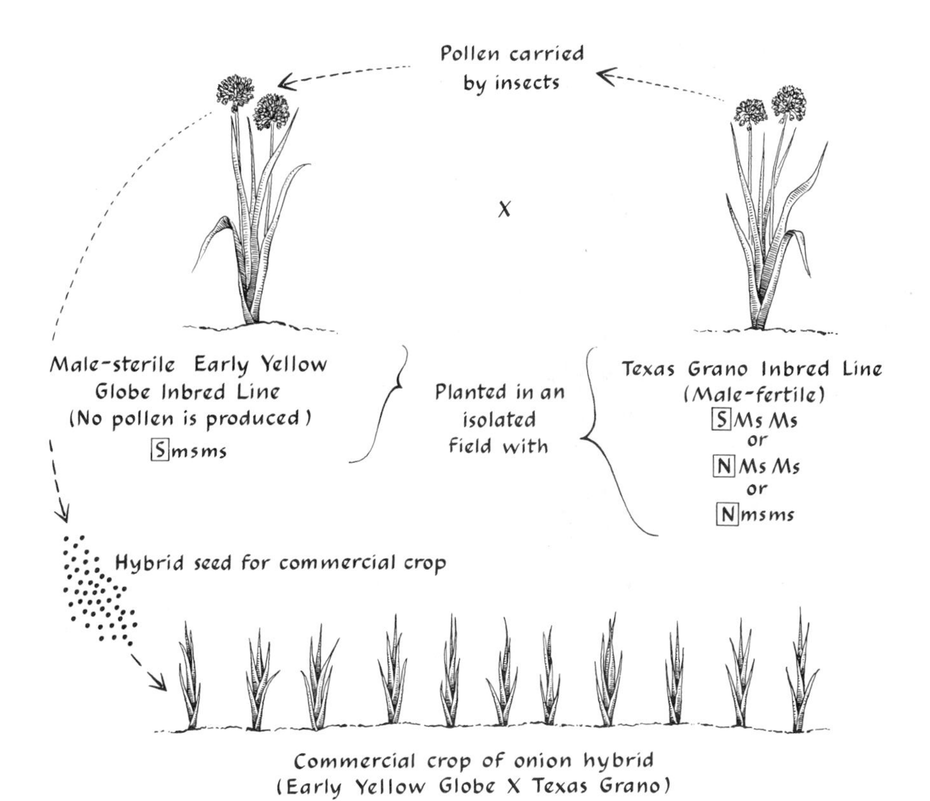

FIGURE 17-17. Controlled hybridization in the onion. Genetically determined male sterility transforms one inbred to a female line. Male sterility in onions results from interaction between a cytoplasmic factor [S] and recessive nuclear gene *ms* in the homozygous state. Plants with [N] (normal) cytoplasm or with gene *Ms*, either homozygous or heterozygous, produce viable pollen. [Adapted from A. M. Srb, R. D. Owen, and R. S. Edgar, *General Genetics*, 2d ed., W. H. Freeman and Company, copyright © 1965.]

dragon, and zinnia. Vigorous hybrid genotypes can be selected from segregating populations instead of creating them from inbred lines (Figure 17-18).

As heterosis appears to be involved to a lesser extent in self-pollinated crops, the creation of F_1 hybrids (actually cultivar crosses) is not usually an important breeding method for these plants. However, since cultivar crosses offer a quick means of producing a particular combination, this method has found a limited place in some self-pollinated crops, such as greenhouse tomatoes.

Although the maximum amount of heterosis is obtained by crossing two inbreds, a number of other, less heterotic combinations may be made. The various kinds of crosses that are referred to as hybrids in the trade are designated as follows.

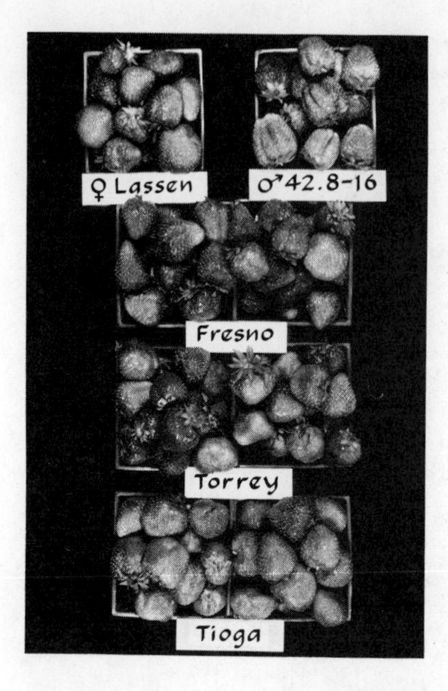

FIGURE 17-18. 'Fresno,' 'Tioga,' and 'Torrey' are three strawberry cultivars obtained from crossing 'Lassen' with a California selection 42.8-16. [Photograph courtesy Royce S. Bringhurst and Victor Voth.]

Type of cross	Name of hybrid
Inbred × inbred	Single cross
F_1 hybrid × inbred	Three-way cross
F_1 hybrid × F_1 hybrid	Double cross
Inbred × cultivar	Top cross

Since many inbreds are relatively weak, the double cross is an attempt to produce seed on a vigorous plant. It has been used very widely in field corn (Figure 17-19). The top cross has been used in spinach, and the three-way cross is popular with sweet corn.

The genetic improvement of inbreds is a specialized type of breeding. Basically, they are treated as self-pollinated plants. However, the success of an inbred is often difficult to determine phenotypically but is related to its success as a parent. This ability of an inbred to produce good hybrids is known as **combining ability.** Combining ability is inherited, but the only way to select for it is to test it directly. It may be selected for early in the inbreeding process by crossing the plant to a tester at the same time it is selfed. The selfed seed is set aside for further inbreeding until after the results of the crossing trial are evaluated.

The Backcross Method

The backcross has been defined as a cross of an F_1 plant with one of its parents. The method is a breeding technique for transferring single characters, readily identifiable, from one cultivar to another. Characters controlled by single genes, such as certain types of disease resistance, male sterility, and so on, are most easily transferred by this method. By repeated backcrossing of the hybrid to the parent cultivar that carries the most desirable characters (the **recurrent parent**), but selecting in each generation for a single character from the other parent (the **nonrecurrent parent**), a genotype will be eventually obtained that has all the genes of the recurrent parent except those affecting the selected character from the nonrecurrent parent. In the following explanatory discussion, assume that the "cultivars" are of self-pollinated plants.

In regard to consequences of the backcross method, two sets of genes must be considered. One set is small, consisting only of the genes from the nonrecurrent parent that are to be transferred. The other set consists of a large group of genes from the recurrent parent, which we do not wish to lose. The genes of the recurrent parent can be shown to be transferred by an increment of 50 percent in each generation of backcrossing. By selecting plants at random in each generation, we can expect the genes of the recurrent parent not only to be incorporated in the hybrid but also to become homozygous. This can be shown easily with respect to a single gene.

Assume a cross involving a single gene pair, $AA \times aa \rightarrow Aa$. The two types of backcrosses are

$$Aa \times AA \longrightarrow 1Aa:1AA$$
$$Aa \times aa \longrightarrow 1Aa:1aa$$

Compare these backcrosses with selfing:

$$Aa \times Aa \longrightarrow 1AA:2Aa:1aa$$

Note that, in backcrossing and selfing, half the genes become homozygous. This result can be generalized for any number of genes. The rate of return to homozygosity in backcrossing is the same as in selfing, but *all the homozygous genes resemble the recurrent parent.* If backcrossing to the same homozygous type is continued for 6 to 10 generations, more than 95 percent of genes of the hybrid will be identical to those of the recurrent parent and will be in the homozygous condition.

If the gene to be transferred is dominant, the procedure is straightforward. After the last backcross, the gene may be made homozygous by selfing. If the character to be transferred is recessive, special testing must be carried on in each generation to be sure the plant selected for backcrossing is heterozygous and thus contains the desired recessive.

The backcross method has found its greatest usefulness in self-pollinated crops and in improving inbred lines of cross-pollinated plants. It is not so useful in cross-pollinated crops because backcrossing is equivalent to selfing in achieving homozygosity. This may be overcome, however, by using a large number of selections of the recurrent line. In this way, inbreeding may be avoided (Figure 17-20).

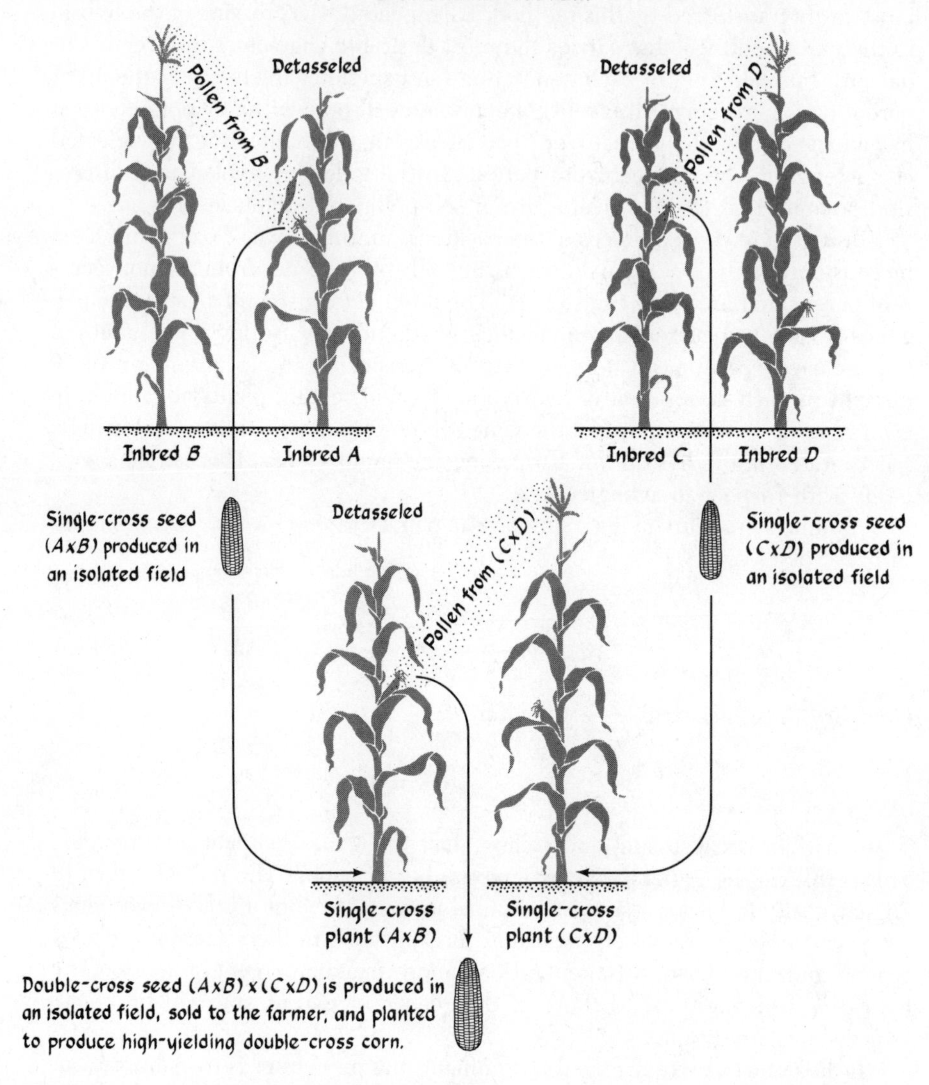

FIGURE 17-19. The production of single-cross and double-cross hybrid corn. The detasseling process (*A*) can be eliminated by manipulating male sterility (*B*) determined by an interaction between a cytoplasmic factor [S] and a double recessive nuclear gene *rf*. Only the combination [S] *rf rf* is male-sterile. Plants with [N] (normal) cytoplasm or with the pollen restoring allele *Rf*, either homozygous or heterozygous, produce viable pollen. The cytoplasmic factor passes only through the egg. Appropriate genotypes are shown in shaded rectangles. Unfortunately,

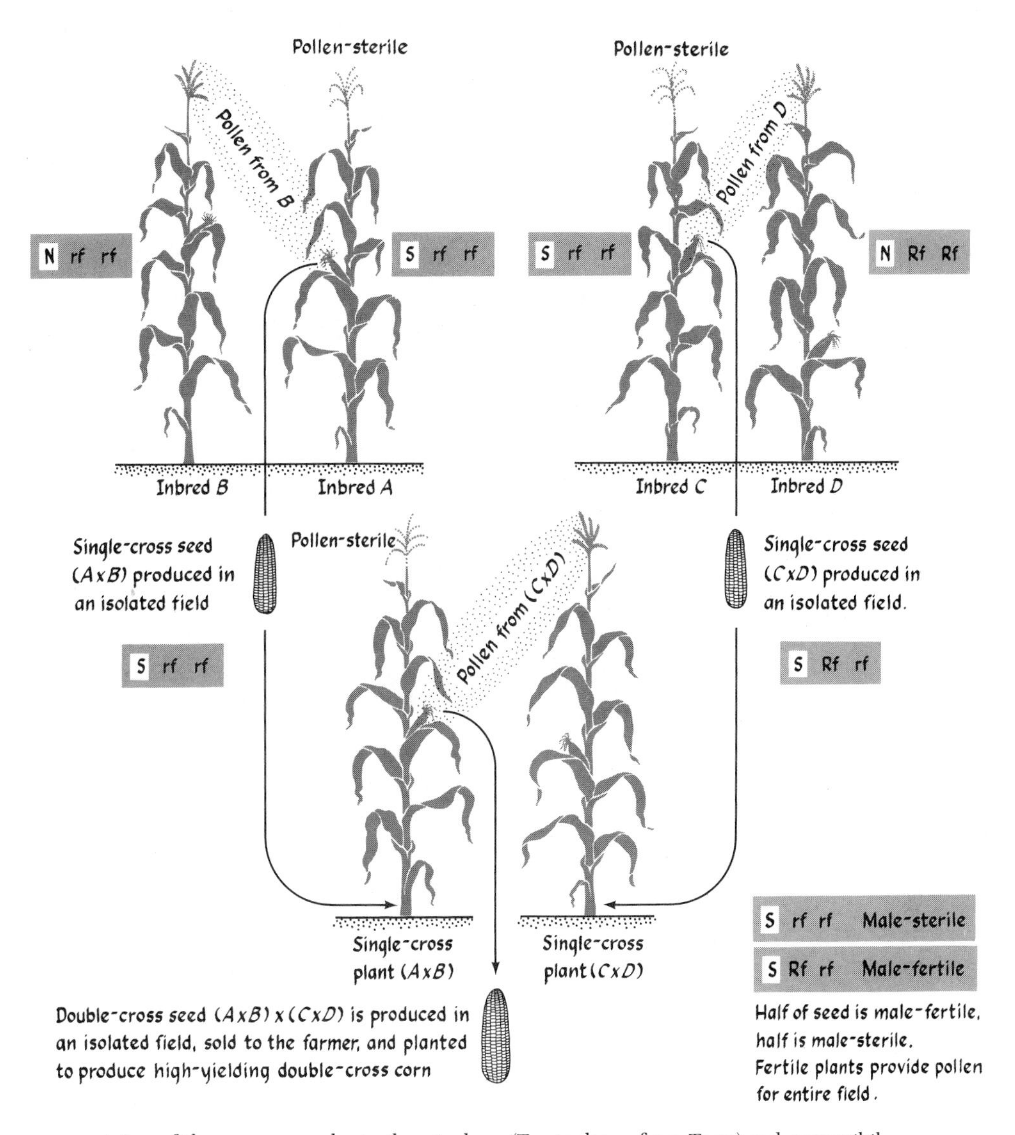

an association of the common male-sterile cytoplasm (T-cytoplasm, from Texas) and susceptibility to Southern corn blight caused by *Helminthosporium maydis* has almost eliminated the commercial use of the cytoplasmic male-sterile system after the disastrous epidemic of 1970. [From J. Janick et al., *Plant Science,* 2d ed., W. H. Freeman and Company, copyright © 1974.]

A
B
C
D
E
F

FIGURE 17-20. Backcross bleeding for scab resistance in the apple. (*A*) The fruits of *Malus floribunda* 821, which is heterozygous for a dominant gene that confers resistance to apple scab. (*B*) Selection derived from cross of *M. floribunda* 821 × 'Rome Beauty.' Subsequent backcrosses to different large-fruited types (*C* and *D*) produce resistant seedlings with increased fruit size and quality. The fruits in the basket (*E*) are the 'Prima' apple, the first release of the scab-resistant apple breeding program cooperatively conducted by Purdue University, Rutgers University, and the University of Illinois. 'Prima' has been successfully introduced in the United States as an early apple both for home gardens and for commercial use. Its use is expanding in Eastern and Western Europe, with the most extensive plantings in Yugoslavia. (*F*) Resistant seedlings are selected in the seedling stage. Susceptible seedlings are killed by the disease.

The main advantage of the backcross method is its predictability. The improved cultivar is often indistinguishable from the old one with the exception of the added character. As a result, extensive testing may be avoided. But the backcross method cannot be expected to improve the cultivar any more than the addition of the single selected character. Moreover, unless this character is inherited relatively simply, the method is difficult to use.

Biotechnology, Genetic Engineering, and Crop Improvement

Conventional plant breeding refers to a system for the selection of superior genotypes from genetically variable populations. It produces superior individuals, lines, or populations largely by using the sexual process to recombine favorable genes. The system is powerful because it is evolutionary: Improved individuals or populations continually serve as parents for ever more improved offspring. The limitations of conventional plant breeding are the inability of the sexual system to incorporate variation from nonrelated species, the reliance on naturally occurring or randomly induced mutations, the inability to obtain specific genetic changes, and finally the difficulty in detecting infrequent or rare recombinants. Furthermore, even when genetic variation is available in usable form, progress depends upon time to generate cycles of recombinants, space to grow the necessary populations to recover superior recombinants, and resources to be able to select, identify, and propagate desirable combinations.

The new biotechnological revolution includes novel techniques based on microbial and molecular genetics that can overcome the limitations of conventional plant breeding. These techniques include cell culture, cell selection, protoplast fusion, and recombinant DNA. Recombinant DNA technology is also dependent upon cell-culture techniques. Sometimes called genetic engineering, it is the most powerful and revolutionary of the new genetic techniques.

Cell Selection and Somaclonal Variation

A shortcut to releasing variability is to impose selection on populations of cells rather than on populations of whole plants. For the technique to be successful, it must be possible to regenerate plants from individual cells. This has been achieved with a great number of horticultural plants (see pages 346–351). The advantage of this system is that extremely large populations of cells (literally in the billions) can be grown in a very small space, such as a few culture tubes or flasks. The problem is the difficulty of selecting for variants at the cellular level that will be expressed as desirable characters in the growing plant.

Variation in cell populations occurs naturally as a result of spontaneous mutation, and it can also be induced by means of various mutagenic agents. Interestingly, the culture system itself appears to generate variants with a higher frequency than would be expected as a result of spontaneous gene mutation. Changes induced by cell and tissue culture have been called **somaclonal variants.** Some somaclonal variation has been traced to gene and chromosomal changes and some may be nongenetic, adaptive changes, but the precise cause of the high frequency is still unknown. Somaclonal variation induced by cell and tissue culture appears to be a new source of variation available to plant breeders.

Protoplast Fusion

Plant-cell hybridization has been shown to be feasible with protoplasts. Protoplasts can be isolated from plant cells by digesting cell walls with appropriate enzymes. A small but significant percentage of protoplasts grown in suspension culture show genetic fusion (Figure 17-21). This means that if protoplasts of two genetic backgrounds are mixed, some of the fused protoplasts will be nuclear hybrids, some will have mixed cytoplasms (identified by a new term as **cybrids**), and some may be both.

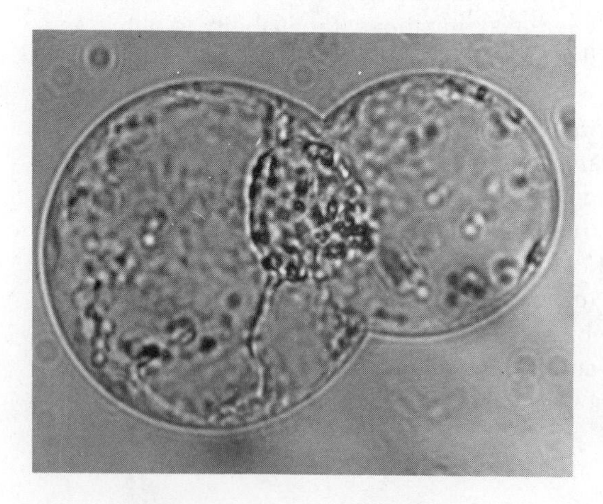

FIGURE 17-21. Initial stage of fusion between two protoplasts of tobacco. [From T. Murashige, *HortScience* 12:127–130, 1977.]

Hybrid protoplasts must be isolated from those that are not fused and from those that are fused with protoplasts of the same genotype. One method is to use a medium in which only hybrid products will be able to grow. Mature plant hybrids have been produced from cellular hybridization in tobacco and petunia. The refinement of this process may make it possible to achieve wide crosses that are impossible with traditional methods of sexual hybridization. Various organelles that are capable of regeneration with their own DNA (such as chloroplasts and mitochondria) may also be transferred from one species to another in this way.

Anther Culture

Pollen of various plants can be cultured to produce haploid cells, or even haploid "embryoids." Haploid cell lines of plants are extremely valuable for mutation breeding because, since dominance is eliminated, all mutated recessive genes will be immediately expressed. Furthermore, doubling such haploids produces "instant" homozygous lines, a process that may require many years with ordinary sexual inbreeding. Haploid cells might also be used to produce "hybrids" directly by protoplast fusion.

Recombinant DNA

The basis of recombinant DNA technology is the ability to isolate DNA fragments containing one or more genes and replicate them in bacteria. Enzymes called **restriction endonucleases** permit DNA to be cleaved into desired sizes that will hybridize, that is, will attach, to other DNA fragments. The vector used is an extrachromosonal, circular, bacterial DNA called a **plasmid.** These plasmids are modified to accept foreign DNA and labeled with biochemical markers to allow their identification after replication in the bacteria. The replication and subsequent selection of these bacteria containing the marked DNA is termed **gene cloning.**

In order for recombinant DNA to be useful for crop improvement, a number of key steps must be executed successfully:

1. Precise characterization of a gene product with a successful trait and isolation of DNA (or messenger RNA) associated with the gene (**gene identification**)
2. Replication of the DNA containing the gene in bacteria (**gene cloning**)
3. Transfer and identification of DNA to host recipient cells (**gene transformation**)
4. Stable replication of DNA in the host DNA (**gene replication**)
5. Regeneration of transformed cells into plants with expression of the new DNA (**plant regeneration**)
6. Predictable transmission of the new trait in the transformed plant (**gene transmission**)

Successful expression in a whole plant of a foreign gene transferred by recombinant DNA technology is the subject of intense effort.

The successful use of genetic engineering to improve crops appears to hinge on the identification and isolation of agriculturally important genes for cloning. These are genes in which messenger RNA can be identified. The traits most amenable to being reproduced have been genes whose protein product is known, such as storage proteins, and traits whose mutant phenotypes result from an altered enzyme. However, most improvements made by conventional plant breeders are due to complex quantitative genes in which the gene product is unknown. Even most genes associated with disease resistance cannot yet be cloned because, although they can be identified by conventional genetic tests their gene product is unknown. Because of the difficulty of agriculturally important genes capable of being transferred and expressed, genetic engineering will most likely be an adjunct to rather than a replacement of conventional breeding systems.

The transfer of a DNA sequence into a plant cell is being attempted by various techniques. The most promising one uses the tumor-inducing (Ti) plasmid of *Agrobacterium tumefaciens,* the bacteria that cause a cancerlike condition in plants called crown gall. (The tumors produce a protein that the bacteria require.) The Ti plasmid of the crown gall bacteria inserts its genes into the plant's chromosomal DNA where it is expressed to form the tumor even when the bacteria is not present. (Even genetic engineering is not really new!) The Ti plasmid can transmit foreign DNA if it is altered to prevent its becoming pathogenic (Figure 17-22).

Recombinant DNA technology makes it possible to clone individual specified genes from one species and insert them in another. This technique has enormous implications for agriculture. For example, it allows genes from plants or animals to be inserted into bacteria so that plant or animal gene products can be manufactured by microorganisms. It further makes it possible to insert genes from one plant species into cells from another—an achievement that has been unreachable by conventional sexual recombination.

The recombinant DNA technique is powerful because *specific* genes, *if identifiable,* can be transposed. Genetic engineering has transferred the human insulin gene to bacteria, producing a human insulin called humulin, and it has transferred and expressed genes from one species (bean) to cell cultures of another (sunflower). The use of genetic engineering to improve plants is still far off although it is within sight.

Controlled Hybridization

Plant breeding is concerned largely with the control of the sexual process in plants. Before two plants can be crossed, they must be induced to flower simultaneously. Because this may not always occur naturally, various environmental factors, such as photoperiod, temperature, and nutrition, may have to be ma-

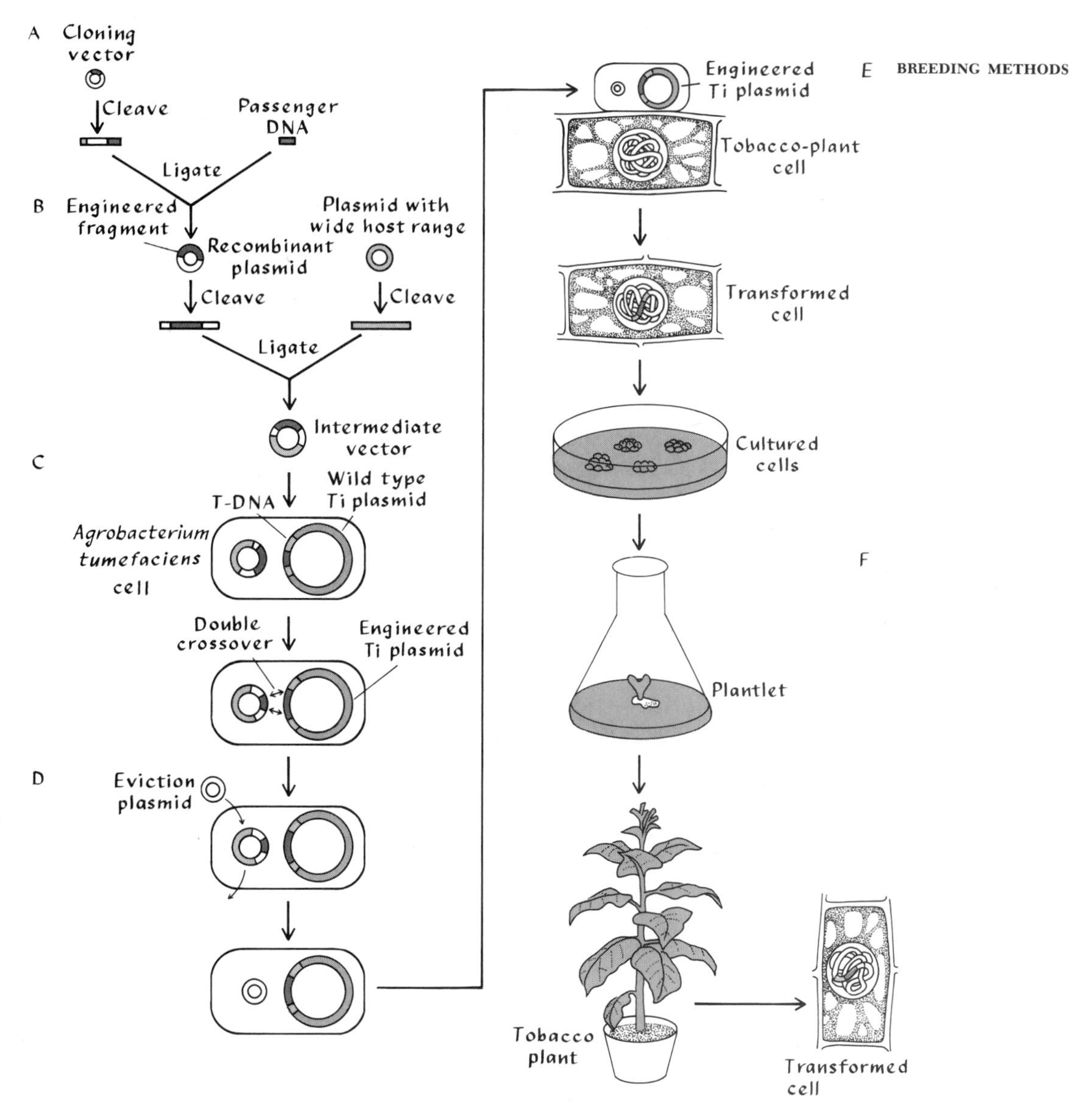

FIGURE 17-22. A technique for inserting foreign genes into plants. (*A*) A small plasmid of a bacteria (*Escherichia coli*) suitable for cloning is cut open with an endonuclease. It contains a fragment of a "disarmed" tumor-inducing gene from *Agrobacterium tumefaciens*. (*B*) A donor gene (passenger DNA) that also contains a genetic marker (a gene for antibiotic resistance) is ligated to the cleaved fragment. (*C*) The recombinant plasmid is attached to a plasmid with a wide host range for insertion into a tumor-inducing (Ti) plasmid of *A. tumefaciens*. (*D*) The vector is evicted by introducing an incompatible plasmid; the engineered, disarmed Ti plasmid is selected by antibiotic resistance. (*E*) The genetically-engineered *A. tumefaciens* containing passenger DNA is used to transform a cell of a dicotyledonous plant. (*F*) The transformed plant cell regenerates healthy plants carrying (and hopefully expressing) the passenger gene. [From M.-D. Chilton, *Scientific American* 248(6):50–59. Copyright © 1983 by Scientific American, Inc. All rights reserved.]

nipulated to achieve synchronization of flowering. In some cultivars of sweetpotato, flowering can be induced only with difficulty; therefore, they must be grafted to species of morning glory and then subjected to specialized photoperiods and temperatures.

The storage and shipping of pollen facilitate artificial hybridization and, in some cases, save many years. Pollen is quite variable in its ability to remain viable. For example, the pollen of cucurbits remains viable only for about 3 hours. Apple pollen stored under low humidity and cool temperatures (30 to 36°F, or −1 to 2°C) may retain its viability for more than 2 years. A new technique of freeze-dehydration is a promising method for the indefinite storage of pollen.

The basic procedure in making artificial crosses involves the following practices: (1) avoidance of contamination before artificial pollination, (2) application of the pollen, and (3) protection of the pollinated flower from subsequent contamination.

In perfect-flowered plants, contamination by selfing is avoided by the process of **emasculation.** This method consists in removing the anthers before they have begun to shed pollen. It is done before the flower opens, and the technique varies with the flower structure (Figure 17-23). In monoecious plants (those with separate pistillate and staminate flowers), the pistillate flowers must be protected before pollination from wind or insect pollination by suitable protective devices, such as paper bags, glassine bags, or cheesecloth nets. In insect-pollinated plants, the petals are often removed to discourage insect visitation.

The pollen may be applied by one of several methods. Often the anthers are collected the day before pollination and dried. The pollen is then applied, preferably in large amounts, with a camel's-hair brush, the fingers, a blackened matchstick, or the dried flower itself used as a brush. In the greenhouse, the pollination of dioecious plants, such as spinach, may be controlled by isolating the pistillate and staminate plants to be crossed. The pollination of onions is

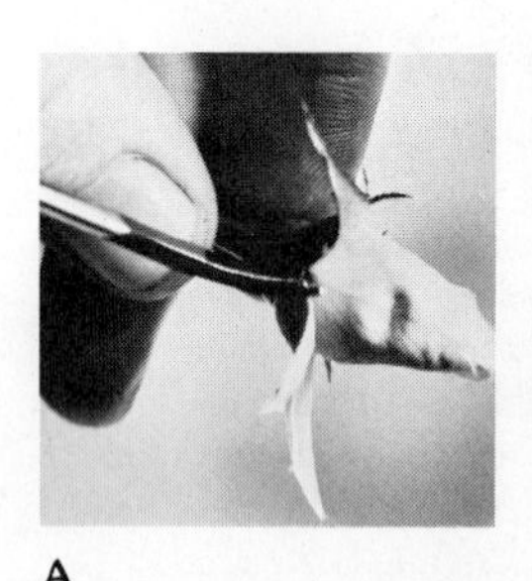
A

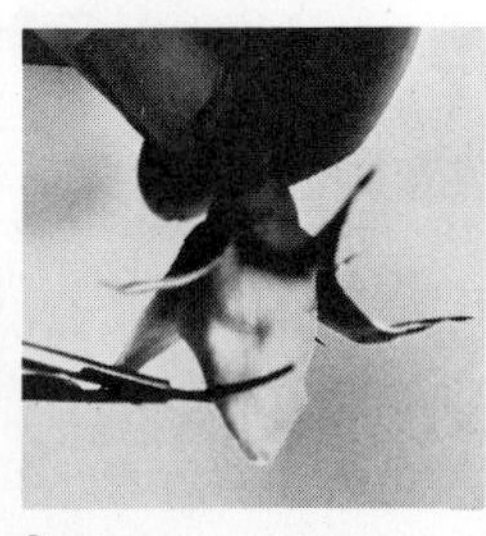
B

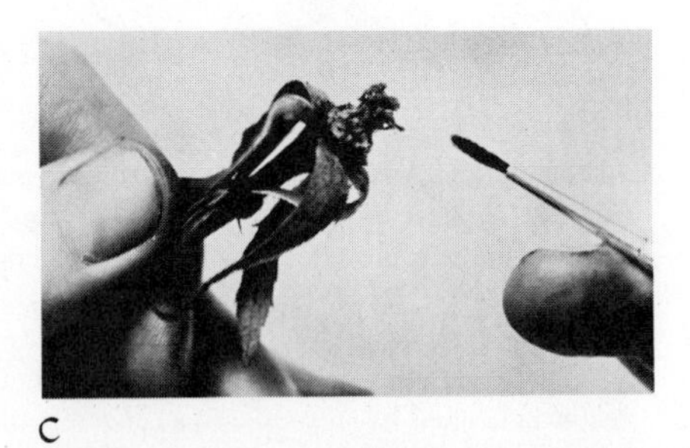
C

FIGURE 17-23. Preparing a rose bud for cross-pollination. The petals are cut at the base (*A*) and at the tip (*B*). Pollen is being supplied to an emasculated flower with a camel-hair brush (*C*). [Photographs courtesy USDA.]

FIGURE 17-24. Pollen transfer in cross of male-fertile × male-sterile onions is achieved by caged blow flies. [Photograph courtesy K. W. Johnson.]

carried out by enclosing their flower heads in mesh cages containing blow flies (Figure 17-24).

After pollination, the flower may need to be protected from contamination by cross-pollination. This is usually accomplished by enclosing the flower in a paper or glassine bag (Figure 17-25). If the petals have been left, as in peas, they may by themselves serve as the protective device with the aid of cellophane tape. When it is necessary only to keep away insects, cheesecloth is usually sufficient. Care must be used in selecting protective devices that do not act as heat traps and thus prevent successful fruit set. In citrus and in apples, insects will not visit the flowers if the petals are removed; protection of emasculated flowers is therefore not necessary.

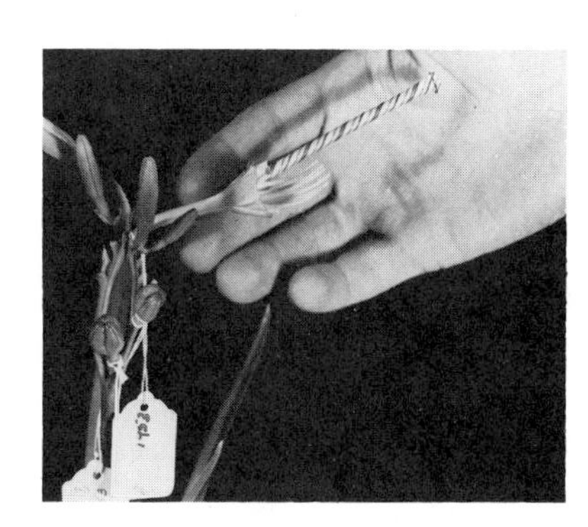

FIGURE 17-25. A piece of ordinary soda straw protects pistil of daylily from contamination before and after pollination. [Photograph courtesy G. M. Fosler.]

Maintaining Genetic Improvement

Once a genetic gain has been achieved, vigilance is still required to maintain improvement. Mutation, natural crossing, contamination, and, in vegetatively propagated material, diseases, especially those caused by viruses, tend to cause the deterioration or "running out" of cultivars. Genetic deterioration may be reduced to a minimum by continued selection and careful propagation. Seed companies and nurseries must strive constantly to keep their stocks free of contamination in order to protect their reputations. In an effort to control the purity of seed, many semipublic organizations and seed associations have been formed.

It has always been possible to protect seed-propagated crops indirectly through copyright of the name, but this did not prevent renaming. However, seed-propagated hybrids can be biologically controlled by the breeder. For example, control of the unique sets of inbreds protects the hybrid even more effectively than does legal control. In hybrids between cultivars of self-pollinated plants, such as tomato, merely keeping secret the names of the two cultivars that are used to make the hybrid from the hundreds available effectively controls its production.

Cultivar Protection

The United States Plant Patent Act of 1930 protects the rights of the originators of certain vegetatively propagated materials to control their increase. This act as passed by Congress excluded plants found in the uncultivated or wild state, seed-propagated plants, and tuber-propagated plants, now interpreted to cover only potato and Jerusalem artichoke. Thus, only vegetative propagation was protected by the plant patent law; propagation by seed was not. Many ornamental and fruit cultivars, such as gladiolus, rose, apple, peach, and pear, are patented under this act. Breeders, whether private individuals, a group, or public agency, may justly profit from their art, just as the inventor or chemist may be rewarded by a discovery, invention, or new process.

Although seed-propagated crops were not covered by the Plant Patent Act of 1930, they are now protected in the United States by the Plant Variety Protection Act of 1970, which now includes all crop species. The establishment of "breeder's rights" has had important implications for the seed industry. However, the Plant Variety Protection Act is national law and has no legal force outside the United States.

The 1980 Chakrabarty decision of the Supreme Court granted a regular patent (under Title 35 of the United States Code) to a genetically engineered strain of bacteria that was used to clean up oil spills. This decision affirms that organisms produced by human ingenuity that have a distinctive name, character, and use may be protected by regular patent. This decision suggests that genetically engineered plants, at least, can be covered by United States patent.

International protection of new cultivars is based on a treaty entitled the Paris Convention of 1961 establishing an International Union for the Protection of New Varieties of Plants (UPOV). The union is now composed of Belgium, Denmark, France, the Federal Republic of Germany, Ireland, Israel, Italy, The Netherlands, New Zealand, South Africa, Spain, Sweden, Switzerland, the United Kingdom, and the United States.

Selected References

Allard, R. W., 1960. *Principles of Plant Breeding*. New York: Wiley. (An outstanding presentation of the genetic foundation of plant breeding.)

Ferwerda, F., and F. Wit (eds.), 1969. *Outlines of Perennial Crop Breeding in the Tropics*. (Miscellaneous Papers 4, Landbouwhogeschool). Wageningen, The Netherlands: H. Veenman. (A review of tropical tree breeding including many tropical fruits.)

Frey, F. J. (ed.), 1981. *Plant Breeding II*. Ames, Iowa: Iowa State University Press. (A collection of articles on plant breeding with emphasis on general methods rather than specific crops. An earlier volume, entitled *Plant Breeding*, still useful, was published in 1966.)

Janick, J., and J. N. Moore (eds.), 1975. *Advances in Fruit Breeding*. West Lafayette, Ind.: Purdue University Press. (A monograph on fruit breeding by crop, each chapter written by a distinguished fruit breeder. Crops covered include apples, pears, strawberries, brambles, grapes, blueberries and cranberries, currants and gooseberries, minor temperate fruits, peaches, plums, cherries, apricots, almonds, pecans and hickories, walnuts, filberts, chestnuts, citrus, avocados, and figs.)

Moore, J. N., and J. Janick, 1983. *Methods in Fruit Breeding*. West Lafayette, Ind.: Purdue University Press. (A companion volume to *Advances in Fruit Breeding*, treating fruit breeding from a procedural and theoretical point of view.)

National Research Council, 1984. *Genetic Engineering of Plants*. Washington, D.C.: National Academy Press. (Research opportunities and policy concerns for molecular biology and genetic engineering relating to crops.)

North, C., 1979. *Plant Breeding and Genetics in Horticulture*. New York: Halsted Press. (A plant breeding text devoted to horticultural crops.)

Simmonds, N. W., 1979. *Principles of Crop Improvement*. London: Longman. (A useful introduction to plant breeding.)

U.S. Department of Agriculture, 1937. *Better Plants and Animals II*. (USDA Yearbook, 1937). (A historical reference for horticultural plant improvement. Together with the 1936 yearbook, this work is a landmark in genetics and breeding of agricultural crops in the United States.)

18 *Marketing*

HORTICULTURE AFTER HARVEST

The path of horticultural products from the growing plant to the consumer is complex. **Marketing** may be defined broadly as the activities that direct the flow of goods from producer to consumer—that is, the operations and transactions involved in their movement, storage, processing, and distribution. These operations have been developed into a highly specialized technology. Many production operations are a direct part of marketing. The grower, when selecting a particular cultivar to plant, is very often making a marketing decision. Production decisions that determine ultimate quality, such as methods of disease control, may ultimately become as much a part of marketing as decisions to store or to sell, or to prepack or to bulk ship. Many production and marketing operations are connected and interrelated. For example, the harvesting, grading, and packing of lettuce is combined in one field operation.

The unique aspects of marketing of horticultural products are due to their perishable nature. The horticultural product is still alive after harvest, but subject to deterioration. The ultimate quality of the product for the consumer, which is what determines its real economic value, hinges upon how it is treated after harvest.

MARKETING FUNCTIONS

Marketing increases the value of horticultural products through the application of **marketing functions.** These functions have been categorized as exchange, physical, and facilitating. The **exchange functions** are those activities related to the transfer of title to goods—that is, buying and selling. Buying in its broadest sense includes the seeking out and assembling of sources of supply. This may involve many transactions and many people in various marketing operations. Similarly, selling includes not only the transfer of goods; it also includes merchandising (packaging, advertising, and promotion). **Physical functions** include storage, transportation, and processing. **Facilitating functions,** which make possible the orderly performance of the exchange and physical functions, include grading, financing, risk bearing, and communications (market information). The total cost of the marketing functions, what is known as the **marketing margin,** may greatly exceed the entire cost of production (Figure 18-1). These are the charges of the much maligned middlemen of agriculture.

This chapter is largely devoted to a discussion of technological problems associated with marketing operations. The detailed structure and economic ramifications of marketing, however, are beyond the scope of this book.

HARVESTING

The close relationship of harvesting to subsequent marketing operations makes its discussion a proper introduction to marketing technology. Harvesting is one of the crucial features in the horticultural operation. It is often the costliest part of production, and its timing has a direct bearing on the final quality of the product. For example, maturity has a great influence on the subsequent storage behavior of many crops. Technological advances in harvesting have become an outstanding example of the current trend of substituting capital for labor in horticultural operations.

Predicting Harvest Dates

A number of factors must be considered with regard to the timing of harvesting: maintaining orderly production operations that make possible the maximum utilization of equipment and labor; setting up an orderly marketing sequence; and the ultimate quality and appearance of the product. For many crops, harvesting must proceed within a certain narrow time interval, for quality is a fleeting and elusive factor. Often there are a number of conflicting factors to be considered. For example, in apple harvest, storage quality is adversely affected by delaying maturity, yet, on the other hand, red color tends to increase with

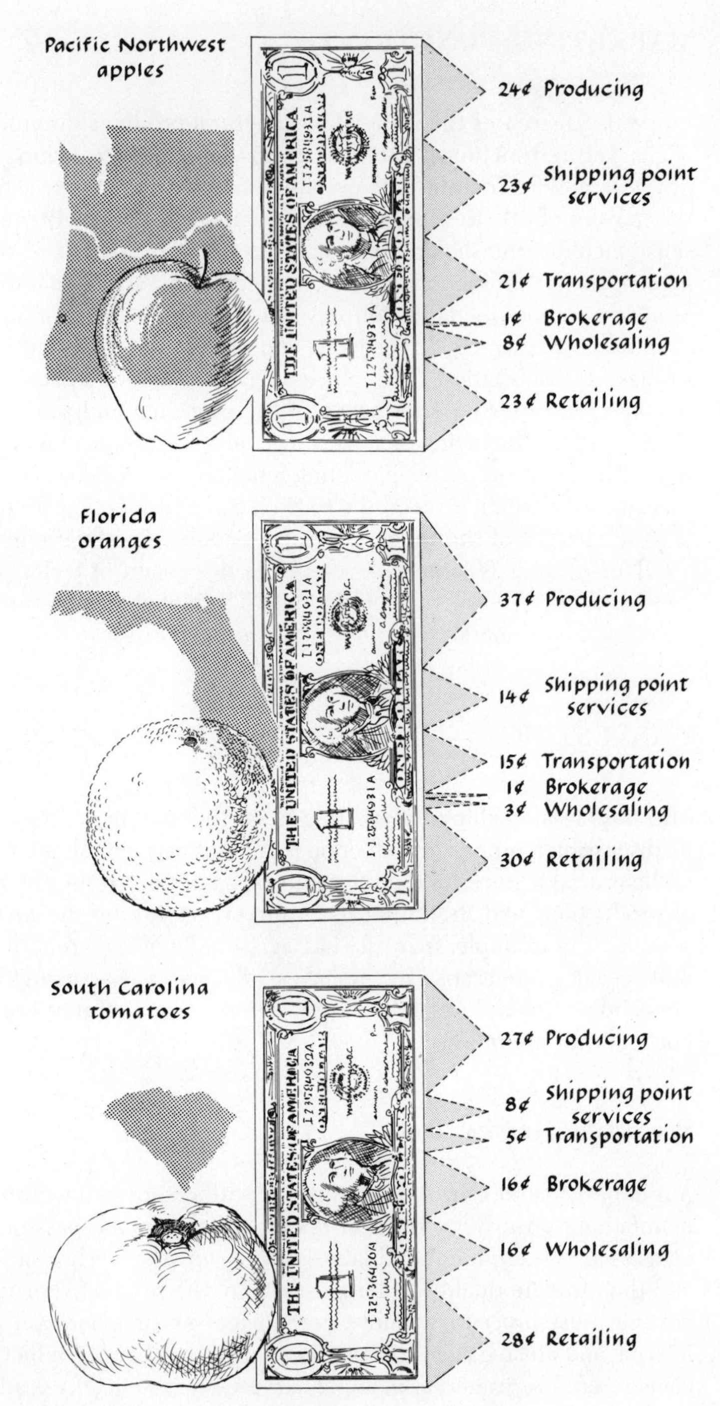

FIGURE 18-1. Marketing margins for apples, oranges, and tomatoes. Note that two-thirds to three-quarters of the consumer's dollar is spent for marketing. [Data from USDA.]

time. Determining the optimum time of harvest is not a simple, straightforward decision.

The factors affecting the harvest date of a crop depend upon the genetic nature of the cultivar, its planting or blooming date, and the environmental factors that exist during the growing season. Successive harvest dates of sweet corn may be obtained by planting cultivars with different maturity dates. Similarly, in tree crops, there are different cultivars that mature at different dates, and planting a well-chosen assortment of such cultivars can make for an orderly succession of crops throughout the season. In the past, tremendous market gluts of peaches have been produced as a result of extensive plantings of one cultivar, 'Elberta'. This has been alleviated by breeding programs that have produced groups of cultivars that ripen in an orderly sequence over almost a 3-month period.

The harvest dates of annual crops may be altered by changing the sequence of planting dates. However, a delay in planting date does not necessarily cause the same delay in harvest date. The time required to reach the harvestable stage varies widely from crop to crop and is based largely on the crop's temperature requirements for growth. When temperatures are below the minimum required for growth, a long delay in planting may not affect the harvest date.

In some crops the time required to reach the harvestable stage may be expressed with temperature-time values called **heat units** by calculating time in relation to temperatures above a certain minimum. For example, if the minimum temperature for growth of a particular crop is 50°F (10°C), then a day with an average temperature of 68°F (20°C) would provide 18 degree-days F (10 degree-days C) of heat units. A day with an average temperature of 41°F (5°C) would provide 0 degree-days of heat units. The harvest date can be ascertained by an accounting of accumulated heat units. In Wisconsin, it has been shown that 1200 to 1250 degree-days F are required for the 'Alaska' cultivar of pea to mature if planted in the early spring. Under increasing spring temperatures it takes longer to accumulate heat units in the early part of the season than later. Assuming that all temperatures above a minimum have similar effects on growth, there would be a decreasing interval between planting and harvest dates as the season progresses.

The heat-unit system has a number of limitations. For example, soil temperatures more accurately indicate early growth than do air temperatures. Differences in day–night temperature shifts, day-length effects, and the differential effects of temperature on various stages of plant growth also affect the results. In addition, temperatures above a minimum may not have a similar effect on growth but, within limits, may act exponentially, approximately doubling many physiological processes with every 18°F (10°C) rise in temperature. The precise determination of harvest date by the accumulation of temperature data depends upon a knowledge of the general climate of an area and upon experience with harvest dates based upon planting dates for each crop.

For many long-season fruit crops, such as apples, the most reliable index for calculating maturity is the number of days from flowering. This value, determined from an analysis of previous data, is surprisingly consistent. It indicates that other environmental factors besides temperature are important in determining fruit maturity.

A number of physiological criteria are used to determine harvest date. In pears the proper harvest date for each cultivar is determined by a combination of criteria, among which are pressure testing of the fruit, ground color of the skin, seed color, and percentage of soluble solids (basically sugar) in the flesh. In some crops, other criteria are also utilized. These include the formation of abscission zones separating fruit from stem (muskmelons), the visible stage of development (such as degree of bud tightness in roses), color (tomatoes), sugar-acid balance (blueberries), and even the way the fruit sounds when thumped (watermelons). A combination of methods is often used to determine the correct degree of maturity.

Harvesting Methods

Hand harvesting is still the only practical method for many high-value crops that either are sensitive to bruising or must be selectively picked (strawberries and apples, for example). The mechanization of harvesting has proceeded in stages. Thus, we may speak of completely mechanized, semimechanized, and nonmechanized harvesting. For example, in Arizona muskmelons are harvested by hand, but they are then placed directly on conveyors, where they are graded, and they are packed in the field on moving machines. Similar operations are used for cabbage, peppers, cauliflower, and broccoli.

The harvesting of many crops is completely mechanized, although it is done in a number of separate steps. For example, when grown for processing, pumpkins and squash are raked into windrows in one operation and are loaded by conveyor in another.

For other crops, harvesting is completely mechanized, and all operations are performed by a single machine in a single step. Thus, potatoes are harvested by a combine that digs, removes vines, and loads tubers. A somewhat similar operation is used to harvest radishes, beets, carrots, and floral bulb crops. Asparagus, pickling cucumbers, snap beans (Figure 18-2), sweet corn (Figure 18-3), and tomatoes are mechanically harvested; canning peas are now picked, vined, and shelled in the field.

An evolutionary pattern is evident in the development of mechanical harvesters and associated growing systems. For example, although late-maturing large-vined cultivars of vegetable crops were planted once in wide rows and hand-harvested several times, early-maturing dwarf cultivars are grown now in close spaced rows for a single destructive harvest. Similarly, harvesting methods proceed from hand harvesting, to mechanical aids for hand harvesting (con-

A

FIGURE 18-2. Mechanical harvest of snap beans in Wisconsin. (*A*) Each picker harvests 8 to 10 acres of snap beans per day. (*B*) A pallet box containing about 700 pounds of snap beans being lowered to the ground. (*C*) Tractors with hydraulic lifts loading pallets onto a truck, which is dispatched to the canning plant. [Photographs courtesy National Canner's Association.]

B

C

veyors), to destructive machine harvest following hand harvest, to a single destructive machine harvest. When machine harvesting is first implemented, the crop is usually collected in the same small containers previously used for hand harvesting; sometimes packing and sorting operations are an integral part of the harvest. In more advanced operations the crop is handled in bulk and transported to a central location for washing, grading, and packing or processing.

Fruit crops present a somewhat different problem in that the plant itself must not be destroyed. Cranberries were the first fruits to be harvested by machine; the plants are essentially combed to remove the berries. Equipment

A

B

FIGURE 18-3. Mechanical harvest of potato (*A*) and sweet corn (*B*). [Photograph at bottom by J. C. Allen & Son; photograph at top courtesy Purdue University.]

has been developed for shaking plums, clingstone peaches, blueberries, and nuts onto canvas sheets. Mechanical harvesters for grape, raspberry, blackberry, and strawberry are also commercially available.

In general, machine harvesting must proceed hand in hand with breeding to develop machine-compatible plants. Such plants must usually have fruits that are highly uniform, that ripen all at the same time, and that are firm enough to resist excessive bruising. When fruit-harvesting machines are perfected, however, they often do a more satisfactory job than hand harvesting. Machine-harvested potatoes show a decrease in defects due to bruising. In many vegetable crops, machine harvesting is facilitated by the planting of F_1 hybrids, which tend to be more uniform in size and shape.

Postharvest Alterations

The desirability, or quality, of horticultural products may be determined by many different things, depending on the product. For processing tomatoes, desirability implies freedom from cracking and good internal red color. In the

'Delicious' apple, however, desirability connotes a particular flavor, shape, and pattern of external red color. Desirability of nursery stock refers to viability, form, and size, plus the proper proportion and packaging of roots and soil.

For the sake of precision, the general term **quality** is best divorced from the meanings suggested by the terms **condition** and **appearance.** In food crops, the word *quality* is best used with reference to palatability. This term implies a pleasing combination of flavor and texture; flavor resulting from taste and smell, and texture perceived as "mouth feel." Quality can also properly be used to describe nonedible products such as seed or nursery stock. Used in this sense, the term refers to the physiological state of the material, connoting high viability and trueness to type. **Condition** is the presence of, or freedom from, disease, injury, or physiological disorders. Although we associate good condition with high quality, one does not always accompany the other. **Appearance** refers to the visible attributes of the product. It includes color, conformation, and size. Unfortunately, appearance is not always a reliable index of quality.

A number of physiological and biochemical processes occur in the harvested, nonprocessed horticultural product that contribute to change (Table 18-1). This may result in the deterioration of quality, condition, and appearance in some crops, and in the improvement of quality and appearance in those crops that complete ripening after harvest. For most commodities, the objec-

TABLE 18.1 Changes that occur in harvested produce.

Change	Process	Examples and significance
Water loss	Transpiration; evaporation	Unattractive appearance, texture changes, weight loss, shriveling
Carbohydrate conversion	Enzymatic	Starch to sugar: detrimental in potatoes, beneficial in bananas and pears
		Sugar to starch: detrimental in sweet corn and most edible crops
Changes in flavor	Enzymatic	Usually detrimental, but beneficial in persimmons, pears, and bananas
Softening	Pectic enzymes	Usually detrimental
	Water loss	Beneficial in pears, bananas
Change in color	Pigment synthesis or destruction	May be detrimental or beneficial
Toughening	Fiber development	Detrimental in celery
Change in vitamin content	Enzymatic	May be gain (vitamin A) or loss (vitamin C)
Sprouting, rooting, or elongation	Growth and development	Detrimental in potatoes, onions, and asparagus
Decay and rot	Pathological; physiological	Detrimental

tive is to maintain the product as close to harvest condition as possible. The product must be maintained in the living state because death causes irreversible biochemical changes. These changes may involve gross deterioration and drastic differences in flavor, texture, and appearance.

GRADING

The inherent variability of horticultural products at harvest and their differences in value make it necessary to grade them according to some objective standard. Grading is the basis of long-distance trade. It permits the description of products in terms that are understandable both to buyer and to seller. Without a system of grading, all products would have to be individually inspected. Grading thus adds tangible value to horticultural products.

Grading has two distinct functions. The first is to eliminate completely all obviously unsatisfactory items. This exclusion is extremely important in packaging, inasmuch as diseases spread rapidly in the enclosed environments afforded by packages. Furthermore, one poor item visibly detracts, out of all proportion to its bulk, from the appearance of a large sample. The second function of grading is consolidation. Products may be consolidated by cultivar, size, appearance, defects, and, where possible, quality. Because of inherent differences, certain crops (for example, apple and pear) are grouped and identified by cultivar. Within a cultivar, size is the most obvious gradable factor (Figure 18-4). Various size characteristics may be used: stem length (roses), stem diameter (trees), spread (shrubs), weight (watermelons), and diameter (most fruits). Grading for appearance may be based on the absence of defects, conformation, and the amount and intensity of color. Various techniques have been used for the objective evaluation of palatability, including determinations of oil content (avocados), firmness (tomatoes), sugar-acid ratio (oranges), specific gravity (potatoes), and various maturity criteria.

One of the factors that determines the quality grade of a processed product is the quality of the raw product. Raw product grades for processing must take into account particular demands of the industry. Some factors assume greater importance, whereas others assume less. For example, apple skin color is of small importance, since the skins are removed in processing, but large fruit size is extremely significant because of the limitations of mechanical peelers. A number of quality features have special importance to processing—for example, total solids (tomatoes), tenderness (peas and sweet corn), and sugar content (grapes for wine). Defects, texture, and internal color become outstanding factors.

Inspection and grading are backed by Federal law. The basic piece of legis-

FIGURE 18-4. This fruit grader sorts fruit by weight. The fiber cartons are tray-packed. [Photograph courtesy FMC Corp.]

lation relating to inspection and grading is Title II of Public Law 733, known as the Agricultural Marketing Act of 1946, which directs the Secretary of Agriculture

> to develop and improve standards of quality. . . . To inspect, certify, and identify the class, quality, quantity, and condition of agricultural products when shipped or received in interstate commerce, under such rules and regulations as the Secretary of Agriculture may prescribe . . . to the end that agricultural products may be marketed, to the best advantage, that trading may be facilitated, and that consumers may be able to obtain the quality product which they desire. . . .

Most federal standards for horticultural products are permissive; that is, they are officially recommended but are not compulsory. But if federal grades are used to describe a product, they must be complied with. Mandatory grades do exist for some crops under certain conditions, however, as in apples and pears intended for export. Most of the grade standards for fruits and vegetables are designed for wholesale trading and are not directly carried over into retail trading.

Federal grading regulations are complemented by state laws. This has created confusion and disorder where state regulations differ widely from one another and from federal standards. The inspection of fresh fruits and vegetables at shipping points is accomplished through the combined efforts of federal and state agencies.

MARKET PREPARATION

Most horticultural crops require special preparation after harvest. They are usually cleaned, trimmed, or specially treated in some manner. Root crops must be cleaned to remove adhering soil and debris. Potatoes grown in muck soil are at a serious disadvantage unless the extremely fine black soil is removed. Washing or brushing of fruit is done to enhance its appearance. The outer leaves of lettuce and cabbage are routinely removed, as are the tops of beets and carrots. Leaves of many florist crops, such as chrysanthemums and snapdragons, are stripped by the grower at harvest. Strawberry plants are "cleaned" by removing the dead leaves and runners.

Many products are waxed to prevent water loss and to improve appearance (turnips, citrus fruits, and cucumbers). Waxing is now a standard practice for dormant nursery stock, especially rose plants. Bananas are dipped in solutions of copper sulfate to control storage and shipping rots.

Conserving Life of Cut Flowers

The life of cut flowers is markedly shorter than that of the same flower left on the growing plant. The reasons for the short "vase life" of most flowers are not completely understood, but water stress accentuated by vascular blockage, the high respiratory activity of petals and the subsequent depletion of metabolites, and the extreme sensitivity of flowers to ethylene are key factors.

The traditional way to conserve the life of cut flowers has been through refrigeration, which slows down metabolism and respiration and minimizes transpiration. Unfortunately, the vase life of flowers after cold storage may be very brief, and in many cases refrigerated flowers do not undergo normal development, especially if harvested too early. Because of the high perishability of flowers and their high value, much effort has been expended recently to find ways to maximize cut-flower longevity by means of postharvest treatments.

Pulsing

The use of immature flowers or unopened buds greatly facilitates packing and shipment, but they must retain the ability to open normally. Pulsing (or "loading") refers to short-term treatment of cut flowers with sugar solutions before shipment. This treatment makes it possible to ship immature gladiolus, carnation, and chrysanthemum flowers and have them open normally. Typically, the stems of the cut flowers are immersed in solutions of sucrose (5 to 40 percent) for about 20 hours. The sugar accumulates in the cells of the petals, enhancing their water-absorbing capacity and allowing them, therefore, to maintain their turgidity even under subsequent water stress. The increased sugar content of pulsed flowers also allows for the maintenance of the high respiratory activity of the petal cells.

Improving Water Relations

Various treatments that maximize water uptake and reduce excess transpiration will extend the vase life of cut flowers. Maximum water uptake may be increased by sharp, diagonal cuts (to avoid crushing) of the stem each time plants are removed from the holding solutions; by using warm solutions (110°F, or about 43°C) when the stems are first immersed; and by the avoidance of microbial buildup in solutions through the use of microbicides or acidifying agents. Diagonal cuts increase the surface area of the stem and prevent the cut end from resting on the bottom of the container. The warm water has a lower viscosity, which increases its uptake, and it is better able to dissolve air bubbles inside the water-conducting tissues. Water loss may be minimized by removing excessive foliage, by maintaining high humidity in shipping containers, and by avoiding high temperatures and excessive air movement.

Vascular blockage has been identified as a major cause of reduced water uptake in cut flowers. Cut roses often die prematurely, because of "bent neck," or water deficiency exacerbated by vascular blockage. The precise cause of vascular blockage is unclear. Because it seems to be increased by bacterial toxins, bactericides such as HQC (8-hydroxyquinoline citrate) are added to cut-flower solutions to reduce the disorder. A combination of HQC and sugar is the basis of most conserving solutions.

Elimination of Contaminants

Flowers in general are extremely sensitive to various gaseous toxins and water pollutants. Flowers are exquisitely affected by ethylene, for example, and should not be stored with fruits or vegetables that release it. A consumer who stores a floral bouquet in a refrigerator containing fruit may reduce, instead of lengthen, the life of the flowers. Recently, silver ions have been shown to counteract the effect of ethylene, and silver nitrate is therefore often added to cut-flower solutions. The use of filters to remove ethylene and other contaminants from the air in flower-storage areas can increase floral life considerably. Water quality is also important: Water high in salts, fluorides, and other pollutants must be avoided when preparing cut-flower solutions.

Components of Conserving Solutions

For prolonging the vase life of flowers, growers, wholesalers, retailers, and consumers may use cut-flower solutions. These solutions increase the availability of energy, reduce microbial build-up and vascular blockage, and provide for maximum water uptake by providing the flowers with increased water-absorbing capacity. The development of cut-flower solutions is a relatively recent innovation in horticulture. Although there are numerous formulations and many physiological differences between species, the following are classes of ingredients common to most solutions.

ENERGY SOURCE. Energy may be supplied to the flowers in the form of a metabolizable sugar, such as sucrose. Most solutions contain 1 to 5 percent sugar. Another effect of sugar is to improve the flower's ability to withstand water stress by increasing the water-absorbing capacity (osmotic potential) of its petal cells.

MICROBICIDE. HQC at 200 to 600 ppm is commonly added to most keeping solutions. Sodium benzoate at 100 ppm may also be used as a microbicide, and citric acid at 75 to 200 ppm may be used to impart antibacterial properties to the solution by lowering its pH.

ANTI-ETHYLENE AGENT. Silver ions, which help to protect flowers against the effects of ethylene, may be supplied by adding silver nitrate to the solution at the rate of 25 to 50 ppm.

WATER. Water of high quality has been shown to be essential to prolonging the vase life of cut flowers. Total dissolved solids of more than 100 ppm, or fluoride levels in excess of 3 or 4 ppm, will shorten flower life. For this reason, distilled or deionized water may be better for cut flowers than tap water.

Curing

Curing is a postharvest treatment used to prolong the storage life of certain commodities, such as potatoes, sweetpotatoes, and bulb crops. Curing involves exposing the product to particular temperature or humidity conditions. Potatoes are cured at 55 to 60°F (about 13 to 16°C) during the first two or three weeks of storage. The humidity is kept very high during this period to prevent shrinkage due to the loss of water. Because the tubers are dormant, they do not sprout. Curing brings the healing of cuts and bruises incurred at harvest. Two physiological processes are involved: **suberization,** the deposition of a waxy material by the tuber, which produces a fairly effective barrier to the loss of moisture; and **active cell division** of the periderm, which produces a new "hide." After the curing period, temperatures are lowered for prolonged storage to prevent sprouting. A similar treatment is undertaken in the processing of sweetpotatoes for storage, although higher curing temperatures are used (around 80°F, or 27°C).

Bulb curing involves the removal of water from the outer scales and neck. This was formerly done in the field, but now the trend is toward the use of artificial drying in storage. Proper curing prevents rot by eliminating from the surface of the bulb those environmental factors favorable to the growth of harmful microorganisms. Unlike the curing of potatoes, the curing of bulbs involves no active cell division.

Packaging

Packaging affords protection, convenience, economy, and appeal. The changes in packaging that have occurred in this century have had a great impact on the horticultural industry. Before World War I, a tremendous variety of wooden containers was used (barrels, boxes, crates, and baskets of various shapes and dimensions). By World War II, a certain amount of standardization had been achieved, and the wooden bushel basket and box became the principal packages used for many kinds of vegetables and fruit. These containers originated as a grower pack, but they were also used by the retailer. The wooden bushel basket has proved unsuitable for many crops. It is bulky, relatively heavy, and costly. It is also expensive to fill, since it must be faced. Nor is its shape particularly suitable for packing: Stacking them wastes space and badly bruises the contents, especially the top layer. Furthermore, the bushel is no longer a convenient retail package, since its use necessitates completely repacking the contents into the refrigerated display cases now commonly in use.

The materials used for packing differ with the product. Lightweight fiber cartons have replaced wood for most packages (Figure 18-5); some are specially coated to give them wet strength. Plastics, paper, and cloth are now commonly accepted as packaging materials.

The term **prepackaging** has been used for the process of putting produce in a consumer unit package at some point before it is put on display in the retail store. The rise of the self-service supermarkets has irrevocably altered packag-

FIGURE 18-5. Different kinds of packages for apples. [Photograph courtesy Purdue University.]

ing of horticultural commodities, although perishable commodities were the last to be prepackaged. By 1970, almost one-half of all fruits and vegetables were prepackaged before they reached the retail store.

The prepackaging of fruit and vegetables has been made possible by the creation of satisfactory films. Polyethylene has proved to be extremely versatile. It is strong, transparent, moisture-resistant, and can be made permeable for gas exchange. The natural appeal of products in transparent bags with a minimum of printing has proved to be a selling attraction. At present, carrots and radishes are nearly always prepackaged. Apples are commonly retailed in 5-pound "poly" bags packaged in fiber cartons. Polyethylene films have had a great influence on the marketing of perennial plants by eliminating the need for heavy moisture carriers such as peat. In general, specialized films are being created to suit individual products. Other materials, such as cardboard, paper, cloth, and plastics, have proven useful for particular commodities.

The prepackaging of produce has put a premium on uniformity. This practice has eliminated customer handling and prevented much in-store damage. Prepackaging has not in any way eliminated refrigeration, though it has in many cases increased shelf life by improving moisture retention. In general, prepackaging has improved quality levels. Many quality-conscious customers, however, still prefer to select individual items. Thus, bulk display of unpackaged fruits and vegetables continues in retail stores.

The question of who should do the prepackaging has not yet been settled. Controversy exists over whether prepackaging is best handled at the shipping point or at the distribution point. Although different crops have been handled in different ways, there is a trend with many commodities for grower packaging. As producers have increased in size, associations of buyers have dealt directly with growers and have specified package requirements. The producer, if large enough, often finds it profitable to assume the packaging operations.

PRECOOLING

Precooling refers to the rapid removal of heat from freshly harvested fruits and vegetables in order to slow ripening and reduce deterioration prior to storage or shipment. The rate of deterioration depends on many factors: temperature, the natural respiration rate of the crop, the moisture content, the presence of natural protective barriers to water loss, and the presence of decay organisms. The major effect of precooling consists in reducing the respiration rate. It also slows deterioration and rot by retarding the growth of decay organisms; and it reduces wilting and shriveling since transpiration and evaporation occur more slowly at low temperatures. The internal temperature of a horticultural product (such as a peach) harvested on a hot day may be 20°F (11°C) higher than the air temperature. The removal of field heat, to reduce the temperature of the harvested product to 32 to 40°F (roughly 0 to 4°C), must be as rapid as possible; consequently, a great deal of energy is required. Hence, the harvesting of many perishable crops is now done at night or early in the morning to avoid excessive field heat. With the field heat removed, considerably less energy is required to maintain low temperatures, inasmuch as the respiration rate at temperatures of 32 to 40°F is relatively slight. The special techniques developed to precool vegetables and fruits are contact icing, hydrocooling, vacuum cooling, and air cooling.

Contact icing designates the use of crushed ice placed in or on the package to effect cooling. Its major advantage is that it does not dry out the food in cooling it. Another advantage is that produce may be shipped immediately after treatment, since cooling takes place in transit. Although icing requires relatively small outlays of special equipment, a large weight of ice must be shipped. Furthermore, after the ice has melted, the package is left only partly full ("slack pack"). When lettuce was precooled by icing, a packed crate would contain 60 pounds of lettuce and 30 pounds of ice. The use of ice spread mechanically over the produce after loading (**top icing**) has eliminated the slack pack, but cooling is not so efficient. Contact icing is now being replaced by hydrocooling and vacuum cooling.

Hydrocooling is the cooling of fruits and vegetables with water (Figure 18-6). The water (usually iced) flows through the packed containers and absorbs heat directly from the produce. The high efficiency of the system is due to the large heat capacity and high rate of heat transfer of water. The time required is related to the thickness of the individual product, as well as the internal temperature at the beginning of the cooling process, in relation to the desired temperature drop. Fungicides, such as calcium hypochlorite at concentrations of 50 to 100 ppm, are used to prevent the spread of decay organisms. One of the chief advantages of hydrocooling is that it prevents the loss of moisture during the process. The crops most commonly hydrocooled are peaches, sweet corn, and celery. The system requires application to large volumes of produce in order to operate efficiently.

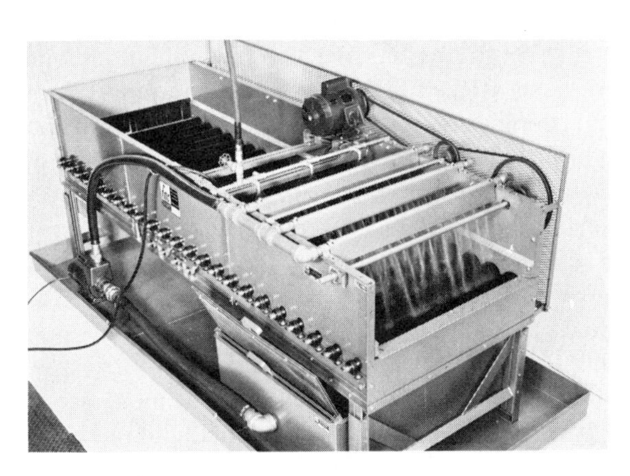

FIGURE 18-6. A hydrocooler in which the fruit may be brushed as it is cooled. [Photograph courtesy FMC Corp.]

Vacuum cooling utilizes the rapid evaporation (actually boiling) of water at reduced pressures to effect cooling. At roughly 0.089 pounds per square inch (which is 1/165 of the normal atmospheric pressure of 14.7 pounds per square inch) water will boil at 32°F (0°C). This rapid evaporation takes up heat directly from the crop. The crop must have a large surface area in proportion to its volume to cool quickly without excessive drying of the surface layers. Lettuce and other leafy crops are ideally suited to vacuum cooling (Figure 18-7). Sweet corn may be vacuum cooled if it is moistened beforehand. The expensive equipment required makes vacuum cooling feasible only for large growers or organizations, although portable vacuum coolers are in use.

Air cooling, the basis of most refrigeration systems, depends upon the movement of cold air. For rapid cooling, special equipment is needed to effect the high-volume air circulation required. The air must be as cold as possible, but it must not freeze the produce. To prevent drying, produce must be removed from the air blast as soon as it is cooled. Solanaceous fruit crops, beans, and berries can be efficiently air-cooled. Leafy crops, however, cannot be satisfactorily precooled by this method because of desiccation.

TRANSPORTATION

Marketing depends upon short-haul movement and handling by the growers and upon large-scale, long-haul transportation facilities. One of the great revolutions in horticultural marketing has been the bulk handling of produce. Great savings are made possible by the synchronization of harvesting and hauling operations to minimize handling. The harvesting container has increased in size to conform to machine power rather than muscle power. For example, the bushel field crate is being replaced as the harvesting unit. Automatic potato and onion harvesters unload directly into truckbeds. In hand-picking operations, as in those for apple and pear, the pickers unload into 30-bushel pallet boxes, which may be stored directly or shipped to processing plants, where they are automatically unloaded (Figure 18-8). Industrial management techniques in packing and storage layout have been utilized to facilitate bulk handling. They have made centralized grading, storage, and packing operations particularly efficient. Thus, a large portion of the fruit production the American Northwest is handled through the centralized facilities of grower organizations.

Long-haul shipment of horticultural products is effected principally by truck, railroad, and boat. Special railroad cars have been constructed to transport potatoes in bulk directly from the harvesting operation. This has materially decreased bruising, an important cause of trouble in potato storage. Air shipment of certain high-value perishables, such as flowers and strawberries, is increasing. Long-haul transportation contributes storage as well as movement, and it necessitates specialized packaging, packing arrangements, and environmental controls for the reduction of in-transit spoilage.

A

B

C

D

FIGURE 18-7. Lettuce harvested in the Salinas Valley of California is picked directly into cartons (*A*), vacuum cooled (*B*), and transferred in pallets (*C*, *D*). [Photograph courtesy T. W. Whitaker, USDA.]

FIGURE 18-8. An automatic pallet unloader in a canning factory. [Photograph courtesy Gerber Products Co.]

STORAGE AND PRESERVATION

The demand for most horticultural products is continuous. Their seasonal production and their rapid deterioration after harvest make preservation and storage essential in order to ensure an extended supply. Extending the supply of any horticultural commodity requires retarding the natural physiological deterioration that occurs inherently in living systems as well as preventing decay by microorganisms. The method to choose depends on the product, its use, and the required storage time. The imposed limitation is that the product be acceptable after storage. If stored plant parts are to be used for reproduction (seeds, bulbs, whole plants), the overriding factor is viability. For fresh fruits, vegetables, and flowers, the maintenance of acceptable quality depends upon preserving the natural, living state, although reproductive viability is not necessary. For the extended storage of certain food crops, various processes have been developed that stop the life functions but maintain edible quality, even though the product may be materially altered from its harvest condition. Changes in form brought about by food processing may result in new products that effectively extend the utility of the original product, rather than simply preserve it. Changing tomatoes to ketchup is an example of this.

Storage of Perishables

The storage of perishable plant products in their natural state may be accomplished by means of environmental control. The ways of prolonging the life of a product are the slowing of respiration in order to retard microbial activity and the prevention of excessive water loss. Respiration and microbial activity may be regulated by the regulation of temperature and the control of oxygen and carbon dioxide levels. Water loss may be prevented by controlling humidity.

The storage life of produce depends not only on storage conditions but on the natural rate of respiration. This rate in turn depends upon the species, the plant part, the plant's maturity level, and its degree of dormancy. Bruising and decay must be considered because of their adverse pathological and physiological effects (including increased respiration and discoloration).

Temperature Control

For best results the storage temperature must be held constant, since the optimum temperature range for many products is narrow. For example, most apple cultivars keep best and longest when held at a constant temperature of 30 to 32°F (−1 to 0°C). Temperature increases of as little as 5°F (3°C) will hasten ripening in proportion to the duration of the increase.

The evolution of heat by respiration must be considered as a factor in storage-temperature control. The heat evolved may be calculated from the respiration rate, which increases with temperature (Table 18-2). The respiration rate of produce varies considerably. The respiration rate of spinach is high enough that, even if it were stored at 40°F (4.4°C), its temperature could increase to more than 100°F (38°C) in 5 days if no heat were allowed to escape. This can be a real problem in shipment if shipping packages are not adequately ventilated.

The lower limits of storage temperature must be determined individually for each crop. In general, most products suffer some injury when frozen, although the degree of injury differs with the product. Some commodities are severely damaged by even slight freezing, whereas others less susceptible may undergo a number of successive freezes and thaws without perceptible permanent injury. Storage temperatures are usually kept above the point at which the commodity freezes, although strawberry plants are often stored frozen.

TABLE 18-2. The relation between temperature and heat evolved due to increasing respiration.

Temperature		Heat (in Btu) evolved per ton of fruit per 24 hours	
(°F)	(°C)	Lemons	Strawberries
32	0.0	480–900	2730–3800
40	4.4	620–1890	3610–6750
50	10.0	1610–3670	7480–13,090
60	15.6	2310–4950	15,640–20,280
70	21.1	4050–5570	22,510–30,160
80	26.7	4530–5490	37,220–64,440

SOURCE: *USDA Agricultural Handbook 66.*

In many commodities, injury can occur at temperatures considerably above freezing. This injury is known as **chilling injury** in contrast to **freezing injury.** Chilling injury interferes with the ripening sequence in tomatoes and bananas. Chilling injury affects the appearance of produce, and it may result in actual breakdown of tissues.

Oxygen and Carbon Dioxide Levels

In addition to being temperature-dependent, respiration is also directly affected by the oxygen and carbon dioxide levels. The atmosphere normally contains about 78 percent nitrogen, 21 percent oxygen, 0.03 percent carbon dioxide, and small percentages of several inert gases. Since respiration is an oxidation process, a reduction in the amount of oxygen reduces the respiration rate. Although slight variations in the amount of carbon dioxide show little effect on respiration, high concentrations may inhibit or prevent respiration and may also act to inhibit the ethylene production that stimulates ripening (as will be discussed in a subsequent section on organic volatiles).

The effect of oxygen and carbon dioxide levels on respiration is utilized in the storage of fruit. Carbon dioxide may be increased in storages by carbon dioxide generators. However, in a gas-tight room filled with freshly harvested "living" apples, the respiration of the fruit will consume the oxygen and at the same time give up equal concentrations of carbon dioxide. When the oxygen is exhausted, the carbon dioxide level will have reached the original level of the oxygen—that is, 21 percent. At this point anaerobic respiration begins, and alcohol is formed in the fruit. However, if fresh air is introduced before this occurs, the amounts of oxygen and carbon dioxide may be kept at compensating levels. The carbon dioxide concentration may be further decreased by passing storage air through a "lime scrubber," usually freshly hydrated lime. This treatment may eliminate the detrimental effect of the high concentration of carbon dioxide observed with some apple cultivars.

If respiration rate is reduced through the control of oxygen and carbon dioxide levels, then storage temperatures may be kept higher than normally required. In the "controlled-atmosphere" storage of apples, storage temperatures may be maintained at 37 to 45°F (3 to 7°C) rather than 30 to 32°F (−1 to 0°C), eliminating disorders of certain kinds associated with low temperatures.

The principle of modified atmosphere may be utilized by using sealed film liners that are differentially permeable to carbon dioxide and oxygen. These materials create a microenvironment one bushel in size (Figure 18-9). Polyethylene film is about five times more permeable to carbon dioxide than to oxygen. When apples are placed in sealed polyethylene film liners, the oxygen in the liner is reduced and the carbon dioxide is increased as a result of respiration. The final concentration depends on the storage temperature and the permeability of the film. In general, increasing the film thickness decreases permeability, which results in lower oxygen and higher carbon dioxide concentrations.

FIGURE 18-9. Polyethylene film liners create a modified-atmosphere storage having a 1-bushel capacity. Film liners are available also for 30-bushel pallet boxes. [Photograph courtesy Purdue University.]

Film density also influences permeability; for example, a film of density 0.928 transmits oxygen and carbon dioxide at one-half the rate of film of density 0.910.

Film liners have made possible an improved method of storing some cultivars of apple. 'Golden Delicious' shows excellent results, although 'Grimes Golden' does very poorly. The high humidity inside the liner is an asset, but temperature must still be closely controlled.

The principle of a modified atmosphere is now extensively used in commercial packaging of fruits, both for home consumption and for shipping by rail, air, truck, and boat.

Organic Volatiles

A number of organic volatiles are produced in ripening fruit. Among them are acids, alcohols, esters, aldehydes, and ketones. The relative proportion of these substances varies with the species and with the degree of maturity, and is reflected in differences in aroma and flavor. In general, the influence of these compounds on respiration is slight. An exception to this statement is ethylene, an organic volatile produced by many fruits during the ripening phase. Ethylene is an unsaturated hydrocarbon (C_2H_6) that is nonpoisonous and has a sweetish odor. It is usually produced in greater amount than the other organic volatiles, usually accounting for two-thirds of the total carbon lost in volatile form. In apples it may be present in an amount 10 to 50 times that of other volatiles.

Ethylene gas has a profound effect on the ripening phase of many fruits, although its exact physiological role is not clear. Ethylene applied externally to immature fruits influences their respiration and ripening, but it has no effect on ripe fruits. A certain minimum concentration and minimum time of exposure trigger an irreversible stimulation of respiration. In 'Honey Dew' melon, 40 ppm is sufficient to stimulate ripening. A 24-hour exposure is long enough if temperatures are suitably high.

Ethylene gas has been used commercially to effect other processes. For example, oranges may be "degreened" by ethylene gas. Its use involves destruction of chlorophyll in the peel. (This effect was first observed when kerosene heaters were used in shipment. Apparently, some ethylene is released by the incomplete combustion of kerosene.) Ethylene gas is not produced naturally by citrus fruits, but it may be produced from a *Penicillium* mold that is often found with them. It is used commercially in banana-ripening rooms to produce uniform ripening or to accelerate ripening. At high concentrations, the gas has been used to defoliate rose bushes. It is also used incidentally in dehusking walnuts, inhibiting potato sprouting, and inducing flowering in pineapple.

Excessive ethylene gas adversely affects growing plants. Thus, neither plants nor cut flowers can be stored with apples; as mentioned previously, scion wood stored together with apples may show severe damage as a result of bark peeling and splitting.

Humidity Control

The control of humidity is directly related to the keeping quality of many horticultural products. In general, low humidity is likely to result in desiccation and wilting. On the other hand, high humidity favors the development of decay, especially if temperatures are too high. Humidity control has become an important feature of modern storage facilities.

The amount of moisture in the air can be expressed as absolute or relative humidity. **Absolute humidity** is the amount of moisture per cubic foot of air. It is expressed as grains of water (one grain equals 1/7000 pound). However, the amount of water vapor that can be held in a given space decreases with falling temperature. **Relative humidity** is the water-vapor content of the air expressed as a percentage of the amount it is capable of absorbing at the same temperature. The ability of a storage to dry out the products it contains is related to relative humidity and temperature. This "drying power" of air is proportional to the water-vapor deficit below saturation. At high temperatures, small differences in the relative humidity represent large differences in drying power. At low temperatures, the reverse is true: Relative humidity may be increased in storage either by adding moisture, as by the use of fine mist, or by lowering the temperature. However, in refrigerated storages, if the difference in temperature between the refrigeration coils and the room is very large, water will condense as ice on the coils and will effectively lower the humidity. This result may be avoided by keeping the coil temperature within 2 to 4°F (1 to 2°C) of the room temperature. The refrigerant system must therefore be large enough to maintain proper temperature. Humidity control becomes more difficult as the rate of air circulation increases.

For the storage of leafy vegetables and root crops, the optimum relative humidity is 90 to 95 percent, but for most fruits and vegetables a relative humidity of 85 to 90 percent is best. Most seed is best stored at relative humidities of 4 to 8 percent.

Ripening

Although most horticultural products ripen on the plant, a number of commodities ripen to optimum quality only when detached from the plant (for example, bananas, pears, and avocados). Pears are picked green and, although stored at 30 to 31°F (about 1°C), they must be ripened at 60 to 65°F (about 16 to 18°C) for optimum quality. The length of the storage period depends on the cultivar. When ripened on the plant, bananas become mealy, lack flavor, and are subject to splitting and subsequent decay. When green, they can be stored at 56°F (about 13°C). They are best ripened at temperatures of about 64°F (18°C) with 90 to 95 percent relative humidity until the fruit is colored. At this point the humidity can be reduced to 85 percent. Ripening can be hastened by holding initial temperatures at not more than 70°F (21°C) for the first 18 to 24 hours and at 66°F (19°C) thereafter. Prolonged high temperatures increase deterioration and decay.

Tomatoes ripen both on and off the plant, but since they do not ship or store satisfactorily when ripe, the fruit may be picked in the "mature" green state and ripened artificially. They will ripen at temperatures above 55°F (13°C), preferably at 60 to 65°F (roughly 16 to 18°C), although the ultimate quality is often less than that of vine-ripened fruit.

Storage Types

In **common storage** the temperature of the atmosphere is utilized, which makes this kind of facility suitable only where temperatures are naturally low enough to benefit the stored produce. Temperature is regulated by insulation and natural circulation. The most primitive type takes advantage of the reduced temperature fluctuations of the soil. Thus, in the fall, trenches or mounds of earth may be used for storing vegetables or plant material. Caves and unheated cellars provide more usable room, but above-ground structures, properly insulated and provided with sufficient ventilation, may be satisfactory in cooler climates. During warm weather, cooling is accomplished by the intake and circulation of the cool night air. Humidity may be kept high with earthen floors. Although common storage is cheap, the lack of precise temperature and humidity control often makes it economically unsound for many horticultural crops.

In **cold storage**, temperatures and humidity are regulated by refrigeration.

A few of the present structures are converted common storages, but large structures with better insulation and convenience features are now built especially for storage purposes. The basic refrigeration and ventilation system involves forced air circulation. The structure must be sufficiently insulated to conserve power.

Controlled atmosphere storage involves the regulation of oxygen and carbon dioxide levels as well as the regulation of temperature. These storages are divided into rooms that are sealed in order that all gaseous exchange can be controlled. The rooms are closed after fruit is stored and remain sealed until it is removed. Temperature, humidity, and gas concentrations are controlled automatically.

A new system called **low-pressure storage** (LPS), also known as **hypobaric storage,** is now under development. This system uses low temperature and subatmospheric pressure (about one-fifth normal), which produce a low partial pressure of oxygen and the other gases in the atmosphere. The system also employs rapid air exchange. Volatiles (such as ethylene) are removed by the air exchange and the reduced pressure. Relative humidity can be controlled within the system.

LPS, by reducing the partial pressure of oxygen as well as ethylene, acts as a modified-atmosphere storage in the control of respiration, ripening, and senescence. Because this control is accomplished by reducing pressure, the system does not need to be continually sealed. It is therefore feasible for portable truck-trailer units as well as large stationary storages. At present, LPS seems most useful for ornamentals because of their high price per volume of storage space required, their sensitivity to ethylene, and their amenability to being stored in atmospheres with low partial pressures of oxygen. The use of LPS vans to transport ornamentals could have important effects on the floriculture industry.

Food Processing

Relatively long-term preservation of food may be achieved by physical and chemical processes that sterilize the food or render it incapable of supporting the growth of microorganisms. These processes include drying, canning, freezing, fermentation and pickling, raising the sugar concentration, and irradiation.

Drying

Drying is one of the most ancient methods of food preservation. The process consists of removing water from the tissues, thus producing a highly concentrated material of enduring quality (Figure 18-10). The natural deterioration of the product by respiration is stopped because of enzyme inactivation. The lack of free water protects the dried products from decay by microorganisms.

FIGURE 18-10. Dehydrated snap beans and the reconstituted product. [Photograph courtesy Quartermaster Food and Container Institute.]

Horticultural products may be naturally dried (**sun drying**) or artificially dried (**dehydration**). Sun drying is relatively inexpensive in locations where summers are sufficiently warm and dry. Dehydration, although a more expensive process, has a number of advantages: The process can be carried on independent of climate, drying time is reduced, and quality may be improved. The yield of dried fruit is slightly higher from a dehydrator than from sun drying because sugar is not lost as a result of continued respiration and yeast fermentation. Furthermore, sun drying requires considerable land and presents sanitation problems.

Dehydration is typically accomplished by hot-air drying. The air conducts the heat to the food and carries away the liberated water vapor. Many types of equipment are used for fruits and vegetables. After being sorted, washed, peeled, and trimmed, fruits to be dehydrated may be treated with sulfur dioxide fumes, which act as a bleaching agent in lighter colored fruits and as a chemical aid to preservation. Safe drying temperatures are near 140°F (60°C). The moisture content of fruits is reduced to 15 to 25 percent. In the dehydration of vegetables, enzyme systems are first inactivated by heating in boiling water or steam (**blanching**). Many vegetables are also more stable if given a sulfur dioxide treatment. For satisfactory storage, the moisture content must be reduced to 4 percent because of the lower sugar content of vegetables as compared with fruits.

A greater quantity of fruit is preserved in the world by drying than by any other method of preservation. Among the important dried fruits are raisin (Figure 18-11), prune, apricot, date, fig, banana, peach, apple, and pear. In contrast, the quantity of dried vegetables on the market is relatively small. Potatoes are the chief dried vegetable. Most successful dried vegetable products are used as flavoring ingredients (for example, onion, celery, parsley, and their powders). Some dehydrated vegetables are sold in soup mixes; others are used in remanufacturing canned products, such as juices, soups, and stews.

A technological development called **freeze-dehydration** may increase the use of drying preservation. Through the use of a high vacuum, quick-frozen

FIGURE 18-11. Sun drying grapes in California. [Photograph courtesy USDA.]

food can be dehydrated. The quality of the reconstituted product is much higher than that of foods dehydrated by ordinary methods. The storage period of dried materials is extended at cool temperatures. At high humidities, mold growth may be a problem.

Thermal Preservation (Canning)

Thermal preservation is a method of preservation that consists in sterilizing food by heating it in an air-tight container. Heating destroys the human-pathogenic and food-spoilage microorganisms, and it inactivates the enzymes that would otherwise decompose the food during storage (Figure 18-12). The sealed container prevents reinfection of the food after it has been sterilized, and it prevents gaseous exchange. There is almost no limit to the size of the container (Figure 18-13).

The application of sufficient heat to sterilize food and inactivate enzymes results in alterations in the color, flavor, texture, and nutritive value of foods. Quality is therefore the limiting factor in thermal preservation. Product quality may be improved by prompt dispatch of high-quality raw products through the processing plant and by proper attention to processing procedures. These measures involve an understanding of the precise relationship between processing time and temperature control. In general, the reduction in processing time, brought about by increasing temperatures, raises the quality of the product. The precise time and temperature required for each commodity is based largely on the natural acidity of the food. There are basically two groups: a **low-acid** group (pH 4.5 to 7.0), which includes most vegetables, and an **acid** group (pH less than 4.5), which includes fruits, berries, tomatoes, and fer-

FIGURE 18-12. Canning cherries. The fruit is being cooked in stainless steel tanks under a constant spray of syrup and color ingredients. The fruit is artificially colored and flavored to make a new product, Maraschino cherries. [Photograph courtesy S&W Corp.]

FIGURE 18-13. Eight 40,000-gallon tanks holding aseptically stored tomato puree in Geneva, Indiana. Bulk storage of processed foods in large containers allows processors to reduce problems associated with seasonal processing, especially seasonal gluts. The partially processed food will be repacked into smaller containers at the convenience of the processor and according to the demand from the marketplace. [Photograph courtesy P. E. Nelson, Purdue University.]

mented and pickled foods. Low-acid foods require relatively severe heat treatment, since they can support the growth of *Clostridium botulinum*, a bacterium that causes a serious form of food poisoning. A millionth of a gram of the toxin produced by this organism is sufficient to kill a normal adult. Thus, all foods capable of sustaining growth of this organism are processed on the assumption that the organism is present and must be destroyed. Since it is extremely heat-resistant, the high temperatures required often reduce the quality of the food.

The storage life of the canned product decreases as the temperature rises. For extended storage (more than 5 years), storage temperatures should be below 50°F (10°C). At storage temperatures above 120°F (roughly 50°C), certain heat-loving bacteria (**thermophiles**), which are not ordinarily all killed by the sterilization process, will continue to grow, causing spoilage. In humid regions, and especially in coastal areas where air-borne salt particles are plentiful, storage life is limited by the life of the container. The salt particles shorten container life by inducing corrosion of metal cans or of the metal lids on glass containers.

A new technological advance in thermal processing is the substitution of plastic envelopes for metal or glass containers. The familiar "tin" can (actually zinc-plated steel) and glass container are tried-and-true packages for food, but usually their cost greatly exceeds the cost of the food they contain as well as the cost of processing. The use of plastic envelopes is made possible by advances in **aseptic packaging,** where the food is first thermally processed (outside the container) and then aseptically poured into plastic and sealed. In traditional canning the heat-sterilization step takes place after the container is filled.

Freezing

Freezing protects food from spoiling because most microorganisms cannot grow at temperatures below 32°F (0°C). The freezing process stops must enzymatic activity and is not in itself destructive to nutrients. Some enzymatic activity in certain products must be stopped by heat treatment (blanching) in order to keep the full flavor and color intact during storage.

The rate of freezing is an important factor in the quality of the thawed product. With slow freezing, relatively large ice crystals develop, fracturing the tissue cells; then, upon thawing, the foods lose cellular fluid, which gives them a soft texture. If freezing is rapid, many small, fine crystals are formed (about one-hundredth the size of those formed during slow freezing). Because these crystals are tightly packed, fewer cells are ruptured.

The basic principle of rapid freezing is the speedy removal of heat from food by various methods, including cold air blasts, direct immersion in a cooling medium, contact with refrigerated plates, and liquid air, nitrogen, or carbon dioxide. Freezing in still air is the slowest method. As living plant cells contain much water, most plant foods freeze between 25 and 31°F (−3.9 and

−0.6°C). The temperature of the food undergoing freezing remains relatively constant at its freezing point until it is almost completely frozen. Quick freezing is a process in which the water in food passes through the zone of maximum ice-crystal formation in 30 minutes or less. This rapidity usually requires refrigerant temperatures of −20 to −40°F (−29 to −14°C).

The success of freezing as a method of preservation depends on the continuous application of the process. Some nutrient loss and deterioration in color, texture, and flavor may occur during frozen storage, depending on the temperature. Best results are obtained at a temperature of −10°F (−23°C), although most storages are kept near 0°F (−18°C). Temperatures must be kept uniform. Frozen foods must be packaged to protect them from dehydration during freezing and subsequent storage. Sublimation of ice occurs in unprotected food, resulting in a freezing disorder called **freezer burn,** which irreversibly alters the color, texture, flavor, and acceptability of frozen foods.

The microwave oven has given a tremendous lift to freezing as a method of preservation because the thawing process is greatly simplified. Quick-frozen meals have become standard fare for airline travelers, as well as for households that do not have the time, desire, or skills to prepare cooked meals from scratch.

Fermentation and Pickling

Although microorganisms are commonly associated with decay, microbial action may be utilized in food preservation. The action of certain bacteria and yeasts in decomposing carbohydrates anaerobically is known as **fermentation.** The decomposition may be accomplished by a number of different organisms, the end products varying with the particular organism. These end products include carbon dioxide and water (complete oxidation), acids (partial oxidation), alcohols (alcoholic fermentation), lactic acid (lactic fermentation), and others. When built up to sufficient concentrations, some of these fermentation products create conditions unfavorable to microorganisms, including the original one. They act as preservatives by retarding enzymatic deterioration, and they impart flavors that are regarded as desirable. Fermentation may be controlled by conditions that favor the growth of one type of organism. This result is achieved through the regulation of pH, oxygen availability, and temperature, and through the use of salt.

Fermentation is an important processing method for some horticultural crops. The fermentation of the juice of grapes or other fruits produces wines. The further fermentation of alcohol to acetic acid is the basis of vinegar production. When used in combination with salting, fermentation is called **pickling.** This term is used especially for cucumbers, but it also applies to other commodities, such as olives and many vegetables (including onions, tomatoes, beans, cauliflower, cabbage, and watermelon rind). Fermented cabbage (sauerkraut) is the result of a number of distinct fermentation processes. Pickling may be accomplished without the direct use of microorganisms by placing food in organic acids (for example, vinegar or citric acid). For extended storage of pickled products, the enzyme systems must be inactivated. This is usually done by canning.

Sugar Concentrates

Acid fruits, concentrated until at least 65 percent of their weight is soluble solids, may be preserved with mild heat treatment if protected from air. The high concentration of sugars and low water content preserve the food. Depending on the recipe used to make it, the product may be called jelly (made from fruit juice), jam (made from concentrated fruit), preserves (made from whole fruit), fruit butter (a semisolid, smooth product made from high concentrations of fruit), or marmalade (made from citrus fruit and rind). The making of candied or glacéed fruits involves the slow impregnation of tissues with sugar. Storage of these products at high temperatures reduces their quality: Their appearance is spoiled by nonenzymatic browning caused by reactions of organic acids with reducing sugars.

Chemical Preservation

In addition to such natural preservatives as salt, vinegar, and spices, a number of chemicals, when added to food, prevent or retard deterioration. They are usually used in conjunction with other methods of preservation. Some of the chemical preservatives used in food preparation are inorganic agents, such as sulfur dioxide and chlorine; others are organic agents, such as benzoic acid, certain fatty acids (including sorbic acid), ethylene and propylene oxides (fumigants), and various antibiotics. Each has a special use in the preservation of fruits and vegetables. For example, sulfur dioxide is widely used in the drying of fruits and vegetables. Chlorine compounds are helpful in hydrocooling and in processing. Potassium sorbate and sodium benzoate are useful in preventing growth of yeasts and molds in such fruit products as fresh cider. Ethylene oxide has been used in the sterilization of spices and flavoring compounds.

Chemical preservatives have a legitimate place in food processing. They are not, however, in the best interests of the public if they are used to deceive or if any danger is inherent in their use.

Radiation Preservation

Radiation is not presently used to preserve horticultural foods, the early enthusiasm with regard to its possible value notwithstanding (Table 18-3). Some of the uses that had been suggested were the extension of storage life of perishables (for example, strawberries) by low-level irradiation, the inhibition of such growth processes as sprouting in potatoes (Figure 18-14) and disinfestation of

TABLE 18-3. Radiation effects on fruits and vegetables.

Dose absorbed (rad)*	Chemical and physiological alterations	Potential horticultural applications
10^1	Interferes with sensitive enzyme systems (auxin, DNA)	
10^2		Treat seed potatoes for storage at room temperature
10^3	Injures many growth processes	Inhibit sprout of potatoes
10^4	Terminates growth processes	Inhibit sprout of bulbs
		Accelerate pear ripening
10^5	Destroys 90–99% of microorganisms	Sterilize fruit and vegetable surfaces to increase shelf life (lengthens storage of strawberries and soft fruits)
10^6	Hydrolysis of carbohydrates	Sterilize fruits
	Termination of respiration	Sterilize vegetables
	Destruction of enzymes, proteins, viruses	Sweeten peas
		Tenderize asparagus
10^7	Complete tissue deterioration	

SOURCE: Data courtesy N. R. Desrosier.

*1 rad = 100 ergs/g of moist tissue. The energy of radiation supplied must be below 2.2 mev (million electron volts) to prevent induced radiation. Cobalt 60 emits radiation below this energy level.

FIGURE 18-14. Inhibition of potato sprouting with irradiation. A dose of 7500 rads permits some cultivars of potatoes to be stored at 50°F (10°C) for 2 years. [Photograph courtesy Quartermaster Food and Container Institute.]

Hawaiian fruit fly in papaya. However, the results of extensive studies indicate that irradiation has adverse effects, including the production of off-flavors and severe injury to delicate tissues. Further, the beneficial effects expected can be obtained by cheaper, more effective means.

Radiation has also been suggested to have potential use in certain unit operations in the food industry, such as the aging of wine, sweetening of peas, and tenderizing of asparagus. None of these uses has proved practical. While no direct evidence of toxic effects has been noted, there is evidence for destruction of vitamins and other nutrients similar to that produced by heat treatments.

Irradiation as a method of food preservation has not lived up to early expectations. The early hopes of irradiation as a substitute for refrigeration, especially in developing countries, has been dispelled. Irradiation cannot substitute for refrigeration and at best can only supplement it.

DISTRIBUTION

Marketing channels are those agencies that handle a commodity along its course from producer to consumer—the actual physical route, as well as the route of title. In general, horticultural commodities travel from producer to wholesaler to retailer to consumer. The concentration of goods by wholesalers and the corresponding dispersion into retail outlets are diagrammed in Figure 18-15. Storage and preservation tend to equalize uneven production with relatively constant demand. The exact marketing channels differ with each commodity and change in pattern over the years.

An interesting phenomenon in horticultural marketing is the resurgence of direct grower sales to customers, especially the trend to organized customer harvesting of crops often labeled as "Pick-Your-Own" or "U-Pick." Many consumers will willingly harvest fruit and vegetables at prices only slightly less than retail because not only is tree- or vine-ripened produce farm-fresh and high quality, but the activity is considered enjoyable and family-oriented. Many "Pick-Your-Own" customers may travel long distances (100 miles round trip is not unusual) for the privilege of harvesting such speciality fruit crops as blueberries or strawberries.

Wholesaling

The wholesaling of horticultural crops varies with the commodity. The transactions involved in wholesaling are performed by intermediaries (commonly called **middlemen**) who specialize in the sale of goods, moving the commodity from producer to consumer.

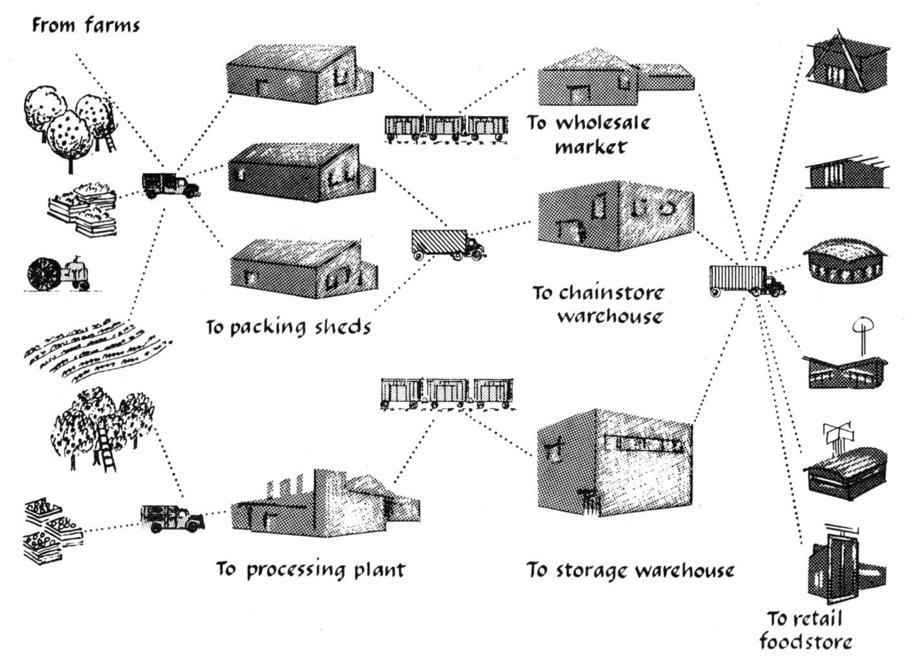

FIGURE 18-15. Principal marketing channels for fruits and vegetables. [Adapted from USDA.]

Merchant middlemen act as representatives for buyers or sellers; they are called **commission men** when they actually handle the product, and **brokers** when their relation to the product is less direct.

Speculators take title to products but merely attempt to profit from market fluctuations. Their effect on the market is mainly on pricing structure. Potato futures and frozen orange-juice concentrate are the chief horticultural commodities traded by speculators in the United States.

A number of organizations, such as **flower exchanges** and **fruit auctions,** are involved in the wholesaling process. They establish procedures at the market and may contribute such services as storage or grading. They operate on fees and assessments paid by those who use the facilities.

Some processors of horticultural products also perform wholesaling functions by acting as their own buying and selling agents. Large processors advertise their own products to build up brand loyalties among consumers. On the other hand, smaller processors distribute their food through food brokers and commission agents. At the present time, processors are large customers for other forms of processed food products. For example, canners are becoming large buyers of frozen foods. Thus, the assortment of marketing channels utilized by processors is becoming quite complex.

Retailing and Merchandising

Retailing, the final outlet for horticultural goods, is the most expensive marketing operation. It accounts for almost half the total marketing cost for fruit and vegetables. In recent years, great innovations at the retail level have created a revolution in marketing that has been felt throughout the horticultural industry—namely, the nearly complete shift to the self-service supermarket. Mass selling, although adaptable to processed products, was not possible with the older methods of retailing fresh fruits and vegetables. The self-service idea has created the packaging concept, which has emphasized better grading and increased standardization. This change has also affected ornamentals. Currently, nursery stock and perennial plants are being sold in this manner.

The self-service trend, and the wide variety and form of products available to the consumer, have resulted in competition that is influenced greatly by promotion and advertising. Different commodities compete for quantity and quality of space in the supermarket. Moreover, certain commodities from one region compete with the same commodities from other regions. This struggle has encouraged producers to organize on the basis of crop and region to facilitate the marketing of their products. Probably the best example of this kind of organization is the California Fruit Growers Exchange, which markets citrus fruits under the Sunkist label. Smaller groups, such as the Michigan Blueberry Association, the Idaho Potato Growers, and Ocean Spray, Inc. (a cranberry-growers cooperative), follow this form. The combined resources of many producers enable them to promote their crop and, at the same time, guarantee a more standardized product.

The marketing of flowers is an exception to this retailing trend. Because of the services they provide (mainly in the form of arrangements), florists have retained their identify as retail units. However, retail florists have organized through such groups as Florist's Telegraph Delivery, Telegraph Delivery Service, and Teleflorists.

Integration

The marketing channels may be integrated horizontally or vertically. **Vertical integration** is the combination under one management of firms that control two or more steps in the production, handling, processing, distribution, or sales of a commodity. A large canning company, such as Dole Pineapple, is a good example of vertical integration. **Horizontal integration** is the combination of firms for the performance of a similar function; a co-op that markets fruits or vegetables from a given area is a good example. Chain stores illustrate both types. They are vertically integrated in that they control their own wholesaling organization—they are not as yet involved in production—and their organization of retail units constitutes horizontal integration. The degree of integration in marketing does not eliminate the main thoroughfares of the marketing channels, although it may streamline them. All marketing operations, such as grading, storage, and packaging, must still be performed.

Selected References

Beringer, L., 1978. *Profitable Garden Center Management.* Reston, Va.: Reston. (A practical guide.)

Fennema, O., 1976. *Principles of Food Science.* New York: Dekker. (An authoritative text.)

Holdsworth, S. O., 1983. *The Preservation of Fruit and Vegetable Food Products.* London: Macmillan. (An English textbook on preservation technology.)

Kohls, R. C., and J. L. Uhl, 1985. *Marketing of Agricultural Products*, 6th ed. New York: Macmillan. (An introduction to agricultural marketing.)

O'Brien, M., and B. F. Cargill, 1983. *Principles and Practices for Harvesting and Handling Fruits and Nuts.* Westport, Conn.: AVI. (Mechanization technology of fruit harvest and handling largely from an engineering standpoint.)

Pfahl, P. B., and P. B. Pfahl, Jr., 1983. *The Retail Florist Business.* Danville, Ill.: Interstate. (A text and manual for the retail florist operator.)

Ryall, A. L., and W. J. Lyston, 1979. *Handling, Transportation, and Storage of Fruits and Vegetables.* Vol. 1, *Vegetables and Melons.* Westport, Conn.: AVI.

Ryall, A. L., and W. T. Penzer, 1982. *Handling, Transportation, and Storage of Fruits and Vegetables.* Vol. 2, *Fruits and Tree Nuts.* Westport, Conn.: AVI. (A two-volume manual for postharvest handling and storage of fresh fruits and vegetables.)

Sullivan, G. H., J. L. Robertson, and G. L. Staby, 1980. *Management for Retail Florists with Application to Nurseries and Garden Centers.* New York: W. H. Freeman and Company. (The fundamentals of management practices for the florist industry.)

USDA Agricultural Marketing Service. *United States Standards.* (Standards and grades are issued and revised periodically by the USDA.)

Wells, R. H., T. H. Lee, D. Graham, W. B. McGlasson, and E. G. Hall, 1981. *Postharvest: An Introduction to the Physiology and Handling of Fruits and Vegetables.* Westport, Conn.: AVI. (An introductory text on postharvest handling of fruits and vegetables.)

HORTICULTURAL INDUSTRY

IV

19 *Horticultural Geography*

The commercial production of horticultural crops is not evenly distributed over agricultural regions, but tends to be concentrated in limited areas of the world. Horticultural geography deals with the distribution of the industry and is concerned with the environmental, economic, and social factors that determine its location and development.

CLIMATE

The physical environment of plants includes many factors whose actions and interactions must be considered. Climate, the summation of an area's weather, includes temperature, moisture, and light effects. It largely determines which plants can be grown in the region and where and when they should be planted. Thus, vegetation is one of the obvious determinants of the differences in appearance between climatic regions. A map of the earth's climatic regions is also an approximation of natural vegetation. Climate, as the fundamental force in the environment, also shapes the soil and, to a lesser extent, the configuration of the earth's surface.

The pattern of climate is a result of the circulation of the atmosphere. Solar radiation, falling more directly on tropical than on polar regions, warms the equatorial air, causing it to flow poleward. The resultant pressure produces a return ground flow of cold polar air. The flow pattern does not follow a simple direct line from the pole to the equator but is deflected as a consequence of (1) the easterly rotational spin of the earth, (2) the seasonal effect, (3) the differential cooling of land and water masses, (4) the altitude and the configuration of the land, and (5) the storms and winds resulting from the interactions of cold and warm air masses. Other extraterrestrial factors, such as sun-spot activity, may affect the weather, but they are poorly understood.

MICROCLIMATE

Microclimate refers to the climate of a "small area." The climate at the ground may differ considerably from that at 30 feet above the ground. These climatic differences are of vital importance to people and their agriculture. We become aware of the microclimate as we drive in and out of pockets of fog on a chilly morning. The orchardists who lose the crop on the lower half of their trees as a result of frost become painfully aware of microclimate.

Whereas **location** identifies a geographic and climatic area, the term **site** implies microclimatic influences within a specific location. The ultimate success of horticultural enterprises depends to a great extent on proper location and site. Microclimatic variations are due to exposure, slope, vegetation, and the thermal capacity and conductive characteristics of the soil. These will be discussed along with the principal climatic elements: temperature, moisture, and light.

CLIMATIC ELEMENTS

Temperature

The temperature at any point on the earth's surface depends on the geographic ordinates of latitude and altitude, season and time of day, and the mediating influence of microclimate. The major factor that determines temperature is the amount of solar radiation received, which depends upon both its intensity and its duration. The more vertical the sun's rays, the less atmosphere they must penetrate. In addition, vertical rays provide a greater concentration of energy per unit area than do the oblique rays that reach the poles. The duration of solar radiation is determined both by day length, which imposes an absolute limit on the amount of solar energy received, and by the variable effects of cloud cover. Furthermore, there is a decrease in temperature with an increase in elevation. This decline averages 3.6°F (2.0°C) for every 1000 feet of elevation, approximately 1000 times the rate of temperature change with latitude. The reason for this fact is that much of the atmospheric thermal energy is received directly from the earth's surface and only indirectly from the sun. In addition, the lower tropospheric air contains more water vapor and dust; it is therefore a more efficient absorber of terrestrial radiation, which explains the presence of snow-capped mountains, such as the famous Mt. Kilimanjaro, near the equator.

The variation in temperature reported at the earth's surface is enormous, with a range of more than 200°F (111°C), from a record −96°F (−71°C) at Verkhoyansk, Siberia, to 136°F (58°C) at Azizia, Libya. The mean annual temperatures range from an estimated −22°F (−30°C) at the South Pole (elevation 8000 feet) to a record 86°F (30°C) at Massawa, Eritrea, Africa. In annual crops the important temperature values are the mean and extreme temperatures as well as the duration of temperatures conducive to plant growth (the length of the "growing season"). Perennial plants are affected by the temperature values during the whole year, and frost (see Chapter 10) is critical, especially in fruit crops. Both seasonal variation and average temperature must be considered in relation to plant growth. Peaches, for example, require a long growing season and warm summer temperatures. Their northern distribution is limited by their degree of hardiness; temperatures below −12°F (−24°C) cause flower-bud injury. Their southern distribution is limited by their chilling requirements. A map of the expected minimum temperatures of the horticulturally important areas of the continental United States and Canada is presented in Figure 19-1.

Moisture

In discussions of climate, **moisture** refers both to precipitation (rain and snow) and to atmospheric humidity. Rainfall is directly related to the circulation pattern of the atmosphere. It results from the cooling of warm, humid air forced upward by the convergence of air currents. Total annual precipitation may vary enormously from one place to another. Because of topographical variation, a marked difference in rainfall may occur between points that are relatively close together. This variation may result in the close proximity of desert and rain forest. A map of average annual precipitation is shown in Figure 19-2.

Of greater agricultural significance than average precipitation is the **effective precipitation**, the water that is not lost by runoff or evaporation and that is consequently available to plants. The percentage of precipitation that is "effec-

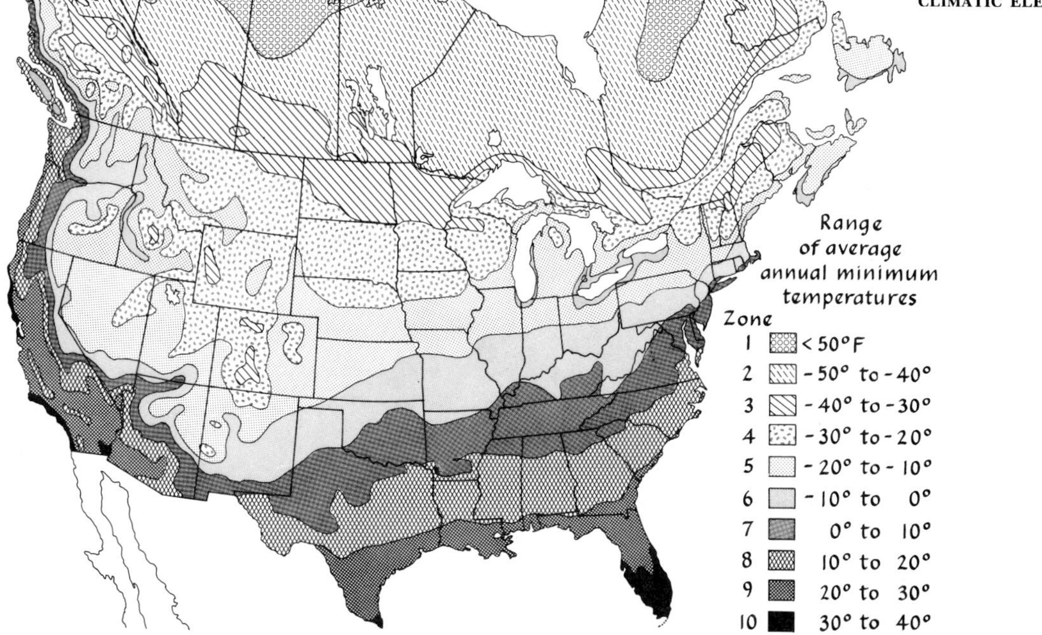

FIGURE 19-1. Zones of plant hardiness in the United States and Canada. [From USDA Miscellaneous Publication 814.]

tive" is higher where temperatures are low. Thus, cool areas require less rain for plants than do warm areas. The natural vegetation of an area is the most satisfactory measure of effective precipitation.

The extremes of precipitation result in drought and flood. Drought may be predictable in some areas on a seasonal basis; in others, it appears to recur on a longer cycle. Unless irrigation is practical, horticultural crop production is effectively restricted in drought areas. Although methods exist for the efficient utilization of existing water, there is no substitute for sufficient water in intensive crop production. Horticultural crops require a plentiful supply of water as compared to agronomic crops. Flood, on the other hand, is equally injurious (Figure 19-3). Except for such water-loving plants as cranberries, which are grown in boggy areas, an excess of water results in extensive damage to horticultural crops. Much of this injury is due to root damage from lack of soil oxygen. Excessive moisture also results in disease problems since high humidity favors the growth of many pathogenic fungi. Many crops (including tomato, peach, apple, and cherry) show abnormal splitting and cracking of the fruit as a consequence of excess moisture during periods of fruit ripening. Flooding can be overcome by proper drainage in some locations. Tiling and proper soil management do much to alleviate the problem.

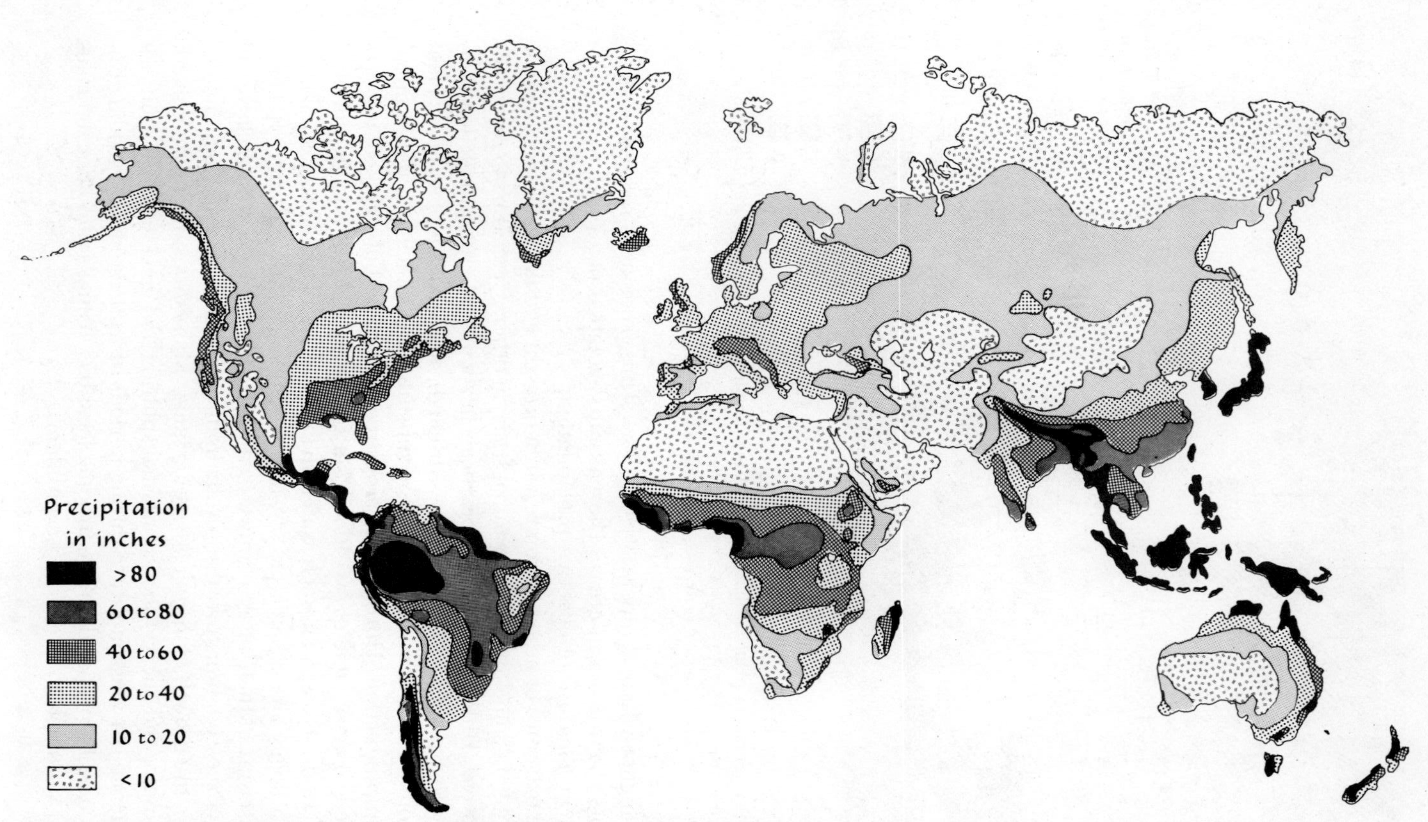

FIGURE 19-2. Average annual precipitation throughout the world. [Courtesy USDA.]

FIGURE 19-3. Spring flooding and erosion. A folk proverb states: "In dry years farmers complain of starvation; in wet years farmers starve." [Photograph by J. C. Allen & Son.]

Light

The quantity of light is a critical climatic feature and a significant part of the plant environment. Day length is the most obvious difference between climates. In the tropics, the day is close to 12 hours long throughout the year. At the polar regions, the summer day length goes up to 24 hours. The world distribution of plant species is greatly determined by their photoperiodic response. It is useless to attempt to grow long-day plants for flowers or fruit in the tropics, for they will remain vegetative indefinitely. In addition to day length, the quantity of light is affected by such atmospheric conditions as the number of sunny versus cloudy days during the growing season. The amount of sunlight greatly affects quality in many crops (for example, grapes and apples).

SOIL AND CLIMATE

The chemical, physical, and biological changes that determine the soil profile are all affected by long-term climatic conditions. The major differences among soils are the result of climate operating through soil-forming processes and vegetation. Thus, climate affects soil formation directly through weathering processes and indirectly through vegetation.

The distribution of the major soil groups can be interpreted through broad climatic types, based on weathering and vegetation (Figure 19-4). The major zonal soil types, grouped on a climatic basis, are (1) the strongly leached red

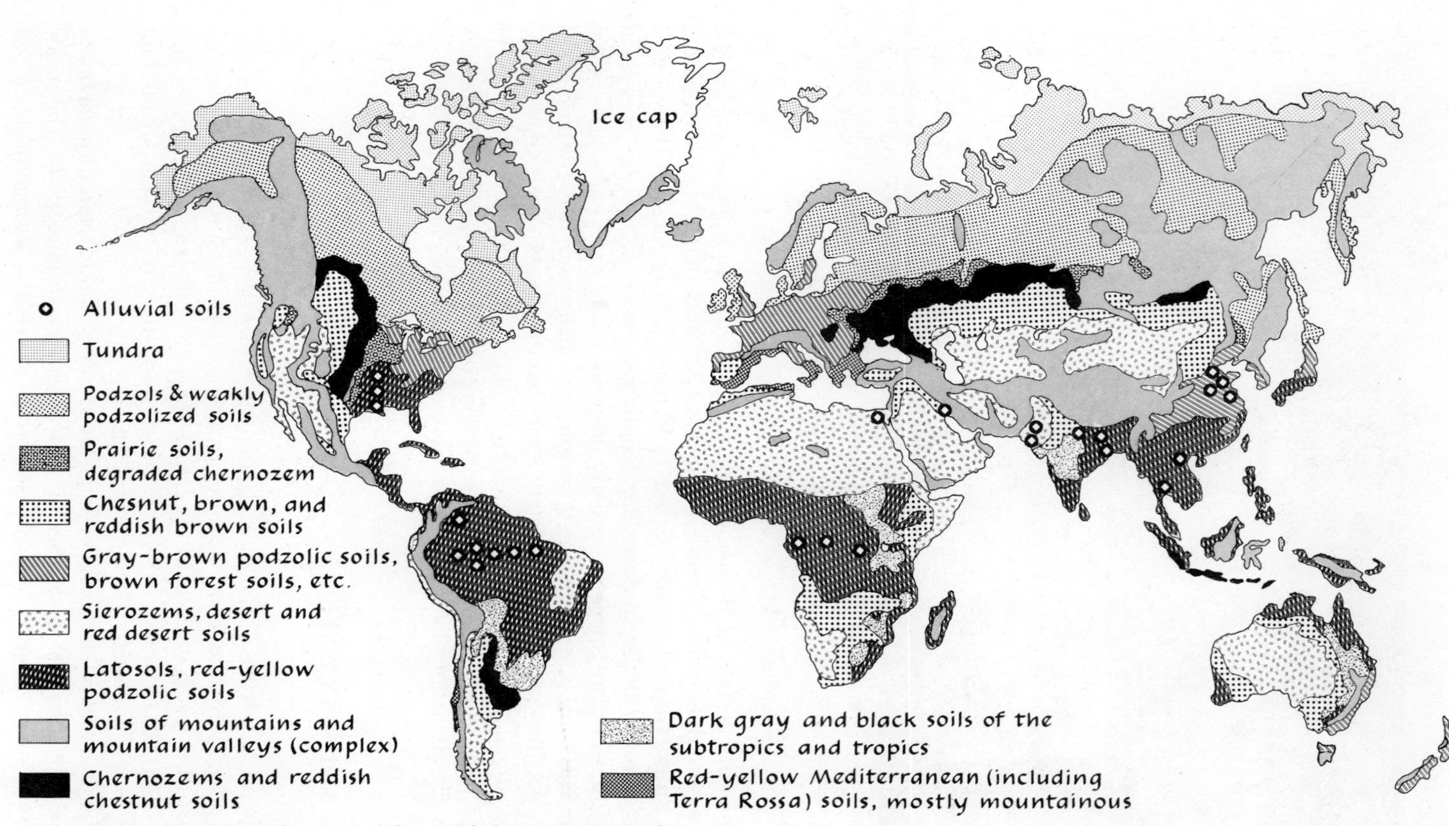

FIGURE 19-4. The major soil groups of the world. [Courtesy USDA.]

soils supporting tropical rain forests in areas of hot, humid climate; (2) the unleached light-colored soils in areas of hot desert climate; (3) the dark-colored soils supporting native tall grasses in areas of subhumid, temperate climate; and (4) the acid, light-colored soils supporting coniferous forests in areas of cool, moist climate. Unfortunately, the climates best suited for plant growth do not always coincide with the areas of naturally fertile soils.

CROP ECOLOGY

Ecology is the study of life forms in relation to their physical and biological environment. Plant ecology deals largely with natural plant communities, and is concerned with the causes responsible for the course and pattern of plant development, succession, and distribution. It involves the relations between climate, soil, and biological interaction.

By definition, horticulture deals with cultivated plants. However, the crop-plant community is subject to many of the same ecological responses as is the natural-plant community. These plant responses determine to a great extent the ability of a region to support successfully a particular crop, and they define the specific problems of land use and crop management.

The climatic environment is a powerful determinant of plant development. It is the extremes of temperature (Table 19-1), rainfall, and light, rather than their yearly means, that determine the status and define the limitations of agriculture. The inappropriate utilization of agriculture in areas of marginal climate results in more "poor years" than "good years." Unfortunately, the occurrence of unusual periods of "good" weather often results in overextension of an unsuited agriculture, with disastrous consequences when the more normal pattern resumes.

The effect of climate upon quality and appearance of horticultural products plays an important role in the location of the horticultural industry. For example, high light intensity favors maximum development of red color in apples. The prominence of the central valleys of Washington State as apple-producing areas is due to their dry climate, which is brought about by a favorable combination of altitude and sheltering mountain ranges that results in bright, cloudless summers.

CLIMATIC REGIONS

A variety of ways exist for classifying climatic regions. We are all familiar with the climatic classification by temperature into tropical, temperate, and polar zones. A more useful classification includes both temperature and moisture to

TABLE 19-1. Classification of common fruit crops by temperature requirements.* Cold-tolerance increases as the crop listing descends.

Tropical	Subtropical		Temperate: Mild winter	Temperate: Severe winter
Coconut				
Banana				
Cacao				
Mango				
Pineapple				
Papaya				
	Coffee			
	Date			
	Fig			
	Avocado			
		Citrus		
		Olive		
		Pomegranate		
			Almond	
			Blackberry	
			Grape (European)	
			Persimmon (Japanese)	
			Quince	
			Peach	
			Cherry	
			Apricot (blossoms tender)	
			Strawberry (very hardy under snow)	
			Blueberry (very hardy under snow)	
			Raspberry	
			Cranberry	
				Pear
				Plum
				Grape (American)
				Currant
				Apple
Low-temperature sensitive		Slightly frost tolerant	Tender	Winter-hardy
Noncold requiring			Cold requiring	

*Variation in tolerance depends to a large extent upon species, variety, plant part, and stage of growth.

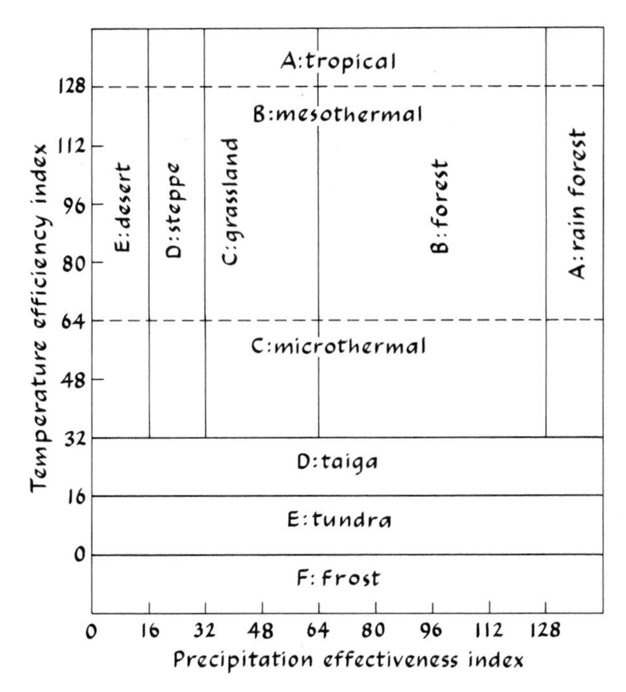

FIGURE 19-5. Temperature and humidity provinces based on natural vegetation. [After C. W. Thornthwaite, *Geogr. Rev.* 21:649, 1931.]

account for seasonal patterns. In this regard the natural vegetation may be used as a meteorological instrument to integrate the climatic elements (Figure 19-5). Using precipitation effectiveness and temperature efficiency,[1] Thornthwaite has divided climate into regions associated with a characteristic vegetation—for example, grassland, forest, and rain forest (Figure 19-6).

A widely used method of classifying climates is the Köppen system, which is based on annual and monthly means of temperature and precipitation. Their boundary lines are selected by natural vegetation and crop responses rather than by atmospheric circulation patterns (Figure 19-7). The classification discussed here is derived from Trewartha's modification of the Köppen classification.

Tropical Rainy Climates

The region of tropical rainy climates lies in an irregular belt 20 to 40° wide that straddles the equator. Its most typical feature is the absence of winter. The difference between the average daytime temperature and the average night-

[1]Precipitation effectiveness is the summation of monthly precipitation divided by evaporation. Temperature efficiency is the summation of monthly mean temperature minus 32°F divided by 4.

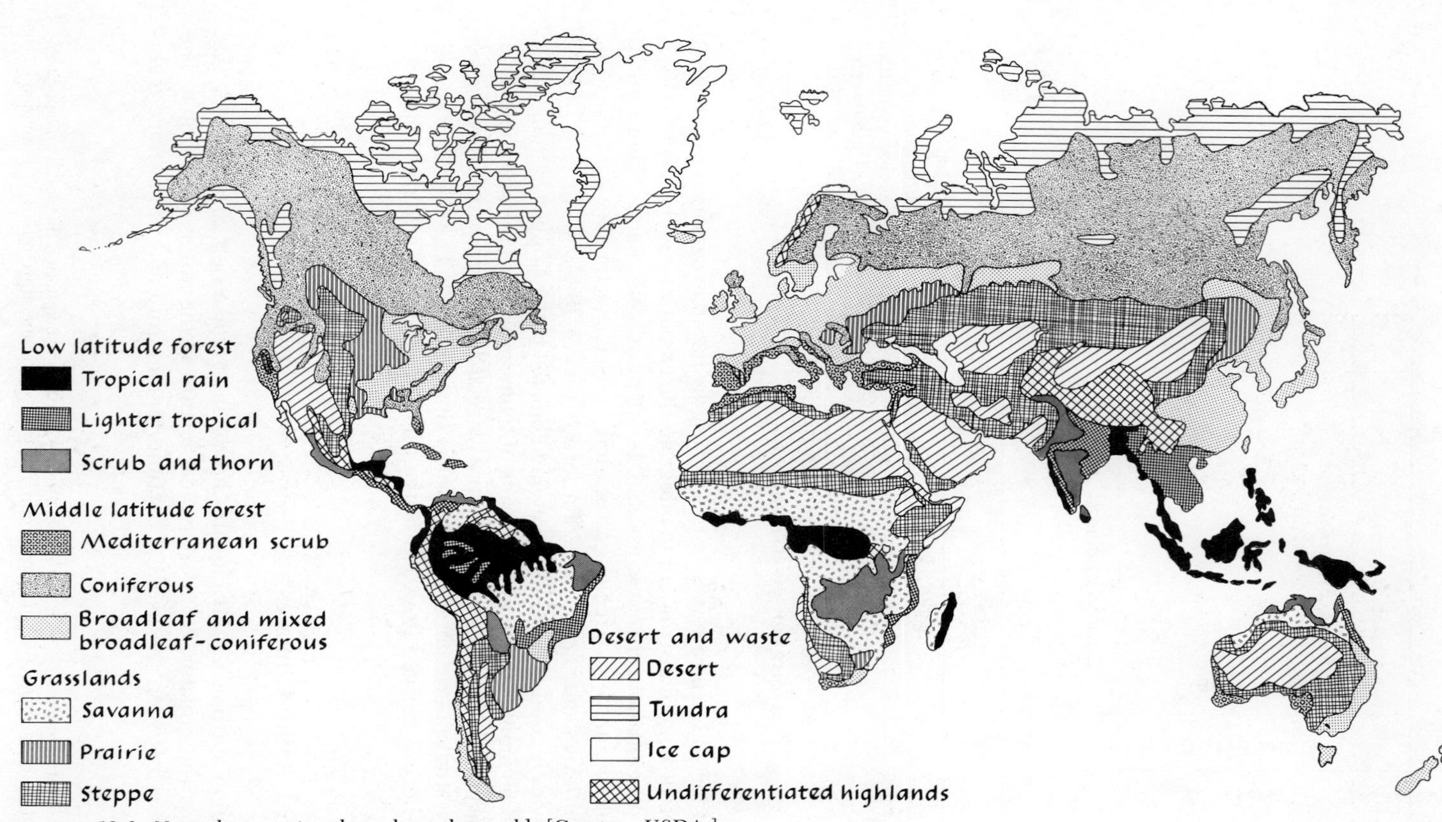

FIGURE 19-6. Natural vegetation throughout the world. [Courtesy USDA.]

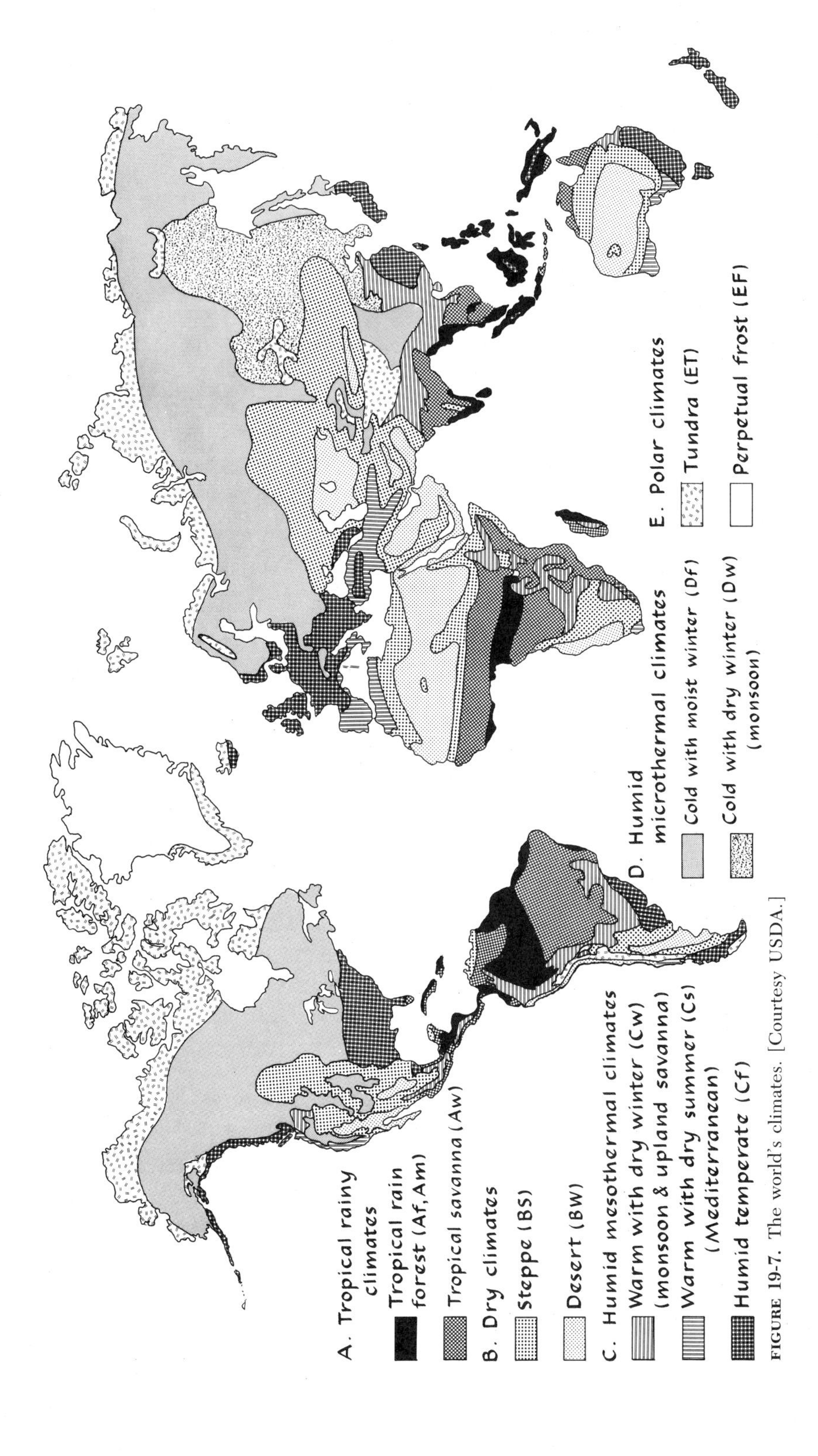

FIGURE 19-7. The world's climates. [Courtesy USDA.]

time temperature exceeds the annual range of 24-hour temperature averages. The temperature boundaries of this region have been arbitrarily placed at the average minimum monthly temperature of 64°F (18°C). This limitation coincides with the poleward limit of plants requiring continuing high temperature, as various tropical palms. Although rainfall is abundant (rarely less than 30 inches per year), its distribution serves to produce two subclimatic types. One type has ample rainfall throughout the year, and although there are seasonal differences, no less than 2½ inches of ran fall in any month. The other subclimate is distinguished by distinct wet and dry seasons.

Tropical soils, which are chiefly **lateritic** soils (red soils rich in iron and aluminum hydroxides), are notoriously unproductive when continually cultivated. The rapid destruction of organic matter in such soils results in a soil structure that deteriorates under intensive crop use. Their low native fertility, a result of constant leaching, makes them responsive to fertilizer application.

In the continually rainy areas of the tropics, the natural vegetation is lush and varied. No other area can compare with the tropics in the diversity of plant species. The natural vegetation is broad-leaved and evergreen and has no dormancy requirement. In the seasonally wet and dry areas of the tropics, the dominant vegetation is coarse grass, which goes dormant under dry periods. A tropical grassland is called a **savanna,** and the climate that produces it is called a **savanna climate.** The number of trees to be found in such places is determined by the duration of the wet season.

The chief horticultural plants of the tropics grow continuously and have no dormancy requirement. As a group, they are quite sensitive to any extended periods without moisture. However, agriculture is generally more extensive in the wet-dry areas of the tropics because continuous rain is unsuitable for many crops. Wet-dry areas are used widely, without irrigation, for grazing. Although many tropical horticultural crops are known principally in their native areas—mango, yam (*Dioscorea*), papaya—many important crops of worldwide significance must be grown in tropical or subtropical climates exclusively. The separation of plant agriculture into horticulture and agronomy is less meaningful in the tropics. Although sugar and rice are considered agronomic crops, the distinction of such crops as palms (for oils), jute, and rubber as horticultural is justifiable because of their intensive culture.

Dry Climates

Dry climates are those in which the potential for evaporation from the soil surface and for vegetative transpiration exceeds annual rainfall. Rainfall is scarce and unpredictable, and water is often deficient. Because evaporation and transpiration increase with temperature, the boundaries of dry climates are affected both by rainfall and by temperature. The dry climates are roughly distributed on either side of the tropical rainy climates.

The dry climates are subdivided on the basis of moisture and temperature into **arid (desert)** and **semiarid (steppe).** Steppe is a transitional zone between humid and desert climates. These dry climates may also be subdivided into tropical and subtropical regions on the basis of minimum temperatures. Thus, there are four general regions—hot and cold desert, and hot and cold steppe.

Although natural vegetation is meager in areas with dry climates, the abundance of such areas (26 percent of the earth's surface) makes them an important agricultural resource. Because the steppe areas support low-growing, shallow-rooted grasses, their chief agricultural use is for grazing.

Since the soils of dry climates are subject to minimum leaching, they are high in nutrients. Although desert soils tend to be too sandy and otherwise unsuitable for agriculture, steppe soils are superior. The grass roots provide considerable organic matter. On steppe soils, irrigation can transform the land, making it ideal for horticulture. The irrigated lands of our Southwest provide a splendid example of this change. There, such subtropical horticultural crops as melons, figs, and citrus are produced in the warmer season, and such temperate-climate vegetables as carrots and lettuce are grown in the cooler season. The agriculture of this area, dependent as it is on irrigation, resembles the oasis agriculture of naturally moist areas (deltas, flood plains, and alluvial fans) within dry climates.

Humid Temperate Climates

The climate of the middle latitudes is characterized by a distinct seasonal rhythm; winter, spring, summer, and autumn have real meaning. The chief factor influencing plant dormancy is low temperature, rather than drought. The humid temperate climates are subdivided into mild-winter (mesothermal) and severe-winter (microthermal) climates. This distinction is based on an average minimum monthly temperature of 32°F (0°C). This condition can be empirically found on the basis of a durable winter snow cover.

Mild-Winter Climates

The mild-wintered portions of the temperate zones are found in the lower latitudes and in marine locations on the westerly side of continents. Within the mild-wintered temperate zone, there are three important climates: (1) Mediterranean, (2) humid subtropical, and (3) marine.

MEDITERRANEAN CLIMATE. The **Mediterranean climate** with its subtropical, dry-summer represents a most important horticultural region. It is characterized by dry, warm-to-hot summers and mild winters with adequate rainfall. Fog is common and provides some summer moisture. The inland areas have hotter summers than the coastal areas. Because of its delightfully mild winters, bright sunny weather, and strong horticultural associations, it is one of the best known climates, found in the Mediterranean region, central and coastal California, central Chile, the southern tip of South Africa, and parts of southern Australia. Although these areas constitute less than 2 percent of the land, they provide a large portion of the world's horticultural products, especially subtropical fruits (citrus, figs, dates, grapes, and olives).

The Mediterranean soils are variable. The natural vegetation consists of mixed trees and stunted woody shrubs, with leaves that are typically thick and glossy, designed to prevent excessive transpiration. The growing season lasts practically the whole year. However, frosts do occur during the winter months. These may be extremely dangerous to horticultural crops that are marginal for the area, such as out-of-season vegetables and citrus. Although the annual rainfall is moderate (15 to 20 inches), the effective precipitation is high, since the bulk of the rain comes in the cool winter, making much of it available for plant growth. Summers are distinctly dry, thus permitting a double dormancy in native vegetation: one due to low winter temperatures, the other to summer droughts. The dry heat of the summer is ideal for fruit drying. It is here that this method of fruit preservation originated, and this area still provides dried figs, dates, plums (prunes), and grapes (raisins).

Flower-seed production is a major part of the horticultural industry in this climate in California. Under irrigation, an extensive fruit and vegetable canning industry has developed. Cool-season crops are shipped almost exclusively from California during the winter months. For example, the Imperial Valley provides most of the United States' winter supply of lettuce.

HUMID SUBTROPICAL CLIMATES. The **humid subtropical climates** are typically on the eastern side of continents (for example, the cotton belt of the United States). The annual precipitation is usually abundant (30 to 65 inches) and well distributed over the entire year. The summers are hot and muggy. Although winters are generally mild, freezing temperatures occasionally occur. The variation, however, is considerable: In the United States, the South experiences more severe winters than do the humid subtropical areas of South America, Australia, or China.

The natural vegetation includes forest (mixed conifers and broad-leaved trees, both deciduous and evergreen) and grassland, depending on precipitation. The soils are deep, owing to weathering, but they are typically low in fertility from constant leaching. The grassland soils are more productive.

The humid subtropics are rich vegetable-producing lands. For example, the long growing season and high summer temperatures of the southeastern United States are suited for the growth of such crops as sweetpotatoes, dried beans, and melons. The relatively mild spring and fall allow two crops of such garden vegetables as spinach, mustard greens, radishes, and snap beans. Tomatoes, peppers, and celery are grown as spring crops. A large industry based on the growing of vegetables and flower transplants for shipment to northern states is located in this area. Only those fruit crops having a low winter chilling requirement can be grown. Thus, this area is more satisfactory for growing peaches than it is for apples. In sites where the danger of spring frost is minimal, peaches are an important crop—particularly true of the Gulf states. Strawberries and blueberries, which have a wide range of adaptation, are common horticultural crops. The nursery industry is expanding along the northern edge of this area.

MARINE. **Marine climates** are found on the western side of middle latitude continents and extend, owing to the ocean's influence, into the high latitudes. The extent to which these climates penetrate depends on topographical features. In North and South America and in Scandinavia, areas with marine climates are narrow because mountain ranges closely parallel the coast. In western Europe, extensive lowlands serve to carry the ocean's influence inland.

In contrast to Mediterranean and humid subtropical climates, the marine summers are cool. The warmest months typically average 65 to 70°F (18 to 21°C). Winters are exceptionally mild considering the latitude, the average minimum monthly temperature being above freezing. The growing season is long, although frosts are frequent and winter is generally long enough to produce a dormant season.

There is adequate rainfall in all seasons, although precipitation varies greatly, from more than 100 inches in the Pacific Northwest to 20 to 30 inches in parts of western Europe, which is surprising considering the many rainy days. Sunshine is limited; fog and mist are abundant. About half the days each year are cloudy.

The natural vegetation is forest, deciduous in Europe and coniferous in the Pacific Northwest. The soils are variable though generally deep. Gray forest soils of this area have excellent structure, and although their natural fertility is moderate, it is higher than that of other forest soils. In addition, there are areas of organic soils (peat-bogs), thin, stony, glacially deposited soils (Scandinavia), and sandy coastal plains (Germany and Denmark).

Horticulture is not the predominant agriculture in marine climates, but it is extensive. Apples and pears are well suited to marine climates, as is the ubiquitous strawberry. Cool-season vegetables—peas, lettuce, and crucifers—do especially well. There is not enough heat, however, for outdoor plantings of some crops, such as melons. For example, cucumbers and grapes are commonly grown under glass in Holland, Belgium, and England. The long spring extends the flowering season. Bulbous flowers (such as tulips, daffodils, and hyacinth) grow well.

Severe-Winter Climates

The severe-winter climates of the humid temperate regions are characterized by summers that are distinct but shorter than those of the mild-winter climates.

These climates, which are located in the interior and poleward areas of North America and Eurasia, are dominated by large land masses. They are bounded in the north by **polar** climates. The climate is divided into areas of warm summers (the United States corn belt), cool summers (the Great Lakes area, western and central Russia), and subarctic (northern Canada, Alaska, Siberia). The insulation provided by permanent deep snow cover serves to prevent excessively low soil temperatures (although winter air temperatures become extremely low in the higher latitudes and vary greatly). Precipitation is generally heaviest in the summer, owing to the influx of warm, humid air from the south; the winters, by contrast, are influenced by cold polar air. This variation is significant in regard to plant growth. Although yearly precipitation is moderate, ample moisture is usually available during the growing season. Largely because of this relatively high effective precipitation, the cool- and warm-summer climates support an abundant agriculture.

The natural vegetation consists of forests (in which conifers predominate) at the northernmost margins, and tall-grass prairies in the subhumid interiors. The gray-brown soils of deciduous forests are good, but the highly acid **podzol** soils of the cooler, more northerly coniferous forests are poor. The prairie soils are excellent in structure and in fertility, which makes these lands particularly suitable for corn and wheat despite the somewhat lower and less reliable rainfall. Although prairie soils are limited in extent, they make up the finest agricultural lands in North America. Glaciation has molded many of the topographical features in this climate (for example, the rocky New England hills).

The warm-summer climates are ideal for such horticultural crops as melons and tomatoes, which require high temperatures. The cool-summer climates more typically support extensive plantings of hardy fruits, such as apples and pears, although cherries and peaches can be produced in protected sites, such as those found in the Great Lakes area. These climates are ideal for cool-season vegetables, such as potatoes and peas. Sweet corn is an important canning crop in the northern part of the United States, where cool summer temperatures tend to prolong the period of good quality for harvesting. The nursery industry has taken advantage of the deep prairie soils; bulb crops, strawberry plants, and nursery shrubs are grown extensively in this region. For example, Iowa, located in the heart of the corn belt, is an important center for mail-order nursery businesses specializing in deciduous plants, which ship well when fully dormant.

The subarctic regions of the severe-winter temperate climates support only a limited horticulture. The long summer days compensate for the cool temperatures and lower light intensity. Here, summer frosts may be a serious hazard. The frost-free season may last only from 50 to 75 days. The main horticultural crops are root crops (such as potatoes, turnips, beets, carrots, and parsnips) and crucifers (including cabbage and cauliflower). As a result of breeding programs in Alaska, some strawberry cultivars adapted to subarctic climates are now available.

The polar climates are found in the high latitudes and at high altitudes. The boundary is set by a mean annual temperature of 32°F (0°C), with the average warmest monthly temperature being 50°F (10°C). There is no horticultural activity in polar climates.

ECONOMIC FACTORS

In addition to such environmental factors as climate and soil, a number of economic considerations strongly affect the complexion of the horticultural industry. They include land costs, availability of labor, distance to market, and the nature of available transportation facilities. The general level of the economy is also an important factor. A highly developed industrial economy is able to afford an abundance of horticultural products. In the United States, an increase in the standard of living results in an increase in the consumption of fresh fruits and leafy vegetables. In addition, the perishable nature of many horticultural crops necessitates advanced technology for movement, storage, and processing. In agricultural societies, most horticultural products are grown for local consumption, and a significant proportion of the products are actually home grown. As a result, the total variety and value of horticultural products are less than in highly developed economies.

Horticultural crops differ as to their adaptability. In general, long-season crops, biennials, and perennials must be restricted to areas where weather remains favorable for extended periods of time; many months are required for long-season annuals, and many years are required for tree-fruit production. This factor greatly restricts their location, especially when markets are highly competitive. On the other hand, annual crops that mature in a comparatively short time often appear to have a wider range of adaptability. This is also true for biennial plants (such as carrots, onions, and cabbage) that are grown as annual crops. Actually, it is their adaptability *as crops* that is large: They may produce a marketable crop in a short time in any one of a number of diverse climates, but it may be necessary for seed to be produced elsewhere under more favorable conditions.

The wide adaptability of annual horticultural crops makes them responsive to local economic advantages. Increased quality and local market preference account for some of this interrelationship. With technological advances in the long-distance movement of horticultural crops, these local advantages decrease in importance. The trend is for annual crop production to be situated in optimum climatic locations, especially those favorable to off-season production. Nevertheless, the great cities are still ringed by relatively small market gardens that grow fresh produce for local consumption, although their number is decreasing. With the increases in processing and in centralization of the marketing operations of chain stores, the demand for the products of local growers has dwindled to an insignificant level in many areas.

Land Costs

Land cost includes not only land prices but local taxes as well. Horticultural crop production is often intensive enough to justify the use of expensive agricultural land, but it cannot long survive in urban areas. The horticultural enterprises originating on the outskirts of large cities may be literally overrun by urban expansion. High urban taxes soon make any agricultural operation unprofitable. Although "selling out" may result in an immediate profit to individual landowners, the horticultural industry is destroyed, and valuable agricultural land is often lost forever. The disappearance of potato farming on Long Island, New York, and of the citrus industry around Los Angeles, California, are examples of such losses.

A movement to preserve prime agricultural land—especially those unique horticultural areas threatened by urbanization—is underway under the theory that the permanent loss of this basic resource is detrimental to society as a whole and will be bitterly regretted by our descendants.

Agricultural land preservation has both an economic and an ethical dimension. Some landowners may lose substantial capital gains as a result of restrictions on the conversion of land from agricultural to nonagricultural use; for others, land preservation enhances present income by protecting farm operations. Some voices argue that agricultural land preservation transcends short-term economic considerations because the land is irreplaceable and is essential to sustain us and our descendants. Under this ethic, land use is considered in terms of stewardship rather than ownership. Limits on absolute freedom of action are considered a social obligation and an extension of the concept that private property—only possible with the creation of a state and a system of laws—may be regulated (by zoning), taxed, and appropriated subject to due process and just compensation.

While reasonable people may agree on the goal of sustaining agricultural productivity, there is no consensus on the means to attain this goal. Our society needs to find a combination of incentives and policy that will conserve the land while protecting individual rights and maintaining the ability to profit in farming both now and in the future.

Labor Supply

Many horticultural operations require an abundant supply of labor at some point, usually during harvest. Because of the extensive need for hand work in the past, the horticultural industries have been dependent upon a plentiful supply of low-cost labor. In many parts of the United States, migrant workers move from one horticultural area to another, harvesting crop after crop as the season progresses. In some areas, workers are brought in from other countries. The organization of this labor, or federal and state legislation requiring minimum wages and standards, increases labor costs and, therefore, the cost of production. This need for transient labor has created severe problems and has encouraged the automation of such operations as transplanting, weed control, and harvesting. In 2 years, for example, the California tomato-processing industry changed from total hand picking to 85 percent mechanized harvesting. Although there is a definite trend toward mechanization, many horticultural operations will continue to depend upon large seasonal labor resources, at least for the foreseeable future.

The recent move of the greenhouse industry away from large centers of population to more rural locations is due to rising labor costs as well as to increased taxation. Cheap labor, however, is not a dependable resource. The solution to extensive labor requirements undoubtedly lies in labor-saving devices and technological innovations.

Market Advantage

Historically, the commercial horticultural industry originated close to the large population centers. The perishable nature of most horticultural products gave a distinct advantage to market proximity by virtue of the close relation of proximity to quality. As transportation and storage facilities have improved, this advantage has steadily diminished. Rapid, refrigerated transportation facilities operating over great distances have equalized the quality differential. The integration of railroad and motor-truck systems have further increased transportation efficiency. Recent advances in the air shipment of such high-priced horticultural products as flowers and strawberries have reduced the advantages of market proximity to growers of these highly perishable commodities.

Transportation costs have remained relatively constant over the years, and do not, therefore, account for the decrease in the industry's dependence on market proximity. Other factors that determine the distribution pattern are market price, season of shipment, and cost of production. Transportation costs,

TABLE 19-2. Transportation costs for 'Delicious' apples, 1955–1957.

State of origin	City marketed	Retail price per pound (cents)	Transportation cost per pound (cents)	Transportation cost as a percentage of retail price
Washington	Los Angeles	20.5	1.4	7
Washington	Chicago	20.6	2.3	11
Washington	New York	20.0	2.8	14
New York (Hudson Valley)	New York	15.7	0.5	3

SOURCE: Data from Agricultural Marketing Service, USDA.

however, are still a significant cost factor (Table 19-2); but when the reduced cost of production of a different region offsets excessive transportation costs, the industry is bound to move.

One of the chief factors in market advantage has been out-of-season production. Thus, early strawberry production in Louisiana and winter production of vegetables in warmer southern and western climates have created important horticultural industries, despite poor soils and the danger of frosts.

The consolidation of marketing operations by large food chains, which deal directly with growers, favors large operations located in areas with the best climates for production. For example, fewer than 200 buying concerns handle 60 percent of the nation's fruit. The trend is toward the centralization of the industry in response to climatic factors.

CULTURAL FACTORS

In addition to environmental and economic factors, a number of cultural factors have played a significant role in the distribution of the horticultural industry. In many cases, the development of the industry in the United States can be traced along with the history of certain national groups. For example, the greenhouse industry was brought to America primarily by the Dutch immigrants, a factor that is reflected today in the large percentage of people of Dutch descent who still engage in the commercial production of vegetables under glass in the Midwest. This explanation is also true of the onion-set industry located around Chicago. Similarly, the Japanese flower and vegetable growers are prominent in West Coast horticulture.

Many local horticultural enterprises of considerable scale relate to a particular family or firm. A single grower in Wisconsin produces a significant percentage of the horseradish used in the United States. The success of the Hill Nurseries, the largest rose center in the United States, has brought a large floricultural industry to Indiana, although there are better adapted areas in terms of light and temperature. The carnation industry in New England traces its origin to a few private conservatories on wealthy estates.

Market preferences have greatly affected the horticultural industry. For example, collards and mustard greens are typical Southern dishes. The migration of people from the South to Chicago has resulted in these crops being grown by market gardeners located in the Chicago area. Similarly, the large number of Chinese restaurants in the East has stimulated specialized vegetable enterprises in Long Island to serve this trade. In general, these unique markets tend to decrease in importance in the United States as the mobility of the population increases. However, national habits and food preferences change relatively slowly.

Selected References

Cox, G. W., and M. D. Atkins, 1979. *Agricultural Ecology: An Analysis of World Food Production Systems.* San Francisco: W. H. Freeman and Company. (An authoritative but readable treatment of the subject as it applies to agricultural systems.)

Geiger, R., 1965. *The Climate Near the Ground.* Cambridge, Mass.: Harvard University Press. (A revised translation—from the fourth German edition—of an outstanding work on microclimate.)

Gleason, H. A., and A. Cronquist, 1964. *The Natural Geography of Plants.* New York: Columbia University Press. (An interesting text on plant geography.)

Klages, K. H. W., 1949. *Ecological Crop Geography.* New York: Macmillan. (A discussion of the factors determining the distribution of crop plants.)

Trewartha, G. T., 1968. *An Introduction to Climate,* 4th ed. New York: McGraw-Hill. (An elementary but comprehensive text.)

U.S. Department of Agriculture, 1941. *Climate and Man.* (USDA Yearbook, 1941). (A broad, popular treatment that is still useful.)

Wilsie, C. P., 1962. *Crop Adaptation and Distribution.* San Francisco: W. H. Freeman and Company. (A general treatment of the principles of crop ecology.)

20 *Horticultural Production Systems*

A list of the horticultural production systems existing in the world today is also a list of all the production systems that have been developed in the course of the history of plant agriculture. Which of these systems is utilized in a particular area is determined by the crop, the population density, the cultural and technological levels of the society, and the forces of tradition. The variation is very great—from primitive planting of a mixed garden to intensive cultivation of a single crop. In New Guinea, primitive gardens are cultivated by isolated tribal people with a Stone Age culture, who, primitive in the use of tools yet sophisticated in the lore of plant material, practice a lengthy rotating system of production in the natural forest based on clearing, burning, production, and abandonment, uninfluenced by the money economy. Banana production in Central America constitutes a highly capitalized plantation system that is in the hands of two large North American corporations whose land holdings are extensive and who have their own internal transportation systems and shipping facilities; the technical personnel come from the United States, but the labor force is made up of local people, many of whom live in company towns. The greenhouse vegetable industry in Holland is based on two crops, tomatoes and cucumbers; it is highly mechanized and the growing operations are conducted by small family businesses whose marketing throughout northern Europe is handled by relatively large cooperatives.

The economic complexity of horticultural crop production is in large part related to the general agricultural situation. At present, the limit to agricultural progress has been set by the slow rate of economic development in many parts of the world; the striking increase in productivity that has been achieved through advances in agricultural science has not benefited all countries. In much of the world, such increases will depend not so much upon improved technical knowledge as upon elimination of both rural poverty and the insecurity of farm tenure, and upon increased availability of reasonable credit. In many areas, farm prices are so unstable that they discourage the investment required for intensive mechanization of agriculture. In others, marketing channels are inadequate for handling perishable crops.

Within the past decades, however, the production of many horticultural crops has continued to show an upward climb (Table 20-1). This reflects the rises not only in world population but also in world trade and the expansion of processing industries. Lengthening the marketing season by cultivar breeding and selection and by improving the technology of distribution has expanded the availability of horticultural crops throughout the year. In contrast to the world prices of many of the staple agricultural crops, the world prices of nearly all horticultural crops are free from artificial controls. However, in many countries there are stringent import controls and production quotas that provide protection for domestic horticultural production.

SUBSISTENCE HORTICULTURE: SHIFTING CULTIVATION

A primitive production system known as shifting cultivation or swidden is practiced by relatively isolated tribal people in many areas of the humid tropics, particularly the tropical highlands of Africa, Indonesia, and the Americas. **Shifting cultivation** is basically a system of subsistence agriculture characterized by cutting and burning the natural forest, long periods of fallow, lack of tools and draft animals, low population density, and low consumption. This type of cultivation is usually considered to represent an early stage in the development of cultivation systems—between food gathering and permanent cultivation—but it is often practiced on land with fragile soils ill-suited to permanent intensive agriculture; under such conditions it has become a highly developed (though still primitive) system.

The system requires a minimum of work: After the forest has been cleared and burned (hence the sobriquet **slash-and-burn**), little soil preparation is required, although in some locations fences may be necessary. The "garden" is a mixed planting of a diversity of species, including annuals and perennials, but generally only such short-term crops as yam, taro, banana, sweetpotato, and cassava. A garden in full growth appears to be a random assortment of crops of

TABLE 20-1. World production totals for vegetables, fruits, and nuts (in thousands of metric tons), 1972–1982.

Economy, region	1972	1974	1976	1978	1980	1982
Vegetables						
Developed	87,186	91,437	86,251	94,330	93,416	98,190
North America	23,469	25,049	25,305	27,406	26,327	29,189
Western Europe	45,318	48,610	42,805	47,741	48,275	49,444
Oceania	1,220	1,300	1,282	1,503	1,421	1,447
Other	17,179	16,479	16,859	17,679	17,393	18,111
Developing	94,420	98,898	109,446	120,069	127,544	133,981
Africa	9,315	9,870	11,119	12,245	13,408	14,199
Latin America	12,846	13,542	14,199	16,335	17,294	17,624
Near East	24,126	25,379	29,145	30,662	33,613	36,151
Asia and Far East	47,877	49,839	54,705	60,541	62,927	65,693
Other	256	269	277	286	302	314
Centrally planned	109,405	114,031	113,968	126,383	128,720	135,570
Asia	64,627	68,031	71,912	80,107	85,296	88,587
USSR and Eastern Europe	44,778	46,936	42,056	46,276	43,424	46,983
World totals	291,010	305,303	309,664	340,782	349,680	367,741
Fruits						
Developed	89,342	100,278	99,466	96,518	108,576	106,907
North America	20,343	23,733	25,372	25,425	29,805	25,585
Western Europe	55,205	63,065	60,963	58,016	64,659	67,155
Oceania	2,503	2,113	2,205	2,191	2,584	2,538
Other	11,292	11,367	10,929	10,887	11,529	11,629
Developing	113,109	122,831	126,728	133,723	143,790	150,235
Africa	21,358	22,787	24,848	25,658	25,977	26,533
Latin America	47,941	52,374	51,368	52,617	57,239	60,451
Near East	14,221	15,141	15,973	17,703	18,693	19,318
Asia and Far East	28,550	31,448	33,584	36,607	40,691	42,695
Other	1,039	1,081	1,104	1,139	1,192	1,238
Centrally planned	25,681	28,817	34,042	34,478	37,095	44,598
Asia	7,319	7,842	8,259	10,254	12,085	13,620
USSR and Eastern Europe	18,362	20,976	25,783	24,225	25,011	30,978
World totals	228,132	251,927	260,236	264,720	289,462	301,740

various heights and shapes. Typically, it is kept free of weeds. As crops are harvested, they are replanted until, after 2 or 3 years, the garden is abandoned and the area reverts to forest.

Attitudes toward shifting agriculture vary. Many agriculturists view the system negatively, describing it as wasteful, extravagant, and an impediment to development. Their reasons are the obvious waste of timber, the use of burning (a distrusted technique in many parts of the world), the extensive use of land, and a skeptical attitude about the effectiveness of a system that makes little use of technology and is employed by primitive people. On the other hand, anthro-

TABLE 20-1. Continued.

Economy, region	1972	1974	1976	1978	1980	1982
Nuts						
Developed	1,339	1,458	1,546	1,519	1,503	1,659
North America	318	389	440	411	543	607
Western Europe	960	1,005	1,035	1,038	905	995
Oceania	2	2	2	3	3	3
Other	60	62	69	67	52	54
Developing	1,438	1,528	1,362	1,536	1,467	1,508
Africa	478	551	360	360	332	333
Latin America	150	120	142	180	179	162
Near East	544	591	649	737	654	694
Asia and Far East	263	263	208	256	297	315
Other	3	3	4	4	4	4
Centrally planned	540	559	451	477	590	622
Asia	265	276	293	312	422	447
USSR and Eastern Europe	276	283	157	165	168	175
World totals	3,317	3,545	3,558	3,352	3,560	3,790

SOURCE: FAO Production Yearbooks. Vol. 30 (1976) for 1972 and 1974; Vol. 32 (1978) for 1976; Vol. 34 (1980) for 1978; Vol. 36 (1982) for 1980 and 1982.

pologists interested in agriculture have romanticized shifting agriculture in ecological terms. Indeed, at its optimum development, shifting cultivation is remarkably in tune with the environment: The lack of tools and draft animals eliminates soil compaction, the mixed planting and permanent cover reduce erosion, the gardens mimic the natural tropical forest in their multistoried complexity and diversity. The nutrients locked in the organic-matter cycle are temporarily released by burning, and the relatively short cropping period allows the land to revert to forest even on steep slopes. The soil structure is restored during the long fallow, which may last as long as 25 years.

The different points of view of agriculturists and anthropologists reflect their different disciplines. The anthropologist is likely to be impressed by a system in which a highly evolved social structure with complex rituals and taboos stabilizes the relation of people to their environment. For instance, in New Guinea some societies are kept in balance by being engaged in symbolic warfare that rarely takes human life. There, shifting cultivation, combined with pig raising, which serves both as a means of stockpiling protein and as a symbol for status and ritual, is a remarkable example of adaptation to the forest environment. The natural forest is made to work for people rather than against them, being transformed into a forest-garden. If energy input rather than yield per acre is measured, the system is remarkably efficient. (The division of labor in which men occasionally clear and burn while the women garden may be another feature of no little attraction to the men of the tribe!)

The carrying capacity of shifting cultivation may be only 100 people per square mile. A breakdown of the ecologically balanced subsistence economy is brought about if population is increased beyond this limit. But such increases often result as new people migrate into an area and impose a civilization that, by comparison, is characterized by less self-sufficiency, a greater number of necessities, and the replacement of society-stabilizing traditions with alternating periods of tranquillity and upheaval, often with attendant violence. To feed the increased population the fallow interval must be decreased and shifting cultivation must be expanded to less humid areas where the deciduous forest recovers slowly from cutting and fire is more difficult to control. An increase in intensive use of land, which is usually accompanied by a shift to monoculture, destroys the ecological equilibrium and results in large-scale deforestation and greater erosion. The equilibrium of the tropical environment is particularly delicate and fragile, as the damage done to the forests of Vietnam during the recent war has clearly shown. Deforested hills in Indonesia, now taken over by cogon grass (*Imperata cylindrica*) and likened to a green desert, attest to this. Under such conditions, a highly developed, sophisticated, primitive society becomes a poorly developed, impoverished parody of civilization. At this point the agriculturist becomes disenchanted with the remnant of shifting cultivation—and with ample justification. The anthropologists also lose interest, their enthusiasm being inversely proportional to the similarity of any culture to their own!

Because of its low carrying capacity, shifting cultivation is incompatible with the civilizations of the world's developed areas. It is probable that this system in its pure form will fade into extinction. Yet, shifting cultivation has contributed much to the developed world in terms of germplasm preservation, the discovery of crop plants (many yet unappreciated), and, most of all, the conservation of natural areas. Further, the system has suggested ways to utilize the humid tropics wisely through fallowing and the development of a natural forest horticulture. Agricultural schemes for these areas have often been carried out in ways best suited to temperate areas (clean cultivation is an example), often with disastrous results.

PLANTATION HORTICULTURE

Plantation horticulture, which is practiced almost exclusively in tropical areas, is characterized by large-scale, centrally managed, commercial production of export crops. At least partial processing accompanies production (for example, oil extraction, drying, or canning). The plantation is a combination agricultural-industrial enterprise tied to the industrial economics of temperate areas rather than to local conditions. Labor, usually hired, is intensive, although this practice is changing as the new agricultural revolution replaces hands and backs with machines and chemicals. The plantation system tends to be monocultural with continuous year-round production. Management is usually by foreign experts, in the past often North American or European. Typically, the plantation is a corporate structure oriented to the creation of profits. High capitalization is a distinguishing feature, and it often involves highly developed internal and external transportation systems (ranging from plantation railroads to refrigerated steamship lines).

The origins of the plantation system can be found in European incursions into tropical America in the sixteenth century. Large-scale production of crops in the tropics for export to temperate countries reached an early pinnacle in the sugar trade from the West Indies. The system ran on a combination of black slavery and extravagant debt financed by London merchants and speculators, and in its heyday in the seventeenth century a saying went: "As rich as a West Indian planter." Croesus had been replaced! The success of the plantation system made the sugar islands of the Caribbean extremely valuable pieces of real estate. Although the boom collapsed, in part because of its own weight but also because of the replacement of tropical cane sugar by temperate beet sugar, the system had proved its capacity for great profits. In the subsequent centuries, the plantation system rapidly spread to the rest of the tropical world with other tropical crops, most of which fit a horticultural definition by most standards. These include palm oil in Africa, rubber in Sumatra and Malaya, tea in Ceylon, coconut (for copra) in the Philippines, bananas in Central America, and pineapples in Hawaii.

Under the colonial system, in which "flag followed trade," the plantations were protected by the military and legal apparatus of the mother country, which profited in a number of ways: Profit from the enterprise filtered back to the investors in the form of dividends; the mother country profited by supplying manufactured goods, and—more important—controlled the trade processes both ways. One of the causes of the American Revolution was the British legislation that inhibited the growth of American industry and ensured that all goods would travel to the colonies in British ships.

Protection of plantation property led to the development of colonial empires; puppet regimes passed quickly to outright foreign rule, a political pattern that had deleterious effects on tropical regions. The stress placed on monoculture of export crops retarded the establishment of diversified agriculture as well as industrial development, and resulted in many agricultural countries' lacking self-sufficiency in food production. The Caribbean sugar islands at the height of sugar production in the seventeenth century depended on North American flour and Newfoundland codfish for food, and that condition still exists. In addition, the colonial superstructure purposefully refused to develop a native managerial class. These restraints were found especially in the Dutch East Indies (now Indonesia), where two parts of a dual economy—plantations based on sugar and rubber, and subsistence farming based on wet (paddy) rice—operated almost independently.

At its best, the plantation system is technically efficient and makes good use of resources. However, because of the high capitalization required and the general poverty of the tropical world, plantations typically have been financed and managed by aliens, sometimes with extreme disregard of the welfare of the local people. Even when management is benign, a good case can be made to show that the development of the tropical country is retarded or hampered by external control.

In spite of the added wealth that it provided to the tropics, the plantation system ran counter to national purpose and pride. In the rush toward self-government after World War II, the system was allowed to deteriorate or was junked outright. It is unfortunate that its social deficiencies brought about the disintegration of what was, in large part, an efficient production system. In many areas, plantation agriculture represents the best use of the land. Where the idea that plantations are no more than manifestations of colonialism can be dispelled, the system will likely be maintained. It can be continued either through training local people for executive positions, or through changes in land tenure and decentralization of processing and shipping operations.

In response to the rise of nationalism and the popular appeal of "agrarian reform," in some areas the former land holdings revert to local ownership or lease while the industrial sector of the operation remains centrally managed. Technical assistance then must be provided by agricultural "extension" agents for the system to remain efficient.

The plantation system is by no means a dead institution. The economies of many developing countries are dependent upon tropical agriculture for foreign exchange revenues, and in many, a single tropical tree crop constitutes more than half the total commercial exports. Indications are that the plantation system is in fact moving into temperate areas, where it is often known as **corporate farming.** Spurred by tax advantages, many corporations in the United States have acquired large holdings of nut and citrus orchards. Mechanization and the unionization of farm labor are rapidly removing the differences between urban and rural industry.

The pineapple industry in Hawaii is an example of a developed modern plantation system in transition (Figure 20-1). All operations from growing to processing are highly mechanized, and many machines have been developed expressly for the industry. Labor is unionized and paid on industry scales. Although increasing labor costs in Hawaii prompted the establishment of plantations in the Philippines, technological progress continues to reduce the labor input per ton of processed fruit. The fruit is still hand-harvested, but self-propelled conveyors now carry it to the packing shed, and the use of growth regulators to cause all the fruit to ripen at the same time will soon make possible complete mechanical harvesting that should maintain the competitive position of Hawaiian pineapple.

Although the structure of banana production varies with location, large corporate plantation systems predominate, especially in Central America.

FIGURE 20-1. Pineapple production in Hawaii. (*A*) A plantation on Oahu, with Diamond Head crater and Honolulu in the distance. (*B*) Harvesting with a conveyor system. (*C*) Unloading and washing fruit before processing. [Courtesy Dole Corp.]

Medium-sized holdings (estates) are typical in Ecuador, Colombia, and the Dominican Republic, with small holdings predominant in Jamaica, Mexico, the Caribbean, the Canary Islands, West Africa, and the Pacific Islands.

Typically, the banana-plantation systems of Central America are vertically integrated. Private firms own the fruit from planting to its sale in the distribution channels of the consuming country. Production is typically organized in divisions ranging in size from 10,000 to 30,000 acres; these are divided into unit farms of 800 to 1000 acres, each with its own operating facilities and labor force. The operations are heavily capitalized, and some include internal railroads. The farms are irrigated, and pests are controlled by airplane application of pesticides. Shipping and marketing facilities are company-operated, as are plants for the manufacture of corrugated boxes. Harvested stems are transported by overhead monorail through the washing area, and selected fruit is cut from the stem in clusters (hands). The hands are graded, packed in trays or corrugated boxes, transported to the docks on the company-owned railroad, and loaded on refrigerated ships either owned or chartered by the company.

COMMERCIAL HORTICULTURE

The commercial production of fruit and vegetables in both temperate and tropical regions is largely conducted by relatively small units, usually family-operated enterprises. The intensive nature of horticultural production results in less extensive land holdings but higher capitalization than in other segments of plant agriculture. The trend, however, is toward larger units and even greater capitalization as mechanization increases.

Because of the differences in planting, growing, and harvesting different kinds of crops, commercial horticulture is best discussed in traditional commodity groupings.

Fruit and Nut Growing

By weight, world production of fruit approaches that of some of the world's staple agricultural crops (see Table 21-1, page 621). Fruit production for human consumption began with the harvesting of wild stands, a procedure still followed with a few crops in some localities. The casual backyard cultivation of trees and vines, which is still widely practiced throughout the world, is the forerunner of the modern fruit industry. Today the commercial production of fruit is a specialized portion of the horticultural industry known as **orcharding** (Figure 20-2). Orcharding is typically based on long-lived perennials, many of which do not bear fruit until several years after they are planted, and in consequence orcharding has a number of unique characteristics. Large-scale com-

A

B

FIGURE 20-2. Apple orchards in the United States. (*A*) An apple orchard in Oregon's Hood River Valley. (*B*) Apples and peaches in Vincennes, Indiana. [Photograph (*A*) courtesy Oregon State Highway Travel Division; photograph (*B*) courtesy U.S. Soil Conservation Service.]

mercial fruit production is concentrated in those areas best adapted to the growth of a given crop. Thus the major production areas for apples and pears in North America are to be found in only a few favorable locations, notably the Pacific Northwest and the Great Lakes region. Similarly, citrus production is heavily concentrated in California and especially in Florida (where Polk County produces as much as all of California). Such concentration in the production of orchard crops is found throughout the world.

Typically, it is not easy to get into or out of the fruit business. Heavy capital expenditures are required, as are financial reserves for the long periods during which the orchardist must wait for newly planted trees to come into bearing. In addition to the costs of ordinary production equipment, great expenditures are necessary for grading, packaging, and storage facilities. Thus, many commercial orchards have been family-owned for generations. In contrast to temperate vegetable production, orcharding is a full-year operation; the nongrowing season is a time of considerable activity, particularly pruning and training.

Although the details of fruit growing differ with species and climates, they share a number of common problems and practices. There must be careful selection of a succession of cultivars suitable both for adaptability and marketing. With some fruits (apples and pears, for example) the choice of cultivars must also take cross-pollination into consideration. Clonal propagation is the rule for almost all fruit crops, and most producing units for tree crops are compound plants consisting of at least two genetically different parts, rootstock and scion. Choice of geographic location and site are important to avoid critical temperatures at flowering and low-temperature injury in winter. Soil type, suitability of drainage and irrigation, and natural soil fertility must also be considered.

The particular planting system is usually a compromise between plant requirements for light (in relation to quality and yield) and management needs (movement throughout the orchard for spraying and harvesting). Most fruit crops require extensive pruning and training to keep plants within bounds, to maintain structural strength, and to influence growth and fruitfulness. Extensive labor for pruning ensures that the fruit growing operation will be a year-round endeavor. Fertilization must maintain a balance between tree growth, productiveness, and fruit quality. Thinning of flowers or fruits is often necessary to control fruit size and the overbearing that may result in alternate bearing, poor quality, and even winter injury.

Because orchards take a long time to become established, pests also have ample time in which to become established. Pest control, therefore, is a critical part of the operation. For apple production in the humid areas, it is not uncommon to have from 12 to 15, even as many as 20, sprays for insect and disease control; even in winter, the grower must contend with pests such as mice and deer.

Finally, harvesting procedures assume key importance in crops that have a short storage life. Mechanical harvesting devices and bulk handling are causing great changes in an industry traditionally dependent on high labor inputs during harvest (Figure 20-3). The life of perishable fruits can be extended through controlled temperature, humidity, and atmosphere, and storage is now an integral part of many orchard operations. Grading and packaging operations may be handled either at the orchard or at some central location.

With the trend toward concentration of the industry in fewer locations, greater reliance on long-term storage, and more exacting marketing requirements, there is a trend also toward larger marketing associations. Thus, most of the avocados in California are sold as Calavo brand, the name of a cooperative marketing association; much of California's citrus is sold with the Sunkist label; and Washington apples move through a number of cooperative packing and marketing associations. In the Northeast many fruit growers produce almost exclusively under processing contracts; Concord grapes are handled almost entirely through grower-processor cooperatives (such as Welch's). In Michigan, the large increase in capitalization available for controlled-atmosphere storage has encouraged the formation of small associations of growers around storage and marketing facilities.

A

B

FIGURE 20-3. Mechanical harvesting: (*A*) wine grapes; (*B*) dates. [Photograph (*A*) courtesy Ontario Ministry of Agriculture and Food; photograph (*B*) courtesy USDA.]

At the same time, small growers outside the major fruit-production areas who have withstood the exodus to more favored locations still find local markets a profitable means of selling their crops—sales on the farm, at roadside markets, and increasingly through "pick-your-own" operations.

Vegetable Growing

The vegetable industry is characterized by its flexibility. Because most vegetables are grown as annuals, shifts in cultivar and crop can be readily made. In the past, a large part of the vegetable enterprise was diversified, and no great long-term investments were required. It was always relatively easy to go into or out of the vegetable business. These moves are less simple as irrigation, specialized equipment, and storage and packing facilities become integral parts of commercial vegetable production. The vegetable industry in the United States may be subdivided into three main categories: home gardening, market gardening, and truck gardening. In addition to these, there are several small, specialized parts of the vegetable industry, including plant growing, greenhouse forcing, seed production, and mushroom culture.

Home gardening, which involves the production of vegetables for home consumption, is still the most important source of vegetables in many countries. It is still a considerable source of vegetable production in the rural United States, although gardening at home appears to be almost more valuable as an outlet for recreation than as a source of food. In times of national peril, home gardens may become a major part of a country's food supply. **Market gardening** developed from local gardens and now involves intensive production of many kinds of vegetables near large centers of population (Figure 20-4). This source is disappearing in the United States as a result of increasing land costs and improvement in food distribution. The large-scale production of vegetables, commonly less diversified than market-garden production, is known as **truck farming** (from the French *troquer*, to barter). Truck farming, which is based on suitable season, climate, and soil rather than market proximity, has become the most important kind of modern vegetable industry. The produce of truck farming is used either for fresh market consumption or for processing. The rise of the processing industry has made many individual vegetable crops part of the general rotation of farm crops. For example, tomatoes and sweet corn have become valuable cash crops in many midwestern general or grain farms.

FIGURE 20-4. An irrigated lettuce farm in Hawaii. [Photograph by J. C. Allen & Son.]

Commercial vegetable production in temperate climates is based on **single cropping,** in which one species is planted *en masse* with only yearly rotations. In tropical areas, however, the traditional pattern of vegetable gardening is **multiple cropping,** which is characterized by simultaneous or relay planting of two or more crops in the same field. The association of beans and maize in South American agriculture is an ancient example of multiple cropping that is still practiced (Figure 20-5). Multiple cropping predominates on small farms, which often produce cereal grains, legumes, starchy roots, and plantains. The system relies on diversity of plant species, closed cycling of nutrients, reduced pest incidence, erosion control, low but stable yield, and high inputs of human energy for cultivation and weed control. With proper management, however, total yields per unit area can be very high.

Multiple cropping provides small farmers with extra sources of income and greater diversity in their diets. However, the intensity of production requires much hand labor, which is not compatible with many modern farming practices. Recently, multiple cropping has received more attention from horticultural and agricultural researchers, and studies suggest that multiple-cropping systems can be compatible with increased technological inputs to exploit the greater productivity inherent in such systems.

Of the several specialty operations in vegetable production, greenhouse production of vegetable crops is perhaps the most important. It is a minor but

FIGURE 20-5. Association of climbing beans and maize in field plantings. Beans and maize also are dietary complements because maize protein is deficient in lysine while bean protein is deficient in the sulfur-containing amino acids cystine and methionine. In Latin America, maize (in the form of tortillas) and beans are staple foods that are often eaten together, thus providing better nutrition than either component provides alone. [Photograph courtesy C. A. Francis.]

progressive part of the vegetable industry. In the United States, greenhouse vegetable production based on tomatoes and lettuce is centered in the Midwest. In the Low Countries of Europe, a relatively large greenhouse industry exists with tomatoes and cucumbers as the major crops.

Another specialty industry is **mushroom culture.** The industry was originally developed in areas with natural caves. Mushrooms are saprophytic and therefore do not need light, but they do require strict temperature and humidity control. At present, the production of mushrooms is a highly capitalized operation utilizing specially constructed houses, although natural caves are still used in some areas. Mushrooms have traditionally been grown on composted horse manure, although synthetic compost formulae have been developed from corn cobs, hay, and inorganic fertilizer. In the United States, cultivation is almost completely confined to *Agaricus brunnescans* (syn. *Agaricus campestris*), a Basidiomycete, whereas a number of other species belonging to the classes Basidiomycetes and Ascomycetes are commercially produced in various parts of the world.

Ornamentals

The production of ornamental plants may be divided on the basis of crop into **floriculture** (cut flowers, potted plants, and greenery), the growing of **landscape plants** (bulbs, rose plants, ground cover, trees, and shrubs), and the growing of **turf** (seed and sod). Although flowering bulbs and flower seed are valuable components of agricultural production in the Low Countries of Europe, ornamentals are relatively insignificant in world trade. World statistics on production are unavailable, but in the United States complete statistics have been obtained in connection with the 1890, 1930, 1950, 1960, and 1970 agricultural censuses.

Floriculture

Floriculture has long been a substantial part of the horticulture industry, especially in Europe and Japan, and it accounts for about half the nonfood horticultural industry in the United States. The wholesale value of florist crops in 1981 was over $1 billion, the retail value was estimated to be $3.6 billion. Because cut flowers and potted plants are mainly grown in plant-growing structures in temperate climates, floriculture is largely a greenhouse industry (Figure 20-6), but there is also outdoor culture of many flowers. Floriculture is a competitive and highly technical business requiring knowledge, skill, and large amounts of capital. Greenhouse floriculture is, in many respects, the most sophisticated kind of plant agriculture. The industry is usually very specialized by crop, and the grower must provide precise environmental control and be vigilant in the constant struggle against pests and other calamities. Exact scheduling must be

FIGURE 20-6. Greenhouse carnation production. Most of the carnations produced in the United States are grown under glass. [Photograph by J. C. Allen & Son.]

arranged because most floral crops are subject to seasonal demand (poinsettias at Christmas, lilies at Easter, roses and carnations on Mother's Day, and so on). Because the product is perishable, market channels must function smoothly to avoid losses.

The floriculture industry consists mainly of the grower, who typically mass-produces flowers for the wholesale market, and the retail florist, who markets to the public and contributes such services as arrangements and delivery. The growing operation is typically family-owned and highly intensive. The average outdoor producing unit in the middle 1960s was less than 5 acres, and greenhouse production per farm averaged 13,000 square feet. Nevertheless, as in all modern agriculture, the size of the growing unit is increasing. There is a movement away from eastern urban areas of high taxes and labor costs to locations having lower tax rates and a rural labor pool as well as more favorable climate (milder temperatures and more sunlight). As a result the United States floriculture industry is moving west and south. The development of air freight has emphasized interregional and international competition. Flowers can now be shipped great distances by air and arrive in fresh condition to compete with locally grown products. International air-cargo shipment of flowers is rising.

Flower growing is based on precise environmental control. Such control involves structures incorporating glass, plastic, lath or shade cloth, and artificial heating and cooling devices. The rising costs of energy have placed heavy burdens on greenhouse floriculture. For example, the prices of oil, gas, coal, and electricity have risen enormously since 1973. Higher fuel costs have been somewhat offset by a rise in the prices that the growers receive, but there has been concern that greenhouse owners may price themselves out of the market.

More dangerous has been the threat individual growers have faced from reduced fuel allotments and even curtailments during recurring energy crises. As a result, greater attention has been paid to energy conservation in the greenhouse industry. Alternatives include the use of greenhouses that are more efficient in capturing solar heat, the use of geothermal heat and waste heat from power plants, and the implementation of energy conservation through such greenhouse modifications as double-wall coverings, thermal blankets (plastic coverings over benches to reduce the volume heated at night), and more attention to heating and cooling practices.

The production of plants under protected cultivation is constantly becoming technically more complex. Greenhouse soils are usually compounded from a variety of ingredients, sterilized to control weeds, disease organisms, and insects and nematodes. Fertility and pH are precisely controlled through the addition of the appropriate chemicals and amendments. Some growers are enriching the atmosphere in their greenhouses with carbon dioxide. Water regimes are automatically regulated, even with pot culture. In greenhouses, temperatures are regulated by steam heating and evaporative cooling (mist and fan-and-pad). Day length is regulated by a combination of shade cloth and illumination to extend the day or interrupt the dark period. The use of artificial light to increase photosynthesis is usually not economically feasible except in special situations with seedlings. Pest control relies on a combination of practices involving sanitation, the use of disease-free plants, and an arsenal of organic pesticides. The use of chemicals to control growth is expanding rapidly. The growth retardants in particular are finding much promise in the industry.

Flower growing is highly competitive. With erosion of profit margins, only the large, well-managed organizations can survive economically, and they keep up with technical complexity and high capital requirements by developing highly specialized businesses (Figure 20-7). The successful grower must be a highly trained, skilled entrepreneur.

The retail florist usually, but not always, operates a business separate from that of the grower. Wholesale cut flowers, plants, greenery and associated products, ribbon, and potware can be transformed into floral arrangements of various types—corsages, table decorations, wreaths, set pieces, and even floral floats. The telephone (and the telegraph before it) enables the retail florist industry to organize to provide nationwide delivery of flowers. New communication systems have linked the florist with a number of broad associations that provide technical information as well as trade channels. Thus the florist is most typically a designer and businessperson rather than a grower.

Landscape Horticulture

The industry of landscape horticulture is divided into growing, maintenance, and design. The growing of plants for the landscape is referred to as the nursery business, although, more broadly, a nursery is the place where any young plant

FIGURE 20-7. A container-filling machine designed for use in the floriculture and landscape industries.

is grown or maintained before permanent planting. There are about 25,000 nurseries of all types in the United States, and together they occupy almost 200,000 acres.

The nursery industry involves the production and distribution of woody and herbaceous plants and includes ornamental "bulb" crops—corms, tubers, rhizomes, and swollen roots as well as true bulbs. The production of cuttings to be grown in greenhouses or for indoor use (foliage plants), as well as the production of bedding plants, is usually considered part of floriculture, but this distinction is fading. While most nursery crops are ornamental, the nursery business also includes fruit plants and certain perennial vegetables used in home gardens—asparagus and rhubarb, for example. Christmas-tree production is also associated with the nursery industry; however, though the operation is similar, the growth of seedling trees for forest plantings is usually considered along with forestry.

The nursery industry in the United States may be divided into wholesale, retail, and mail-order enterprises, although there may be some overlapping. Typical wholesale nurseries specialize in relatively few crops and deal largely in plant propagation, supplying only retail nurseries or florists. They sell young "lining out" stock of woody material to the retail nursery, which then performs two basic functions: care of the plant until growth is complete, and resale to the public. Often, though not always, the retail nursery provides planting and maintenance service (Figure 20-8). Many nurseries also execute the design of plantings in addition to furnishing the plants. This service is analogous to the provision of architectural services by some lumber yards. This activity is a complex one, since some states license landscape architects as members of a distinct profession.

Mail-order nursery companies usually grow only a small proportion of their catalogue listings, and subcontract the rest from wholesale houses. Although some of the largest and most reputable businesses in the country deal only in

A

B

FIGURE 20-8. Installation and maintenance are important components of the nursery industry. (*A*) Landscaping is an integral part of the home construction business. (*B*) Moving large trees in winter. [Photograph (*A*) courtesy Ford Motor Co.; photograph (*B*) courtesy G. Kessler.]

mail order, very often fly-by-night operators are attracted to businesses of this sort.

Ornamental trees and shrubs represent the most important of the nursery crops (ornamental shrubs are the backbone of the nursery trade); next are fruit plants and then "bulb" crops. The most popular single plant grown for outdoor cultivation is the rose. The type of plants grown by a nursery depends on location; in general, the northern nurseries provide deciduous trees and shrubs and coniferous evergreens, whereas the southern nurseries offer tender broad-leaved evergreens.

The nursery industry is especially prominent in the Great Lakes and Middle Atlantic states, and in Florida and California. In some of these areas, nursery sales (including sales of roses, shade trees, and small fruits for the home) account for nearly 40 percent of the retail trade in horticultural specialties. The nursery business is about equally divided among the production of (1) coniferous evergreens such as yew, juniper, spruce, and pine; (2) broad-leaved evergreens such as rhododendron, camellia, holly, and boxwood; (3) deciduous plants including *Forsythia, Viburnum,* barberries, privets, lilacs, and flowering vines; and (4) roses.

Fields of specialization have evolved within the ornamental-shrub industry. Some firms confine their activities mostly to the production of "lining out" stock—seedlings and rooted cuttings for sale to growers for field planting that must be tended for several years before reaching salable size. Such firms will generally sell not only the bare-root liners themselves, but also small plants started in plant bands or slightly larger ones in containers. They may also offer small grafted plants of expensive or rare cultivars. Lining-out stock of a typical evergreen, such as the Pfitzer juniper (*Juniperus chinensis* 'Pfitzeriana'), is relatively inexpensive, usually much less than a dollar. It is more economical for growers to purchase these ready to plant in the field than to undertake the specialized propagation themselves.

The field grower may, in turn, specialize in mass growing for the wholesale trade only. The field plantings are tended until they attain marketable size, having been shaped and sheared as necessary, provided with pest control, irrigated, culled, and so on. Because of the time required to produce a marketable crop, and because of appreciating labor costs, this phase of the nursery industry can easily be caught in a profit squeeze, especially if landscaping preferences change and no quick market for a field of mature shrubs exists. In contrast, wholesale growing escapes the high overhead of retail marketing in urban areas and, although many growers do raise retail stock at their nurseries, they generally try to avoid the expensive merchandising required of the typical urban-area garden center. Growers are particularly interested in new technology that will lessen labor requirements, and they are turning to herbicidal control of weeds (as with dichlobenil) and short-cut methods for transplanting (as substitutes for hand digging and burlapping).

Most shrubs are usually balled and burlapped ("B and B") for market. The plants are dug with as much soil as possible clinging to the roots, and the root ball is then wrapped with burlap or other suitable material. Recently a trend has developed toward container-grown stock—nursery stock grown in the container (Figure 20-9) in which it is to be sold. This method allows year-round sales of plant material. Deciduous shrubs may be marketed with bare roots. There are obvious economies in not having to dig and wrap the plant carefully, thereby handling added weight and soil. Fairly satisfactory techniques have been developed for holding bare-root ornamental shrubs in viable condition until marketed. Usually this involves digging after leaf fall in autumn and stor-

FIGURE 20-9. A field of container-grown evergreens. [Photograph courtesy D. Hill Nursery Co., Dundee, Illinois.]

ing the plants through winter in cold cellars with regulated humidity for spring sale. Unfortunately, the plants are often handled carelessly. The chief cause of loss during storage is desiccation, but respiration also plays a part when temperatures rise. But, even if the plants are properly cared for during cold storage, they often suffer from warming and drying at the sales outlet. Bare-root plants ordered by mail frequently arrive in poor condition because of warming and drying in transit. The great effort spent to control humidity and temperature during winter storage apparently can be defeated if there are no cold-storage facilities at retail outlets.

In recent years, a breed of "nurseryman" who is more merchandiser than planter has taken over much of the retail trade. Exemplifying this change are the numerous diversified garden centers that sell all sorts of horticultural merchandise and maintain display yards of shrubs, trees, and roses (Figure 20-10). Usually these are purchased wholesale from large growers. Many such firms have landscaping services and offer counsel, product, and service to the homeowner.

Turf

The turf-grass industry includes the production and maintenance of specialized grasses and other ground covers for utility, recreation, and beautification. Grass is ubiquitous in many urban and suburban facilities. In the United States, maintenance expenditures for turf run to more than $5 billion, of which 70 percent is spent on residential lawns, 11 percent on highway plantings, 8 percent on cemeteries, and 6 percent on golf courses. About 10 million acres of land is

FIGURE 20-10. Entrance to the garden center of D. Hill Nursery Company, Dundee, Illinois.

tended by about 50 million homeowners in the United States, and much of it is lawn.

The two basic commodities of the turf industry are seed and sod. Associated products include fertilizer, pesticides, and equipment such as lawn mowers. The turf-grass industry reaches a tremendous market throughout the affluent countries of the world. Commercial home lawn care (fertilization and weed control) is an expanding new industry in the United States. The maritime climate of the British Isles is particularly suited to beautiful turf. Although lawns are usually not the major landscape feature of Britain's homes as in the United States, lawns are more prominent in public areas in Great Britain.

The major turf species in Europe is *Poa annua*. The major kinds grown in the northern two-thirds of the United States are three cool-season grasses, Kentucky bluegrass (*P. pratensis*), red fescue (*Festuca rubra*), and bentgrass (*Agrostis* species). A number of grass species are adapted to the South: Bermuda grass (*Cynodon dactylon*), bahia grass (*Paspalum notatum*), carpet grass (*Axonopus compressus*), centipede grass (*Eremochloa ophiuroides*), *Zoysia* species, and St. Augustine grass (*Stenotaphrum secundatum*).

Seed Production

Seed production is a relatively small but essential part of the horticultural industry. The seed business involves not only a great many crops but literally thousands of cultivars and hybrids, and not only plant growing but manufacturing and processing (milling, cleaning, packaging, and storing of seeds).

The principal world areas of horticultural seed production are Holland, Denmark, England, Japan, and the United States. In the United States, the chief seed-producing areas are in the western states. California produces practically all the flower seeds, but a good share of the vegetable seed, especially of large-seeded crops such as beans, peas, and corn, is produced in Idaho. The actual growing of plants is often contracted out to growers, although large seed houses may produce some of their own specialties directly. Also, a number of specialty growers confine their production to a few crops, such as pansies or sweet corn.

The seed business itself is divided into wholesale, retail, and mail-order houses. To the majority of the American public, the commission packages are the most familiar wares of the seed industry, but these represent only a small percentage of seed sales.

An obvious relationship exists between seed growing and seed improvement. The progressive seed houses have initiated the breeding of many crops, especially flowers. The Department of Agriculture and State Agricultural Experiment Stations have also supported much of the breeding programs, and, as a result, many experiment stations have sponsored seed associations. At present, horticultural crops account for only a minor portion of this activity.

Selected References

Arthur, H. B., J. P. Houck, and G. L. Beckford, 1968. *Tropical Agribusiness Structures and Adjustments—Bananas.* Boston: Division of Research, Harvard Business School. (An economic analysis of the world banana industry.)

Childers, N. F., 1983. *Modern Fruit Science: Orchard and Small Fruit Culture,* 9th ed. New Brunswick, N.J.: Horticultural Publications. (A text on temperate pomology emphasizing the production practices for apple.)

Copeland, L. O., 1976. *Principles of Seed Science and Technology.* Minneapolis: Burgess. (A basic text on seed technology.)

Dalrymple, D. G., 1973. *Controlled Environment Agriculture: A Global Review of Greenhouse Food Production.* (Foreign Agricultural Economic Report 89). Washington, D.C.: Economic Research Service, U.S. Department of Agriculture. (A world survey of food production under glass.)

Davidson, H., and R. Mecklenberg, 1981. *Nursery Management, Administration and Culture.* Englewood Cliffs, N.J.: Prentice-Hall. (A text covering cultural practices and business management.)

Geertz, C., 1963. *Agricultural Involution: The Process of Ecological Change in Indonesia.* Berkeley and Los Angeles: University of California Press. (See especially Chapter 2 for a comparison of shifting cultivation with sedentary agriculture.)

Gourou, P., 1966. *The Tropical World: Its Social and Economic Conditions and Its Future Status,* 4th ed. New York: Wiley. (An introduction to the tropics.)

Hanan, J. J., W. D. Holley, and K. L. Goldsberry, 1978. *Greenhouse Management.* New York: Springer-Verlag. (The mechanical and engineering aspects of greenhouse construction and management.)

Hanson, A. A., and F. V. Juska (eds.), 1969. *Turfgrass Science.* Madison, Wisc.: American Society of Agronomy. (A monograph on turf with each chapter contributed by a specialist.)

Harris, R. W., 1983. *Arboriculture: Care of Trees, Shrubs, and Vines in the Landscape.* Englewood Cliffs, N.J.: Prentice-Hall. (A comprehensive text on the care of woody plants in the landscape.)

Hartmann, H. T., W. J. Flocker, and A. M. Kofranek, 1981. *Plant Science: Growth, Development, and Utilization of Cultivated Plants.* Englewood Cliffs, N.J.: Prentice-Hall. (A survey of economic plant production.)

Hebblethwaite, P. D. (ed.), 1978. *Seed Production.* London: Buttersworth. (A collection of 43 chapters on all aspects of seed production.)

Janick, J., R. W. Schery, F. W. Woods, and V. W. Ruttan, 1981. *Plant Science: An Introduction to World Crops,* 3d ed. San Francisco: W. H. Freeman and Company. (An introduction to crop science. See Chapter 15, "Cropping Systems and Practices.")

Jaynes, R. A. (ed.), 1979. *Nut Tree Culture in North America.* Hamden, Conn.: Northern Nut Growers Association. (The culture of edible nuts.)

Joiner, J. N. (ed.), 1981. *Foliage Plant Production.* Englewood Cliffs, N.J.: Prentice-Hall. (Comprehensive coverage of the foliage-plant industry.)

Laurie, A., D. C. Kiplinger, and K. S. Nelson, 1979. *Commercial Flower Forcing*, 8th ed. New York: McGraw-Hill. (A textbook on greenhouse florist crops with emphasis on practices.)

Lorenz, O. A., and D. N. Maynard, 1980. *Knott's Handbook for Vegetable Growers*, 2d ed. New York: Wiley. (A concise ready reference for the vegetable industry.)

Manshard, W., 1974. *Tropical Agriculture: A Geographical Introduction and Appraisal.* London: Longman. (An introduction to agricultural systems of the tropics.)

Mastalerz, J. W., 1977. *The Greenhouse Environment: The Effect of Environmental Factors on Flower Crops.* New York: Wiley. (An advanced review of factors affecting greenhouse production.)

Nelson, P. V., 1981. *Greenhouse Operation and Management*, 2d ed. Reston, Va.: Reston. (A basic text on floriculture production in greenhouses.)

Ochse, J. J., M. J. Soule, Jr., M. J. Dijkman, and C. Wehlburn, 1961. *Tropical and Sub-tropical Agriculture.* New York: Macmillan. (A huge two-volume work dealing with tropical plantation agriculture. The first volume deals with general cultural practices, the second with crops.)

Splittstoesser, W. E., 1984. *Vegetable Growing Handbook*, 2d ed. Westport, Conn.: AVI. (A text on vegetable-crop gardening emphasizing organic and traditional methods.)

Turgeon, A. J., 1980. *Turfgrass Management.* Reston, Va.: Reston. (An introductory text.)

Ware, G. W., and J. D. McCollum, 1980. *Producing Vegetable Crops*, 3d ed. Danville, Ill.: Interstate. (A practical book on vegetable-crop production.)

Westwood, M. N., 1978. *Temperate-Zone Pomology.* San Francisco: W. H. Freeman and Company. (An excellent, up-to-date text on temperate pomology.)

Woodruff, J. G., 1979. *Tree Nuts: Production, Processing, Products*, 2d ed. Westport, Conn.: AVI. (Nut culture with emphasis on processing.)

21 *Horticultural Crops*

Horticulture ultimately reduces to the study of individual crops, the basis of the industry. Notwithstanding the diversity of horticulture, most temperate horticultural crops can be grouped into the traditional categories: fruits, vegetables, and ornamentals. However, for a world view of horticulture, additional categories are required to include beverage crops, spice and drug crops, and others.

Accurate world production figures for all horticultural crops are not available. This lack is due in part to inconsistent international communication about such data and in part to poor statistical records. Production statistics may be incomplete or may be computed differently in different countries. For example, some countries report only commercial production; others report total production. The lack of reliable data is especially noticeable for crops that do not enter significantly into world trade. As a result, the highly perishable horticultural crops are underestimated even though they may be grown extensively in home or local gardens. Production figures of the major crops, shown in Table 21-1, indicate the worldwide importance of horticulture. World production statistics for selected horticultural crops are presented in Table 21-2.

TABLE 21-1. Comparison of 1982 world production of some major crops.

Crop	Production (millions of metric tons)	
Cereals		1695.1
Wheat	481.0	
Rice (paddy)	411.9	
Maize	455.4	
Barley	160.3	
Root crops		556.4
Potato	254.9	
Pulses (legumes)		44.7
Vegetables and melons		367.7
Fruits		301.7
Grape	70.6	
Citrus	53.8	
Banana	41.2	
Apple	39.4	
Nuts		3.8
Vegetable oils		57.0
Sugar (centrifugal, raw)		101.4
Cocoa beans		1.6
Coffee (green)		4.9
Tea		1.9
Tobacco		6.1
Rubber (natural)		3.9

SOURCE: *FAO Production Yearbook 1982*, Vol. 36, 1983.

FRUIT CROPS

The fruits of certain perennial plants have long been prized as sources of refreshment, for their delightful flavors and aromas, and as nourishing foods. Throughout the ages, these fruits have been regarded as both luxuries and necessities, and the combination of utility and beauty has earned them a special place in the relationship of people to plants:

> Sustain me with raisins, refresh me with apples;
> For I am sick with love.
>
> SONG OF SOLOMON 2:5

Although there are exceptions—banana and coconut, for example—fruits are not a prime source of calories, minerals, or protein; nevertheless, considered together, fruit crops compare favorably with the world's staple agricultural

TABLE 21-2. World production figures (1000 metric tons) for selected horticultural crops, 1982. (SOURCE: *FAO Production Yearbook 1982*, Vol. 36, 1983.)

Crop	World	Africa	North America	South America	Asia	Europe	Oceania	USSR
Fruits								
Grape								
Table	70,605	2,313	6,383	5,564	7,557	40,551	937	7,300
Wine	37,155	1,161	2,047	3,728	249	26,124	445	3,400
Raisin	973	31	244	8	438	153	100	
Citrus								
Orange	35,952	3,120	9,565	11,966	6,690	3,965	375	270
Tangerine and clementine	7,218	549	708	943	3,724	1,255	40	
Lemon and lime	5,234	248	1,587	751	1,306	1,300	41	
Grapefruit and pumello	4,484	299	2,931	304	895	22	34	
Other citrus	960	374	114	11	449	2	11	
Tropical and subtropical fruits								
Banana	41,224	4,445	7,421	12,232	15,513	489	1,125	
Plantain	22,582	13,652	1,637	4,664	2,625		4	
Papaya	1,869	220	347	538	747		17	
Pineapple	8,864	1,258	1,355	1,142	4,974	2	133	
Mango	13,508	844	1,371	857	10,426		9	
Date	2,630	978	24		1,615	12		
Olive	10,577	855	164	128	1,932	7,496	2	
Avocado	1,520	126	971	340	78	2	3	
Pome fruits								
Apple	39,391	502	4,375	1,446	8,151	16,995	522	7,400
Pear	8,915	239	803	234	2,914	3,897	128	700
Stone fruits								
Plum and prune	6,227	66	612	99	912	3,517	21	1,000
Apricot	1,658	177	114	34	440	676	38	180
Peach and nectarine	7,107	244	1,425	551	1,187	3,185	94	420
Berries								
Strawberry	1,863		497	15	314	928	8	100
Raspberry	247		26			129	2	90
Currant	511					439	2	70
Crucifers								
Cabbage	36,225	731	1,831	601	15,587	7,767	110	9,600
Cauliflower	4,752	168	325	71	1,851	2,209	117	10

Nuts								
Almond	1,071	52	261		217	532	1	7
Hazelnut	424		17		257	146		4
Chestnut	541			11	377	147		6
Walnut	792	6	208	7	267	250		53
Pistachio	93		20		70	3		
Coconut (whole nut)	36,530	1,532	1,456	604	30,632		2,307	
Cashew	461	181	3	76	200			
Beverage crops								
Coffee	4,934	1,206	923	2,096	654		55	
Cocoa	1,587	918	95	473	66		35	
Tea	1,925	221		49	1,501		8	145
Roots, tubers, and bulbs								
Potato	254,861	5,472	19,949	10,067	38,882	101,319	1,173	78,000
Sweetpotato	140,186	5,164	1,429	1,483	131,353	134	624	
Cassava	128,944	48,758	942	3,220	48,793		231	
Onion, dry	22,397	1,725	2,090	1,603	10,293	4,394	191	2,100
Carrot	11,176	444	1,430	451	2,802	3,500	148	2,400
Garlic	2,315	177	126	143	1,407	435	3	22
Legumes								
Bean, dry	14,195	1,497	2,666	3,578	5,751	624	3	75
Bean, green	2,421	193	214	112	685	1,175	42	
Broad bean, dry	4,224	868	83	119	2,690	448	15	
Pea, dry	9,807	387	290	92	2,778	941	120	5,200
Pea, green	4,769	126	1,311	117	622	2,201	152	240
Chick pea	6,158	301	260	260	5,492	85		
Lentil	1,292	81	173	38	935	59		8
Peanut (shell)	18,580	4,553	1,725	638	11,573	23	66	1
Cucurbits								
Watermelon	26,076	2,145	1,908	1,004	14,899	3,069	51	3,000
Muskmelon and others	7,328	439	1,204	329	3,622	1,734		
Cucumber	11,093	343	1,080	46	5,491	2,517	16	1,600
Pumpkin, squash, and gourd	5,388	1,055	181	731	2,276	1,063	83	
Solanaceous fruits								
Chili pepper	7,209	1,141	728	169	2,963	2,207	1	
Eggplant	4,802	466	69	34	3,697	562		
Tomato	53,892	5,380	10,502	3,179	12,996	14,269	266	7,300
Miscellaneous vegetables/flavorings								
Artichoke	1,245	100	51	71	14	1,009		
Hops	132		36		3	82	2	8

crops on a tonnage basis. Orcharding is now a specialized and highly technical part of the horticultural industry.

Fruits may be classified in a number of ways. However, to the horticulturist, the basic distinction is a climatic one. Fruits are called temperate, subtropical, or tropical. Unlike vegetables, of which many tropical perennial species grow well as annuals in temperate areas (tomato, for example), long-lived fruit crops cannot be transplanted to alien conditions.

There is a tremendous diversity of tropical fruits, many of which are little known. Tropical fruit plants are evergreen and extremely sensitive to low-temperature injury. Although there are fewer temperate fruits, they have received much attention. Virtually all temperate fruit crops are deciduous and require a period of cold. Subtropical crops may be either deciduous or evergreen, and they are usually able to withstand light frost; some require a period of cool weather for maximum fruiting.

Most fruits can be consumed fresh when they contain large amounts of water. They are generally extremely perishable once the climacteric has been reached. The proper stage for consumption may therefore be very brief and transitory. For this reason, some delicious tropical fruits, such as the cherimoya, are virtually unknown beyond their area of production. As a result, most fruits that enter commerce either ship well in the fresh state, as do citrus and apple, or adapt well to preservation through drying (date and fig), the use of perservatives (as in fruit jellies and jams), canning (pineapple), or freezing (strawberry). The advent of air transport, refrigerated storage, and new packaging techniques has recently made it possible for a tremendous variety of fruits from all parts of the world to be available fresh in the largest metropolitan centers throughout the year. Yet at the same time, in smaller cities of the United States, common fruits that were formerly plentiful at least once a year from widespread local production (raspberries, for example) have virtually disappeared and have been relegated to the status of luxury products.

Temperate Fruits

Grape

Grapes (Figure 21-1) are the most important fruit crop, accounting for about one-quarter of the fruit production of the world. They are grown throughout the temperate regions, especially in warm, sunny climates with mild winters and dry periods during fruit ripening. In northern Europe, particularly in Belgium and the Netherlands, some grapes are still produced under glass. In the United States, production is located largely in California, although a sizable grape industry exists in the Great Lakes region.

About 78 percent of the world's crop is pressed into wine, 8 percent is

FIGURE 21-1. Heavy fruit production in the grape. (*Inset*) 'Fiesta,' a new early-ripening seedless grape, has 'Thompson Seedless' in its pedigree. [Photographs courtesy USDA.]

consumed fresh, and 13 percent is dried. Specific cultivars are usually grown for each purpose; 'Thompson Seedless' ('Sultanina') can be used for all purposes, however. And, in California, the European grape, *Vitis vinifera* (Vitaceae), is grown for wine, fresh fruit, and raisins. Various combinations of soil type, climate, grape cultivar, and production technique account for the differences in wines produced throughout the world. Of the entire California crop, 59 percent of the grapes are grown for wine, 28 percent for table use, and 13 percent for raisins. In the past, raisins and other dried fruit made up the bulk of United States fruit exports, but this market has been decreasing.

Vitis vinifera grapes are unadapted to most areas of the United States because of their lack of resistance to native diseases and soil pests and their lack of cold-hardiness. Therefore, the grape industry in the Great Lakes region has been based almost exclusively on a single cultivar, 'Concord,' an interspecific hybrid between *V. vinifera* and the native American fox grape, *V. labrusca*. The foxy flavor of 'Concord' is prized for nonfermented grape juice and sweet dessert wines. The grape juice is usually canned or bottled, although frozen concentrate has achieved market acceptance.

In the United States, increasing consumption of dry or table wines (wines made from grapes with high acidity and low sugar) has resulted in a wine-grape boom that has been going on since the middle 1960s. As prices have climbed, grape acreage and wine production have risen. Much of the increase in wine production has resulted from expansion of plantings of *Vitis vinifera* grapes in California, especially such famous cultivars as 'White Riesling,' 'Chardonnay,' 'Cabernet Sauvignon,' and 'Pinot Noir.'

Because the foxy flavor of the American-type slip-skin grape cultivars derived from *Vitis labrusca* is considered undesirable for table wines, alternative wine grapes have been eagerly sought for wine production in the United States outside of California. In France, hybridization of *V. vinifera* grapes with other native American species, including *V. rupestris* and *V. lincecumii,* has long been underway. These American species were introduced to France as rootstocks because of their resistance to *Phylloxera vitifoliae,* an insect carried into France from the eastern and central United States before 1860 that threatened to wipe out the French grape industry. A series of breeding programs, conducted mainly by private French breeders (Seibel, Ravat, Seyve-Villard, Seyne, and others), has been going on for three-quarters of a century and has resulted in a new class of wine grapes known as "French hybrids."

These new grapes have not only achieved prominence in France (where they account for one-quarter of the present wine acreage) but have been found adapted to many areas of the United States, where they are now being widely planted. French hybrid cultivars adapted to the Midwest include 'Aurore' (a cross between 'Seibel 788' and 'Seibel 29') and 'De Chaunac' ('Seibel 9549').

Kiwifruit

The fruit crop to be domesticated most recently is the kiwifruit, *Actinidia chinensis* (Actinidiaceae), also known as Chinese gooseberry, mihoutao, and yantao. For centuries, however, the fruit has been collected from wild plants in its native land, China. It was introduced to Europe, the United States, and New Zealand at the beginning of the twentieth century but was only seriously cultivated as a crop in the 1930s in New Zealand, where it was established as an export crop in the 1970s. Recent interest in kiwifruit has encouraged widespread trials throughout the temperate world, and it is now grown as a crop in California.

The plant is a perennial, twining vine, grown on trellises, that produces ovoid-shaped fruit the size of a large hen's egg. Because it is dioecious, both staminate and pistillate clones must be planted to ensure pollination and fruit set. The pubescent, brown-colored fruit is unattractive on the outside, but the inner flesh has fine flavor and a brilliant bright green color interspersed with 600 to 1000 tiny edible seeds. The fruit is very high in vitamin C, having twice that of oranges and 10 times that of apples. Kiwifruit can be stored for as long as 6 to 8 months.

Apple and Pear

Apple and pear, the **pome fruits,** are major tree fruits of the temperate climates. They are closely related botanically, and their culture is similar. Both crops are often grown in the same orchard.

The apple, *Malus domestica* (Rosaceae), is the most important single species of tree fruit in the world. It is widely adapted and is grown in temperate regions that have a distinct cold period. Although the United States is the largest single producer and the place of origin of most of the world's most popular cultivars, the bulk of the world's crop is grown in Europe. Many European cultivars are cooking and cider apples. Cider, which is sold with an alcoholic content of about 6 to 8 percent, is quite popular in England and northern Europe. Since World War II, there has been a trend away from cooking apples to dessert types, with increased emphasis on the more attractive red cultivars.

Although apples are grown throughout the United States, the largest concentrations are on the West Coast (Washington and California), in the Great Lakes area (New York and Michigan), and in the Appalachian area (Virginia, West Virginia, and Pennsylvania). The western orchards must be irrigated, but the low humidity reduces the disease problem, and the climate is uniquely adapted for the production of fruit of high "finish" and attractiveness.

About 10 cultivars account for more than 90 percent of the United States crop. The trend over the past 60 years has been toward a great reduction in the number of cultivars and a change to the red sports of the more popular cultivars, such as 'Delicious,' 'Rome,' 'Jonathan,' and 'McIntosh.' 'Golden Delicious' is the most important yellow apple and 'Granny Smith,' a tart green apple from Australia, is gaining in popularity. The small "farm orchards" have disap-

TABLE 21-3. World production of major apple cultivars; all originated in the United States except 'Granny Smith,' which comes from Australia.

Cultivar	World Production*	
	Thousands of metric tons	Percent of total
'Delicious'	5143	26.8
'Golden Delicious'	4457	23.2
'Jonathan'	1162	6.1
'McIntosh'	762	4.0
'Granny Smith'	705	3.7
'Rome Beauty'	666	3.5
Total apple	19,165	100.0

SOURCE: *USDA-FAS 1979 International Apple Institute Reference Book and Directory.*

*Includes 32 countries but not USSR, China, and some of Eastern Europe.

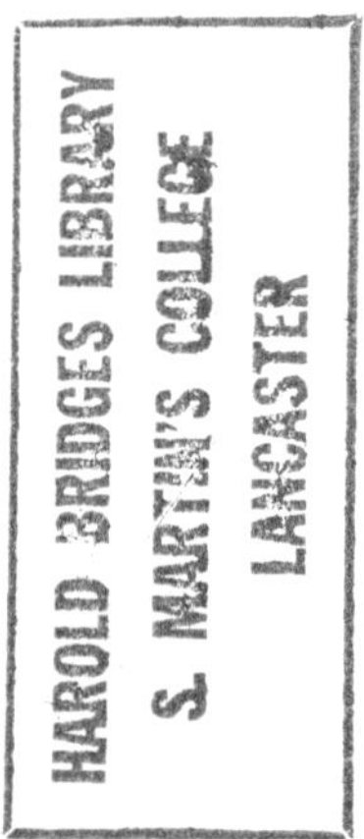

peared, and production has concentrated in more adapted locations. Although the number of trees has gradually declined until recently, yields are going up sharply. The export market, which has dropped from pre-World War II levels, is now increasing. The rise in population and the decline in consumption of apples per capita has stabilized the total apple consumption in the United States over the past two decades.

Production technology in the apple industry has shown great changes. In the United States, there is a trend toward greater use of dwarfing rootstocks, particularly of the "semidwarf" types, such as EM 2 and EM 7, but seedling rootstocks are still the most commonly used. The use of controlled-atmosphere storage makes it possible to obtain high-quality apples the year round. As this technological advance has increased in use, it has brought significant changes to the pattern of apple marketing. It has already interfered with the marketing of early-summer cultivars. Probably the most significant advance in production technology has been the greater use of organic fungicides and insecticides in the spray schedule. Most northeastern orchards use high-concentrate speed sprayers, but dusting is more common in the West.

The world's production of pear, *Pyrus communis* and other species (Rosaceae), is about a third of the world's apple production. The major production areas are located in Western Europe. About a third of the European crop goes into production of perry ("pear cider"); most of the rest is consumed fresh. In the United States, pears are relatively less in demand, and total production represents only about one-fourth that of apples. The United States industry is based largely on the 'Bartlett' cultivar, which accounts for about 80 percent of the total production. 'Bartlett' is used both for fresh consumption and for processing. It is also one of the most important cultivars in Europe, where it is known as 'Williams' or 'William Bon Crétien.' Almost 90 percent of the United States production is located in the states of Washington, Oregon, and California. Of the total crop, about half is processed by canning.

Although the pear is not quite as hardy as the apple, the limiting factor to its production has been the bacterial disease **fireblight,** which has eliminated commercial pear production from the warmer humid regions of the United States. The disease had been confined to the Western Hemisphere, but it is now established in Northern Europe. Various breeding programs are underway in the United States and Europe to combine fireblight resistance and high fruit quality. 'Magness,' a highly fireblight-resistant cultivar introduced by the U.S. Department of Agriculture, produces fruit of extremely high quality but its commercial potential is limited by low productivity.

Stone Fruits

Species of the genus *Prunus* (Rosaceae) constitute the so-called **stone fruits.** These include the plum, peach (Figure 21-2), cherry, and apricot. The stone fruits require a cold period to break their rest period, but they are subject to

FIGURE 21-2. 'Flamecrest,' a new yellow-fleshed freestone peach produced as a shipping type for California. [Photograph courtesy USDA.]

winter killing and frost injury and consequently can be grown profitably only in restricted locations.

Plums and peaches are the principal stone fruits. There are a great number of plum cultivars and species, with a correspondingly wide range of adaptability. Plums that have a high sugar content and can be dried are known as prune-type plums, or simply as prunes. The great bulk of the world's plums are grown in Europe, chiefly in Yugoslavia, Rumania, and Germany. In the United States, the West Coast accounts for almost the entire commercial crop, of which a large proportion is dried and sold as prunes.

The United States is the leading peach-producing country. About half the crop is grown in California. Two-thirds of the California crop are clingstone peaches. These are "rubbery fleshed" peaches, and virtually all of them are processed by canning. The remaining California production, and the rest of the United States production, is devoted to freestone peaches. These have "melting flesh," and the pit more or less separates from the flesh when ripe. Probably the most famous freestone peach is the 'Elberta,' which is still used as a basis for comparison in the industry. A number of breeding programs have produced peaches that ripen over a 2-month period at most locations. The early red-colored 'Redhaven' is now the most widely planted cultivar.

Nectarines are smooth-skinned peaches; this character is caused by a single recessive gene. Nectarine production, once confined almost exclusively to California, is now expanding to other peach-growing areas with the introduction of adapted cultivars.

Cherries are an important European fruit. They are relatively less popular in the United States. The European crop consists mainly of sweet cherries (*Prunus avium*), whereas more than half the United States production consists of tart cherries (*P. cerasus*). The cultivar 'Montmorency' is the basis of the tart-cherry industry in the United States. The production of sweet cherries is

FIGURE 21-3. The blueberry has become an important small fruit crop, transforming many of our supposedly worthless acid soils into valuable cropland. [Photograph by J. C. Allen & Son.]

concentrated on the West Coast, whereas tart cherries are grown most abundantly in the Great Lakes region.

The United States commercial production of apricots is almost completely confined to California. The bulk of the crop is processed by canning or drying.

Small Fruits

The **small fruits** grown on plants of small stature are also known as **soft** or **berry fruits.** They include the grape (usually considered separately, as here); the strawberry, *Fragaria* × *ananassa* (Rosaceae); the brambles, all *Rubus* species (Rosaceae), such as red and black raspberry, blackberry, dewberry, and loganberry; currant and gooseberry, *Ribes* species (Saxifragaceae); cranberry, *Vaccinium macrocarpon* (Ericaceae); blueberry, *Vaccinium* species (Ericaceae), as in Figure 21-3; and various others. Except for grape, the most important small fruit is the strawberry (Figure 21-4), followed by the raspberry (Figure 2-2). Although these crops are widely adaptable to other areas, the West Coast strawberry industry has expanded to the point where California accounts for

FIGURE 21-4. The strawberry is one of the most widely adapted fruit crops. Present-day cultivars descend from hybrids of *Fragaria virginiana*, native to the East Coast, and *F. chiloensis*, native to the Pacific coasts of North America and South America. The 'Salinas' cultivar shown above is a California introduction produced by Royce Bringhurst and Victor Voth. [Photograph courtesy Royce Bringhurst.]

more than half the United States crop. In California the acreage devoted to this crop increases and decreases in response to price changes.

The high labor costs associated with small fruits have limited the expansion of these crops in the United States. Although production has stabilized, yields per acre have steadily gone up with the increased care of improved cultivars and with the adoption of better cultural practices. The use of virus-free plants has increased the productivity of many small-fruit cultivars. With strawberries, irrigation has played an important role in improving performance. Irrigation is also widely used to control damage by spring frosts. The development of the frozen-food industry and the success of harvesting machines with bramble crops indicate that future expansion of the industry is likely.

Subtropical and Tropical Fruits

Citrus Fruits

The citrus group (Rutaceae) of evergreen fruits are native to the subtropical regions of eastern Asia. They are now the major fruit crop of the subtropical climates, and they rank in commercial importance with apple and pear. Although citrus fruit can be grown in the tropics, fruit quality there is inferior. Main centers of world production include Brazil, the United States, Japan, Spain, Italy, and Mexico. In the United States, Florida is the main producing state, followed by California. Some devastating winter freezes have stabilized Florida citrus production. Brazil has taken over world leadership in citrus production and even exports citrus concentrate to the United States.

The genus *Citrus* contains a great number of edible species. Of the commercially grown types, the sweet orange, *C. sinensis,* is the most important. The California orange industry is based primarily on two cultivars, 'Washington Navel' (the 'Bahia' cultivar of Brazil), which is relatively seedless, and 'Valencia' (Figure 21-5). The Florida industry is based on various thin-skinned cultivars, of which the chief ones are 'Valencia' and 'Pineapple.' Mandarin oranges, *C. reticulata,* are typically rough-skinned and have easily separated segments. (Deep orange-red mandarins are called tangerines.) Mandarins are widely popular in Japan and have recently stimulated considerable demand in the United States.

A number of artificial hybrids within the citrus group have been produced. Among them are the tangor (tangerine × sweet orange) and the tangelo (tangerine × grapefruit). The chief one of these hybrids is the 'Temple' orange, which is believed to be a tangor. The 'Temple' orange has achieved considerable importance in Florida. Grapefruits (*C. paradisi*), lemons (*C. limonia*), and limes (*C. aurantifolia*) are the other major citrus crops.

The high concentration of citrus production in the United States has led to grove management on a contract basis by cooperative or caretaking organizations. The size of individually owned groves is commonly as small as 20 acres,

FIGURE 21-5. 'Valencia' oranges in California. [Photograph by J. C. Allen & Son.]

although the Florida enterprises tend to be larger. The great increase in the citrus industry in the United States is due to a number of factors, including low production costs (low as compared with apples, for example), efficient marketing, joint industry advertising and promotion, and the rise of citrus as a valuable processing crop. A major part of the Florida citrus crop is utilized as frozen concentrate.

Olive

The olive, *Olea europea* (Oleaceae), is native to the Mediterranean region, where it has been cultivated for many centuries. World production is principally in Spain, Italy, Greece, Turkey, and Morocco. Most of the world's olive crop is produced for oil, but ripe (black) and immature (green) olives are pickled for table use. The California crop is used largely for pickling and canning. Olive fruits contain a bitter glycoside that must be removed by hydrolysis with sodium or potassium hydroxide. Bitter Greek olives are cured in a highly concentrated salt solution without hydrolysis.

The small, tough evergreen tree begins production only 6 years after propagation, but several decades are required for full production. Although production declines after 50 years, the trees live for centuries. Biennial bearing is often a problem.

The tree is subtropical and is killed outright at 14°F (9°C), although it survives frost well. Slight winter chilling, not lower than 43 to 46°F (6 to 8°C), promotes fruiting. Because olive trees do best with low atmospheric humidity, they are typically grown in arid regions. Spacing varies with water supply.

Propagation is by cutting and grafting. Hand harvesting is general throughout the world, although mechanical shakers are being used in California. Olives for pickling must be picked carefully to prevent bruising.

Date

The date palm, *Phoenix dactylifera* (Palmaceae), is an ancient crop plant in arid regions (Figure 2-7). Production is centered in Asia Minor and North Africa (Iran, Iraq, Algeria, Egypt, and Morocco), with small acreages in Spain, Mexico, and the United States. Although the tree is able to withstand light frost, it requires an average temperature of 86°F (30°C) for proper fruit ripening. Low humidity during fruit maturation is essential. Because the trees need a good supply of moisture, date palms are usually planted in oases or where irrigation is possible.

The date is dioecious. Select pistillate clones are propagated through offshoots that grow from the base of the trunk, but rapid propagation by tissue culture via asexual embryos appears feasible. To improve fruit set, flower clusters are usually pollinated by hand with pollen from staminate trees grown in separate plantings. Pollen affects tissues of the fruit outside the embryo sac (**metaxenia**); thus the source of pollen influences the form, size, and rate of ripening of the fruit.

Flowering may begin in 3-year-old trees of precocious cultivars. The developing fruit is often enclosed in paper or net bags to protect it from birds and other pests. A mature tree may yield more than 100 pounds of fruit yearly. Harvested dates are first fumigated and then allowed to ripen to increase sugars and precipitate astringent components. The partially mature astringent fruits are eaten in the Middle East, but those that enter commerce are ripened and dried.

Fig

The fig, *Ficus carica* (Moraceae), a small deciduous tree, has been cultivated since antiquity in the eastern Mediterranean region. It was known to the ancient Egyptians as early as 2700 B.C.E., and numerous references to it can be found in the Old Testament and in ancient Greek and Roman writings. World production totals about 2 million tons on a fresh-weight basis. About two-thirds of the crop is dried, and much of the dried fruit is treated with sulfur dioxide to maintain a light color. Major production is in Portugal, Italy, Greece, Turkey, and the Near East; U.S. production is chiefly in California. Fig trees thrive best in hot, dry summers and cool, moist winters. Mature trees will withstand temperatures of 14 to 23°F (−10 to −5°C) if completely dormant. Uniform soil moisture is essential.

The unisexual flowers are borne inside a pear-shaped, nearly closed hollow peduncle. The entire peduncle becomes fleshy and forms the fruit, which is called a **syconium** (Figure 21-6). Skin color may be shades of green, yellow, pink, purple, brown, or black. Figs are propagated by means of cuttings.

The many major cultivars have been divided on the basis of pollination requirements into four types: caprifigs, Smyrna figs, common figs, and San Pedro figs. **Caprifigs** are primitive cultivated types with short-styled pistillate flowers and functional staminate flowers. Most caprifigs are not edible but are grown because they harbor the *Blastophaga* fig wasp, which develops in the fruit. The winged female escapes from the fruit of the caprifig in the spring covered with pollen and seeks developing figs in order to lay her eggs. She may enter the **Smyrna** fig, which contains only pistillate flowers. The pollen she carries with her fertilizes these flowers and thus allows the fruit to mature. Caprifigs are therefore essential to the cultivation of Smyrna figs, such as the large-fruited 'Calimyrna' cultivar, which will not mature if unpollinated. The wasp is prevented from laying eggs in the young synconia of the Smyrna fig because her ovipositor is too short to penetrate the long-styled flowers; the wasp either dies there or flies away. The process of growing caprifigs to achieve pollination is known as **caprification.** Fruit set can be induced with growth

FIGURE 21-6. Fig synconia at pollination time. The characteristic inflorescence of *Carica* species, the synconum, is a complex, enlarged, fleshy, hollow flowering stem or peduncle bearing massed flowers on the entire inner wall. (*Left*) A monoecious caprifig with staminate flowers in anthesis; (*right*) pistillate fig with flowers in anthesis.

regulators, but such fruits are seedless and lack the characteristic grittiness of figs produced by caprification. **Common figs,** such as the 'Kadota' and 'Mission' cultivars, set fruit parthenocarpically without the stimulus of caprification and seed development. **San Pedro** types set the first crop (**brebas**) parthenocarpically from buds initiated early in the spring, but they require caprification for the second crop.

Avocado

The avocado, *Persea americana* (Lauraceae), is of New World origin, as the specific name implies. Although still not widely known in the Eastern Hemisphere, avocados have been planted commercially in South Africa and Israel. Avocado is a popular local fruit in tropical and subtropical America, and there is considerable production in California and Florida.

The fruit is extremely rich in oil—fruits of some cultivars contain as much as 30 percent. However, the oil content varies with location, so 'Fuerte' avocados contain about 25 percent in California but only 13 to 15 percent in Florida. In rural areas of Brazil, the avocado is used to make soap.

The many cultivars of avocado have been divided into three "horticultural" races, as shown in Table 21-4. Fruit size varies greatly, as do other characteristics. Many cultivars are hybrids between these races—'Fuerte,' for example, is considered to be a natural Guatemalan–Mexican hybrid.

TABLE 21-4. Distinguishing characteristics of the cultivated avocado.

Characteristic	Mexican	Guatemalan	West Indian
Native habitat and elevation	Highlands of Mexico (also Ecuador, Peru, and Chile), 7900–9200 ft	Highlands of Central America (also Mexico), 2600–7900 ft	Lowlands of Central and South America, below 2600 ft
Climatic zone	Semitropical	Subtropical	Tropical
Hardiness (of dormant trees)	Slight injury at 21°F (−6°C)	Severe injury at 24°F (−4°C)	Severe injury at 28°F (−2°C)
Season of fruit maturity	Summer	Winter and spring	Summer and fall
Foliage scent	Anise odor	No anise odor	No anise odor
Fruit size	½ lb	½–5 lb	1–5 lb
Fruit stem length	Short	Long (3 inches)	Short (1–2 inches)
Skin texture and thickness	Smooth, papery (0.8 mm thick)	Rough, woody, brittle (3–6 mm thick)	Smooth, leathery (1.5–3 mm thick)
Seed size and fit	Large and tight in seed cavity	Small in proportion to fruit size and usually tight	Large and usually loose

SOURCE: Ochse, Soule, Dijkman, and Wehlburg, 1961, *Tropical and Subtropical Agriculture* (Macmillan).

Planting usually requires more than one cultivar in order to synchronize pollen shedding and stigma receptivity. Flowers of some cultivars (in Class A) open in the morning with receptive stigmas but shed no pollen; the flower closes at noon and reopens the afternoon of the following day to shed pollen, at which time the stigmas are nonreceptive. Flowers of other cultivars (Class B) open with receptive stigmas in the afternoon of the first day and reopen for pollen shed the next morning or the morning of the third day.

	First day		Second day	
	A.M.	P.M.	A.M.	P.M.
Class A	♀			♂
Class B		♀	♂	

In Florida, cultivars of classes A and B are interplanted to assure adequate fruit set.

Banana

Bananas, *Musa* species (Musaceae), originated in the tropical regions of southern Asia, and are among the oldest cultivated plants (Figure 21-7). They are true tropical fruits and require temperatures that do not fall below 50°F (10°C) or rise above 105°F (41°C). Although many species are edible, the chief ones are the seedless selections of *M. sapientum*, exemplified by the cultivar 'Gros Michel.' In Asia, there are many starchy types of bananas, known as plantains (*M. paradisica*), that must be cooked before eating. Bananas have become one of the most widely used fruits of the world, and they are the best known of the tropical fruits.

The banana plant is a giant herb with an underground stem (rhizome). The above-ground portion, the **pseudostem,** is formed from a compact mass of leaf stalks through which the blossom emerges. Although each pseudostem produces only one fruiting stalk, the underground rhizome produces numerous suckers. Thus, although each pseudostem is an annual, one planting is capable of producing continuous crops.

Bananas are propagated by planting rhizome pieces or young suckers. When a fruit bunch is harvested, the pseudostem from which it is produced is cut down and a sucker from the rhizome grows to replace it. If not shortened by disease, production cycles often continue for as long as 20 years before another rhizome must be planted.

The leading banana-producing countries in the Western Hemisphere are Brazil, Equador, and Mexico, whereas the leading Asian producers are India and the Philippines. The present export industry is located largely in the Caribbean countries and is controlled by a few companies, of which United Fruit is the largest. The main limitations to production in Central America have been

FIGURE 21-7. The banana plant. As the stem elongates, the blossoms unfold. The fruited stalk, or bunch, consist of many clusters of flowers called hands. Each pistallate flower developed into a fruit—the "finger." Staminate flowers are produced toward the apex end of the stem. [Photograph by J. C. Allen & Son.]

two devastating diseases, **Panama disease** (a root wilt caused by the fungus *Fusarium oxysporum* f. *cubense*) and **sigatoka** (a leaf spot caused by the fungus *Cercospora musae*). In the past, the only economical way of controlling Panama disease was to move plantings to new, wilt-free locations. Since 1963, however, a new cultivar, 'Valery,' which is resistant to Panama disase, has been used to replace the susceptible 'Gros Michel.' 'Valery' is thin-skinned and must be shipped in cartons rather than on stems.

Pineapple

Pineapple, *Ananas comosus* (Bromeliaceae), also originated in the New World. The fruit vaguely resembles a pine cone, and the Spanish explorers called it "Piña de las Indias." Imported to Europe and raised in greenhouses in the seventeenth and eighteenth centuries, this tropical fruit became a gourmet delicacy and a symbol of hospitality. The development of canning established pineapple as a commercial plantation crop at the close of the nineteenth century. The dominance of Hawaii as a center of pineapple production has deteriorated, and the major producing countries are now Sri Lanka (formerly Ceylon), the Philippines, Brazil, and Mexico.

The family of Bromeliaceae consists largely of epiphytes, or "air-growing" plants (Spanish moss is an example). Although pineapple is terrestrial, it has many epiphytic adaptations, including a short stem and coarse, stiff, narrow, spiny leaves arranged in the form of a rosette. The pineapple is well adapted to dry habitats; the close-fitting basal leaves form a funnel capable of holding and absorbing water. The multiple fruit is usually seedless. Therefore, propagation is from the **crown** (the leafy shoot arising from the top of the fruit), from **slips** (shoots growing below the fruit) or from **suckers** (offshoots). These propagules are remarkably resistant to desiccation.

In Hawaii, planting materials are set out through plastic mulch in strips of fumigated soil. The plants normally flower 20 to 24 months after being planted (usually in the summer) and bear one large fruit. A number of growth regulators are used to initiate flowering out of season, and this, in turn, ensures a continuous supply of pineapple for the canning factories. After the first crop ("plant crop") is harvested, the growth of basal branches called suckers yields smaller fruits the following year (a yield known as the "first ratoon crop"). The plants bear still smaller fruits in the "second ratoon crop," after which the field is disked and the enormous amount of plant residue is burned. Then the land is prepared and the cycle is repeated. The crop is harvested by hand in Hawaii; conveyors move the fruits through the field for quick dispatch by truck or barge to a centralized canning factory on the island of Oahu. The Hawaiian industry is based on the 'Smooth Cayenne' cultivar, thought to have originated in French Guinea but described also at Versailles in 1841. The fruit of 'Smooth Cayenne' is tough and well suited in size and shape to the Ginaca machine, a remarkable device that trims and cores the pineapple at a rate of up to 120 fruits per minute. Pineapple by-products include bran, syrup, and alcohol.

Mango

Native to Southeast Asia, the mango, *Mangifera indica* (Anacardiaceae), has been cultivated since prehistoric times in India, where it is today the most popular fruit. The mango is grown now in practically all tropical areas. Although it is not a major export fruit, it has very great local use, being planted widely in small groves and in backyards. The tree, magnificent and glossy-leaved, bears juicy, aromatic, one-seeded (or drupe) fruits on a terminal panicle (Figure 21-8). The single large seed has fibers that extend into the flesh. Some select cultivars are smooth-fleshed, however, and have few noticeable fibers. The flesh has a characteristic aroma, which ranges from mild (peachlike) to very rich and strong (turpentinelike). Fruits range in color from green through yellow to bright red, and they vary in weight from ¼ to 2 pounds. Many mangos are apomictic, with adventitious embryos produced from nucellus tissue, as in citrus. Thus, mangos may breed true from seed. Often, mango trees may bear only in alternate years.

A

B

FIGURE 21-8. (*A*) 'Tommy Atkins' mango is widely planted in Florida because of its high productivity and its attractive appearance. The fruit is a bright reddish orange. (*B*) Mango inflorescences are large panicles containing many hundred individual flowers with a diameter twice the size of the head of a pin (4 mm). [Photograph courtesy S. E. Malo.]

Other members of the family Anacardiaceae are the cashew (*Anacardium occidentale*), which is grown for its nut as well as for its edible, pear-shaped peduncle, and the pistachio (*Pistacia vera*). It is interesting to note that poison oak and poison ivy are also members of this family. To one degree or another, all members of the family contain toxic substances that may be irritating to the skin.

Papaya

The papaya, *Carica papaya* (Caricaceae), is indigenous to tropical America but is now grown throughout the tropics. It is known by a number of other descriptive names: for example, papaw (New Zealand), mamão (Brazil), lechosa (Venezuela), and fruita-bomba (Cuba). The tree—fast-growing, short-lived, and herb-like—is normally single-stemmed, but occasionally lateral branches result from wounding. Fruit is borne on the axils of the large leaves that open in sequence up the stem as the tree grows. The crop is propagated by seed.

The fruit, which may be round, pyriform, or oval, ranges in weight from ½ to 15 pounds. The skin color is orange; inside, the flesh is pale orange or light red.

Papaya has both staminate and pistillate flowers and a great array of bisexual flowers. On the basis of their flowers, the trees are classified as staminate (all staminate flowers), pistillate (all pistillate flowers), or hermaphroditic (hermaphroditic and staminate). Hermaphroditic trees may be further divided according to the type of hermaphroditic flowers they produce. Moreover, some papaya trees are continuously fertile and others are sterile in the summer.

Most cultivars are dioecious with staminate and pistillate plants produced in a 1:1 ratio, but some are gynodioecious with hermaphroditic and pistillate plants. Sex determination seems to be controlled by a single gene with three alleles, M, M_1, and m, in accordance with this scheme:

$$Mm = \text{staminate}$$
$$M_1m = \text{hermaphroditic}$$
$$mm = \text{pistillate}$$

The genotypes MM, M_1M, and M_1M_1 are lethal and give rise to nonviable seeds. Dioecious cultivars produce staminate and pistillate plants in equal ratios:

$$mm\,♀ \times Mm\,♂ \longrightarrow \underset{\text{Staminate}}{1\,Mm} : \underset{\text{Pistillate}}{1\,mm}$$

In general, only a few staminate plants are needed for pollination. Usually, three seedlings per hill are allowed to grow until flowering, and staminate plants are removed as they flower. At present, there is no way to distinguish the sex of the trees before flowering. In seven-eighths of the hills there will be at least one pistillate plant. In one-eighth of the hills all three seedlings will be staminate, of which one is kept to provide pollination.

In Hawaii the standard papaya cultivar is 'Solo' (Figure 2-6), a very high-quality small-fruited type composed of hermaphroditic and pistillate plants. Hermaphroditic flowers yield pyriform fruit; pistillate flowers yield round fruit that is discriminated against (apparently without reason). Seed saved from hermaphroditic fruit yields two new hermaphroditic plants for each new pistillate plant:

$$M_1m\,⚥ \times M_1m\,⚥ \longrightarrow \underset{\text{Lethal}}{1\,M_1M_1} : \underset{\text{Hermaphroditic}}{2\,M_1m} : \underset{\text{Pistillate}}{1\,mm}$$

Thus, growing three seedlings per hill with subsequent thinning to one hermaphroditic plant per hill yields a crop in which only 1⁄27 (3 percent) of the spaces grow the unpopular pistillate plants.

In addition to providing fresh fruit, papaya is the source of papain, a proteolytic enzyme that is used as a beer clarifier and in meat tenderizers. Papain is obtained by collecting and drying the latex exuded from scratches in the surfaces of slightly immature fruit. It also has medicinal uses, such as to pretreat red blood cells prior to cross-matching and to dissolve cartilagelike substances that often develop in disks between vertebrae.

NUT CROPS

> I had a little nut tree, nothing would it bear
> But a silver nutmeg and a golden pear.
> The King of Spain's daughter came to visit me,
> All for the sake of my little nut tree.
>
> ENGLISH NURSERY RHYME

The edible nuts—seeds of woody plants with firm shells that separate from an inner kernel—are often considered along with fruit crops, even in nursery rhymes. However, although nuts are tree crops, their cultivation is typically more casual; many nuts are still gathered from wild stands. Nut production has been suggested as a use for marginal lands, a proposal that undoubtedly has great potential through genetic improvement and better cultural practices; however, those who champion the raising of nut crops have not received the attention they deserve.

Characteristically, nuts are rich in oil; indeed, the inedible tung nut, *Aleurites fordii* (Euphorbiaceae), is grown to supply oil for paint, varnishes, and other uses. Edible nuts have been cultivated, in general, as an expensive luxury food, a special delicacy consumed often with candy or sweetmeats. Important temperate nut crops include almond, chestnut, hazelnut, pecan, pistachio, and walnut; among tropical nuts are the cashew, Brazil nut, and macadamia.

Temperate Nuts

The almond, *Prunus amygdalus* (Rosaceae), is closely related to the peach. The outer flesh of the almond is astringent and tough, however, splitting as the fruit tipens to expose the pit—the almond shell. The edible sweet almond (*P. amygdalus* var. *dulcis*) is the source of confectionary nuts (Figure 21-9); the bitter almond (*P. amygdalus* var. *amara*) contains a poison, prussic acid, and is grown only for the extraction of almond oil for use as a food flavoring. Almond crops are produced in Spain, Italy, and the United States, where California harvests the largest crop.

Almond cropland in California hovered around 100,000 acres from 1930 to 1970 but more than tripled to over 300,000 acres in 1980, while production increased four to four-and-a-half fold. In 1982, total acreage of bearing and

FIGURE 21-9. 'Solo,' a new almond cultivar from Israel. [Photograph courtesy P. Spiegel-Roy and J. Kockba.]

nonbearing plantings was 420,000 acres with total production of 350 to 400 million pounds, two-thirds of the world supply. 'Non-pareil' is the predominant cultivar in California. The almond is self-incompatible, so different cultivars are planted in adjoining rows.

Pecan, *Carya illinoensis* (Juglandaceae), is the chief native American nut crop. Originating in the southeastern United States and Mexico, it is still harvested from wild and seedling trees, called **natives,** in Texas, Oklahoma, and Louisiana. In 1982, cultivated pecans (Figure 21-10) accounted for 41 percent of total acreage but over 61 percent of total production in the United States. The industry is increasing rapidly in Georgia, Texas, Alabama, and Mississippi, with selected cultivars propagated by patch budding or grafting to seedling rootstocks. Selection favors thin shells ("papershells") and large size. Most of the crop is sold to bakers and processors, with the poor-quality, unshelled nuts being sold to less knowledgeable buyers—usually tourists. Efficient cracking and shelling machines can now produce a high percentage of unbroken halves.

The pistachio, *Pistacia vera* (Anacardiaceae), is a dioecious, slow-growing tree that is capable of attaining great age. The pistachio fruit has a pale red or yellow husk enclosing a small, smooth, thin, hard-shelled nut (Figure 21-11) with a richly flavored, fine-textured green kernel. The tree is native to western Asia, and the greatest production areas are Iran, Turkey, and Syria. Pistachio is presently one of California's newest and most rapidly expanding industries—from 300 acres in 1968 to more than 47,000 acres in 1985. The first commercial crop of significance, 4.5 million pounds (2000 metric tons), was harvested in 1977; 43 million pounds (20,000 metric tons) were harvested in 1982, and 80 million tons (36,000 metric tons) are expected in 1990. Annual U.S. consumption in the 1970s was only about 22 million pounds (10,000 metric tons). The nuts traditionally imported from the Middle East were usually dyed red to obscure fungal-stained shells, but the pistachios produced in California are

FIGURE 21-10. In-shell and shelled nuts of 'Shoshoni,' an improved pecan cultivar obtained by breeding. [Photograph courtesy USDA.]

stain-free and sold in their natural straw color. California production is based on the nut-producing 'Kermen' and pollen-producing 'Peters.' 'Jolly,' a new cultivar released in 1979 by the University of California, is promising.

The chestnut is important in Europe and Asia as both a livestock feed and an edible nut. The European or Spanish chestnut, *Castanea sativa* (Fagaceae), is a semicultivated tree of southern Europe, and it produces the nut so often found "roasting by an open fire" during the Christmas season. The native North American chestnut, *C. dentata*, has been eliminated by chestnut blight. Both the Chinese chestnut, *C. mollissima*, and the Japanese chestnut, *C. crenata*, are resistant to chestnut blight but are of inferior quality. Perhaps hybrids between *Castanea* species that are resistant to blight and those that produce nuts of high quality can restore chestnut cultivation to the United States.

Hazelnuts, various species of *Corylus* (Betulaceae), grow wild in most of the world's northern temperate areas. The cultivated types are known as filberts, cobnuts, Barcelona nuts, and Turkish nuts. Hazelnuts are grown extensively in northern Turkey, Italy, and Spain. United States production is mostly in Oregon. The native American species *C. americana* and *C. rostrata* are pleasantly flavored but are considered too small for commercial use.

Although commercially the most important walnut with an edible nut is the Persian, Regia, or English walnut, *Juglans regia* (Juglandaceae), many species of walnut produce edible nuts, including the black walnut (*J. nigra*) and the

FIGURE 21-11. The pistachio bears its fruits laterally on wood produced the previous season. Nut-bearing branches bear in alternate years because most of the inflorescence buds abscise on the year of fruiting.

butternut or white walnut (*J. cinerea*). More than 200,000 tons of in-shell English walnuts (Figure 21-12) are produced in the United States each year, mainly in California and Oregon. By contrast, Turkey produces 125,000 tons, China 115,000 tons, and Portugal 40,000 tons. Unlike the husk of the black walnut, the husk of the Persian walnut readily separates from the shell. After being gathered, washed, and bleached, nuts are often individually labeled when they are to be sold in the shell for table use.

Tropical Nuts

Coconut, *Cocos nucifera* (Palmaceae), which is grown throughout the tropics, is the source of **copra** (dried coconut meat that is pressed to produce oil) and **coir** (the husk fiber). Although the stately trees require 60 inches of rainfall evenly distributed throughout the year, they can easily withstand brackish water. They are typically found along tropical sandy shorelines. Important cen-

FIGURE 21-12. Fruit of the English walnut, showing the outer husk and the shell-covered nut. [Photograph courtesy USDA.]

ters of production lie in a zone no wider than 15° of latitude. Annual world production of copra approaches 5 million tons, most of which comes from the Phillipines and Indonesia. However, much coconut is consumed locally and does not enter into world trade.

Propagation is from seed; the young seedlings are planted about 40 or 50 to the acre. Such a planting may yield 900 to 1100 pounds of copra per acre, with yields per tree of up to 88 pounds. Production varies from 50 to 200 nuts per tree. Approximately 5500 nuts will yield 1 metric ton of copra (of which 61.5 percent is oil). Trees bear from 6 to 10 years, though full production is not reached until the trees are 15 to 20 years old. Fruits require about one year to develop. The 'Malay' tree starts to bear fruit close to the ground, but eventually these apparent dwarf trees reach great heights.

The cashew, *Anacardium occidentale* (Anacardiaceae), is native to Brazil but now grows in many tropical areas (Figure 21-13). In temperate countries, it is well known for its high-quality edible nut, but in tropical areas the fleshy astringent pedicel or stalk (the cashew "apple"), which bears the nut-containing fruit, is also eaten fresh or made into juice. Moreover, the cashew shell contains a liquid—the nut-shell "oil"—that is used as a resin. Most cashews consumed

FIGURE 21-13. Fresh cashews as harvested in Guatemala. The true fruit at the tip contains the cashew nut; the fleshy pear-shaped pedicel and receptacle can be eaten fresh or preserved. [Photograph courtesy USDA.]

in the United States are grown in India. Cashew nuts, like their pedicels, are astringent; unlike the pedicels, the nuts are toxic when raw and are roasted before export.

The large Brazil nut, *Bertholletia excelcea* (Lecythidaceae), is borne in a natural "jumble pack" in urnlike ligneous fruits on great trees of the Amazon rain forest. The nuts are gathered from natural stands, and about 30,000 tons are exported annually to the United States and Europe.

Macadamia ternifolia (Proteaceae), although native to Australia, is cultivated chiefly on the stony volcanic soils of the island of Hawaii—the "big" island—where its nuts are rapidly becoming a valuable crop. In the economy of the state of Hawaii, it is still a crop of minor value compared to sugar and pineapple. The spherical nut, enclosed in a fleshy husk, is extremely tough, but it cracks easily between rotating drums to yield an unusually rich kernel (70 percent oil) of excellent quality. The nuts fall naturally onto the smooth, stony orchard floor and are vacuumed up by special harvesting machines. In processing, the entire crop is first dried, then cracked, graded, cooked slightly in oil, salted, and vacuum-sealed in glasses or cans. Imperfect nuts are used in a number of other products, chiefly brittle and ice cream. At present, the demand for macadamia nuts outstrips the supply, and production is increasing.

BEVERAGE CROPS

Tropical plants are the source of the world's three most popular nonalcoholic beverages—coffee, tea, and cocoa. The original basis of their popularity was the "lift" provided by the caffeine (or caffeinelike alkaloids) that they contain, though their consumption with sugar and milk products also makes these products refreshing foods. As currently consumed, cocoa contains little stimulant and is more widely known for its confectionary uses. World demand has created huge industries in connection with growing, processing, and transporting the bulky materials from which these beverages are brewed. The crops represent a large proportion of the tropical world's agricultural exports; in many tropical countries, their production is the major industry.

Coffee

Coffee is made by steeping ground roasted seeds (or "beans") of *Coffea* species (Rubiaceae) in hot water. Most of the world's production is of *C. arabica* (Arabian coffee), with small quantities of *C. conephora* (Robusta coffee), *C. liberica* (Liberian coffee), *C. excelsa* (Excelsa coffee), and others. Coffee was being grown in Arabia as early as the seventh century, and it reached Europe in the sixteenth and seventeenth centuries with the opening of the East by the Dutch and English. The City of Mocha, Arabia, a city immortalized in the name still used for one type of coffee, remained the center of the trade, but the Dutch introduced *C. arabica* into Ceylon in 1658, into Java in 1699, and, about 1619, into Martinique, whence it spread into the New World.

Coffee now has its major production centers in the Americas. Brazil accounts for about one-third of the world's annual production of more than 4 million metric tons. Unlike many plantation crops, coffee tends to be produced by local entrepreneurs. Marketing is characterized by periods of glut and shortage, with corresponding fluctuations in price.

The coffee plant is a glossy-leaved shrub that flowers seasonally, usually at the onset of wet weather. The fruit ("cherry") is a two-seeded drupe. Seedlings are commonly transplanted to a density of approximately 500 plants per acre. Fields may be planted in shade, but the highest yields are achieved in full sunlight with large amounts of fertilizer. Trees are pruned to establish strong frameworks, with single or multiple-stem systems. Renewal pruning is practiced to maintain a good quantity of fruiting wood less than 5 years of age. Harvesting is generally done by hand, so plants are pruned to control height. Experimental machines that either shake the fruit into catch-frames or comb the limbs are now under development.

Processing requires removal of the fruit skin (exocarp), the mucilaginous flesh (mesocarp), a parchment coat (endocarp), and the delicate silverskin (seed coat). Fruit may be **dry-processed:** After drying, the pulp is separated mechanically from the seed. More commonly, the fruit is **wet-processed:** Defective fruit is first floated off in water, then the pulp is separated from the seed by machine. Brief fermentation then removes the remaining flesh adhering to the parchment, and the beans are dried and cured (through further milling and polishing that removes parchment and silverskin). Roasting for about 5 minutes at 500°F (about 280°C) is the final step in either process.

Coffee requires a dry season for fruit maturation; a monsoon climate with a double dry season is preferred to obtain two crops per year. The best coffee is grown at elevations of 4000 to 5500 feet with rainfall between 80 and 120 inches and mean temperatures of 60 to 72°F (15 to 22°C).

Cacao

Although cacao, *Theobroma cacao* (Sterculiaceae), is a native of the New World, its greatest production is in Africa; Brazil now ranks a close second. In some areas of pre-Columbian America, the cacao "bean" was used as currency. The Aztecs also greated prized a bitter, nourishing concoction made of cacao, vanilla, corn meal, and chilis. To please the Spanish palate, sugar replaced the corn meal and cinnamon substituted for the chilis. The new mixture made chocolate (from the Aztec *chocolatl*) a popular beverage in Europe by the mid-seventeenth century. Consumption increased rapidly when, in 1828, a Dutch manufacturer perfected a method for separating the fat (cocoa butter) to produce the versatile residue, cocoa powder. The Swiss invention of milk chocolate in 1876 further increased the demand for cacao.

Cacao cultivars are usually divided by "race": The Forastero has bitter purple seeds and the Criollo has sweet white or violet seeds. Hybrids between them are known as Trinitario. A mutant with white cotyledons produces a lighter chocolate. Football-sized fruits are borne directly on trunks and large branches (Figure 21-14). Cacao trees are related to *Cola acuminata*, the source of the cola nuts used in soft drinks.

The fruit is hand-harvested and split open with a machete. The seeds are removed and fermented to remove the sweet white pulp and to improve their flavor. After fermentation, the beans are dried, sacked, and shipped.

Propagation is commonly from seed. Recently improved "hybrid seed" has been produced by crossing selected clones. Clones may be propagated from cuttings or bud grafts. Unless a self-fertile clone is chosen, a number of clones must be planted to ensure pollination.

Cacao is a true tropical crop. Production is concentrated in lowlands in a zone that extends no more than 10° on either side of the equator. Because of its sensitivity to drought, cacao requires an even distribution of rainfall, and it is grown under the shade of a tree canopy.

Tea

Tea, an infusion produced from the dried tip leaves of *Thea sinensis* (Theaceae), is the most widely used beverage in the world. Of Chinese origin, it was introduced to Europe by the Dutch in the seventeenth century. In England, tea replaced coffee in popularity, a preference still maintained in British Commonwealth countries.

A

B

FIGURE 21-14. (*A*) A cacao tree bearing fruits (commonly called "pods") on the trunk and large branches. (*B*) The sweet pulp surrounding each seed can be saved and made into a jelly, but it is usually removed by fermentation. [Photographs courtesy USDA.]

The glossy leaves of the tea plant contain numerous glands that produce an essential oil. The terminal leaf or two are plucked and either dried immediately to produce green tea or processed by fermentation to produce black tea, the main tea of commerce in the West.

Tea can be cultivated in subtropical areas or in mountainous regions of the tropics with even distribution of rainfall. The tea plant is an evergreen shrub propagated from seeds, with some clonal propagation by cutting or bud grafting. Plants are pruned to establish a low, flat top, to keep plants within bounds, and to renew old plantings. Leguminous shade trees may be used at first, but they are removed as the plantings age. Most harvesting is done by hand, although mechanical harvesters are now extensively used in the Soviet Union.

SPICE AND DRUG CROPS

> [Joseph and his brothers] . . . saw a caravan of Ishmaelites coming from Gilead with their camels bearing gum, balm, and myrrh, on their way to carry it down into Egypt.
>
> GENESIS 37:25

Plants with strong aromas and flavors, spices very early attracted human attention, probably at first for magical rites, spells, purification ceremonies, and embalming, and then also for fragrances and perfumes, flavorings and condiments, food preservatives, curatives, aphrodisiacs, and poisons. The distinctive odors of many of these plants (essences) are due in large part to **essential oils,** benzene or terpene derivatives that are oily, vaporizing, and flammable. Although often associated with the flower, these oils may be found in all plant parts, usually as secretions from specialized glands. Because of these oils, the plants have been widely used with a variety of effects—some real, some fanciful. For example, clove is valued for flavoring and is often tried as a remedy for stomach complaints.

Although spices and medicinal plants overlap, those whose primary value is therapeutic will be classified here as **drug crops.** Many (but not all) the curative powers of drug crops are derived from **alkaloids,** a class of nitrogenous bases that cause profound physiological responses in animal systems (see Table 4-5, page 135). The high value placed on spices and medicinals in the ancient world attracted traders and merchants who traversed well-established routes from sources in the East to the metropolitan centers of the West. The "spice trade" developed the first contacts between East and West.

The traffic in spices centered early in Arabia, a crossroads leading East to the Malabar coast of India by water (via the Red Sea and the Persian Gulf) and in other directions by overland caravan. However, with the rise of the West, control over the spice trade passed from Arabia to Venice and Constantinople, two cities so situated that they were effectively like bottlenecks to trade entering the Mediterranean from the Far East. The prize passed next to Portugal: Although the explorations of Columbus failed to discover a competitive all-water route to the spice islands of the East, Vasco Da Gama in 1497 skirted Africa and found a direct route to Calicut, the Indian center of the land of cinnamon, ginger, and black pepper. By combining their navigational skill and daring with greed and ruthlessness, the Portuguese soon monopolized the spice trade of the East Indies until they were overwhelmed by the Dutch in the seventeenth century. Spices continued to influence the course of empire and history as power in the East Indies was contested by the French and finally (successfully) by the English. Today, the monopoly is long broken and the spices that once lured the great powers no longer have great importance in world trade, even though the mischief created in the era of spice colonialism is with us still in the form of political instability and underdevelopment. Still, spices claim their places the world over as silent partners to cooks and as essential luxuries even to jaded palates.

Today there are more than 200 species of spice and drug plants collected for use in the United States alone. Many of them are slow-growing tropical vines, shrubs, or trees—black pepper, cinnamon, clove, allspice, vanilla, nutmeg, and mace; and opium, quinine, rotenone, and strychnine. Temperate condiment plants, often known as **savory herbs,** include mustard, cayenne and paprika, sage, and caraway; temperate drug plants include belladonna, digitalis, henbane, and stramonium. In the following discussion, only a few of the more important spice and drug crops are reviewed.

Cinnamon is made from the inner bark of the small, bushy cinnamon-laurel tree, *Cinnamomum zeylanicum* (Lauraceae). The bark is stripped from second-year wood, dried, and then peeled of epidermis particles. These are formed into cinnamon **quills** (hollow, tubelike "pipes"), or are ground to yield powdered cinnamon. Oil of cinnamon is prepared from the leaves, fruit, or root. The highest-quality cinnamon is produced in southern Sri Lanka. Camphor is derived from wood of the closely related *C. camphora*.

Cloves are the dried unexpanded flower buds of a small tropical tree, *Eugenia aromatica* (Myrtaceae). The Dutch at one time obtained a complete monopoly in the clove trade and confined production to a single island of the Moluccas, uprooting all other native clove trees. The French, however, introduced the clove into Mauritius in 1770, and from there production spread to Brazil, the West Indies, and Zanzibar. The ripe, bright red flower buds, when sun dried, turn dark brown. The resulting spice has a powerful fragrance and a hot, acrid taste. Oil of cloves is extracted by distillation and consists mainly of eugenol ($C_{10}H_{12}O_2$), the raw material for the synthesis of vanillin.

Nutmeg and mace, two fairly well-known kitchen spices, are both derived from the fruit of the nutmeg tree, *Myristica fragrans* (Myristicaceae). Nutmeg is the seed, and mace is obtained from the aril, the bright red membrane surrounding the seed. "Oil of mace" is expressed or distilled from the seed, rather than from the aril. The nutmeg tree is native to the Moluccas, but it is now cultivated also in both the East and West Indies.

Black pepper, *Piper nigrum* (Piperaceae), should not be confused with chili peppers, *Capsicum* spp. (Solanaceae). Black pepper, native to India, is one of the oldest and most important of all the spices. At one time it was also used as a medicine. The fruits are red when ripe, but they darken when dried to produce black peppercorns. These are ground to produce the familiar spice in pepper shakers. Whole peppercorns are widely used in meat flavoring and preservation (of pepperoni and salami). White pepper is the kernel of the peppercorn, from which the outer skin and pulp (mesocarp) have been removed after soaking in water. The pepper plant is a perennial climbing vine usually grown on trellises. Propagation is typically by cuttings. Production is centered in India and Southeast Asia, with small quantities produced in Brazil. World export ranges between 50,000 and 75,000 tons annually. The United States consumes about one-third of the world supply at the rate of about 100 g (4 ounces) per capita each year.

Peppermint oil, one of the best known and most ancient of the essential oils, is a distillation production of *Mentha piperita* (Figure 21-15), a member of the family Labiatae. Menthol, the chief ingredient, is used as a flavoring in many products—chewing gum, confections, dentifrices, medicines, and tobacco products. The oil of the closely related spearmint, *M. spicata,* is used similarly. The mints are perennial herbs, well adapted to moist areas of temperate climates; oil production requires long days. Peppermint and—to a lesser extent—spearmint are cultivated on a commercial scale on muck soils in the United States (Michigan, Indiana, and the Columbia River basin of Oregon and Washington), as well as in Japan, England, and continental Europe. The plant is propagated from stolons in rows the first year (row mint); in subsequent years, the plants cover the whole field (meadow mint). The crop is mowed by machine, field-dried, and steam-distilled. Oil yields average about 30 pounds per acre. Verticillium wilt, one of the major diseases, has been controlled by genetic resistance obtained via mutation breeding.

Vanilla is produced from the dried, fermented seed pods of a climbing orchid *Vanilla planifolia* (Orchidaceae). The name *vanilla* is Spanish, a diminu-

FIGURE 21-15. Peppermint (*Mentha piperita*) contains an essential oil in its foliage. The oil, which is produced when the plant is subjected to long days, is extracted for flavors and medicinals.

tive of *vaina*, a pod. Vanilla is indigenous to tropical America and, as noted earlier, was used by the Aztecs in compounding the beverage *chocolatl*. Vanilla beans were highly prized and were exacted as tribute. The marvelously fragrant pods yield an essential oil, which can be extracted with alcohol. The active substance vanillin ($C_8H_8O_3$) is now widely synthesized with other materials, but the natural flavor is far superior to the synthetics. The vanilla plant is grown on supports, under shade. Flowers are often hand-pollinated to ensure set, and the fruits may be thinned. The plant requires a hot, moist climate. Most of the world's supply is produced in Madagascar.

Many spices are products of plants already discussed as fruits and vegetables. For example, seeds of celery, *Apium graveolens* (Umbelliferae), are used as a spice (the savory seeds are used whole or ground) or for the production of essential oil. Other species of the Umbelliferae used as spices include anise (*Pimpinella anisum*), caraway (*Carum carvi*), coriander (*Coriandrum sativum*), cumin (*Cuminum cyminum*), fennel (*Foeniculum vulgare*), and parsley (*Petroselinum sativum*). Similarly, many *Citrus* species (lemon, grapefruit, orange, lime) are used for their essential oils. A number of crucifers are grown as spices—horseradish (*Armoracia rusticana*) and mustard (*Brassica nigra*) are the most notable examples.

VEGETABLE CROPS

The "vegetable" food plants of horticultural interest include many species and many families, of which Solanaceae, Cucurbitaceae, Cruciferae, and Leguminosae are particularly prominent. Some of the starchy crops, such as potato, sweetpotato, cassava, and yam, are valuable calorie sources, and rival the cereals as nourishment for people. Many of the edible legumes (peas and beans) are important protein sources. However, many garden vegetables are classed as **protective food,** supplying vitamins and minerals to the diet, and are not major sources of energy or protein.

Although many vegetable crops are merely of local interest, some have assumed tremendous economic importance. Thus, lettuce consumption in the United States in 1985 is almost 10 million heads per day, with an annual farm value of almost $1 billion. Because of the diversity of vegetables (they are much more numerous than fruit crops), the following discussion is by no means complete.

Salad Plants

Salad plants are leafy vegetables usually consumed raw, although this distinction is arbitrary. As a group, they are fast-growing cool-season plants. The major salad crops are species of three families: Compositae—lettuce (*Lactuca sativa*), endive (*Cichorium endivia*), chicory (*C. intybus*), and dandelion (*Taraxacum officinale*) (Figure 21-16); Umbelliferae—celery (*Apium*

FIGURE 21-16. Cultivated dandelion. [Photograph courtesy USDA.]

graveolens), parsley (*Petroselinum sativum*), and chervil (*Anthriscus cerefolium*); and Criciferae—watercress (*Nasturtium officinale*) and garden cress (*Lepidium sativum*). A number of other crucifers that may be considered as salad crops, such as Chinese cabbage, are discussed briefly with the cole crops in a subsequent section.

Lettuce is the world's most important salad crop. It appears to have been domesticated within historic times, probably derived from the wild *Lactuca serriola* native to Asia Minor. The numerous cultivars are usually classified into five types: **crisphead,** firm heads with brittle texture; **butterhead,** soft heads with pliable leaves and delicate flavor; **cos (romaine),** upright, loaf-shaped heads with long narrow leaves; **looseleaf (bunching),** nonheading leaf lettuce; and **stem,** with enlarged edible seedstalks (Figure 21-17). Lettuce is an annual well adapted to cool climates; temperatures that are too high result in seedstalk elongation (bolting). Most of the United States lettuce crop is produced in California, and production is confined to crisphead types, which are typically vacuum-cooled and shipped in ice or in refrigerated transport. The butterhead, cos, and looseleaf types ship poorly and are usually grown only for local markets. Breeding programs have produced heat-tolerant strains with resistance to downy mildew, a disease that can cause severe crop losses. Mosaic, a virus disease, also a severe problem in lettuce-producing areas, can be avoided by using only virus-free seed.

Although celery (Figure 21-18) is native to marsh lands from Sweden to northern Africa and from southern Asia to the Caucasus, it is most popular in the United States and Canada, where it is the second most important salad crop after lettuce. Celery is a moisture-loving, long-seasoned annual usually grown under cool conditions; however, exposure of transplants to temperatures as low as 35 to 50°F (2 to 10°C) increases bolting in some strains. In Europe, celery is commonly blanched by covering the leaf stalks with soil or paper for about 3 weeks before harvest to produce white tender stalks.

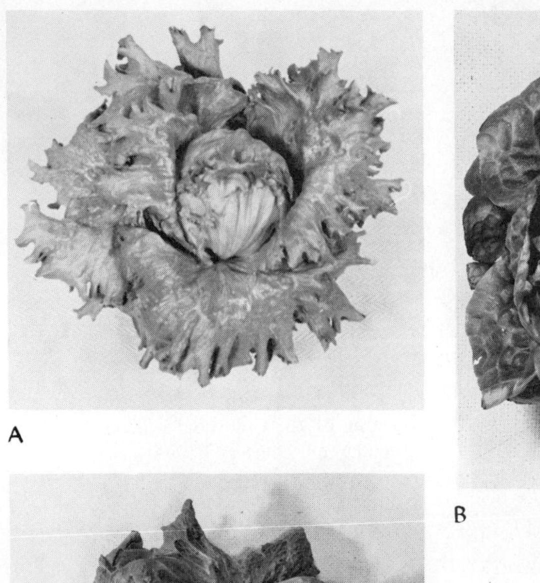

FIGURE 21-17. Three lettuce types: (*A*) 'Merit,' a widely used crisphead cultivar; (*B*) 'Big Boston,' the most extensively used butterhead cultivar; and (*C*) 'Vanguard,' a distinctive winter cultivar used in desert areas that has genes from a wild lettuce, *Lactuca virosa*. [Photographs courtesy T. W. Whitaker, USDA.]

Solanaceous Fruits

The family of which the potato is a member, the Solanaceae, contains three important members grown for their fruit: the tomato, eggplant, and capsicum pepper.

The tomato (*Lycopersicon esculentum*), a native of South America, was introduced into Europe before 1544 by the Spanish conquistadores. Superstitions concerning its presumed poisonous qualities discouraged its use in many European countries, with the exception of the Mediterranean regions, until the late eighteenth century. Tomatoes did not become popular in the United States

FIGURE 21-18. Celery before and after trimming for shipment. [Photograph courtesy USDA.]

until the latter part of the nineteenth century, although they had been reintroduced from Europe more than 100 years earlier.

Although it is frost-susceptible, the tomato is one of the most widely cultivated plants: It is grown from the equator to as far north as 65° latitude (Fort Norman, Canada). It is also grown (largely under glass) in northern Europe. In the United States, it is the most important greenhouse vegetable and ranks as the most popular home-garden food plant. The leading countries in tomato production are the United States, China, Italy, Turkey, Egypt, and Spain; the United States accounts for over one-seventh of the world's production. Florida and California are the leading states for fresh market production. Tomatoes destined for fresh consumption are picked when pink and ripened naturally off the plant. They can also be picked green, stored, and artificially ripened. They are then referred to as "green wraps." California is far and away the leading tomato-processing state. Tomatoes are canned whole and also as juice, puree, sauce, catsup, and paste (Figure 21-19).

Although the tomato is a perennial plant, it is treated as an annual. The usual method of planting is by transplants, but direct seeding is increasing in popularity because of the effectiveness of chemical weed control. The mechanical harvesting of tomatoes is well developed and has resulted in great changes in the industry.

The eggplant (*Solanum melongena*) is assumed to have been domesticated in tropical India and secondarily in China. It is a staple vegetable in many oriental countries (in Japan, much of the crop is pickled). Although featured prominently in the various Mediterranean cuisines, the eggplant is only in minor demand in the United States. The round, oblong, or pear-shaped fruits vary greatly in size; their color ranges from purplish black, through light violet

A

B

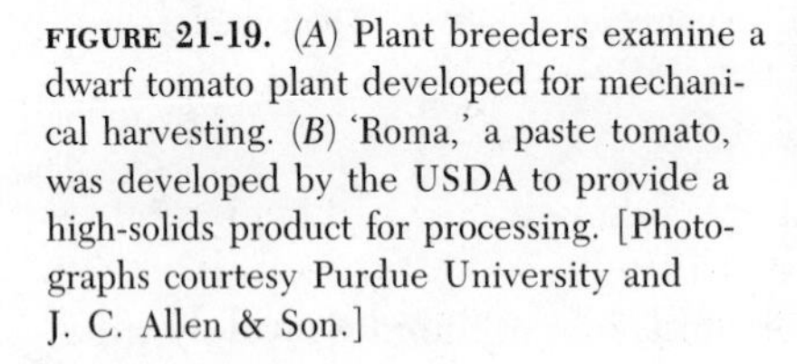

FIGURE 21-19. (*A*) Plant breeders examine a dwarf tomato plant developed for mechanical harvesting. (*B*) 'Roma,' a paste tomato, was developed by the USDA to provide a high-solids product for processing. [Photographs courtesy Purdue University and J. C. Allen & Son.]

and white, to green. The plant is very sensitive to frost. The crop is usually started from transplants.

The capsicum peppers are native to tropical America, where they have long been a much-used condiment and food. "Pepper" is actually a misnomer deriving from a pungency in some species that is reminiscent of the Old World black pepper, *Piper nigrum* (Piperaceae). To avoid confusion with this totally unrelated plant, members of the genus *Capsicum* have been given the names capsicum or chili peppers. The pungent chilis were brought to Europe on Columbus's first voyage; spread first by the Portuguese, they have since gained worldwide popularity. They have become important spices in Asia and vital ingredients of such delicacies as the powerful Korean staple kimchee (a fermented cabbage relish) and Indian curry (made from roasted dried chilis mixed with turmeric, coriander, cumin, and other spices).

The capsicum peppers are an extremely variable group, and their nomenclature is confusing. Most cultivars are contained in two species: *Capsicum annuum* (annual plant, fruit borne singly) and *C. frutescens* (perennial, fruit borne in groups). *C. annuum* is unknown in the wild state. Horticulturally, capsicum peppers may be divided between "sweet" and "hot" peppers. The **sweet pepper,** also known a bell pepper and—most confusingly—as "mango" in the southern United States, is relatively nonpungent, with thick flesh; the fist-sized green fruit ripens to red or yellow. It is produced mainly in the United States. Sweet peppers are eaten either raw, cooked (often stuffed with meat), or pickled, and are a rich source of vitamin C. There are a number of cultivars and types of sweet peppers. The red "pimiento" is also used for stuffing olives or combined with processed meat or cheese. Other sweet types are dried and ground to produce the red condiment "paprika," largely in Spain and Hungary, where it is used as one of the ingredients of goulash. The **hot** or **spicy peppers,** when dried, are also known as "red" or "cayenne" peppers. The pungency is due to the volatile phenol capsicin ($C_{18}H_{27}NO_3$). Hot sauces, such as Tabasco, are made by pickling the pungent pulp in vinegar or brine. The largest exporter of pungent capsicum peppers is India, followed by Thailand.

Edible Legumes

The edible **legumes,** or **pulse crops,** which include many genera of the Leguminosae, have long been used as food (Figure 21-20). There have been other uses for legumes: The expression "blackball," for example, comes from the ancient Greek and Roman practice of using beans for voting, a white bean signifying acceptance and a black bean for rejection. And today, soybeans, although edible as a vegetable, are grown in the United States principally as an oil and feed crop. Edible legumes are good sources of protein, although they are deficient in the sulfur-containing amino acids, cystine and methionine, and they are especially well-known in Asia and South America, where they constitute a key part of the diet. In the United States, the most important of the edible legumes are the common bean and pea.

The common bean (*Phaseolus vulgaris*) is a warm-season annual quite sensitive to frost. It and the lima or butter bean (*P. lunatus*) are both native to South America, although they are now extensively grown throughout the world. The bean is edible in various stages of its growth, but distinct cultivars are now grown to provide each type of food. The edible-podded types are called snap or stringless beans (formerly string beans!) and may be either the climbing (pole beans) or nonclimbing (bush beans). Beans are also harvested when mature but still green (shell beans) or in the dry-seeded stage. Dried beans are a substantially larger crop than snap beans and account for about 10 times the acreage. Among the various types of dried beans grown in the United States are the navy, red kidney, pinto, great northern, marrow, and yellow-eye beans. The mung bean (*P. aureus*) is a small-seeded Asiatic species commonly consumed after germination, the bean sprout being ubiquitous in Chinese cookery.

In contrast, the garden pea (*Pisum sativum*), also native to Asia, is a cool-season crop. It is harvested for use in either the mature green stage or the dry-seeded stage, although some cultivars are edible in the podded stage (sugar peas, or snow peas). Almost the entire commercial green-pea crop is now pro-

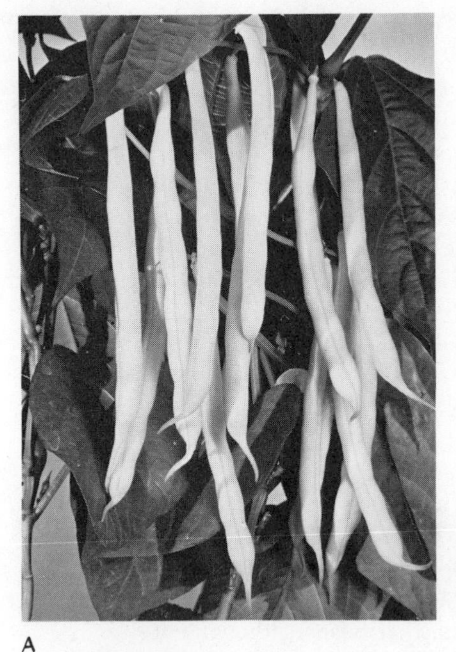

A

B

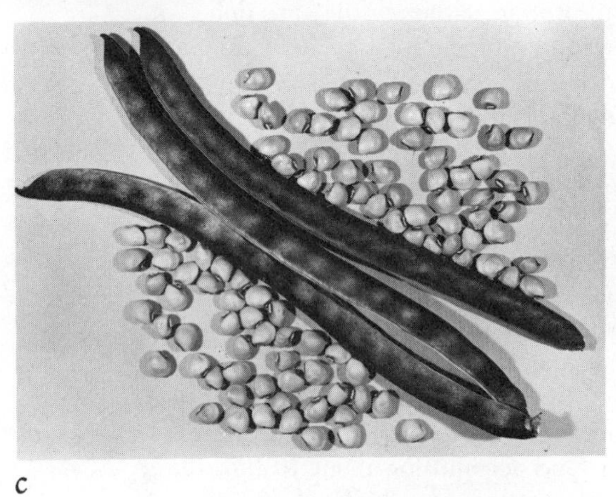

C

D

FIGURE 21-20. Some examples of edible legumes. (*A*) 'Goldcrop' wax bean (*Phaseolus vulgaris*). (*B*) 'Oregon Sugarpod' pea (*Pisum sativum*), which has an edible pod. (*C*) 'Worthmore' southernpea (*Vigna unguiculata*). (*D*) Pigeon pea (*Cajanus cajan*), a tropical bush legume high in protein. [Photographs courtesy All-American Selections (*A*), J. R. Baggett (*B*), J. Dan Gay (*C*), Julia F. Morton (*D*).]

duced for processing. At present, these peas are one of the most popular frozen vegetables; however the bulk of the crop is still canned. The major areas producing green peas are the northern states of Wisconsin, Washington, Minnesota, and Oregon. Dried peas, almost as large a crop as green peas, are produced mainly in Washington and Oregon.

The peanut (*Arachis hypogaea*), indigenous to Brazil, is well known in the United States in the form of peanut butter, salted table nuts, and, when roasted in the shell, as standard fare at sporting events. However, peanuts are used chiefly as an oil crop, and the by-product, the protein-rich presscake, is used as a livestock feed. Flowers are formed above ground, but after pollination they are gradually pushed beneath the ground where the shell of the peanut matures; hence the English name "groundnut."

Other edible legumes include the broad bean (*Vicia faba*), an annual with large thick pods. Native to the Old World, the broad bean is widely cultivated in Europe and North Africa. The chick-pea, also known as the gram pea or garbanzo (*Cicer arietinum*), is well adapted to arid regions and is an important crop in India, North Africa, Spain, and Portugal. The black-eyed pea, also known as southernpea or cowpea (*Vigna unguiculata*), a useful fodder plant, is grown chiefly in Asia. A great number of edible legumes are popular in Asia, including lentils (*Lens esculenta*), the Congo bean or pigeon pea (*Cajanus cajan*), and the hyacinth bean or lablab (*Dolichos lablab*).

Starchy Root Vegetables

The major starchy vegetable crops of the world are underground crops grown for their tubers, corms, or roots. They include potato (tuber), cassava (root), sweetpotato (root), yam (tuber), and taro (corm). Fleshy roots and underground stems are less compact than cereal grains and legume seeds; consequently, they are relatively more difficult to store and transport. Root crops are high in carbohydrates and are therefore prized as energy foods. However, the low concentration of protein makes them inadequate for sustenance if not supplemented with protein-rich foods.

Potato

The potato, *Solanum tuberosum* (Solanaceae), although a New World crop, now has its world center of production in Europe where it is a staple for livestock as well as for humans. It is now one of the outstanding food crops of the world, especially adapted to northern Europe, the northern United States, and southern Canada. However, because of their bulk, potatoes have never become significant in international trade. They are still cultivated as a mainstay in the Andes, where small colorful tubers are made into a flourlike product called chuño by alternately freezing and trampling them, then thawing and drying them.

A temperature of 70°F (21°C) is optimum for tuber formation. At higher temperatures, the increased respiration rate reduces the amount of stored carbohydrates and consequently reduces yield and quality. Tuberization is generally accelerated by short days, although some cultivars produce tubers over a wide range of photoperiods. Although adequate rainfall is essential for potatoes, excess rain has been feared because it can result in severe infestations of the late-blight disease (caused by the fungus *Phytophthora infestans*), the same disease responsible for the potato famines in Ireland in the 1840s.

Potatoes store well at 39°F (4°C). Above 50°F (10°C), sprouting is stimulated, although this can be inhibited by chemical treatment, such as with maleic hydrazide. To avoid an undesirable brown discoloration, tubers are held for a few days or weeks at 60 to 70°F (16 to 21°C) before processing. During this **reconditioning** treatment, the concentration of reducing sugars (glucose and fructose) is decreased by respiration and by conversion to starch.

In the United States, potato production is concentrated in the North—Idaho and Maine are the largest producers—although early-maturing cultivars are widely grown as fall and spring crops in southern latitudes. Even though total acreage has sharply decreased, average yields have almost tripled between 1840 and 1980. As a result, the United States production has stabilized, whereas consumption per capita showed a steady decrease from the 1900s until recently, when this trend was reversed. The recent increase is due in large part to the use of processed potato products, mainly chips, frozen french fries, and other convenience products.

Cassava

Cassava, *Manihot esculenta* (Euphorbiaceae), another New World crop, is grown extensively in tropical America and also in tropical Africa and Indonesia. The plant adapts to poor soils and casual cultivation and has become a staple in many of the poor and less developed parts of the tropics. In the temperate areas of the world, cassava is known through a dessert pudding made from **tapioca,** heated cassava starch that agglutinates into small round pellets. Tapioca is also used for biscuits and confections. In Brazil, cassava is ground into a bland-tasting meal called **farinha** that has a coarse sawdustlike appearance. In addition to food uses, cassava starch is utilized in sizings, adhesives, and the production of alcohol by fermentation.

The cassava is a short-lived shrub whose tuberous roots resemble the sweetpotato. The plant is deciduous and sheds its leaves during prolonged periods of drought. The many cultivars are roughly divided into sweet and bitter types. **Sweet cassavas** have a low percentage of hydrocyanic glucoside that is generally confined to the outer layers of the tuberous root. Sweet cassavas generally have light green leaves and stems. **Bitter cassavas** have a high percentage of hydrocyanic glucoside distributed throughout the tuber. They generally have dark green or reddish leaves and stems. It is poisonous unless cooked, and its use is confined largely to industrial manufacture of starch, alcohol, and acetone.

Sweetpotato

The sweetpotato, *Ipomoea batatas* (Convolvulaceae), is known only in cultivation (Figure 2-4). It is one of the few New World crops that reached Polynesia before Columbus's "discovery" of America.[1] Its presence in America and Polynesia suggests pre-Columbian contacts, a theory popularized by the adventures described by Thor Heyerdahl in *Kon-Tiki*, but the precise route of the wandering sweetpotato remains a mystery. Easter Island, where it was found by the first European settlers, may have been a possible transfer point in the trans-Pacific migration. However, some botanists do not favor this "Pacific regatta" theory and suggest that sweetpotato seed capsules, possibly attached to a floating log, may have been carried by ocean currents. Seeds will germinate after immersion in ocean water.

Sweetpotatoes are now grown throughout the tropics, principally in Asia, with 55 percent of world production grown in China. The edible portion of this vinelike trailing perennial is the tuberlike root. The crop is typically produced as an annual from stem cuttings. Cultivars are of two fleshy types: dry and mealy or moist and soft. (The moist, soft types are also known as "yams" in the southern United States.) Although yellow- and orange-fleshed cultivars are rich sources of vitamin A, many tropical cultivars have white flesh.

Yam

Although moist-fleshed cultivars of sweetpotatoes (which are dicots) are sometimes referred to as yams, the true yams are monocots of the genus *Dioscorea* (Dioscoreaceae). There are hundreds of species, many still not fully described. The most commonly cultivated species is *D. alata* (water yam or white yam), native to southeast Asia but now grown throughout the tropics with significant cultivation in West Africa, Asia, and the Caribbean (Figure 21-21). *D. trifida* (yempi yam) is a New World species. Yams are a prestigious food in the Pacific Islands and individuals vie to provide large tubers for traditional feasts.

The yam is a dioecious perennial vine often grown on a trellis. The top of the tuber, where the vine is attached, is planted. The crop takes 8 to 10 months to mature. Tubers range from the size of small potatoes to more than 100 pounds each. Wild yams offer a standby food in times of food scarcity. They are also gathered as a source of steroidal sapogenins for the production of cortisone.

Taro

Taros, species of the family Araceae, produce edible corms from soft-stemmed plants that grow in shady marsh areas (Figure 21-22). All cultivated aroids except *Xanthosoma* probably originated in Southeast Asia and Indonesia. Taros

[1]Other crops known in both the New and Old Worlds include the bottle gourd (*Lagenaria siceraria*), coconut (*Cocos nucifera*), and one species of cotton (*Gossypium herbaceum*).

FIGURE 21-21. Typical tuber of 'Morado' yam (*Dioscorea alata*). [Photograph courtesy F. W. Martin.]

are grown commonly in the Pacific Islands as a source of food. Sweet taro, *Colocasia esculenta*, is the best-tasting species, and its leaves are also edible. The famous Hawaiian poi, a purplish, sticky, highly digestible food, is a fermented product made from crushed corms of sweet taro. In tropical America, the plant is known as dasheen (perhaps from the French *de Chine*, from China); in Africa, as cocoyam. The wild taro, *Alocasia macrorrhiza*, is the most common species and produces corms above ground. These are used for pig feed because the high content of calcium oxalate crystals renders them virtually unfit

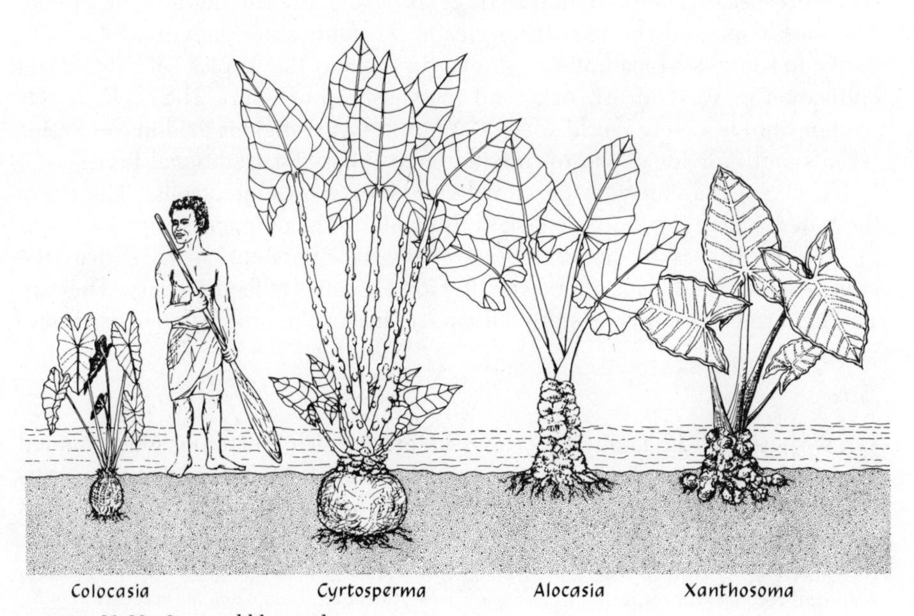

FIGURE 21-22. Some edible aroids.

for human consumption. Giant swamp taro, *Crytosperma chamissonis*, the taro of Micronesia, is the largest of the taros. After 10 to 15 years of growth, its corms each weigh between 100 and 200 pounds. *Xanthosoma sagittifolium* is native to the New World and is called yautias or American taro. Under the name "new cocoyam," *Xanthosoma* is spreading into Africa and Asia, where it is relatively free of pests. In New Guinea, it is confusingly called taro cong, or Chinese taro.

Other Root Vegetables

In addition to those just described, many other root crops are grown as vegetables throughout the world. They include the table beet, *Beta vulgaris* (Chenopodiaceae), a biennial plant whose swollen red root is eaten fresh, canned, or pickled, and is the main ingredient of borscht, a delicious soup eaten hot or cold with sour cream. The sugar beet, an important source of sugar in temperate climates, is considered to be the same species. Carrot, *Daucus carota* (Figure 21-23), and parsnip, *Pastinaca sativa*, are well-known root crops of the family Umbelliferae. Arrachacha (*Arracacia xanthorrhiza*), another interesting carrotlike vegetable, is just one of the little-known root crops with

A

B

FIGURE 21-23. (*A*) 'Imperator 58,' a popular carrot cultivar for the fresh market. (*B*) A carrot field at harvest time. [Photographs courtesy of T. W. Whitaker, USDA.]

local importance in South America. Jerusalem artichoke, *Helianthus tuberosus* (Compositae), is of interest because the reserve carbohydrate in the potatolike rhizome is inulin, a polymer of fructose. Inulin, but not glucose-containing carbohydrates, can be metabolized by diabetics. Salsify, *Tragopogon possifolius* (Compositae), has fleshy roots resembling parsnip. In the family Cruciferae, radish (*Raphanus sativus*), rutabaga (*Brassica napobrassica*), and turnip (*B. rapa*) are often classed as root crops, along with horseradish (*Armoracia rusticana*), whose powerfully pungent roots are used as a condiment.

The Onion and Its Pungent Allies

The onion and the other pungent species of the genus *Allium* (leek, garlic, and chive) are among the most ancient of cultivated vegetables. The strong odors and flavors are due to sulfur compounds, mostly *n*-propyl disulfide in onion and methylallyl disulfide in garlic. These species are all biennial bulb-forming plants of the lily family (Liliaceae). The onion, the most important of this group, is grown in all temperate climates (Figure 21-24). The chief onion-producing countries are China, India, and the United States. In the United States, the crop is widely grown, with California, New York, Texas, and Michigan being the chief producing states. Onions require cool temperatures during early growth and long days and high temperatures for bulb formation. Green onions, or scallions (the nonbulbous stage of onion), are a significant winter crop in the southern United States.

In the northern states onions are commonly planted from seed in the spring; in the southern states plantings are made in the fall or winter from seed, seedlings, or onion sets. The production of onion sets, which are small bulbs produced by crowding, is a specialized industry in Michigan, Illinois, and Wisconsin. Fall onions are stored for sale in the winter. A relatively small percentage of the crop is processed as flavorings for use in other products.

FIGURE 21-24. Onions are among the most ancient of vegetable crops. [Photograph by J. C. Allen & Son.]

Cole Crops

The leading vegetables of the family Cruciferae include cabbage, cauliflower, broccoli, Brussel sprouts, Chinese cabbage, kale, collards, and mustard. These are all species of the genus *Brassica,* and are known collectively as **cole crops** (Figure 21-25). The radish (*Raphanus sativus*) is also a crucifer, but it belongs to a different genus. The cole crops are cool-season, hardy plants. Except for cauliflower and broccoli, which are annuals, they are biennials, requiring a cold treatment to flower.

The cole crops are grown widely in Europe, where they constitute the major "green" vegetables. In the United States, cabbage is the most important

FIGURE 21-25. The All-America hybrid crucifers for 1969: (*A*) 'Tokyo Cross' turnip; (*B*) 'Stonehead' cabbage; (*C*) 'Green Comet' broccoli; (*D*) 'Snow King' cauliflower.

cole crop, and, in the past, was a common winter vegetable, since it is easily stored. The production of cabbage has sharply declined, however, reflecting a general consumer shift to other vegetables. About 10 percent of the present acreage is grown for the production of sauerkraut. Broccoli, although a minor crop compared with cabbage, has increased in demand and has become very popular as a frozen vegetable.

The cole crops are usually set out as transplants. In the North, they are planted in the spring; in the warmer climate of the South, they are planted in the fall or winter. Insect control is particularly essential. In general, soils with a pH greater than 7 are used to control club root, a serious root disease caused by *Plasmodiophora brassicae*. However, resistant cultivars are now commercially available. Hybrid crucifers are a recent innovation and have made possible the production of uniform plantings for mechanical harvesting.

Vine Crops

Crops of the curcurbit family (Cucurbitaceae) are known as **vine crops** (Figure 21-26). They include the cucumber, muskmelon, watermelon, pumpkin, squash, and the chayote. All are warm-season crops and very susceptible to cold injury.

The cucumber, *Cucumis sativus*, probably native to Asia, has been in cultivation for thousands of years. In Europe it is a common greenhouse crop: The extremely long fruits, grown in the absence of pollinating insects, develop without seeds. In the United States, most of the cucumber acreage is devoted to the production of pickling cucumbers. The crop is widely planted, but Michigan and Wisconsin lead the other states in production. Very successful mechanical harvesting machines have been developed.

Muskmelons, *Cucumis melo*, native to Iran, are a relatively recent crop in the United States, where they are often erroneously called cantaloupes. (True cantaloupes are hard-shelled melons grown only rarely in the United States.) California, Texas, and Arizona are the chief producing states. There are many types of melons in addition to the netted, musky type, with the leading one the late-ripening winter melon, of which the most common cultivar is the 'Honey Dew,' a white-skinned, large-fruited type.

Watermelon, *Citrullus lanatus*, is native to Africa, but has achieved its greatest popularity in the United States. The bulk of the planting is in the southern states, although there are important areas as far north as Iowa. The types range from large 30-pound fruits to small round "icebox types" weighing between 5 and 10 pounds. There are also yellow-fleshed types, but the red-fleshed kind is the most popular. The seedless watermelon, a triploid fruit first created in Japan in the 1940s from crossing tetraploids and diploids, has found its greatest success in Taiwan, but it is still just a novelty in the United States.

Pumpkin and squash are native to the New World. The distinction between

A

B

C

FIGURE 21-26. Vine crops: (*A*) various patterns and shapes of the watermelon; (*B*) hybrid cucumbers trained on a snow fence; (*C*) winter squash.

the two is a loose one. Some would use the word *pumpkin* only for species of *Cucurbita pepo*, referring to *C. maxima* and *C. moschata* as squash. However, the fast-growing types of *C. pepo* are commonly called summer squash. It seems that the orange, large-fruited, smooth types will continue to be called pumpkins regardless of species. Squash represents a major food crop in South America, ranking next to maize and beans. The flowers, flesh, and seeds may be eaten. The chayote (*Sechium edule*), practically unknown in the United States, is also an important South American plant. It is customarily trained on trellises.

Sweet Corn

Sweet corn, *Zea mays* (Gramineae), a sugary-seeded maize native to South America, has been popular only since the last century. Although field corn is a well-known crop in many parts of the world, the popularity of sweet corn is confined to the United States.

Sweet corn is a warm-season crop, but it is produced mainly in the northern states. This is because of the short period over which quality can be maintained in extremely hot weather. In Wisconsin, Minnesota, and Illinois, about two-thirds of the sweet corn is grown for processing, but sizable acreage for fresh-market production has built up in Florida and Texas.

The transitory nature of quality in sweet corn is due to the rapid conversion of sugar to starch. Experimental combinations of other genes, in addition to the common sugary factor, produce not only a higher sugar content but a less rapid conversion to starch. These types would no doubt extend commercial sweet corn production to areas of hotter climates, but germination in these types is irregular. Two genes, *brittle (bt)* and *shrunken-2 (sh2)*, have demonstrated potential over *sugary (su)* in improving sweetness.

ORNAMENTAL CROPS

Ornamentals are commonly divided on the basis of the industry into **florist crops** (flower and foliage plants) and **landscape crops** (nursery plants, including turf, ground covers, deciduous and evergreen shrubs, and trees). A tremendous number of species may be considered as ornamentals. They include many of our food plants, especially fruit trees, which are also greatly prized for their decorative value. A complete list of species used as ornamentals must therefore be encyclopedic in scope.

This available diversity notwithstanding, the commercial ornamental industry consists of relatively few species. The bulk of the United States cut-flower industry is based on less than a half-dozen flowers: chrysanthemum, rose, carnation, orchid, snapdragon, and gladiolus. One annual, the petunia, accounts for 50 percent of the sales of bedding plants. The turf industry involves relatively few grasses, as was discussed in Chapter 20. Even the most popular deciduous and evergreen trees and shrubs make up a modest list: yew, juniper, boxwood, privet, forsythia, dogwood, magnolia, crabapple, oak, and maple.

Although the bulk of sales is concentrated in a manageable inventory, nevertheless the total number of species in the trade is impressive. Even more astonishing are the countless cultivars and variants of each, with intense competition continually bringing out more. Often trivial variations are given the special status of an appealing name. Double naming is not uncommon. The plethora of names greatly magnifies the available variability. The registers of most ornamentals are endless; the rose fancier is bombarded with scores of new cultivars each year. In contrast, variation in vegetables and fruits is minimized at the retail level. The public is thought to be confused by cultivar names. Therefore, only about five or six names of apples are used, even fewer for oranges; vegetables are almost never sold by their cultivar names in the marketplace.

Food-crop species make up a relatively stable list, and the introduction of new ones is very difficult because of people's conservative food habits. This is not true of ornamentals; apparently people are more innovative with plants they do not have to eat. One may expect that the number of species in the trade will increase as new ornamentals are introduced and domesticated. However, it is still difficult for a particular ornamental crop to become established as something other than a novelty. Recently, anthurium has gained popularity as a cut flower as it has been made more easily obtainable through increasing air shipments from Hawaii. Proteas (Figure 21-27) are other tropical flowers with commercial possibilities. Both anthuriums and proteas have extremely long shelf lives measured in weeks rather than days. With the trend toward suburban living, people show greater awareness and interest in new plants, and the industry is expanding to meet this fresh challenge.

Chrysanthemum

The chrysanthemum, *Chrysanthemum morifolium* (Compositae), long cultivated in Asia, has become the largest cut-flower crop in the United States. It is considered a warm-temperature crop, performing best at about 65°F (18°C). The numerous cultivars are usually classified into two types: standards, or large-flowered types, and Pompons, or sprays. Pompons, small-flowered types, may be globular or daisylike and are used largely as fillers in floral arrangements.

The main stimulus to chrysanthemum production has been the possibility of producing continuous flowering by the manipulation of photoperiod. The

FIGURE 21-27. Ornamental proteas: (*A*) *Banksia bastinia;* (*B*) *Banksia prionotes;* (*C*) *Protea barbigera;* (*D*) *Leucospermum cordifolium* (pincushion protea). [Photographs courtesy D. P. Watson (*A*), and E. Memmler (*B, C, D*).]

plant is a perennial and will bloom for many years, but it is grown commercially as an annual crop. Chrysanthemums are grown under glass in the northern and midwestern states, under plastic in California (the center of production), and under cloth in Florida.

Rose

Roses, members of the genus *Rosa* (Rosaceae), have long been one of the major cut-flower crops. The industry is based largely on the hybrid tea rose, a perpetually blooming, long-stemmed rose whose pedigree incorporates many species. In greenhouse culture, the plants are budded or grafted onto *Rosa manetti*, a vigorous rootstock. If properly managed, the plants bloom continuously, although the intensity of production is affected by temperature and fertility. Production can be manipulated by pruning (known as pinching) and temperature control to take advantage of favorable marketing periods. Producing plants are usually kept from 4 to 7 years, depending on their vigor.

The ideal growing temperature is 60°F (16°C). Commercial rose growing has been restricted, for the most part, to areas north of Lexington, Kentucky, but some establishments are located in the higher elevations of North Carolina. Evaporative cooling is being utilized to reduce daytime temperatures farther south, but, in general, areas that have night temperatures above 65°F (18°C) for extensive periods of the year are not suitable for the production of roses. Most of the rose production is carried out under glass, although a trend is developing toward semipermanent construction utilizing lath and plastic, as in the San Francisco area.

The high capitalization required for rose culture has generally limited production to establishments not smaller than 10,000 square feet. The importance of novelties, as well as the better keeping qualities in the hybrid teas, has made hybridization a significant part of many rose-growing establishments. The most widely grown forcing cultivar, however, is 'Better Times,' originated in 1931 by Hills Brothers Nursery of Richmond, Indiana.

The production of rose plants for garden use is a very specialized operation. Most roses are budded onto *Rosa multiflora* rootstocks to achieve rapid and economical increase to meet market demands. Large-scale production of roses for the garden has tended to center in the southwest (from Texas to southern California), where the nearly year-long growing season makes rapid production possible, but it is also carried on near large consuming areas, such as northern Ohio.

Partly because of widespread promotion, and partly because of the great popular interest in the rose, the marketing of roses is nearly as important a facet of the nursery industry as is the production of other classes of shrubs. Mail-order sales from attractive catalogs that picture the roses in color account for an appreciable part of the market. Enthusiasm for new selections is heightened by an elaborate scheme for testing and selecting "All-America" roses (only a few roses receive the coveted award each year), and by a strong national rose society with headquarters in Columbus, Ohio, that encourages rose clubs and sponsors rose shows throughout the nation. Each year accredited judges compile and issue a rating sheet that ranks all available named cultivars on a numerical performance scale. Thus even the neophyte can determine readily the named roses that have been the most successful through the years, and purchase a 'Peace' or 'Tropicana' on the basis of authorative testing rather than extravagantly worded catalog descriptions.

The production of roses requires skilled hand labor in the budding operation and expensive field maintenance. Recently, techniques have been developed to improve the distribution of roses. In the mail-order handling of bareroot roses, coated paper or plastic bags are now used to retain humidity, in contrast to the former practice of wrapping the roots in damp moss. For mass marketing through chain outlets, the stems of dormant stock are coated with wax to prevent desiccation during prolonged storage and handling, and the root system is "potted" in a moss-filled carton on which the name and a picture of the rose appear. Although much stock held dormant for long periods is not of first quality, this method of penetrating the mass market is now well established in the nursery trade.

Carnation

Carnations, *Dianthus caryophyllus* (Caryophyllaceae), require cool temperatures (48 to 50°F, or 9 to 10°C) and high light intensities for maximum-quality production. The industry in carnations grown for cut flowers is based on perpetual-flowering types, and production proceeds throughout the year. Plantings are established from cuttings and are maintained from 1 to 2 years. Production is greatest in California and Colorado, although a sizable industry exists in Massachusetts.

Although there are many carnation cultivars in production today the 'Red Sim' and 'White Sim' are the most popular (Figure 21-28). These cultivars and their sports are characterized by vigorous growth, heavy production, and long stems. Their chief disadvantage is a tendency toward calyx splitting, a disorder not well understood but believed to be accentuated by large differences between night and day temperatures. Recently, artificial dyeing of carnations has become popular. The dye is taken up from the cut end of the stem and is absorbed into the veins of the petals. This process permits a wide variety of pastel shades to be achieved while only the white-flowered cultivars need be grown.

FIGURE 21-28. The 'White Sim' carnation is a periclinal (hand-in-glove) chimeral sport of the 'Red Sim.' The internal tissues carry genetic factors for red color. The 'White Sim' carnation can be identified by flecks of red tissue in the flower. [Photograph by J. C. Allen & Son.]

Orchid

The culture of orchids (members of the family Orchidaceae) is a specialized greenhouse industry. The main type is the showy *Cattleya* (Figure 21-29) but *Cymbidium* is becoming increasingly popular. Orchids are a warm-season crop with optimum growth at 65 to 70°F (18 to 21°C). Shade is necessary during the summer, but it must be removed to maintain adequate light intensity during the winter. The plants are commonly propagated by divisions. Most orchids are epiphytes, and obtain nutrients from decomposing organic matter. They are grown either in **osmunda fiber** (the chopped roots of a kind of fern) or in shredded white-fir bark. Since repotting, one of the costliest items of culture, is difficult with osmunda, shredded bark is more widely used by commercial growers. The plants bloom once a year. The harvested flowers are usually placed in tubes of water and are relatively long-lasting.

FIGURE 21-29. The *Cattleya* orchid. [Photograph by J. C. Allen & Son.]

Breeding is an important part of orchid culture, for new cultivars may become quite profitable. The time-consuming and exacting techniques involved in growing the plants from seed have encouraged specialized enterprises that grow seedlings on a commission basis. Successful breeding methods include hybridization of diploid and tetraploid types, since triploid orchids have proven to be extremely vigorous. Many new orchids are increased rapidly by tissue culture.

Snapdragon

Snapdragons, *Antirrhinum majus* (Scrophulariaceae), although perennial, are grown as annuals and are planted from seed. Formerly grown only under relatively cool temperatures, selection for cultivars that perform well under hot temperatures has created a year-round program for this crop. In the Midwest, snapdragons appear to fit into a profitable rotation with greenhouse-forcing tomatoes. Snapdragons tolerate low light if temperatures are cool, and they can be grown during the winter when conditions are unfavorable for tomatoes.

At present, practically the entire collection of commercial cultivars consists of F_1 hybrids, which have uniformity and vigor. Artificially created tetraploid cultivars have achieved some success because of their larger flowers.

Bulbous Crops

The "bulb" crops include plants produced from true bulbs, corms, rhizomes, tubers, or tuberous roots. Examples are tulip, hyacinth, narcissus (true bulb), crocus (corms), iris (bulbs or rhizomes), daylily (tuberous roots), and dahlia (tubers). Among bulbous crops are nonhardy bulbs for sale as indoor potted plants and summer outdoor plantings, such as amaryllis (*Hippeastrum*), anemones, various tuberous begonias, *Caladium*, cannas, *Clivia*, dahlias, freesias, gladiolus, spiderlilies, montbretias, ranunculus, *Tigrida*, tuberoses, *Zephyranthes*, and others. Hardy bulbs, those left in the soil through the winter, include various crocuses (*Colchicum* and *Crocus*), fritillaries, snowdrops, hyacinths, lilies, daffodils, *Lycoris*, *Scilla*, tulips, irises, and others (Figure 21-30).

Many of these plants are of Old World origin, introduced into horticulture long ago and subjected to selection and crossing through the years to yield many modern cultivars. One of the most popular is the tulip, *Tulipa* (Liliaceae). Various botanical species are grown in gardens; but tulips are especially prized in select forms of the garden tulip (which arose from crosses between thousands of cultivars representing several species). Garden tulips are roughly grouped as early tulips, breeder's tulips, cottage tulips, Darwin tulips, lily-flowered tulips, triumph tulips, Mendel tulips, parrot tulips, and so on. The garden tulips seem to have been developed first in Turkey and then were spread throughout Europe and were adopted enthusiastically by the Dutch. Holland has been the center of tulip breeding ever since the eighteenth century, when interest in the tulip was so intense that single bulbs of a select type were sometimes valued at thousands of dollars. The collapse of that "tulipomania" left economic scars for many decades. Holland remains the chief source of tulip bulbs planted in the

FIGURE 21-30. Some hardy bulb crops: (*A*) tulip; (*B*) daffodil; (*C*) hyacinth; (*D*) bulbous iris. [Photographs courtesy Netherlands Flower Bulb Institute.]

United States; over 150 million are imported annually. Holland has also specialized in the production of related bulbs in the lily family, and provides narcissus, crocus, and others to the United States market annually. The Dutch finance extensive promotion of their bulbs in the United States to support their market. Years of meticulous growing are required to yield a commercial tulip bulb from seed. Bulbs sent to market meet specifications as to size and quality, which assure at least one year's bloom even if the bulb (having been subjected to cold induction) is supplied nothing more than warmth and moisture. The inflorescence is already initiated, and the necessary food is stored in the bulb. Under less favorable maintenance than prevails in Holland, a subsequent year's bloom may be smaller and less reliable. For featured display, such as at botanical gardens, bulbs are discarded after one year's performance although homeowners usually derive several years of satisfaction from a single bulb planting if maturation of the tulip foliage is allowed after flowering.

Gladiolus species (Iridaceae) have practically disappeared as a greenhouse forcing crop. Now almost all are produced in outdoor culture, mainly in Florida, California, and North Carolina. A great variety of colors and types exists (Figure 21-31). Owing to the nature of the plant, bulb and flower production are part of the same business.

Bulbs generally grow best when planted fairly deeply (several times the vertical dimension of the bulb itself) in friable, loamy soil well supplied with

FIGURE 21-31. All-America gladiolus selections: (*left to right*) 'Thunderbird,' 'Horizon,' and 'Ben-Hur.'

phosphatic fertilizer. Autumn, a few weeks before freezing weather, is the recommended planting time for tulips, hyacinths, daffodils, and crocus. The bulbs have a chance to root and become established, and will flower in spring as soon as the soil warms. After the foliage has died back it is often recommended that the bulbs be dug up and reset, a practice that is of some value for rearranging the planting pattern and determining which bulbs are in good condition if nothing else. Some bulbs, such as daffodils and hyacinths, proliferate; digging them up provides additional stock.

Selected References

GENERAL

Berrie, A. M. M., 1977. *An Introduction to the Botany of the Major Crop Plants.* London: Heyden. (A botanical description of the major crop families.)

Cobley, L. S., 1977. *An Introduction to the Botany of Tropical Crops,* 2d ed. Revised by W. M. Steele. London: Longman. (A survey of the botanical features of plants grown in the tropics.)

Harrison, S. G., G. B. Masefield, and M. Wallis, 1969. *The Oxford Book of Food Plants.* London: Oxford University Press. (A description of important food plants with remarkably beautiful illustrations by B. E. Nicholson.)

FAO Production Yearbook. Rome: Food and Agriculture Organization. (An invaluable compilation on world food production; published yearly.)

Martin, F. W. (ed.), 1984. *Handbook of Tropical Food Crops.* Boca Raton, Fla.: CRC Press. (A useful guide to edible plant species of the tropics including many minor crops.)

Purseglove, J. W., 1968, 1972. *Tropical Crops: Dicotyledons* (2 vols., 1968); *Tropical Crops: Monocotyledons* (2 vols., 1972). New York: Wiley. (The complete botany and culture of tropical plants; an invaluable compilation.)

Schery, R. W., 1972. *Plants for Man,* 2d ed. Englewood Cliffs, N.J.: Prentice-Hall. (A very comprehensive text on economic botany.)

FRUIT, NUT, AND BEVERAGE CROPS

Chandler, W. H., 1950. *Deciduous Orchards* and *Evergreen Orchards.* Philadelphia: Lea & Febiger. (These classics are companion volumes on temperate and subtropical fruit.)

Collins, J. L., 1960. *The Pineapple: Botany, Cultivation, and Utilization.* New York: Wiley. (An excellent treatment.)

Condit, I. J., 1969. *Ficus: The Exotic Species.* Berkeley: University of California Press. (A thorough technical review of the fig and its relatives.)

Darrow, G. M., et al., 1966. *The Strawberry: History, Breeding and Physiology.* New

York: Holt, Rinehart and Winston. (A readable review covering the history, breeding, and culture of this small fruit.)

Dawson, V. H. W., and A. Aten, 1962. *Dates—Handling, Processing and Packing.* Rome: Food and Agriculture Organization. (The United Nations' overall look at the date industry.)

Ecke, P., and N. F. Childers (eds.), 1966. *Blueberry Culture.* New Brunswick, N.J.: Horticultural Publications. (A broad compilation of information on this special crop, including its botany, breeding, and growing.)

Eden, T., 1976. *Tea,* 3d ed. London: Longman. (Discussion of history, botany, planting, care, harvest, and processing of this important beverage crop, with references, by the Director of the Tea Research Institute in Sri Lanka and Africa.)

Haarer, A. E., 1962. *Modern Coffee Production.* London: Leonard Hill. (On the history, genetics, physiology, and practical growing of coffee by an English expert with wide practical experience in East Africa.)

Reuther, W., L. D. Batchelor, and H. J. Webber (eds.), 1967, 1968, 1972, 1978. *The Citrus Industry,* rev. ed. Berkeley: University of California Press. (A monumental work on citrus in four volumes. Vol. I covers history, world distribution, botany, and varieties; Vol. II, anatomy, physiology, mineral nutrition, seed reproduction, genetics, and growth regulators; Vol. III, propagation, planting, weed control, soils, fertilizing, pruning, irrigating, climate, and frost control; and Vol. IV, biology and control of pests and diseases.)

Shoemaker, J. S., 1978. *Small-fruit Culture,* 5th ed. Philadelphia: Blakiston. (A standard reference through the years.)

Simmonds, N. W., 1959. *Bananas.* (Tropical Agriculture Series). London: Longman. (An authoritative monograph.)

Singh, L. B., 1960. *The Mango: Botany, Cultivation, and Utilization.* New York: Wiley. (A treatise on one of the most popular and luscious of tropical fruits.)

Tellez, M. R., 1978. *Modern Olive-growing.* New York: Unipub. (The techniques of modern olive production.)

Weaver, R. J., 1976. *Grape Growing.* New York: Wiley. (Emphasizes cultural factors of California grape production.)

Wood, G. A. R., 1975. *Cocoa,* 3d ed. New York: Longman. (A thorough review of cacao worldwide, with special attention to its growth in different regions.)

Woodroof, J. G., 1979. *Coconuts: Production, Processing, Products,* 2d ed. Westport, Conn.: AVI. (The story of coconuts, with emphasis on processing.)

VEGETABLE AND SPICE CROPS

Change, S. T., and W. A. Hayes (eds.), 1978. *The Biology and Cultivation of Edible Mushrooms.* New York: Academic Press. (A collection of articles reviewing the science of mushroom culture.)

Cousey, D. G., 1977. *Yams,* 2d ed. New York: Humanities Press. (A review of *Dioscorea.*)

Edmond, J. B., and G. R. Ammerman, 1971. *Sweet Potato: Production, Processing, Marketing.* Westport, Conn.: AVI. (A comprehensive coverage.)

Heiser, C. B., Jr., 1969. *Nightshades: The Paradoxical Plants.* San Francisco: W. H. Freeman and Company. (A charming book on solanaceous plants.)

Jones, H. A., and K. L. Mann, 1963. *Onions and Their Allies: Botany, Cultivation, and Utilization.* New York: Wiley. (A monograph on the cultivated *Allium* species.)

National Academy of Sciences, 1979. *Tropical Legumes: Resources for the Future.* Washington, D.C. (An analysis of useful underexploited plants of the Leguminosae.)

Nieuwhof, M., 1969. *Cole Crops.* London: Leonard Hill. (An excellent review of the production, harvesting, storage, conservation, and utilization of the cabbage family.)

Onwaene, I. C., 1978. *The Tropical Tuber Crops: Yams, Cassava, Sweet Potatoes, Cocoyams.* New York: Wiley. (A review of four of the tropical world's underground staple food crops.)

Parry, J. W., 1969. *Spices.* New York: Chemical Publishing Company. (A 2-volume work on the history and botany of spices.)

Rosengarten, F., 1981. *The Book of Spices.* New York: Jove. (A popular but authoritative book on spices and herbs.)

Ryder, E. J., 1979. *Leafy Salad Vegetables.* Westport, Conn.: AVI. (A discussion of lettuce, celery, cabbage, endive, chicory, Chinese cabbage, mustard, watercress, parsley, dandelion, cress, sea kale, chervil, and corn salad.)

Salaman, R. N., 1949. *The History and Social Influence of the Potato.* Cambridge, Eng.: Cambridge University Press. (The potato throughout its history—a delightful treatment.)

Smartt, J., 1976. *Tropical Pulses.* New York: Longman. (The botany, physiology, and culture of tropical legumes.)

Tindall, H. D., 1983. *Vegetables in the Tropics.* Westport, Conn.: AVI. (A handbook of vegetable-crop species.)

Wang, J-K. (ed.), 1983. *Taro: A Review of* Colocasia Esculenta *and Its Potentials.* Honolulu: University of Hawaii Press. (A collection of papers on taro covering history, production technology, utilization, and crop future.)

Whitaker, T. W., and G. N. Davis, 1962. *Cucurbits: Botany, Cultivation, and Utilization.* New York: Wiley. (A comprehensive treatment of the cultivated cucurbits.)

Yamaguchi, M., 1983. *World Vegetables: Principles, Production, and Nutritive Values.* Westport, Conn.: AVI. (Traditional and nontraditional vegetable-crop coverage.)

ORNAMENTAL CROPS

Bectel, H., 1981. *The Manual of Cultivated Orchid Species.* Cambridge, Mass.: MIT Press. (A comprehensive description of orchid species beautifully illustrated, including cultural data.)

Carpenter, P. L., T. D. Walker, and F. O. Lanphear, 1975. *Plants in the Landscape.* San Francisco: W. H. Freeman and Company. (An excellent treatment of the landscape industry and landscape plants.)

De Hertogh, A. A., 1985. *Holland Bulb Forcer's Guide*, 3d ed. Hillgom, The Netherlands: International Flower Bulb Center. (A grower's manual.)

De Hertogh, A. A., L. H. Aung, and M. Benschop, 1983. The tulip: botany, usage, growth, and development. *Horticultural Reviews* 5:45–125. (The story of tulips and tulip research.)

Ecke, P., and O. A. Matkin (eds.), 1976. *The Poinsettia Manual.* Encinitas, Calif.: Paul Ecke Poinsettias. (A grower's manual.)

Flint, H. L., 1983. *Landscape Plants for Eastern North America.* New York: Wiley. (A reference work on the landscape use of 1500 plants with the emphasis on woody trees, shrubs, and groundcovers.)

Graf, A. B., 1963. *Exotica 3: Pictorial Cyclopedia of Exotic Plants.* Rutherford, N.J.: Roehrs. (A comprehensive pictorial record of ornamentals, mostly tropical.)

Harrison, C. R., 1975. *Ornamental Conifers.* New York: Hafner Press. (Conifers of the world illustrated in color.)

Holley, W. D., and R. Baker, 1963. *Carnation Production.* Duguque, Iowa: Wm. C. Brown. (A grower's manual.)

Kofranek, A. M., and R. A. Larson (eds.), 1975. *Growing Azaleas Commercially.* Richmond: University of California Field Station. (A grower's manual.)

Langhans, R. W. (ed.), 1962. *Snapdragons: A Manual of the Culture, Insects and Diseases, and Economics of Snapdragons.* Ithaca, N.Y.: New York State Extension Service and New York State Flower Growers Association. (A grower's manual.)

Langhans, R. W., 1964. *Chrysanthemums: A Manual of the Culture, Diseases, Insects, and Economics of Chrysanthemums.* Ithaca, N.Y.: New York State Extension Service and New York State Flower Growers Association. (A grower's manual.)

Langhans, R. W., and D. C. Kiplinger (eds.), 1967. *Easter Lilies: The Culture, Diseases, Insects, and Economics of Easter Lilies.* Ithaca, N.Y.: New York State Extension Service; Columbus: Ohio State University. (A grower's manual.)

Larson, R. A. (ed.), 1980. *Introduction to Floriculture.* New York: Academic Press. (Twenty-one chapters on commercial floricultural crops.)

Mastalerz, J. W. (ed.), 1976. *Bedding Plants.* University Park, Pa.: Pennsylvania Flower Growers. (A grower's manual.)

Mastalerz, J. W., and E. J. Holcomb, 1982. *Geraniums: A Manual on the Greenhouse Crop.* University Park, Pa.: Pennsylvania Flower Growers. (The sourcebook of geranium culture.)

Mastalerz, J. W., and R. W. Langhans (eds.), 1969. *Roses: A Manual on the Culture, Management, Diseases, Insects, Economics, and Breeding of Greenhouse Roses.* University Park, Pa.: Pennsylvania Flower Growers; Ithaca, N.Y.: New York State Flower Growers Association; Haslett, Mich.: Roses, Inc. (A grower's manual.)

Ouden, P. den, and B. K. Bloom, 1965. *Manual of Cultivated Conifers Hardy in the Cold- and Warm-Temperate Zone.* The Hague: Martinus Nyhoff. (The complete book of conifers.)

Zucker, I., 1966. *Flowering Shrubs.* Princeton, N.J.: Van Nostrand. (A source book for ornamental plantings.)

Esthetics of Horticulture 22

ESTHETIC VALUES

In addition to their utility, plants have esthetic value. Owing to particular qualities that collectively we call beauty, certain plants provide us with pleasure. Beauty is not a tangible quality that can be measured or weighed, it is a value judgment. A thing is beautiful when someone decides that it is. The artist is one who can make this judgment and communicate the experience. This judgment is a reflection of cultural tradition. People of widely different heritages will have quite different opinions about what is beautiful and what is ugly.

Experiencing visual beauty depends upon our response to things sensed visually. Although a certain amount of our perception is innate, many perceptual responses are learned. To a great extent we are aware only of what we are able to interpret. For example, upon hearing a foreign language, we do not actually perceive most of the nuances of sound and inflection until we have learned to imitate them; yet even a newborn baby is aware of a sudden loud noise and can distinguish between gentle and disapproving tones of voice. Similarly, the botanist learns to discern small differences in plants that may be all but invisible to the layperson. So it is with beauty: We must learn to recognize it.

With reference to the concept of beauty, it is difficult to determine what part the innate psychological stimulation plays in the learned response. If any generalization can be made, it is that we tend to enjoy the full exercise of our perceptive facilities. Consider, for instance, the universal preferences for color, depth, and contrast for our visual experiences. Nevertheless, experiencing beauty is basically a learned response. This fact explains the underlying conservatism concerning beauty. We prefer what we are used to and tend to reject the completely strange and new. Yet, we learn to enjoy small, subtle differences and can be "trained" to expect them, as the automobile manufacturers and flower breeders have discovered.

If we accept that beauty is relative, it is apparent that we cannot arbitrarily define it. We cannot say absolutely whether a particular object, or arrangement of objects, is beautiful or not. We must suspend judgment until we have examined the object in relation to its beholder. Snakes, spiders, and worms are considered ugly by many people but beautiful by others. It is no coincidence that they are feared objects in our culture and that a certain fear of them is passed down by each generation. Thus, our perception of beauty is strongly affected by our emotional feelings and by our cultural attitudes toward objects. This is to say that the standards of one culture cannot in time be applied in all cases to another, for our method of evaluation—our yardstick—has been molded by the culture in which it developed. Generally, the things that have been accepted as beautiful for long periods of time, and which are more or less universally admired, have a basic simplicity and harmony of form and function. In conclusion, our concept of beauty is made up of two parts: (1) sensory stimulation and (2) our responses to this stimulation, which have strong personal and cultural components.

Most plants have an inherent capacity to stimulate visually. Their most obvious feature is their coloring; not only the brilliant hues of flowers, fruits, and (in some plants) leaves, but the muted tones of stem and bark. Green, of course, is the most common color, and our response is probably more than coincidental since it is also psychologically the most restful. The stimulation that plant color provides is enhanced by contrast and texture.

Also significant with respect to visual effects are the plant's structure and shape; that is, its form. Form can be seen not only in the plant as a whole but in its parts as well. The forms of plants are infinitely varied. But the same could be said of random stones, which are considerably less interesting. The perpetual interest in plants is a result of their ordered arrangements of parts, which involves symmetry, the repetition of parts on either side of the axis.

Symmetry can make any random shape an orderly one. The psychological satisfaction experienced in viewing symmetrical objects is probably due to their inherent order. The human being exhibits a universal awareness of symmetry, which is not strange considering its common occurrence in biological forms (Figure 22-1). Although all plants show some types of symmetry, the

FIGURE 22-1. Symmetry in the rose. [Photograph by J. C. Allen & Son.]

growth of many plants produces asymmetrical patterns. It is this deviation from symmetry that makes for visual interest. The basis of contemporary design is to achieve balance and harmony without the monotony of perfect symmetry.

With the possible exception of arctic peoples, human cultures have developed in plant-dominated environments. Plants provide food for people and their animals, as well as fiber, shelter, and shade. Our dependence upon plants has influenced and molded our esthetic consideration of them. We need plants, and no doubt plants have been culturally accepted as beautiful partially because they are useful. In our present American culture, in which only a relatively few people are directly involved with the growing of plants (although we still all depend on them), all of us have traditional attachments to plant material. Horticulture has a place in all our lives.

DESIGN

Design refers to the manner in which objects are artificially arranged in order to achieve a particular objective. Usually, but not always, this objective involves both a functional and a visibly pleasing arrangement. Designs are evaluated esthetically with regard to their elements of color, texture, form, and line by long-established, human value judgments called design principles: balance, rhythm, emphasis, and harmony. The importance attributed to each of these will vary with the objective of the design. When a design is successful, it is usually considered appropriate, functional, and beautiful.

Elements of Design

The design elements are visible features of all objects.

Color is the visual sensation produced by different wavelengths of light. It may be described in terms of its **hue** (red, blue, yellow), **value** (light versus dark), and **intensity,** or **chroma** (saturation or brilliance).

Texture in design refers to the visual effect of tactile surface qualities. Consider, for instance, the visual difference between burlap and silk or between the surface of a pineapple and that of a rose petal.

Form refers to the shape and structure of a three-dimensional object (sphere, cube, pyramid). However, when we view these forms in a plane, they may appear to be two-dimensional (as a circle, a square, and a triangle, respectively). In design we are concerned not only with the form of the individual objects but with the larger forms made up by their arrangement.

Line delimits shape and structure. The concept of line in design involves the means by which form guides the eye. Line becomes a one-dimensional interpretation of form. Emotional significance has been attributed to line, as shown in Figure 22-2.

Principles of Design

The principles of design (balance, rhythm, emphasis, and harmony) apply to each of the design elements as well as to their interrelations. Thus, we speak of balanced color as well as of balanced form. The artistic application of these principles is the basis of esthetic success as measured by beauty and expressiveness.

Balance implies stability. Our eye becomes accustomed to material balance, and as a result we grow uneasy about objects that appear unstable or ready to topple over. This concept is carried over to arrangements in which balance applies to the illusion of equilibrium around a vertical axis. It is achieved automatically by the symmetrical placement of objects around a central vertical axis. Balance is also achieved in nonsymmetrical arrangements (utilizing the lever principle) by the coordination of mass, distance, and space (Figure 22-3). Emphasis should be placed on the fact that, in design, we are concerned with the illusion of balance rather than with actual physical balance.

Rhythm, in the auditory sense, refers to a pulsating beat. Similarly, rhythm in the visual context lies in the pattern of "spatial" beats that our eye follows in any arrangement of objects. Rhythm leads and directs the eye through the design. It suggests movement; design without rhythm becomes uninteresting. Its proper use makes for expression and excitement.

Emphasis in design serves to lead the eye and focus its attention on some

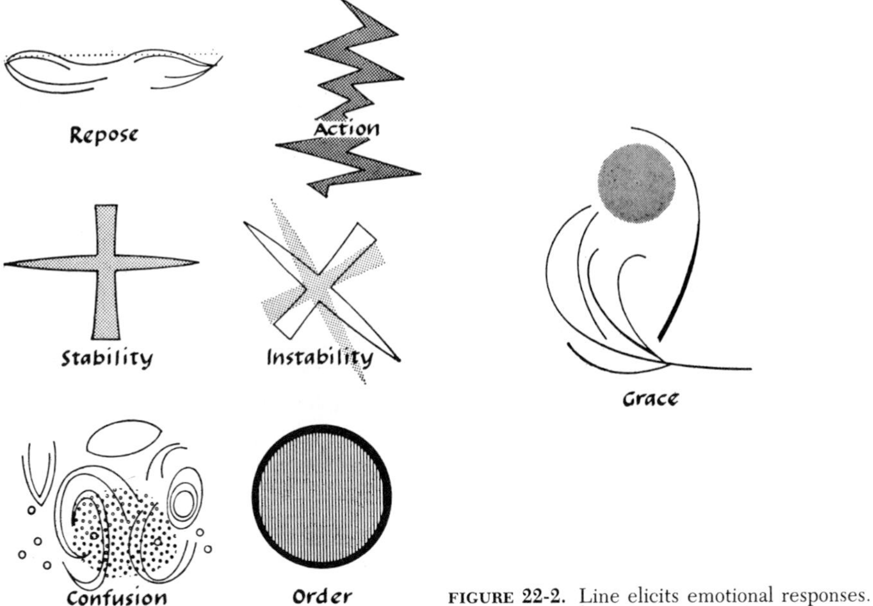

FIGURE 22-2. Line elicits emotional responses.

dominant aspect of the design. By accenting and emphasizing various elements (for example, a particular form, a strong horizontal line, or a brilliant color), the separate parts of the design may be drawn together. Emphasis, properly made, coordinates the design elements and creates an orderly and simplified arrangement.

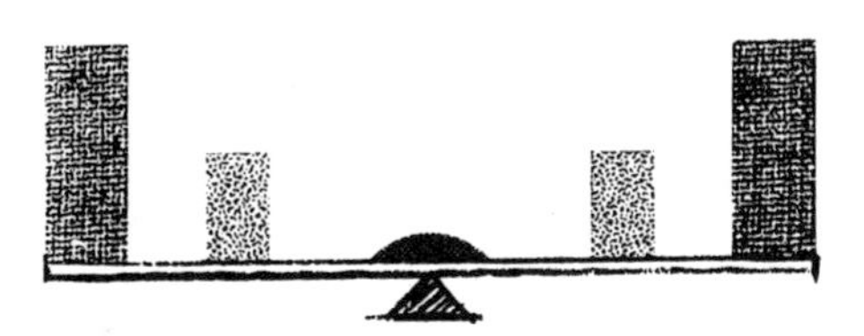

FIGURE 22-3. Graphic representation of symmetrical and asymmetrical balance.

Harmony qualifies the unity and completeness of the design. This quality is seldom achieved except by proper planning and organization. It relies principally on scale and proportion, the pleasing relationship of size and shape. The separate components lose their identity to become part of an idea, the basis of design.

GARDENS

The origin of the garden is rooted in the human desire to be surrounded by plants. The great variety of plants permits us to select, in addition to the useful, the esthetically pleasing. Their pleasant fragrance, as well as their beauty, plays a role in this selection. With respect to plants that become culturally significant, as fruit trees have, it is often difficult to separate the purely functional from the esthetic.

The first gardens in recorded history were those of the ancient cultures of Egypt and China. It was in these cultures that the two opposing traditions in gardens originated, namely, **formalism** and **naturalism.**

Formalism and the Western Tradition

The Egyptian gardens (Figure 22-4), which were developed at the edge of the deserts where the natural vegetation was sparse, represented the development of an artificial oasis. The enclosed garden is cool and leafy, typified by water and shade. The dry climate demands irrigation, which results in small, formal,

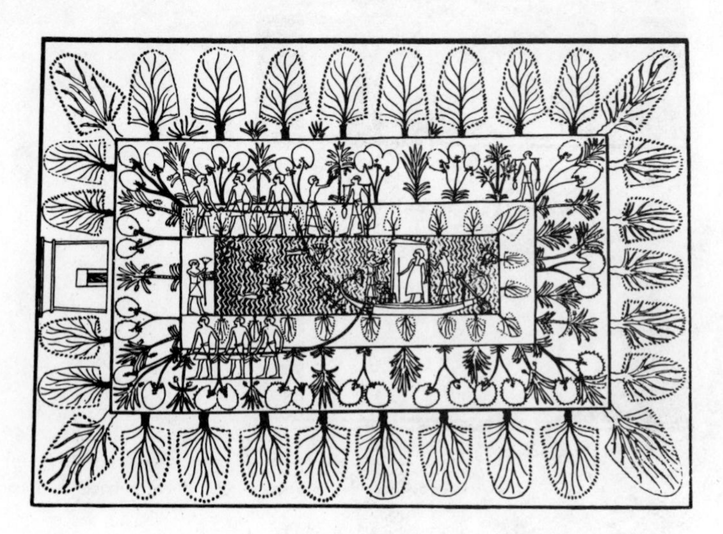

FIGURE 22-4. A portrayal of a formal Egyptian garden from a tomb at Thebes, about 1450 B.C.E. The garden contains doum palms, date palms, acacias, and other trees and shrubs. A statue is being towed in the lotus pool toward a pavilion. [From C. Singer, *History of Technology*, vol. 1, Oxford University Press, 1954.]

orderly, arranged plantings. The garden became a human triumph, as it were, over nature.

The Egyptian garden, copied everywhere though changed by local variation in land, plants, and climate, spread to Syria, Persia, and India, and ultimately to Rome. It can be traced through the remains of Egyptian paintings, Persian rugs, and Roman frescoes. This concept of the garden as a separate outdoor room is by no means outdated. The tradition has remained relatively unbroken in the Western world through the courtyard gardens of the Arab cultures and medieval cloister gardens to present-day patio gardens.

During the Renaissance, the grand period of the West's cultural revival, the concept of the garden was transformed from relative insignificance to a magnificent splendor befitting the age. The grounds design became the important concept, whereas the plant was treated rather impersonally as merely an architectural material. The plant was pruned, clipped, and trained to conform to the plan. Even architecture became subservient to the landscape plan, the garden engulfing and dominating the stately palace. The resultant "noble symmetry" included courtyards, terraces, statuary, staircases, cascades, and fountains. The emphasis was on long symmetrical vistas and promenades. The small, enclosed garden remained, but only within the walls of the buildings, as a component part of the grand plan. Formalism reached its peak in the age of Louis XIV. The master architectural gardens of André Le Nôtre still remain unsurpassed examples of this concept of design predominant over nature (Figure 22-5).

A

B

FIGURE 22-5. Formal French gardens by André Le Nôtre, near Paris. (*A*) Vaux-le-Vicomte. (*B*) Versailles; foreground, the garden called the Parterre du Midi.

Naturalism and the East

Naturalism as a concept in gardens can be interpreted as an attempt to live with nature rather than to dominate it. The desired effect is the appearance of a "happy accident of nature," although it is achieved through methods fully as artificial as those of formalism. Although the separation between garden and landscape in formalism is severe, in naturalism it is vague and indistinct. The landscape blends into the garden. If formalism is the severe line of geometry, naturalism is the free curve.

The concept of naturalism has been traced to China, but it has reached its highest development in Japan (Figure 22-6). It has also developed independently in the West. In the Eastern tradition, plants have symbolic significance. This concept is carried over to the arrangement of plants and miniature landscapes and to the development of the whole landscape. The fusion of Eastern naturalism and Western formalism took place in eighteenth-century England, where the influence of Chinese culture coincided with a movement away from formalism. The marriage was not always a happy one. Some English gardens became interspersed with Chinese pagodas amid fake antique Gothic ruins. This idea of naturalism survives today in the use of curved walks, artificial wishing wells, and herbaceous borders (Figure 22-7).

The contemporary trend in gardens is to develop a meaningful design for

FIGURE 22-6. A naturalistic Japanese garden, Rengei Temple, Kyoto. [Photograph courtesy Consulate General of Japan.]

FIGURE 22-7. A naturalistic Chinese garden in Wisconsin. The ivy-covered moon gate will seem either charming or inappropriately "cute," depending on the eye of the beholder. [Photograph courtesy Burlington Industries.]

living. Freed from the confines of "naturalism" or "formalism," the modernist strives to reach esthetic expression through the capacity for abstraction. Plants and people, as in the past, make good companions. We have turned full circle to the concept of the "garden" as a vital need in our society and not merely as an esthetic mix.

LANDSCAPE ARCHITECTURE

Landscape architecture in its broadest sense is concerned with the relationship between people and the landscape and, as such, is involved in all aspects of **land use.** The profession deals with site development, building arrangement, grading, paving, plantings and gardens, playgrounds, and pools. It is concerned with the individual home and the whole community, with parks and parkways. If the landscape architect must be first an artist, he or she must also be a horticulturist and a civil engineer. Although landscape architecture was in the past intimately associated with architecture, the two have become—unfortunately, perhaps—distinct professions. The objectives of the landscape architect have been to integrate, functionally and esthetically, people, building, and site. The main tools at the professional's disposal are plants and space (Figure 22-8).

FIGURE 22-8. The landscape architect integrates plants, structures, and space for human use. [Photograph courtesy T. D. Church.]

Plant Materials and Design

Although the materials of the landscape architect also include stone, mortar, and wood, the main ingredients are the plants of the horticulturist, who propagates, grows, and maintains them. Unlike the materials of the painter or sculptor, plants are not static but change seasonally and with time. The design is in living material; the composition is a growing one. These plants, along with the structural components (paving, walls, and buildings), form the elements of the landscape design. Let us consider plant material in relation to the design elements.

Color

The changing colors of plant material—foliage and bark as well as flower and fruit—must be considered through the seasons in relation to the landscape. Although the material is stationary, the patterns and color contrasts are transient. Compare, for instance, the way birch trees look in the winter with the way they appear in the spring, or compare the spring brilliance of flowering crabapple trees with their scarlet October fruit. The horticultural palette is a rich one, and variety is available for many tastes and effects. The delights of color can be planned by the allocation of space for herbaceous material (for example, bulbs, annuals, and succulents). Home landscape planning makes room for gardens to provide cut flowers for decoration indoors.

Texture

Although the variation in textural quality of plants is large, the use of plants next to structural forms (brick, stone, and wood) achieves striking contrasts. Think of trees in front of brick; grass together with flagstones; ground cover against paving; or flowering stocks within stone walls. Deciduous trees and shrubs offer changing textural patterns from winter to summer.

Form

Plant material naturally duplicates most of the common forms of solid geometry as well as the more complicated and more interesting forms. Ground covers can be made to form any two-dimensional pattern. Forms can be created by massing plants. Care must be taken in planting, however, to consider plants in terms of their mature forms.

Line

Line is created, as is form, by the arrangement. The dominant vertical lines are created with trees; the horizontal lines, for the most part, by grading—using the ground itself. The use of individual "specimen" plants, by virtue of their own interesting "line," creates focal points—visual interest centers—within the landscape design.

Functional Use of Plant Materials

Terrain without plants is not a fit place for people (Figure 22-9). The owner of a new home located on a raw piece of ground will readily testify that there are primary problems in mud and erosion, lack of privacy, and a need for protection from sun and wind. Equally dissatisfying is the uninteresting outlook on the wounds of construction. These problems can be beautifully solved with plants.

Ground Cover

The surface of the ground may be successfully protected from erosion by covering it with plants. Living **ground cover** serves to disperse the force of the driving rain, but, more important, it entangles and holds the soil with roots. Grass, shrubs, trees, and vines all act as successful erosion controls, even on steeply sloping terrain. Perhaps too much cannot be said to extol the virtues of grass sod as a landscape material (Figure 22-10). It makes an ideal surface for recreation; it is cool and free of dust and glare. Grass, as a living floor, is beautiful, remarkably efficient, and relatively easy to maintain, although it must be cut. In areas where less maintenance is desired, such as on slopes or in

FIGURE 22.9. A home before and after landscaping. [Courtesy Ford Motor Co., Tractor and Implement Division.]

inaccessible areas, vines and other spreading materials (myrtle, Japanese spurge, ivy, and so forth) make excellent ground covers. Herbaceous annual plants will also serve as surfacing materials, but they are efficient only for half the year in temperate climates.

Enclosure

Shrubby plant material that is high and dense ensures privacy by restricting movement and vision. Where space is limited, plants may be used in combination with structural fences to restrict intrusion (Figure 22-11). By screening unsightly areas with plants, the view may be controlled. Screening thus permits a planned vista. Objectional features of the landscape (garbage cans, incin-

FIGURE 22-10. Sodding a lawn. [Courtesy Ford Motor Co., Tractor and Implement Division.]

FIGURE 22-11. Structural materials and plants may be combined to achieve enclosure and privacy. [Photograph courtesy American Association of Nurserymen.]

erators, clothes lines, parking areas) may be successfully blocked from view by plant material. Screening also serves as protection from the elements. Windbreaks are important for unprotected areas in colder regions. Garden enclosures are the sides of the landscape and, as such, create the feeling of space. Depending on their composition and arrangement, they can produce feelings of seclusion or grandeur.

Shade

Protection from the heat and glare of the sun becomes a vital function of plants. Shade trees are not only important to outdoor living but greatly affect the indoor temperature. Care must be taken that trees do not offer potential hazards to buildings. Well-placed, they provide a ceiling to the outdoor room. Deciduous trees further offer the advantage of shade in summer and sun in winter.

Home Landscape Design

The well-designed home landscape is planned to meet the needs and desires of the people who live there (Figure 22-12). It is concerned with establishing a functional and esthetic relationship between building and site. It is more than the placing of shrubs around the building, although foundation planting is certainly a part of it. In the successful design, the house and the surrounding area

FIGURE 22-12. Backyards are areas children use. The circular course makes this backyard an attractive place for children to play. [Photograph courtesy T. D. Church.]

are treated as a single home unit—as "two sides of the same door." The practical plan must by its very nature be a compromise on a number of conflicting desires. This requirement involves consideration of the distribution of space, maintenance, and, for most people, expense.

Houses of simple form naturally subdivide a rectangular lot into areas—front, back, and side yards. The problem of home landscaping is to transform these spaces into usable areas. How many uses must be met depends upon the people who live there. However, most homes need at least three general areas: (1) for public access, (2) for service and work, and (3) for living.

The area of public access is usually the front yard. This is the portion of the house that is on view to the public, the passerby as well as the welcome guest. It creates the setting and tone of the home, and should appear large enough to set off the house from the street (Figure 22-13); it may need to allow for safe access by any automobile; and it should be hospitable.

The service area accommodates places for garbage and trash disposal, clothes drying, vegetable gardens, compost or mulch piles, dog houses, sandboxes, lawn mowers, storage facilities, and other unsightly items. The problem is one of screening and yet maintaining accessibility. In this regard, the service area can often be separated into two locations, and a side yard usually accommodates one of them. The disappearance of back alleys and the change in garbage collections from the back of the lot to the front have complicated the once-simple arrangement of service areas in home landscape planning.

To be most functional, the outdoor living area should be an extension of the

FIGURE 22-13. A handsome lawn and a generous setback make an attractive setting for this home. [Photograph courtesy Purdue University.]

interior of the dwelling—in other words, accessibility from the inside living area. It should be level, or should consist of at least a series of level areas, sloping only enough to provide proper drainage. This area should be sufficiently screened and sheltered to assure privacy and comfort, taking full advantage of shade and cooling breezes. All elements of this area should contribute to a total effect of pleasant and beautiful living space (Figure 22-14).

FIGURE 22-14. The outdoor living area of a well-landscaped home. [Photograph courtesy Purdue University.]

Prominent in the outdoor living area may be a patio, a paved area attached to the house, with or without a structural ceiling. Through the utilization of paving, plant enclosures, and at least one wall of the building, the patio can be made to serve as an outdoor room (Figure 22-15). The current use of glass to separate the patio from the house helps to bring garden and house into closer harmony. A private patio that is close to (or at least on the same level as) the kitchen is ideal for dining and relaxation. Additional access from the indoor living area is desirable, however. When space permits, this area may be separated from other areas used for more active recreational purposes.

The relative size of the outdoor areas is a matter of need, preference, and availability of space. Where space is limited, it would seem unnecessary to waste it on the public area. On the other hand, where space is ample, and only maintenance is the limiting factor, there is no reason to skimp. Nothing is more inviting than a generous setback for the house, but few things are as valuable as adequate living space. If large enough, the outdoor living area may be divided to include both privacy for relaxation and spaciousness for more vigorous recreation.

That the landscaped home must be maintained is a horticultural truism. The completely paved lot is quite unappealing. If no-maintenance living is desired, the only real alternative is an urban apartment. However, the amount of maintenance required by a landscaped home may be modified with the plan. This goal entails minimizing lawn areas, annual plants, trimmed shrubs, and

FIGURE 22-15. A patio is enclosed by two sides of a house and a bamboo wall. Although there are relatively few plants, they are effectively placed. [Photograph courtesy T. D. Church.]

any plants that require special care. It involves increasing paved areas and planning for efficiency. The wise selection of hardy, easily grown plants of the proper mature size is necessary. The design of lawns to accommodate power mowers efficiently is an important consideration.

Urban Planning

Communities in the United States with populations of more than 5000 contain about 60 percent of the population. Ours is indeed an urban society. Yet, at the same time, our cities, sprawling at the edges and decaying in the center, appear to be committing suicide by strangulation. This danger is reflected in the rush, by those who can afford it, to the outskirts of the city. The suburban movement is often self-defeating, since shops and businesses soon follow, and the increased stress on already inadequate public transportation further increases congestion.

Cities need not be ugly, dirty, or congested. The problems of rehabilitating the expanding modern city can be solved by planning and subsequent action. The aim is to shape the physical development of the city in harmony with its social and economic needs. This relationship involves land-use and transportation control, architectural design, recreation facilities, and sometimes harbor development also.

Cities have a strong social need to be beautiful, to contain fine buildings, monuments, and parks. Beauty is a positive aim, and in this respect, green living plants are a necessary part of the urban pattern. This is not to say that nature must dominate the city but only that the city should develop such that nature provides contrast and relief. Just as trees are practical for the home, they are also practical for the city: to provide shade, to act as noise buffers, and to serve as windbreaks. When properly used, they contribute to the architecture and provide visual contrast with the drab colors of construction. The provision of open space in the form of parks and gardens fills a real need of the urbanite, not only for recreational use but for esthetic fulfillment (Figure 22-16). They are not meant as an escape from the city but as an integral part of it. Finally, open green space may be considered an economic necessity. Blighted, congested cities lead to grave and expensive social problems.

Parks

> Moreover, he hath left you all his walks,
> His private arbours, and new-planted orchards
> On this side Tiber; he hath left them you,
> And to your heirs for ever—common pleasures,
> To walk abroad and recreate yourselves.
>
> SHAKESPEARE, *Julius Caesar* [III. 2]

FIGURE 22-16. Garden design in a backyard in New York City. Remarkable transformations are possible through imaginative planning. [Photograph courtesy J. Rose and The New York Times.]

Although open spaces in cities for the use of all citizens date from earliest times, the urban system of parks and playgrounds is a recent development. The industrial revolution has made them a pressing need. **Parks** and **recreational areas** are classified on the basis of size and use. **Squares** and **plazas** are ornamental, restful areas, the smallest units in a park system. They are expensive, for many small units of green space create maintenance problems. **Playgrounds** (Figure 22-17) and **athletic fields** are primarily recreational areas, and may or may not be part of a connected park system. When properly designed, golf courses and athletic fields not only accommodate recreational activities but offer all the benefits of scenic interest. **Neighborhood parks** serve areas inaccessible to large parks or are located to take advantage of local scenery. An innovation for highly developed urban areas is the development of small green spaces popularly called **vest pocket parks** or **miniparks.** They are frequently less than an acre in size and are often developed on sites formerly occupied by buildings.

Large parks (more than 100 acres), on the other hand, are designed to serve the city as a whole. They may provide facilities for such special recreational activities as horseback riding and golf, and, where water or beach is available, swimming and boating. In addition, they often include museums, memorials, zoos, and botanical gardens (which are usually horticultural-

FIGURE 22-17. A well-landscaped playground combines play activity with landscape elements. [Photograph courtesy George Patton, Landscape Architect.].

ly oriented). **State** and **national parks,** although serving a greater area, become important as recreational centers for nearby cities.

The way in which parks are landscaped determines their use. At one end of the spectrum is the **wild park,** which is left almost completely natural except for the creation of a few roads or trails. This limits its uses. Such parks are not suitable for large crowds unless they are unusually vast, such as national parks and forests. The **developed park** combines increased landscape development with natural effects and allows for more intensive use. Historic Central Park in New York City is an outstanding example. Finally, **formal parks** and **gardens** (Figure 22-18) offer the most intensive park use. Zoological and botanical parks are prime examples. When properly planned and designed, they are able to handle large crowds gracefully.

The midway atmosphere of commercial amusement parks deteriorates the landscape. These visual vulgarities are distinguished by their lack of landscaping and their general unsightliness. A new trend is the large thematic amusement park (for example, Disneyland), which is landscaped and designed to give a park effect. It is difficult to determine whether the effect of these places will be to raise the level of amusement parks in general or merely to stimulate a host of cheap imitations.

FIGURE 22-18. A formal park thoroughfare at the University of Pennsylvania that makes good use of common materials. [Photograph courtesy George Patton, Landscape Architect.]

Subdivisions

The design for **subdivisions** for residences and businesses (such as shopping centers) is an important area for the skills of the landscape architect. The problem is to integrate great numbers of structures into a pattern that maintains function, esthetic appearance, and safety. Well-planned subdivisions preserve the character of the landscape and avoid monotonous repetition. Streets are designed for efficient traffic flow and the avoidance of congestion and dangerous conditions. The architect, when specifying the building locations to conform with zoning regulations, may also specify the character of the individual units. Too often poor planning or the lack of any planning creates the stereotyped development (Figure 22-19) described by the phrase "ticky-tacky." Such neighborhoods lack individual identity and become, in effect, suburban slums.

Landscaping and Public Buildings

Public buildings include schools, hospitals, and museums in addition to national, state, and municipal buildings. (In the broadest sense, the term must also include churches.) Such buildings are often expected to be formal, even monumental, in their design and setting as compared with most commercial buildings. Representing the spirit and ideals of a municipality, they require

FIGURE 22-19. Poor planning creates a suburbia ill-at-ease with the natural environment. [Photograph courtesy USDA.]

special esthetic treatment. They should be set off so that they can be seen to best advantage. The landscaping should be spacious, dignified, and distinctive.

Schools, rural or urban, deserve special consideration—the best efforts of the community. An ideal school grounds plan could create a spacious and extensively planted park. Landscaped areas might provide adequate room for free play and supervised recreation, sports events, and outdoor ceremonies. In many cases, rural schools, possessing adequate space and good soil, have greater possibilities for development than urban schools have. The development of the rural school grounds is often accomplished with community help. Natural plant materials are sometimes used to supplement nursery-grown stock.

Industrial Landscaping

Factories and industrial plants are moving out of heavily congested urban areas. Many companies have begun to pay attention to esthetic considerations as part of their obligations as corporate citizens. The appearance of factories becomes a major factor in both public and labor relations. A well-landscaped industrial plant can be very attractive (Figure 22-20).

FIGURE 22-20. A splendid example of industrial landscaping. Note the contrast created by the exciting use of water, the paving, and the plant materials. [Photograph courtesy General Motors Corp.]

Highways and Roadside Development

Highway design must satisfy the requirements of utility and safety, but it should also satisfy esthetic considerations. The landscaping objectives include the utilization of existing scenic advantages in the proposed routes: Although bridges and pavement may need replacement eventually, scenic values can be permanently enhanced through careful planning. Details of subsequent roadside development—outlooks, picnic areas, parks, and the like—become important in planning highway routes.

Highway design should harmonize with the natural topography. Existing trees and lesser vegetation should be conserved. Plantings to control erosion, if in harmony with the natural surroundings, will accomplish a natural transition between construction and landscape. To be effective, such plantings require the use of extensive land adjacent to the right of way. Zoning is essential for control and regulation of outdoor advertising and commercial structures along the highway. Safety considerations dictate roadside development that will not be monotonous but that also will not distract the attention of drivers. Interesting scenery, long sites, and gentle curves and grades all help to create these results.

Landscape objectives and engineering objectives alike include erosion control, economical maintenance, safety, sound construction, and conservation of natural beauty. That the public desires landscaped, well-designed highways is evident. Toll roads and freeways have become increasingly utilized in interstate travel. Tremendous amounts of tax money are spent for safe, restful, scenic driving. The **parkway,** a highway for noncommercial traffic located on a strip of parkland with limited access, although only a generation ago considered an extravagant form of highway construction, is today considered an essential form.

Floral Design

Floral design bears about the same relationship to landscape design as a string quartet does to a symphony orchestra. The principles of design are the same; only the scale is reduced. Floral design is one of the decorative arts, along with painting and sculpture.

Although arranging flowers[1] is a means of individual artistic expression, it is also the basis of commercial floriculture, which constitutes a large segment of the horticultural industry. All cut flowers are for ultimate use in some sort of arrangement. It may consist of the simple placement of a dozen roses in a vase, or it may involve the creation of a large floral float.

Planters

Planters, large containers for growing plants, are popular for indoor and outdoor use (Figure 22-21). Planters are well suited in outdoor courts and plazas. Depending on the season, a wide variety of plants may be used, including evergreen shrubs, bulbs, and annuals. Growing plants contribute to interior decoration just as they do to architecture. Because of the unavailability of indoor light, foliage plants having low light requirements (such as *Philodendron*, *Sansevieria*, and *Ficus*) are usually grown. Planters have become especially prominent as interior decor in lobbies, offices, and restaurants. The use of artificial foliage plants is also increasing, but these imitations have little esthetic appeal.

Flower Arrangement

Arrangements of flowers, plants, and plant parts have long been used for decoration. In Japan, flower arrangement, or **ikebana,** is a continuing tradition that has been an integral part of cultural life for thirteen centuries. Its significance in Japanese home life was established in the fifteenth century, along with the

---

[1]The term *flower* is used in its broadest sense to include all decorative plants and plant parts, especially the morphological flower.

FIGURE 22-21. A window box of succulents is a part of balcony gardening. [Courtesy Longwood Gardens.]

tea ceremony. In its conception, ikebana symbolized certain philosophical and religious concepts. Today, much of the religious connotation has been lost, but the symbolism still remains a key part of the arrangement. Thus, the expression of seasonal change and the passage of time are vital parts of all arrangements, as are appropriate representations of traditional holidays and festivals. Unlike the Western concepts of floral design, the Oriental tradition emphasizes the element of line over form and color. In the classical concept, line is symbolically partitioned into the representation of heaven (vertical), earth (horizontal), and humanity (diagonal and intermediate). The chief aim is to achieve a beautiful flowing line; to accomplish this end, the most ordinary materials may be used. The concept of naturalism is expressed throughout. All symmetrical effects are avoided.

Flower arrangement is still an important part of Japanese life. There are many different styles and schools: **ikenobo,** classical arrangements; **rikka,** larger, ornate, upright reproductions of the landscape by means of flowers and plants; **nageire,** simple, naturalistic arrangements; and **moribana,** expressive, scenic arrangements with greater use of foliage and flowers. These schools of flower arrangement differ in opinion and conception, but the basic principles of the art are preserved in common.

Other typically Oriental types of artistic expression involve growing plants. **Bonsai,** the culture of miniature potted trees, dwarfed by pruning and controlled nutrition, is a spectacular example of the horticultural art. Living trees, some of them hundreds of years old and yet only a few feet high, are grown in containers arranged to resemble a portion of the natural landscape (Figure 22-22). **Bonseki** is the construction of a miniature landscape out of stone, sand, and living vegetation.

In Europe, flowers are readily purchased in the market and are in common use as part of normal living. In the United States, on the other hand, the use of

FIGURE 22-22. Bonsai, or tray culture, is an oriental art achieved through pruning and controlled nutrition. The four examples shown are part of a gift to the United States from Japan and are on permanent display at the U.S. National Arboretum in Washington, D.C. (*A*) Japanese white pine (*Pinus pentaphylla* var. *himekomatsu*), 350 years old. (*B*) Thorny elaeagnus (*Elaeagnus pungens*), 150 years old. (*C*) Azalea (*Rhododendron kiushianum*), 150 years old. (*D*) Japanese beech (*Fagus crenata*), 50 years old.

flowers is usually limited to special occasions such as formal dining, decoration at religious holidays, appropriate remembrances (especially birthdays, anniversaries, and Mother's Day), and as "get well" gifts. The ceremonial use of flowers in weddings and funerals is the backbone of the florist business. Bouquets and wreaths are standard fare for concert artists, beauty queens, openings, and derby winners. Corsages and boutonnieres become a significant part of the costume at dances and other formal occasions.

HORTICULTURE AS RECREATION

Horticulture has long been, and will continue to be, an outlet for recreation and pleasure. Gardening is probably the true national pastime, for horticulture may be actively enjoyed at many levels. It provides either vigorous or sedentary activity that may be pursued on any scale. Both the young and strong and the aged and infirm may enjoy its respite. There is room for the innovator, the gadget-minded, the artist, and the faddist. The joys of solitude are available, as is the bustle of organizational activity. Horticulture yields the sweets of anticipation along with the bitterness of disappointment. Tangible rewards are available for a minimum of effort, and they increase in proportion to skill and persistence. All who partake of it soon acquire a keener awareness of the mysteries of life, growth, and death. The beautiful as well as the delicious are readily available to those who seek them.

Selected References

Berrall, J. S., 1966. *The Garden: An Illustrated History*. New York: Viking. (A beautiful and well-written history of gardens.)

Booth, N. K., 1983. *Basic Elements of Landscape Architectural Design*. New York: Elsevier. (The theory of landscape design.)

Brooks, J., 1978. *The Small Garden*. New York: Macmillan. (Landscape architecture on a small scale.)

Carpenter, P. L., T. D. Walker, and T. O. Lanphear, 1975. *Plants in the Landscape*. San Francisco: W. H. Freeman and Company. (An excellent treatment of the landscape industry and landscape plants.)

Church, T. D., 1983. *Gardens Are for People*, 2d ed. New York: McGraw-Hill. (A monumental book on gardens, bringing together the work and theory of one of the greatest of landscape architects: Thomas D. Church, 1902–1978.)

Hannebaum, L., 1981. *Landscape Design: A Practical Approach*. Reston, Va.: Reston. (An introduction to the landscape design process.)

Huxley, A., 1978. *An Illustrated History of Gardening*. New York: Paddington Press. (The world of garden history, profusely illustrated.)

Hyams, E., 1971. *A History of Gardens and Gardening*. New York: Praeger. (A splendid history; highly recommended.)

Manaker, G. M., 1981. *Interior Plantscapes: Installation, Maintenance, and Management*. Englewood Cliffs, N.J.: Prentice-Hall. (The use and culture of plants for interiorscaping.)

Pierceall, G. M., 1984. *Residential Landscapes: Graphics, Planning, and Design*. Reston, Va.: Reston. (An introduction to landscape design and residential site development.)

Plants & Gardens. New York: Brooklyn Botanic Gardens. (A series of useful handbooks for advanced gardeners. Handbook 51, *Bonsai: Special Techniques*, and Handbook 2, *Bonsai for Indoors*, are recommended for bonsai fanciers.)

Riester, D. W., 1958. *Design for Flower Arrangers*. Princeton, N.J.: Van Nostrand. (A beautiful book.)

Rose, J., 1958. *Creative Gardens*. New York: Reinhold. (Well-illustrated book of beautifully conceived landscapes showing unique combinations of natural and structural materials.)

Rutledge, A. J., 1981. *A Visual Approach to Park Design*. New York: Garland. (Park design with a focus on social and behavioral patterns.)

Simmonds, J. O., 1978. *Earthscape: A Manual of Environmental Planning*. New York: McGraw-Hill. (A comprehensive approach to landscape architecture and environmental quality.)

Simonds, J. O., 1983. *Landscape Architecture*, 2d ed. New York: F. W. Dodge. (Landscape planning and site development.)

CONVERSION TABLES

CELSIUS–FAHRENHEIT

To convert a temperature in either Celsius or Fahrenheit to the other scale, find that temperature in the center column, and then find the equivalent temperature in the other scale in either the Celsius column to the left or in the Fahrenheit column to the right.

On the Celsius scale the temperature of melting ice is 0° and the temperature of boiling water is 100° at normal atmospheric pressure. On the Fahrenheit scale the equivalent temperatures are 32° and 212° respectively. The formula for converting Celsius to Fahrenheit is $°F = \frac{9}{5}°C + 32$, and the formula for converting Fahrenheit to Celsius is $°C = \frac{5}{9}(°F - 32)$.

C	C or F	F	C	C or F	F	C	C or F	F
−73.33	**−100**	−148.0	−6.67	**20**	68.0	15.6	**60**	140.0
−70.56	**−95**	−139.0	−6.11	**21**	69.8	16.1	**61**	141.8
−67.78	**−90**	−130.0	−5.56	**22**	71.6	16.7	**62**	143.6
−65.00	**−85**	−121.0	−5.00	**23**	73.4	17.2	**63**	145.4
−62.22	**−80**	−112.0	−4.44	**24**	75.2	17.8	**64**	147.2
−59.45	**−75**	−103.0	−3.89	**25**	77.0	18.3	**65**	149.0
−56.67	**−70**	−94.0	−3.33	**26**	78.8	18.9	**66**	150.8
−53.89	**−65**	−85.0	−2.78	**27**	80.6	19.4	**67**	152.6
−51.11	**−60**	−76.0	−2.22	**28**	82.4	20.0	**68**	154.4
−48.34	**−55**	−67.0	−1.67	**29**	84.2	20.6	**69**	156.2
−45.56	**−50**	−58.0	−1.11	**30**	86.0	21.1	**70**	158.0
−42.78	**−45**	−49.0	−0.56	**31**	87.8	21.7	**71**	159.8
−40.0	**−40**	−40.0	0	**32**	89.6	22.2	**72**	161.6
−37.23	**−35**	−31.0	0.56	**33**	91.4	22.8	**73**	163.4
−34.44	**−30**	−22.0	1.11	**34**	93.2	23.3	**74**	165.2
−31.67	**−25**	−13.0	1.67	**35**	95.0	23.9	**75**	167.0
−28.89	**−20**	−4.0	2.22	**36**	96.8	24.4	**76**	168.8
−26.12	**−15**	5.0	2.78	**37**	98.6	25.0	**77**	170.6
−23.33	**−10**	14.0	3.33	**38**	100.4	25.6	**78**	172.4
−20.56	**−5**	23.0	3.89	**39**	102.2	26.1	**79**	174.2
−17.8	**0**	32.0	4.44	**40**	104.0	26.7	**80**	176.0
−17.2	**1**	33.8	5.00	**41**	105.8	27.2	**81**	177.8
−16.7	**2**	35.6	5.56	**42**	107.6	27.8	**82**	179.6
−16.1	**3**	37.4	6.11	**43**	109.4	28.3	**83**	181.4
−15.6	**4**	39.2	6.67	**44**	111.2	28.9	**84**	183.2
−15.0	**5**	41.0	7.22	**45**	113.0	29.4	**85**	185.0
−14.4	**6**	42.8	7.78	**46**	114.8	30.0	**86**	186.8
−13.9	**7**	44.6	8.33	**47**	116.6	30.6	**87**	188.6
−13.3	**8**	46.4	8.89	**48**	118.4	31.1	**88**	190.4
−12.8	**9**	48.2	9.44	**49**	120.2	31.7	**89**	192.2
−12.2	**10**	50.0	10.0	**50**	122.0	32.2	**90**	194.0
−11.7	**11**	51.8	10.6	**51**	123.8	32.8	**91**	195.8
−11.1	**12**	53.6	11.1	**52**	125.6	33.3	**92**	197.6
−10.6	**13**	55.4	11.7	**53**	127.4	33.9	**93**	199.4
−10.0	**14**	57.2	12.2	**54**	129.2	34.4	**94**	201.2
−9.44	**15**	59.0	12.8	**55**	131.0	35.0	**95**	203.0
−8.89	**16**	60.8	13.3	**56**	132.8	35.6	**96**	204.8
−8.33	**17**	62.6	13.9	**57**	134.6	36.1	**97**	206.6
−7.78	**18**	64.4	14.4	**58**	136.4	36.7	**98**	208.4
−7.22	**19**	66.2	15.0	**59**	138.2	37.2	**99**	210.2
						37.8	**100**	212.0

AREA

Metric

1 square centimeter	=	0.155 sq inch
	=	100 sq millimeters
1 square meter	=	1550 sq inches
	=	10.764 sq feet
	=	1.196 sq yards
	=	10,000 sq centimeters
1 square kilometer	=	0.3861 sq mile
	=	1,000,000 sq meters
1 hectare	=	2.471 acres
	=	10,000 sq meters

Imperial

1 square inch	=	6.452 sq centimeters
	=	1/144 sq foot
	=	1/1296 sq yard
1 square foot	=	929.088 sq centimeters
	=	0.0929 sq meter
1 square yard	=	8361.3 sq centimeters
	=	0.8361 sq meter
	=	1296 sq inches
	=	9 sq feet
1 square mile	=	2.59 sq kilometers
	=	640 acres
1 acre	=	0.4047 hectare
	=	43,560 sq feet
	=	4840 sq yards
	=	4046.87 sq meters

LENGTH

Metric

1 nanometer	=	0.001 micrometer
1 micrometer	=	0.001 millimeter
1 millimeter	=	0.001 meter
	=	0.0394 inch
1 centimeter	=	10 millimeters
	=	0.3937 inch
	=	0.01 meter
1 meter	=	39.37 inches
	=	3.281 feet
	=	1000 millimeters
	=	100 centimeters
1 kilometer	=	3281 feet
	=	1094 yards
	=	0.621 mile
	=	1000 meters

Imperial

1 inch	=	25.4 millimeters
	=	2.54 centimeters
1 foot	=	30.48 centimeters
	=	0.3048 meter
	=	12 inches
1 yard	=	0.9144 meter
	=	91.44 centimeters
	=	3 feet
1 mile	=	1609.347 meters
	=	1.609 kilometers
	=	5280 feet
	=	1760 yards

WEIGHT

Metric

1 milligram	=	0.001 gram
	=	0.0154 grain
1 gram	=	0.0353 avoirdupois ounce
	=	15.4324 grains
1 kilogram	=	1000 grams
	=	353 avoirdupois ounces
	=	2.2046 avoirdupois pounds
1 metric ton	=	1000 kilograms
	=	2204.6 pounds
	=	1.102 short tons
	=	0.984 long ton

Imperial

1 grain	=	1/7000 avoirdupois pound
	=	0.064799 gram
1 ounce (avoirdupois)	=	28.3496 grams
	=	437.5 grains
	=	1/16 pound
1 pound (avoirdupois)	=	453.593 grams
	=	0.45369 kilograms
	=	16 ounces
1 short ton	=	907.184 kilograms
	=	0.9072 metric tons
	=	2000 pounds

YIELD

Metric

1 kilogram per hectare	=	0.89 pound per acre
1 cubic meter per hectare	=	14.2916 cubic feet per acre

Imperial

1 pound per acre	=	1.121 kilograms per hectare
1 ton (2000 lb) per acre	=	2.42 metric tons per hectare
1 cubic foot per acre	=	0.0699 cubic meter per hectare
1 bushel (60 lb) per acre	=	67.26 kilograms per hectare

VOLUME

Metric

1 liter	=	1.057 U.S. quarts (liquid)
	=	0.9081 quart (dry)
	=	0.2642 U.S. gallon
	=	0.221 Imperial gallon
	=	1000 milliliters
	=	0.0353 cubic foot
	=	61.02 cubic inches
	=	0.001 cubic meter
1 cubic meter	=	61,023.38 cubic inches
	=	35.314 cubic feet
	=	1.308 cubic yards
	=	264.17 U.S. gallons
	=	1000 liters
	=	28.38 U.S. bushels
	=	1,000,000 cubic centimeters
	=	1,000,000,000 cubic millimeters

Imperial

1 fluid ounce	=	1/128 gallon
	=	29.57 cubic centimeters
	=	29.562 milliliters
	=	1.805 cubic inches
	=	0.0625 U.S. pint (liquid)
1 U.S. quart (liquid)	=	946.3 milliliters
	=	57.75 cubic inches
	=	32 fluid ounces
	=	4 cups
	=	¼ gallon
	=	2 U.S. pints (liquid)
	=	0.946 liter
1 quart (dry)	=	1.1012 liters
	=	67.20 cubic inches
	=	2 pints (dry)
	=	0.125 peck
	=	1/32 bushel
1 cubic inch	=	16.387 cubic centimeters
1 cubic foot	=	28,317 cubic centimeters
	=	0.0283 cubic meter
	=	28.316 liters
	=	7.481 U.S. gallons
	=	1728 cubic inches
1 U.S. gallon	=	16 cups
	=	3785 liters
	=	231 cubic inches
	=	4 U.S. quarts (liquid)
	=	8 U.S. pints (liquid)
	=	8.3453 pounds of water
	=	128 fluid ounces
	=	0.8327 British Imperial gallon
1 British Imperial gallon	=	4.546 liters
	=	1.201 U.S. gallons
	=	227.42 cubic inches
1 U.S. bushel	=	35.24 liters
	=	2,150.42 cubic inches
	=	1.2444 cubic feet
	=	0.03524 cubic meter
	=	2 pecks
	=	32 quarts (dry)
	=	64 pints (dry)

INDEX

Page numbers in *italics* indicate illustrations.
Page numbers followed by *t* indicate tables.

A

Index

B

C

D

E

F

G

H

I

J

K

L

M

N

O

P

Q

R

S

T

U

V

W

X

Y

Z

KU-005-796

EMINENT VICTORIANS

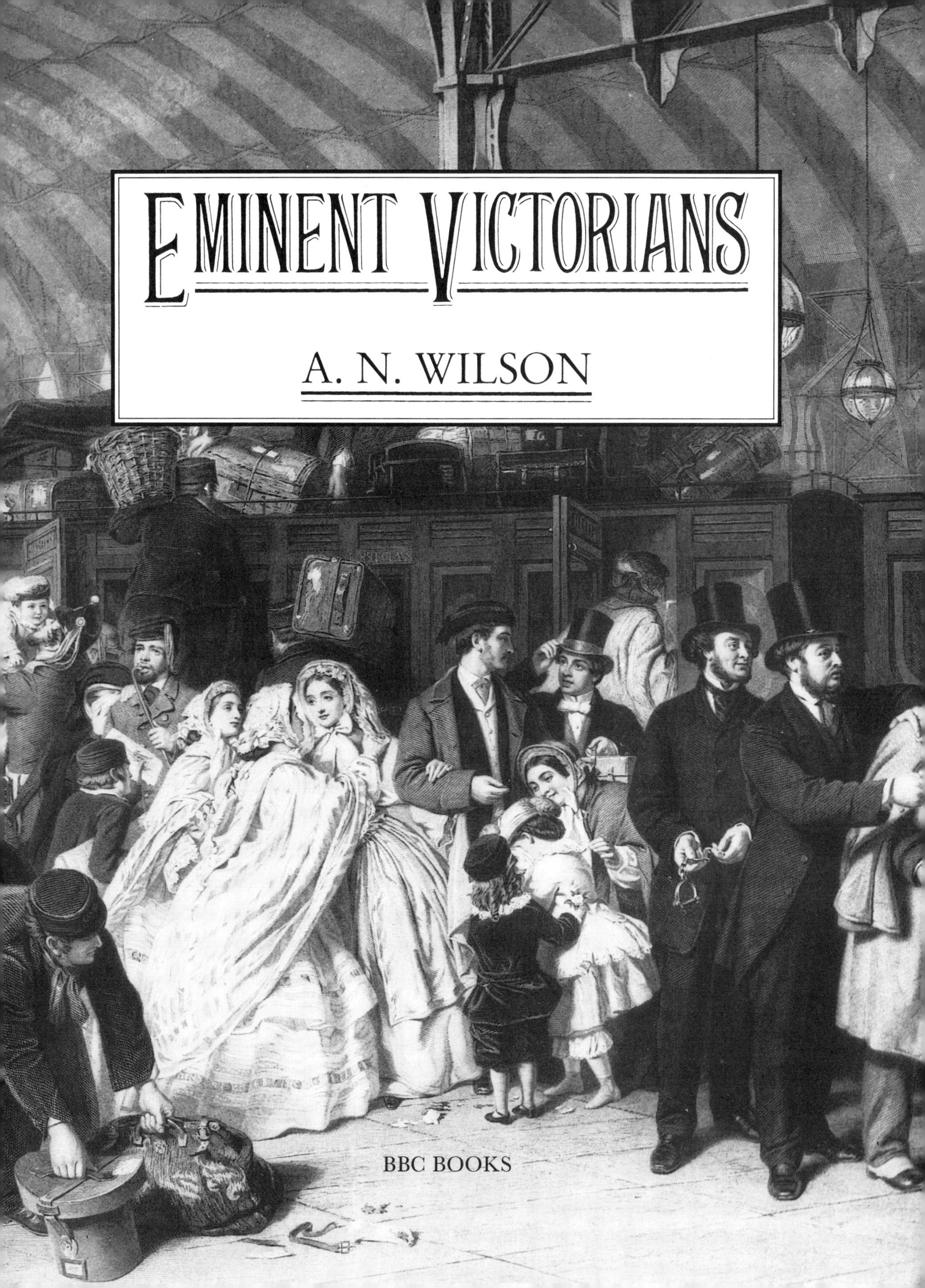
EMINENT VICTORIANS
A. N. WILSON
BBC BOOKS

for John Lucas

Frith's crowd scenes give vivid glimpses of Victorian life. His famous painting of Paddington station seethes with human activity and reminds us of the revolutionary importance of the railways.

Published by BBC Books,
a division of BBC Enterprises Limited,
Woodlands, 80 Wood Lane, London W12 0TT
First published 1989

ISBN 0 563 20719 1

Set in 12 on 14pt Baskerville by Phoenix Photosetting, Chatham
and printed and bound in Great Britain by Mackays of Chatham PLC, Chatham, Kent
Colour originated by Technik Limited, Berkhamsted
and printed by Chorley and Pickersgill Limited, Leeds
Jacket printed by Belmont Press, Northampton

'THE HISTORY OF THE VICTORIAN AGE will never be written: we know too much about it.' Seventy years have elapsed since Lytton Strachey wrote those words as the preface to his hilarious volume of essays, *Eminent Victorians*. He added, with his tongue so firmly in his cheek that it is a wonder the words came forth at all, 'Je n'impose rien; je ne propose rien; j'expose.' In fact, his highly contentious 'expositions' of four much-revered Victorian figures are full of imposition and proposition. Many of the more grotesque 'facts' he unearthed – such as the assertion that one of Dr Arnold's legs was much shorter than the other – were invented because Strachey thought they made his subjects more ridiculous. Rendering his four subjects absurd – and by implication all the values they represented – was the purpose of his volume. It was a book which had a profound effect on the minds of the whole post-1918 generation. The values of public school and imperial militarism, embodied in Dr Arnold and General Gordon, were cut down to size by contemplating the supposedly ludicrous characters of Arnold and Gordon themselves. Religion, in the person of Strachey's caricature of Cardinal Manning, was thrown overboard with equal glee. As for the opening sentence, intended ironically at the time, it has been overtaken by the more gargantuan joke of Victorian Studies – masses and masses of material about every aspect of life in the last century. Not only do we 'know too much' about the Victorian Age; we have also been treated to a series of extraordinary Victorian revivals, both in taste and in morals. Even in my boyhood (born 1950) 'Victorian' was a term of aesthetic abuse when applied to architecture or painting. John Betjeman could see beauty in George Gilbert Scott's St Pancras Hotel or the Albert Memorial. But nobody could be quite sure whether or not he was joking. His friend Evelyn Waugh, meanwhile, bought up a huge collection of Victorian 'narrative' paintings of 'The Last Night in the Old Home', 'And When Did You Last See Your Father?' genre; almost every canvas must have come to him cheap.

Then, fashions changed. As the developers moved in to every provincial town in England and began to destroy what had been regarded as

'Victorian monstrosities', people began to take John Betjeman seriously. Gothic revival architects like George Gilbert Scott, William Butterfield, or G. F. Bodley came to be the subjects of serious study. Faced with the brutalism of post-Bauhaus architects, persons of taste and refinement began to see the glories of George Edmund Street's Law Courts or even Alfred Waterhouse's Manchester Town Hall.

Victorian paintings which only a few decades earlier had been the subject of laughter suddenly became valuable. The Pre-Raphaelites now command enormous prices in the sale rooms, and even the lesser Victorians are beyond the most avaricious dreams of the private collector. Victorian furniture which had been lumpy and inelegant when it was in grandma's attic suddenly seemed full of character under the auctioneer's hammer. Button-backed chairs and chaises-longues returned to the drawing-room. Repro-Victorian wallpaper from Sandersons or Laura

Alfred Waterhouse's Manchester Town Hall is one of the most magnificent examples of civic architecture in the Gothic manner, bringing 'the last enchantments of the Middle Ages' to the industrial North.

Ashley began to transform every household in the land. And as house prices shot up, and the Georgian rectories became the preserve of millionaires, the rest of us started to discover that Charles and Carrie Pooter, of The Laurels, Brickfield Terrace, Holloway, had in fact been arbiters of the greatest good taste. Those tiles round the fireplace, which ten years before we would have paid a builder to hack out, are seen to be of the greatest beauty. Thank goodness for that little piece of stained glass window over the front door! How glad we are to have those wrought-iron railings in the garden fence.

Charles Pooter is of course the fictitious narrator of *The Diary of a Nobody* by George and Weedon Grossmith, first published in 1892 and evergreen to this day, not only as the best comic novel in the language, but also as offering a glimpse of life in an 'ordinary' household towards the close of Queen Victoria's reign. The suburbs of England were full of Pooters, and in launching his attack on the 'Eminent Victorians' it is not surprising that Lytton Strachey did not even give them a glance. Because he was born into one of the great intellectual dynasties of the Victorian Age, Strachey, and his more serious Bloomsbury or Fabian coevals, took it for granted that they could overthrow the systems erected by the patricians like Arnold and Manning. It seemed to them as though the great Victorian institutions, the public schools, the Army, the Church, had had their day; likewise, probably, the monarchy. The new civilisation, based on the values of the *New Statesman*, would create an England that was altogether different. This, to a large extent, took place – particularly in the 1945 Parliament of Clement Attlee, who, though he looked like a Pooter, was really a Fabian patrician at heart.

This reading of events failed to see that the Pooters were the really triumphant class of the Victorian period. Karl Marx, surveying England in the 1840s, became convinced that it was only a matter of time before the proletarian masses rose up against their capitalist masters. It did not happen, because almost every member of the proletarian masses was dreaming of that cosy domestic respectability which Charles Pooter, the clerk in the City, had established for his wife Carrie and his raffish son Lupin. This huge class of person, who took their holidays at Broadstairs and followed new crazes like the purchase of Christmas cards, had no interest in Marxist revolutionary programmes. But that is not to say that they were not revolutionaries. Their kind had not existed in eighteenth-century England. Once they came into being, they were an unshakeable force. They wanted the best for themselves within their modest, hard-earned incomes and carefully-put-by savings. It was the Pooters who supported, or dismissed, the governments of Gladstone, Disraeli, Salisbury and Rosebery. It was they who wanted, and obtained, the vote. It was they who educated their children. In turn, they would start to enter the civil service and the business community. Because of the extraordinary class system in England, they would have to wait eighty years before they truly obtained a political voice. No wonder, that once a Pooter was established in Downing Street (and you can tell from *The Diary of a Nobody* that it would be Carrie not Charles who had the will-power for that) there was much talk of a revival in Victorian values.

While the sale-rooms paid prices for the paintings of William Holman Hunt which in a sane world would be reserved for Titians, and while Laura Ashley made her fortune, Mrs Thatcher reintroduced simple Gladstonian economics: you don't spend money you don't possess. Instead of systems, she stood for the little family with its savings, and its decent values. The result was the most successful political career in British twentieth-century history. The last laugh was on Pooter.

In the course of the Victorian revival which we have so surprisingly witnessed in our own times, there has been a lot of talk about 'Victorian values'. At the time of the Suez crisis in 1956, the British nation was quite largely divided between those who still clung to some quasi-imperial sense that 'we' had a right to invade Egypt if Johnny Egyptian was misbehaving himself, and others who thought that the whole enterprise was mad, anachronistic, bloodthirsty. There was indeed such divide in 1982. The surprising thing was that when General Galtieri invaded the Falkland Islands he provided the Prime Minister, whose political fortunes until that juncture had seemed to be flagging, with the chance to summon up reserves of 'jingoism' in the British people that most *bien-pensant* observers would have imagined to be long since dead. The rights and the wrongs of the Falklands campaign will probably continue to be debated. What was so arresting at the time was the atmosphere in Britain. People said that it was a revival of the Dunkirk spirit; but it wasn't. At Dunkirk, there was an imminent likelihood of invasion by the most dangerous army in Europe. The Falklands War was far away, and, at least until the sinking of HMS *Sheffield*, it felt like a carnival. If it re-awoke the spirit of an earlier war, it was that of the Crimean, when the Black Sea was invaded not only by the 'task force' of British troop ships, but also by sightseers, excited by a dizzy sense of vicarious national heroism. It was a strange example of the way that our Victorian legacy has the power to return and haunt us.

Unlike Strachey, I am not close enough to the Victorians to feel superior to them. A child of my time, I have gone through periods of 'nostalgia' for an age I never knew, and a feeling that the Victorians built buildings, followed creeds, or made achievements which it was sad to discard. Viewed in another light, however, they seem so different from us that it seems farcical to hope that we could ever hope to understand them, let alone imitate them. The passage of time, which seems so cruel when one is in nostalgic or elegiac mood, is never without mercies. We might regret steam trains, but who would choose to live in a Victorian slum? Oxford was more beautiful in the days before Lord Nuffield built the Cowley works, but for me the beauty would have been hard to appreciate if, like Dr Pusey, I had believed that the majority of the human race was doomed to hell-fire throughout eternity. Men's clothes were nicer, but children were starving, and worse (see my chapter on Josephine Butler) on the streets of London. On the whole, I have come to echo Lucky Jim in his definitive lecture on the Middle Ages, 'Thank God for the twentieth century.'

My relationship with Strachey, like my relationship with the Victorians, has had its ups and downs, too. As a stylist, I used to love him too much. Set pieces, like his death of Queen Victoria (in his biography of the Queen) got into my bloodstream, with the result that for a period I rejected him as nothing but a mannerist. The Bloomsbury air of sniggering superiority to the Victorians wore very thin by the time I was about twenty-five. After all, the great Victorians have a sheer bigness about them which is impossible to laugh off. No English poet of the twentieth century matches Tennyson, even though any schoolboy can point out Tennyson's faults. For sheer improbability of plots, repetition, sloppy writing, it would be hard to find a writer as bad as Dickens on a bad day. But the achievement of Dickens remains something which makes many 'better' novelists of our century seem like pigmies.

My selection of 'Eminent Victorians' has not been arbitrary, but inevitably it fails to be comprehensive. What? No Dickens? No Darwin? No Gilbert and Sullivan? No William Morris? No Butterfield? No Lord Leighton? What kind of nineteenth century is this? I have selected the Prince Consort partly because of the multifarious range of his interests,

The nineteenth century saw a vast expansion of the lower middle classes of whom Charles and Carrie Pooter in The Diary of a Nobody *are the archetypal examples. Streets like this one in London are typical of those built to accommodate them.*

but partly as a symptom of that 'Pooterite' revolution at which I hinted in an earlier paragraph. Gladstone, as well as being the greatest statesman of the age, was also a man who contained many characteristically Victorian contradictions within his nature – the Tractarian recluse at war with the man of action; the old Tory at war with 'the people's William'. Charlotte Brontë could, like many of the great Victorian writers, perhaps be described as a Late Romantic as much as she was a Victorian. Her story is perpetually haunting, just as *Jane Eyre* is endlessly re-readable, and her presence here reminds us that there is nobody who is typical, everyone is an oddity, if few odder than she. Few, that is, unless it be Mrs Cameron the photographer who although she could not be described as 'eminent' in the sense that Newman or Gladstone were eminent, was one of the most stylishly accomplished practitioners of that art-form or species of chicanery which was first invented and propagated during Victoria's reign. Josephine Butler, on the other hand, on the face of it the most typical of the group, showed exceptional courage and honesty in the way she challenged the accepted sexual mores of the day. Like my readers, I could have chosen a different six, though I would always be sad to leave out Newman who seems, the more one thinks about him, to have been one of the most remarkable people who ever lived.

Like Strachey, I think it is possible to learn more from short biographies of a roughly contemporaneous nature than from some generalised know-all survey. Also like Strachey (only in my case quite sincerely), I have not tried to make a point. It has seemed to me, though, the more I have dwelt upon these six characters, that any attempt to revive 'Victorian values' in our own day would be a precarious exercise. The greatest Victorians were the ones who were on the move. 'To live is to change,' said Newman. All six of my subjects were aware of it. Newman himself did not only change his own religion. He radically changed both the Church which he left and the Church which he joined. By the very fact of his having lived, it would be impossible to return to 'Victorian values' either in the Roman or the English Churches because they are both almost unrecognisably different from their Victorian manifestations. Likewise, in the political sphere, the programme of reform and change presided over by Gladstone has meant that, except in the broadest outlines, it is impossible to think of how one might return to Gladstonian politics. Besides, having rumbled 'Victorian values', both Newman and Gladstone in their different ways were intent on destroying them, rooting them out. Mrs Butler was the same. The double standard by which a household was kept 'respectable' by the husband being tacitly allowed to visit the stews is arguably the safety-valve on which much British life depended. Once that valve was burst by Mrs Butler, it could be said that all kinds of other changes were inevitable, changes not only to the lives of poor women, but to women in general. Once women were liberated from the thraldom in which Charlotte Brontë and her sisters grew up – in which the choice of career was confined to marriage or becoming a governess, or some equally old-maidish occupation – the world was indeed going to change.

In the course of writing this book, my admiration for Strachey himself has revived. His lightness of touch, his pure elegance, is not something which I could imitate. We live in different ages, and that is the point. My initial purpose in stealing his title was to cut him down to size, to remind his ghost that there really *were* eminent Victorians, giants in those days. Now that the book is finished and laid before the public, I can only repeat his words, 'Je n'impose rien; je ne propose rien. . . .' His book is a thousand times better than mine, and it will still be read when mine is forgotten. But when both Strachey and I are forgotten, they will still be reading *Jane Eyre* and Newman's *Apologia*.

PRINCE ALBERT

1819-61

Father of the modern British monarchy

By the end of his days no one looked more respectably middle class than the Prince Consort, photographed here by Mayall.

When Queen Victoria died, on 22 January 1901, few people can have felt a heartier relief than her personal physician, Sir James Reid, whom she had led a sorry dance for the previous twenty years. Deprived of the power to be a tyrant in a political capacity, Victoria as an old lady had enjoyed keeping all her courtiers on their toes. Sir James was even dragged away from his own father's death-bed to attend to the Queen's favourite, John Brown. And there was no moment of Sir James's private life which his monarch considered too private that she could not intrude upon it with bulletins of her own digestive or bowel processes.

Demanding to the last, she left Sir James with rigorous instructions for her own burial. It was, in fact, probably the first time that Sir James had ever seen the body of his patient, since in life she had a strong sense of her own dignity and only allowed him to discover the state of her health through verbal inquiry. Now, if he chose, he could look at her. His wife considered that in death 'her face' was 'like a lovely marble statue, no sign of illness or age, and she still looked "the Queen", her veil over her face and a few loose flowers on her bed – all so simple and grand.'

There was to be no embalming and no Lying in State. The Queen, as head of the Army, had asked for a military funeral, and she was to lie in her coffin wearing – as presumably heads of the Army do in death – a simple white silk dress and the Order of the Garter. What was less simple was her request that Sir James should order her coffin to be filled with mementoes. Though the Queen may very well have been in law the Head of the Established Church of Scotland and the Established Church of England, she seemed in death to be casting back to the chieftains of the Viking Age, who set sail on their funeral barges laden with the jewels, wives, armour and hens which they had managed to accumulate during a life of plunder; or perhaps she harked back to the Pharaohs, who liked to hoard in their pyramid tombs the food, the libations and the ornaments which would accompany them for their long trek through eternity. So, Sir James recorded in his journal, he was instructed to place into Queen Victoria's coffin favourite photographs of Prince Albert, and of her children; a garment worked upon by Princess Alice; keepsakes from all her favourite servants and relations – chains, bracelets, lockets, shawls, handkerchiefs. Then there were the plaster casts which she had had taken of her children's little limbs. Heaven was to be cluttered with *objets* and mementoes, just like one of the Queen's drawing-rooms at Osborne or Balmoral.

The funeral took place on 22 February 1901 and – as has very often been observed – it was one of the last great occasions when all the crowned heads of Europe were assembled together: all those kings and grand dukes and emperors whose power and manner of life, within a couple of decades, would have been blown from the earth by war and revolution. Of all the dead little mementoes which Sir James Reid had placed in Queen Victoria's coffin, surely none could have been deader, in 1901, than the concept of monarchy itself.

As we watch them today on that faded old film, in their stiff uniforms and cocked hats, the monarchs seem almost as pathetic as those other sad films of fifteen years later – the wounded returning from the Front, their eyes bandaged, their sight destroyed by the gas. Both are processions of the blind. The twentieth century was soon to be called the century of the Common Man – but it was a title more properly owing to the Victorian Age. The century which produced Karl Marx and Mr Pooter could not for long contain these medalled chests, these fluttering feathers, these spiked helmets, these proud empty heads. Monarchies are expressions of national mythology. They survive in societies which, willingly or unwillingly, place all power in the hands of one individual and their circle. It usually requires perpetual vigilance by the military, and a preparedness to put down dissent with the utmost ruthlessness for modern (post-Napoleonic) monarchies to survive. In the seventeenth century, monarchs survived by claiming that they had been put on their thrones by God. But in the nineteenth century, when God himself was dethroned, where did that leave the monarchs? Even in countries like Russia – where the news of God's death (proclaimed by Nietzsche) was slow in arriving – the divinely-appointed Emperors needed to protect themselves with savage laws and cruel intolerance. It was of no avail to them when the time came. The Romanovs, like the Bourbons and the Hapsburgs and the other great European dynasts who assembled behind that well-laden coffin of Queen Victoria, were all doomed to be scrubbed out of the history books. They were as obsolete as their clothes.

Victoria's son, Edward VII, was a popular king in England – largely perhaps because people felt sorry for him having to wait so long to inherit. In many ways, he was unlike his parents, a determined throwback to the Hanoverian days of dissipation when kings were really kings – their favourite occupations being to interfere in international politics and to sleep with other men's wives. Edward VII's love of Paris provided ample opportunity for both activities, and the establishment of the famous Entente Cordiale. But had he lived longer into the twentieth century, I wonder whether the British would have continued to like him, or even to keep him on at Buckingham Palace? The long reign of King George V and Queen Mary was a reversion to something very different. With the brief and unhappy interlude of the reign of Edward VIII, the British royal family has kept going where many another ancient dynasty has collapsed. Whereas all the famous kings and queens of Europe have been famous precisely because of their political influence, the British have decided that their monarchs 'traditionally' have no political bias or indeed interest whatsoever. How can this convincingly be argued on behalf of a family who, merely on a private level, must own half the wealth of Scotland and Norfolk, and who, in their ritualistic public display – in their State Openings of Parliament, their glass coaches, their crowns and uniforms – hark back to an infinitely obsolete hierarchical view of society?

The answer, surely, partly resides in the fact that the British royal family is *déclassé*. They certainly are not aristocratic, though they have married into the aristocracy. The Queen Mother, for example, is very evidently a Scottish aristocrat when she is met in a private capacity. But her public role is something rather different. We would not use phrases like 'the Queen Mum' about a duchess or some terrifyingly upper-class lady from the pages of a gossip column. Her Majesty Queen Elizabeth the Queen Mother has become a fantasy grandmother for everyone – for members of the working class as much as for the rich and grand. Hence her extraordinary clothes, which are less those of an aristocrat than of a charwoman who has won the football pools.

When you come to examine the irrational basis by which newspapers assess the popularity or otherwise of members of the royal family, you find that it resides precisely in the extent to which they are usable as vehicles of such fantasy. Those who conduct themselves like their dethroned royal cousins in Europe, who flaunt their wealth, or stamp their feet, or 'come the royal personage' are on the whole detested by the British public, though they provide amusement for the readers of American or European newspapers and magazines. Those on the other hand who appear like mega-versions of us, the middle or lower-middle classes, are adored. It does not really matter if the activities themselves – four-in-hand racing, shooting, wild-life preservation – would be beyond the means of many of us. Prince Philip seems essentially like Dad with his hobbies. And though it may be the case that members of the royal family have gradually begun to allow into their midst divorced persons – and even on occasion to get divorced themselves – they have maintained, throughout the century, the sort of level of morality which would have been expected not of an aristocratic family but of a family like mine – a provincial middle-class to lower-middle-class family which had never, until the 1970s, known a divorce.

How did this strange institution, the modern British monarchy develop, with its apolitical constitutional position, its fuddy-duddy morals, its concepts of public service, and its rather quaint notion – particularly among the males of its number – that they should be 'good all-rounders'? It is not in the least a natural idea, and it certainly has very little to do with any other type of monarchy which you might encounter in the history books. The answer, I suspect, is to be found in almost every big town in Britain. Either consciously or unconsciously, nations betray themselves by their monuments. When we visit the Soviet Union, we perhaps feel that the Russians are trying too hard to persuade themselves that Lenin was responsible for all the good things in the world when he was, in fact, as absolute a monarch as the Tsars, and ruthlessly imposed a political system on the people from above.

The British abundance of statuary, dating from the 1860s onwards, portrays a man whose achievement was subtler than that. If the favourite comic book of the latter end of Queen Victoria's reign was *The Diary of a Nobody*, then their favourite statue was of a prince who, by all historical rights, should have been a nobody, but whose peculiarly small-scale qualities matched the moment and created something which was to survive and outlast all the most illustrious dynasties of Europe. For the modern British monarchy is, without question, the invention of Prince Albert.

Prince Albert was born on 26 August 1819 in the tiny German Duchy of Saxe-Coburg, the second son of preposterously ill-matched parents. His father Ernest, the Duke of this diminutive domain, was an old roué whose protruding lower teeth and baggy, bloodshot eyes gave him the appearance of a lecherous bulldog. He had innumerable illegitimate children by more or less any woman who was to hand but he had chosen as his bride a girl of only sixteen years old. Not surprisingly perhaps, the Duchess Louise abandoned him at the earliest possible opportunity, eloping to Paris with an officer in the Coburg army. She died in 1831, when Prince Albert was only twelve. He did not see much of her after the age of five.

None of this really mattered, because the truly important characters in Prince Albert's story were not his parents but three other people in every way more robustly ambitious on the family's behalf – his Uncle Leopold,

The Schloss Rosenau was the country retreat in Saxe-Coburg where Prince Albert spent his childhood. It always occupied a special place in his heart and he recreated much of its atmosphere at his Scottish seat, Balmoral.

his Aunt Victoire, and above all the family doctor, tutor and general adviser, the Baron Christian Frederick Stockmar.

First – Uncle Leopold, Duke Ernest's younger brother. It did not worry him that, as a younger son, he stood no chance of inheriting an obscure German duchy. He managed to marry, in 1816, the only child of George IV – and the heiress to the throne of England and Scotland. When Leopold married Princess Charlotte, he took the title of Prince Consort and Parliament voted him a stipend of £50,000 per annum. It was very clear from the outset that Leopold did not intend to be a mere stooge. He was determined to interfere in English political life as much as he could, and to take upon himself the power as well as the state of a monarch.

Strangely, this was not merely acceptable but positively appropriate in the entrepreneurial world of early nineteenth-century England. Just as the old Whig aristocratic dynasties were going to have to stand aside for the prosperous members of the middle class, so the British were ripe to be governed by a self-made monarch.

There was no historic or dynastic link between the throne of England and the Duchy of Saxe-Coburg. Leopold and Stockmar forged the links which were eventually to sweep away the old Hanoverian inheritance of George IV and his brothers, replacing it with what they hoped would be a Germanic autocracy of the kind favoured by Frederick the Great of Prussia. All their ambitions were fulfilled, but fulfilled in ways they did not in the least expect. Leopold, who looked so likely to produce a new royal line in England, was rather to live to see the progeny of his brother and sister ruling there in his stead.

The first blow to Leopold's ambition fell when Princess Charlotte died in childbirth after he had been married to her for only eighteen months.

Albert's Aunt Victoire had also married into the British royal family. In the scramble, not without comedy, which followed the death of Princess Charlotte, the Royal Dukes, none of them thin, none of them young, all attempted to provide themselves with the dignity of siring a future king or queen of England. To this end, Edward, Duke of Kent, third son of George III, who had previously shown no predilection for marriage, chose as his wife Leopold's sister. She had been married before to Emich Charles, Prince of Leiningen, who died in 1814. By her first husband, Victoire had produced two children – Charles, Prince of

Leiningen, and Feodora, who was doomed to marry the infinitely obscure Prince Ernest of Hohenlohe-Langenburg. Victoire's second husband was the only surviving son of George III who managed to produce a legitimate heir which is why the first-born of Victoire of Saxe-Coburg should have died as a German princeling of infinite obscurity, but her third child, by virtue of being the daughter of the Duke of Kent, should have been Queen Victoria, Defender of the Faith, Queen of the most prosperous and expansive commercial power in Europe and – in the fullness of time, through the fancy of her most fanciful and favourite Prime Minister – the very Empress of India. Not bad going for the House of Saxe-Coburg – a duchy rather smaller in size than the English county of Dorset.

But the glorification of the House of Saxe-Coburg, the unification of its two scions – Victoria and Albert – would perhaps never have happened, or never have had such momentous consequences, had it not been for the efficient stage management of Leopold's indispensable right-hand man, Baron Christian Frederick Stockmar.

Stockmar – described by Mr Gladstone in splenetic mood as 'a mischievous old prig' – was not only Prince Albert's tutor and mentor, but a perpetual presence in his life until the middle of the century. It was Stockmar who recognised that Albert's good looks and intelligence marked him out as something rather special – anyway as far as the house of Saxe-Coburg was concerned. It was Stockmar who saw that the boy received a good education – a grounding in languages, mathematics, and literature, but also an appreciation of the manner in which school-room subjects might affect the world. From an early age, Albert was trained in economics, agriculture, architecture. He learnt to paint. He was keenly musical. Stockmar made sure that Albert, together with his far less promising elder brother Ernest, travelled to Italy to study art and music; and that the boys attended Bonn University. It is quite unthinkable that any of George III's sons might have gone to Oxford or Cambridge, let alone submitted themselves to the wide-ranging academic disciplines which were expected of German students of the time. Stockmar was busy creating a monarch for the modern age – a figure who could preside with plausibility over an expanding industrial and imperial power.

Ever since it had become clear that none of Victoria's uncles would produce an heir, Leopold had been using his influence with his sister to groom the young princess of Kent for the office of monarch of England.

The Duchess of Kent was a silly, somewhat tyrannical woman, under the sway of her devious secretary and adviser Sir John Conroy. Throughout Victoria's minority, he (together with her governess Baroness Lehzen) brought appalling pressure on her to behave as they thought fit. On more than one occasion – sometimes choosing moments when Victoria was seriously ill – Conroy tried to force her to sign pledges that she would appoint him as her secretary, and in effect the Regent, if she were to succeed before her majority. Common sense, plain stubbornness and her Uncle Leopold – who at one point sent over Baron Stockmar to advise her – saved her. By the time she did succeed, at the age of eighteen, Victoria had formed her first important political alliance – with Lord Melbourne – and she was able to send Conroy, Lehzen and her mother about their business. It is one of the most liberating stories in history when their spell is broken and Victoria reigns alone.

The marriage between Victoria and Albert, the two cousins, which took place on 10 February 1840, was something which had been mooted since their childhood. It was fortunate for all concerned that Victoria, a self-willed individual, who was, in the early years of her reign, very happy in her freedom, very much enjoying power, and totally besotted with her Prime Minister, Lord Melbourne, should have fallen in so readily with the plan to get married. After all, she was only twenty and Albert just a few months younger.

It was physical infatuation which decided her, as her Journal entry for 15 October 1839 makes clear:

> I said to him that I thought he must be aware *why* I wished him to come here, and that it would make me too *happy* if he would consent to what I wished (to marry me); we embraced each other over and over again, and he was *so* kind, *so* affectionate; Oh! to *feel* I was and am loved by *such* an Angel as Albert was too *great a delight* to describe! He is *perfection*; perfection in every way – in beauty – in everything!

It has the breathless enthusiasm of a schoolgirl who has been given a new puppy. For the next twenty-two years, Albert and Victoria were to provide an ideal royal partnership and, through their nine children, exercise a vast influence in almost all the royal families of Europe. By the close of the century the blood of Albert and Victoria had provided kaisers, tsars,

This revealing photograph by Roger Fenton captures the Queen and the Prince Consort in a thoughtful mood. 'I am only the husband,' Albert confided to a friend, 'and not the master in the house.'

grand dukes, kings and queens for Germany, Russia, Schleswig Holstein, Hesse, Norway, Romania, Spain, Sweden, Greece and Denmark.

But, beyond providing a sort of royal stud farm from the comparatively untried stables of Saxe-Coburg, what was the role of the Prince Consort to be? What was Albert's function? To ask a more fundamental question – what was the function of the monarchy itself?

It should not be supposed that the overweening ambition of the Saxe-Coburgs – of Leopold, of the Duchess of Kent, of Stockmar – had gone unnoticed among the Whig aristocratic families, nor among the rising new Tories who held the key positions of power in London political circles. By the same token, Albert had not been educated by Stockmar to think of himself either as an upper-class roué like his father, nor as a stuffed dummy to stand behind the Queen on ceremonial occasions.

The determination of Parliament to show the new arrival that he must not overstep the mark *politically* was clear very early on when the House of Commons was discussing the question of Albert's allowance. Deep resentment was still felt against King Leopold, who had by then taken some million pounds out of the Exchequer. Led by the Tory Opposition of Sir Robert Peel, the Commons voted overwhelmingly that the new

Prince should only be granted £30,000. It was, of course, enough to live on. But a message had been sent by Parliament to the Court, and it was one which was profoundly resented.

Victoria had grown up with the English tradition that the monarch, like most of the good families in London, was a Whig. Her atavistic distaste for the Tories turned to active dislike when she saw the way they intended to treat her beloved Albert.

But Albert himself was to change all this. When it came to the point, it was discovered that his own political views coincided much more closely with those of Sir Robert Peel than with the old Whigs. But this was not why he urged the Queen to abandon her early partisanship. Taught by Stockmar, he believed that a sovereign should be above party politics. Stockmar's reasoning was that the sovereign could thereby continue to influence events whichever party was in power. It was because he wished to extend his power, not reduce it, that Albert took the Olympian non-party view. By paradox, however, it was this which was to enable the British monarchy to survive when so many others went under. Even when the last vestiges of political power had, in effect, slipped through their fingers, the British royal family had – thanks to Prince Albert – established a role for themselves in national life beyond their political function as constitutional figureheads.

This is not for a moment to suggest that Albert wanted the monarchy to become politically emasculated. He made frequent efforts to interfere in the government of the realm. The earliest rebuttals he received were from Victoria herself. She refused him, in the initial stages of their marriage, permission to see any of the official State papers which came to the palace in boxes to be signed and approved.

'In my home life I am contented,' Albert confided to a boyhood friend, Prince William of Lowenstein, 'but the difficulty of filling my place with proper dignity is that I am only the husband, and not the master in the house.' The faithful Lord Melbourne thought that it was foolish of the Queen to be so cagey with the boxes. 'The Queen has not started on a right principle – she should by degrees impart everything to him, but there is a danger in his wishing it all at once.' In fact, Victoria did consent to allow Albert to handle some of the State papers during her first confinement barely nine months after their marriage and, in the event of her death and the survival of an infant king or queen, it was agreed that Albert should be the Prince Regent.

Little by little, Victoria learnt to rely on Albert's political judgement. For example, her violent hostility to the Tories became modified and she came to share his admiration for Sir Robert Peel. The principle of the sovereign being above party rivalry began to develop.

Naturally, given his birth and origins, Albert took a keener interest in foreign affairs than Victoria did at first, and it was usually in this area that he clashed with the politicians. Palmerston was his greatest sparring partner, and there were some embarrassingly public differences between the Foreign Secretary and the Prince Consort. For example, by the end of 1846 the Portuguese were in a state of virtual civil war. The clash was between the ultra-reactionary royal family, whose Queen Donna Maria was being, as Victoria put it, 'as foolish as ever' and a revolutionary military junta. Albert was anxious that there should be a commonsense constitutional solution, with concessions made by the Queen to the demands of the libertarians. A representative was sent in the person of one Colonel Wylde to see if there was any possibility of compromise. Wylde was sent by the Government but Albert felt that he was very much the representative of the royal family, and it infuriated him to learn that, while Wylde's private negotiations were in progress in Lisbon, Palmerston was openly instructing the British diplomats in Lisbon to support the rebels. 'The Queen of Portugal had told Wylde that the belief that England wished well to the cause of the rebels was one of the chief causes of their strength,' wrote the furiously sensible twenty-six-year-old Albert to the cheerfully belligerent sexagenarian Palmerston.

Both Victoria and Albert profoundly disliked the cynicism with which, after the 1848 revolutions in Europe, Palmerston was prepared to treat with the enemies of monarchy. When Louis Philippe was thrown off his throne in France, for example, the Queen was scandalised that Palmerston could write to her, in his bluff way, that the adventures of the royal refugees were 'like one of Walter Scott's best tales'. (Like many girls of her generation, Victoria had a passion for Scott.) Similarly, when the King of Sardinia led an army of liberation against the Austrians, the British Crown and the British Foreign Office were once more at variance. Britain, urged Albert, by which he meant himself and his wife, could not endure the creation of an independent Italian state.

But – this is where the Darwinian powers of survival by the Victoria and Albert team show themselves so strongly – they did, of course tolerate something which was totally contrary to their original wishes. In a similar way, having expressed horror at the expulsion of Louis Philippe, they were perfectly happy to befriend Napoleon III and to accord him full imperial honours. Consistency is the enemy of survival in political life. Albert showed the powers of adaptation which (as Darwin was demonstrating with the lower species of the Galapagos islands) is the essential ingredient in ensuring genetic success. Figures like Queen Donna Maria of Portugal were well on their way to becoming political dinosaurs.

One sees this most glaringly in the most exciting of Palmerston's political adventures. Throughout his career, Palmerston had been a famous advocate of gun-boat diplomacy. Never send an ambassador if a destroyer will do the job, seemed to be his approach. When trouble blew up in the Black Sea, and the Russians occupied Balkan territories which by right belonged to the Ottoman Turks, there was an international outcry. The Cabinet in London was in general in favour of finding some

"DANGEROUS!"

This Punch *cartoon of 7 January 1854 makes play with Albert's known love of skating and shows the suspicion with which his desire to intervene in foreign affairs was regarded.*

peaceable solution to the problem. However, egged on by Palmerston, the Prime Minister of the day, Lord Aberdeen, prepared for the British Fleet to enter the Black Sea. They did so without consulting the Queen, who was on holiday at Balmoral. Albert was furious. He wrote to the Cabinet that it was 'morally and constitutionally wrong' for the Cabinet to act without consulting the Sovereign. 'We ought to be quite sure', he wrote that 'they do not drive at war while we aim at peace.'

But war fever and anti-Russian feeling swept the country. The news that Albert was opposed to the war was carefully leaked to the Press by his political enemies – very likely by Palmerston himself. By the end of 1853, the papers were full of suggestions that Albert was a pro-Russian. The *Daily News* complained that 'The nation distrusts the politics, however much they may admire the taste, of a Prince who had breathed from childhood the air of courts tainted by the imaginative servility of Goethe.' *Punch* carried a cartoon of the Prince nervously skating by a board labelled 'Foreign Affairs, very dangerous'. (The point of the drawing was that Albert was a keen skater, who helped to popularise the sport in England.)

But it is entirely typical of this adaptable monarchical species that Albert, far from maintaining an anti-war line, became, once the Crimean hostilities had broken out, a keen patriot.

Moreover, the war became a magnificent excuse for him to exercise his uncontrollable urge to suggest reforms and reorganisations of inefficient British institutions. None was worse organised than the Army itself. He had a chance to see this at first hand by observing the antics of his own regiment, the Eleventh, or Prince Albert's Own Hussars – better known in history as the Light Brigade. Lord Cardigan, who had bought his command of the regiment for £40,000, had transformed, at huge personal cost to himself, the Hussar blue of the uniform to crimson, the colour of the Saxe-Coburg livery; the motto of the regiment was changed to Prince Albert's own – *Treu und Fest*.

Not all Albert's attempted Army reforms were particularly happy. *Punch* described his own design for new helmets for infantrymen as 'a cross between a muff, a coalscuttle, and a slop pail'.

Albert watched the spectacle of the Crimean War with a mixture of excitement and horror. Sixteen thousand British troops died of dysentery and disease before a single shot was fired. By the time the Russians had withdrawn from the disputed territories in the Balkans, there was no reasonable ground for the war to continue. But the newspapers at home were crying for blood. They wanted their army to 'lick the Russians'. The Prince was gravely suspicious of the Press. He had suffered enough at their hands, and he felt that far too much about the war was let out by the gossips who masqueraded as 'war correspondents'. 'The pen and ink of one miserable scribbler', he wrote, 'is despoiling the country of all the advantages which the hearts blood of 20,000 of its noblest sons should have earned.'

Both he and the Queen were profoundly impressed by the work of Florence Nightingale at Scutari. 'Such a head!' said Victoria when she heard of the organisational skills of the Lady with the Lamp. 'I wish we had her at the War Office.'

Perhaps the most famous moment in the war, however, occurred when, as a result of misheard misdirections, the Noble Six Hundred of the Light Brigade charged to their deaths, the inevitable victims of the Russian guns. Tennyson's famous lyric sanctified the blunder:

Honour the charge they made! . . . Noble six hundred! . . .

Prince Albert's Colonelcy-in-Chief of the regiment helped to dignify it.

At Deene Park in Northamptonshire, the seat of Lord Cardigan, there hangs a famous picture which depicts Lord Cardigan at Windsor Castle explaining the Charge to Prince Albert and his family. It is a strange icon, which speaks volumes. On the one hand we observe that Cardigan, who was a bad man and a bad soldier, has been transformed into a military hero. He had returned to England after the Crimea to the strains of 'See the Conquering Hero comes!' And already, by the time this picture was painted, all over Britain, streets and squares and villas were being named after him. So it is that we find Prince Albert by his side – and with him, the royal children. But there is one very noticeable absentee from the canvas – someone we should expect to have been beside Prince Albert's side when the heroic story of Balaclava and the Light Brigade was related: Queen Victoria herself. The apocryphal story, which I rather like, is that the Queen was in the original painting but that she had herself painted out when she discovered the profligacy of Lord Cardigan's lifestyle: that Deene had for years been the scene of wild parties, attended by London prostitutes, and that Cardigan himself had for years openly lived with a woman half his age before making her his second bride.

As the *Punch* critic was supposed to have remarked when attending Shakespeare's *Antony and Cleopatra*, 'How unlike the homelife of our own dear Queen.'

Albert, from the moment he arrived in this country, brought with him an atmosphere of respectability and 'family values' which – like his political views – owe far more to Baron Stockmar than to his own dissolute parents. It may be that both Albert and Victoria were consciously repelled by 'aristocratic' dissipation because of the lifestyles of their parents and uncles. It may be that their reaction against all 'that kind of thing' was instinctive, each generation naturally behaving differently from its parents.

Albert was a keen huntsman and showed courage in the saddle which impressed his early hosts in England who had not expected him to be equal to riding to hounds in the Quorn hunt. He is pictured here with his beagles.

Deene Park is a good place to reflect on that fact. Cardigan had been one of the first English noblemen to welcome Albert to England and the Queen and Prince Consort had visited Deene Park in the very early days of their married life. The Queen was acutely conscious of the fact that the English aristocracy wanted to put Albert to the test, and one obvious way of discovering the man's mettle was to see whether he was brave enough to hunt.

He followed the Belvoir Hunt in 1843 at a meet when the going was so rough that even his secretary, George Anson, fell off. 'Albert's riding so well and so boldly, and so hard has made such a sensation that it has been written all over the country,' said the Queen in an enthusiastic letter to Uncle Leopold. It was true. Whatever else he was, the man was no milk-sop. 'It has put an end to all sneering for the future about Albert's riding.'

But unlike her father, and all her royal uncles, Victoria was never to move freely among the aristocracy, nor to belong to any fast set. Both lonely in childhood, both naturally demure, both passionately fond of one another, Victoria and Albert wanted to be homemakers rather than guests in the great houses of England.

Albert was a punctilious administrator, and he soon set to work reorganising the royal households. He thoroughly overhauled the chaotic administration of Buckingham Palace. At Windsor Castle, which had lately been refurbished by George IV, Albert made a true home, adorning the walls with paintings which he had discovered in the vaults of Hampton Court, and with modern masterpieces by his favourite painters Landseer and Winterhalter. At Windsor, too, he experimented in the establishment of a model village on the estate, and a dairy was set up according to his own plans and run according to his own specifications. Back at the Castle, he could watch his children grow, and indulge his inherent love of family which, throughout his own childhood, had been so much frustrated. It was he who, in 1840, first popularised Christmas trees here. If it was Charles Dickens who might be said to have invented the English Christmas three years earlier in *The Pickwick Papers*, it was surely Prince Albert who gave the English Christmas some of its more distinctive features.

But tireless as he was at making sure that Windsor and Buckingham Palace were organised as befitted royal residences, Albert was really always hankering after a *schloss* on the German pattern. He had enormously enjoyed his time spent in childhood among the quiet simplicities of Schloss Rosenau, a neo-Gothic country retreat, near Coburg, the place where for five brief years – before she was banished – Albert had known the company of his mother. It was the atmosphere of the Rosenau which he and Victoria wished to recreate for themselves in their two family residences in Scotland and in the Isle of Wight.

As lovers of late Romantic art and literature – the poetry of Sir Walter Scott, the paintings of Landseer, the music of Mendelssohn – it is no sur-

The royal family at Balmoral, 29 September 1855. Left to right, Prince Alfred, Prince Frederick William of Prussia, Princess Alice, the Prince of Wales, Queen Victoria, Prince Albert, the Princess Royal (Vicky) who on this day became engaged to Frederick William (Fritz), the first of the great dynastic matches forged by Queen Victoria.

prise that Victoria and Albert should have fallen in love with the Highlands of Scotland. Their initial intention was to find or build a Highland home on the West Coast. But although they were in raptures at the beauty of Argyll – of the Crinan canal in particular and the seaviews of the Hebrides – they concluded that the west was too wet and too fly-blown. After several delightful visits to Scotland they finally, in 1848, alighted upon Deeside, one of the driest and sandiest parts of the north-east. Balmoral Castle had been rebuilt in the 1830s by Sir Robert Gordon in the Scottish Baronial manner. Victoria and Albert greatly added to it. A ballroom, where they could both learn to dance reels, was to be one of the necessities of life here: as well as dozens of extra rooms where guests, relations, and even members of the Government could be accommodated.

The royal link with Deeside continues to this day unabated. It was not merely the hunting, the painting in water-colours, the invigorating walks, the peace which the Queen and the Prince Consort enjoyed when they came to Balmoral. That, after all, could have been had in Wales or Yorkshire. As the yards of tartan carpets in Balmoral testify, the Queen and Prince Albert actually wanted to be Scottish. They wanted to dance reels, to wear the tartan, to attend the Kirk – thereby causing considerable offence to those members of the Church of England who took seriously her titular headship of that body – to hear the pipes played in the heather, and to sing old Scottish airs. Balmorality, as it came to be known, had an enormous effect on nineteenth-century taste.

Not on aristocratic taste, so much as on that of the middle and the lower middle classes. The Scottish Baronial style crops up in the unlikeliest settings. Purely suburban places like Great Malvern and Leamington Spa can boast their turreted middle-class versions of Balmoral. Even the thistle-festooned pubs and tartan biscuit tins which became a part of any Victorian man and woman's life owe their existence to Prince Albert's Deeside dreams.

The Scottish theme is carried on, too, in one of the most distinctive domestic interiors in the South of England – in the Horn Room of Osborne House on the Isle of Wight. This was Victoria and Albert's English country retreat, and once again, as you look about Osborne, you see how it is merely a big version of a style of house which might have accommodated not a duke or an earl but perhaps a doctor or a bank manager. Osborne was paid for by the sale of the Brighton Pavilion.

Victoria had refused to use her uncle's seaside fantasy house as her country residence. Ostensibly, there were good reasons for this. The Pavilion was on a public road, and there had been attempts on her life. More deep seated reasons must have had to do with her feeling about her uncles. The Chinoiserie and exotic interiors of the Brighton Pavilion had been the backcloth for a lifestyle which Victoria and Albert would assiduously avoid imitating.

Osborne was very dear to them both. When she purchased the estate in 1845, the Queen wrote to her Uncle Leopold: 'It sounds so snug and nice to have a place of one's own, quiet and retired.' That is what it felt like – a place of her own. She who ruled over the British Empire wrote with all the eagerness of a young wife whose husband – a bank clerk, perhaps, or a jobber on the Stock Exchange – having been compelled to share their early days with relations, had at last found an independent abode in the suburbs.

It was fitting that the Queen was to die here – for she had spent many of her happiest hours at Osborne. Her very accomplished water-colours of the place breathe lightness and happiness. The bright splash of the 'Hay Harvest at Barton Farm' – on the Osborne estate – is captured with all the colours in the royal paint-box. The seaviews of the Solent could be used for a holiday poster. How could she be anything but happy on the Isle of Wight? For this is where she came to be with her family, and it was very much the creation of Prince Albert.

Osborne reflected Albert's good taste and modernity. Since, for many years, Victoria lived there as a widow, we need to discount her additions to the place if we are to catch a glimpse of Albert's original conception. Clutter, which she loved so much as an older woman, did not fill the rooms as Albert conceived them. The beautiful corridor, leading to the Audience Chamber, houses his collection of fine sculpture. At the same time, the clear, stately interiors planned by this Enlightenment Prince, contain jarring surprises, explicable in terms of his homesick childish self which he never outgrew. The chandelier in the Audience Chamber, florid with glass convolvulus, exactly echoes the motif of the bedroom in the Schloss Rosenau where his mother gave birth to him. And the spectacularly hideous Horn Room, in which every single item of furniture – sofas, tables, lamp brackets – is made from antlers, would be more in place in some German hunting lodge than in the house of a modern British prince. The imperial exotica which Victoria added in her

LEFT The Horn Room at Osborne. All the furniture, lamp fittings and other features are made out of antlers and horns, as in some German hunting-lodge. Some of the more aesthetically surprising interiors reflect Albert's profound nostalgia for his German childhood.

RIGHT Osborne House on the Isle of Wight, designed to Prince Albert's specifications. Its Italianate style reflects Albert's love of the Renaissance.

widowhood – in particular, the splendid Durbar Room, which feels like the state apartments of an Indian nabob – ironically suggest a throwback to her uncle the Prince Regent's love of the oriental. One wonders whether Albert would have liked it.

When Osborne was still a dream in his mind, Albert planned out every bit of the place. He discarded all the architect's plans for the conversion of the original house and chose a master builder – Thomas Cubitt who had lately impressed him by his grand reconstructions of Belgravia, not far from Buckingham Palace. He planned the towers, the asymmetrical skyline, the bow windows with their superb views of the sea. He also planned that the place should have every modern comfort. Most unusually for the 1840s, Osborne had hot running water in the bathrooms. The Prince Consort's bath is presided over by two vast naked figures – is the man Albert or is it Hercules? The nudes which look down from the walls of Osborne with such uninhibited voluptuousness reflect the early married happiness of the royal pair.

If their parents had been strange and remote, Victoria and Albert took a close interest in the lives of their own children from an early age. Prince Albert appears at his most attractive as the young father of young children. When he had died, Victoria remembered for example how 'If he was not ready, Baby [Princess Beatrice] generally went into his dressing room and stopped with him until he followed with her at his hand

coming along the passage with his dear heavenly face. . . . Poor darling little Beatrice used to be so delighted to see him dress and when she arrived and he was dressed she made dearest Albert laugh so, by saying, "What a pity!".'

Stockmar's nursery prodigy from the Schloss Rosenau became, very easily and happily, the president of the nursery at Osborne. It is nowadays taken for granted, not only by the British public but by a much wider public in the world at large, that the royal family is a sort of emblem of family happiness, a picture of what it should be like in all well-conducted households. If this is so, it is the legacy of Prince Albert. No one held up the Prince Regent or George III as ideal fathers. Most British monarchs have had private lives which did not bear too close a scrutiny.

A royal group on the terrace at Osborne. From the first, Albert planned Osborne as a family house, where his children could grow up with the wholesome influence which had been largely lacking in his own childhood.

There is a paradox here with which the modern royal family is still living. It is frequently insisted upon that they have a private life which deserves to be respected. True enough – and all persons of good taste are offended when stories, usually untrue or wildly exaggerated, about the marriages and love lives of the royal family become the subject for splashy headlines in the newspapers. But the matter is complicated by the fact that, on one level, it is for their private lives that we now value the royal family. Royal weddings, which used to happen in private, are now big public events. Royal films give us tantalising glimpses of life behind the scenes at Balmoral or Windsor.

Such phrases as 'The Royal Soap Opera' reflect the actual function which the royal family occupies in the imagination, as well as in the newspapers, of the world. This is very much the legacy of Prince Albert, who longed to be a king, and a great political figure, but provided the British people instead with an image of family life.

This, incidentally, explains – I think – why so many British people feel comfortable with a matriarchal ruler. Their favourite monarchs – certainly their most long-lasting ones – have been women: two Elizabeths and Victoria. As for politicians – well, in the history of female politicians, it is early days. . . .

Strange psychosexually determined prejudices underlie our attitudes to the archetypal family-projection which the British have set up in Buckingham Palace. Instinct appears to tell many people that it is right and proper to have a Queen. But where does that leave the male members of the royal family? In asking this question, we are not merely asking what is the function of princes, but what is the function of males. When they have done their work as studs, and supplied the world with the next generation, is there not an element of pathos about most men as they sink into middle age? Few of them do 'work' which really needs doing. Are all the great schemes promoted by royal men – the concern for good causes, the sporting life, the interests in art and science – just a glorified version of Dad and his hobbies? The royal equivalent of the pigeon hutch at the end of the back yard?

If so, Prince Albert led the way with some very distinguished pigeon hutches indeed. The most splendid of them all was the Crystal Palace, the site of the Great Exhibition in Hyde Park in 1851. The idea originated with Henry Cole, a leading light of the Society of Arts, of which Prince Albert was the patron. Cole had visited the Paris Exhibition of 1849, and fired Prince Albert with the idea that London could go one better – hold a vast international exhibition, displaying all the best achievements of modern art, science and technology, and at the same time advertising British supremacy and boosting British trade. Launching the appeal for subscriptions to pay for the exhibition, Prince Albert said, 'The distances which separated the nations are rapidly vanishing before the achievements of modern invention, and we can traverse them with incredible ease. . . . The publicity of the present day causes that no sooner is a discovery made than it is already improved upon and surpassed by competing efforts. The products of all quarters of the globe are placed at our disposal and we have only to choose which is best and cheapest for our purposes and the powers of production are intrusted [*sic*] to the stimulus of *competition and capital*.' (His italics.) He could have been a modern Prince speaking of the competition from Japanese technology in the age of the micro-chip.

London had never seen anything like the Great Exhibition of 1851. There were 13,937 exhibitors – 6556 of them foreign, the rest British. There were over 100,000 exhibits, and all housed in the vast, airy spaces of what came to be called the Crystal Palace, a huge glass construction, designed by Joseph Paxton with many hints from Prince Albert himself. Everyone remembers the crisis when the birds inhabiting the elm trees of Hyde Park encased within the glass began to cause havoc to the exhibits below. What could be done? The Queen consulted the aged Duke of Wellington, who knew the answer to everything. 'Sparrowhawks, ma'am', was his solution.

But it is sometimes easy for us to forget the extraordinary visual and emotional impact of the Crystal Palace. Now the name just refers to a football team, for the reason that after the exhibition, the Palace was reconstructed in the suburbs of South London on Sydenham Hill. The local team took its name from the Palace which was destroyed by fire in 1936. Now that it's no longer visible, we easily forget how magnificent it was. Queen Victoria recorded, 'The sight as we came to the centre where the steps and chair was [*sic*] placed, facing the beautiful fountain was magic and impressive. The tremendous cheering, the joy expressed in every face, the vastness of the building, with all its decorations and exhibits, the sound of the organ (with 200 instruments and voices, which seemed nothing) and my beloved husband, the creator of this peace festival "uniting the art and industry of all nations of the earth", all this was moving, and a day to live forever. God bless my dear Albert, and my dear country, which has shown itself so great today.'

On the day it was opened, 700,000 people lined the route from Buckingham Palace to Hyde Park. Over 30,000 got inside the Crystal Palace to see magnificent achievements of art and science – the statue of the Queen made out of zinc, the group of stuffed frogs, one of them holding up an umbrella, the patent pulpit fitted with gutta-percha tubes to the pews of the deaf, or the doctor's walking stick which contained an enema. Charles Kingsley, author of *The Water Babies*, wept when he stepped over the threshold to see so united a display of human ingenuity. So did the Prime Minister, Lord John Russell, and the Home Secretary. And, when it was all over, nobody wept to discover that the Royal Commissioners who had patronised the exhibition had made a profit of £180,000, which Albert could use to erect a permanent exhibition-centre in South Kensington.

The Great Exhibition of 1851 was one of Prince Albert's finest achievements. The exhibition was housed in the Crystal Palace, the first prefabricated building in history, and was the brainchild of Joseph Paxton who had designed the glasshouses at Chatsworth.

He wanted his Albertopolis, as it came to be known, to be a vast structure with four institutions devoted to raw materials, machinery, manufacture, and the plastic arts. Modern industry should be shown to be as important as Geology and Zoology, which both had societies in London patronised by royalty. In the event, the Albertopolis did not quite get off the ground in the form that the Prince wished, but the South Kensington museums – particularly the V and A – survive as a permanent reminder of his patronage of the arts and sciences.

His artistic taste is interesting. We tend to associate him exclusively with the glossy, tranquil canvases of Winterhalter, and the set-piece doggy or hunting scenes of Landseer. Certainly these represent one side of Albert's nature. But the paintings which he actually bought with his own money reveal a broader range of sympathies than that. There is comedy (inextricably connected in our minds with Dickens whose novels he illustrated) in George Cruikshank's *Disturbing the Congregation*, bought by the Prince in 1850. There is the stagey but decidedly erotic passion of Augustus Egg's *L'Amante*, which Albert hung in his dressing-room at Windsor. Most striking of all, perhaps, is Albert's taste for the work of John Martin, the visionary landscape painter whose *Eve of the Deluge* Albert purchased in 1840.

But it would be paradoxical not to recognise the German-born Winterhalter as the Prince's great *trouvail* and creation. As so often happens with the relationship between an artist and a painter, Albert himself became Winterhalter's greatest work, whether as the paterfamilias, surrounded by adoring and adorable children who appear to have been moulded – like something in the Brothers Grimm – out of candy or coloured marzipan, or as the serious, slightly tubby, slightly balding uniformed statesman, arrayed in the uniform of the Colonel of the Rifle Brigade.

In Winterhalter's most revealing royal canvas – *The First of May 1851* – we see Albert's creation, the Crystal Palace, in the background. But the foreground is uneasily reminiscent of the iconography of Christian art. The subject is notionally the Duke of Wellington who is kneeling to present a gift to his godson and namesake Prince Arthur (later to become the Duke of Connaught). The baby holds out to the grand old soldier a bunch of lilies of the valley, emblems – like the Palace in the background of the picture – of universal peace. Anyone who has ever visited a picture gallery cannot fail to be reminded of an Adoration of the Magi by one of the Italian masters. No wonder Albert looks, somewhat pensively, out of the canvas and away from the Mother and Child. For what part does St Joseph have in such a scene, except to stand politely, or reverently, in the background? Winterhalter did not wish us to identify the infant Prince Arthur with the Incarnate. But he was making a rather wistful point about Prince Albert himself.

What was his role to be, as the children grew up? Like many parents who are besottedly fond of their children as young infants, Albert rather

A sketch of Prince Albert by the German artist, Franz Winterhalter.

resented their reaching maturity; and he and the Queen had, in particular, an abhorrence of the Prince of Wales growing up like any normal young man. There was so much 'bad blood' in the family. They so desperately did not wish their first-born son to take after the dissolute bulldog Duke Ernest of Saxe-Coburg, or after any of Victoria's fat, lecherous, greedy old Hanoverian uncles. The programme drawn up by Prince Albert for the instruction of the Prince of Wales was dauntingly worthy. This was how the young Prince had to divide his day when he was seven years old.

RELIGION	half an hour at a stretch
ENGLISH	one hour
WRITING	half an hour
FRENCH	one hour
MUSIC	half an hour
CALCULATING	half an hour
GERMAN	one hour
DRAWING	half an hour
GEOGRAPHY	half an hour

No wonder, by the time that he grew up, that 'Bertie', as they called the boy, was determined to stretch his wings and enjoy himself. Unlike his father he was not clever, and he did not enjoy his enforced attendance at the Universities. They tried sending him to Oxford which he did not like. Then they tried sending him to Cambridge, which he liked even less. This was an embarrassment to Albert who had allowed himself to stand for election as the Chancellor of Cambridge University, and who took his position there extremely seriously.

The more they watched him grow, the more the Prince of Wales appeared to his parents like a reproachful throwback to the bad old days. Prince Albert's elaborate programme of education seemed to have no discernibly beneficial effects whatsoever on the young man. Unsure what to do with him next, they packed him off (like so many royal princes after him) to do a stint in the Army.

In the summer of 1861 Lord Torrington, 'that arch gossip of all gossips' as he was called by the Prince Consort, brought to Albert's attention what was being said in the London clubs. The Prince of Wales had, it seemed, formed a liaison with . . . an actress! What was more, he had brought her over . . . to Windsor Castle itself!

Albert was the father of nine children and, as the Queen's journals make abundantly clear, the royal pair continued to enjoy conjugal relations to the end. At the time when the Prince of Wales's liaison with the actress, Nellie Clifden, was uncovered, Albert was only forty-two years old. And yet he wrote to his son in terms which would have seemed prissy coming from a maiden aunt of seventy-five. The strongest passion of Albert's life, one feels on reading this letter, was a passion for respectability. Why should he have felt that royal families must be respectable? They never had been before. But the spectre of Prince Albert's puritanism has hovered over them ever since. It was Albert, through old Queen Mary and Stanley Baldwin, who put paid to Mrs Simpson's chances of becoming Queen of England. To George IV, she might have seemed an admirable bride. It was surely Albert's ghost who made it impossible for Princess Margaret to marry Group-Captain Townsend, at a period when large numbers of 'respectable' people did in fact condone divorce. It is surely the puritanical shade of Prince Albert who inspires the prurient and vulgar gutter-press of today in their constant vigilance to guard the moral standards of the royal house.

Bertie, his father wrote, might try to deny any association with Nellie Clifden:

> 'If you were to try to deny it, she can drag you into a Court of Law to force you to own it and there with you (the Prince of Wales) in the witness box, she will be able to give before a greedy multitude disgusting details of your profligacy for the sake of convincing the Jury, yourself cross-examined by a railing indecent attorney and hooted and yelled at by a Lawless Mob!! Oh horrible prospect, which this person has in her power, any day to realise! and to break your poor parents' hearts!

It is an extraordinary letter to have penned to an officer in the Grenadier Guards, as Bertie was at the time. Probably there was not a fellow officer in his entire mess who had not from time to time had some association with the likes of Nellie Clifden, and the idea of anyone being shocked by it is the purest Pooterism.

It was a sign of Albert's failing powers that he wrote in quite such immoderate terms. There is no doubt that the Prince Consort was under stress. In the spring of that year he had had to cope with Victoria's excessive grief on the death of her mother, the Duchess of Kent. His tireless overwork, his neuralgia, and his toothache had all helped to age him prematurely. The Adonis of Landseer's early canvases had been cruelly transformed by the passage of years into the paunchy balding figure of the photographs and the last Winterhalter renderings.

But it was not overwork which killed Prince Albert. Like so many of his wife's subjects, he was killed by typhoid, probably bred in the drains of Windsor Castle. It is astonishing to us, who are aware of the dangers of typhoid, cholera, polio and the like, to see how long it took our ancestors to guess the connection between water-supplies and disease. Prince Albert had encouraged Sir Edwin Chadwick in his life-work of providing England with proper drains. He did not realise that from the point of view of sanitation, Windsor Castle was as dangerous as a slum in one of the northern cities. His already overstrained constitution gave in to the fever, but its early symptoms went unrecognised. Albert was too busy. His preoccupation with his son's moral welfare, and his concern for the administration of government policies – Army reform, a crisis in British relations with the United States – continued unabated. But suddenly there entered in a new and premonitory note of resignation. 'I do not cling to life,' he had once said to the Queen. 'You do, but I set no store by it. I am sure if I had a severe illness I should give up at once, I should not struggle for life.' Isn't this a strange thing for a man hardly in early middle age to say to his wife? It surely reveals unhappiness, a sense that for all his business, his organising skills, his so-called 'interests' in art, music, the state of industry, he really felt his life to be totally pointless.

On 22 November 1861, Albert inspected the newly-opened Staff College and Royal Military Academy at Sandhurst, and took the opportunity to give a further dressing down to the Prince of Wales. It was a pouring wet day and Albert returned to Windsor drenched and feverish.

Anxiety about the American situation seemed to make his condition worse. The Civil War there had been going on for nearly a year. The Northern States had seized a British ship, the *Trent*, on the high seas and arrested two Confederate diplomats who were travelling under British protection. This act of aggression was just the kind of thing which delighted Lord Palmerston, the Prime Minister, and arrangements were made to withdraw the British Ambassador from Washington. Albert, shivering on his bed of sickness, was appalled by the intemperance of the diplomatic despatches and did in fact intervene. He couldn't eat. He was hardly strong enough to hold a pen. But he insisted on toning down the despatches sent by Palmerston and Lord John Russell and to withdraw Great Britain from the very brink of war with the Northern States. It was his last political piece of interference, but it was very effective.

It was against this background that Palmerston's medical advice to his sovereign must be seen. The Queen, and both her doctors, Sir James Clark and William Jenner, insisted that Albert was suffering from no more than a feverish cold. When the Prime Minister urged Victoria to take another doctor's opinion, she rejected his proposal as mere impertinence. On 3 December she wrote, 'Dreadfully annoyed at a letter from Lord Palmerston suggesting Dr Ferguson should be called in as he heard Albert could not sleep and eat. Very angry about it. In an agony of despair about my dearest Albert and crying much, for saw no improvement and my dearest Albert was so listless and took so little notice. Good kind old Sir James . . . reassured me and explained to Dr Jenner too that there was no cause whatever for alarm.'

As the illness progressed, Albert's diarrhoea got a little better, but his temper worse. Although still calling her Weibchen (little wife) he started to display violent anger against her. 'He is so kind calling me gutes Weibchen [excellent little wife] and liking me to hold his dear hand.' Nevertheless, if he felt that she was bossing him, he lost his temper again. It is surely not fanciful to see in these outbursts of rage a crucial element in their whole relationship, and indeed in Albert's life. He was a frustrated man, a man whose cleverness and talents were all made subservient to his rather absurd role as the Queen's toy man. With a part of himself, he must have hated being Prince Consort, and hated his wife for forcing him into this ignominious role. Perhaps if he had not hated either, with a part of himself, he would not have been so eager to lay down the burden of life. No wonder, when he died, that her happiest relationships were with male servants. They obsequiously provided what Albert was not temperamentally equipped to give.

By the middle of December, the Queen at last began to realise what was happening. On the morning of Friday 13 December, she found him awake, with blank, staring eyes, and his breath coming in shallow gasps. When the Prince of Wales arrived, summoned from Cambridge by his sister Princess Alice, he found his father almost dead and his mother

Queen Victoria is pictured here perusing State papers in 1893. Her beloved friend the Munshi stands in attendance. He was an adviser in whom she reposed absolute confidence, even though he was subsequently unmasked as a fraud.

distraught. 'I prayed and cried as if I should go mad,' the Queen confided in one of her ladies in waiting.

Albert's symptoms were too obvious to ignore – a brown tongue, shivering fits, extreme weakness, diarrhoea. This did not prevent Dr Watson, the specialist sent in at Palmerston's insistence, from telling Victoria that 'I never despair with fever.'

The next day, 14 December, Dr Brown, another of the Prince's medical advisers, woke her at 6 am with the good news: 'I've no hesitation in saying . . . that I think there is ground to hope the crisis is over.'

An hour later, the Queen entered the Blue Room to find Albert awake, his eyes unusually bright and staring. In the course of the morning, Albert was got out of bed, and wheeled out for a change of air in another room while they changed his bedding. The Queen stayed with him, though from time to time she walked out on to the terrace, but the sound of military music, playing in the distance, reduced her to tears. She noticed that Albert's face had a 'dusky hue'.

Gradually, throughout the day, a succession of visitors trooped through the Blue Room. The Queen was afraid that it would agitate the patient to see Bertie, but the Prince of Wales, followed by the other children, all came in to take their leave. Their father was too weak to say anything to them. The Keeper of the Privy Purse came, and the Master of the Household.

When, yet again, they had to change his bedding, the Queen went next door and lay down, exhausted. Then she heard heavy breathing from the Blue Room, and ran back in. 'Es ist die kleine Frauchen,' she said – it is your little wife. She asked him for 'ein Kuss' and with her hands she moved his head so that he could kiss her. Then, in anguish, she ran from the room again.

The agony was lingered out a little more, as with almost instinctive formality, the royal household gathered around the bed: Princess Alice and the Prince of Wales kneeling. Princess Helena was there, Dean Wellesley and General Bruce. The gentlemen of the household stood in the corridor. The next time the Queen entered the room, the obvious fact, which she had not quite been able to face until then, dawned on her with hideous clarity. 'Oh, this is death,' she murmured. 'I know this, I have seen it before.'

'Two or three long but perfectly gentle breaths were drawn, the hand clasping mine, & (oh! it turns me sick to write it) *all all* was over. . . . I stood up, kissing his dear heavenly forehead & called out in a bitter agonising cry, "Oh my dear Darling!" & then dropped on my knees in mute, distracted despair, unable to utter a word or shed a tear.'

Victoria's life was shattered by the removal of Albert from the scene. She was only forty-two years old, and her reign had another forty years to run. There can be no doubt that, on her side at least, the marriage had been a passionate love affair. In Albert, she lost her lover, her friend, the father of her children, and an invaluable political adviser. It is very unlikely, for example, had Albert lived (he the great Peelite and sympathiser with many views which were to be promoted by Gladstone) that the Queen's feud with the great Liberal leader would ever have happened. It is certainly unlikely, had Albert been alive, that she would have had quite so close or so flirtatious a relationship with Disraeli. Disraeli's worldly glossiness, his grandiose forms of flattery – such as getting her to call herself the Empress of India (a name somehow better suited to a pig or a railway engine than to a person) – would presumably have had no chances of success had Albert's moderating influence been to hand.

Victoria's cult of Albert has something ghoulish about it, no doubt, but even in death, Albert managed unintentionally to be a useful projection of national fantasy. He had failed to become the modern dictatorial monarch which Stockmar and Uncle Leopold had wanted him to be. But he had become, with his essentially meritocratic views, his interest in industry, his cultivation of home virtues, the soap opera monarch of modern times. If the middle classes, and even the lower classes, could somehow identify with Albert the good, the hen-pecked family man, could they not also see that he, like so many of them, was subject to avoidable death by something so unmonarchical as the drains? These were times when in some northern towns, the average age of death was twenty-six. To us, with antibiotics and indoor toilets, the young family who has suffered the death of a child or a young parent is a tragic rarity. It happens, because we are all mortal. But it is unusual. For the Victorians, it was the norm. Though the long seclusion of the Widow Queen made her increasingly unpopular, the initial bereavement linked her to her people far more intimately than anything which she would have been able to do in her own rather haughty Hanoverian person.

It is not entirely fanciful, therefore, to see the abundance of Albert memorials which sprang up all over Britain as monuments not simply to a very remarkable man, but also to one with whom they could identify. You do not much come across the Christian name Albert among the upper classes, the upper ten thousand with whom Albert in life never had much to do. Miners, station masters and foremen were called Albert.

As the forty-year mourning proceeded, Albert became a sort of divinity to the Queen. But, as I have said, she was not without consolation. Her happiest friendship was with the Highland ghillie, John Brown, with whom she could spend days on the moors near Balmoral, or walking in the grounds of Osborne. The exact nature of their happy relationship will probably never be known. They liked drinking whisky together. Brown was one of those who got on well with royalty because he could be natural – even brusque – in the Queen's company without for a second surrendering the sense that he was her inferior, her servant. When he died, she was desolated. Her Indian servant, the Munshi, provided some substitute for Brown's company, and indeed she allowed the Munshi more confidential knowledge of her State papers and the mysterious red boxes than she had allowed to Albert. Her doctors and advisers had great difficulty in persuading her of the (regrettably true) reports that he was a con-man.

It was appropriate, then, that when she came to die, Victoria should have been placed in her coffin with mementoes not only of her children and friends, and not only of the Prince Consort, but of her beloved servant John Brown, whose photograph, if we are to believe Sir James, was put into her left hand before the coffin lid was placed over the curious collection of bric-à-brac with which she chose to be laid in Frogmore. There she lies, beside Albert, to this day.

Charlotte Brontë
1816-55

Romantic novelist

George Richmond sketched and prettified nearly every famous person of the mid-nineteenth century in England. His portrait of Charlotte Brontë was especially dear to her husband, Arthur Bell Nicholls. When he died, in 1906, his coffin rested the night beneath Richmond's portrait, as he had asked.

In 1847, Messrs Smith Elder and Co. published a novel by a new author of whom no one had heard. The author's name was Currer Bell. The book was called *Jane Eyre*. It caused an instant sensation, not least because some of the reviewers considered it so extremely wicked. Queen Victoria herself was thrilled by the novel, and recorded in her Journal:

> Finished *Jane Eyre*, which is really a wonderful book, very peculiar in parts, but so powerfully and admirably written, such a fine tone it is, such fine religious feeling, and such beautiful writings. The description of the mysterious maniac's nightly appearances awfully thrilling, Mr Rochester's character a very remarkable one, and Jane Eyre's herself a beautiful one . . .

People have been raving about *Jane Eyre* ever since. It is one of the most readable and exciting novels in the language. It is the story of how a tiny, outwardly unremarkable girl (the daughter of a clergyman) suffers the indignity of an early nineteenth-century education, the cruelty of a bad school and of insensitive religious grown-ups, the loss of dear childhood friends through death, and is rendered thereby fit only for the life of the school-ma'am or the governess. She seems to be doomed to become, like Miss Prism in *The Importance of Being Earnest*, 'a female person of repellent aspect remotely connected with education'. But then the real drama and excitement of the book begins. Jane goes as a governess to Thornfield Hall, the seat of the mysterious Mr Rochester. In spite of being a man of the world, with a little string of mistresses in his past, and the rich daughters of the local gentry wanting his hand in marriage, Rochester falls in love with 'Our Jane'. It is the ultimate fulfilment of romantic fantasy – popular novelettes, from Smith Elder and Co. to Mills and Boon in our own day, have been feeding the public ever since with fantasies of this kind.

Into *Jane Eyre*, however, there creeps a feeling of real tragedy and darkness. Queen Victoria, who doubtless responded warmly to a story in which the diminutive, plain heroine is found so alluring by the hero, was right to point to the passages where the maniac appears as the most gripping in the tale. For, as everyone knows, the mysterious Mr Rochester is not in a position to marry. Upstairs in the attic is his mad Creole wife Bertha, whom he had married for mercenary reasons before having time to discover that she was a dissolute person with madness in her blood. All the subsequent passages in the book – the destruction of Thornfield by fire, the maiming and temporary blinding of Rochester, the mysterious moment when Jane, who has run away from him, hears his voice in the winds, lifts us out of the world of novelettes – romances with a small 'r' – and into the world of Romance, as in Romantic movement, as in the paintings of John Martin, as in Goethe, as in Byron. The themes of love and death – the two being so closely linked with inescapable pain and religious awe, which the novel-reading public of the 1840s enjoyed in the work of Currer Bell – were to delight the opera-goers at the close of the century in the works of Wagner.

But who was Currer Bell? The most famous English novelist of the age, William Makepeace Thackeray, wished that he knew. He also wished that the publisher had not sent him the book: 'It interested me so much that I have lost (or won if you like) a whole day in reading it at the

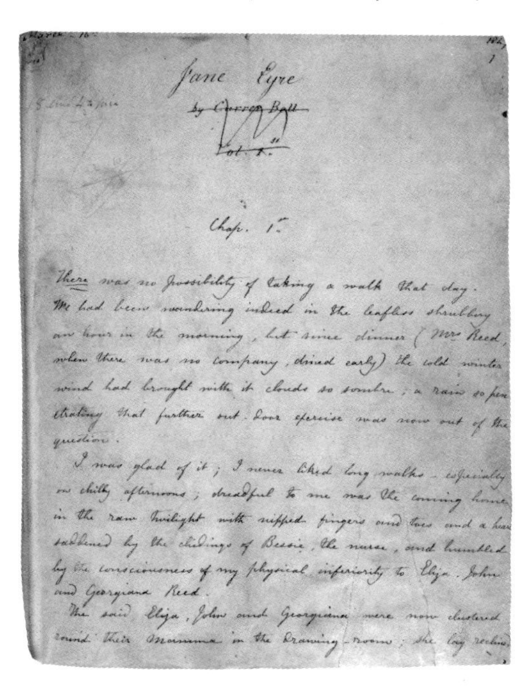

Jane Eyre

~~by Currer Bell~~

~~Vol. I.~~

Chap. 1st

There was no possibility of taking a walk that day. We had been wandering indeed in the leafless shrubbery an hour in the morning, but since dinner (Mrs Reed, when there was no company, dined early) the cold winter wind had brought with it clouds so sombre, a rain so penetrating that further out-door exercise was now out of the question.

I was glad of it; I never liked long walks – especially on chilly afternoons; dreadful to me was the coming home in the raw twilight with nipped fingers and toes and a heart saddened by the chidings of Bessie, the nurse, and humbled by the consciousness of my physical inferiority to Eliza, John and Georgiana Reed.

The said Eliza, John and Georgiana were now clustered round their Mamma in the drawing-room; she lay reclin

The first page of Jane Eyre *in Charlotte Brontë's own handwriting. Most readers of the novel shared Queen Victoria's sense that it was 'a really wonderful' book.*

busiest period with the printers I know waiting for copy. . . . I don't know why I tell you this but I have been exceedingly moved and pleased by "Jane Eyre". It is a woman's writing, but whose? Give my respects and thanks to the author, whose novel is the first English one . . . that I've been able to read for many a day.'

William Smith Williams, Smith Elder and Co.'s reader, passed on this highly complimentary report to the unknown author. As it happened, Currer Bell was a devoted admirer of Thackeray's and asked if the second edition, published in 1848, might be dedicated to the great man, and this was done.

What an excitement that caused in the gossipy world of literary London! Currer Bell, whoever she or he was, was living in the provinces. Mr Williams corresponded with Mr Bell by writing to a post office in an obscure Yorkshire mill town. How could Currer Bell know all the London gossip about Thackeray, how he himself was married to a lunatic whom, in the most painful circumstances, he had been obliged to put away? Moreover, he was notoriously a highly susceptible man where the opposite sex were concerned, and those engaged as nurses or governesses for his two daughters were known to be capable of exciting his devotion. Before long, the rumours were flying around. *Jane Eyre* had been written by a woman who had worked as a governess in Thackeray's house, become his mistress, and been discarded.

It is a curious episode in literary history. Almost everything in *Jane Eyre* – the size and appearance of the heroine, her educational history, her hopeless love for a married man older than herself, her decision (before her reunion with Rochester) to find happiness, if at all, within the innocent circle of her own family, in running a village school, all these things were directly based on 'real life'. The only thing which was purest melodrama and fabrication was the mad wife. This one fantasy was what the knowing public seized upon as the most flagrant example of putting real life into fiction!

Currer Bell wrote in some distress to Mr Williams of Smith Elder, regretting the embarrassment which the whole episode had caused Thackeray:

> It appears that his private position is in some points similar to that I have ascribed to Mr Rochester, that thence arose a report that 'Jane Eyre' had been written by a governess in his family, and that the

> dedication coming now has confirmed everybody in the surmise. . . . I am very, very sorry that my inadvertent blunder should have made his name and affairs a subject for common gossip.

But now, another drama blew up in the life and career of the novelist Currer Bell. Not long before, in 1846, there had been published a slim volume entitled *Poems by Currer, Ellis and Acton Bell.* It had made no impact whatsoever, even though, if readers had had the wit to spot the fact, one of the poets, Ellis Bell, was a person of unique genius. The poems had been printed by a religious publisher called Aylott and Jones. As was often the case in the nineteenth century, the cost of publication was borne by the author. There was therefore no loss for Aylott and Jones when the poems failed to make any impact whatsoever. In fact, though Currer Bell had unwisely ordered a print run of a thousand, only two copies of the poems were sold. Amazingly, almost none of the thousand copies survive. The volume which one can see today at the Parsonage in Haworth is a priceless bibliographical rarity. Aylott and Jones could not foresee that, and when the Bell brothers offered them novels, they were able to decline on the grounds that, as religious publishers, they did not deal in anything so frivolous. But, business and principle do not always marry, even in the most religious souls. When Smith Elder and Co., a much cannier publisher, had accepted *Jane Eyre* Aylott and Jones saw their chance to make a killing. They could sell the *poems* of the three brothers by putting it about that in fact there was only one author, masquerading as three, and that was Currer Bell, the author of *Jane Eyre.* It was this confusion, and the same claim being made by an American publisher, which made George Smith write to Currer Bell and insist that 'he' came clean. Smith himself was one of the great Victorian publishers. We can now see, more clearly than a generation ago, since the unearthing of Smith's letters, what care he took with his author, advising against the publication of *The Professor* and encouraging, as a good editor should, the writing of more mature novels. He was one of the first publishers who made a figure in society, the forerunner of many a social-climbing modern counterpart.

Wuthering Heights by Ellis Bell was published (by yet another publisher, T. C. Newby) in the same year as *Jane Eyre* and, very surprisingly to us, it made very little impression on the public. It was reprinted in 1850 because Smith Elder had been persuaded to take over

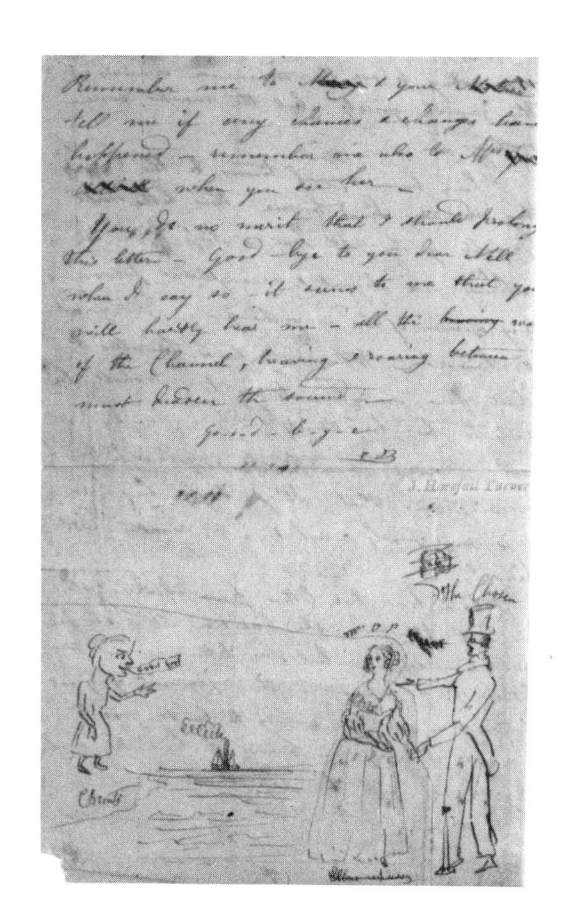

Remember me to Mary & your Mother
tell me if any chances & changes have
happened — remember me also to Mrs [illegible]
when you see her —
You do not want that I should prolong
this letter — Good-bye to you dear Nell
when I say so it seems to me that you
will hardly hear me — all the [illegible]
of the Channel, heaving & roaring between
must deaden the sound —
good-bye
CB

An extract from a letter written by Charlotte to her life-long friend, Ellen Nussey. The sketches at the foot of the letter are particularly revealing of Charlotte's rather poor image of herself as opposed to that of her more glamorous friend.

publication of 'Acton' and 'Ellis' Bell's books – Acton Bell's *Agnes Grey* had also been published by Newby in 1847 – but its great qualities seemed to have passed everyone by in the midst of the *Jane Eyre* mania which swept London in 1848.

Who on earth *were* this mysterious family? Acton, Ellis and Currer Bell were obviously some sort of pseudonym. The scandalous rumours about Thackeray and Mrs Rochester only fuelled speculations about the identity of the authors – or author. Yes, author. Wasn't it really only too probable that Acton, Ellis and Currer Bell were truly one and the same person? Matters came to a head in July 1848. *Jane Eyre* had been hugely successful in the United States, and Newby had seized the opportunity to publish Acton's second novel – *The Tenant of Wildfell Hall* – there, claiming it was an early work of Currer Bell, in order to cash in on this success. George Smith wrote a hurt letter to Currer asking if it was true that he had allowed another publisher to publish his work.

He could have no idea of the flurry and the anguish his letter was to cause his correspondent in Yorkshire. But it was prompted by understandable anxiety. Not only did he have a business to run; but, as a publisher who had brought before the world the novel which everyone was talking about, it was only natural that he should want to be assured about its author.

A little while later, George Smith was busy at his desk one Saturday morning in the bookshop and publishing company at 65 Cornhill, London, which he ran on behalf of his father. Work was piling up, and he did not particularly welcome an interruption from his clerk who informed him that he had two visitors – ladies who had come on private business, but refused to give their names.

A little unwillingly, he asked that they should be shown into his office. 'Two rather quaintly dressed little ladies, pale-faced and anxious-looking, walked into my room.' Others who knew them have remarked on the fact that they were really very small indeed – under five feet in height. One of the ladies (we learn from other sources that she had a poor complexion, and was acutely conscious of being spotty and plain) held out a letter to George Smith, written in his own handwriting. It was addressed to 'Currer Bell Esq'.

With some sharpness, he asked, 'Where did you get this from?'

'From the post office,' was the reply. It was a striking accent. Years before, one of her school friends had been struck by her blighting timidity and her Irish brogue. But it was a Northern Irish voice, almost like a Scottish Lowland accent. She continued, reverting to the letter, 'It was addressed to me. We have both come that you might have ocular proof that there are at least two of us.'

With extraordinary excitement, and some incredulity, George Smith realised that he was in the presence of his best-selling author. She then presented herself with her real name, Charlotte Brontë. Her companion was her sister Anne.

Mr Smith immediately proposed that they should come to stay with his mother and sister in Bayswater. The shy sisters had already established themselves in a hotel near St Paul's Cathedral. Their father had taken Charlotte and Emily there before, *en route* to Belgium six years earlier. Emily had so hated the earlier journey to Belgium that thereafter she refused to leave Yorkshire; and this was the reason that Charlotte and Anne were in London without her. They excused themselves Mr Smith's kind hospitality, but they could not escape altogether. Although Charlotte was exhausted by her journey south, and by a blinding sick headache, she consented to the suggestion that Mr Smith and his partner

When this carte de visite, *dated 1854, was discovered in the 1970s there were those who doubted that it was, as is stated on its reverse side, Charlotte Brontë at the time of her marriage, but the resemblance to existing portraits seems unmistakable. It is the only photograph of her to survive and probably the only one to have been taken.*

Mr Williams – 'a pale strapping man of fifty' – should take the sisters to the Opera. Charlotte and Anne went back to the hotel to be sick, to lie down, and to take a strong dose of sal volatile. Towards evening, Messrs Smith and Williams arrived, with George Smith's mother and sister in tow, all elegantly clad in smart evening dress. The Brontë sisters had only their travelling clothes, and felt self-consciously dowdy as they were led into the splendours of the Royal Opera House. Charlotte bravely decided to 'put her headache in her pocket', and gave herself up to the excitement of the occasion. The opera – Rossini's *The Barber of Seville* – was 'very brilliant, though I fancy there are things I should like better. We got home after one o'clock; we had never been in bed the night before and had been in constant excitement for twenty-four hours. You may imagine we were tired.'

During the evening, Anne Brontë had expressed an interest in hearing the famous Dr Croly preach, so next morning, Mr Williams arrived at their hotel and took the two sisters off to Morning Prayer at the lovely Wren church of St Stephen Walbrook. Dr Croly, the rector, was not there that morning. The next item on their programme was dinner in Bayswater with the Smiths – just a family affair with Mr Smith's mother, sisters and little brother. 'We had a fine dinner which neither Anne nor I had appetite to eat, and were very glad when it was over,' Charlotte recorded. The next day was devoted to sight-seeing and shopping. They saw the National Gallery, the Royal Academy, and bought as many books as they could carry. In the course of these days, Mr Smith and Mr Williams were able to discover that Charlotte Brontë, this infinitely shy, tiny woman, was by no means a weakling. She had a strong, even forceful personality, a good sense of humour, and high intelligence. She also had, without being a flirt, the gift of sympathetic intimacy. It is apparent in the thank-you letter which she wrote to William Smith Williams, whom, having met on the Saturday, she left on the Tuesday to return to her home in Yorkshire. It is the letter of a vigorous human being who has struck up a true friendship with another human being.

> It was a somewhat hasty step to hurry up to town as we did, but I do not regret having taken it. In the first place mystery is irksome, and I was glad to shake it off with you and Mr Smith and to show myself to you for what I am, neither more nor less – thus removing any false expectations that may have arisen under the idea that Currer Bell had a just claim to the masculine cognomen he, perhaps somewhat presumptuously, adopted – that he was, in short, of the nobler sex.

This humorous, but easily intimate tone, is highly characteristic of Charlotte's letters. As well as being one of the great novelists of the nineteenth century, she is also one of its great letter-writers. And it was in her letters that Williams began to learn some of the extraordinary circumstances of Charlotte Brontë's life.

At this date, she was thirty-two years old, by nineteenth-century standards a spinster who was almost certain to remain an old maid for the rest of her life. And what sort of life had that been? By almost any standards, it had been quite staggeringly uneventful; and yet, as generations of readers have demonstrated, the lives of Charlotte and her sisters are, in their way, quite as fascinating as their work.

As Charlotte left London, on that hot July day in 1848, she was returning to a Yorkshire parsonage where she had lived ever since she was four. Her father, the Reverend Patrick Brontë, was a self-made man of unusual accomplishments. His birthplace in County Down, in the North of Ireland, is today a heap of stones, but one can still make out its tiny dimensions. Twelve hens, if they inhabited the space, might well feel constricted. Hugh Brunty (Patrick's father) filled the space with twelve human beings – his wife and ten children. Purely by his own wits, Patrick

Brunty emerged from this peasant background, became a schoolmaster, and eventually went to St John's College, Cambridge. People have mocked him for dandifying the ordinary name of Brunty into Brontë. The spelling is taken from one of Lord Nelson's titles, the Duke of Brontë – just such a vein of fantasy was to lead Charlotte, when a child, to people her imaginary world of Angria with the Duke of Wellington and the Marquess of Douro. But Patrick Brunty was probably the first member of his family who was able or desirous to write the name down; as a pioneer in the art of handwriting, why should he not have spelt his own name as he chose? He, and probably his children, would have continued to pronounce the name in the Irish way, *Brunty*.

It is a very important ingredient in Charlotte's emotional history that her father was purely the product of education. If it had not been possible through cleverness to rise through the ranks of society in this way, Patrick Brontë would have lived, as most of his relations did and had done, the life of an agricultural labourer. Cambridge, ordination as a clergyman in the Church of England, the eventual cure of souls which came his way at Haworth in Yorkshire – all these advantages (regardless of their social and religious significance) released him and his family to live the life of the mind, to inhabit the world of books. This was something which he had done, from the moment circumstances allowed it. The fact that he was not a good writer does not detract from the important point – that he assumed that one might want to be a writer. Those 'thoughts which wander through eternity' – the evanescent fancies of an average brain, are things which the writer (unlike the generality of people) wishes to immortalise or hold. Most people are story-tellers on some rudimentary level; most, presumably, have had some consciousness of the inner life, have fallen in love, have confronted the awfulness of nature. The unusual thing is to confront these experiences with pen in hand, and this was what Patrick Brontë had done in his pamphlets *The Cottage in the Wood: or the Art of becoming Rich and Happy* and *The Maid of Killarney*. In his own life he never displayed the art of becoming either rich or happy, but he passed on to his children the habit of getting through life by means of literature-as-drug.

He married a Cornishwoman called Maria Branwell. They had six children in fairly swift succession (Maria born 1814, Elizabeth born 1815, Charlotte born 1816, Patrick – called Branwell – 1817, Emily Jane 1818, and Anne 1820). Their mother Maria Brontë died of cancer in

Haworth Parsonage. Note the preponderance of flat, imperfectly drained tombs: the place stank habitually of the dead.

1821, the year after they had all moved to Haworth. Her sister, a somewhat austere lady of Calvinistic outlook, came to live in the Parsonage to look after the motherless children.

Haworth in the first half of the nineteenth century was a semi-industrial village with a handful of mills, built on a steep gradient on the edge of the moors. When Patrick Brontë and his young family went to live there it must have been very remote. The railway had not yet been built – that would come in 1867. All the village's water supply flowed through the overcrowded graveyard, contaminating every mouthful imbibed. A report by a public health inspector in 1850 considered Haworth to be one of the least sanitary villages in England. There were no water closets in the place in the Brontës' day. Half the population died before they were six years old, and the average age of death was twenty-six.

Not only was it unhealthy. It was also a place where congenial society for a man such as Patrick Brontë would have been hard to find. There were not many reading men thereabouts, nor many churchmen. Brontë himself was only just a churchman, as far as his theology went. His wife Maria had been a Wesleyan Methodist in Cornwall, and Patrick himself, when his curate was there to substitute for him at the service in Haworth parish church, used sometimes to attend the local Methodist chapel. One of his own predecessors at Haworth parish church had been a famous friend of the Wesleys called William Grimshaw. He was noteworthy for the

William Grimshaw, a former vicar of Haworth, was a friend of Wesley's and a leading evangelical. His teapot, emblazoned with the words of St Paul, was one of the Brontës' most treasured possessions; it is preserved at the Parsonage.

manner of his dying. He left strict instructions in his will that however wild the weather, they were to carry his dead body ten miles across the moors to a place called Luddenden to be buried beside his wife. The rather ghoulish tale was a favourite of Tabitha Aykroyd, the Brontës' family servant, who was doubtless the source of that scene which was eventually, when filtered through the imagination of Emily Brontë, to become the story of Heathcliff digging up the corpse of his beloved Cathy. A teapot belonging to William Grimshaw was kept at the Parsonage (it is there today). Emblazoned with the legend, *To live is Christ, To die is Gain* – a text which however uplifting in its proper place must have been disconcerting to observe as the brown steaming fluid poured from the spout into the teacup – it was no doubt used by Aunt Branwell, and suggests something of her relentlessly theological preoccupations.

To be a churchman in such a Methodistical heartland was only one of the many things which isolated Patrick Brontë. He did loyally uphold church principles against the chapel – even though he did so with a strong sense of how attractive the chapel was. Within his own church, he was entirely unsympathetic either to the Calvinism of the Evangelical party or to the crypto-Catholicism of the Puseyites. Likewise in politics, he was in favour of moderate reform, and had supported the Reform Bill of 1831/32. But as the local representative of Church and State he had felt obliged to side with the mill-owners against the rioting workers at the time of the Luddite protests against the introduction of machinery.

When these violent bands descended on the mill towns to destroy the new mechanical looms (scenes memorably evoked in Charlotte's novel *Shirley*) it was said that Patrick Brontë had been armed with a gun, and that his own life was threatened (nevertheless he buried Luddite rioters in secret). All this was scarcely the way for a parish priest to make himself feel at home among his people. Add to this the fact that he continued to speak with a strong Ulster brogue, and that he suffered from failing eyesight, and the isolation of Patrick Brontë at Haworth becomes the more conspicuous. The small, neat Parsonage at the top of the village was destined to become an isolated imaginative world.

They were an extraordinarily close family. Strangers would observe how, in the presence of visitors, these nervous, small people would hug one another like timorous animals huddling together against predators. The children inhabited a curious imaginative world of their own creating, partly culled from their wide and miscellaneous reading and partly from dreams. Emily's imagined world was called Gondal. It appears to have been a more passionate mystical place than Charlotte's imagined kingdom of Angria (the crossness suggested by the homophone is surely revealing) which was a strange blend of fantasy about high society with erotic and emotional dreamings. Tens of thousands of words were, over the years, composed in stories and poems based on these secret kingdoms. As she grew up, Charlotte began to be suspicious of her own need to create such dreams. Her hope that she might be adored by some demonic figure like Zamorna ('his face whitened more and more, something like foam became apparent on his lip – and he knitted his brow convulsively'), a figure who had started life in her Angria tales as the Marquess of Douro, developed into more ordinary erotic fantasies about the men she actually knew. Emily, by contrast, as she grew up, inhabited a world which had less and less in common with her discernible surroundings. She became ever more inward-looking and mysterious. This difference is seen very strongly when Charlotte's very patchy, uneven achievement as a novelist, with her gossipy and frequently comic interest in human character and eccentricity, is compared with the white-hot intensity and self-contained brilliance of Emily's masterpiece, *Wuthering Heights*.

Another difference was that Charlotte's spirit was always contriving ways to escape Haworth, whereas Emily's spirit never left the place even when, as inevitably happened, she was compelled to go away to seek

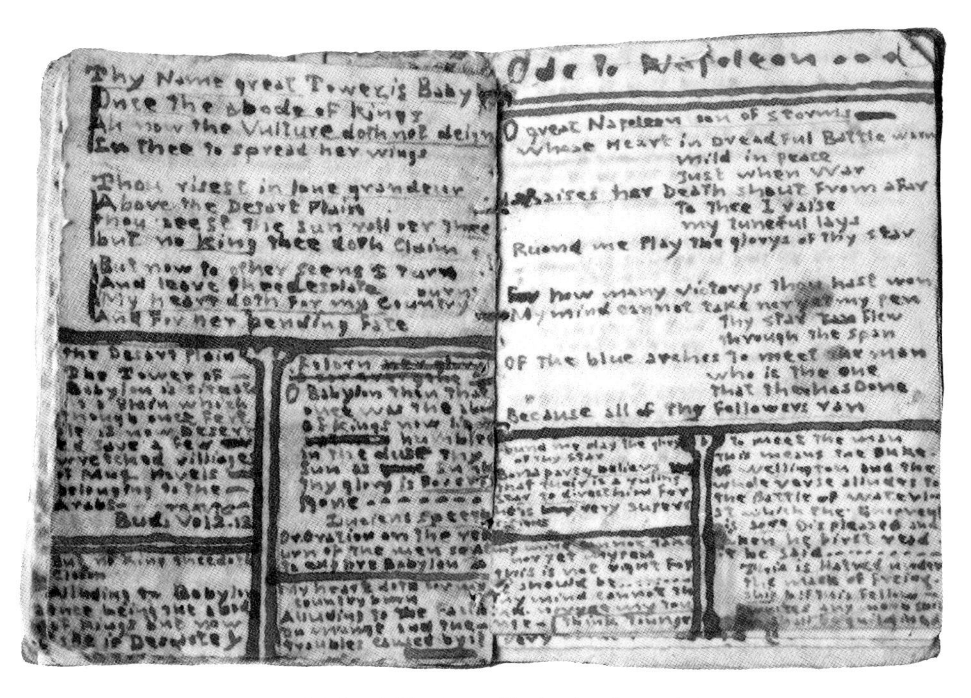

Thy Name great Tower is Babylon
Once the abode of Kings
Ah now the Vulture doth not deign
In thee to spread her wings

Thou risest in lone grandeur
Above the Desert Plain
Thou seest the sun roll over thee
but no King thee doth Claim

But now to other scenes I turn
And leave thee desolate
My heart doth for my Country
And for her pending fate

Ode to Napoleon

O great Napoleon son of storms
Whose heart in dreadful Battle warm
Mild in peace
Just when War
Raises her Death shout from afar
To thee I raise
my tuneful lays
Round me play the glorys of thy star

For how many victorys thou hast won
My mind cannot take nor yet my pen
Thy star has flew
through the span
Of the blue arches to meet the man
who is the one
that thee has done
Because all of thy followers ran

In her childhood Charlotte, like her brother and sisters,
wrote many poems and stories
in tiny books like this one.

education or employment. The moorland itself is the first thing to confront us in Lockwood's difficult walk to Wuthering Heights. It is the place which gives life and freedom to the anarchic, quasi-incestuous love of Cathy and Heathcliff. It inspires much of Emily's greatest lyric poetry. Though, as has been said by her best modern biographer, Charlotte was less inclined than Emily to spend as many of her waking hours as possible outdoors, the moorland around Haworth became a perpetual and important part of her inner landscape. The almost Wordsworthian moral significance the landscape held for all the Brontës is thinly fictionalised in *Jane Eyre* when Jane seems to be on the point of settling with her cousins Diana and Mary Rivers.

> They loved their sequestered home. I, too, in the grey, small, antique structure, with its low roof, its latticed casements, its mouldering walls, its avenue of aged firs – all grown aslant under the stress of mountain winds; its garden dark with yew and holly – and

> where no flowers but of the hardiest species would bloom – found a charm both potent and permanent. They clung to the purple moors behind and around their dwelling – to the hollow vale into which the pebbly bridlepath leading from their gate descended; and which wound between fern-banks first, and then amongst a few of the wildest little pasture-fields that ever bordered a wilderness of heath, or gave sustenance to a flock of grey moorland sheep, with their little mossy-faced lambs: they clung to this scene, I say, with a perfect enthusiasm of attachment. I could comprehend the feeling, and share both its strength and sweep – on the wild colouring communicated to ridge and dell, by moss, by heath bell, by flower-sprinkled turf, by brilliant bracken and mellow granite crag. These details were just to me what they were to them – so many pure and sweet sources of pleasure. The strong blast and the soft breeze; the rough and the halcyon day; the hours of sunrise and sunset; the moonlight and the clouded night, developed, for me, in these regions, the same attraction as for them – wound round my faculties the same spell that entranced hers.

And yet, to judge both from her fiction and her letters, Charlotte seems to have felt from the beginning that her childhood was a dream from which she would one day wake up. Her siblings never came to terms with this, and some of them, sadly, never needed to do so.

Charlotte wrote in 1835:

We wove a web in childhood,
A web of sunny air;
We dug a spring in infancy
Of water pure and fair;

We sowed in youth a mustard seed
We cut an almond rod;
We are grown-up to riper age –
Are they withered in the sod?

Although the poem sets out to deny any such mortality to the workings of the imagination, Charlotte shared with her siblings the knowledge that something died forever when they grew up; and since she was destined to be the last survivor of this ill-starred dynasty, she also had a painful sense of how frail was their physical hold on existence itself.

Education, which had been the making of the Brontë father, was to be the undoing of the Brontë children. Charlotte, at the age of eight, was sent to the Clergy Daughters' School at Cowan Bridge, near Tunstall in Lancashire. Her sisters Maria and Elizabeth were already in residence at this recently founded establishment, and little Emily, not seven years old, was due to follow by the end of 1824. This school was the 'Lowood' of *Jane Eyre*, a place of great physical austerity, run by an Evangelical clergyman called the Reverend Carus Wilson, the model for Mr Brocklehurst in the novel. One can tell instantly, when reading *Jane Eyre*, that Brocklehurst is based on a figure in real life. Just as Mr Rochester has all the solid reality of a perfectly realised fantasy figure – he is so well-rounded because he has become so fully what his creatrix wanted him to be – so Brocklehurst, like a photograph not quite in focus, comes before

Cowan Bridge School was the model for Lowood in Jane Eyre, *where the conditions were insanitary and the regimen cruel. All the Brontë sisters attended this place until the deaths of Maria and Elizabeth, aged 10 and 11 respectively, as a result of tuberculosis contracted here.*

us distorted by the author's hatred. His cruelty and his religious humbug seem too extreme to be real. It is only when we read about the real Mr Wilson that we discover, of course, that 'Brocklehurst' is really rather a benign portrait of a man who was far more monstrous than any novelist could paint. Not only did he force the children to live in insanitary conditions (one privy for over seventy people); not only were they cold, hungry, exhausted much of the time. But Wilson's morbid belief in a Calvinistic system of predestination made him believe that death itself, if it overtook the children in his care, was to be seen as a positive blessing. He was one of those Christians who believed that only a very few have been designated for Heaven in God's scheme. Most – and these unfortunately included Charlotte Brontë – were in the dubious position of not having been elected to glory. It gives force to the joke in *Jane Eyre* when the infant is asked how she is to avoid the pains of hell. 'I must keep in good health, and not die.' This really was the only way of avoiding hell in Mr Wilson's school, and if he thought that you were elected to glory, then he was quite cheerful about the prospect of your early demise, even if you were only eleven. Readers of Wilson's monthly magazine *The Children's Friend* were able to peruse many an edifying yarn about little children happily going to their maker or – if they belonged to the infamous majority – being snatched by such playground misfortunes as skating on thin ice or being bitten by a dog into the flames of immortal perdition. But this was not just written for show. Wilson really meant it. One of Charlotte's companions, Sarah Bicker, died at the age of eleven. She made a pious end; though suffering from the last agonies of typhoid, with inflammation of the bowel, she was compelled to spend her last minutes being catechised by Mr Wilson to make quite certain that Christ had died for her sins. 'And He will save all men?' asked Wilson. 'No, Sir,' said Sarah, 'only those that trust in him.' 'I bless God,' Wilson wrote in *The Children's Friend*, 'that he has taken from us the child of whose salvation we have the best hope and may her death be the means of rousing many of her schoolfellows to seek the Lord while He may be found.'

Not all parents would find this a reassuring item in a school curriculum. The combination of insanitary conditions and crazed Calvinistic theology were to carry off two of Charlotte's sisters. Maria was taken home to Haworth 'in ill health' in February 1825. She was the model for Helen Burns in *Jane Eyre*. She died of consumption on 6 May 1825, aged eleven. Her sister Elizabeth died about a fortnight later.

Mr Brontë immediately removed Charlotte and Emily from Cowan Bridge, but the terrible experience of being there was something from which the girls never escaped. The pure anger which she felt about the deaths of Maria and Elizabeth sharpens every word of the opening chapters of *Jane Eyre*. But, as happens to Jane Eyre at Lowood, Charlotte had begun to guess at Cowan Bridge that life for such as her, at that particular date, offered an alarmingly restricted range of options. The sickly daughter of an impoverished clergyman had only the option, if she failed to secure a husband, of staying at home or entering the world of education. And not education as we understand the phrase – not colleges, not institutes, not universities. The best she could hope for was a post as governess in a pleasant household, or as an assistant mistress in some such school as Cowan Bridge. It was with this in mind that Mr Brontë decided, after a delicious period of five years in which Charlotte and Emily did not go to school, that they should once more be sent off to the classroom – this time to the much more benign establishment at Roe Head, presided over by Miss Wooler. Charlotte was evidently a born academic. She flourished at Miss Wooler's and had the kind of brain which would have benefited from further education at a university – quite out of the question at this date, of course. Emily, who had cocooned herself in the world of her own private dreams, was not happy at Miss Wooler's and soon enough left. Anne, the youngest child, also came to Roe Head while Charlotte was there and seems to have flourished.

The two most important things which happened to Charlotte at Roe Head were the forming of friendships and the improvement of her French. The two friends she made there – Mary Taylor and Ellen Nussey – remained friends for life, and the letters Charlotte wrote to them for the rest of her days are among the most valuable sources for her biography. In particular, I think, one is interested in Mary Taylor whose family had a genuinely modern outlook on things, and whose feminism was a noticeable influence on Charlotte's thinking. Just as Charlotte, while unmarried, found it difficult to find a place in English society, so did the independent-minded Mary Taylor, who ultimately emigrated to New Zealand rather than follow the male-dominated rules set down for the way Englishwomen should behave.

The French – begun at home and brushed up at Miss Wooler's – enabled Charlotte, in her late twenties, to have her only sustained experience of life outside an English parsonage, schoolroom or governess's

Charlotte was a ready and fluent letter writer and formed an intimate pen-friendship with her publishers. One of her letters is shown here.

some proportionate addition to the remuneration you at present offer. On this ground of confidence in your generosity and honour, I accept your conditions.

I shall be glad to know when the work will appear. I shall be happy also to receive any advice you can give me as to choice of subject or style of treatment in my next effort – and if you can point out any works peculiarly remarkable for the qualities in which I am deficient, I would study them carefully and endeavour to remedy my errors.

Allow me in conclusion to express my sense of the punctuality, straight-forwardness and intelligence which have hitherto marked your dealings with me

parlour. The experience of being a governess – not quite a servant, but certainly not the equal of the family for whom she worked – was mortifying to one of her sensitivity. When she left one family for whom she worked, the Sidgwicks, Charlotte wrote that 'I was never so glad to get out of a house in my life'; and she told her future biographer the novelist Mrs Gaskell that, 'None but those who had been in the position of a governess could ever realise the dark side of "respectable" human nature.' Anne, her sister, was having an almost equally miserable time with a family called Ingham, near Mirfield, and collecting the experience which would enable her to write *Agnes Grey*.

But between these unpleasant experiences, all four surviving Brontë siblings – Branwell, Charlotte, Emily and Anne – were much together, much at home. Their teenage years were perhaps their happiest: by day, happily absorbed in the routines of the Parsonage (they were passionately house-proud, so that polishing and cleaning took up much of their time), the livestock (two geese called Victoria and Adelaide, and a succession of dogs) absorbed much of their emotional energy as well as the delights of Nature and Art. All were accomplished sketchers and painters. All four were musical. Emily was a brilliant pianist – the piano which Mr Brontë bought for her still survives at the Parsonage – and Branwell developed a particular facility on the church organ. And all four were writers. At the end of an average day, Mr Brontë, his sight already fading, would have retreated to his study, an austerely furnished room immediately to the right of the front door, while, a few feet across the passage-way, his children would assemble to read aloud from their voluminous written works, the chronicles of Angria and Gondal.

Charlotte's memory was of their pacing about the room with excitement as they spoke of their creations. When you see the room, which is really quite small, this circumambulation seems all the more peculiar. But then you remember two things. One is that all the siblings but Emily (who was five feet six) were really very tiny indeed. Peering into their dining room, which itself has a somewhat doll's house quality, you have to imagine people of almost doll-like size themselves. The second thing to recall is how still you were expected to sit, right up to the period of the 1950s, in genteel English society – at least in the sort of provincial society to which Charlotte Brontë (or I) belonged. Children were taught to sit and sit and sit. One was meant to sit with a straight back, too. I imagine that the Brontë children, when they were not scrambling over the moor, or completing the household tasks they were set from an early age, were expected to sit still for most of the day. They would certainly not pace about the room in the presence of a servant, nor of their father or aunt. So, when they were on their own, this slightly frenetic movement began.

The three girls' devotion to their brother Branwell has been made the subject of endless speculation. In particular, it has been hinted that Emily's passionate nature nursed incestuous feelings for him, or that the story of Heathcliff and Cathy Earnshaw might conceal an actual incestuous affair between the two. Of course, no one can ever know the truth of these rather wild speculations, but I should be amazed if any such thing ever existed at the conscious level. The success of the Brontë novels, and in particular of *Wuthering Heights*, derives from Emily's and Charlotte's instinctive knowledge that sex is something in the head. Such knowledge, when translated into art, very rarely derives from experience. For some reason or another, real sexual experiences or feelings produce a kind of imaginative taboo in all but the most relentlessly autobiographical of pens. Novelists tend to write about the experiences they did not have, not the ones which they had. Besides, Branwell Brontë appears to have been an altogether jollier and more genial companion than Heathcliff.

The wonder is, when we consider how much trouble was devoted to the girls' education, that he received so little. Probably his poor father depended too much on his society. The loneliness of Mr Brontë is one of the underlying sadnesses of the whole story. He evidently taught Branwell, in his fashion, to a much higher level than many schools of the time would have done. Branwell's translation of *The Odes of Horace* (Book One) for example, has been thought worth reprinting in this century. It is true that Mr Brontë had very little money. But that is more than *no* money, which his own father had had. He, with no money or position, had managed to get to St John's College, Cambridge. Branwell never came near such heights, though it was not until the 1840s that he formed that attachment to alcohol and laudanum which was to have such a disastrous consequence.

His chosen field, after much family discussion, was painting. He was sent off to the Royal Academy in 1835, but we don't even know if he got there. When he came home again, however, he made an occasional living as a portrait painter. The Bradford worthies whom Branwell commemorated cannot have been particularly pleased by his amateurish daubs. Only occasionally did he rise to a sort of brilliance with the brush. The most noticeable example of his talent is his most famous picture, the depiction of his three sisters which now hangs in the National Portrait Gallery in London. It is a haunting picture, one of the most inescapably emotional portraits in the whole gallery. Even if we knew nothing about their work, we should know as we stood in front of it that an extraordinary story lay behind the picture. On the extreme left is Anne – she looks as if she is about to say something, but beginning to think better of it. Her eyes are particularly well painted, the charming, very faint suggestion of a squint is immediately recognisable as authentic. Next to her is oval-faced Emily. Her overpowering strength of personality, and her beauty, strike us. So does the difference between Emily and the other two sisters. If they might conceivably be about to entertain us with a pot of tea, Emily's features suggest something altogether more mystical and arcane. The smile playing on her lips is not sociable; it hints of ecstasy.

She is staring beyond us to some unseen Presence. Because Emily is there, the group ceases to appear domestic; we start to think of three Norns, three sybils – even semi-mythical figures like three Furies. There is no nonsense about the figure on the extreme right, the figure of Charlotte. If Anne looks as if she is about to say something, Charlotte looks as though she just *has* said something, and is now wondering whether it was somewhat too acerbic. It is the most Irish of the three faces, and in the curl of the lips we discern the humour that could give us wonderful sentences like the opening passage of *Shirley*: 'Of late years, an abundant shower of curates has fallen upon the north of England: they lie very thick on the hills; every parish has one or more of them . . .'

Perhaps the most alarming face on the canvas is the fourth, the one which is not there. Between Emily and Charlotte, Branwell painted himself. Then, with a gesture of self-abnegation, if not self-hatred, he painted himself out. With similar despondency in life, he drifted from one position to another, usually tight, and always short of funds. Once a life has started to go downhill in the way that, from an early stage, it was indicated that Branwell's should go, it is difficult to single out any particular moment as the lowest ebb. In terms of embarrassment, it would have been hard to match the time that he lost his job as station master at Luddenden Foot railway station for an alleged irregularity with the petty cash; or the sad culmination of his career as a private family tutor to the Robinsons at Thorp Green. He was passionately in love with his employer's wife. Charlotte persisted in believing (perhaps rightly) that it was Mrs Robinson who had made the running in the affair. Twenty years Branwell's senior, she had flirted with the young man 'in the very presence of her children, fast approaching to maturity; and they would threaten her that if she did not grant them such and such indulgences, they would tell their bed-ridden father "how she went on with Mr Brontë".' It was from this incident, Charlotte averred to her friend Ellen Nussey, that Branwell's real decline could be dated, when 'he began his career as an habitual drunkard to drown remorse'. When one visits the Parsonage at Haworth today, one wonders whether it was as simple as that. One sees the small bedroom which Branwell shared with his myopic father; the tiny dining-room where he sat with his loving sisters. Anything he did would be forgiven. Their love for him was unconditional and, one suspects, smothering, emasculating. When short of funds and low in spirits Branwell would threaten suicide. It would usually produce a sovereign from his father. The local pub, the Black Bull, was only a few yards from the house, and the apothecary – ready supplier of opium (in the form of laudanum) without prescription – was just opposite in the village square.

One of the things which makes Charlotte remarkable is that one suspects she shared much of Branwell's pessimism of outlook. It was not given to her to be consoled, as Emily was, purely by dreams and the inner life. While Emily continued to weave the tales of Gondal in her head throughout her life, Charlotte firmly brought an end to her Angria fantasies when she entered adulthood (though she allowed herself to rework them in *The Professor*). She saw as clearly as Branwell did that life offered very little for such as them. She also knew at first-hand that poverty and boredom, and the humiliation of the near-servant status 'enjoyed' by tutors and governesses in those days, were not the worst that life could offer. Like Branwell, she was to discover the misery of an impossible love for a married person. But unlike Branwell, she stared all these bleak realities in the face. One sees it so clearly in his painting of them all. He has taken the easy way, by scrubbing his face out of the picture altogether. Charlotte is doggedly staring, her jaw set, her courage and her sense of humour diminished but not extinguished by the reality of things.

The love-misery came to her in Belgium where, very enterprisingly, she had gone with Emily to escape the tedium of governessdom and curates falling on the Yorkshire hills, thick as snow. Mr Brontë accompanied Emily and Charlotte to Brussels in February 1842. Her school friend Mary Taylor, and her brother James Taylor, went with them. The two sisters attached themselves to the Pensionnat Heger, a girls' school in a handsome seventeenth-century building in a quiet part of the Belgian capital. Emily hated the school. She felt uncomfortable with her fellow-pupils. She had a profound horror of Roman Catholicism. As always, when separated from Haworth – the warmth of the family circle, the freedom of the moor – Emily was homesick. When they came home for Aunt Branwell's funeral, it was decided that Emily should remain at Haworth to take care of her father. Charlotte, however, went back to Belgium for another year, this time not as a student, but as a teacher at M. Heger's Pensionnat: or rather, as a sort of student teacher, for she was paid a salary of £16 per annum, out of which money was deducted for her German lessons and her laundry.

In 1842 Charlotte and Emily went to Brussels to study, subsequently to teach at the Pensionnat Heger. Charlotte fell in love with M. Heger, the Principal of this establishment, who is photographed here many years later in the 1870s with his pupils.

Whereas her first year in Brussels had been dominated by looking after Emily's needs, and by something like a friendship which could develop between the two sisters and Monsieur and Madame Heger, the second year was very different. Madame Heger could not fail to be aware that the young English *gouvernante* had fallen helplessly in love with her husband, *le professeur*. Seven years older than Charlotte, and five years younger than his (second) wife, Constantine George Romain Heger was a man with a melodramatic history. He had fought in the Belgian Revolution on the side of the nationalists. His brother-in-law had been shot, at his side, on the barricades. His first wife had died of cholera, as had his first child. In temperament, he had much in common with Mr Rochester and even more with Paul Emmanuel, the fascinating hero of *Villette*, Charlotte's Belgian novel. He had a furious temper ('he fumed like a bottled storm', Charlotte lovingly remembered), but he could simmer down and be kind and generous. He was an inspired and brilliant teacher. He possessed (as an obituary notice said of him) 'a kind of intellectual magnetism with children'. Like Charlotte, he combined an emotional intensity and an intellectual rigour with a deeply religious temperament.

Love for him was to become the dominant and searing emotional fact in Charlotte's life – at least for the next few years, perhaps forever. It is quite obvious that he did not return her love, although he found her companionship entertainingly intelligent. As soon as the situation became apparent, his wife jealously kept the pair apart. It meant that there were many moments of desperate loneliness, such as the occasion in the summer of 1843, when the Hegers took their children on holiday to Blankenburg, and left Charlotte behind at the *pensionnat* – all the children being also on vacation – with no companionship but the domestics. In her misery, she found herself going into the cathedral and muttering her secret love to a priest through the grille of one of the confessionals: an incident vividly written up, first in a letter to Emily and then, more famously, in one of the most poignant scenes in *Villette*. 'Of course the adventure stops there,' she wrote to Emily, 'and I hope I shall never see the priest again. I think you had better not tell papa of this. He will not understand that it was only a freak, and will perhaps think I am going to turn Catholic.'

The sheer emotional and intellectual impossibility of this is apparent to anyone who has read even briefly in Charlotte's correspondence. The moment in the confessional was nothing to do with religion, it was a desperate cry for help, like a modern person ringing up the Samaritans. It is true that living in Brussels had its compensations. She came to like the place. And there were amusing interludes, such as Queen Victoria's visit to her Uncle Leopold. Charlotte went out into the street to see the royal personage 'flashing through the Rue Royale in a carriage and six . . . a little stout vivacious lady, very plainly dressed, not much dignity or pretension about her'. But the horror of her situation deepened. Rather cruelly, M. Heger valued her as a teacher – and, very likely, as a companion – and would not hear of her leaving. It was only the forceful advice, given by post by her friend Mary Taylor, which finally enabled Charlotte to do the sensible thing. When she was back at Haworth, she inscribed a book called *Les Fleurs de la Poésie Française depuis le Commencement du XVIe Siècle*. She wrote, 'Given to me by Monsieur Heger on the 1st January 1844, the morning I left Brussels.'

Dotted through the stout volumes of Brontë correspondence and memorabilia are the four surviving letters which Charlotte wrote to M.

Heger. It was said by his children that he was in the habit of destroying her letters as soon as he received them, and these four were retrieved from a waste-paper basket and stuck together by his wife. They are the most painful things Charlotte ever wrote. Yet, without the sheer misery of unrequited love which Charlotte suffered, it is doubtful whether we should have had her masterpieces, *Jane Eyre* and *Villette*. The first attempt she made to come to terms with her pitiful situation was in writing *The Professor*. But it is a dull book, and Smith was right to reject it. She skirts round her unhappiness, writes well about children and classroom life, but gives the teacher in love with the Professor too easy a ride. It is in *Jane Eyre* that she comes to terms with (among so many other things such as the unrelieved awfulness of her schooling at Cowan Bridge) the difficulty of M. Heger's character. But the ending of that novel – hence part of its enduring appeal – is sentimental. It is the ending we all want. Only in *Villette* does she dare to face up to the fact that life does not offer consolation or happy endings. We can weave a happy ending for Lucy Snowe and Paul Emmanuel if we wish, but Charlotte Brontë will not do it for us.

> That storm roared frenzied for seven days. It did not cease till the Atlantic was strewn with wrecks: it did not lull till the deeps had gorged their full of sustenance. Not till the destroying angel of tempest had achieved his perfect work would he fold the wings whose waft was hunger – the tremor of whose plumes was storm.
>
> Peace, be still! Oh! a thousand weepers, praying in agony on waiting shores, listened for that voice, but it was not uttered – not uttered till, when the hush came, some could not feel it: till, when the sun returned, his light was night to some!
>
> Here pause: pause at once. There is enough said. Trouble no quiet, kind love; leave sunny imaginations hope. Let it be theirs to conceive the delight of joy born again fresh out of great terror, the rapture of rescue from peril, the wondrous reprieve from dread, the fruition of return . . .

It may have been the case that when Messrs Williams and Smith first met 'Acton' and 'Currer Bell', some such happy ending to their troubles was writing itself in all their minds. The success of *Jane Eyre* could hardly fail to bring delight. Charlotte even confided in her father that she had written a book. Thinking she merely meant written one out in her own hand, he declined at first to read it, knowing that his eyesight would not be strong enough to decipher a manuscript. But then she presented him with the bound, finished copy. When he next appeared at tea, Mr Brontë remarked to Anne and Emily, 'Girls, do you know Charlotte has been writing a book, and it is much better than likely?'

It was a comparatively happy time in the life of the household. None of them could possibly guess how short a period it was to be. All four suffered from tuberculosis and Branwell, weakened by drink, was the first to be killed by it.

In their deaths, Charlotte's three siblings all displayed something of their essential character. Branwell, though looking terrible with unkempt hair ('the thin white lips not trembling but shaking', a friend remembered) had been out and about in the village until a few days before the end. He had looked in on friends at the Black Bull, drunk some brandy, and even managed a bite to eat. Then he took to his bed, and by Sunday morning, 24 September 1848, he realised that the end had come. It was rumoured in the village that he insisted on standing up to die, his pockets stuffed with letters from Mrs Robinson, to whom he had remained besottedly attached.

Mr Brontë's grief was extreme: 'he cried out for his loss like David for that of Absalom,' Charlotte tells us. Charlotte herself, always the victim of nervous stomach disorders, took to her bed with 'bilious fever'. The day of the funeral was raw, and Emily caught cold at it. The persistent cough, the loss of all appetite, the feverish, sleepless nights which she endured soon filled Charlotte and Anne with alarm. Charlotte poured out her anxieties in her letters. She had begun to treat Mr Williams, her publisher, as a confidant. Williams, a fervent follower of the fad for homeopathic medicine, urged them to bring a homeopathic physician to 'Ellis Bell''s sofa. But Emily refused to see a doctor of any kind.

At this point, two further reviews of *Wuthering Heights* appeared (it had been published the previous year). The *Spectator* was particularly hostile, driving Charlotte into outraged fury on her sister's behalf. 'Blind is he as any bat,' Charlotte wrote, 'insensate as any stone to the merits of Ellis.' The *North American Review* was even harsher. Emily greeted it with a smile. She was already preparing herself for a sphere where they neither review nor give in reviews.

But (no coward soul, hers) there was nothing resigned about the manner of her death. She refused to go to bed. She refused even to acknowledge that she was ill. It has been several times noticed that in her death, Emily Brontë showed kinship with the animals with whom she so profoundly empathised. She resisted death for as long as she needed to do so. Then she went, without fuss.

She died on 19 December 1848. On Christmas Day, Charlotte wrote to Mr Williams:

> Emily is nowhere here now. Her wasted mortal remains are taken out of the house. We have laid her cherished head under the church aisle beside my mother's, my two sisters' – dead long ago – and my poor hapless brother's . . . Well, the loss is ours, not hers, and some sad comfort I take, as I hear the wind blow and feel the cutting keenness of the frost in knowing that the elements bring her no more suffering; their severity cannot reach her grave; her fever is quieted, her restlessness soothed, her deep hollow cough is hushed forever; we do not hear it in the night nor listen for it in the morning; we have not the conflict of the strangely strong spirit and the fragile frame before us – relentless conflict – once seen, never to be forgotten.

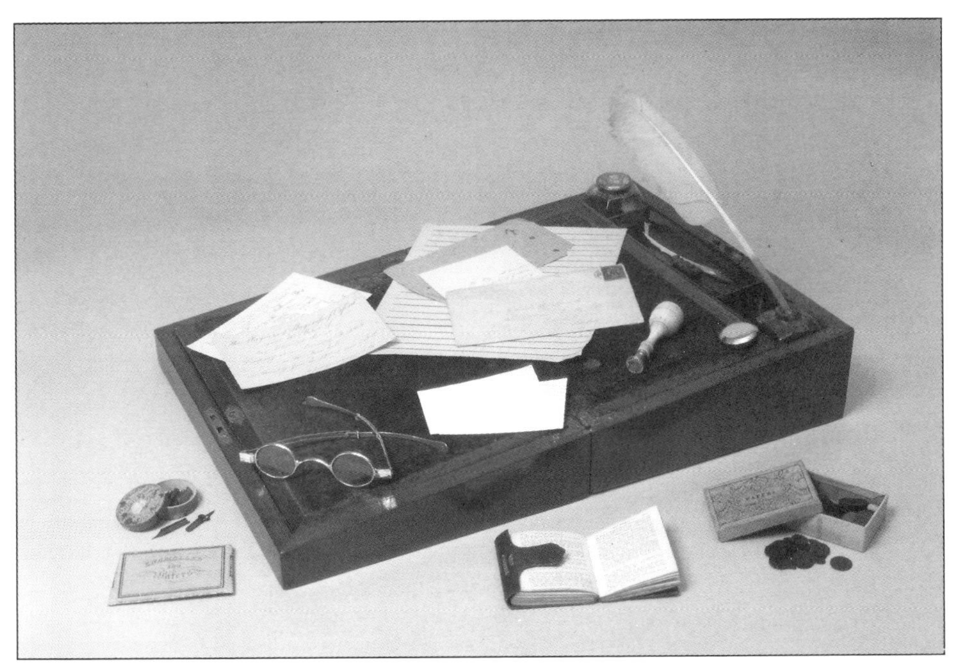

Charlotte Brontë's desk at Haworth parsonage.

Charlotte was surely right to emphasise that the loss was ours, not Emily's. The lives of artists, whether cut short by death or merely fizzling out in youth, do not have to be long, and there is nothing necessarily 'tragic' about the death of someone who has achieved in the course of their life a corpus of highly distinctive poems and a novel upon which it would be impossible to improve. The same could be said of Lermontov, who was even younger than Emily Brontë when he died, and who had also, interestingly enough, developed the narrative experiment, as she had done, of a story told by several different first-person narrators.

Charlotte's destiny was very different. She was marked out as the survivor. 'My father says to me almost hourly, "Charlotte, you must bear up, I shall sink if you fail me"; these words you can conceive, are a stimulus to nature. The sight too of my sister Anne's very still but deep sorrow, awakens in me such fear for her that I dare not falter.' The fears were fully justified. By January 1849, Anne had developed the ominous honking cough which was indelibly associated in their minds with the last days of Emily. Unlike Emily, she declined slowly, quietly, one might almost say modestly. It was hoped, when spring came, that a period by the sea would improve her condition. She died there, at Scarborough, 'a quiet Christian death' which, Charlotte told Mr Williams, 'did not rend my heart as Emily's stern, simple, undemonstrative end had done.' To spare her father the agony of a third family burial service at Haworth in so short a space of time, Charlotte arranged for Anne to be buried at Scarborough. Her grave can still be seen there, the only one of the immediate Brontë family not to be buried in the family vault at Haworth. 'Papa has now me only – the weakest, puniest, least promising of his six children.'

Perhaps in the lives of all great writers, there comes a moment when they feel compelled to put their childhood behind them, to try something new. For some, such as Proust, this is the whole business of their writing lives. For others, the psychological readjustment is more compartmentalised. Tolstoy, having written his pseudo-autobiographical pieces, *Childhood, Boyhood* and *Youth,* is prepared for the work of his maturity. Dickens, with *David Copperfield* behind him, sets forth with a new voice to write the masterpieces of the later period.

Charlotte's severance with her past could not have been more violent or savage. Indeed, for a great many people with only a smattering of knowledge about the Brontës, it is the most important fact about them –

that they all died prematurely. It has given the surely false sense that life in the Parsonage was gloomy, miserable, a reflection of the stormier passages in Emily's poems and fiction. In fact, for much of the time, it was idyllically happy and – like the lives of many largish families in remote or provincial places – entirely self-contained. All through her adult life Charlotte had struggled to make an independent life for them all since they would be left with nothing – not even a house to live in – when their father died. Now, without Anne, Emily and even Branwell to consider, she was left to confront the great question alone – what did she intend to *do* with her life?

It all came at a time when the success of *Jane Eyre* enabled her to hope that the pattern of her life would be shaped by her celebrity as an author.

In July 1849, Charlotte wrote to W. S. Williams:

> The hush and gloom of our house would be more oppressive to a buoyant than to a subdued spirit. The fact is, work is my best companion – hereafter I look for no great earthly comfort except what congenial occupation can give – for society – long seclusion has in a great measure unfitted me – I doubt whether I should enjoy it if I might have it. Sometimes I think I should, and I thirst for it – but at other times I doubt my capability of pleasing or deriving pleasure. The prisoner in solitary confinement – the toad in the block of marble – all in time shape themselves to their lot.

It was in this frame of mind that she took up again the writing of *Shirley*, begun before the deaths of her siblings, a book which, if you come to it fresh from *Jane Eyre*, is bound to disappoint, but which has merits of its own. It has comedy. And in its accounts of the industrial troubles of the earlier part of the century, it has a much broader and more self-consciously serious subject. While waiting for proofs Charlotte read the new Dickens, *David Copperfield*. People were already talking about the affinity between Dickens's novel of childhood and *Jane Eyre*. It is indeed unlikely that the one would have been written without the other. She saw this, but added – 'only what an advantage has Dickens in his varied knowledge of men and things!' Novelists need material to feed on. Unlike poets, they should not have an undiluted diet of solitude. Charlotte was acutely conscious of the danger that if she wrote too much, the ore would run out.

But there is a paradox about her reverence for Dickens's 'knowledge of men and things'. Dickens's characters – and even his scenes – may be regarded as creatures of pure fantasy. It may be the case that he needed the stimulus of society and London life to have invented them, but nothing in literature is more obviously the product of the imagination than the works of Dickens. The people in his books are the creatures of his brain. Charlotte's characters in the fiction of her maturity are very recognisably more realistic than anything Dickens attempted. Realism was not his mode, as it was to become hers. Indeed, one could almost say that Dickens would have written better novels, very likely, if he had been compelled to live in Haworth and look after Mr Brontë, while Charlotte, with her acute eye and her commonsense vision of current affairs (ever the avid reader of contemporary periodicals) might have profited much more than Dickens did from prolonged exposure to London. Still, lives of artists as of others, are what they are, and not what they should be, and we must take the nineteenth-century writers as we find them.

One issue which begins to become very apparent in *Shirley* and which probably owes something to Charlotte's friendship with Mary Taylor, is that of feminism. The naive conversations between Caroline Helstone and Shirley Keeldar must echo sentiments which had passed between Charlotte Brontë and Mary Taylor:

> "Men and women, husbands and wives quarrel horribly, Shirley".
> "Poor things! – poor, fallen degenerate things! God made them for another lot, for other feelings".
> "But are we men's equals, or are we not?"

In an extremely gentle, low-key way, the novel challenges the conventional, Biblical view of women as expressed in the story of Eve, or Milton's great interpretation of that myth. Shirley says that Milton tried to see the first woman, but 'it was his cook that he saw; or it was Mrs Gill, as I have seen her, making custards in the heat of summer.' Shirley's vision of woman is something different – of 'a woman Titan . . . Jehovah's daughter'.

There is a conscious farewell to the *Jane Eyre* style of writing in the novel's conclusion, when Martha (the servant who is clearly based on Charlotte's own Tabitha Aykroyd) describes the district where they live.

Lytton Strachey, whose book Eminent Victorians, *published in 1919, did so much to puncture the inflated regard in which 'Victorian values' were held.*

Franz Winterhalter, the German artist who painted so many of the royal heads of Europe in the middle years of the nineteenth century, had a particular feeling for Prince Albert.

ABOVE 'The First of May, 1851', by Winterhalter. It is an Adoration scene. A Wise Man (the Duke of Wellington) presents his gift to his godson, Arthur. The Mother is the Queen. Albert, cast in the St Joseph role, stares away from the scene with some wistfulness.

RIGHT Lord Cardigan explains the Battle of Balaclava to the royal family. It is said that the Queen had herself painted out when she discovered Cardigan's dissolute manner of life.

ABOVE A watercolour by Anne Brontë of her dog, Flossy. All the Brontës were devoted to animals and birds, and many of their paintings have animals as their subject.

RIGHT Branwell Brontë's famous portrait of his sisters in the National Portrait Gallery. Left to right they are Anne, Emily and Charlotte. The blank smudge in the centre of the canvas is where Branwell attempted a self-portrait, and then scrubbed himself out, a potent image of his urge for self-destruction.

her father and
mother was very
rich Mr and Mrs
Wood were there
names and she
was there only
child but she h
was not too mu
indulged once
little Ann and
her mother
went to see
a fine castle
near
London about
ten miles

from it Ann
was very mu
ch pleased with

ABOVE An extract from Charlotte Brontë's Stories of Angria, the imagined world she invented during her childhood.

RIGHT An alcove at the top of the stairs in Haworth Parsonage, in which may be seen some of Charlotte's possessions, including the trunk she took to Brussels.

Below Sydney Prior Hall's portrait of Gladstone shows him at the end of his life in his study, still mulling over literary texts and ruminating on the problems of the day.

ABOVE Christ Church Hall at Oxford, where generations of undergraduates have dined. A high proportion of Christ Church men have always been aristocrats and many young men who sat here went on to become bishops, judges or high-ranking statesmen. Wealth and the Church of England here reside side by side; it is appropriate that Gladstone should have been educated here.

RIGHT Charles Stewart Parnell in a caricature for Vanity Fair.

A triumphal icon (from the Illustrated London News*) of the advances in sea and land transport, and in street lighting, that marked the 60 years of Queen Victoria's reign. The 'Whig' view of history, and the Darwinian view of nature, alike persuaded the Victorians that history had been a perpetual process of improvement, each age better than the last.*

'This world has queer changes.' Martha can remember not only the early industrial days of Yorkshire, but also the days before that, the days when such supernatural happenings as occur in *Wuthering Heights* or *Jane Eyre* are part of the common folklore. 'I can tell, one summer evening, fifty years syne, my mother coming running in just at the edge of dark, almost fleyed out of her wits, saying she had seen a fairish (fairy) in Fieldhead Hollow; and that was the last fairish that ever was seen on this countryside . . . A lonesome spot it was – and a bonnie spot – full of oak trees and nut trees. It is altered now.'

If the young Brontës were the heirs of the Romantic movement, Charlotte in her maturity was preparing to be something different: a modern realist. The subject of her developed fiction is what women are meant to *do* in a world whose limited tedium was determined for them by men. The 'abundant shower of curates' at whom she pokes such fun in *Shirley* were Charlotte Brontë's only real hope of escape from her clergyman father, who in any case required her increasing attention as his sight failed. As for love, and the bleak prospects for those who fall into it, *Villette* is her soberest and saddest development of that theme. All the wretchedness she had felt in Brussels, and the hopelessness of her attachment to M. Heger was poured into the experiences of Lucy Snowe, and her love for M. Paul Emmanuel. '*Villette* touches on no matter of public interest. I cannot write books handling the topics of the day; it is no use trying,' she wrote to her publisher George Smith in October 1852. Like John Keats, she was 'certain of nothing but the holiness of the heart's affections' which, of course, is why her public loved her.

Her fame could not console her for the loss of those she best loved – either by death or unrequited affection. Nevertheless, it provided her with welcome moments of diversion and new friendships, it offered more excuses to escape Haworth and the difficult companionship of her father, and it introduced her to one of the most important figures in her life, her biographer. The novelist Mrs Gaskell, the wife of a Manchester Unitarian minister, wrote to express her admiration of *Shirley*. It is clear from Mrs Gaskell's letters that she was fascinated by the idea of Currer Bell before ever meeting Charlotte. The high-minded agnostic feminist Harriet Martineau (the author of *Deerbrook* whom Charlotte had long admired and had met in London) had, Mrs Gaskell wrote to a friend, 'sworn an eternal friendship with the authoress of "Shirley".' When they did meet Mrs Gaskell was surprised to find Currer Bell 'a little, very little bright-haired sprite, looking not above 15, very unsophisticated neat and tidy'. From the earliest, Mrs Gaskell adopted Charlotte as a 'character' whom she would feel compelled to write up – whether in a work of fiction or as the subject of a biography was not yet clear.

It was in the August of 1850 that the two women got to know one another while staying at the house of a mutual acquaintance near Lake Windermere. Mrs Gaskell was struck by Charlotte's tininess, by her plainness ('large mouth and many teeth gone') and by her social awkwardness. While in the district, the two women went over to Fox House, the home of Dr Arnold, the famous headmaster of Rugby, and still inhabited by his widow. Charlotte, who had been profoundly affected by Arnold's biography by Dean Stanley ('where can we find justice, firmness, independence, earnestness, sincerity, fuller and purer than in him?'), was impressed by Mrs and Miss Arnold. To her son Matthew (already the published author of two volumes of poems), she talked about French novels and her father's curates. He found her 'past thirty and plain'. His manner displeased her by its 'seeming foppery' and his 'assumed conceit' but she was prepared to believe those who told her that 'Mr Arnold improved upon acquaintance'.

It was open to Charlotte, from this point in her life, to have entered the world of Victorian intellectual society. James Taylor, a partner in Smith Elder, fell in love with her and repeatedly proposed marriage. But as she confided in Ellen Nussey, 'if Mr Taylor be the only husband fate offers to me, single I must remain'. She did not love him; in fact, she found him physically repugnant. She went on loving M. Heger. Perhaps more importantly than love, the world offered by a London publisher was not one where she could have survived. The crushing loneliness of the Parsonage, her father's difficult character, the provincialism of Haworth – it was by now too late for her to escape these things. Her nature had grown used to them. And, as the lighter passages of *Shirley* show, she did not entirely hate them.

Fate, in the event, offered her more than one partner to choose from. He was the latest in a line of her father's curates, an Irishman called Arthur Bell Nicholls. A follower of Dr Pusey in doctrine, his manner of reading the services annoyed Charlotte's father. Charlotte found him 'good, mild, uncontentious' as he came in and out of the Parsonage to report on the doings of the parish. Then, just before Christmas 1852, he proposed.

This rather ineffectual young man had been in the parish for eight years now; beyond providing Charlotte with a comic vignette for *Shirley*, he had hardly intruded at all upon her consciousness. Everything that had happened – her falling in love with M. Heger, the deaths of Emily, Branwell and Anne, the illness of her father, the discovery of her fame – had been so much stronger experiences than anything which Mr Nicholls seemed to be offering. Now he revealed, in a potentially embarrassing outburst, that he had been in love with Charlotte 'for months', and that he 'could endure no longer'. Very bravely, Charlotte went into the study to inform her father of the conversation which had just taken place. There was a characteristic explosion of Ulster rage. In Nicholls, Mr Brontë now saw the man who proposed to take away from him his last surviving child and he resisted with ferocity. How could they marry on Mr Nicholls's paltry £100 a year? The upshot of the quarrel was that Mr Nicholls was asked to leave the parish and seek alternative employment.

But this dull, amiable man had offered Charlotte something which she was not prepared to forgo. Perhaps with the love of her sisters and brother around her, she would have been content to sit at home, and dream of fantasy figures like Mr Rochester, and write painful love letters to M. Heger. But now she was alone – painfully lonely and depressed. She was not in love with Mr Nicholls, but she desperately needed what he offered to her.

Another curate was engaged – a Mr De Renzy – and they began to count the days before Nicholls's departure. The last time he celebrated the Communion, on Whitsunday, Mr Brontë stayed away, even though it is one of the Sundays in the year when members of the Church of England are obliged to receive the sacrament. When Nicholls approached Charlotte, kneeling at the altar rails to receive, he 'struggled, faltered, and then lost command over himself, stood before my eyes and in the sight of all the communicants, white, shaking, voiceless!'

This was a story which was gradually unfolded to Mrs Gaskell when she came to stay in the Parsonage later that year. Mr Brontë took a liking to Mrs Gaskell and reminisced expansively. 'He had a sort of grand and stately way of describing past times,' she recollected, noting that 'he never seemed quite to have lost the feeling that Charlotte was a child to be guided and ruled when she was present.'

It was undoubtedly Mrs Gaskell's goodness of heart and practical common sense which determined her to do something for Mr Nicholls and Charlotte. But perhaps, too, instinct was already prompting her that she should write Charlotte's story, and that this story should have (as seemed possible then) a happy ending. Though no Anglican, Mrs Gaskell worked out how it was possible to get some extra money from the Additional Curates Society to supplement a curate's salary were Nicholls to be allowed back to Haworth by Mr Brontë. Then at least, the financial objections could be overruled.

It would appear that Nicholls's devotion to Charlotte was unwavering, and by early 1854, she confessed to her father that she had been continuing to correspond with the man. As she admitted to Mrs Gaskell, her father's were not the only objections to the match which had to be overcome. 'I cannot deny that I had a battle to fight with myself: I am not sure that I have even yet conquered certain inward combatants.' Mr Nicholls's apparent religious bigotry, his devotion to Puseyism, his lack of interest in literature or the arts were all handicaps. It had been said that when he was a curate in Haworth he would not speak to Non-conformists. What would he make of Charlotte's friendship with the Unitarian Mrs Gaskell and the lapsed Unitarian, now agnostic, Harriet Martineau?

But she had time to think it over. She was thirty-eight years old (Mr Nicholls two years younger). Mr Brontë's objections were at length over-come. A back room at the Parsonage was refitted as a small study. Charlotte, who had a domestic genius as well as a genius for writing novels, saw how to make the room attractive, by knocking through a fireplace and papering it with a light floral pattern. The wedding was fixed for 29 June.

Mr Brontë stayed indoors and did not accompany them to the church. The ceremony was conducted by the vicar of nearby Hebden Bridge. Her faithful friends Ellen Nussey and Margaret Wooler had arrived to stay the day before; and, in the absence of any suitable male, Miss Wooler gave away the bride in marriage. There was something purely apposite about this. For Miss Wooler, who had been her mentor at Roe Head, represented all that was best about her spinsterhood, and its one hope of usefulness, the educational profession. There was no one in the church at 8 a.m. on that day to represent the 'Currer Bell' side of Char-lotte's life – no fellow-authors, no publishers, no London literary folk.

1854. Marriage solemnized at Haworth in the Parish of Bradford in the County of York

No.	When Married.	Name and Surname.	Age.	Condition.	Rank or Profession.	Residence at the Time of Marriage.	Father's Name and Surname.	Rank or Profession of Father.
346	June 29th	Arthur Bell Nicholls	full age	Bachelor	Clerk	Kirk Smeaton	William Nicholls	Farmer
		Charlotte Brontë	full age	Spinster	—	Haworth	Patrick Brontë	Clerk.

Married in the Church of Haworth according to the Rites and Ceremonies of the Established Church by Licence by me. Sutcliffe Sowden

This Marriage was solemnized between us, Arthur Bell Nicholls, Charlotte Brontë

in the Presence of us, Ellen Nussey, Margaret Wooler

Charlotte Brontë's marriage lines. She and Arthur Bell Nicholls were, somewhat to her surprise, extremely happy together, but she died after only nine months of marriage as a result of excessive morning sickness.

Charlotte wore a white muslin dress with a motif of ivy-leaves, and a bonnet which is still preserved at the Parsonage. When the little party returned for the wedding-breakfast, they found that Mr Brontë had recovered his spirits and he was 'the life and soul of the party'. When Charlotte had changed into her shot-silk travelling dress – it still survives in her bedroom – she and her bridegroom planted evergreen trees at the bottom of the garden.

The honeymoon was spent in Ireland, and it was a great success, although it began with Charlotte nursing a nasty cold. The despised Mr Nicholls turned out to have rather grand relations – the Bells of Cuba House; and Charlotte enjoyed staying there, as well as seeing the haunts of his student life at Trinity College, Dublin. It is obvious, too, that the physical aspects of marriage brought unexpected delight. The next six months found Charlotte happy in a way that she had never been happy before. Although when they returned from Ireland, Mr and Mrs Nicholls found Mr Brontë rather unwell, their lives were unclouded. It appeared that the two men, who had quarrelled so much about the prospect of Charlotte's marriage, settled down together perfectly happily once it had happened. Mr Nicholls was now 'my dear Arthur' and 'my dear boy'; and Charlotte was pregnant.

None of them realised how debilitating the pregnancy would be to Charlotte; she was thirty-nine and very small. Her lingering 'cold' was never quite shaken off, and she made it worse in January 1855 by walking on to the moors in thin shoes and coming home with wet feet. Her usually minimal appetite diminished to pure anorexia. 'A wren could have starved on what she ate during those last six weeks,' wrote one observer.

On 21 February, her much-loved servant Tabby Aykroyd died, and was buried just beyond the garden gate in the churchyard. Mr Nicholls took the service. Charlotte was by then confined to her bed. When, some weeks later, she realised quite how seriously ill she was, she exclaimed to her husband: 'Oh, I'm not going to die, am I? He will not separate us – we have been so happy.' She died on 31 March 1855. Her father lived for another six years, Mr Nicholls remaining with him as his faithful nurse and companion. When Mr Brontë died, Nicholls returned to Ireland where he married again – to a cousin. He did not die until 1906. A photograph of him in old age survives. It is clear that he never quite recovered from the loss of Charlotte. The lively Richmond drawing of her always hung on the wall of his dining-room at Hill House, Banagher. When he died, his wife had the coffin carried so that it was placed beneath Charlotte's face. It was said by his nephews and nieces that often, in those last few days before he died he would fall into abstracted thought, meditating upon the reunion with Charlotte: 'I wonder how it will be?'

By then, the Brontë cult was flourishing. Only four months after Charlotte's death, Mr Brontë had written to Mrs Gaskell, asking her to Haworth to discuss the possibility of a biography. *The Life of Charlotte Brontë* is one of the great biographies of the nineteenth century, as well as being the basis for all subsequent studies of the family. Mrs Gaskell was criticised for sensationalism; and inevitably, she got some details wrong. She blamed Charlotte's near-anorexic eating habits, as she had observed them at first-hand, on a supposed regimen of old Mr Brontë's, claiming that he had deprived his daughters of meat. But she was handsome enough to correct this and other mistakes in her second edition. She was punctilious in her research, and she interviewed everyone who might have known her subject – everyone, that is to say, except the husband. Mr Nicholls (very understandably) refused to co-operate in the enterprise. Otherwise, villagers, schoolfriends, pupils in Yorkshire and

This studio portrait of Arthur Bell Nicholls was perhaps taken at the same time as Charlotte's carte de visite *(page 57). He was Patrick Brontë's curate and, after initial difficulties, was allowed to marry Charlotte Brontë.*

Belgium, publishers – all were quizzed for their evidence of this remarkable genius.

But the *Life* which emerged probably had an effect which Mr Brontë did not intend. The fact that a memorial was written helped to console him for the appalling misfortune of losing all six of his offspring. Something was salvaged from the destruction wrought by time. He was vividly aware, as such an intelligent man could not fail to have been aware, that his children had been exceptional people – at least one and possibly two geniuses, and all four who survived childhood of remarkable talents. To preserve their memory, he was bravely prepared to have the sad story told of his son's alcoholism. There were, of course, great protests from the Clergy Daughters' school – now moved from Cowan Bridge to Casterton – at the revelation that Charlotte had based Lowood in *Jane Eyre* so closely on Mr Wilson's educational mode. But the protest was not truly justified, even though it was generally recognised that, since the move to Casterton, the Clergy Daughters' school had 'improved'. A year after Charlotte's death, a Miss Dorothea Beale taught there. She did not find it as monstrous as Lowood, but she left, and went on to become the first great headmistress of Cheltenham Ladies' College. It is a symptomatic little link between the hopeless absence of opportunity for women which existed at the beginning of the century and the very considerably improved standards at the end. Figures like Miss Beale, and her counterpart at the North London Collegiate School, Miss Buss, provided as high a standard of education for girls as was then available for boys. They were closely involved with the foundation of women's colleges at the Universities under women like Elizabeth Wordsworth.

It was all a far cry from the kitchen at Haworth, where Emily tried to teach herself German from a grammar propped up among the bread-dough; or from the miserable 'daughters of the clergy' school from which the Brontë sisters could only hope to emerge as potential governesses or, if they were lucky, as curates' wives. The pupils of Miss Beale and Miss Buss could hope to become doctors, scientists, lawyers, civil servants, engineers. There was of course the inevitable taunt that such ideals were not 'feminine':

Miss Buss and Miss Beale
Cupid's darts do not feel.
They are not like us
Miss Beale and Miss Buss.

But once started, the movement for women's education became a self-perpetuating process.

Mrs Gaskell's *Life of Charlotte Brontë* was a study of people of the highest talents and capability who were compelled, merely by virtue of their sex, to live in a sort of protestant purdah. The genius of the Brontës set them apart from humanity. Their lives, in many particulars, were like the lives of thousands of provincial men and women throughout England, Scotland, Wales and Ireland. Those who read Mrs Gaskell's book could discover a mirror-version of their own tedious, limited existence. That is part of its intense fascination.

But only part. Though many provincial people – no telephone, no radio, no electricity, miles from anywhere and with few, if any, congenial neighbours – led lives which by modern standards would be considered intolerable, one can confidently say that none were quite like the Brontës. The social isolation of their lives, as well as the shocking brevity, make them all the more dramatic. But we would not be so interested if Haworth Parsonage had not produced *Jane Eyre* and *Wuthering Heights*.

Nor – it has to be said – would all the advantages of an enlightened modern society have necessarily done the Brontë sisters much good. Because they were people of high imaginative genius, they would doubtless be producing wonderful books; but they would certainly not have written *Wuthering Heights* or *Jane Eyre*, books which depend for their huge success on the isolation and innerness of their creators. Charlotte was sad that no one liked her later books as well as they had liked *Jane Eyre*. It is the element of fantasy, of suppressed eroticism, of isolation from humanity, which gives the great scenes of that book their power.

That is why people all over the world continue to read it. It is also the reason, presumably, why millions of tourists or pilgrims flock each year to Haworth and stream through the little Parsonage house which, since 1928, has been owned by the Brontë Society and run as a museum. Here may be seen Charlotte's tiny gloves and shoes; the daubs which unhappy Branwell made; the piano where Emily played, the sofa where she died. It is one of the best-run museums of its kind in the world, rivalled only perhaps by the Tolstoy museum in Moscow or the Carlyle House in Cheyne Row, London. In spite of the huge crowds flocking through it all the time, the house retains its private feeling. Have all the millions who go there read *Jane Eyre*? Many have, one suspects, though more have seen the film. Is it the thought of the premature deaths of the Brontës which makes them such haunting figures? Or is it that their books are so inescapably linked to the place? Step out above the village, with all its souvenir shops, its Villette tea room and Heathcliff Café, and you will very quickly be exposed to the same wild and rather alarming moorland which confronts us in *Wuthering Heights*.

If people suppose that the early deaths of the Brontës are *tragic*, however, they are wrong. Certainly, from the purely selfish point of view of the reader who admires their work, there is no reason to suppose that they would have produced much more of the same quality had they lived longer. Charlotte was writing more *maturely* – but that was just the trouble! And Emily might well have fizzled out altogether. To have written *Wuthering Heights* and some of the best poems in the language is enough. It doesn't matter that she was only thirty when she died. Her life was in a sense complete.

Charlotte's was not. From a literary point of view, had she lived, we would, I suspect, have witnessed her getting further and further away from the things she was truly good at. There would have been plenty of good characterisation, social comedy, bleak observation about human life; but the intense drama of Bertha Rochester bursting from her locked-up room was never to be repeated. It is not because we want to read any more of her unsatisfactory fragment *Emma* that we wish Charlotte had not died so soon. It is because, as she said herself, she had just begun to be happy. It is because we wish that she had been able to have her baby, and had lived to enjoy it. Such a happy eventuality would have ruined Mrs Gaskell's biography, and would, I suspect, have reduced the future number of pilgrims to Haworth by a thousandfold.

William Ewart Gladstone

1809-98

The great Liberal Prime Minister

William Ewart Gladstone, photographed in old age. By then he was the Grand Old Man, the 'People's William'. His handsome face is marked by the moral toughness with which he confronted his essentially divided nature.

When I think of Victorian England, I think of energy: irrepressible physical energy, intellectual, industrial, moral energy. I think of a place where machines are perpetually turning, where factories belch smoke, where canals and railways, laden with produce, carry freight to warehouses and ports. I think, too, of the great ships, setting out from Liverpool and Hull and London to destinations all over the world. I think of the merchants, the explorers, the colonisers, the evangelists and the engineers all self-confidently taking abroad their skills and prejudices and calling the result of their endeavours the British Empire.

At home, I think of the stupendous engineering achievements of Brunel; I think of the literary fertility of Carlyle, Ruskin or Browning, filling volume after volume of the library shelf. I think of all the movement and life of the Victorian city – the crowds, the street-cries, the clatter of wheels on cobblestones, the plight of the poor, and the adventures of the criminal. I think of the London of Mayhew, Dickens, and Sherlock Holmes.

In the very heart of London's Kensington Gardens, G. F. Watts's statue of *Physical Energy* (1904) is a reflective meditation on a prosperous industrial and imperial power at its zenith. The horse, with brawny flanks and upraised hoof, tugs at the reins in the direction of Kensington Palace, where Queen Victoria herself was born. The barefoot rider, rather more wistfully, looks down the avenue of trees towards the Albert Memorial. Is he wondering what all this expenditure of energy was for? Has the horse got out of control, and taken him into a world he does not either trust or understand? Spirit and matter, morality and desire, money and romance are torn in this divided emblem.

Of all the great Victorians, no one demonstrates this sort of ambivalence more dramatically than the greatest of their Prime Ministers. He spans the century like a colossus, never quite knowing what he makes of it, nor what it makes of him. Like Watts's statue, Gladstone's was an essentially divided nature. Few men have been more intimately involved in the world. Yet his deepest longing was unworldly. Though one of the

G. F. Watts's statue, 'Physical Energy' in Kensington Gardens, London. Spirit and matter, morality and desire, money and romance are torn in this divided emblem. The rider gazes in one direction, while the horse strains in another.

longest-serving of all Cabinet ministers, he had originally yearned to be a minister of religion. Moreover, the shift in his political views – one of the strangest stories in the history of statecraft – follows a most unusual pattern.

When he was an old man of eighty-two, a friend remarked to him, 'You know the saying that nobody is worth much who has not been a bit of a radical in his youth, and a bit of a Tory in his fuller age?' Gladstone laughed and replied, 'Ah, I'm afraid that hits me rather hard. But for myself I think I can truly put all the change that has come into my politics into a sentence; I was brought up to distrust and to dislike liberty, and I learned to believe in it. That is the key to all my change.'

He began his political career, in Lord Macaulay's famous words, as 'the rising hope of those stern and unbending Tories'. He ended his days as 'the people's William', dreaded by the Queen, and by the landed classes as the father of radicalism and the fomenter of social discord at home and in Ireland. But for others, there was something more appropriate in his journalistic nickname – the G.O.M. He was the grand old man whose sheer bigness, and grandeur, and moral weightiness was never to be repeated on the political scene.

William Ewart Gladstone was born on 29 December 1809. The place and timing of his birth are both important. He was born in Rodney Street, Liverpool. His father, John Gladstone, was a self-made Scottish merchant, who had come to Liverpool to further his fortune in trade. And much of that trade was conducted with the West Indies, and much of the trade in the West Indies was not in sugar or rum, but in human lives, in slaves. In 1795, John Gladstone was worth £15,000 – a considerable sum in today's money. By 1799, he had more than doubled it and was worth £40,700. By 1820, he was worth £333,600 and by 1828, £502,550. This was serious money.

Like many devout Evangelical families, the Gladstones attributed the steady accumulation of their capital to the loving favour of the Lord, and they built two churches as a testimony to His goodness. Anglican churches, of course, for in coming south and becoming a gentleman, John Gladstone had abandoned the Presbyterianism of his youth. William Gladstone, therefore, together with his five brothers and sisters (he was the fourth son), was born into an inheritance of vast prosperity; but he was never to forget from what that prosperity derived. He was too complex, and ultimately too Protestant a figure, to be able to separate the guilt from the blessing. You might have thought that a man who felt awkwardness about inheriting a fortune from money made in the slave trade would have supported the Yankees in the American Civil War, but Gladstone was a firm, and indeed tactless, supporter of the Confederation, his outspoken utterances on the subject at the time of the American Civil War leading to an international incident. Though he never appears to have had the 'conversion experience' which characterises so many Evangelical lives (such as that of Cardinal Newman) Gladstone believed in justification by faith. Guilt was an essential condition of life, washed by Grace, but always there. Money and Anglicanism are Mr

Gladstone's two great wet-nurses, sustaining him throughout his long moral adventure in the 'vale of soul-making'.

He was sent to Eton in 1821. His elder brothers were already there. Nothing but the best for the new money of the northern merchant. By the mysterious alchemy of money, Gladstone, whose grandparents had been peasant-proprietors in the Scottish borders, was instantaneously propelled into a world where he mixed as an equal with the sons of great scholars, churchmen and aristocrats. Being large and athletic, boarding school held no terror for Gladstone. It was here that many of his most distinctive habits and talents were developed: at the Eton Society (later in the century known as 'Pop') he revealed himself as a powerful debater, always arguing from the 'stern unbending Tory' point of view.

No less remarkable than his energy in debate was his academic prowess. It was at Eton that he developed his lifelong addiction to Homer. The *Iliad* remained Gladstone's favourite book, with Dante's *Divine Comedy* running it a close second. As the years of his grown-up life passed, with such time-consuming distractions as being Prime Minister four times, he was always trying to spare time for Homer. He wanted to prove Homer's essential kinship with the Divine revelation in the Hebrew scriptures. In the 1860s, for example, when he was Chancellor of the Exchequer, we find him laying aside the Budget to snatch a few hours with the manuscript of his book, *Studies in Homer*. 'If it were even tolerably done', he wrote, 'it would be a good service to religion as to literature and I mistrustfully offer it up to God.' At the same period, the Chancellor of the Exchequer was observed, having changed trains at midnight, sitting in the station waiting-room at Crewe, struggling with his translation of the *Iliad* into English verse.

At Eton, too, Gladstone was confirmed in the Church of England, and there began that lifelong pattern of extreme piety which was so marked a feature of his character. He wanted to be a parson, but his father who had got himself elected as a Tory MP for a pocket borough, wanted the monied interest of the Gladstones to continue to be represented in the House of Commons. William's powers of oratory, meant by Nature, if not by God, for the pulpit, were destined to echo in the secular chamber, and in the drawing-room of his hostile monarch. Meanwhile, ever 'mistrustfully' wondering whether he should not have given his life completely to God, Gladstone attended church every day of his life – sometimes twice, sometimes three times. In all the multifarious business

As a priggish High Church young man, Gladstone was described as the rising hope 'of those stern unbending Tories'.

of his existence, church business occupied the most emotionally exciting role – whether it was the establishment in London of the Margaret Street Chapel, or the use of Gladstone money for the founding of the Episcopalian seminary at Glenalmond in Scotland. However busy he was, Gladstone usually found time to write a sermon each week which he declaimed to his servants at family prayers on Sunday morning.

At Eton, Gladstone also began his habit of a daily diary, or journal as he himself termed it. He kept it up for seventy more years – over 25,000 entries with hardly a day missed. There is a certain degree of soul-searching in Gladstone's diaries, at key moments in the year, when he reviews his sins and accomplishments. But for the most part, the volume after volume of Gladstone's diary, the gradual publication of which is even now keeping an Oxford academic in full-time employment, is more a diary of doing than one of being. It is a chronicle of prodigious, one might almost say manic effort and activity, an extraordinary compilation of books read, committees attended, journeys made, speeches declaimed (often of several hours duration), of money invested, land cultivated, arguments won or lost. He was a giant – a Titan of emotional, physical and mental energy.

As a schoolboy, he does not appear to have had much homosexuality in his make-up; but he was highly-sexed and given to profound emotional attachments. His Eton friendships meant a lot to him. Greatest among the Eton friends was Arthur Hallam. This figure, destined to become legendary in our literature, was evidently, emotionally speaking, a flirt. In 1826, Hallam told his sister that he had been 'walking out a great deal, and running the changes on Rogers, Gladstone, Farr and Hanmer'. While Hallam 'ran the changes', his friends agonised for the chief place in his affections. Later, when he was an undergraduate, Gladstone wrote down a brief history of his friendship with Hallam. It is typical of the two sides of his nature – the efficient administrator who wanted to keep minutes of everything, even the changes of the heart's affections; and the figure who was all passion, the man who underscored all the most fervently emotional passages in Dante's *Paradiso*, and who really did believe that it was Love which moved the 'sun and other stars'.

Here is the friendship with Hallam, as minuted by Gladstone:

> It began late in 1824, more at his seeking than mine.
> It slackened soon, more on my account than his.
> It recommenced in 1825, late, more at my seeking than his.
> It ripened much from the early part of 1826 to the middle. In the middle Farr rather took my place . . .

And so on, until the last sad stage: 'At present, almost in uncertainty, very painful, whether I may call Hallam my friend or not.'

What appears to have made Arthur Hallam so attractive to his friends was a combination of innocence and intellectual intensity. He was a Whig, but he did not – in life at least – manage to shake Gladstone's Toryism. When they left Eton, the two boys went their separate ways. Hallam went to Cambridge, where he befriended a boy at Trinity called Alfred Tennyson. When Hallam died at the age of twenty-two, he left a whole generation bereft. Tennyson, in a long cycle of lyrics, mourned:

> *My Arthur, whom I shall not see*
> *Till all my widow'd race be run.*

Arthur Hallam was one of Gladstone's closest friends at Eton. At Cambridge he befriended Tennyson and after Hallam's death at the age of 22 the poet mourned him in the elegy In Memoriam.

These lines prompted Dickens, who read the anonymous poem, to speculate that it was written by the young widow of a naval officer, lost at sea. But the elegy, with its haunting sense that young death makes life meaningless, with its tormented religious doubts, its fascination with and fear of science was to be of huge importance. It was called *In Memoriam*, one of those poems, which – like *The Waste Land* of T. S. Eliot – seems to speak for all its generation. The Queen, deep in grief for the Prince Consort when her friendship with Tennyson was at its height, found that his sonorous expressions of despair had 'helped her much'.

Gladstone, however, was not destined to go to Cambridge – Cambridge of poets, Cambridge of speculative meditative scholars, Cambridge of the inner life. Instead he went to politicking, pious, public-life Oxford. And, being John Gladstone's son, he went to the best that money could buy, the aristocratic foundation of Christ Church. Nearly all the Prime Ministers of modern times went to Oxford, rather than to Cambridge, and this is no accident. In the hall at Christ Church today you see, beside the portrait of Gladstone, a host of lord chancellors, archbishops and elder statesmen. The place feels like the very embodiment in stone and wood and canvas and marble of the Establishment. In splendidly palatial quadrangles and rooms around, the young gentlemen were accommodated. It feels closer to one of the great London houses or one of the big stately homes than a place of scholarship and learning. And in the middle of it all, instead of a mere college chapel, there is a cathedral, with a dean and an archdeacon and a bishop's throne. What a perfect allegory of Gladstone's early life! What a parable of the way that the young Gladstone saw England. He was a Tory, not because he believed in privilege, but because he believed in duty – the God-given duty of the few to preserve the Crown and the Church for the many.

To us, in the last decades of the twentieth century, the Church of England may be one of many things. Larkin called it

That vast moth-eaten musical brocade
Invented to pretend we do not die.

When the poet died in 1985, rather than denouncing these lines, the Sub-dean of Westminster Abbey quoted them with reverence at Larkin's memorial service. It seemed the ultimate expression of how 'marginalised' the Church had become: no longer even one Christian denomination among many, but a fusty piece of old furniture of which even its paid-up custodians seemed to have forgotten the significance.

For Gladstone's generation, the Church of England was something quite different; particularly for those, like Gladstone, who were Tories. Nineteenth-century England is difficult enough to understand. But we will never begin to do so unless we grasp the importance, in the Victorians' scheme of things, of the Church of England. When Gladstone came up to Oxford in 1828, he was obliged to subscribe to the Thirty-Nine Articles – the rule-book or party manifesto of the National Church. Those who were unable to subscribe – Jews, Catholics, Methodists – were not admitted to Oxford. As Samuel Johnson had remarked when this was deemed unjust: 'Sir, we permit cows in a meadow but we drive them out of the garden.'

Was this merely Privilege defending Privilege? Was it merely a spiritual department of state keeping the riff-raff out of the Church just as their hopelessly corrupt and antiquated political system, with its rotten boroughs and limited franchise kept the *hoi polloi* out of Parliament, and the system of purchase made it impossible for any but the rich to serve as officers in the Army? Or was it, as Oxford Tories believed, something quite other? Was the Church not merely a national institution founded by Henry VIII, but a divinely founded society, purged but not destroyed at the Reformation, and protected by the Holy Spirit to enshrine the

Truth? If this was the case, then anything which smacked of pluralism, of toleration, would imply that the truth of God was not true. If you *tolerated* the heresies of the Non-conforming sects, such as Roman Catholicism or the Countess of Huntingdon's Connection, were you not implying that the Thirty-Nine Articles were wrong?

This argument – esoteric by today's standards – first came Gladstone's way in a manner which was to be full of portent for his future career – in the national debate concerning Ireland and the Roman Catholic Church.

The Duke of Wellington, who grew up in Ireland and was in any case a pragmatist, knew that the Irish would go on practising their faith, Thirty-Nine Articles or no Thirty-Nine Articles. There was about as much chance of getting them to become Anglicans as there was of making them into Muslims. Liberals of various complexions felt that it was a simple injustice to deny franchise to fellow-Christians who could not, for reasons of conscience, support the National Church. And not just the *vote* – that was a privilege reserved for few enough in any case. Before 1829, Catholics could not practise at the Bar, could not attend the Universities, and could not really play any serious part in the practical running of their nation's affairs.

The Catholic Emancipation Act of 1829 went some way to remedying this state of things. It was the first stage in a monumental change which was overtaking the whole country, and which would culminate in the Great Reform Bill of 1832.

But for the Church Tories of Oxford, all this constituted, in the famous words of their leader John Keble, a 'national apostasy'. Keble and his followers were only an inch away from believing that God had appointed the Tories and the Landed Class to govern England in perpetuity. In a series of pamphlets, known as *Tracts for the Times*, they poured out their worry about the increasing secularisation of the age and their insistence that the Church by Law Established was divine – the same Holy Catholic and Apostolic body which had sustained the early fathers of Christendom. It was only the second generation of 'Tractarians', as believers in these Tracts came to be called, who saw that Catholicism insists not only upon truth, but upon justice – a distinction which enabled Tractarians like Gladstone to shake loose from the Tory party.

But throughout his early years, Gladstone believed – or sort-of-believed, for his religious beliefs are never crude, however peculiar they may seem to us – in John Keble's Church Toryism. With the advanced Catholicism of its theology, Gladstone, in adopting this creed, had come a long way from his boyhood Evangelicalism.

Nothing before or since has ever remotely resembled the Oxford Movement. The medieval revival which, in the previous Romantic generation, had been primarily an aesthetic matter, became cerebral and political. Sir Walter Scott had created a vastly popular medieval fantasy world in novels like *Ivanhoe* and *The Talisman* which were adapted in endless operas, circus performances and paintings. Keats and Coleridge had sung of La Belle Dame Sans Merci and Christabel. Architects had rediscovered the beauties of Gothick. But it was only really in Oxford, and particularly in Oriel College, that men like John Keble had begun to think that the Catholic theology of the Middle Ages might actually be true. Until Newman, some decades later, took the logical step of joining the Church of Rome (the church of St Thomas Aquinas, St Albert the Great and the other medieval schoolmen) there was no thorough-going attempt to undermine the Reformation. Rather, Keble, Pusey and his friends believed that the truths which had been enshrined in the medieval Church had not been lost at the time of Henry VIII's divorce from Catherine of Aragon. The Church of England, far from being a botched-together quasi-political compromise, was in Keble's eyes the very body of Christ Himself, a holy society of Divine foundation. It followed that when Church matters were discussed in Parliament, far more than politics were at stake. Since only the Ascendancy families in Ireland were adherents of the Anglican Church there, it made sense to the Whig Parliament in London to reduce the number of Irish bishoprics. But from the perspective of Oriel, where the fellows had refashioned the established Church in a stained glass medieval glow of their own imagining, this essentially practical question was full of religious significance. Members of Parliament had no authority in the Body of Christ. To reduce the Irish bishoprics was to imply an unbelief in the full, catholic and apostolic character of the National Church. So issues which, with the perspective of history, seem almost inconceivably petty and unimportant, were test cases to show forth the truth that God Himself was directly involved with all the issues of the hour.

Within a generation, all the best minds in Oxford had rejected these High Church preoccupations. Some had gone off to join the Church of Rome – among them, Gladstone's close Oxford friend Henry Manning, destined to become the Cardinal Archbishop of Westminster. Others – the greater part – such as Mark Pattison, J. A. Froude, Benjamin Jowett, Matthew Arnold – had drifted into some form of 'liberal' Christianity, or rejected Christianity altogether. It was mere chance, or destiny, which determined that Gladstone was at Oxford when the religious revival in the Church was just about to begin. It was to colour the way he viewed the world, and his own political career, for the rest of his days.

Gladstone entered the House of Commons in 1833, the very year that Keble first denounced national apostasy in his famous Assize Sermon preached from the pulpit of St Mary's, Oxford, the University church. By 1841, when Peel was Prime Minister for another term, Gladstone, aged thirty-two, felt aggrieved not to be offered a Cabinet post. There was never any fake humility about him. He knew himself to be remarkable, and he put his talents at the service of the nation. It was therefore remiss of the Prime Minister not to offer him what God had obviously intended. But he had not long to wait; appointed Deputy-President of the Board of Trade, he had soon become President, and was then transferred to the Colonial Office.

This photograph was taken during Gladstone's second term as Chancellor of the Exchequer, when his political ambitions and his psychosexual tensions were at their height.

From the first, Gladstone was a tireless administrator. He had a tremendous eye for detail. Rather like Mrs Thatcher, he always knew – throughout his life – exactly what everything cost, and while having largeness of vision he retained the ability, essential in any government minister, to concern himself with small things, too.

Throughout this period, when he was finding his feet as a Cabinet minister, all his Church Toryism was being undermined by the pressure of events. Keble's vision of a theocracy – of not just the Church of England, but England itself, being the Tory Party at Prayer – did not correspond to the facts of things as they actually were and, as Gladstone's favourite philosopher Bishop Butler observed, 'things and actions are what they are, and the consequences of them will be what they will be.' The Tractarian position was considerably weakened in 1845 when its most brilliant exponent, John Henry Newman, became a Roman Catholic. In Gladstone's undergraduate days, Newman had been regarded as rather suspiciously Protestant and Low Church. How fast things were changing. It was not long before Henry Manning, perhaps Gladstone's closest Oxford friend, was to follow Newman into the Roman Catholic Church. It was a crushing emotional blow to Gladstone, made worse when his own sister Helen was also converted to that creed. Like many sick people in that period, Helen Gladstone also became addicted to laudanum. In his hectoring disapproval of her, it is sometimes hard to know whether Gladstone thought her drug addiction or her Romanism the more insufferable deviation.

As a young Cabinet minister, he had tried to keep alive his old Anglo-Catholic or High Church view derived from Keble. When the Government increased its grant to the Irish seminary of Maynooth – more or less the only place of higher education available to the Irish – Gladstone resigned from the Cabinet. The absurd 'logic' behind this resignation was the idea that the young Irish seminarians, instead of preparing for the Catholic priesthood, should all become Anglicans. But as a practical young Peelite economist, Gladstone the MP was a different being from Gladstone the minister. Though resigning on principle in protest against the Maynooth grant, he *voted for it in the House of Commons.* This was because common sense and common humanity told him that you could not continue to lay burdens on the Irish without some form of explosion. The famines which had wiped out so many Irish people in the 1840s were beginning to create the inevitable political backlash.

Violence was threatened. For where people are repressed and justice is violated, violence would inevitably break out. It was the matter of Ireland which broke Gladstone's Toryism. Though he remained a Peelite Tory until such a thing became obsolete, he knew, after 1845, that his position was untenable.

There was, besides, another side to Gladstone which quite differed from the parochialism of his Oxford Tory self. There was Gladstone the lover of Italy. His first visit to Rome in 1831 had made an overwhelming impression upon him. The sight of St Peter's had made him see what 'unity in the Church' might mean. 'May God bind up the wounds of his bleeding church,' he wrote in his diary on that occasion, adding, of course, an explanation to God, that the blame for schism rested entirely upon 'Rome itself'. This lofty sense of a truly universal or catholic church never left Gladstone. Paradoxically, it was what fired his violent hostility to the Papacy. When, later in the century, Pius IX declared himself to be infallible, Gladstone furiously denounced the measure as an encouragement to unbelief.

In 1850, Gladstone went to Naples and was able to see for himself what religious absolutism was like when given its head. It completely changed his own rigidly Tory attitude to political authority. Here was a Catholic kingdom which, *mutatis mutandis*, was not so very unlike what Gladstone and his young Oxford friends had hoped to establish in England. Gladstone visited Poerio, the recently deposed Liberal leader in Naples, and found that he had been sentenced to twenty-four years in irons. The prisoners were chained two and two in double irons with murderers and violent thieves.

There followed Gladstone's written outburst to the British Prime Minister of the day, the famous *Letters to Lord Aberdeen* in which he denounced the Bourbon kingdom of Naples as 'the negation of God erected into a system of Government'.

Gladstone was never to be the same man again. Both political parties were fractured in the middle of the last century – old Whigs divided from the new radicals; old Tories from Peelites and the rising new Conservatives who looked to Disraeli as their leader. It was inevitable in the midst of these new alignments that Gladstone would ally himself with the party which would eventually take the name Liberal, the party which identified itself with freedom and independence. Free Trade was the watchword economically – a market unshackled by tariffs or government interference. Minimal taxation was necessary in those days. As Chancellor of the Exchequer in Aberdeen's ministry, Gladstone would have liked to abolish income tax altogether. As it was he was never able to reduce it below the scandalously high figure of 2.5 per cent – sixpence in the pound. From these freedoms – and from the natural consequence of such economic freedom, the growth in power and importance of the urban trading classes – Gladstone learnt to believe in other freedoms. By the end of the century the young Tory die-hard had become a believer in the ballot box; in one man one vote; even, most extraordinary of all, in religious freedom, and the disestablishment of the Church in Wales and Scotland.

Freedom of morals, in sexual matters, was something rather different. Gladstone was happy in the choice of his wife. Like many Victorians, he married the sister of a great friend. And, like many rich mercantile families, he married into the aristocracy. Catherine Glynne was the sister of Sir Stephen Glynne, whose country seat, Hawarden Castle in Cheshire, became Gladstone's most treasured refuge. Here, even when Prime Minister, he would retreat for anything up to four months of the year. Forestry was his passion. When he was not digging or cutting down trees, he was going for long walks and ruminating on the rights and wrongs of things. And one of the most notorious wrongs of Victorian England, one with which Gladstone was obsessed, was prostitution.

It is symptomatic of the extraordinary double standards of the age that, only a stone's throw from the front door of Gladstone's palatial town house – 11 Carlton House Terrace – was the Haymarket. Today, it contains nothing more racy than shops and theatres. In the days of Mr Gladstone, the Haymarket was swarming with prostitutes, plying their trade. Most nineteenth-century men took their existence for granted. In a world where divorce was almost unthinkable, and sex had to be buried, a case could be made for believing prostitution to be a social necessity. The Russian novelist Tolstoy, no stranger in his youth to the pleasures of the *bordello*, defended the existence of prostitution even after his own conversion to a form of ethical Christianity. 'Imagine London without its 80,000 Magdalens!' he wrote. 'What would become of families? How many wives or daughters would remain chaste? What would become of the laws of morality which people so love to observe? It seems to me that this class of woman is *essential* to the family under the present complex forms of life.'

ABOVE Gladstone's wife was Catherine Glynne, a merry, intelligent Welsh aristocrat who was his devoted companion for more than half a century.

RIGHT The Glynne family seat in North Wales, Hawarden Castle, was Gladstone's most cherished retreat and refuge. Here he could escape for four months of each year from the hurly-burly of London to read, to pray, to write and to chop down trees.

Gladstone took a more austere view. In his encounters with dozens of women, and men, who undertook this form of work, he endeavoured to persuade them to desist. Together with other High Church friends, he set up the House of St Barnabas in Soho as a refuge for prostitutes as early as the 1840s. His obsession with prostitutes continued throughout the coming decades; and there can be no doubt that he was deeply attracted to the women whom he attempted to save. In order to curb his passions he procured a special whip, donated for the purpose by Dr Pusey at Oxford who was supplied with these disciplines by a continental convent. From the 1860s onwards, Gladstone made a habit of scourging himself with the discipline after conversing with a prostitute.

Prostitutes were to discover what the Queen, and the Gladstone family, and the House of Commons had known for some time – that when Gladstone spoke, he spoke at length, adumbrating points under different headings, weighing moral as well as practical considerations, outlining the problem, and suggesting the solution. Sometimes his discourses with these women lasted over two hours. In a profession where time is money, his interruptions to their business cannot always have been welcome. Some of those whom he rescued were not always grateful, like the woman he persuaded to abandon 'the game' in favour of life in the Anglican convent at Clewer, near Windsor. After six months, she apologetically confided in Gladstone that she would prefer the risk of being beaten by her pimp or infected by her clients to the insufferable atmosphere of this worthy Tractarian house.

Did the ladies ever oblige Gladstone by giving him the discipline of flagellation themselves? Perhaps. Not long before he died, Gladstone emphasised to his clergyman son that he had never 'been guilty of the act which is known as that of infidelity to the marriage bed'. It is a typically Gladstonian and long-winded way of saying that he had never committed adultery. Whatever else he was, he was no liar. It would seem likely that even the flagellation happened on his own, in private, and that the encounters with these ladies were purely conversational. Though they exposed him to torments of lust, we may believe that he never compromised himself. Like everything else he ever did, his hundreds of chronicled encounters with prostitutes were motivated by a sense of Christian obligation. He had as much duty, having been enlightened with the truth, to persuade prostitutes to abandon the pavements as he had to persuade 'the West End' and the 'top ten thousand' of the population – groups whose antagonism he risked and courted – of the injustice of their privileges.

Needless to say, the hazards of his interest in the sexual underworld of Victorian London were considerable. His changing allegiance politically brought him many enemies. Just before he presented his 1853 Budget, the Chancellor of the Exchequer was seen holding an animated and extended conversation with a woman of doubtful appearance on the corner of Haymarket. No doubt he was expounding to her the eighth chapter of St John where the Saviour tells a woman to 'go and sin no more'. A Scotsman by the name of Wilson unwisely attempted to blackmail Gladstone, who instantly handed him over to the police.

Gladstone undertook the rescue of prostitutes as a charitable work in his early twenties. This picture shows girls bribing the Beadle of the Burlington Arcade, off London's Piccadilly, so that they could ply their trade in this popular shopping place.

Is it too neat to attribute, in part, Gladstone's extraordinary capacity for hard work – those endless parliamentary speeches, those long hours in committee, those thousands of words spoken, minuted and recycled – to pent-up sexual energy? In the early years of his marriage, Gladstone's wife was almost perpetually pregnant and spent much of her time in the country while he was in London. We can never imagine Gladstone relaxing. Even in extreme old age, when holidaying at Biarritz with his biographer John Morley, we find him going into shops to inquire the prices of everything; we find him mulling over textual cruces in Homer; we find him ruminating on the problems of the hour. In addition to the nation's economy, he was an efficient manager not only of his own money but that of his wife's family. His preparedness to go through everything with a tooth-comb was largely responsible for his ability, not without effort, to rescue his brother-in-law from bankruptcy. There was never a period when he was still, unless it was when he was on his knees, and even then it is all too easy to imagine the Almighty, like Queen Victoria, feeling that He was being addressed as if He were a public meeting.

Public meetings were Gladstone's forte and – in the form of them at which he excelled – it could be said that they were his invention. The Conservatives were not all wrong to think of him as a demagogue. He knew how to manage big crowds and to use the power of the crowd as an extra-parliamentary weapon.

During the election of 1868, Gladstone returned again and again to the theme of Ireland in his public speeches. He spoke of the Protestant Ascendancy there as:

> some tall tree of noxious growth, lifting its head to Heaven and poisoning the atmosphere of the land so far as its shadow can extend. It is still there, gentlemen, but now at last the day has come when, as we hope, the axe has been laid to its root. (Loud cheers) It is deeply cut round and round. It nods and quivers from top to base (Cheers). There lacks, gentlemen, but one stroke more – the stroke of these elections (Loud cheers) It will then, once for all, topple to its fall, and on that day, the heart of Ireland will leap for joy, and the mind and conscience of England and Scotland will repose with thankful satisfaction upon the thought that something has been done towards the discharge of national duty, and towards the deepening and widening the foundations of public strength, security and peace. (Loud and sustained applause).

It was a triumphant election for Gladstone. The Liberals were returned with a majority in the House of Commons of 112 seats. Gladstone (who had previously represented Newark and Oxford University) now found himself sitting for Greenwich. At the beginning of December he was at Hawarden, cutting down a tree in the park, when a telegram arrived, informing him that General Grey, the Queen's Secretary, would arrive that evening. Fully aware of the drama of the situation, Gladstone merely said, 'Very significant.' Then he resumed his work with the axe. After a few minutes he paused and said to his companion (the Hon. Evelyn Ashley, a son of Lord Shaftesbury), with a voice of deep earnestness and with an intense expression, 'My mission is to pacify Ireland.'

But how was he to do it? At nearly sixty, he could not have dreamed that he had a quarter of a century of active political life ahead of him. By the time he died, Gladstone had done all that was humanly possible to avert an Irish calamity, but he had failed; and to this hour, the people of Ireland live with Gladstone's failure.

As well as being the most dramatic arena of his political failing, Ireland was also a mirror to Gladstone's extraordinary mind. In examining his changing attitudes to Ireland, we see more clearly than anywhere else the kinds of thing which motivated him, as a politician and as a man. In the beginning, as a High Anglican bigot, he had hated the idea of Catholic Emancipation, and had felt as a matter of principle that the Maynooth seminary should not have an increase in its grant. It was, for the young Tory Cabinet minister, a resigning matter. As the years went by, however, he realised that it was not possible, nor was it even desirable, for the governing class of a country to make all its citizens follow one creed or submit blindly to a single authority. The sight of the repressive regime in Naples made him particularly sensitive to the position of the Irish. Here were citizens of the United Kingdom who had suffered religious persecution, political repression, and starvation. Not surprisingly, there had grown up a movement desiring self-government for the Irish, a movement inevitably attracting to itself hotheads, bomb-throwers, and murderers.

Justice demanded that the Irish case be met. Gladstone did what he could, in his first administration as Prime Minister, to appease the Irish, to make it clear to them that he was on their side. The trouble was to find a solution which would work, and one which would be acceptable on both sides of the Irish Sea. The crucial factor in the whole case – the position of the Ulster Protestants – was one which Gladstone ignored totally. Indeed, having only visited Ireland once for a month in the 1840s, he did not really know about it. For him, the issue was simply one of justice to the Catholics, and the pacification of the land of Ireland as a whole. So he, the High Anglican, presided over the disestablishment of the Church of Ireland in 1869. Henceforth Roman Catholics in Ireland were not in the anomalous position of having to be described as Non-conformists to a national church in which none of them believed. A year later, the Irish Land Act gave a measure of protection to small tenant farmers against the rapacity of Ascendancy landlords, to avert a repetition of the conditions which had led to the famines of the 1840s.

Gladstone had a profound personal and political antipathy to Disraeli, whose genius was the exact opposite of his own. The rivalry between the two greatest statesmen of the age was the subject of many cartoons, such as this one from Punch *in 1868.*

But Ireland was not the only thing on the political agenda in the early 1870s, and much else intervened to guarantee that by 1874 Gladstone's great adversary Disraeli would win the election and throw the Liberals out of office. By then, Gladstone was sixty-five years old. The side of his nature which craved prayer and scholarship (he used to pray that he would die in church 'though not at a time to disturb worshippers') wanted to quit the political scene altogether. He sold up his big London house, moving his vast collection of porcelain and pictures to a smaller house in Harley Street. And he retired to Hawarden. The leadership of the Liberal party passed to Lord Hartington, the heir to the Duke of Devonshire. It looked as if Gladstone's spell as Liberal Prime Minister was to have been dramatic, but brief.

His enemies, observing the huge contradictions in Gladstone's nature, dismissed him as a humbug and a hypocrite. But this is unfair. He was not one man pretending to be another. By contrast, he was many powerfully different personalities all waging war with one another. It is typical of the man that, having abandoned politics for ever, he should have been back on the hustings, more loquacious and energetic than ever, within a year.

It was a time of some of the worst trouble in the Balkans. Disraeli's government, enthusiastically supported by the Queen who pined persistently for a repeat performance of the Crimean War, supported the Turks in their hostility to their Christian minorities, and to the threatened eruption of nationalism among the small Balkan states against their Ottoman masters. The Russians, a great Christian Empire, who in Gladstone's eyes had the supreme advantage of being non-Roman but highly Catholic, protested at the treatment of Christians under Turkish rule.

Gladstone's term as Chancellor of the Exchequer in Aberdeen's Cabinet during the Crimean War had left a residual guilt about the folly and pointlessness of that conflict. Conviction told him that Christian nations should be allies and that the British should not have lined up with the Turks against the Russians. Then, in 1875, there began to emerge the terrible stories of Turkish atrocities against the Bulgarians – babies, women and children butchered in their thousands. For the Queen and Disraeli, it remained an imperial question. Which European power should dominate the Mediterranean? Should it be Russia, with its known desire to reoccupy the Constantinople patriarchate? Or should it be the weakened Turks, who would of course look to the British as their protectors? The actual fate of the Bulgarians meant less to Disraeli and to Queen Victoria than the shape of the map of Europe. How much, in the power game, did the British control?

Gladstone, who had been tormented for ten years now by thoughts of Ireland, saw this as another case of a small nation struggling for independence against the overwhelming might of ignorant imperial armies. The sufferings of the Bulgarians were, in any case, intolerable for their own sake. It sickened him that the British could regard as their allies a nation who could behave as the Ottoman Turks were behaving.

With one of the most mammoth displays of political energy of his whole life, the Grand Old Man emerged from retirement at Hawarden and made a series of speeches across the north of England. By 1879, his mind was drawn not only to the Bulgarian atrocities but also to what the British were doing, or not doing, in Afghanistan, and in South Africa. Everywhere, it seemed, there was this conflict between the desire of small nations to determine their own destiny, and the desire of Empires to crush them. This was the era when the word 'jingoism' came into the English language. The crude music-hall song was the perfect expression of that vulgar attitude of mind displayed by the Queen, the Conservative party, and the Press:

> *We don't want to fight but by jingo if we do*
> *We've got the ships, we've got the men, we've got the money too!*

For Gladstone, the world was not a playground in which powerful nations could tolerably behave like the school bully.

In 1879 he decided to offer himself at the next election to the voters of Scotland, the land of his high-minded Presbyterian ancestors. In the first of his legendary Midlothian campaigns he made a series of magnificent and hugely long speeches – one of the great ones was in the Corn Exchange in Edinburgh, but he followed it up, speaking to a crowd of thousands in the open air. Standing in the frosty Scottish air, he told his audience:

> Remember that the sanctity of life in the hill villages of Afghanistan is as inviolable in the eye of Almighty God as can be your own. Remember that He who has united you as human beings in the same flesh and blood has bound you by the law of mutual love; that that mutual love is not limited by the shores of this island, is not limited by the boundaries of Christian civilisation, that it passes over the whole surface of the earth and embraces the meanest along with the greatest in its unmeasured scope.

In place of the commercial rapacity and sheer political arrogance of the imperial ideal, Gladstone held out a vision of something very different.

When, in 1880, the Conservative government appealed unsuccessfully to the country, they made Irish Home Rule the main issue. The Liberals won the election and Gladstone won his seat. The Queen asked Lord Hartington to form an administration. He pointed out to her – and how little she liked it – that it would be unthinkable to form a Liberal government of which Gladstone was not a member. Gladstone was approached and asked if he would serve. He replied, with his usual absence of nonsense when political power was in the offing, that he would serve in no other capacity than that of Prime Minister.

The Midlothian campaign was therefore an extraordinary demonstration of popular politicking, of demagoguery. It sits oddly beside the reclusive and quiet side of Gladstone's nature. It is hard to see the old

The most exciting moments of Gladstone's later political life were spent on the hustings in the Midlothian Campaigns where he addressed and enthralled crowds of thousands, sometimes speaking for as long as five hours at a stretch.

man, almost drunk with the sound of his magnificently resonant deep voice and with the cheers of the Midlothian crowds, as the same person as the scholar on his knees in the chapel at Hawarden.

The Midlothian campaign – like much else in Gladstone's life – changed the face of British politics. It was the first election campaign fought across the country by a senior politician. Just how strange it was may be realised at once if we try to picture Lords Palmerston or Aberdeen 'touring the country' in the manner of a modern democratic leader. In so far as it was an exercise in attracting political support outside Parliament by a man who at that point held no office under the Crown, it was not surprising that the Queen viewed the Midlothian campaign with distaste and alarm. The waves of radicalism were riding high. The new generation of English Liberals contained many, like Joseph Chamberlain in Birmingham, who were anti-monarchical, and even republican. The Queen did nothing to counteract her growing unpopularity with many sections of the populace who resented her refusal to take part in public life while busily hoarding away her Civil List payment from public funds. The private fortune of the present royal family in part owes its origins to the miserliness of the Widow of Windsor who for years after her husband's death never made public appearances, never attempted to meet her people, and really failed to do her job. It would not have been at all difficult for Gladstone in the 1880s to bring the British monarchy to an end altogether. As on the last occasion when it had happened, it is improbable that there would have been much, if any, protest. The popularity with which Victoria was regarded when she was a very old woman had not yet begun. Her refusal to come out of mourning and her unenlightened political views made her much hated. It was very largely Gladstone's personal devotion to the monarchy, based on his reading of the Book of Common Prayer rather than on the Queen's intolerable behaviour towards him, which saved the monarchy from extinction.

The Midlothian campaign of 1879–80 is also the key to Gladstone's Irish policy. From now onwards, it was common humanity which counted for most with Gladstone, whatever his own religious or personal preferences. Since 1875, there had been an Irish nationalist party in the House of Commons in Westminster, led by a Protestant squire by the name of Charles Parnell.

One of the greatest tragedies in the Irish situation is that Parnell was never able to trust Gladstone. As in so many of Max Beerbohm's cartoons, there is truth in his depiction of Parnell meeting Gladstone in Heaven. Parnell has his fists raised. Gladstone, with ineffable loftiness, is saying, 'If you could but give me three hours and a half of your time, I could explain *everything*!' Parnell was never able to recognise how much Gladstone, even before his mind cleared and he saw how things truly stood in Ireland, was anxious to do the right thing by that country. In exchange for some big concessions in a reworking of the Land Act – in particular, releasing tenants from arrears in their rent – Gladstone extracted from Parnell a promise that he would modify the violence of his Sinn Fein supporters. Inevitably, there had been bomb outrages not only in Ireland but in England – in London, Manchester and Liverpool. In May 1882, Gladstone sent Lord Spencer – the great-great-great grandfather of the present Princess of Wales – to be Viceroy of Ireland, and with him a nephew of Gladstone's marriage, Lord Frederick Cavendish, the brother of Lord Hartington. On 6 May, only the day after his arrival, Cavendish was walking with Burke, the head of the Irish executive, in

Above Charles Stewart Parnell, the Irish Nationalist leader.

Right It was the Clerkenwell bombings, when some Fenians blew up a London prison to rescue their imprisoned comrades, that panicked Gladstone into the belief that it was his mission to pacify Ireland.

Phoenix Park in Dublin when the two men were stabbed to death in broad daylight under the eyes of the Viceroy. For the rest of that summer there were incidents of murder throughout the Irish countryside. On 17 August 1882, a family in Connemara was butchered. Unlike the murderers of Burke and Cavendish, the men responsible for this crime were caught and brought to justice. At the trial, however, it was discovered that the defendants either could not or would not speak English. For the Irish, the hanging of these men, who did not even understand what was being said to them by their accusers, brought home the cruel implacability of British justice. For the English, the Irish situation called for a solution. But what was to be done? The Conservatives, and a high proportion of the Liberal party, believed that the answer was greater and greater repression of the unruly elements in Irish society. Violence must be met with violence. As for the yearning of the Irish people for nationhood, it did not *do*! The Irish were being as tiresome as those Bulgars, those Zulus, those Greeks – all those little peoples who wanted self-government. To give the Irish self-government would be to dissolve the imperial ideal, to remove a brick which might begin a process by which the whole edifice of Empire would crumble! Ireland was as much a part of the United Kingdom as was Wales or Scotland.

Gladstone, little by little, learnt that this was *not* the solution. Many things delayed his conversion; and after the conversion, there were many things which contributed to the Irish distrust of Gladstone – not least his mounting hostility to the Vaticanism, the spiritual imperialism, as he saw it, of the Papacy, rampant since the Vatican Council of 1870 of which his old friend Manning was chief architect, and which had declared the Pope, in certain circumstances, to be infallible. But since the Irish were now consistently sending back a majority of republican members to Westminster, and since all the violent agitation in town and country up and down the length of Ireland made it clear that they wanted self-government, Gladstone saw no alternative which did not ultimately have self-government as its end. He was converted at last to the view that what he believed held true for the Neapolitans and the Afghans and the Zulus was true too for the Irish. He became a convert to Irish Home Rule.

The first Home Rule Bill of 1886 was defeated in the Commons and lost Gladstone the election of that year. In the years that intervened before Gladstone was able to present the matter to the Commons a second time, a number of disasters had occurred that doomed the Irish cause to be lost for ever.

Not least of these was the rather ludicrous fact that Parnell, the leader of the Irish nationalists, had, like so many Victorian gentlemen, been leading a double life. By his enemies, Parnell was accused, quite falsely, of colluding in murder and condoning the atrocities which each month were committed in the name of Irish nationalism. A man in this position could not afford to lose the support of his friends. But this is what happened – and for the most trivial of reasons. From 1880 onwards, Parnell had been in love with Katherine O'Shea, the wife of an Irish political colleague. Her husband was prepared to turn a blind eye to what was in effect her quasi-marriage to Parnell. She shared a house with him in Surrey and bore him three children. The situation continued equably because the O'Sheas stood to inherit a small fortune from an aunt, a

Gladstone's peroration in the House of Commons, defending against all odds his belief that the Irish people should be allowed to determine their own affairs. This Home Rule Bill, like its successors in Gladstone's lifetime, was defeated.

respectable old woman who would have cut Mrs O'Shea out of her will had the truth been known. Had that aunt been either more broad-minded, or longer lived, it is conceivable that we should have peace on the streets of Belfast today.

In 1890, the aunt died. Anxious to get his share of the inheritance, O'Shea sued his wife for divorce and cited Parnell as co-respondent. With a rashness which seems almost as extraordinary to us as Oscar Wilde's determination to sue Lord Queensberry, Mrs O'Shea contested her husband's action, and thus, in a single stroke, the cause of peaceful Irish nationalism was destroyed. Parnell was, in the eyes of his own side, disgraced. Many of his supporters were anxious to hold to him in disaster as in triumph. But the priests of the Roman Church, who had been happy enough to condone the murders done in the republican cause, were unable to stomach Parnell's more venial sins of the flesh. The fact that he was a Protestant made it all the worse.

In 1891, Parnell died suddenly. By then, the forces of Irish nationalism were hopelessly divided among themselves. Gladstone, aged eighty, fought the election of the following year on the Irish question. Ever a pragmatist – what he called 'a good parliamentary hand' – he knew that he could not force a measure like Home Rule through the House of Commons on his own. He was therefore prepared to throw in his support with the English political wing who were prepared to support it: the extreme radicals. The man who had first entered Parliament as a young prig opposing the measures of the Great Reform Act found himself with some very different friends at the age of eighty. As he stood on the hustings in Midlothian he was calling for the principle of one man one vote; he had accepted the idea of disestablishing the Churches of Wales and Scotland; local government was to be reformed to allow elections to parish councils, and – for the dreary teetotallers formed a powerful group among his supporters – local councils were to be given the power to regulate and even to forbid the sale of alcoholic drink in this country. All this – much of which went against the grain – was worth it, as far as Gladstone was concerned, if it would pacify Ireland. It was not a matter of appeasement or cowardice. Gladstone did not need to be fighting an election as he entered his ninth decade of life. No one could ever accuse him of cowardice. It was a simple matter of justice. He wanted freedom for the Irish – freedom to live as they wished, freedom to live without the terror of madmen and murderers on the one hand and the oppression of Protestant landlords on the other.

It would have been so much easier for the old man to leave the problem to his successors. As it was, it propelled the break-up of the Liberal party. The most brilliant and likely of Gladstone's successors as leader, Joseph Chamberlain, joined the Conservatives, while many of the more genial souls in the Liberal party, like Lord Rosebery and Herbert Asquith, repelled by the pro-Irishry and anti-imperialism of some of Gladstone's new friends, were really preparing themselves for a split in their party. The Liberals remained, under Campbell-Bannerman and Asquith, as an uneasy coalition between those who were in favour of Empire and the Irish Union and those who were much more politically and economically radical and would one day themselves split – some behind Lloyd George, others to join the expanding ranks of the Labour Party. In this matter of Ireland, in fact, we see the whole future of the English Left in the melting-pot, and there is a strange paradox that it should have been presided over by this ancient Victorian gentleman, who went to church every day, and who would have felt no more at home in the twentieth century than would Duns Scotus or Alfred the Great.

The Queen dismissed him as 'an old wild, incomprehensible old man'. Lord Randolph Churchill in the election campaign spoke for many when he said:

> Mr Gladstone has reserved for his closing days a conspiracy against the honour of Britain and the welfare of Ireland more startlingly base and nefarious than any of those other numerous designs and plots which, during the last quarter of a century, have occupied his imagination . . . all useful and desired reforms are to be indefinitely postponed, the British Constitution is to be torn up, the Liberal party shivered into fragments. And why? For this reason and no other: to gratify the ambition of an old man in a hurry.

In February 1893, a much modified Home Rule Bill for Ireland was introduced into the House of Commons. It was merely proposed that there should be an independent legislative assembly at Dublin and that the Irish should continue to send their eighty MPs to Westminster. The debate occupied eighty-five days of parliamentary time, and the bill was *just* passed in its third reading by thirty-four votes. Needless to say, when it went to the House of Lords, it was defeated by 419 votes to 41.

Far from seeing Gladstone as an ambitious old man in a hurry, we now must see his great political failure as a piece of extraordinary quixotry. Better perhaps than anyone in England, he knew the political make-up of the House of Lords. He knew that Irish Home Rule had no chance of being allowed by a Westminster Parliament.

But, like the Trollope hero, 'he knew he was right'. He had begun to see with prophetic clarity what would happen in Ireland if the Irish were not allowed to govern themselves. He knew how much notice was taken of the House of Lords in Connemara, in Cork, and even in Dublin. His failure on a political level ws not his idealistic promotion of Home Rule. It was his complete inability to see into the minds of Ulster Protestants.

There was a sort of Olympian, donnish loftiness in Mr Gladstone which did not grow less with the years. Whether as a young Tory or an old radical, he tended to assess questions of the day against the high standards of religion and absolute morality rather than by the vulgar enthusiasms of the hour. He lacked any of Joseph Chamberlain's or Margaret Thatcher's political 'nose' – what some would call vulgarity. One sees this in the notorious incident of General Gordon's death, when Gladstone's failure to read the 'mind of the British people' cost his party an election.

Gordon, it will be remembered, was stuck in Khartoum in 1884 without reinforcement against the overwhelmingly superior opposition of the Mahdi, the self-proclaimed religious leader of a revolt against British rule in Egypt. It was the Government's indecision and prevarication that delayed the sending of a relieving expedition. When one was dispatched, under Sir Garnet Wolseley, it arrived too late. Gordon had been killed two days before. Gladstone afterwards declared himself to have been

This painting of Gordon's death at Khartoum was executed by G. W. Joy. It was fast to become an icon, reproduced in thousands of prints and engravings, hung in parlours and school-rooms, a set-piece of Victorian heroism.

'tortured' by the affair. He had never met Gordon. He had been reluctant to send him out to Khartoum in the first place. But while the Queen and the British public, with that irrational fervour that sometimes grips imperial powers (compare the popularity of Colonel Oliver North), were only able to see Gordon as a hero, Gladstone's more detached mind saw it all differently. He had the greatest objection to British soldiers being in Egypt in the first place. He did not see it as the British vocation to be the policemen to the rest of the world. Ridden with guilt as he was about his father's fortune deriving from the slave trade, he had a lifelong suspicion of interference for ultimately commercial ends in the political life of other nations. So, while the jingoistic Press proclaimed Gordon as a hero, Gladstone regarded him as a disobedient fanatic, a menace who was making a difficult situation worse.

At length, the Grand Old Man did retire from public life, and as a very old man he lived almost entirely at Hawarden, seldom venturing on to a public rostrum. But there was an exception. In 1896, when news came of yet more appalling massacres of Armenians by the Turks, Gladstone, whose deep voice was still resonant at the age of eighty-five, made a speech in Liverpool in which he described the massacres as 'the most terrible and the most monstrous series of proceedings that has ever been recorded in the dismal and deplorable history of human crime.'

And then he gave utterance to that marvellous sentence which could be said to be his creed: 'The ground on which we stand is not British, nor European, but it is human.'

It is that which coloured Gladstone's largeness of vision. Nourished by the great classics of European literature and by an intelligent exercise of religion, Gladstone was never petty-minded, never parochial, never mean.

He died at Hawarden, aged eighty-nine, on a bright May morning – Ascension Day – on 19 May 1898. All his family was kneeling at the bedside, a scene which was quintessentially Victorian. His death was felt as much in Europe as it had been in England: a sensation, it was said, as great as the death of that other great lover of liberty, Garibaldi.

His legacy to the nation is Hawarden itself, St Deiniol's Library where people may come to do what Gladstone best loved to do himself – to pray in the chapel and to read. It is partly an Oxford college and partly a Tractarian retreat house. Is Gladstone himself as quaintly irrelevant to the concerns of the generality of modern people as this place seems today? Or do his great rallying-cries, for decency in public life, and above all for liberty, have greater force now than they ever did? He wanted economic freedom – an ending of income tax was his dream. He believed passionately in the freedom of the Christian gospel against what he saw as the 'Asiatic monarchy' of the Vatican. He believed in self-determining nationhood. An enemy of Gladstone's could say that this was symptomatic of his blindness – his inability to see that these lofty-sounding ideas would turn into the rampant and selfishly destructive forces of a free market economy, into religious sectarianism and, most dangerous of all, into the rise of nationalism, the fatal effects of which in the forty years after Gladstone's death – particularly in Germany – he seems to have been entirely unable to predict. He quite failed to see the importance of Bismarck.

Certainly it is easy to see much that was unintentionally comic about a man who chewed every mouthful of food thirty-two times, and who was so prolix in conversation that he could indulge in monologues of two and a half hours with prostitutes, crowned heads, clergymen or parliamentary colleagues. The Beerbohm cartoons of Gladstone's reception into Heaven, which now adorn the walls of the Carlton Club, are the Conservative answer to the icons of Gladstone in the National Liberal Club. They hint amusingly at the sort of Gladstone we should have seen had he appeared as a character in Strachey's *Eminent Victorians*. We, with a bigger perspective than Strachey's – seventy years more perspective – may laugh at Gladstone's eccentricities but perhaps be better able to see him for what he was, just as, the further one drives away from a mountain, the better one is able to make out its mighty outline against the sky.

John Henry Newman

1801-90

Poet and priest

This picture of Newman as a Roman Catholic priest dates from the time when he wrote his famous autobiography, the Apologia pro Vita Sua, *which gives a moving account of his decision to leave the Church of England.*

Three miles outside Oxford, in a village called Littlemore, there is a row of simple cottages which date back to the old coaching days, before the coming of the railways. Then the buildings used to serve as a staging-post to allow the coaches that made the east-west journey across England from Cambridge to Oxford to change horses.

This old staging-post was also the scene of a more momentous change. For it was here, on 9 October 1845, that the eloquent vicar of St Mary's Oxford, John Henry Newman, by far the most famous and influential intellectual leader of his generation, shocked the world by becoming a Roman Catholic.

Nowadays in England, the Roman Catholic Church numbers more regular adherents than any other Christian body; there are more Roman Catholic churchgoers than there are practising members of the Church of England. Though not all Christians accept their doctrines, the Roman Catholic Church is by far the most vigorous of all the denominations. Though pockets of prejudice against it remain, and presumably always will remain, it is now a fully accepted part of the British religious scene. As a result of the Act of Settlement at the close of the seventeenth century which determined that the British monarchy would be Protestant, the Sovereign is still forbidden either to be or to marry a Roman Catholic. But it is now a familiar part of public ritual that the Cardinal Archbishop of Westminster, or some other public dignitary of the Roman Catholic Church, will take part in such events as royal weddings, along with representatives from the Non-conformist churches.

The Roman Catholics have arrived.

So much do we take this for granted that it is almost impossible to imagine that only 140 years ago, the Roman Catholic Church in England barely existed. A very few families, mostly in the north of England, had clung to their old religion ever since the Reformation, and since the Irish famines in the early decades of the century, there had been an influx of Irish workers into the great industrial cities. But in so far as it actually affected national life in the 1830s and 1840s, the Roman Catholic Church hardly counted in England. In a town like Oxford, there might be one small Roman Catholic chapel to cater for the tiny number of local Papists who were at least (since 1778) allowed to practise their religion without infringing the law. But the scenes which are familiar today were quite alien then: several large Roman Catholic churches in every town, attracting good congregations each Sunday; Roman Catholic schools, Roman Catholic cathedrals and bishops; nuns glimpsed on a bus; monsignors, or ex-monsignors, leading anti-nuclear demonstrations.

If you had suggested to an Oxford don of the 1820s that England in the 1980s might contain so many Roman Catholics, he would have considered the proposition as fantastic. As fantastic, perhaps, as the wild notion that there might one day be a mosque in London's Regent's Park.

Ever since the Reformation, English people had nursed a dread of Roman Catholicism which might at any moment flare up in displays of irrational fear. One thinks, for example, of the Gordon riots in 1780 (described so vividly by Dickens in *Barnaby Rudge*) when London mobs, egged on by the demagogue Lord George Gordon, looted and burnt the house of supposed Roman Catholics in protest against the lifting of the legal ban against their religion. Or one thinks of the controversies which seized Oxford fifty years later in the time of Newman.

Most people arriving in Oxford for the first time are struck by the spindly grace of the Martyrs' Memorial, standing at the bottom of St Giles's. When I first saw it as a boy, and made up my mind that Oxford was the place for me, the Martyrs' Memorial struck me as the quintessence of the Middle Ages. In fact, the Memorial was erected between 1841 and 1843 to the designs of George Gilbert Scott. It commemorates the brave death, not far from this spot, of Thomas Cranmer, author of the English Prayer Book, and of Bishops Ridley and Latimer, all of whom were burnt at the stake in the reign of Bloody Mary for their refusal to accept the Roman Catholic religion and the authority of the Pope.

> Be of good comfort Master Ridley! and play the man. We shall this day light such a candle by God's grace in England as (I trust) shall never be put out.

That candle was the Protestant religion. And why did they wait until 1841 before they thought of building a memorial to the Protestant martyrs? Because in 1841 the dons of Oxford were panicked into believing that the unthinkable was about to happen. The candle lit by Ridley and Latimer was guttering in its socket. And the man who with devastating logic and sleight of hand was going to snuff it out was John Henry Newman.

The nerve-centre of the controversy was Oriel College, of which Newman was a fellow. It was here that he and his friends – John Keble, Edward Bouverie Pusey, Hurrell Froude and others – began their analysis of the Christian religion and its history which was to cause so much upset.

Very nearly all their deliberations, when read today, seem to contain about as much matter for intellectual excitement as the controversies of Tweedledum and Tweedledee in the works of that other Oxford clergyman Charles Dodgson, otherwise known as Lewis Carroll. Their *Tracts for the Times*, published at irregular intervals throughout the 1830s, were on such esoteric subjects as Fasting in the Early Church or the need for a new translation of the Psalter.

But these tracts sold in enormous numbers up and down England. Not only did parsons in their country rectories devour them; but all thinking

Oriel College, Oxford, of which Newman became a fellow in 1827, was at that period a centre of religious debate. Newman occupied rooms just to the right of the oriel window.

people in London rushed to buy them on their appearance, rather in the way that, in a later generation, they might have subscribed to the Left Book Club. The reason for this is partly to be found in the quite extraordinary degree of interest which early nineteenth-century men and women took in the minutiae of religious questions. But, as the Tracts continued to appear and the Tractarians, as their authors and admirers came to be known, unfolded their position, the nature of what they were saying became direly clear – clear and revolutionary.

You have to imagine a world where certain propositions were taken for granted as intellectual certainties by believing Christians: that there was a God who made the world, and declared His laws to that created world by means of a revealed religion. The first part of the revelation was through Moses and the Ten Commandments. Then, in the fullness of time, He wound up the first Testament and made a New Testament with a New Israel, the Church. He came in His own person, clothed in human flesh but fully divine, to found this society, to establish the sacraments of baptism and holy communion and to give His Church authority to teach and baptise all people in the world.

This is why people felt so passionately about religious controversies. What was at issue was the will of God Himself, who would reward those who were on His side with eternal bliss, but – as most Victorian Christians appeared to have no difficulty in believing – would consign those who got this question wrong to an eternity of damnation.

The Church of England view, in so far as it had been thought out at all, was that God, having founded His Divine society the Church, remained within it as a sacramental presence. In other words, he was there in the holy communion and in baptism. But this had not stopped the Church, over the years, becoming hopelessly corrupted, and all kinds of new doctrines, particularly doctrines about the Virgin Mary, being adopted into its system. Therefore, in the sixteenth century, when the time was ripe, it was perfectly legitimate for a more purified branch of the Church to break away from the Pope and set up on its own. The Church of England was, by this view, much purer and closer to the original Church which Christ had founded than was the Church of Rome from which it had broken away.

It would be largely true to say that for most of the eighteenth century, these questions were a dead issue in the Church of England. The origins of the Church were much less interesting to people than whether the

Church was alive at all. The controversy then was largely between those who were caught up by the fervour of the Wesleys into a lively personal faith in Christ, and those like Bishop Butler who told Wesley, 'Sir, pretending to gifts of the Holy Ghost is a horrid thing, a very horrid thing.'

A rather different sort of religious revival was to emerge through the influence of the dons of Oriel College, Oxford in the 1830s. From 1817 to 1820, Newman had been a very shy undergraduate at Trinity College, loathing the raucous behaviour of his fellow-undergraduates, and spending much of his time alone in his rooms there, playing the fiddle. All his life he was a passionate violinist. He was one of those very clever people who could not take examinations. He overworked for the Schools (as the Final exams are called at Oxford). In Mathematics, at which he was rather good, he failed, while in Classics he was placed in the lower division of the second class.

This was a shattering blow to Newman, who had no mean estimate of his own capabilities. He had set his heart on becoming a Fellow of Oriel College, for that was the place where all the cleverest men in Oxford were believed to be assembled. Different colleges acquire these reputations in different generations. When Oxford, and intellectuals generally, began to react against the interests and ideas fostered at Oriel, it was the turn of another college, Balliol – broadminded almost to the point of free-thinking and politically radical – to claim the predominance in the university.

In spite of his failure in the Schools, and in spite of the fact that he was desperately nervous about the Oriel fellowship examination, Newman managed to do justice to himself in the papers. On 12 April 1822, as he sat playing his violin in his rooms at Trinity, there was a knock at the door. It was the Provost of Oriel's butler, who said that he had 'disagreeable news to announce. Mr Newman was elected Fellow of Oriel and that his immediate presence was required there.'

The Common Room at Oriel 'stank of logic' as a contemporary remarked, but it was not what modern philosophers would mean by the word logic. Not long after his election, Newman, like the greater proportion of Oxford dons, was ordained as an Anglican clergyman, and although he was painfully shy in their presence, he settled down to get to know his colleagues. A great debate was taking place at Oriel, and neither side of the debate would immediately seem attractive to

Newman's priggish evangelical soul. On the one hand, there were men like Thomas Arnold, the future headmaster of Rugby, or Richard Whately, later Archbishop of Dublin. They were what was called liberals, arguing that if Christianity was to survive the assaults of scepticism, it was essential that it accommodate itself to the findings of modern thinking and scholarship; otherwise, Christianity would become purely obscurantist, and ultimately impossible for intelligent people to believe. These men tended to take a 'liberal' view in politics also, supporting in the fullness of time such measures as the Great Reform Bill of 1832. They were sometimes called 'latitudinarians', and they had a benign, tolerant view, for example, of Non-conformist Christians.

On the other side of the argument, there were those men who grouped themselves behind the saintly figure of the poet John Keble. These included such brilliant Hebraists as Edward Bouverie Pusey and historians like Hurrell Froude. They argued that if you took this 'liberal' view of religion, you would ultimately end up with no religion at all, since each doctrine of the Church could be scrapped if it were found to be incompatible with modern wisdom. The Church, Keble and his friends argued, was something different. It was a Divine society, instituted by Christ Himself during the days of His earthly ministry. It had been corrupted by medieval Roman Catholicism, but it was in essence Catholic, sacramental and apostolic. Indeed, it would seem that there was no purer form of Christianity, closer to the customs of the primitive Church or the intentions of its Founder, than that to be found in the Prayer Book of the Church of England.

This point of view took some arguing. It involved ignoring the historical evidence about the origins of the Church of England, and putting an imaginative interpretation upon the origins of Christianity itself. But – perhaps precisely because it was a paradoxical and difficult belief – it was the side of the argument which ultimately appealed strongly to Newman. For it was the Oriel men of the 1820s and 1830s who began to look again at the historical origins as well as the theological justification of the Church of England's existence. What became very clear, as the examination proceeded, piece by piece in the *Tracts for the Times*, was that the Church of England did not come out of it very well.

After all, no historian could claim that the Reformation had happened because of the fervour of Protestant theologians. It had happened because Henry VIII had wanted to marry his pregnant mistress, Anne

Boleyn. And as for continuity between the old Church of the first few centuries of Christendom and the modern Church, where did that lead you? However much you deplored the corruption of Popes, the superstition of individual Roman Catholics, the inadmissibility of certain modern Roman Catholic beliefs, there was only one Church which could be seen to stretch back in an unbroken line to the time of the Apostles and to Christ himself. If God was God, and if God founded a visible Church on earth, and if he promised that this Church would never be divided, all the other churches are, by this definition, only sects, who have deliberately separated themselves from the parent stem.

It was this line of logic which the Tractarians started up, and which Newman was brave enough to follow to its conclusion.

No doubt, if the question had occurred in other minds, at other times, or in other places, it would have had different effects, or no effects at all. It has to be seen against the background of precisely the kind of religious enthusiasm which the Methodist John Wesley had stirred up in the eighteenth century. A few months before coming up to Trinity College at the age of fifteen, Newman had undergone an Evangelical conversion-experience. He knew that he was 'elected to eternal glory' and for the rest of his life, as he tells us in his autobiography, he rested in the thought 'of two and two only supreme and luminously self-evident beings, myself and my Creator'.

It was in such a glow of pious enthusiasm that this profoundly impressionable youth was exposed to Oxford. He was what we would now call an aesthete. His great passion was for music, and he was a brilliant violinist. But he had also, like most readers of his generation, wallowed in the poetry of Sir Walter Scott and discovered, in *The Lay of the Last Minstrel* and *The Lady of the Lake*, a vision of the past which was to transform the look of the British Isles.

Gothic architecture, which to earlier generations had seemed quaint, or rude, or at best picturesque, now appeared to be the quintessence of beauty. 'I think the Gothic style', Newman once told a lecture audience, 'is endowed with a profound and commanding beauty such as no other style possesses with which we are acquainted and which probably the Church will not see surpassed till it attain to the Celestial city.'

Many of Newman's contemporaries were unable to wait until they reached the Celestial city, and zealously got to work, transforming the secular cities of Manchester and Birmingham and London into neo-

Gothic paradises, with Gothic hotels, Gothic town-halls and Gothic post offices.

At Oxford, Newman found himself surrounded by genuine Gothic of intoxicating beauty. 'Beautiful city!' the poet Matthew Arnold recorded, 'so venerable, so lovely, so unravaged by the fierce intellectual life of our century, so serene! spreading her gardens to the moonlight, and whispering from her towers the last enchantments of the Middle Ages.'

St Mary's Church, Oxford, where Newman was vicar and preached his greatest sermons. 'Who could resist the charm of that spiritual apparition,' asked the poet Matthew Arnold, 'gliding through the afternoon light of St Mary's?'

Few individuals have ever been more enchanted by the Middle Ages than Hurrell Froude, the beautiful young man who became a fellow of Oriel College in 1827, and, the year following, became an intimate friend of Newman's. Froude, good-looking and aristocratic, exercised a spell over Newman by the extremism of his romantic religious faith. He practised somewhat bizarre forms of mortification, sleeping on the floor, and castigating himself with a whip for lustful thoughts or even for such heinous sins as eating too much buttered toast. He burned with the zeal of a medieval monk. His diaries are full of his hatred of the Reformation and everything which the Protestant reformers stood for. Newman, who was profoundly impressionable to male charm, lapped all this up.

He and Froude travelled in Europe together. Newman always said that he loved Hurrell Froude as a man, and that he believed he owed him his Catholic soul. They saw Rome together. 'It is the first of cities,' Newman exclaimed. 'And all I ever saw are but as dust (even dear Oxford inclusive) compared with its majesty and glory.' Newman tried to persuade himself that he still detested the *Roman* Catholic system while becoming more and more attached to Catholicism. 'I quite love the little monks and seminarists,' he wrote, 'they look so innocent and bright, poor boys!'

The year was 1833. Back in Oxford, John Keble preached his sermon in the University Church that recalled England from her apostasy. It was the day which Newman thought marked the beginning of the Oxford Movement. Significantly, he was far away at the time, in Europe, and ill. In Rome, he and Froude had gone their separate ways, and Newman had travelled down to Sicily where he became very ill and suffered for three weeks from a dangerous fever. But as he recovered and took the boat for home, he had a renewed sense of God leading him on to some task he knew not what. In the Straits of Bonifacio, his ship was becalmed, and it was there that he wrote his famous lyric, 'Lead kindly Light'. At fifteen, there had been for him only two luminously self-evident beings, himself and his Creator. Now, in the last verse, there are others – himself, yes, led on 'o'er moor and fen, o'er crag and torrent' by that unseen hand. But also, as he awaited his reunion with Hurrell Froude and his other Oxford friends, it seemed to him like an image of heaven.

And with the morn those angel faces smile
Which I have loved long since and lost awhile.

This caricature of Newman in 1841 still hangs in the vestry of his old church, St Mary's. It was in that year that he wrote his Tract No. 90 and realised that his Oxford career was in ruins. 'I fear that I am clean dished.'

In the decade which followed, Newman, who had by now been appointed vicar of the University Church, St Mary's, became the object of a besotted cult-following among the undergraduates. The poet Matthew Arnold explained his magic as a kind of silvery aesthetic charm:

> Who could resist the charm of that spiritual apparition, gliding in the dim afternoon light through the aisles of St Mary's rising into the pulpit and then, in the most entrancing of voices, breaking the silence with words and thoughts which were of a religious music – subtle, sweet, mournful?

Arnold, like so many of that generation, was bowled over by Newman's *charm*, but in later life felt quite unable to go along with his religious views – he became, in fact, an agnostic. So too did Hurrell Froude's brother James, the famous historian who in his day – like Mark Pattison and so many of the other famous unbelievers of the nineteenth century – had been drunk on what they jokingly called Newmania. James Froude attributed his attraction partly to Newman's appearance – 'his head was large, the face remarkably like that of Julius Caesar' – and partly to the bigness of his mind. Whereas most of the dons then – like most of the dons now – were content to burrow down like White Rabbits into a Lewis Carroll world of footling research and ludicrously parochial quarrels, Newman held out to the young a vision of something bigger. James Froude wrote in after-years:

> Newman's mind was world-wide. He was interested in everything that was going on, in science, in politics, in literature. Nothing was too large for him, nothing too trivial, if it threw light upon the central question, what man really was and what was his destiny.

For a few extraordinary years – agonising years as they turned out to be – this vast question of what man really was and what was his destiny, became connected in the minds of those involved in the debate with the – to us semi-comic – question of what the Church of England really was. Hurrell Froude – as beautiful young geniuses have a way of doing – died young, at the age of thirty-three, in 1836. His death left a great gap and a great sadness in Newman's heart which was not to be filled for a number of years. In devotion to his memory, Newman arranged for Froude's papers and diaries – his Remains as his friends called them – to be printed. Their publication in 1837, the very year when the Protestant succession passed into the hands of the impeccably Hanoverian Queen Victoria, caused a storm of protest.

What the world had begun to suspect was here made clear to the public! These so-called Tractarians were little better than crypto-Roman Catholics. Here were phrases out of Froude's own mouth: 'Odious Protestantism sticks in people's gizzards' and 'The Reformation was a limb badly set – it must be broken again in order to be righted.' And these sentiments were expressed by a young man who quite plainly, both in his callow pieties and his fervent devotions to his friends, was not exactly what one might call manly.

After the death of Hurrell Froude and the publication of the *Remains*, Newman became more and more addicted to exciting the young with his extreme and paradoxical religious opinions. It made him feared and hated by other dons, and he began to plan for himself some form of rear-guard action. He did not yet quite know what it should be, but in 1840 he bought the little row of cottages in Littlemore with a view to living there for periods of prayer and retreat.

Littlemore had long since been attached to the living of St Mary's.

Newman had built the church there; his mother laid the foundation stone of it in 1837. As his position as a cult idol among the undergraduates became more and more heady, and his position in the Church of England more and more doubtful, Newman found great refreshment in the simple pastoral work of a country parish priest. He was an assiduous visitor of the old, the sick, and the housebound. But although Littlemore provided him with respite, the controversies surrounding his position were not going to go away.

In 1841, Newman published the ninetieth of the Tracts. Though he claimed afterwards that he had not expected to attract any particular notice, he must have been deceiving himself. For it was in this tract that he claimed that there was nothing in the doctrinal formularies of the Church of England – in the Thirty-Nine Articles – which could not be believed by a Catholic. It is true that, in his subtle way, Newman qualified this by all kinds of tenuous arguments. It is also true that when you actually examine the contents of Tract No. 90, you realise what a very long way Newman still was from believing in Roman Catholicism. For example, he states that it would be perfectly possible and indeed desir-

LEFT Newman built the parish church at Littlemore and his mother laid the foundation stone. He found pastoral work in a small rural parish a welcome respite from academic controversy.

RIGHT The cottages at Littlemore, near Oxford, converted from an old staging post, where Newman and his friends lived a semi-monastic existence while he decided whether to remain in the Church of England or join the Roman Catholic Church. It was here that he was received into the Roman Catholic Church by an Italian missionary on 9 October 1845.

able to abolish the Papacy; and that the Bread and Wine of the Eucharist remain bread and wine in their natural substance even after being consecrated by a priest.

Such quibbles did not change the effect Tract No. 90 had upon Oxford – and upon the country at large. Here was a man who was meant to be the greatest intellect in the Church of England, which as everyone knew was a Protestant Church, saying that really, in essence, it was a Roman Catholic Church. This time, Newman had gone too far. On 12 March 1841 all the Heads of Houses in Oxford, save two, assembled in the Sheldonian theatre and condemned Tract No. 90. 'I fear,' Newman wrote to one of his beloved sisters, 'I am clean dished.'

The Bishop of Oxford was bombarded with requests from all over England to suspend this undercover Papist firebrand. He did not do so. Newman was left to work out his own salvation. By the end of the year, he had retreated to live in the cottages at Littlemore, and the world waited, nearly four years, to see what he would do.

Newman could see that the logic of his position was leading him inexorably towards Rome, but it was more than mere cowardice which held him back. Not only was he in love with the Church of England; but also, as we have seen, there was not really much of a Church of Rome to go to. In a Church which barely existed in England, what could be the place of a man like Newman, to whom all the religious minds in England looked, either with fear or respect, but with the sense that here was a

most original and important thinker?

In the cottages at Littlemore, he collected about him a group of like-minded friends, and they lived a life of the utmost austerity and simplicity, based on study and prayer. Today, it is a Catholic shrine. To one like myself, who is alas unable to accept the logic of Newman's position, it is a place of peculiar sanctity nonetheless. This meagre little place, little more than a shed, is to me – much more than the grand colleges of Oxford – a monument to intellectual integrity. Newman, like so many other men of the nineteenth century, could so very easily have put the issues of the 1830s out of his mind, and opted for a comfortable life as a much respected Oxford don. Given the quizzical and paradoxical cast of his mind, his ability to see the plausibility of other people's viewpoints, one can easily imagine him, like his friends Pusey and Keble, muddling along, trying to introduce, piecemeal and undercover, Catholic beliefs and practices into a scandalised Protestant National Church.

But that was not Newman's way. He was prepared to put aside his vast following, his glittering career, his comfort and – what perhaps mattered to him more than anything else – the feelings of his friends for the sake of what he believed to be true. In 1843, he resigned the living of St Mary's preaching a sermon in Littlemore which came to be known by the title *The Parting of Friends*. It is said that on that occasion, his own voice broke, but that the sermon was in any case inaudible, because of the sobs of his friends in the congregation. How typical, incidentally, of Newman, that he took leave of the dons not in their own home territory, but in the little country church he had built himself, surrounded by the children he had catechised, who were arrayed for the occasion in frocks and bonnets which he had bought for them himself.

Newman's life had been changed, not only by the great public drama of his resignation but by meeting a young man who to a large extent took the place in his affections of Hurrell Froude. Ambrose St John was a clergyman of twenty-eight years who came to join the community living in the Littlemore *mone* or monastery as it came to be known. 'Dear Ambrose St John,' Newman calls him in his autobiography, 'whom God gave me, when He took every one else away.' The fact that a young male friend was prepared to follow him on his journey gave Newman the strength he needed to take the plunge. For two years more, the world waited to see what Newman would do. Newspapermen hovered about outside the cottages anxious to discover bizarre monkish practices in

Ambrose St John was Newman's devoted companion from the Littlemore years onwards. 'Dear Ambrose St John, whom God gave me, when He took every one else away.'

progress, but were unfortunately unable to discover any practice odder than that Newman and his friends were always praying. Perhaps that is as odd as can be.

At long last, Newman realised that he had come to the end of his journey, the end of the period which he was later to call his Anglican deathbed. Ambrose St John went off to Bath and became a Roman Catholic on 2 October 1845. Newman immediately wrote to the Provost of Oriel to resign his fellowship. St John returned, bringing with him an Italian priest of the Passionist order called Father Dominic Barberi. It was a rainy night – 8 October – and Father Dominic arrived soaked to the skin, having travelled on the top of a coach for five hours. He was drying himself by the fire when Newman burst into his room, knelt at his feet, and begged to be received into the Catholic Church. He began to make his confession.

It went on so long that the priest had to tell him to stop, and continue it in the morning when they were both less tired. The next day they walked into Oxford in pouring rain and heard mass at the little Roman Catholic chapel in St Clements. Then they returned to Littlemore where Father Dominic heard the conclusion of Newman's confession, gave him conditional baptism, and celebrated mass on Newman's writing desk. When the ceremony was over, Newman walked down to his library at the opposite end of the row of cottages and picked up a volume of the Fathers. He kissed it and said, 'Now I belong to you.'

Newman had hardly ever met a Roman Catholic, and his future was now highly uncertain. He was forty-four years old. There was, of course, no possibility of his continuing his career at Oxford where Roman Catholics were still refused admittance. He threw himself on the mercy and guidance of the Roman Catholic authorities, such as they were. After a short period of training in Rome, it was decided that Newman and his friend should not become monks but should set up a religious house in the tradition of St Philip Neri, a sixteenth-century Roman priest who had established Oratories to house groups of secular priests living together for the purposes of instructing the faithful and attracting converts by their splendid musical tradition. As interpreted by Newman at least, the ideals of the Oratory, and the Oratorian way of life, were almost a carbon copy of an Oxford common room of the 1830s or 1840s. And the Oratory which he established with St John and his other convert friends in the Hagley Road in Birmingham has much the feel of Oxford. The room where the Oratorians meet for recreation was consciously designed by Newman to remind him of Oriel. His own library is rather more extensive than the average library in a Roman Catholic presbytery – more like a college library. And his rooms, unchanged since he left them, are the rooms of a Victorian don. It was here that, with a few interruptions, Newman was destined to spend the rest of his days, writing, instructing converts, and ministering to a large parish of poor Irish immigrants.

His conversion changed the religious complexion of England. Neither the Church of England nor the Church of Rome were ever the same again. The Church of England lost thousands of converts to Rome following Newman's agonised decision in Littlemore, and that inevitably changed the nature of the Roman Catholic Church. For instead of being a backwater Church, followed by a few heroically stubborn English families and the Irish section of the working class, it now became a Church to which clever middle-class English people felt attracted. There was no more important convert, after Newman's change of mind, than the Archdeacon of Chichester, Henry Manning, the close friend of Mr Gladstone and the very essence of an Establishment man. Manning began life as a civil servant in the Colonial Office. He always was the

Henry Edward Manning, when Cardinal Archbishop of Westminster, had the nickname of 'the Marble Arch'. His desire for a 'triumphalist' and markedly Italianate form of Catholicism was not in tune with Newman's more subtle intelligence.

brilliant civil servant and the manipulative archdeacon. Whereas Newman was the dreamer, the intellectual, the splitter of hairs, Manning was the brilliant administrator. He was the man who got the show on the road, as far as the Roman Catholic Church was concerned. And what a show it was! Thanks to Manning (and in spite of howls of protest from Gladstone and others) the Roman Catholics soon had cathedrals of their own all over England, and bishops and dioceses as if they were

Anglicans! The Roman Catholic Church as we know it in England today, with its schools, convents and hospitals, largely paid for by the addiction of the faithful to bingo, is the creation of Manning – Cardinal Manning as, needless to say, he soon became.

Nor was he content to create and organise the Roman Catholic Church in England. It was largely through Manning's influence that Pius IX established the First Vatican Council in 1870, that same council which declared that most difficult doctrine of the Pope's infallibility to be an article of faith necessary to salvation.

To some, the growth in power and influence of the Roman Catholic Church as a recognisable material institution will be seen as a sure token of Divine grace. I must confess that I am sadly blind to this fact – if fact it be. The existence of the Roman Catholic Church seems to admit of all kinds of explanations. Its being and essence seems to me no more miraculous than that of any other well-organised human institution, religious or secular. Popes, whatever their own personal qualities of goodness or wisdom, are no more likely to be infallible than any other human being. So, though I would recognise that Newman's most obvious achievement was in persuading a lot of people to become Roman Catholic who would not otherwise have dreamed of doing so, I would not think that was his greatest achievement.

In fact, by a strange sort of paradox, it could be said that after he had joined the Roman Catholic Church most of the issues which so worried him while he was an Anglican seemed insignificant, and he was able to concentrate all the powers of his delicately-honed intelligence on questions which are of momentous importance for all of us, whether we are Catholics, Hindus, or atheists.

It was the agnostic James Froude, rather than his pious brother, who put his finger on what made, and makes, Newman so compelling a writer – his unhesitating preoccupation with the big questions, with 'what man really was and what was his destiny'.

The nineteenth century is the first century in which thinking people in large numbers began openly to entertain the notion that Christianity was untrue, and that God, the all-powerful, all-loving Creator of this unhappy world, was not merely an intellectual improbability, but also a scientific and moral impossibility.

Traditional belief taught that the world could only be explained in terms of a Creator. Matter and animal life in all its varied species could not exist without Someone who made them. Charles Darwin's discoveries showed that this syllogism was simply untrue. It became perfectly possible to demonstrate how species have evolved on this planet without resorting to a theory that there is One who 'gave their glowing colours' and 'made their tiny wings'.

Then again, earlier Christian writers taught that the Bible was an infallible book, every word of which had been dictated by God Himself. This was as true of the passages which contradicted one another as of those which did not. But ever since the Germans, and in particular the Biblical scholars of Tübingen, had begun to investigate the history of how the Bible came to be composed, it had become clear that any such simple-minded fundamentalism was intellectually impossible. Like all other ancient texts, the Bible bore the traces of being the product of human beings in a particular time and place.

With the removal of these two great props – the Creator-God and the Infallible Word of God in Scripture – Christianity in Britain lost many of its adherents not to Rome but to Unbelief. Matthew Arnold, who as a young man had listened so entranced to Newman's rhetoric at Oxford, expressed the mood of a whole generation in his poem 'Dover Beach', when he saw the Sea of Faith ebbing out

> *Retreating to the breath*
> *Of the night-wind, down the vast edges drear*
> *And naked shingles of the world.*

It was to this world of intelligent, questioning people that Newman addressed himself.

> Go to the spring, and draw thence at your pleasure, for your cup or your pitcher, in supply of your wants; you have a ready servant, a domestic ever at hand, in large quantity or in small, to satisfy your thirst, or to purify you from the dust and mire of the world. But go from home, reach the coast: and you will see that same humble element transformed before your eyes. You were equal to it in its condescension, but who shall gaze without astonishment at its vast expanse in the bosom of the ocean? who shall hear without awe the dashing of its mighty billows along the beach? who shall without terror feel it heaving under him, and swelling up and yawning wide, till he, its very sport and mockery, is thrown hither and thither, at

> the mere mercy of a power which was just now his companion and almost his slave?

Newman got small thanks for his endeavours to present the faith in a manner which might be intelligible to intelligent people. The Roman Catholic Church, with its absolutism and dogmatism, failed to appreciate him until he was more or less dead. The only great task which it asked of him was to set up the Catholic University in Dublin. The brilliant lectures which he gave there, and which were published under the title of *The Idea of a University*, filled the Irish clergy with forebodings and suspicion, and before very long, Newman, to his great relief, was sent back to Birmingham. The bishops and monsignors of Catholic Dublin were as little appreciative of the tangential subtlety of Newman's mind as had been the Protestant dons in Oxford.

But we may be thankful that Newman developed an unfair reputation for casuistry and insincerity, for it drew out of him his best-known and most passionately felt book, the *Apologia pro Vita Sua*.

That Newman had become a Roman Catholic was common knowledge. But how and why were private questions to which very few, if any, really knew the answer. It was commonly put about by Protestants that he had secretly been a Roman Catholic all along, and had merely delayed his conversion in order to undermine the Church of England. Charles Kingsley, the robust no-nonsense advocate of muscular Christianity, was one such critic. Better known as the author of *The Water Babies* and *Westward Ho!*, Kingsley, in the Christmas number of *Macmillan's Magazine* in 1863, reviewed J. A. Froude's *The History of England*. In the course of his review, he let fall the remark, 'Truth for its own sake had never been a virtue with the Roman clergy. Father Newman informs us that it need not be, and on the whole ought not to be.' Newman at once replied, demanding that Kingsley withdraw the slander. Kingsley refused. In the correspondence which followed, Kingsley attributed to Newman, while he had still been an Anglican, views which he had only developed later. 'What then,' he asked in exasperation, 'does Dr Newman mean?'

Stung by the accusations of deviousness and double-talk, Newman resolved to tell Kingsley, and the world, how he had arrived at his opinions. The result, written in a few weeks, often in tears, and always standing at his writing-desk in his room in Birmingham, was his famous

Charles Kingsley, best known today as author of The Water Babies, *prompted Newman to write his* Apologia *by his assaults on Newman's good name in* Macmillan's Magazine.

Defence of His Own Life. After twenty years, Newman returned in his mind to the events which had led up to that crisis of conscience in Littlemore. He revisited in his memory all those 'angel faces' which he had 'loved long since and lost awhile' – Pusey, Keble, the Froudes. Though nearly all the issues with which it deals are dead, the *Apologia* remains a fascinating book because its primary subject is how a mind moves and develops. What do we mean when we say that we are changing our mind, and how do we arrive at any opinions whatsoever?

> It was not logic that carried me on; as well might one say that the quicksilver in the barometer changes the weather. It is the concrete being that reasons; pass a number of years, and I find my mind in a new place; how? the whole man moves. Paper logic is but the record of it.

Newman's moving account of how, by painful steps, he came to the conclusion that the Church of England was not a part of the Catholic Church, and that he must therefore leave his friends and all that he loved, had a powerful effect on the reading public. It was immediately recognised as an autobiography of genius, and a testimony of great intellectual honesty. Strangely enough, it probably made him more friends in the Church which he had left behind than in the Church of which he was now a somewhat uneasy member. By many of his co-religionists, Manning at the head of them, Newman was regarded as a dangerous liberal, a compromiser. Before the First Vatican Council, he was quite open about his reservations concerning papal infallibility, stating that he considered it to be 'not a dogma but a theological opinion'. When the Council – the huge proportion of whose members absented themselves rather than vote for it – proclaimed the Pope to be infallible, Newman's more reasonable view of things got pushed to one side. But he had enough, when he was a Protestant, of internal squabbles among the clergy; and although the activities of the First Vatican Council, and the controversies surrounding it, caused him dismay, he managed, at least in his intellectual life, to rise above the immediate concerns of the day. For he had been planning and writing over a good many years what turned out to be his greatest book, *The Grammar of Assent.* Forgetting Roman bishops and Anglican bishops, he returns to the essence of the religious question. In a world in which, as Arnold wrote, we have 'nor certitude, nor peace, nor help for pain', how do some people – Christians – arrive at their strange and, on the the surface, improbable collection of beliefs?

If ever anyone made out a case for belief in God, Newman did so. To paraphrase it would be to make crude what is essentially delicate, to distort the very measured way in which Newman demonstrated that we assert our belief in matters for which there can never be verifiable proof. Of these, only one instance, though a primary one for Newman, is the existence of conscience, our belief in a moral law, our capacity to feel guilt when we know that we have gone against this law and the opposite when we feel that we have obeyed it. '"The wicked flees, when no one pursueth"; then, why does he flee? whence his terror? Who is that sees in solitude, in darkness, in the hidden chambers of his heart?'

These questions of Newman's are similar to the ones that Dostoyevsky was asking in his great novels at exactly this date. If God does not exist, then is anything permitted to us? Since we feel that everything is not permitted, does this not give us pause before dismissing the notion of a Supreme Lawgiver? Newman never loses sight, though, of the extreme mysteriousness of life, the fact that with our present range of limitations and sympathies, we shall never know much.

> Break a ray of light into its constituent colours, each is beautiful, each may be enjoyed; attempt to unite them, and perhaps you pro-

> duce only dirty white. The pure and indivisible Light is seen only by the blessed inhabitants of Heaven; here we have but such faint reflections of it as its diffraction supplies.

It is hardly surprising that the Roman Catholic Church did not know what to do with such a man as this. For years and years, his was a strange position. He enjoyed widespread respect and enormous national popularity among Catholics and Protestants, but no kind of official recognition came to him from Rome. Nor could it be said that he did not mind, for with all Newman's sensitivity to the beauty of words and the subtlety of argument, there went an over-sensitivity in personal matters. He was touchy and awkward. I remember one of the old Oratory fathers in Birmingham telling me of a man who joined the community in the 1870s. On his very first day – and nobody knew how he did it – this unfortunate young man offended Newman. That man lived in the same house as Newman for the next twenty years without Newman addressing a single word to him.

This sort of touchiness led to a rift between Newman and the London Oratorians, and as is notorious, he got on badly with Manning. There are many stories of Manning trying to block Newman's promotion in the Church. If they were true, I would not consider them damaging to Manning, for Newman was not the stuff of which bishops or administrators are made. He was a scholar, a poet, a mystic. The humdrum chores of a diocesan bishop would not have appealed to him. His extreme readiness to take offence might well have been disastrous when trying to deal with parish priests. Newman noted in his diary the anniversaries of quarrels and disputes, the day he stopped speaking to Gladstone or Wilberforce or Pusey. How could an ordinary Irish monsignor of little education have dealt with such a person?

'Poor Newman,' Manning is supposed to have remarked, 'he is a great hater.' I am not sure that this is fair, but he certainly went in for intense feelings about other people which were not always markedly controlled.

In 1875, he lost the best of them, Ambrose St John. 'From the first,' Newman said, 'he loved me with an intensity of love which was unaccountable. . . . The Romans called him my Angel Guardian. . . . He has not intermitted this love for an hour up to his last breath.' When St John's last illness came he hugged Newman so tightly that Newman said jokingly to the doctor, 'He will give me a stiff neck,' but then as his friend's hand was clasped in his own, he realised that this was no joke. The great parting was about to happen. At midnight, Newman was summoned from his room to the sick-bed. Another priest, Father Neville, was trying to give St John some arrowroot. St John rose up on one elbow and then fell back on the pillow and died. Newman, at this date seventy-four years old, broke into a paroxysm of weeping and flung himself on to the bed. He lay there with the corpse, sobbing, for the rest of the night, only consenting to be parted from it in the morning. St John was buried in the Oratorian burial ground at Rednal. 'I trust I may now be allowed to die in peace,' Newman remarked. 'What a dream life is!'

It was 1875, and Newman had another fifteen years to live in this world without his great friend. Recognition from the Pope did come at last in 1879 when he was offered a cardinal's hat. It was a rule of the Church that you could not become a cardinal without residing in Rome unless you were a bishop. Newman was too old and too set in his ways to leave the Oratory, but when permission was granted to reside in Birmingham, he accepted the hat, and with it, two hundred pounds worth of expensive regalia. He seemed to have travelled a long way from that dark wet night in Littlemore of thirty years before.

Thereafter, Newman was a 'character', an institution, an object of curiosity whom just about everyone who was anyone reckoned on coming to hear or see. His frail figure in robes of black and scarlet was still a familiar one in the memories of old people I have met in the Hagley Road, Birmingham. With his rather wispy white hair and his curious attire he more and more came to look like an old woman. While the

Newman's sensitive nature wrestled with the profoundest religious questions. In old age he spoke and wrote less of ecclesiastical controversy and more about fundamental questions of belief.

Sovereign Pontiff heaped honours on his head, and the Church of his adoption at last began to see that he was a religious genius, the boys at the school which he founded displayed that wonderful refusal to take anyone seriously which boys will always display. One of them was the future poet Hilaire Belloc, who remembered that their nickname for the Cardinal was Jack. Every year he used to help them to produce a Latin play. The great joke for the boys consisted in trying to make 'Jack' recite particular lines of Latin poetry which would make him burst into tears.

Tears are never far from the surface in Newman's poetry or prose. It is appropriate, I think, that he is nowadays chiefly loved and remembered for his hymns and his poetry. When one of the Oratorian Fathers, Father Gordon, died in 1882, Newman was moved to write his most famous poem, 'The Dream of Gerontius', the story of a man on his death-bed coming face to face with God. Ever since, as a boy, Newman had come to believe that there were two, and two only, luminously self-evident beings, himself and his Creator, that meeting was one for which he had been preparing, the meeting when he would shake off this mortal coil and meet his Creator face to face.

Throughout his life, Newman was passionately fond of music. He once composed a violin sonata. Like Sherlock Holmes, he spent long periods of solitude in his rooms, sawing away at the fiddle. Above all, he had a taste for the most exquisitely melancholy final quartets of Beethoven – perhaps the most highly charged of all music for strings. There is much in those quartets of Newman's mysticism, of his overwrought emotions, his courage triumphing over depression, loneliness and dejection.

I think that it would have pleased him that his best poem had been transformed, by a devotee and fellow Roman Catholic called Edward Elgar, into possibly the most magnificent piece of English choral music in the repertoire: *The Dream of Gerontius*. It is not a brilliant poem, but it acquires brilliance when married to Elgar's music.

Newman was very nearly ninety years old when he died. By then, he was an eminence indeed, a prince of the Church, a much-loved institution, a name who merited a six-column obituary in *The Times*. They buried him on the outskirts of Birmingham, in the same grave as Ambrose St John.

In his memory, the Oratorian priests erected an imposing church in the Hagley Road in Birmingham, a place to match or rival the Italianate splendours of Father Faber's Oratory church in London. It feels like a

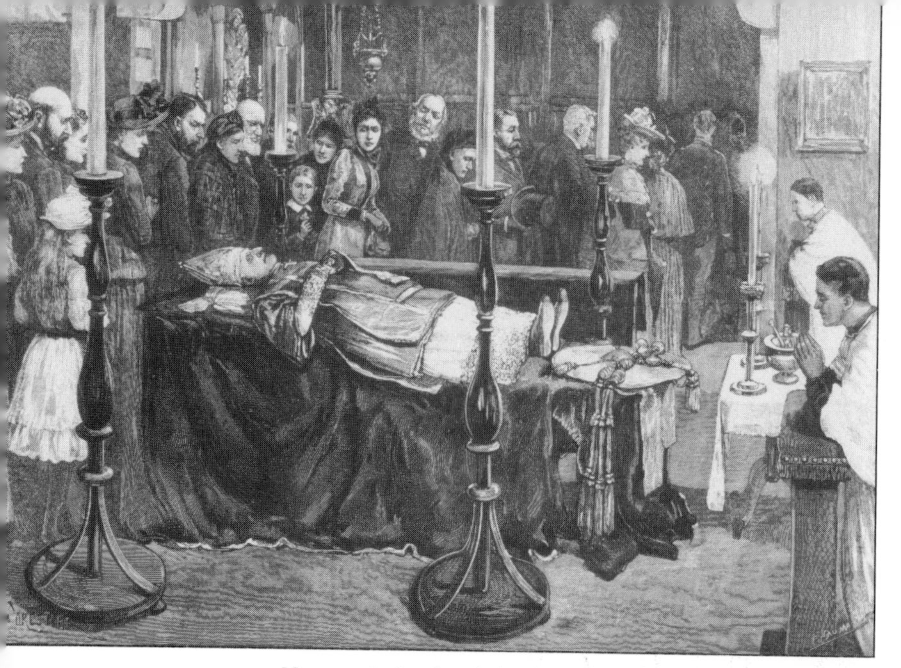

Newman insisted on being buried, not like most Cardinals in a cathedral tomb with his hat hanging above him, but in a dreary Birmingham suburb beside his devoted friend Ambrose St John.

church in Rome. So very successfully have they created a pastiche of a continental church that it is rather a surprise to emerge from the place and see the familiar old Birmingham buses making their way down to Five Ways.

It was really in the twentieth century that Newman came into his own. He has been hailed as the forerunner of modern Catholicism. I am not sure that I fully understand why this is so. But, when Pope John XXIII revolutionised the Roman Catholic Church by setting in train the Second Vatican Council, Newman was called by some the Father of that Council, the genius who was its primary inspiration. He has even been spoken of as the prophet of modern ecumenism and Christian unity, which seems a slightly strange idea to entertain about a man who suffered such pains to demonstrate the falsehood, as he saw it, of Protestantism.

If Newman was the father of modern Catholicism, it is a somewhat dubious accolade. Institutional Christianity in any form, whether in the Church of England or the Church of Rome, attracts fewer and fewer adherents, and it is doubtful whether many Roman Catholics today would share Newman's opinion that salvation outside that Church was an impossibility.

I have lately – over the last year or so – re-read most of what Newman wrote, and found it on the whole a depressing experience. We associate the one-track mind either with fools or psychopaths, and Newman was neither. But when we survey his works on the library shelf – the volumes and volumes of letters and diaries, the two novels, the works of autobiography, the book about university education, the poems, and of course the sermons and theology – we realise that there is hardly a page in the entire oeuvre which is not concerned, obsessively so, with religion. Many of the things which mattered so much to him seem almost pathetically remote. And reading of the controversies and squabbles of the Victorian clergy I feel like Louis MacNeice reading the Ancient Greeks

> *And how one can imagine oneself among them*
> *I do not know;*
> *It was all so unimaginably different*
> *And all so long ago.*

And yet, for a brief period of my life – in my late teens – I was almost in love with Newman, and I read him obsessively. The first thing I did when I arrived at Oxford as an undergraduate was to walk out to Littlemore. Now, my teenage self seems almost as remote to me as the Victorians do. It is not so much that I do not believe in what Newman wrote, so much as that I cannot imagine what it was like to care about – for example – the Arian heresy in the fourth-century Church, or whether it was permissible in the nineteenth century to have a shared Lutheran and Anglican bishopric in Jerusalem. Today, invincibly ignorant as I am, I find it inconceivable that my eternal destiny might depend upon membership of any religious group. The pronouncements of the clergy, whether made as individuals or in collections of synods and councils, seem for the most part remote from the things which concern me, and remote from the New Testament. I cannot even imagine how Newman – let alone I – thought that there could be any connection between the mysterious events and personages of the Gospels and the very far from mysterious set-up in the Vatican.

'There are but two alternatives,' Newman wrote, 'the way to Rome and the way to Atheism: Anglicanism is the halfway house on the one side, and Liberalism is the halfway house on the other.' Was it this sort of extremism which appealed to me once? Peeping gingerly from the windows of my halfway house, I am rather glad not to have opted for either of Newman's bleak alternatives. Viewed from the halfway house, his life seems sad. So many of his sacrifices seem pointless. And if his conversion to Rome has meant that there are two big Churches with bishops in

England rather than one, so what? It wasn't, I now realise, his church interest which really attracted me to Newman. Was it then merely an aesthetic attraction? He still seems to me incomparably the most flawless English stylist. The music of his prose still has the power to turn me to gooseflesh. But is it that alone? 'If I looked into the mirror and did not see my face,' he wrote, 'I should have the sort of feeling which actually comes upon me when I look into this living busy world, and see no reflection of its Creator. . . . Were it not for his voice, speaking so clearly in my conscience and my heart, I should be an atheist.'

Neither as the popular vicar of St Mary's, nor as the – dare one say it – rather prissy dignitary of the Roman Catholic Church does Newman any longer attract me. But the honest man of Littlemore, who was prepared to change the course of his entire life for what he believed, will never fail to move me. And throughout his writings, every so often, one comes across a paragraph of disturbing religious power.

Do the workings of our conscience imply the existence of a moral law, and the moral lawgiver, or are they a meaningless kink in the evolutionary machine? Is our sense of awe in the presence of great art or of Nature something which has no meaning, or are such moments echoes of a homeland from which we are temporarily exiled? Above all, when we die, will the meaningless sum of all our days be blotted out, or will we, like Gerontius, open our eyes in death and wake up as one refreshed?

John Henry Newman was no more able than anyone else to give us infallible answers to such questions. But for anyone who finds it impossible, quite, to put such questions out of their mind, Newman will always be a giant, and a hero.

Josephine Butler
1828-1906

Feminist and social reformer

A photograph of Josephine Butler in middle age. Her keen, spiritual features are marked by sorrow and illness.

Josephine Butler was the woman who was brave enough to defend that unmentionable class of Victorian person, the prostitute. How many prostitutes existed in nineteenth-century Britain, we shall probably never know. In 1797 it was calculated that there were 50,000 such women in London alone. Different observers calculated the figure at around 80,000 in the late 1830s and 1840s. In 1858 Whitehorne opined that one in six unmarried women between the ages of fifteen and fifty earned their living in this manner – some 83,000 women. Police estimates of known prostitutes for the period are much smaller. In 1858, for instance, it was estimated that there were 7194 prostitutes in London and 27,113 in England and Wales as a whole. How these estimates were made forms part of the story of Josephine Butler.

The actual statistics are of enormous interest, but they are not the central point at issue. What the vast number of Victorian prostitutes reveals is an attitude towards women which extended far beyond the brothels of the back street, the fumbled little encounter and the exchange of money before the man went back to his house and family in the suburbs. The existence of these women in such quantities was not, as some respectable observers of the time might have wanted to believe, in violent contrast with the respectable home life of their clients. It was merely the underbelly of the same creature. The strange, sexless heroines of Dickens's novels – sometimes like Little Nell and Little Dorrit, part child, part woman, sometimes, like Agnes Wickfield, more of an angel than a physical being – tell something of the same story. So do the paintings of Lord Leighton, Holman Hunt and Alma-Tadema. It was more than their pubic hair which the artists painted out. It was any sense of women having an independent, let alone equal, status.

The greatest single difference between Victorian times and our own is the difference in the position of women. So great is the change effected by the feminist revolution that we who live in after-times find it difficult to grasp the absolute non-status of nineteenth-century women. As we turn the pages of Jane Austen or Charlotte Brontë we do well to remind ourselves that all women in Europe – and, come to that in America – were

Wentworth Street, Whitechapel, a fairly representative London slum, is here depicted by the French engraver Gustave Doré, famous for his intricate interpretations of spiritual crisis in illustrations of Coleridge's Ancient Mariner *or Dante's* Inferno. *Like Doré, Josephine Butler found more than merely physical squalor in the lives of the poor. Her work for prostitutes was to open up infernal glimpses into the soul.*

leading lives which were every bit as restricted, legally, morally and spiritually, as those of Elizabeth Bennett or Jane Eyre. There really were two choices for women when Victoria came to the throne (unless, of course, you were of monarchical status): marriage or old maidhood.

Education for women in the early decades of Victoria's reign was limited to those few enlightened families where the male relations were prepared to share their reading-matter with their sisters. Those who survived such schools as Cowan Bridge, where the Brontë sisters were 'educated', had no chance of further education or university. It was totally unthinkable for women to enter the professions. It took Elizabeth Garrett Anderson years of heroic struggle to qualify as the first woman doctor, and it was not until the mid-1870s that she was able to set up the London School of Medicine for Women. The number of women doctors, even by the close of the century, was tiny.

It was not a question of women being denied the vote, denied education, and denied professional status, but in all other respects being regarded as human beings. In the eyes of English law married women existed only as wives, or not at all. Again, in an age when feminism has by and large achieved its objectives we can be slow to understand what this meant. The reality only comes to the surface when we read about someone desperate enough to question the system.

Take Caroline Norton, whose case came to trial a year before

Victoria's accession. A society girl, granddaughter of the dramatist Richard Brinsley Sheridan, and a writer of novels herself, she had married at the age of nineteen, the Hon. Richard Norton, who gave her three children with a great number of bruises and unpaid bills along the way. It was a disastrously unhappy marriage. After one particularly bad quarrel, Norton took the three children and placed them under the care of a cousin. It was only then that Caroline became aware of her legal status. She discovered that in the eyes of the law she was her husband's property. She had no right to see her children if he removed them from her; nor, having married the Honourable Richard, did she have any rights over her own property or even the income she derived from her writings. All belonged, in the eyes of the law, to him. He refused to let her see her children, and furthermore, sued Lord Melbourne, Victoria's beloved adviser and mentor – on preposterously unlikely grounds – for 'criminal conversation' with Caroline. Since there was no evidence for this whatsoever, the jury dismissed the case without so much as retiring from court. But the trial revealed to Caroline Norton another fact about her legal non-status. As a woman she was neither allowed to sue, nor to be sued, nor to defend her good name in a case of this nature.

Desperate to have her children returned to her, what could she do? It was only because a family friend, Sergeant Talfourd, happened to be a Member of Parliament that by 1839, there was passed into law, by a bemused House of Commons, the Infants' Custody Act, which laid down that mothers against whom adultery was not proven might be allowed custody of their children under the age of seven but only if the

An illustration to Charles Dickens's Bleak House *(1853). Dickens's novels, like the paintings of Millais or Lord Leighton, delighted in desexing women. His favourite heroines were innocents, or small women who might be mistaken for children.*

woman was of 'good character'. A woman could not sue for divorce. Divorce, which we now take for granted, was still something which women could only obtain if they were in a position to move a special Act of Parliament – in each and every divorce case. Not until 1857 could divorces be obtained though the law courts, and even then, while a man could divorce his wife on grounds of her adultery only, a wife had to prove her husband had additionally committed incest, bigamy, cruelty or desertion. And it was not until 1925 (two years after the grounds for divorce were made the same for women as for men) that women were allowed full equal guardianship of their children under English law.

'I really lost my young children,' Caroline Norton wrote, after she had obtained the right to see her two surviving sons from time to time, 'craved for them, struggled for them, was barred from them, – and came too late to see one that had died . . . except in his coffin.'

It is one of many sad stories in legal history (Annie Besant's – who lost custody of her daughter to her clergyman husband after she published a treatise on birth control with the radical atheist Charles Bradlaugh – is another). But however often a modern person reads it, she or he will still find it difficult to grasp what it means. The difference between women then and women now is as great as the difference between those who are now full citizens of the United States – as it were, senators, surgeons, professors – but whose ancestors were slaves. The difference goes to the very heart of what we mean, and what the Victorians meant, by being human. There is a strange, liberating thread which connects us here; as we see in the remarkable story of Josephine Butler. For she successfully campaigned against the laws themselves that made women the victims of the Victorian 'double standard'. The repeal of the Contagious Diseases Acts and the raising of the age of consent are both owed to her indomitable courage.

Josephine Butler was born, the seventh child of John Grey and Hannah Annett, at Milfield in Northumberland, in 1828, four years before the passing of the Great Reform Bill. She was related to Lord Grey, the Prime Minister of the day. Her cousin was Charles Grey, an equerry of Prince Albert. She grew up with all the advantages of a farflung branch of an aristocratic Whig family. John Grey made sure that all his children, including the girls, were politically educated – indeed, as cousins of the Prime Minister who had brought in the Great Reform Bill, it could

Josephine Grey was born into a land-owning family in Northumberland. Ewart Park, Wooler, was the home of her liberal-minded father.

hardly have been otherwise. As a young girl, as well as the usual accomplishments of a young lady – needlecraft, sketching, music, and a reading knowledge of French – Josephine knew all about the pioneering work of her father's friend Sir Edwin Chadwick who, the year Josephine was born, had drafted reports on the conditions of British cities which led to the passing of the Public Health Bill. It was the first step towards recognising that the State had a responsibility for the health of the nation. By growing up with a knowledge of Chadwick On Drains, Josephine learnt some of her keen sense of responsibility towards the urban poor. Her destiny was to clean up vice as Chadwick cleaned up sewage.

In addition to her political education, or rather as part of it, Josephine Grey learnt the Christian religion. That religion has at its heart the belief that God who is rich for our sakes became poor. Those who respond most vividly or affectionately to the Gospel have almost invariably been drawn to identify with the poor. This has been a fact which has linked many different Christians of widely various ethnic or cultural backgrounds. We see it in the life of Francis of Assisi, in General Booth of the Salvation Army, in Mother Teresa of Calcutta today. Because God loved the world, such Christians have always believed that it was their duty to love the world too; for 'in as much as ye have done it unto the least of these my brethren ye have done it unto me.' The dichotomy in the minds of many people between 'politics' and 'religion' does not exist for these Christians. Probably neither word plays a large part in their vocabulary. But they are living in the world which they believe God has loved and inhabited, they naturally look for justice in society, relief for the poor.

In the early part of the last century this was particularly true of the Evangelical wing of the Church of England. It was from this religious perspective that William Wilberforce brought about the abolition of the slave trade, and that, later in the century, Lord Shaftesbury prevented the use of children as what was no more or no less than a slave labour force in British mines and factories. Josephine Butler was very much the heir to this way of reading the Gospel. What she saw very clearly was that, in spite of the misogynistic traditions of Christianity, the Gospel contains within it the seeds of what was later called feminism. Just as the equality of Jew and Gentile as proclaimed in the Gospel made it ultimately unthinkable for Christians to allow the continuance of the slave-trade (though for 1800 years Christians did not *see* this!) so, for Josephine, the respect shewn by Christ for women, in spite of the social conventions of His day, led inevitably to the conclusion that women – all women, not just the prostitutes with whom her name is associated – should enjoy equal rights with men.

> Among the great typical acts of Christ which were evidently and intentionally for the announcement of a principle for the guidance of Society, none were more markedly so than His acts towards women: and I appeal to the open Book, and to the intelligence of every candid student of Gospel history for the justification of my assertion, that in all important instances of his dealings with women, His dismissal of each case was accompanied by a distinct act of Liberation.

From an early age then, radicalism and evangelical piety were in Josephine's blood. Evangelical piety, for those of us who are not used to it, can be embarrassing, even cloying. But she was never this. Feminist or political radicalism can often be strident. Josephine was never that either. For me, she is one of the most *attractive* people who ever lived: not merely beautiful, but one of those extemely rare people who is good through and through without for one second seeming 'goody-goody'.

She was twenty-two years old when she married George Butler, the son of a headmaster of Harrow, and a clergyman. Butler was very much influenced by another rich strand in the Anglican tradition, exemplified by the figure of F. D. Maurice. A friend of Tennyson's, Maurice was what that generation would have called a Christian socialist, whose faith meant nothing to him if it was not translated into action. He was regarded as a heretic by the Establishment, and sacked from his job as Professor of Theology at King's College London for his doubts about eternal punishment, that favourite doctrine of the Victorian orthodox. (Tennyson wrote a big-hearted poem about the sacking.) Others in roughly the same mould were Charles Kingsley and Thomas Hughes, the author of *Tom Brown's Schooldays*. Their theology was more 'liberal' than the Evangelicals, admitting and indeed welcoming the advances in Biblical scholarship which inevitably modified the way in which the Bible was viewed. But they were no less ardent in their insistence that Christianity involves a social commitment. Josephine with her inquiring mind and profound interest in society found no conflict of views in her marriage to George Butler which was idyllically happy for forty-eight years, until his death in 1890. His words to her, some four years after they were married, are remarkable for the extent to which they recognise his young wife's equality.

> No words can express what you are to me. I hope I may be able to cheer you in moments of gloom and despondency . . . and by means of possessing greater physical strength . . . I may be enabled to help you in the years to come to carry out plans, which may under God's blessing, do some good, and make men speak of us with respect.

One can hardly emphasise too strongly that society in the last century was changed almost entirely by people like the Butlers – the enlightened and educated classes, some aristocratic, mostly middle or upper middle class. Marx could not have been more wrong in his prediction that England would be changed by a vast proletarian upheaval. The enormous changes which took place – in attitudes to the sexes (such as that letter of George Butler reveals), to class, to capitalism, to the conditions of urban life, to public health, to franchise – all stemmed from the educated classes, and in particular from the Christian educated classes. (Of course there were many other educated reformers, from J. S. Mill and Carlyle downwards, who were not Christian, but the point still stands.)

The first five years of Josephine's married life were spent in Oxford, where George had been a fellow of Exeter College. Those were the days when fellows of Oxford colleges (but not professors) were obliged to be

Josephine Grey married George Butler, a clergyman of broad church views who came from a distinguished academic dynasty. Butler believed in marriage as a partnership between equals, and he was fully supportive of Josephine's campaigns for her fellow women. Very few of George's Oxford contemporaries shared his feminist principles. It was in the early years of married life at Oxford that Josephine came face to face with Victorian 'double standards' in sexual morality.

celibate. On his marriage, George resigned his fellowship, and set up a hall of residence for undergraduates, called Butler Hall. He combined this housemasterly duty with trying to set up a School of Geography within the University at Oxford, and hoping for a professorial job. His expertise was as a cartographer, and Josephine helped him with his maps. Oxford was not a bad place for an aspirant feminist to cut her teeth as a young wife for, even in recent years, its attitude towards married women has been not dissimilar to the Shi'ite Persians. It was only in the middle of the last century that Oxford dons were permitted to marry, and their whole lives, until quite lately, have been shaped with the assumption that they are really celibates. No woman dined in Christ Church until 1960, and it was only in that decade that the custom whereby Oxford dons dined each night in the hall of their colleges, leaving their wives behind in their dingy houses in North Oxford to eat baked beans on toast with the children, fell into abeyance.

A sympton of Josephine's quiet determination to live in a world less circumscribed than that of her neighbours was her application for a reader's ticket at the University Library, the Bodleian. It seems more than likely that she was the first woman ever to think of doing this. Certainly, at that period, the Bodleian – like everything else in Oxford, including of course the colleges – was a male preserve. Josephine got her ticket. Life among the dodos of Oxford cannot have been much fun.

An added constraint was the fact that the domestic architecture of the place was suburban and uncongenial, and the climate damp – very far from the clear skies of her native Northumberland, and the 34,356 acres of her father's estate. George's career went poorly in Oxford, too. His plans for a School of Geography there were thwarted by his colleagues, more stick in the mud than he. He applied for the Chair of Latin, and was turned down. It was at that point that he cut his losses and they moved to Cheltenham, where he became Vice-principal of the Boys' College.

Two incidents from the Butlers' Oxford life, however, are worth recording. One was a discussion arising among their limited social circle of Mrs Gaskell's novel *Ruth*. A young man seriously advanced the view that he would not 'allow' his mother to read such a book. It then emerged that the opinion of nearly everyone there was that if a young girl became pregnant, it was she and not her seducer who was to blame. This struck Mrs Butler as odd, to say the least. 'A moral sin in a woman was spoken of as immensely worse than in a man.' It was one of the first chords, very unobtrusively played, in what was to be the grand music of Josephine Butler's life. What may be called the introduction of the theme, albeit in a minor key, came at about the same time. A young woman serving a sentence in Newgate prison for the murder of her child (it is not clear whether it was a case of infanticide or abortion) was brought to the Butlers' attention. As it happened, they knew the father of the child who was a respected member of the University. But it was in keeping with their way that when the girl came out of prison, rather than expose the man as a bounder and a cad, they had her to live with them; she was the first in a succession of so-called 'fallen women' who found a welcome in the Butlers' household.

In 1857, then, they moved to Cheltenham, a place which Josephine found in every way more congenial than Oxford. The Ladies' College was still in its infancy, and its first great Principal, Dorothea Beale, was appointed in the year after the Butlers' arrival in Cheltenham. It was largely through Cheltenham Ladies' College that Josephine Butler became aware of the developments that were taking place in the education of women. Miss Beale set up the Schools Enquiry Commission in 1865, and when the Butlers moved north a few years later, Josephine undertook the same work there. The North of England Council for the Higher Education of Women was to lead to the founding of such pioneer institutions in the older Universities as Newnham College, Cambridge.

George meanwhile was happy teaching at the Boys' College in Cheltenham. By this time their family numbered four – three boys and a little girl, Eva. Had no personal misfortune befallen them, the Butlers might very well have stayed in the comfortable southern spa town for years – their benevolence towards the unfortunate doled out from a position of comfortable security and becoming, perhaps, just a trifle smug.

One evening, however, in 1863, something happened which was to change their lives for ever. Returning home from a drive, the children rushed on to an upstairs landing to greet them as they entered the hall. Little Eva fell over the banister on to the hard, tiled floor below, and lay insensible at her parents' feet. A few hours later, she died.

> Never can I lose that memory – the fall, the sudden cry, and then the silence. It was pitiful to see her, helpless in her father's arms, her little drooping head resting on his shoulder and her beautiful golden hair, all stained with blood, falling over his arm!

The torture of grief for this child was something which Josephine was unable to assuage. Never strong (she had poor lungs) she soon fell seriously ill. The house where the accident had happened became a place of dread to both the parents, and when the chance came to accept a new job in a completely different part of the country, George seized it. He became Headmaster of Liverpool College. The genteel and homogeneously cultivated atmosphere of Cheltenham was exchanged for the altogether rougher climate of a prosperous northern port – one of the biggest and most important in the British Empire. Here, the vast prosperity of Liverpool merchant families (like the Gladstones) stood in stark contrast to the plight of the poor. As in all Victorian ports, where there were families waiting to emigrate, vagrancy and homelessness were endemic. And, with a huge population of the poor, combined with a big migratory population of sailors and travellers passing through, there was a great amount of prostitution. At this date there were probably about 9000 prostitutes working in Liverpool, many of them children (1500 under 15 and 500 under 13, according to one estimate).

Josephine decided that the grief she felt for Eva could find no outlet at home.

> I became possessed with an irresistible desire to go forth, and find some pain keener than my own – to meet with people more unhappy than myself (for I knew there were thousands of such). I did not exaggerate my own trial; I only knew that my heart ached night and day, and that the only solace would seem to be to find other hearts which ached night and day.

In the Liverpool of the 1860s, she did not have far to look. She began visiting the vagrant women who had been rounded up into the notorious Brownlow Hill Workhouse, an establishment which makes the one in *Oliver Twist* seem positively benign. In exchange for a night's lodging and a hunk of bread, the girls had to work in the sheds stripping oakum (tarry hemp) from piles of rope, the same tedious and painful work – it tears all the skin off your fingers – doled out to prisoners serving penal servitude (Oscar Wilde did it in Reading Gaol). Josephine was shocked at her first sight of the oakum sheds, but her immediate and characteristic response was not to enter them as a lady bountiful, dispensing good advice or soup. Instead, she sympathised in the literal sense of the word: she suffered with these women.

> I went into the oakum shed and begged admission. I was taken into an immense, gloomy vault, filled with women and girls – more than two hundred, probably, at that time. I sat on the floor among them and picked oakum. They laughed at me, and told me my fingers were of no use for that work, which was true. But while we laughed we became friends.

The unselfconscious Evangelical felt no difficulty, in these miserable circumstances, in speaking to 'this audience – wretched, draggled, ignorant, criminal' about her Christian faith. She got one girl, tall and dark, standing up amid the heaps of tarred rope, to repeat the words of St John's Gospel – 'Let not your heart be troubled. Neither let it be afraid.' And then, they prayed. 'It was a strange sound, that united wail – continuous, pitiful, strong – like a great sign or murmur of vague desire and hope issuing from the heart of despair.' The scene reminds us of that of the prostitute reading St John's Gospel to Raskolnikov at the end of Dostoyevsky's *Crime and Punishment.*

It was not long before Josephine became aware of how easy it was for a woman to be tempted into taking apparently lighter work than picking oakum. One of the first girls whom Josephine befriended in Liverpool was called Marion. A farmer's daughter, she had been seduced when she was quite young, drifted into Liverpool, and soon been imprisoned in a brothel. By the time she met Josephine Butler, Marion was consumptive and riddled with venereal disease. Since she had nowhere else to live, the Butlers invited her to share their home. The neighbours in their comfortable bourgeois suburb of Sefton Park could hardly believe their eyes when they watched the arrival of this woman at the Butler residence. She was most visibly the sort of person whom no Victorian household would wish to 'receive'. George Butler welcomed her at the front door, showed her his arm, as if she were the grandest lady in the land, and escorted her upstairs. 'I hear that Mrs Butler takes in these creatures,' said a disgruntled neighbour to one of the Butlers' maids. 'Yes, Madam,' said the maid, 'I am proud to say she does.' When Marion died within the year, Josephine had her coffin filled with white camellias, the symbol of purity.

The story of Marion was to be typical of Josephine Butler's whole approach to her great campaigns of social reform. She was never a mere busybody who felt it was her duty to improve people. Still less was she a mere social engineer, interested in changing the condition of some impersonal thing such as 'society'. She was always an imaginative and sympathetic human being, and the end of all her work was to remind her fellow-citizens that you did not stop being human, and deserving of the honour due to all humanity, merely because you were female, or poor, or, by the rather convoluted morality of the times, a sinner.

There were far too many Marions in Liverpool for the Butlers to be able to take them all in. What was worse, the longer Josephine worked among them, the more she discovered that it was not simply a matter of reclaiming individuals. These women and children were the victims of precisely that attitude which she had heard expressed at Oxford when Mrs Gaskell's *Ruth* was under discussion: 'A moral sin in a woman was . . . immensely worse than in a man.' Women are temptresses, hoydens, harbourers of disease and corruption; men on the other hand will be men and must be indulged and forgiven. This was the 'morality' of mid-Victorian England.

This delicate portrait of Newman in extreme old age by Emmeline Deane (1889) shows an ethereal face. At this period, hobbling along the Hagley Road in Birmingham in his cassock, Newman was often mistaken for an old woman.

Right Newman's study at the Birmingham Oratory is today preserved more or less as he left it at his death. It is called 'the Cardinal's room' but it feels less like the room of a prince of the Italian church than the book-lined den of an Oxford don in the 1830s.

Opposite Top Left The church at Littlemore that Newman built, and where he preached his last sermon as an Anglican – 'The parting of friends'. His words were scarcely audible above the sobs of his disconsolate followers.

Opposite Top Right A design for a stained glass window, which was never made, on the theme of the Newmania, reflecting the way the cult of the Middle Ages in Oxford in the 1830s was bound up with the cult of personality surrounding Newman himself.

Opposite Bottom Newman's room at Littlemore was as austere as any monastic cell. It was there that he wrote his book on The Development of Christian Doctrine and prepared himself for the decision to leave the Church of England.

The Newmania

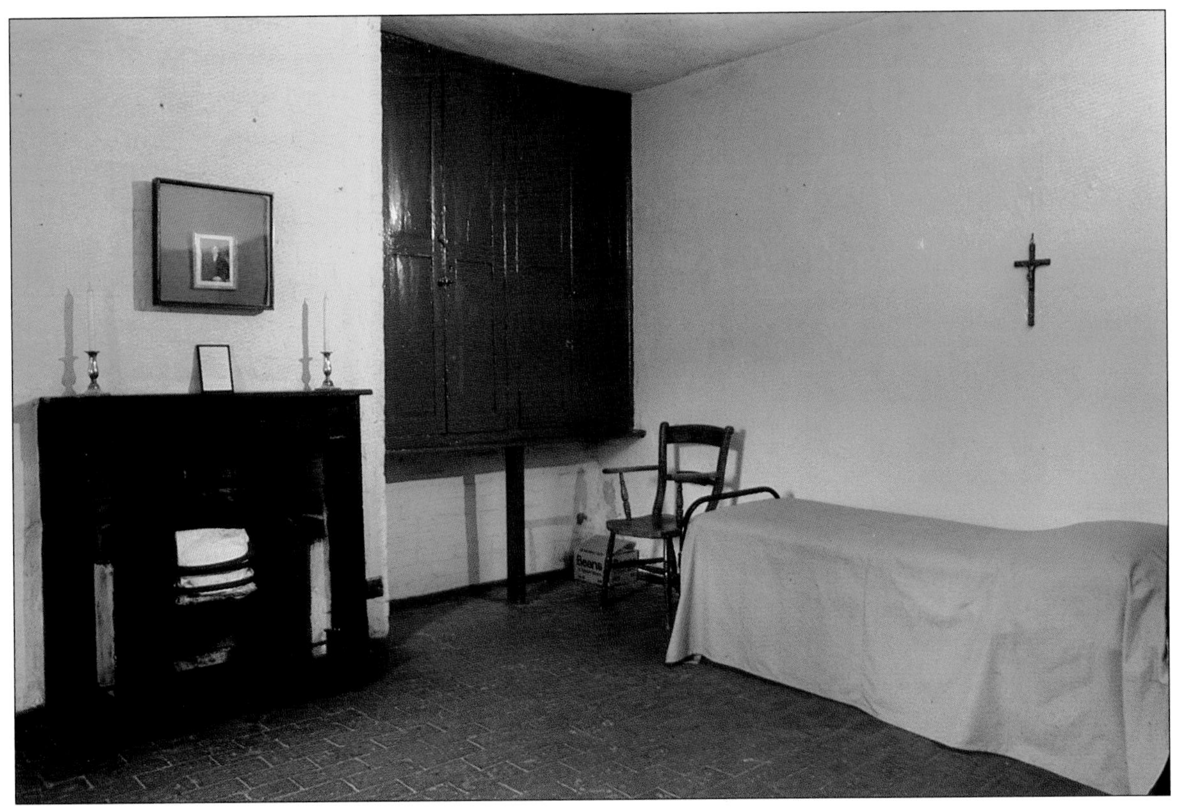

My prayer for my husband.
Sept. 25. 1864

My God, I pray not for myself alone, but for him who is as dear to me as my own soul.
Author of Faith and Source of Strength, give him a living faith in Thee, and make him strong in Thy Might.
Confirm & strengthen him in all goodness.
Give him a selfrenouncing will, bold to take up, and firm to sustain the consecrated Cross, which all must daily bear who follow Thee.
Reveal to his inmost soul the true nature of sin, and the beauty of holiness.
Draw near to him – nearer & nearer. Reveal Thyself to him as his Saviour, Brother, & Friend, The Lamb slain, the immaculate

OPPOSITE G. F. Watts's portrait of Josephine Butler was painted as part of his Hall of Fame series of those who had 'made the century'. She, in some senses, helped to unmake it by exposing the double standards of Victorian sexual morality. When it was finished, she said she felt 'so sorry' for the figure depicted.

LEFT Extract from a prayer written by Josephine Butler to her husband.

BELOW One of Josephine Butler's watercolours, depicting a family picnic in her beloved Northumberland.

RIGHT G. F. Watts's portrait of Julia Margaret Cameron. Watts was one of the many bearded figures on whom Mrs Cameron doted. She called him 'Signor' and arranged, when he was in his forties, for him to marry the sixteen-year-old Ellen Terry. Watts himself sat for many of Mrs Cameron's tableaux.

OPPOSITE 'Paul and Virginia'. Mrs Cameron, like many Victorians, admired Bernardin de St Pierre's 1787 romance of two children leading the life of virtuous savages in Mauritius. The eponymous hero and heroine are here modelled by two Freshwater children, Freddy Gould and Elizabeth Keown. The clothes, and presumably the umbrella, which came from India, were deemed appropriate appurtenances.

LEFT 'King David'. Mrs Cameron greatly admired Henry Taylor's plays, poems and handsome bearded appearance. He is here representing the biblical king and psalmist.

Julia Margaret Cameron's photographic techniques had an influence on contemporary painters. Dante Gabriel Rossetti refused to meet her, but his canvas Beata Beatrix *reveals that he had absorbed many of her ideas. The smudgy edges of the painting, and the semi-focused close-up of the face are directly drawn from Cameron's techniques.*

It was more than a mere piece of convention which you could ignore or follow at your choice. It was written into the law of the land. The first Contagious Diseases Act was passed in 1864, with the aim of reducing the spread of sexually-transmitted disease in the armed forces. In effect it meant the establishment of state brothels for the naval and the military, but it also involved a gross violation of the civil rights not only of prostitutes but, by implication, of every woman in Britain. In 1866 and 1868, with the passing of further Contagious Diseases Acts, its original powers were extended far beyond the confines of the military encampments.

The Contagious Diseases Acts effectively abolished Habeas Corpus in Great Britain. By the provisions of these Acts, special police were empowered to arrest any woman, compel her to submit to an examination for venereal disease, and require that she should present herself to the Justice of the Peace. Her guilt was presumed unless she could prove herself innocent. No witnesses were required, and no evidence on the part of the officer making the arrest. If the woman protested or refused to co-operate with the law, she was liable to a period of penal servitude. If she did submit to the examination, it was in the eyes of many who investigated the matter little better than an 'instrumental rape'. The woman was forced into a straitjacket to prevent her from struggling. Her legs were then forced apart by metal clamps. One girl interviewed by Josephine Butler after such a 'medical' examination had rolled off the couch with a ruptured hymen. She turned out, as it happened, to be a virgin. The police paid her a few shillings' hush money, but she went at once to Josephine Butler. Another woman, walking innocently along one evening with her daughter, was arrested and charged by the special police with being a 'common prostitute'. This was how all women were now defined by the law of England unless they could prove to the contrary. This particular woman committed suicide rather than submit to the horrors of the examination.

These are matters about which many people today still find it difficult to speak in public without embarrassment. How much truer that was in Josephine Butler's day! She was not a strident, foul-mouthed woman who found it easy to mention matters normally only spoken about in the doctor's consulting-rooms (if there). But a vitally important matter of human liberty was at stake, one which would never get reformed unless someone were brave enough to challenge it. To do so would be to risk the charge of prurience and impropriety. When it was further discovered

Josephine Butler's water colours suggest an almost obsessive preoccupation with female suffering. This sketch of a young girl in chains, gazing wistfully towards her prison window, may not have been a conscious self-portrait, but it bears a strong resemblance to the artist. For much of her life, Josephine took opium, in the form of laudanum, as a relief for pain. This possibly inspired some of her stranger artistic fantasies.

that Josephine was protecting 'immoral women', she would obviously be charged with wickedness. She was an upper middle-class woman with a position to maintain. This meant little to her personally, but the reputation of her sons and her husband meant everything. George was a clergyman and a schoolmaster with responsibility for the care of the young. His entire career was put in jeopardy by the very idea of having a wife whom 'he could not control', who was prepared to peer so mercilessly beneath the respectable surface of Victorian life and reveal the cess-pit which lurked there. What should she do?

She put the problem to George. She knew that he was sympathetic to the cause, but was he prepared to risk the obloquy and anger which the campaign against the Contagious Diseases Acts would provoke? Josephine now knew more, from first-hand experience, than any other educated woman in England about the plight of urban prostitutes. The feminist campaigner, Elizabeth Wolstenholme, wrote to her in 1869 and asked her to lead the protest against the Acts. George had no doubt about where her duty lay. Quoting Saul's words to young David when he went forth to fight the giant Goliath, he told her, 'Go, and the Lord go with you.'

The Goliath whom Josephine had to fight represented almost the entire male-dominated British Establishment. This was not just a case (brave as that would have been) of standing up to the pimps, and the brothel-keepers and the madams. She had, for one thing, the medical profession against her. Since thousands of young men in the armed forces were suffering from venereal disease, it was felt that anything justified the halt of it: and for the doctors of Victorian England, who saw things from so one-sided a point of view, that merely meant controlling the prostitutes. 'It is only insofar as a woman exercises trade which is physically dangerous to the community that Government has any right to interfere,' conceded *The Lancet* of 27 November 1869. But the notion that the Contagious Diseases Acts in effect deprived women in garrison towns of civil liberty, whether they were prostitutes or not, and that it had no effect on combating the spread of sexually transmitted diseases did not seem to have occurred to the author of the article. Most doctors reacted as Dr Preston of Plymouth, who wrote on 24 June 1870

> I will pass over Mrs. Josephine Butler's address in public before men . . . because I believe that a very large majority of our sex . . . can only characterise it as the height of indecency to say the least. But it is my opinion that women are ignorant of the subject . . . but not Mrs. Josephine Butler and Company – they know nothing about it . . . Certainly if such women as Mrs. Butler continue to go about addressing public meetings – they may ultimately do so but at present I venture to say that they are ignorant and long may they remain so. No men, whomever they may be, admire women who openly show that they know as much on disgusting subjects as they do themselves, much less so those who are so indelicate as to discuss them in public.

But Josephine risked the extreme ignominy of making speeches about venereal disease because she knew she was right, even though none of the medical profession would support her. The very few women doctors who were struggling into existence at this period were frightened of their positions vis-à-vis their male colleagues. Only Dr Elizabeth Blackwell was brave enough to oppose the Contagious Diseases Acts from the first, followed eventually by Dr Elizabeth Garrett Anderson, who had been the first woman to qualify as a doctor in 1865.

Josephine Butler was eventually to collect some influential male allies – such figures as F. D. Maurice, James Stansfeld (a Liberal MP) and the philosopher John Stuart Mill. But many whom one might have thought would be sympathetic were not. None of her old friends at Oxford would lend their support. Benjamin Jowett, for example, the liberal-minded Master of Balliol, was priggish enough to say, 'Mrs Butler takes an interest in a class of sinners whom she had better left to themselves.' More surprisingly, Gladstone, himself so keen on rescuing prostitutes, was deeply unsympathetic to her cause. He considered it unfortunate that she should try to make it a political issue.

But of course it was a political issue, since her aim was to repeal a series of Acts of Parliament. This could only be done by scaring the Liberal government of the day into some kind of action. Since women did not have the vote, Mrs Butler had to appeal to men – and that meant extensive travelling around the country to speak to frequently bawdy or hostile audiences, as well as a ceaseless stream of written campaign material, largely gathered up in her news-sheet, *The Shield*. In her speeches and articles, she was punctilious in her collection of evidence; and what she began to reveal, with hideous clarity, was not the narrowly important question of the Contagious Diseases Acts and their injustice, but also the much wider question of double standards in Victorian society. This, beyond question, is why her campaigns, from the very first, got so many people on the raw. One of her friends who was sent to prison for soliciting on 2 March 1870 told *The Shield*, 'It did seem hard, ma'am, that the Magistrate on the bench who gave the casting vote for my imprisonment had paid me several shillings, a day or two before, in the street, to go with him.'

This double standard was extremely widespread. In a society where men were supposed to delay marriage until they could afford to maintain a household, and in a society where marital breakdown was not relieved by divorce, prostitutes provided an essential role in keeping the whole facade of 'Victorian values' unscathed. It was, moreover, the prostitute who supposedly made sure that the promiscuous middle-class man did not infect women of his own class – the sort of woman he might dance with, play croquet or bridge with, or escort into dinner.

In an age when there was no cure for the rampant disease of syphilis, it is easy to see how these standards grew up. It is equally easy to see how Josephine Butler's attempt to expose the standards was seen, and

Since no woman could stand for parliament, Josephine and her friends were compelled to make public speeches in their campaign for the repeal of the Contagious Diseases Acts. It required great courage to face audiences of hostile men.

intended, as a political act. She saw the Contagious Diseases Acts as 'a tyranny of the upper classes against the lower classes'. And that is why she got the Liberal Party on the run.

The first major political triumph was the Colchester by-election of 1870 when the Liberal candidate, Sir Henry Storks, was challenged by a Tory candidate who promised to work for the abolition of the Acts. When it was discovered that Josephine Butler was in Colchester, her hotel was stormed, and the mob threatened to set it alight. The manager asked her to leave, very late at night, and she was obliged to slink out with a prostitute friend who found her refuge in a grocer's shop. They hid behind piles of soap and candles until the hooligans dispersed. It was not the first time she was to be confronted with dangerous physical violence. Storks, whom she nicknamed the Bird, had been promised the War Office by Gladstone if he won the by-election. He was an experienced military administrator, and a hero of the Crimea, where he had helped Florence Nightingale set up the military hospitals at Scutari. A few days later, when the votes had been counted at Colchester, Josephine received a telegram which read simply, 'Bird Shot Dead'. Storks had been defeated by 400 votes – a big majority in those days before universal suffrage. It was a sign that the message was getting across, and that the essential injustice of the Contagious Diseases Acts was becoming intolerable to the electorate.

Realising that something had to be done, but wanting to put off the evil day, the Government set up a Royal Commission. Josephine Butler was summoned before it in 1871 to face a panel which was made up of bishops, doctors, naval and military experts and MPs. Every member of this Commission was declaredly opposed to repealing the Acts and when she stood before them, she was made conscious of their hostility. 'It was distressing to me owing to the hard, harsh view which some of these men take of poor women, and the lives of the poor generally. . . . I felt very weak and lonely. But there was One who stood by me.'

Josephine meant this sincerely and literally. She and her husband were both of the view that Christ's morality was simple, and obligatory on all Christians. They were impatient of the doctrinal wranglings which so interested Roman Catholics and High Churchmen. 'I am sure', George once wrote, 'Mary who sat at the feet of Jesus would have been puzzled by the reading over to her of the Athanasian Creed and the injunction to accept it all at the peril of the loss of her soul; but she understood what Jesus meant when He said "One thing is needful".' Josephine felt that 'those who profess the religion of Jesus must bring into public life and into the legislature the stern practical social, real side of the Gospel'. And this in turn brought the realisation that 'economics lie at the root of practical morality'.

It is for these underlying reasons that the Campaign for the Repeal of the Contagious Diseases Acts became so important. Since the Liberal Party and the Church – the two institutions to whom she would have expected to look for support – were so unhelpful, Josephine Butler looked around for any support, however heterogeneous. The friends she made in this way cut across the barriers of class, religious denomination and even of obvious political alignment. But they were united in discovering a great many things about society which had nothing specifically to do with venereal disease. They were discovering how undesirable they found an all-male-dominated, upper or upper-middle class dominated society. They wanted something different. And some of the different things they achieved – including women's suffrage itself – came about as an indirect but very certain consequence of Josephine Butler's campaigns.

The first Royal Commission on the Contagious Diseases Acts reinforced the double standard. It even published, in its report, that odious sentence which, for Josephine Butler, summed up the whole of what she was fighting against: 'There is no comparison to be made between prostitutes and the men who consort with them. With the one sex the offence is committed as a matter of gain, with the other it is an irregular indulgence of a natural impulse.'

By the time the next Royal Commission was set up, things were very different. Mrs Butler had made her point. By then, six of the Commission, out of fifteen, were abolitionists, and they were addressed by John Stuart Mill in forthright terms: 'I do not consider [such legislation] justifiable in principle, because it appears to me to be opposed to one of the greatest principles of legislation, the security of personal liberty. It appears to me that legislation of this sort takes away that security almost entirely from a particular class of woman intentionally but incidentally and unintentionally, one may say, for all women whatever.' At last the point had been made.

When Parliament opened on 28 February 1883, Gladstone was still evasive about the question of abolition, fearing it would have dangerous political repercussions. Nevertheless, the thing had by then gathered its own momentum. The end came on 20 April during a late-night sitting of the House of Commons. A Women's Prayer Meeting was held in the nearby Westminster Palace Hotel, and outside on the pavement middle-class women knelt with poor prostitutes to pray. Josephine Butler herself was inside the House in the Ladies' Gallery as James Stansfeld moved against compulsory examination. The measure was carried – and the Acts in effect neutered – by 182 votes to 110. It was at half past one in the morning that the news came. Josephine was on the terrace of the House of Commons by that time. 'The fog had cleared away and it was very calm under the starlit sky. All the bustle of the city was stilled, and the only sound was that of the dark water lapping against the buttresses of the broad stone terrace . . . it almost seemed like a dream.'

When the Repeal Bill was finally passed through the House on 18 March 1886 the full iniquity and absurdity of the Acts was shown. Before they had been brought in, in 1865, 260 in every thousand serving in the armed forces suffered from venereal disease. When the Acts were repealed, the figure was still 260 per thousand.

'Say unto Jerusalem', exclaimed Josephine Butler, quoting the Prophet, 'that her warfare is accomplished.'

But as she was almost immediately to recognise, this was only partly the case. For, in the course of her campaigns against the abuse of a particular piece of parliamentary nonsense, she had uncovered the much deeper and wider problem of prostitution itself. And this problem would not go away overnight.

From the moment she took up the challenge to fight the Contagious Diseases Acts, Josephine Butler had travelled all over Europe, collecting information about prostitution. In Paris, she saw fully-fledged state-sponsored brothels in operation and they were not very impressive. She has left a very remarkable account of her meeting with C. R. Lecours, the chief of the dreaded Agents de Moeurs. Lecours, well-groomed, harsh and cynical, had, as she pointed out, 'the keys of heaven and hell, the power of life and death for the women of Paris'. Josephine thought that 'Service de Débauche' would have been a better name for his office than 'Service des Moeurs'. The police who were supposed to be the custodians of public morals were, as she quickly found out, doing a roaring trade themselves out of the prostitutes whom they were supposed to discourage. It was Lecours's view, as he explained to her, that vice would go on increasing and increasing, and this was very largely the fault of the prostitutes.

> I interrupted rather abruptly by reminding him that, in this crime – prostitution – which he was denouncing, there were two parties implicated. I asked him if he had been so long at the Préfecture without it occurring to him that the men for whose health he labours – and for whom he enslaves women – are guilty.

At this point, Lecours became very angry and began to pace the room, waving his arms about like a monkey and talking about the pardonable 'carelessness' of, perhaps, a young man who had taken a little too much wine and did not know what he was doing.

Josephine Butler returned to the attack.

> Is it not the case that the woman is poor, for I know that in Paris work is scarcely to be found just now; or else she is a slave in one of your permitted houses and is sent out by her employers on what is their, other than her, business.

By the end of the interview, Lecours was shrugging, and assuring her that he was himself a highly religious man. Yves Guyot, the radical Minister of Public Works, remembered Josephine Butler addressing the Municipal Court of Paris.

> Elegant, refined, of an upright figure; she gave quite an original piquancy. She expressed herself with great simplicity; but by this simplicity she reached irresistible rhetorical effects. . . . Everybody felt moved in his inmost fibres by this pathetic address which owed nothing to rhetoric, but was the spontaneous outcome of an intense feeling of justice and humanity.

She was not a fanatical man-hater, and as the depths of the problem unfolded, she even began to understand its pathos from the point of view of the clients. At a very early stage, as she had told the Royal Commission, she had braved the licensed brothels of Chatham:

> I gathered the [young men] round me, or rather they gathered themselves round me, and I spoke to them as a mother to sons. I did not speak to them altogether in a directly religious manner. I spoke to them lovingly. 'Why do you come here?' I asked one of them.
>
> 'The soldiers all come here, and these are Government women.'
>
> 'Here is a young lad' – he was pushed forward – 'he has just joined. Couldn't you get him out of this? He doesn't understand what it means. It is a shame they don't give us some proper entertainment.'

In the case of these young men, as Josephine Butler saw, there was perhaps 'a natural impulse' – in that odious phrase – finding its 'irregular indulgence'. But as she was to discover, many clients of brothels are on the look-out for something quite other than a 'natural impulse'.

> I doubt if many people have come face to face, as I have sometimes, with 'demoniacal possession' in the form of raging impurity and unconquerable lust. I have, in a sense, looked into hell. I have been filled with a deep pity for many men so possessed, rather than horror, for I felt they were themselves victims dominated and tormented like the Demoniac of Gadara. The faces of some haunt me. You will not misunderstand me when I assure you I have known men of gentle and lovable natures, 'true gentlemen' generous and ever ready to do a kind act and who at the same time have been 'possessed' by a spirit of impurity and lust so that they were driven to deep despair.

Such gentlemen were often in the grip of sado-masochistic or paedophile desires which it was quite unthinkable to gratify within the confines of respectable domestic life. As Josephine Butler discovered, England was the supplier of a vast European chain of vice, not least because it had such a low age of consent (twelve), and so persistent a habit of turning a blind eye to these problems, that it was easy to carry on the abduction, and even the sale of children, either for use in London or for export to the continent.

With the help of Alfred Dyer, a Quaker publisher, Josephine Butler exposed the trade in children which was being carried on by the Belgian Chief of Special Police, who was imprisoned in 1883. But the British Government would take no sort of action. Her son George, enlisted to help the Cause, found brothels in London where there were padded cells in which children could be tortured without their screams being heard. A certain Harley Street physician – and he was not unique – made a brisk trade in procuring under-age children for his patients. In one season, he sold over one hundred.

It was in the face of these horrors, and of an indifferent Government determined not to disturb things by mentioning the unmentionable, that Josephine Butler decided to make the matter public. She enlisted the help of W. T. Stead, the redoubtable editor of the *Pall Mall Gazette*.

Living in her house was a woman called Rebecca Jarrett, a reformed alcoholic ex-brothel keeper to whom Josephine had been introduced by the Salvation Army. She knew the ropes for procuring a child, and undertook to purchase, for the sum of £5, a girl called Eliza Armstrong, and hand her over to Stead. The girl was then smuggled out of England by a member of the Salvation Army to Paris.

As a result of this escapade, Stead was able to demonstrate, quite unequivocally, that it was possible to buy a child in London, for the purposes of sexual abuse, in the year of grace 1885. He published the story, under the title of 'The Maiden Tribute of Modern Babylon' in a sensational series of articles for the *Pall Mall Gazette* and was able to follow up his account with dozens and dozens of further examples of what was going on. 'My object throughout has been to indicate crimes virtually encouraged by the law . . .'. He then introduced his readers to 'a tiny mite' called Annie Bryant, aged five, who was in the care of the Society for the Protection of Children.

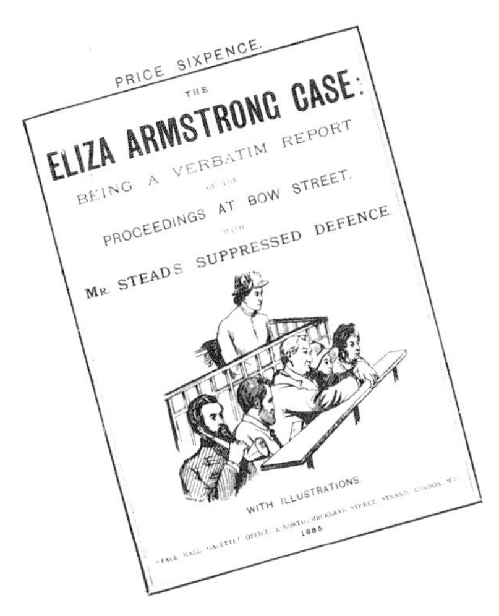

W. T. Stead, the editor of the Pall Mall Gazette. *Stead's articles, 'The Maiden Tribute of Modern Babylon' provoked an outcry, and Stead himself was imprisoned for procuring a child. But the articles led to the eventual raising of the age of consent.*

> She is now just five years old. Yet that baby girl has been the victim of rape. She was enticed together with a companion into a house in the New Cut on May 28 and forcibly outraged, first by a young man named William Hemmings and then by a fellow-lodger. The offence was committed and the poor little child had internal injuries from which it is doubtful whether she will ever entirely recover. The scoundrel is now doing two years' penal servitude, but his accomplice escaped. A penny cake was the lure which enticed the baby to her ruin.

Stead reckoned that there were '10,000 little girls living in sin in Christian England'. It was difficult enough to bring to justice those who abused children under the age of twelve, the legal age of consent. Once over that age, the child was more or less outside the protection of the law. Stead recounted the story of a man who had been sentenced for indecent assault on a child just under the age of twelve. In the course of the trial it became clear that he had 'violated more than a dozen children just over thirteen whom he had enticed into backyards by promises of sweetmeats, but though they did not know what he was doing until they felt the pain, they were over age and so he escaped scot free.'

Many people reacted with stark incredulity to Stead's revelations.

George Bernard Shaw, for example, who knew a certain amount about Victorian low life, dismissed 'The Maiden Tribute of Modern Babylon' article as a put-up job. But those who had dared to explore the underworld knew that Stead was telling the truth.

The *Pall Mall Gazette* articles caused an uproar. It was clearly outrageous that the law of England allowed such abuses. The articles were published at a time when the Government's popularity was shaky as a result of its Irish policy, and it could not afford the risk of appearing to resist reform. As the articles appeared, it was rumoured that Stead knew the names of those in the House of Commons and the House of Lords whose tastes lay in the direction of sadistic paedophilia, and who had themselves been clients at the child-brothels. How many names in 'high society' were connected with these dens of vice is unclear; but Gladstone obviously felt that it was at least possible that members of his Government – certainly Members of Parliament – were implicated. This fact alone speaks volumes.

So, reform came fast. Within the same year, Parliament had rushed through the Criminal Law Amendment Bill which introduced heavy penalties for those guilty of child abuse, and which raised the age of consent from twelve to sixteen.

To this day, journalists still take differing views of Stead. There are those who regard him as the arch humbug, the spiritual father of those newspaper editors who profess to be horrified by what is going on in the world, but merely reveal (or invent) horrors in order to increase the circulation of their paper. Journalists like Stead, it is argued, do not really improve the world, though they may be extremely self-righteous. For others, however, W. T. Stead is a hero, and his association with Josephine Butler and Rebecca Jarrett is seen as a great example (Zola's *J'Accuse* is the greatest of all) of the power of the written word in a democracy to Get Things Done. Probably neither of these extreme positions represents the full truth and Stead was, like many successful journalists after him, a strange mixture of hypocrisy and sensationalism. Whatever the fact of the matter, he had committed a criminal offence by procuring Eliza Armstrong, and he was prosecuted and sentenced to three months imprisonment. Rebecca Jarrett, however, was the one who suffered most. She was regarded as having played the most disgraceful role in the whole affair, and was awarded two years' penal servitude. Josephine Butler felt remorse to the end of her life for the trouble in which she had landed Rebecca. Jarrett herself never felt resentment against her old friend, and always spoke of Josephine Butler with gratitude and affection.

At about this time, Canon Scott Holland, the leader of the Christian Social Union, had a glimpse of Josephine Butler of which he has left a memorable account:

> A face looked at me out of a hurrying hansom, which arrested and frightened me. It was framed on pure and beautiful lines but it was smitten and bitten into as by some East wind that blighted it into grey sadness. It had seen that which took all colour and joy out of it. . . . Shortly after, all European civilisation shook with the horror of Mr. Stead's disclosures . . . I knew I had seen Mrs. Butler in the thick of the terrible work she had undertaken for God. She was passing through her martyrdom. The splendid beauty of her face, so spiritual in its high and clear outlines, bore the mark of that death upon it to which she stood hourly and daily committed. There was no hell on earth into which she would not willingly travel if, by sacrificing herself she could reach a hand of help to those poor children whom nothing short of such a sacrifice could touch. The sorrow of it passed into her being. She had the look of the world's grim tragedies in her eyes. She had dared to take the measure of the black infamy of sin; and the terrible knowledge had left its cruel mark upon a soul of strange and singular purity.

Scott Holland, famous in his day as a great social reformer and thinker in his own right, is today best known for his hymn 'Judge eternal, throned in splendour'. Those who do not know the background to the work which he, and Josephine Butler, undertook, have perhaps smiled at the language of those verses, which seem so quaintly dated today:

Still the weary folk are pining
For the hour which brings release,
And the city's crowded clangour
Cries aloud for sin to cease. . . .

But the words reflect an absolutely hideous reality which few were brave enough to face.

Josephine Butler's career shows how few people it takes to change the world. The numbers who were prepared to commit themselves to her campaigns in their initial stages were tiny yet within twenty years she had managed to organise the repeal of the Contagious Diseases Acts and to raise the age of consent. But these legislative reforms, although they did a certain amount to reduce human misery, had also uncovered an amount of wretchedness and sheer evil which could not be reformed by any Act of Parliament. As a young girl from a prosperous and high-minded Northumberland family, it had seemed to Josephine that England could be changed by extending the suffrage and passing laws to produce a juster society. She had assumed that the level of decency which her own family took for granted was one on which everyone was agreed, even though certain individuals, perhaps through little fault of their own, were not yet in a position to pursue it. Her journeys through the dark places of English cities had uncovered something infinitely more troubling. Her career exemplified in the most painful way a dilemma which was one of general application for the decent, or cultivated, classes of Victorian England. Matthew Arnold, as a schools inspector, had not seen the horrors which confronted Josephine Butler, but he was aware that decent values were not, in fact, accepted, or even known about, by the generality of his fellow countrymen. It was the underlying point of his great essay on *Culture and Anarchy*. With terrifying speed, Great Britain had been transformed in a few generations from a small island with trading interests around the world into a great Empire, sustained and paid for by an ugly and entirely inequitable industrial expansion at home. The imperial glory which haunted Disraeli's fantasy-life, the commercial prosperity on which Gladstone's Liberalism depended, were both alike covering over the grisly reality of what Victorian cities in fact were – seething hotbeds of vice and disease, where the poor had no status whatsoever, and where one half of the population, however rich or comfortable they might be, had no status either in the eyes of politicians or the law. Josephine came to believe, and write:

> Womanhood is *solidaire*. We cannot successfully elevate the standard of public opinion in the matter of justice to woman, and of equality of all in its truest sense, if we are content that a practical, hideous, calculated, manufactured and legally maintained degradation of a portion of womanhood is allowed to go on . . . 'Remember them that are in bonds as being bound with them.'

The generations immediately following Josephine Butler looked to her as a heroine of social reform, and of the feminist movement; both of which she was. Blake's prophetic lines

The harlot's cry, from street to street
Shall weave old England's winding-sheet

was a favourite among early twentieth-century socialists. The capitalist with his big cigar and his top-hat was seen as the creator of the poor harlot in the gutter. But this was not really Josephine Butler's view, even though she had done so much to point out that no woman would *choose* to become a prostitute if all the economic circumstances of her life would allow her to do something different. If it was economic necessity which determined that there would always be fresh supplies of girls for the brothels, it was not economic necessity which meant that these girls would have clients. The existence of the clients said sadder and more far-reaching things about that much cherished Victorian institution, the family. The very values which Josephine had grown up to cherish – chastity and decency among them – were those which had helped to create a climate in which quite a number of men felt driven to some form of release outside the stifling confines of 'decency'.

What singles Josephine out from the run of do-gooders and reformers is that she had such a very human understanding of this. She was a tough fighter, but there was nothing self-righteous about her, nothing strident. And though she was intelligent enough to see the economic causes which led to the abuses she attacked, she had an ultimately larger scale of values by which to assess the whole problem. From the rescue of one

By the end of her life, Josephine Butler had become associated with almost every 'good cause' in the world, from the plight of the Russian Jews to that of Dreyfus in France. She travelled constantly.

Liverpool prostitute, she moved on to the larger question of prostitution, of the Contagious Diseases Acts, and of what that told her about the position of women in society; it led her to a form of socialism, and to a belief that men and women must change their whole attitude to each other. But, ultimately, her feeling that society was out of joint derived from the impulse which had led her, in the first place, to help those less fortunate than herself. That is, it derived from her Christian belief. There were many things, from the death of her own daughter onwards, which contributed to moods of doubt and despair, when she felt tempted to 'a hopeless and final denial of the Divine goodness'. But she repeatedly returned to the central notion that the world fell out of joint when it departed from the purposes of that Divine Goodness itself. At the heart of her practical concerns for others there remained something like a mystical love of God. Although it was refined by much suffering and frequent doubt, she could say,

> as we get nearer to God, all prayer resolves itself into *communion*. To the Holy of Holies, face to face with Jesus, all perplexities vanish. No difficulties can live. If I may dare to tell a little of what He has taught me, even in days and weeks of bodily suffering, it is this; that in prayer I am still, silent, waiting for the Spirit and the Spirit is granted, so that He prompts every request.

The overtly religious tenor of her utterances, and her motivation, is possibly one of the things which enabled Josephine Butler's name to sink into some obscurity after the early years of this century, when other feminist heroines could be less embarrassingly quoted. Among her co-religionists, on the other hand, her name became forgotten for subtly different reasons. Prostitutes and venereal diseases did not cease to be unmentionable subjects. In Anglican folk-memory, however, she has always occupied a much-loved place. When Liverpool Cathedral eventually came to be built, a window was erected there to her memory. Being represented in stained glass is only a short step away from being made a saint. The Church of England does not have a Process, like the Church of Rome, by which individuals can be declared saints, but when the new Prayer Book of 1980 was compiled, provision was made on 30 December, the anniversary of her death, to commemorate Josephine Butler. It is hard to know what she would have made of that.

Josephine Butler and her family. There is no doubt that Josephine Butler's family suffered as a consequence of her campaigns. Her son's career at the Bar became impossible, and George Butler was compelled to resign his post as Principal of Liverpool College. But they all supported her because they knew she was right.

In the notoriety of her campaigns, George Butler had felt it better to resign his position as Principal of Liverpool College. 'He was even more to me in later life than a wise and noble supporter in the work which may be called especially my own', she remembered. Much to the credit of the established Church – which is not always to be relied upon to be unstuffy – they made him a canon of Winchester Cathedral. It was in that beautiful and gentle southern town that the Butlers were able to live for the five years following the *Pall Mall Gazette* furore. When George died in 1890, Josephine's home life, and her happiness, had in some ways been brought to an end.

Though she had nursed George Butler throughout his last illness, she was unable, quite, to remain a purely domestic creature, even in this sorrow. Life had made her into a supporter of Causes, and the remaining sixteen years of life were to see her name being added to those who continued to fight for the reform of vice in European towns. She was pro-Dreyfus, pro-Boer, pro any good cause you could mention until her death. There is a haunting photograph of her taken in old age, a lace cap on her head, a pen in her hand: a strange link between the high-minded generation who grew up after the Great Reform Bill, and the 'modern' women she helped to liberate. She died, where she had begun, in Northumberland, at Wooler, on 30 December 1906.

Some years before, in the 1890s, the prolific painter G. F. Watts was engaged in his great 'Hall of Fame' series of portraits of those who had 'made the century'. He asked Josephine Butler to sit for him. He told her to 'look into Eternity, looking at something no one else sees'.

When he had finished the picture, she told him that she felt inclined to burst into tears.

> I will tell you why. I felt so sorry for her. Your power has brought up out of the depths of the past, the record of a conflict which no one but God knows of. It is written in the eyes and in the whole face. Your picture has brought back to me all that I suffered, and the sorrows through which the Angel of God's presence brought me out alive. I thank you that you have not made that poor woman look severe or bitter but only sad and purposeful.

The sorrow which the portrait reveals was, as she confided, a deeply personal one. But it was also one which her whole class, and all decent-minded people, must have felt when surveying Great Britain and the British Empire. She was one who 'made the century' in one sense. In another, she helped to unmake it, because it did not deserve to be, and in unmaking it, she and her like were to unpick the very fabric of so much that they took for granted and loved. 'God and one woman make a majority', she had once said. It turned out to be true, and it was a devastating combination.

JULIA MARGARET CAMERON
1815-79

Pioneer of portrait photography

This picture of Julia Margaret Cameron was taken by her fellow enthusiast Lewis Carroll. She and Carroll belonged to the small band of Victorian photographers who saw the imaginative and artistic potential of the camera, and stood apart from the dull conventional work of the studio photographers.

Two years after Queen Victoria came to the throne, some small boys were playing near the foot of Charles I's statue at the Trafalgar Square end of Whitehall. Since that statue was first erected close to the site of the king's execution, thousands of children must have sat there in the hot weather to play marbles, spun hoops there in the bracing mists of spring, or scooped up snowballs from its iron railings on winter days. There was nothing different about this particular group of boys – Artful Dodgers in top hats – but there was something exceptional about the moment they had chosen to play there. For on the other side of the road, in Trafalgar Square, a Frenchman called de St Croix was tinkering with a mysterious piece of equipment. He was in the process of taking a daguerreotype on that autumn day in 1839. Those boys must be among the first people in England to have been photographed.

The first real photograph may be said to have been produced in 1827 by Nicéphore Niepce, a Frenchman who pioneered methods of fixing the image obtained by a *camera obscura*, or pin-hole camera. His techniques were developed by his fellow-countryman L. J. M. Daguerre who, as a painter, had been famous since 1823 as the man responsible for the striking topographical views displayed in the Diorama he had opened on the edge of London's Regent's Park. It was de St Croix who first demonstrated Daguerre's discoveries to a London audience at 7 Piccadilly on 13 September 1839. The technique exploited the sensitivity of silver salts to light. A polished and silvered copper plate was made light-sensitive with iodine vapour and, after exposure in the camera, was developed with mercury vapour and fixed into a permanent image with hydrosulphite of soda.

Quacks and mountebanks, showing off their latest invention, were a familiar part of the London scene. Gullible invited audiences must often have been ushered into such rooms as these, to be entertained with the latest developments in phrenology (the science of telling character from the bumps and lumps on your skull), the mysteries of psychic research (as demonstrated by Mr Sludge the Medium), hydropathy (the application of water to cure almost every known ailment, as pioneered by the celebrated Dr Gully and practised by almost every intellectual of note in the mid-nineteenth century) or genetics (as expounded in 1859 by Charles Darwin). As the list suggests, some of these 'discoveries' have affected the way we look at the universe today. Most of them – like the huge majority of scientific 'discoveries' in history – have been shown to be purely bogus.

The discovery of photography, however, changed the world in a manner with which, even now – 150 years later – it is not always easy to come to terms. Questions about the nature of time, and of reality itself, are raised by the figure of M. de St Croix, standing in Trafalgar Square with his little box. Until then, each moment had been essentially evanescent. The only media in which a moment could be 'frozen' or the past be 'captured' were manifestly interpretative – through the written word, or the painted or sculpted image. It was taken for granted that the here and now – the particular minute or two in the gardens of Westminster Abbey, for example, it took for a man in a peaked cap and a frock coat to walk along with a woman who could be Dora Copperfield – would pass by for ever. Men and women no more thought of holding such moments than a butterfly would so think. But there it remains for us, in a famous calotype of around 1845, taken by William Fox Talbot.

The arrival of the camera had an almost immediate effect on the other arts, which by the end of the century had diverged into many fascinating reactions and counter-reactions. It more or less goes without saying that the history of painting after 1840 is haunted by the knowledge that pencils, brushes and crayons are no longer the only means of representing reality on paper. The extreme, detailed 'realism' of Dominique Ingres, or of John Everett Millais or William Holman Hunt, owed something to the existence of M. Daguerre's box. So, obviously, did the opposite development in painting, the discovery that a painter was now liberated from being a mere recorder and could become an interpreter of the world, as all great artists always had been. Impressionists could look at light, not as the camera artificially freezes it, but as the eye sees and feels it. Post-Impressionists could convey, as poets can, what landscape feels like, not just what it looks like. Liberated from representationalism, painters and sculptors could begin to experiment, as Cézanne did in his later canvases, with form which is of interest for its own sake. Cubism is only a step away in one direction, and non-representational painting in another. Without the camera there would be no Picasso, no Braque, no Mondrian, no Miró.

Literature, either consciously or subliminally, is influenced by the discovery of the camera, too. The neo-realism of such childhood-recreations as *Jane Eyre* and *David Copperfield* provides in prose what a family photograph-album was soon to give every middle-class household in England: an exact representation of the self at its most ephemeral period. It was, however, abroad that these tendencies, hinted at by English novelists, were taken to their greatest extreme. Proust is the most obvious example, his great *roman fleuve* attempting to provide a series of endless inner photographs, reclaiming something which, until that point of history, could never, by definition, be reclaimed: time itself.

All this has had a disconcerting effect upon our idea of reality. The audience at 7 Piccadilly on 13 September 1839 were not being tricked. M. de St Croix really had, by employing the ingenious methods of M. Daguerre, produced photographic images of the streets of London. In another sense, however, it could be said that he was unfolding the most fraudulent of all illusions; no greater untruth was ever uttered than the saying that the camera cannot lie.

The nineteenth century, so unlike other eras in history in so many ways, apparently distinguishes itself yet again by being the first which we can apparently *see*. Our vision of the Napoleonic Wars, and of all the previous European conflicts, we believe, would be quite different if, as during the Crimean campaigns, photographers had been present on both sides. If it is arguable that Samuel Johnson, Voltaire or Frederick the Great, familiar to us through the portrait artist's eye, would be discernibly different historical figures if we knew them through photographs, how much more would Mary Queen of Scots, Shakespeare, Francis of Assisi be. Photography increases our sense of life's vividness, but it also diminishes. Scientific perusal of the Turin Shroud prompts the apparently profane, actually religious, question, whether the doctrine of the Incarnation itself could ever have been formulated had the Romans invented photography, and had the *Ecce Homo* been preserved on daguerreotype.

The invention of photography advanced the severance which had already begun between what was 'imaginary' and what was 'real' in the human mind. For many, it confirmed the notion that there was such a thing as 'objective' or 'scientific' reality. Photography became a stan-

The nineteenth century is the first epoch in history which we can see.
This photograph by Roger Fenton shows Lt Gen Sir George Brown and his staff in the Crimea in 1856. It is interesting to reflect on how differently we would regard earlier wars if we had photographs of Marathon, Agincourt or Waterloo.

dard, a criterion of objectivity – something which political propagandists in the early decades of the twentieth century were not slow to exploit. It is surely no accident that most of the footage of film which documents the Russian revolution was not newsreel but posed and invented stuff by Eisenstein. Those crowds sweeping into the Winter Palace, for example, which for many of us are the image of the October Revolution of 1917, were filmed in the 1920s. The reality of the coup was undramatic; no crowds, few shots fired. The film-makers and photographers needed to create something as vivid as, say, Carlyle's *French Revolution.*

The history of photography-as-history is full of such cases: either of photography being used fraudulently to convey reality, or, much more disconcertingly, of people assuming it to be real merely because it is on film, and regardless of the photographer's motive. Photography, as most of the great early examples of men like William Fox Talbot show, is an extremely unrealistic medium. True, on an obvious level, it can record the shapes of buildings, the clothes people wore, the existence or non-existence of a particular landmark (though these can be easily altered in the darkroom). But, in various quite important degrees, common sense shows that photography, unlike painting, can never be completely true. For one thing – to limit our discussion to still photography – the world never has been still. Most Victorian photographs had to be taken on a long exposure, so that the faces of the sitters had to be frozen for anything up to ten minutes or quarter of an hour. No wonder one gets the impression that they were a gloomy crowd (besides, saying 'cheese' was an embarrassment in an age when dentistry simply meant drawing rotten teeth, and when most people aged forty were by modern standards almost toothless). Fox Talbot tells us a certain amount about the look of Victorian England. But the painter William Frith with his crowded *Derby Day* or seaside scenes tells us much more. His canvases move with life and colour.

Eyes, as opposed to cameras, interpret, quickly and changeably, the images which are played upon the retina. No eye has ever seen the same image for ten minutes without movement, which is why slowly-exposed photographs always have an air of unreality. By contrast, no eye has ever frozen one single image from a split second, which is why modern fast-exposed photographs, however vivid or beautiful, also have an air of unreality.

He thought he saw an Elephant,
That practised on a fife:
He looked again, and found it was
A letter from his wife.
"At length I realise", he said,
"The bitterness of life".

Such moments of 'double-take', to which eye and brain constantly readjust in the course of an ordinary day, are impossible for the camera. Filmed from the wrong or right angle, the letter from the wife always looks like an elephant. It was no accident that such an *aperçu* should have come to one of the most distinguished amateur photographers of the Victorian age, Lewis Carroll.

Carroll's photography of children sets up an awkward set of responses in the modern observer. Rather than enjoy the haunting beauty of his frames, we are more inclined to pose as guardians of Carroll's morals, or at least of his *psyche.* A dangerous game, whether you play it with the dead or with the living.

Then they joined and all abused it –
Unrestrainedly abused it –
As 'the worst and ugliest picture
That could possibly be taken . . .'.

Carroll belonged to that early generation of photographers who saw at once that, whether you use a pencil, a brush or a camera, it requires human imagination to make pictures. While the majority of 'studio photographers' in the nineteenth century set out to make their living (quite understandably) from dull posed family groups standing by potted plants, or from cartes de visite, there was a handful of genuine artists who immediately saw the point of photography, its imaginative potential.

To emphasise the contrast between these artists and those who merely took photographs, you only have to turn the pages of some Victorian family album in which the compiler has forgotten to label the images. Nothing is deader than these faces. A silhouette or a sketch of their grandparents, executed in the days of the Regency, has a thousand times more animation! The slow exposure and the bad studio lighting so often contrived to deprive their features of character. This is oblivion indeed; for we really feel that these nameless images could be *anyone.* Far from emphasising the distinctiveness of their subjects, as a true portraitist or artist would do, these depressing productions – like the faces on passports or ID cards today – make one doubt the very existence of human individuality. An album of cats and dogs, if we are being candid, would contain more variety. It is in the light of these pictures, and not with thoughts of modern child pornography, that one should look at Lewis Carroll's photographs, or at those of an even better photographer, Julia Margaret Cameron.

Julia Margaret Cameron's interest in photography was first aroused at the age of twenty when, recuperating from an illness at the Cape of Good Hope, she met the astronomer Sir John Herschel, one of the pioneers of photographic processing. It is to Herschel that we owe the first use of the terms 'negative' and 'positive' as applied to this process, 'snap-shot', and the word 'photography' itself. Mrs Cameron did not, however, take her first picture until she was forty-eight years old. Before that momentous day on the Isle of Wight in January 1864, there stretched 'a life crowded with incident'.

She was born Julia Margaret Pattle, in Calcutta on 11 June 1815, a week before the Battle of Waterloo. It was a big family – ten children, of whom Julia Margaret was the fourth. Her father was a senior official of the East India Company, and the saying out there – originating with Lord Dalhousie – was that society divided into 'men, women and Pattles': not that there was anything androgynous or hermaphroditic about the Pattle girls, but – like many large families – they to a certain extent devised their own patterns of behaviour, often at variance with those of the rest of the world. 'Pattledom' became a term synonymous with oddity. They were a family about whom tall stories were told, and they all appear to have been happy to live up to their reputation as originals.

Julia Margaret Cameron came from a famously peculiar Anglo-Indian family called Pattle. She was unfortunate in being the only Pattle sister who was not beautiful.

When James Pattle died in 1845, his body was preserved in alcohol and put on a ship bound for England. He had expressed a wish for interment in his native soil. Mrs Pattle (the daughter of a man who had been one of Marie-Antoinette's courtiers, the Chevalier de l'Etang) accompanied her husband on his last journey. Rounding the Cape of Good Hope, the ship ran into a storm. The cask of alcohol containing Mr Pattle rolled over and burst, and his widow, having heard the explosion, went next door to see what was happening. The sight of her husband's pickled corpse bursting from its cask is said to have driven her mad. Two months later she too died, 'Raving . . . the shock sent her off her head.'

Her daughter Julia Margaret – never far, it would seem, from the mental condition in which her mother died – had married Charles Hay Cameron when she was twenty-two and he an extremely elderly forty-three. He had a distinguished career in India, rising to become President of the Calcutta Council of Education. His chief passions were Ceylon, where he owned considerable estates, Homer, whose poetry he liked to read aloud, and liberalism. He was a follower of the great utilitarian philosopher, Jeremy Bentham, and a keen supporter of good causes. Despite her Pattle eccentricities, Julia Margaret could contrive great feats of organisation. We have an early glimpse of her energies when news reached Calcutta of the Irish potato famine in 1845. She managed to raise £14,000 for the relief-fund; and she ran the campaign more or less single-handed.

By the time Charles Cameron was fifty-four, he felt ready for retirement and the family (by then there were six children) decided to return to England. Since they had no idea where, eventually, they would settle, the journey assumed an almost Abrahamic feeling of caravan as they loaded on board ship their children, together with trunkfuls of belongings – fabrics, rugs, carvings, spices, curry-powder and coffins (mindful of her parent's unhappy experience at the Cape, Mrs Cameron never went anywhere without taking something to be buried in).

She was a short woman, and unlike her sisters, rather plain. Her olive complexion and dark hair were a throwback to her French ancestors. A pronounced mole or wart dominated the left-hand side of her face just next to the nostril, and her eyes were too small for her brow. Undeterred by these aesthetic disadvantages, she enlivened her appearance with a quantity of brightly coloured shawls, mantillas, and festoons of velvet drapery. Since her husband had shoulder-length, prematurely white hair and a long beard which would have done credit (indeed, *was* to do credit) to Merlin or King Lear, the couple must have made a striking entrance into the drawing-rooms of London.

Arrived in England, Julia Margaret happily falling back into Pattledom, made contact with her sisters. She made the hottest bee-line for Sara, also married to an elderly Indian civil servant, called Thoby Prinsep, and the centre of an artistic coterie which included the Pre-Raphaelite painters John Everett Millais, Dante Gabriel Rossetti, Edward Burne-Jones, William Holman Hunt and G. F. Watts. He was to become Julia Margaret's favourite. Sara Prinsep inhabited a residence known as Little Holland House – rented from Lord Holland, who lived in Italy – and Watts, nicknamed Signor by Sara, who 'came for three days and remained thirty years' had a room there where he could paint, pontificate and entertain his friends. Sara and Julia Margaret were responsible some years later for urging Watts, when quite a greybeard, to marry the sixteen-year-old Ellen Terry. It is one of the more chilling little episodes in the history of 'Pattledom'. Ellen, though beginning to make her way on the stage, was emotionally immature, and even after her marriage liked to play, as one of the children, with Mrs Cameron's offspring. After a year, her unconsummated union to Signor was allowed to be deemed a failure.

As well as painters, Mrs Cameron liked to cultivate poets. The most famous of her poets was Alfred Tennyson: it was because he resided for a while at Putney that the Camerons found themselves taking up brief residence on Putney Heath. But a poet she admired with equal fervour was Henry Taylor. Nowadays, he is little more than a footnote in the history of literature, but in his day he was an enormously prolific playwright and epic poet. Like most of the men Mrs Cameron admired, Taylor had a beard which grew generously over his chest. Taylor and his wife soon learnt what it was to be the devotees of Julia Margaret's generosity. She liked to shower her friends with gifts, and Indian shawls, bracelets, ivory, inlaid portfolios poured on to the doorstep of their house at Mortlake, as did huge letters at least six pages long. On one occasion, Henry Taylor and his wife were sitting in a railway carriage, about to leave Waterloo Station, when a dishevelled lady rushed up at the last moment and flung a Persian rug through the window. Just as the train began to move, Taylor had the presence of mind to roll up the rug and plop it back on to the platform. Refusing her bounty was not always that easy. Mrs Taylor once returned an extravagantly valuable shawl which Mrs Cameron had forced upon her. Many months later, when visiting the hospital for incurables at Putney, she was surprised to see an invalid sofa inscribed as a gift from herself. Mrs Cameron had sold the spurned shawl and given the money to the hospital in Mrs Taylor's name.

All this sounds as though it has something about it of the intrusive; as though Julia Margaret was a tuft-hunter. That is to mistake the exuberance of her generosity altogether. It applied to everyone, not just the famous. Years later, when police in the Isle of Wight informed her that trippers and locals were picking the flowers in a rose hedge which ran along the road at the bottom of her garden, she replied that was why she had planted the hedge in the first place – so that people could pick the flowers.

It is with the Isle of Wight that we most associate Julia Margaret Cameron. After a peripatetic existence, and an attempt to make a life for themselves in Tunbridge Wells, the Camerons heard, in 1860, that the Tennysons had moved to Farringford House on Freshwater Bay at the western end of the island. That summer, Mrs Cameron visited Tennyson. (She was, incidentally, the only woman apart from his wife whom he ever called by her Christian name.) The cliffs, the sea air, the colour, the proximity of a great, heavily bearded poetic genius all determined Mrs Cameron. She immediately bought the two cottages adjoining Tennyson's extensive grounds, erected a tower between them, and named them, after her husband's estates in Ceylon, 'Dimbola'.

The Camerons arrived in Dimbola with their strange caravan of belongings and offspring. A new individual had been added to the throng, an Irish girl called Mary Ryan, one of the most noted beneficiaries of Julia Margaret's strange, swooping possessively exuberant heart. Mary had been a little vagrant girl until her mother, an Irish tinker, had been unguarded enough to knock at the door of Mrs Cameron's cottage in Putney and beg for alms. Mrs Cameron immediately decided that she was not prepared to give mere gifts of money to the pair. They must come and live with her. Unsurprisingly Mrs Ryan beat a hasty retreat, preferring the hardships of hedgerow and heath to the oppressive attentions of Mrs Cameron's good nature. Mary, however, stayed. Not quite a daughter and not quite a parlour-maid, she was kept in Mrs Cameron's household, her hour not yet come. Once adopted, Mary received no particular affection from Mrs Cameron. She

Farringford House, Freshwater Bay – the home of the poet laureate, Alfred Tennyson, and his family.

collected people rather as she collected objects, largely because their looks were appealing. For example, though she had almost no garden during the last years of her life in Ceylon, she employed a gardener there because she liked the shape of his back. Before the arrival of the camera in her life, Julia Margaret was a natural photographer, taking, we may believe, a myriad plates inside her highly distinctive brain.

Mary Ryan was destined to become one of her most remarkable photographic subjects, and with far-reaching consequences for herself.

Dimbola, the Camerons' house on the Isle of Wight, named after Charles Cameron's estates in Ceylon.

Her sad eyes, her full lips, her thick hair, made a lasting impression on one of the visitors to an early Cameron exhibition in London. He was a young Indian civil servant called Henry Cotton. With long hair and wild eyes, he made the journey to Dimbola and, having sent in his card, announced to Mrs Cameron, 'I have come to ask for the hand of your housemaid. I saw her at your exhibition and have all the time kept the bill she wrote out for me next to my heart.'

Objections from the Cotton family were eventually overruled. Only the vagrant Mrs Ryan protested at the unsuitability of the match. Under the benign eye of their patroness, the young people were married at Freshwater in 1867. Within eight years, he was Sir Henry Cotton with a salary of £2400. 'What is more important,' said Mrs Cameron, 'it was a marriage of bliss.' And how is this defined? 'With children worthy of being photographed, as their mother had been, for their beauty.'

Julia Margaret kept tirelessly in touch with her circle of friends and family. She never sent less than six telegrams a day. 'Let out all Henry's throats' was one wire she bafflingly dispatched to the wife of Henry Brotherton, the artist. 'I had a dream my Henry was suffocating,' she explained in a letter which followed, 'and in case my dream were to come true, I telegraphed Mary to make his throats (i.e. shirt collars) bigger.' Increasingly, Mr Cameron found it necessary to attend to his estates in Ceylon, leaving his wife behind on the Isle of Wight. Since their children were now all grown up, or on the verge of being so, she spent much of her time in the post office, sending them letters or telegrams. When two of her sons accompanied her husband on a prolonged visit to Ceylon, she would sit in the post office until the last minute before the departure of the Indian mail, scribbling her interminable missives to them. 'How much longer?' she would ask the clerk. 'Five minutes madam.' Scribble, scribble, scribble. 'How much longer now?' 'Two minutes, madam.' And so on, until the moment that the last mailbag was closed.

Three of her sons eventually settled in Ceylon, to grow coffee on their father's estate. Her daughter Julia, distressed by her mother's loneliness, was aware of how much she minded the boys' absence. 'The more it is prolonged,' Mrs Cameron admitted, 'the more the wound seems to widen.' It was at this point that Julia and her husband Charles Norman thought of a present to revive Julia Margaret's spirits. They gave her a large wooden camera (the plates it took were 9 in × 11 in) and all the necessary dark-room equipment. 'It may amuse you, Mother, to try to photograph during your solitude at Freshwater.'

The coalhouse at Dimbola was instantly transformed into a dark-room, and Mrs Cameron set to work. She made mistakes; but from the first, there was a distinctiveness about her way of looking at people which immediately communicated itself. At first, she concentrated on portraits. She paid 'Farmer Rice' half a crown to pose for her, and it was not a great success. Then in January 1864, she found a little girl called Annie who made a much better subject. 'I was in a transport of delight,' she tells us in *Annals of My Glass House*. 'I ran all over the house to search for gifts for the child. I felt as if she entirely had made the picture. I printed, toned, fixed and framed it, and presented it to her father that same day.'

The result was the picture inscribed, 'Annie – My First Success'. Much of the success, as even Mrs Cameron's gushing prose allows us to recognise, is her competent grasp of the photographic medium. The chemistry of the art was something she mastered at once, and she was to exploit it as she matured. But a mere knowledge of how to develop a picture would not have produced 'Annie'. We can see an understandable wariness in the sitter. Who would not have been wary of this lady, described by William Allingham, at about the same date, as 'shrieking' and wearing 'a funny red openwork shawl'. But beneath all the 'Pattledom' there was an extraordinarily sympathetic imagination. We feel we *know* Annie.

One of the reasons for Mrs Cameron's success as a portraitist is that she boldly employed the use of close-up. She refused to enlarge, using only her 11 in × 9 in plate negatives, going on in 1866 to 15 in × 12 in ones. Often she is so close to her subject that it is out of focus, causing the famous Cameron 'imperfect vision'. But there are no better Victorian portraits than hers. She seems to see into the souls of her subjects. And because of her irrepressible boldness, she managed to capture many of the most eminent Victorians of them all.

Anthony Trollope, the novelist, was an early *trouvail* – he came to the Isle of Wight in 1864. Here, Julia Margaret's 'imperfect vision' has made of the borders of the picture what Sir Joshua Reynolds would have made in one of his oil sketches. They are just a smudge, leading our eye directly into the features of the subject. In her portrait of Benjamin Jowett, the famous Master of Balliol College, taken the following year, she has brilliantly caught the contradictory qualities of that fascinating man. The neat Oxford MA hood has become a fuzz, and the translator of Plato comes before us, as is appropriate, in a kind of spiritual emanation. The face, however, has only half glimpsed the wisdom of the Greeks. The superciliousness of the mouth tells us of Jowett the snob, the social climber, the cultivator of the society of duchesses. The clever eyes are both amused at the absurdity of undergraduates, and meditating upon the mystery of things.

If her Jowett is on the edge of being a comic figure, her Carlyle is pure, unremitting tragedy. The historian of *The French Revolution*, Thomas Carlyle's increasingly desperate vision of the human condition was blackened by a crushingly depressive temperament, exacerbated by the fact that he had no gift for marriage. It was Tennyson's surprising malice which thanked God for making Carlyle marry Mrs Carlyle – thereby making two people unhappy instead of four. In 1866 Carlyle's clever and long-suffering wife Jane died, quite suddenly, of a stroke while in her carriage going round Hyde Park Corner. Mrs Cameron captures Carlyle a year after this melancholy event, his face full of the remorse which he felt for the rest of his days, for the deliberate withholding of affection from a woman to whom he owed so much. He used to go and stand on Hyde Park Corner to do penance for the sufferings he had caused her; and he made sure that his biographer, J. A. Froude, knew the whole miserable story, so that he should not be allowed to escape the judgement of posterity.

The melancholy of Lord Tennyson's face, by comparison, is much more showy. We notice here (particularly in the famous picture which he nicknamed 'The Dirty Monk') how Mrs Cameron liked not merely to take close-ups, but to see her subjects in an informal, intimate way. 'Although I bully you, I have a corner of worship for you in my heart,' she told Tennyson, who for some time had resisted posing before her camera.

The 'bullying' certainly took intrusive forms, even by her bossy standards. Since they were her immediate neighbours at Freshwater, the

Tennyson with his wife Emily and sons Lionel and Hallam in the garden of Farringford House on the Isle of Wight. It was here that Julia Margaret Cameron attempted to photograph Garibaldi and was impatiently dismissed as a beggar.

Tennysons bore the brunt of her generosity. 'She is more wonderful than ever in her wild beaming benevolence,' Tennyson said to his wife about Julia Margaret one night after dining with her. He was less sanguine when she decided that he should undergo a vaccination against small-pox. Tennyson locked himself in his tower at Farringford when he heard that Mrs Cameron was on her way with the doctor. She stood at the bottom of the staircase and shouted, 'You're a coward, Alfred, you're a coward!' until he opened the door and submitted to the needle. The vaccine was past its best, and Tennyson was confined to his bed until the effects had worn off.

On another occasion, long past the ordinary hour when the Tennysons normally dined, Mrs Cameron decided that they should eat a leg of mutton, and pushed her way through to the kitchens, insisting that their cook put the meat in the oven straight away. Again, Emily Tennyson was surprised to come downstairs one morning to find her hall heaped with rolls of fashionable new 'Elgin Marbles' wallpaper which Mrs Cameron – without consulting the Tennysons – had decided would be an appropriate adornment to the Poet Laureate's household. At least she did not hang the paper – unlike the occasion when she stayed with the Henry Taylors and covered their spare bedroom, furniture and walls, with floral transfers during the night.

As well as loving Tennyson as a dear neighbour, Mrs Cameron saw him as an admirable 'bait' to attract subjects for her portraiture. Many of the Poet Laureate's distinguished visitors were resistant to Mrs Cameron, but she was hard to deter. When she heard that Garibaldi, the great hero of the Italian Risorgimento who was being feted wherever he went in England, had arrived at Farringford, she rushed round to persuade him to be photographed. He was sitting in the garden, and was astonished by the advance of someone he took to be a gipsy woman, her hair awry, he hands blackened by silver nitrate, her shawls and mantillas stained and dishevelled. She threw herself at his feet. He dismissed her with a wave of the hand, assuming her to be a beggar.

Another visitor to Tennyson's house who refused to be photographed was Edward Lear, the nonsense poet and watercolourist, no great lover of women. 'Mrs Cameron came in only once – with feminine perception, not delighting in your humble servant,' Lear told Holman Hunt. Knowing Lear's fondness for music, and hearing that the Tennysons' piano was out of tune, Mrs Cameron hit upon what seemed like the per-

Alfred Tennyson, c. 1867. Tennyson was Mrs Cameron's neighbour at Freshwater on the Isle of Wight and he was her most celebrated subject.

fect way of winning his heart – or at least his willingness to sit for his portrait. She ordered eight men to carry her own piano around to Farringford and offer it to Lear for his use while staying on the Isle of Wight. He still refused to have his picture taken.

Nevertheless, she managed in the end to have an impressive gallery of characters in her portrait collection – as well as Tennyson, G. F. Watts, Henry Taylor, Thomas Carlyle, Benjamin Jowett, and Anthony Trollope there were the poets Robert Browning and Henry Longfellow, Charles Darwin, William Holman Hunt, the sculptor Thomas Woolner, and the historian William Lecky, not to speak of Ellen Terry and Alice Liddell of *Through the Looking Glass*. This was surely an impressive collection of 'Eminent Victorians'.

One of the reasons why her portraits are so lively is undoubtedly the simple fact that her sitters were with *her*. You could not be in Mrs Cameron's company and not react to it in some manner or another, even if your reaction – as so many of her sitters' reactions seem to have been – was absolute incredulity that anyone could be quite so extraordinary as the lady behind the camera. She did not need to tell *them* to watch the birdie, as she swirled to and fro between their chair and the camera, ruffling their hair with silver-nitrate-stained fingers, and talking poetic nonsense in a loud voice. Sometimes, as is the case with Trollope I think, her sitters registered intolerance, the sense that being 'a character' was one thing, but that Mrs Cameron went too far. Sometimes, as in Jowett's case, there is barely concealed amusement in the features. What they perhaps did not realise until they saw the results (and not always even then) was that she had an extremely intelligent and penetrating eye. There is no doubt, as we survey her portfolio, that she was in control of her subjects and, when we consider the eminence of the subjects, that is in itself sufficiently remarkable. She was in control because she did not consider herself her subjects' superior. Not in the least. She was a genuinely humble person. When she went down on her knees to Garibaldi, there was nothing false in the gesture. She believed herself to be in the presence of a giant, but when he dismissed her, she was defiant. Waving her filthy hands, she remonstrated, 'This is Art, not Dirt.'

As well as being a portraitist, however, she was also an imaginative artist, a Pre-Raphaelite with a camera. Roger Fry, the Bloomsbury artist and art critic, believed that she was in fact the greatest artist of the Victorian age. 'Mrs Cameron's photographs already bid fair to outlive most of the works of the artists who were her contemporaries,' he wrote. 'One day we may hope that the National Portrait Gallery may be deprived of so large a part of its grant that it will turn to fostering the art of photography.' Fry was being deliberately provocative, and he praised Mrs Cameron partly because he admired her but chiefly as a way of expressing contempt for the productions of Millais, Holman Hunt, Rossetti, Lord Leighton or Alma-Tadema. It is indeed hard to believe that some of her Tennysonian *tableaux*, for example, are meant seriously, rather than being satires on the subject-matter and composition of her contemporaries in the field of painting. But such pictures as her 'Gareth and Lynette' or her 'Passing of Arthur' are offered in the same spirit of artistic seriousness as masterpieces like the portraits of Carlyle. The contrast between the two, however, to a modern eye is so glaring that some comment seems to be necessary.

I think one should not be too beguiled by the word 'serious'. The

Pattles, like most big families until the invention of wireless and television, enjoyed parlour games, charades, and amateur dramatics of a more or less formal character. Julia Margaret had grown up with the idea, now lost to us, that it was fun to dress up, pose and act for a small circle of family and private friends. Even when the subject of the charade or tableau was 'serious', it would be a mistake to think that the game was played with no sense of fun. It is heavy-handed, perhaps, even to say so, but when some modern commentators poke fun at Mrs Cameron's 'Pre-Raphaelite' scenes, they do not give sufficient attention to the fact that these scenes were often executed in a spirit of fun, though not of course meant to be 'funny'. Plenty of people found them unintentionally comic at the time. We need not think that because they raise a smile on our lips, this somehow proves our emotional sophistication or aesthetic superiority to Mrs Cameron and her generation.

The tableau-photographs, however, cannot really be defended on aesthetic grounds, but they do, unintentionally and fascinatingly, raise all sorts of questions about Victorian aesthetics and, indeed, Victorian sexuality. Nor was the influence all one way. Mrs Cameron did not merely draw inspiration from Pre-Raphaelites; she herself created images which the painters tried to copy. Dante Gabriel Rossetti's *Beata Beatrix* in the Tate Gallery, with its smudged background and artificial glow around the head, surely owes much to Mrs Cameron's techniques (though in fact Rossetti refused even to meet her). But you only have to look at a picture where she herself has imitated Rossetti (her 'Too Late! Too Late!' of 1868 owes much to his *Rosa Triplex* of 1867) to see where the difficulties arise. The whole point about Rossetti's maidens (as that great admirer and collector of his work, L. S. Lowry, observed) is that they are not real. 'They are dreams,' as Lowry said. You can paint a dream, but you cannot photograph one. The girls in 'Too Late! Too Late!' just look like real girls who are horribly bored. (For a picture called 'Despair' Mrs Cameron was reputed to have locked the sitter in a cupboard for several hours to obtain the appropriate expression.) Moreover, by taking her camera so close, and by exaggerating the muddiness and tactility of her girls' complexions and expressions, Mrs Cameron vividly presents us with the very thing which mid-to-late Victorian painters all sought so strenuously to exclude from their canvases: female sexuality. No wonder Henry Cotton fell in love with Mary Ryan from her photograph. Here, unlike anything you would see in a gallery of contemporary paintings, is an overpoweringly attractive woman with real skin, real hair. Rossetti, and even more Lord Leighton and Alma-Tadema, removed the least vestige of human texture from their female subjects' limbs. Their nudes have no pubic hair. Their faces are vacuous, purely receptive. The messages given out by these, in many ways repugnant, canvases are quite clear. Women – and by implication sexual relations – have to be presented with the same kind of saccharine unreality as the heroines of Dickens and Thackeray. These pretty-pretty beings are in the greatest possible contrast to *real* girls, like Mrs Cameron's models. Florence Anson in 'Girl Praying' is probably trying to look soulful. She fills the male observer of the photograph with other than soulful thoughts. Mary Hillier in such pictures as 'The Kiss of Peace' or Alice Liddell in young adulthood, are full of fleshy reality.

The critic John Ruskin hated Mrs Cameron's pictures as much as she idolised his writing. Both reactions were quite appropriate. She, as a committed and passionate aesthete, naturally drew inspiration from the

'The Gardener's Daughter,' 1867. This illustrates Tennyson's poem of the same title. It is an unusual Cameron picture, being taken out-of-doors, probably in the garden of Dimbola, her house on the Isle of Wight.

greatest aesthete of the age. Ruskin could plainly see that her photographs gave the lie to many of the modern paintings of which he had been the godfather or inspiration. Ruskin, we remember, was the man who was incapable of consummating his marriage because of his appalled sense of shock, on the wedding night, at the discovery that women had pubic hair. He was to move on, living frequently on the borders of insanity as well as having periods of raving madness, to a besotted devotion to a pubescent girl called Rose La Touche. Yet, the Victorian male devotion to young girls seems vaguely obscene in a picture like Cameron's 'The Whisper of the Muse'. She is trying to convey to us a face of great genius – G. F. Watts – being visited by angel voices. We see only a slightly unsavoury man with a beard being whispered to by a Muse whose name in real life was Lizzie Keown. Freddy Gould, the other muse, 'spoils' the effect of the picture (but actually makes it) by cheekily refusing to act, and staring instead at the photographer. His look speaks volumes about what it was like to know Mrs Cameron.

Victorian women were only meant to have hair, abundant, flowing, hair, growing from the top of their heads. All other parts of the body had to be rendered with marbly smoothness by the painters. The men who feasted on these images, however, sported more facial hair than at almost any other time in post-Merovingian European history. In the eighteenth century, men were clean-shaven unless, like George III and other poor creatures of that kind, they could not be trusted with a razor. By the reign of that bearded monarch, Edward VII, it had once again become normal for men to be clean-shaven; street urchins liked to call out 'Beaver!' in Edwardian England, if they saw a bearded man walking down the street. Forty years earlier, there were so many bearded men about that the boys would have been shouting all day long. Mrs Cameron's gallery of men is a positive Valhalla of Gnomes: Signor (G. F. Watts) himself, Longfellow, Trollope, Darwin, William Michael Rossetti, Tennyson, Henry Taylor – all sprout, not just beards, but flowing beards. Set beside the women (or the faintly androgynous Jowett) these men seem not to be just of a different sex but of a different order of being. Whatever psychosexual explanation (if any) we may wish to produce for the conspicuous Victorian cult of beards, it is certainly something which, on a visual level, distinguishes them from the previous and subsequent generations.

Their predilection for beards is only one of a number of matters – their attitude to the family, to religion, to money are others – on which the

Victorians taken as a whole seem unlike any other cluster of generations in history. That is why 'parallels' between the Victorians and our own generation are so seldom successfully drawn. How can you compare a culture such as our own in which few people have even read the Bible, and one in which the majority of educated people knew much of it by heart and believed it to be not merely an infallible guide to morals, but also to physics and biology? How can a generation such as our own which takes for granted the findings of psychology to such a sophisticated degree that we can afford to be post-Freudian, or post-Jungian – how can such a generation be compared with one which knew almost nothing of such matters, and therefore, in their attitudes to sex, will always seem like innocents? (Nowadays, the Church would feel the need to justify or suppress Newman's wholly 'innocent' expressions of adoration for his friend Ambrose St John; just as the University Proctors would have complaints if Christ Church dons such as Charles Dodgson chose to photograph the children of North Oxford in a state of undress.) How can a western democracy – let alone any other country in the world – begin to understand the political atmosphere of a country where the Prime Minister spent five months of each year outside the political arena, cutting down trees and reading Dante? There are no parallels. Mrs Cameron's photographs, which seem to be depicting a different world, almost a different species, perhaps really were doing so. The world changed deeply, fundamentally, at the time of the First World War, leaving the Victorians to seem as obsolete as the dinosaurs.

Julia Margaret Cameron's true period of activity as an artist was from the end of 1863, when Julia and Charles Norman gave her her first camera, to 1875 when the Camerons finally left Freshwater. During this period, she said herself (it is thought that she exaggerated, but not much) that she produced over 500 plates each year. At first, she exhibited her work widely – in Dublin, Berlin, London and Paris. After 1873, when her daughter Julia, the donor of the camera, died in childbirth, a shattering blow for Mrs Cameron who plunged herself thereafter more and more deeply into work, she tried to develop her charade-style tableaux into book illustrations, abandoning exhibitions. Two volumes of Tennyson's *Idylls of the King* were published with Mrs Cameron's photographs: 'The Passing of Arthur' is typical of these illustrations. Quentin Bell comments on this picture, '"So like a shattered column lay the King".

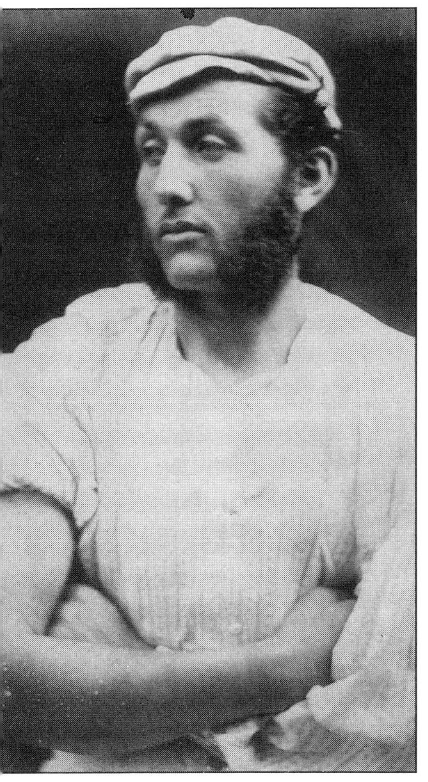

Above 'Connor of the Royal Artillery, winner of prizes at the RA Games at Freshwater, October 1864'.

Right 'Annie, my first success, January 1864'.

He looks as though he might easily be sick. They are going to lose an oar. Something unexpected has happened to the moon, and to the water. In fact Mrs Cameron has more poetry than she can deal with.'

Since love of family was as strong in Mrs Cameron as love of art, it is perhaps not surprising, after half a lifetime in England, that she eventually felt drawn to Ceylon. Her husband was by now an invalid, Ceylon was always his first love, and it was the residence of their three coffee-growing sons (their fourth son, Hardinge, accompanied his parents on the journey). It was no mere matter of loading a few cases into the hold of a ship. There were the coffins to take on board in case any member of the party should die at sea. Then – how to avoid tuberculosis on a sea-voyage? Why, take your own cow, of course, to ensure fresh milk each day. The number of crates, boxes and packing-cases which accompanied the Camerons seemed to be without number, and the long-suffering porters, carrying all the stuff on board, were surprised to discover, when their labours were complete, that they were tipped not with cash but with original Cameron photographs. Those short of money would probably have preferred sixpence to one of Mrs Cameron's bored-looking entourage posing as Tennyson's Princess with a banjo. But those whose families kept these remarkable gratuities may have been grateful: not because of their financial value, but because sixpences come and go, but Mrs Cameron is unique.

They did return to England once, in 1878, 'a visit of turmoil, sickness, marriages and deaths', Mrs Cameron called it. For the rest of her life, however, she was marooned in Ceylon. She liked the inhabitants but, strangely enough, her inspiration seems to have left her under the tropical skies. She took some photographs in Ceylon, but the distinctive Cameron touch is absent in these plates.

In January 1879, when she was staying at the house of her son Henry Herschel Cameron, soon to become a professional photographer himself, she became ill, and took to her bed. Henry arranged it so that his mother could see through her window and across a balcony. One evening, the stars were particularly bright. 'Beautiful', said Mrs Cameron, and died. Surely nobody could possibly have invented for her a more appropriate last word.

Perhaps the most striking faces in Julia Margaret Cameron's plates are those of the children – sometimes bored stiff, sometimes suppressing the giggles, very frequently looking as if they want, by some means or another, to get their own back. Well, they did. One of the pieces of family business which Julia Margaret Cameron wanted to discuss on her last visit back to England in 1878 was the fate of her newly-widowed niece Julia Duckworth, daughter of her sister Mia. Mrs Cameron's idea was that she should marry poor Charles Norman, the widower of her own daughter Julia. But Mrs Duckworth married instead the morosely agnostic ex-clergyman and man of letters Leslie Stephen. When we see Mrs Cameron's statuesque portraits of Mrs Duckworth we do not need

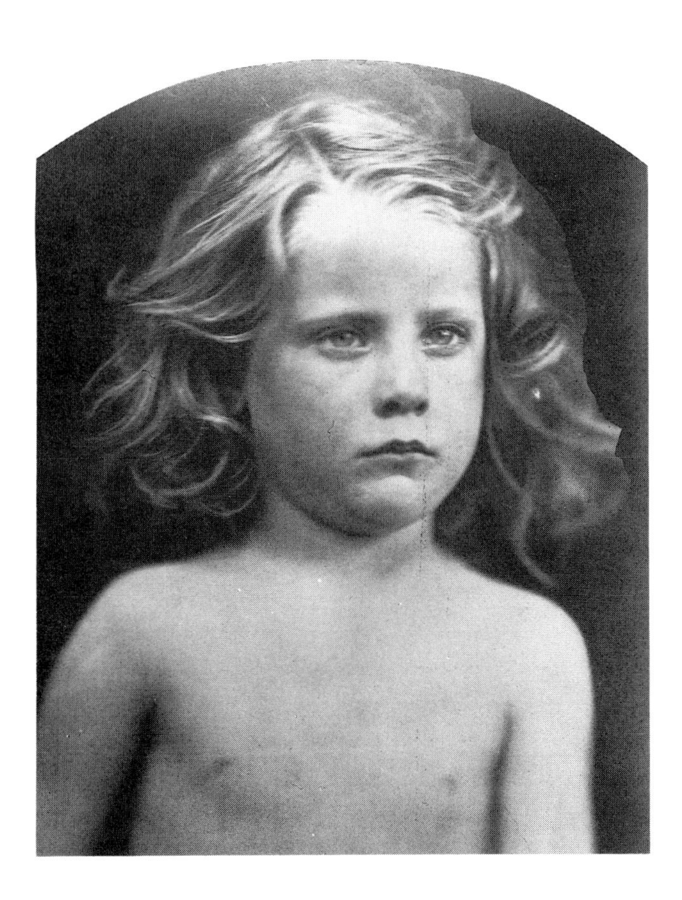

'Young Astyanax', 1866. This is a rare excursion for Mrs Cameron into classical literature. Astyanax, or Scamandrius, was the younger child of King Priam of Troy, who was killed by being thrown over the battlements by Agamemnon. He is here represented by Freddy Gould who looks bored and cold.

to guess whose mother she was destined to become: the resemblance to Virginia Woolf is uncanny.

Within a generation, the sniggering sophisticates of Bloomsbury were gathering in the studio of Vanessa Bell, Julia Stephen's other daughter, to see enacted a play by Virginia, entitled *Freshwater: A comedy*. In Virginia Woolf's casting list, we read that Mrs Cameron was played by Vanessa Bell herself, Charles Hay Cameron by Leonard Woolf, G. F. Watts by Duncan Grant, Alfred Tennyson by Adrian Stephen and the young, exploited Ellen Terry by Angelica Bell, Vanessa's daughter by Duncan Grant. In the light of all that we know about Bloomsbury there seems something almost ghoulish in this *jeu-d'esprit*. The vaguely obscene jokes about Ellen Terry and G. F. Watts, instead of making us titter, as they doubtless made the Bloomsbury set titter over their cigars and lemonade, strike a chilling note: Angelica Bell's own memoirs reveal that the emotional torture of growing up in an emancipated post-Victorian world could be every bit as dreadful as the supposed troubles of Ellen Terry matched to the Signor.

> NELL: *My name is Mrs George Frederick Watts.*
>
> JOHN: *But haven't you got another?*
>
> NELL: *Oh, plenty! Sometimes I'm Modesty. Sometimes I'm Poetry. Sometimes I'm Chastity. Sometimes, generally before breakfast I'm merely Nell.*

Now that we know more than we would ever have asked to know about the Bloomsbury families, we might very well echo the servant Mary's words in the final act of *Freshwater* – 'Gorblime! What a set! What a set!'

Yet Bloomsbury, particularly through the malicious wit of Lytton Strachey, was very largely responsible for shaping twentieth-century attitudes to the Victorian Age. *Freshwater* is a joke, and a private family joke; it would be heavy-handed to build too much upon it. But it reveals in caricatured form the impish, mischievous belief that everything about the Victorians was essentially ridiculous. Beneath the giggling, there are serious political and religious feelings. The generation of which Bloomsbury were the most articulate and amusing mouthpieces had watched 'Victorian values' take their effect, and it is not particularly surprising that they were unimpressed. The combination of unchecked industrial expansion with Gladstonian economics had led to conditions of poverty and squalor which (in numerical terms) were without equal in civilised history. As members of the Fabian Society, or readers of Beatrice Webb's *My Apprenticeship*, could observe, there were children starving on the streets of London at the turn of the century. That was the achievement of Victorian capitalism.

In terms of family life – that other supposed bastion of Victorian virtue – we have only to read the Bloomsbury memoirs themselves, or, say, the novels of Ivy Compton-Burnett, to discover what emotional havoc can be wrought by growing up in a family, particularly in one of those large, prosperous middle or upper-class families which were supposed to be the finest flowering of the English system. The Wilde trial and its aftermath (1895) had revealed the depths of ignorance which existed of human character and variety, just as the Dilke and Parnell scandals had uncovered the extent of English humbug. To the pains of growing up in one of those Victorian families was often added the pure horror of a public school – again the product of Victorian thinking, and Victorian values, particularly those of Dr Thomas Arnold.

Nor did the enlightened reformers of the Victorian age seem to have achieved all that much. The 'liberalism' advocated by such as Matthew Arnold had no effect in dispelling the 'philistinism' which hung over the suburbs and guided the way the British thought about sex, politics, or religion. Though Darwin on one level and the historians of the Bible texts on another had exploded any possibility of clinging to orthodox Christianity in its unreconstituted, fundamentalist form, there seemed no shortage of people who were prepared to assert the truth of things which had been – as it seemed to the next generation – manifestly proved false. Arnold's 'sweetness and light' seemed to make no more headway in the world than Gladstone's innocent nationalism. Nationalism itself, far from being a peaceful desire for self-determination, had taken on Nietzschean, Wagnerian overtones of doom and strength.

'My favourite picture. My niece Julia,' April 1867. Julia Jackson's second husband was Sir Leslie Stephen by whom she bore two daughters – the painter Vanessa Bell and the writer Virginia Woolf. Julia's resemblance to Virginia is very strong as this 'favourite picture' shows.

Between Gladstone and Disraeli the debate had appeared to be between innocent nationalism and jingoistic imperialism. By the end of the century these no longer seemed like alternatives. Instead, they were both capable of fuelling the suicidal urge which overcame the great European powers in the decade after Queen Victoria died. No wonder, as Russia, Germany, France and Britain prepared to fight the most pointless and the most destructive war in history, if the children of the Victorians saw the whole sorry struggle as a tragic enactment of the 'values' cherished and built up by the 'eminent Victorians'. Sending young men to their death in the mud of Flanders was all part of the same silly ideal of personal heroism, detatched from political common sense, which had inspired such famous Victorian set-pieces as the death of General Gordon of Khartoum or the Charge of the Light Brigade.

Those who protested against the First World War found themselves treated with great harshness. Bertrand Russell, as a conscientious objector, was in Brixton Gaol in 1918 when his wife Alys brought him a copy of a recently published book by Lytton Strachey. It was called *Eminent Victorians*. 'It caused me to laugh so loud', Russell remembered, 'that the officer came to my cell, saying I must remember that prison is a place of punishment.'

It is that story, to my mind, which illustrates better than any other why *Eminent Victorians* had such an overpowering effect when it was first published. Strachey had conceived the idea of the book in 1912 when he was thirty-two. The timing of the publication had much to do with its effect. But so, too, did Strachey's manner, the elegance of his satire, the refusal to take his Victorian idols remotely seriously. Just as the Victorians had enjoyed huge stodgy meals, multiplied themselves in vast families (like the Stracheys themselves) and lived in big, lumpy houses, so they had written monumental biographies of their great men and women. Strachey immediately cut them down to size by presenting four such idols – Dr Arnold, Florence Nightingale, Cardinal Manning and General Gordon – in minimal caricature. He does not need to say what he thinks about 'Victorian values' by the time he has finished making these four ridiculous. The release from those values is made by the most liberating means of all: laughter.

In Strachey's dismissal of his subjects, however, as in a rebellious child's rudeness about his parents, there is more than a little love mingled with the hatred. We see this even more strongly in his life of

Queen Victoria, published in 1921, where the affection for his subject – indeed, identification with her – is very marked. We see it, too, perhaps in the rather trivial but possibly significant fact that Strachey, unlike most men of his generation, chose to grow a long flowing beard, so that the famous Henry Lamb portrait, now hanging in the Tate Gallery in London, looks less like a scourge of the Victorians than the last etiolated specimen of their kind.

We are different. We do not need to score points off the Eminent Victorians as if (or as in the case of Bloomsbury, because) they were embarrassing members of our family. From our more distant perspective, we can see that, for all the horrors of life in the nineteenth century, it did throw up figures of a stature not easily rivalled in the post-Strachey years. With the exception of Churchill – and only then in a wartime context – there simply have been no twentieth-century British statesmen to match Peel, Gladstone, Disraeli or Lord Salisbury. There has been no theological figure of the stature of John Henry Newman, either in the Church of Rome or in the Church of England. He outsoars them all. Few modern engineers have ever matched the sheer energy and expertise of Isambard Kingdom Brunel. And think of the great Victorian architects! Where is there in the twentieth century an architect who can come near them for the scale, majesty and self-confidence of their achievement? I am thinking of the front-rank architects – George Gilbert Scott, Augustus Welby Pugin, Charles Barry, William Butterfield, George Edmund Street, G. F. Bodley – but there are a whole host of lesser ones who produced buildings, each one of which, when it has been allowed to survive into our own day, provides us with something like a reproach.

It should not be supposed that I am trying to make some sweeping or indefensible suggestion (except perhaps in the case of architecture where the temptation to do so is almost insuperable) about the superiority of nineteenth over twentieth-century taste or achievement. That is not what I am suggesting at all. In terms of paintings, for example, admirable as individual canvases are, I find my taste for the High Victorians ebbing away to the point where it has almost vanished, whereas the twentieth century seems full of excitement. Likewise, if I am honest, I find the twentieth century (in Britain) a much more exciting one in the history of literature than the nineteenth. What I am talking about is the size and self-confidence of the Victorian achievement. There might, for example, be better historians writing nowadays (though I somewhat

'Faith', 1864. The first of the theological virtues is here represented by Mrs Cameron's long-suffering maid, Mary Ann Hillier, who posed for many of her most successful plates.

doubt it) but there is no one writing on the massive scale and with the grandly self-confident poise of Macaulay, Froude, Lecky.

No one will ever write a funnier book about the Victorians than Lytton Strachey did, nor one more elegant. What I have tried to suggest in this book is that there really were 'giants in those days'. It does not mean that they were without foibles, sometimes hilarious ones. But their stature is unquestionable. Virginia Woolf mocked 'Aunt Julia' Cameron, Lord Tennyson and the others because by implication they embodied a set of 'Victorian values' which the Bloomsburyites were too sophisticated any longer to accept. Strachey even more, in his selected four 'Eminent' Victorians chose figures who may be said to have exemplified values which modern people might hope to have outgrown.

In latter years, we have seen a reaction, and 'Victorian values', far from being things to snigger at, appear to be vote-winning concepts in the minds of politicians. Usually, it turns out on examination that the cherished values concerned are no more 'Victorian' than they are 'modern'. It is simply that, in an age of Laura Ashley wallpaper and successful Dickens movies, we might also be expected to admire politicians and journalists more if they, too, dress up their ideas as 'Victorian'.

My purpose in this book has been much simpler, much less propagandist, much more like that of Julia Margaret Cameron in her close-up portraits. I have done my best to look at six individuals, Mrs Cameron included, with the sole aim of finding out what they were like. I have not been trying to make a point, either about our own age, or about theirs. As it happens, far from being six exemplars of 'Victorian values', I find that I have chosen six people who, in one way or another, turned supposed 'Victorian values' on their head. Prince Albert was a foreigner who never came to terms with England or the English. As it happened, he defended the monarchy, but only by unintentionally making it into something more modern, and wholly unpolitical. Gladstone – a figure synonymous in many ways with the Victorian Church and State – was a man whose mind was awake, and who therefore went on changing, and went on upsetting people, to the end. Newman, too, was brave enough to change his mind: not just before he became a Roman Catholic, but afterwards, too. No one could call him a defender of 'Victorian values'.

The three women in my collection have a more ambivalent relationship with the society in which they found themselves. Charlotte Brontë led a life which was almost buried, hidden from the world. And yet her private fantasy, fashioned into art, became one of the most popular novels of the age. She is a good example of the private and public faces of art. The artist is always a hermit; but unless what she creates is of interest to the outside world, then her labour is vain.

Josephine Butler, beautiful, 'fulfilled', with a good husband and children, devout, might seem at first sight to be a 'typical' Victorian woman. But capitalist society in its most corrupt manifestations had fewer more devastating opponents. In spite of sex, let alone venereal diseases, being unmentionable words, she managed to change the conscience of society. She looked at it without illusions, for what it was. There was tremendous heroism in her opposition to the Contagious Diseases Acts, and in her more generalised defence of feminism and women's suffrage. She is a good example of why Lytton Strachey's satirical approach to the nineteenth century is misleading. She saw more clearly than Strachey did what was wrong with Victorian England. So did Gladstone and Newman. But they all in their vigorous way did something about it.

Mrs Cameron, her hands black with silver nitrate and the other chemicals of her strange alchemy, took an altogether less sociological view. She was through and through an aesthete. But it is thanks to her, more than to her great-niece Virginia Woolf and the Bloomsbury circle, that we have some of the most memorable images of the Victorian age. She believed in divinity. So, strangely, did all the six in my book, and I think this would have been true, even had I chosen another six, with a few carefully selected atheists or agnostics, such as Charles Bradlaugh, J. A. Froude or George Eliot. Long after they had abandoned a belief in a personal Creator, such people continued to believe in a moral law, in duty, and in the divine potential of human beings. It was this divine potential which got lost during the First World War, and led to so much literature of despair, of which Strachey's *Eminent Victorians* is but one, elegant and hilarious example. The world has not got any nicer since Strachey wrote his book – in fact, the reverse. But somehow, it is no longer possible to dismiss anyone, whether dead or alive, in quite the debonair spirit in which he caricatures his subjects. We share a common humanity with people in the past, even when they baffle us, and puzzle us. It is common humanity with the Victorians which we have recovered. In recovering it, we also recover no small sense of their greatness.

Index

Page numbers in *italics* refer to illustrations.

PICTURE CREDITS

Colour Photographs
Page 81 Tate Gallery; 82 National Portrait Gallery; 83 *top* The Royal Collection, *bottom* Deene Park; 84 *top* Brontë Society, *bottom* National Portrait Gallery; 85 *top, bottom left* and *right* Brontë Society; 86 National Portrait Gallery; 87 *top* Andrew Anderson, *bottom* The Mansell Collection; 88 Illustrated London News; 177 National Portrait Gallery; 178 Canon Bartlett; 179 *top left and right* Magdalen College Oxford; 179 *bottom* Canon Bartlett; 180 *top and bottom* Mary Evans Picture Library; 181 and 182 *top* National Portrait Gallery; 182 *bottom* and 183 National Museum of Photography; 184 Tate Gallery.

Black-and-white Photographs
Page 2–3 Hulton Picture Company; 7 Illustrated London News; 11 Greater London Photograph Library; 13 *and* 14 Hulton Picture Company; 19 Her Majesty The Queen; 23 Victoria and Albert Museum; 26 Punch; 29 Hulton Picture Company; 31 Her Majesty The Queen; 34 English Heritage; 35 *and* 36 Her Majesty The Queen; 39 Illustrated London News; 41 Hulton Picture Company; 45 National Portrait Gallery; 49 Hulton Picture Company; 50 Brontë Society; 52 British Library; 55, 57, 60, 61, 63, 65, 68, 73, 77, 93 *and* 95 Brontë Society; 99 *and* 100 Hulton Picture Company; 102 Julian Selmes; 105 Illustrated London News; 107 *and* 111 Hulton Picture Company; 115 *left* Mary Evans Picture Library, *right* Illustrated London News; 117, 120 *and* 123 Mary Evans Picture Library; 125 *left* Hulton Picture Company, *right and* 127 Illustrated London News; 130, 133 *and* 134 Hulton Picture Company; 137 Royal Commission on Historical Monuments; 142 Bodleian Library (G A Oxon a 68, page 17 item 23a.); 144 Hulton Picture Company; 146 *and* 147 Oxfordshire County Libraries; 149 Birmingham Oratory; 151 National Portrait Gallery; 155 Hulton Picture Company; 158 The Mansell Collection; 160 *left* Mary Evans Picture Library, *right* Birmingham Oratory; 163 *and* 164 Fawcett Library; 166 The Mansell Collection; 167 Hulton Picture Company; 169 Mary Evans Picture Library; 172 *left* Illustrated London News, *right and* 186 Fawcett Library; 189 Salvationist Publishing; 195 Hulton Picture Company; 199 Mary Evans Picture Library; 201 Fawcett Library; 203, 204, 208 *and* 211 Hulton Picture Company; 215 *and* 216 Joy Lester; 219 Lincolnshire County Council; 221, 224, 227 *left and right*, 229, 231 *and* 234 National Museum of Photography.

FRONT COVER PHOTOGRAPHS *Top left* Fawcett Library; *all others* Hulton Picture Company

BACK COVER PHOTOGRAPH Tara Heinemann.